Springer
*Berlin
Heidelberg
New York
Barcelona
Budapest
Hong Kong
London
Milan
Santa Clara
Singapore
Paris
Tokyo*

Acoustic Remote Sensing Applications

Editor
S.P. Singal

Springer-Verlag

Narosa Publishing House

Dr. S.P. Singal
Formerly, Deputy Director
National Physical Laboratory
New Delhi-110 012, India

Copyright © 1997 Narosa Publishing House

Cataloging-in-Publication data applied for

All rights reserved. No part of this publication may be reproduced, stored in a retrieval system, or transmitted in any form or by any means electronic, mechanical, photocopying, recording or otherwise, without the prior permission of the publisher.

Exclusive distribution in North America (including Mexico), Canada and Europe by Springer-Verlag Berlin Heidelberg New York. For all other countries, Narosa Publishing House, New Delhi.

All export rights for this book vest exclusively with Narosa Publishing House. Unauthorised export is a violation of Copyright Law and is subject to legal action.

ISBN 3-540-61612-8 Springer-Verlag Berlin Heidelberg New York
ISBN 0-387-61612-8 Springer-Verlag New York Berlin Heidelberg
ISBN 81-7319-110-7 Narosa Publishing House, New Delhi

Printed in India.

Foreword

Acoustic remote sensing is one of the most useful methods in our technological kit of tools for research and modification of our environment. In the ocean *sonars* provide information on depths and bottom profiles, locate shoals of fish, trace motions of pollutants in the benthic boundary layer, and measure the spectrum of wind-driven waves in the oceanic surface layer. In the atmospheric boundary layer where we (other than "frequent fliers" and astronauts) live most of the time, *sodars* (or *echosondes*) provide data on wind shears, turbulence parameters needed for pollution control, and inversion heights. The earliest form of acoustic remote sensing—*range finding*—antedates the use by humans. Some species of bats emit sound pulses with frequencies up to 200 kHz, and determine distances to ground, cave walls, or prey from the time delay of the echoes. Similarly, some dolphin species find ranges to shores, colleagues, or shoals of fish in the same way. For man, the use of sound for remote sensing could not begin until the speed of sound was known. The first measurement of the speed of sound in air probably was that of Mersenne[1] using rather primitive methods: eyes and ears to detect a musket shot at night, with the delay between the two timed by his own pulse or, later, by the swings of a pendulum. A few decades after Mersenne's work, William Derham[2] invented an improved pendulum timer that, together with longer paths, allowed more accurate measurements of sound speed. These results, presented by Derham at the Royal Society, induced the President of the Society, Isaac Newton, to try to improve his incorrect prediction (from assuming an isothermal process) by unconvincing mathematics. The discrepancy remained for almost a hundred years until Laplace[3] obtained the correct value by assuming an adiabatic process.

With a reasonable estimate of ambient sound speed in hand, Derham was able to pursue answers to a number of questions about the nature of sound. Among other results, he proved that wind along the sound path linearly increased or decreased the measured apparent speed, depending on whether the wind was blowing toward or away from the observer. Values of wind were obtained by another of his inventions, throwing up chaff (from grain-not radar chaff) and

---

[1] F.M. Mersenne, **Harmonicorum Liber.,** Paris (1636).
[2] W. Derham, Experimenta et observationes de soni motu, **Phil. Trans. Roy. Soc.** London, **26,** 2–32 (1708).
[3] P.S. Laplace, **Méchanique Céleste,** Paris (1799).

measuring the distance it traveled before hitting the ground. Recently, Delaney[4] plotted Derham's data on sound speed vs wind, clearly showing the linear dependence as claimed. Finally, Derham advanced and explained the echo-delay method for range finding. For example, using a building on the far bank as a reflector, he obtained reasonably accurate measurements of the width of the Thames near London. Plans to make a similar measurement of the width of the English Channel were aborted by the outbreak of war between the Dutch and the French. All in all we should consider Derham as the first pioneer in acoustic remote sensing, that is, the application of sound to measurement of environmental properties.

In the second half of the 19th century, financed by a society for retired seamen, John Tyndall conducted research on sound propagation over the ocean and through fogs, with the hope of improving warning systems, such as fog horns. His results, which were reported in a paper[5] at the Royal Society, showed that sound propagated with smaller loss through fog than through clear air, and that sound projected horizontally over the ocean through clear air returned echoes even though no ships were present to cause reflections. He attributed the echoes to backscatter from "flocculence" in air temperatures and supported this explanation by creating a laboratory model of the process. Later, he turned his apparatus to project pulses vertically upward into the atmosphere and again obtained returning echoes. We could consider this vertical system to be the first monostatic sodar.

For the speed of sound in water, although several sporadic attempts to measure it occurred earlier, according to Hersey[6] the first reliable value—1500 m/s was obtained by Francois Beudant for sea water near Marseilles in 1820. Later, reporting their own measurements in Lake Geneva in 1826, Daniel Colladon and Charles Sturm[7] credited Beudant and his experimental value. Using the observed value, researchers attempted to exploit echo ranging in the ocean for depth sounding, with varying degrees of success. By World War I, however, use of depth sounders to provide single measurements became more common. In 1919, M. Marti invented the *facsimile recorder,* an instrument that—with constant ship motion—provided complete bottom profiles.

Little was done on acoustic remote sensing for the atmosphere in the first half of the 20th century, but in 1969 the development of the *sodar* or *echosonde* by McAllister[8] and his Australian group, together with improved theoretical studies, mainly by Russian physicists, led to a vastly expanded interest in acoustic remote

---

[4] M.E. Delaney, Sound propagation in the atmosphere: an historical review, **Acustica, 38,** 201–203 (1977)

[5] J. Tyndall, On the atmosphere as a vehicle of sound, **Trans. Roy. Soc. London, Ser. A. 164,** 183–244 (1874)

[6] J.B. Hersey, A chronical of man's use of ocean acoustics, **Oceanus, 20,** 8–21 (1977)

[7] J.D. Colladon and J.C.F. Sturm, Memoire on the compression of liquids, **Ann. Chim. Phys., 36,** 225–257 (1827).

[8] L.G. McAllister, J.R. Pollard, A.R. Mahoney, and P.J.R. Shaw, Acoustic sounding—A new approach to the sudy of atmospheric structure, **Proc. IEEE, 57,** 579–587 (1969).

sensing and its applications. Programs set up in the USA by C.G. Little, in France with A. Weill, in Italy with G. Fiocco and G.M. Mastrantonio, in India with S.P. Singal, and in Japan with M. Fukushima, to name only a few, led to substantial improvements in hardware, configurations, data processing and recording, predictions of scattering cross-sections, together with the addition of other measurements, such as winds from Doppler shifts, or temperatures from hybrid radar/acoustic (RASS) systems.

One result of the increase in numbers of research and development programs was a considerable expansion in the number of researchers, of published papers, of improvements in theory, and descriptions of sodar and sonar equipment. Some of the earlier results were covered by Brown and Hall[9], or Brown[10] but no comprehensive or unified survey of the many developments in the over 15 years since the review[9] has appeared. Therefore, the collection of papers being put together by S.P. Singal in the present volume should be most welcome, especially by newcomers to the field, graduate students, and those involved in environmental applications.

EDMUND H. BROWN
Ex-Head, Acoustic Sounding Proprogramme
NOAA, Environmental Technological Laboratory
Boulder, Colorado, USA

---

[9] E.H. Brown and F.F. Hall, Jr., Advances in atmospheric acoustics, **Rev. Geophys. and Space Phys., 16,** 47–110 (1978).

[10] E.H. Brown, Acoustic remote sensing in the air and the sea, **Acoustic Sensing and Probing** (A. Alippi, Editor), pp 41–56, World Scientific Publishing Co., Singapore (1992).

Preface

Acoustic waves while propagating in a medium undergo scattering from the inhomogeneities in the medium. This scattering is very strong compared to the scattering of electromagnetic waves under similar conditions. These scattered signals if received can thus be made use of to characterize the medium itself.

During the last few decades considerable developments have taken place in instrumentation and theory to exploit this property to probe the structure of the lower atmosphere and the oceans. The information is however, hidden, in research papers published in journals or in proceedings. A researcher entering into this area has to search the original papers which many a times do not become available to him. These original research papers have again to be searched if a course on the subject has to be given to the graduate students. The environmentalists undertaking applications of acoustic remote sensing also need easily available but authentic information on the subject.

Considering the above, a great need was being felt to compose or write a book which should acquaint the reader with the subject ab initio and at the same time it should be precise and concise. Theoretical advancements in the subject should be described. Various physical principles and technical details in instrumentation should be discussed in such a minute detail that one may assemble a system if one likes. For the users, the technical details of the various applications should be described and for the researchers the latest developments in the subject (the state of the art) should be covered.

Taking into view the above stated needs, the purpose of the present volume is to give an all round progress of technical developments in instrumentation and of theoretical advancements in the subject of acoustic remote sensing, to explain utilization of acoustic techniques in various atmospheric and ocean applications and to describe the potential and limitations of the techniques for future studies. To fulfil this scope it was proposed to compose a set of selected articles written by experts in their respective areas. Thus original researchers on different aspects of the subject anywhere in the world were requested to contribute articles to this volume stating very clearly its scope.

This volume containing 23 articles is divided into three parts. Part One deals with atmospheric acoustics and instrumentation along with the history of the subject, principle of application, theoretical approach and instrumentation design and fabrication etc. Part Two describes applications of acoustic remote sensing techniques to study the structure of the atmospheric boundary layer, to determine its dynamics and the turbulent structure parameters and to monitor/nowcast hazardous situations in air pollution and microwave propagation. Part Three deals with ocean acoustics which describes studies in ocean tomography, measurements of oceanflows, currents and effluent plume dilutions, ocean-atmosphere interactions and acoustic propagational characteristics.

I am highly thankful to the contributors for their cooperation without which this volume would not have been possible. I am also thankful to Dr. E.H. Brown who readily agreed to write the foreword. Thanks are also due to Dr Ian Bourne for his constant valuable suggestions and to all my well wishers who encouraged me during this endeavour.

S.P. SINGAL

Contents

Foreword   *v*
Preface   *ix*

PART ONE: Atmospheric Acoustics and Instrumentation   1-176

1. Physical Grounds for Acoustic Remote Sensing of the Atmospheric Boundary Layer
   M.A. Kallistratova   **3**

2. An Overview of the Technological Development of Atmospheric Echosounders (SODARS)
   John A. Kleppe   **35**

3. Design of a Tri-Monostatic Doppler Sodar System
   Yoshiki Ito   **85**

4. A Modular PC-Based Multiband Sodar System
   G. Mastrantonio and *S. Argentini*   **105**

5. Mini Acoustic Sounding-A Power Tool for ABL Applications: Recent Advances
   D.N. Asimakopoulos, C.G. Helmis and *M. Petrakis*   **117**

6. Radio-Acoustic Temperature Profiling in the Troposphere
   G. Bonino   **133**

7. Radio Acoustic Sounding system (RASS) for Studying the Lower Atmospher
   S.P. Singal and *Malti Goel*   **142**

PART TWO: Applications in the Atmosphere   177-406

8. Determination of the Turbulent Structure Parameters
   Pan Naixian   **179**

9. Turbulence Variables Derived from Sodar Data
   R.L. Coulter — 191

10. Development of Sodar Detection and Its Application for Studies of Atmospheric Boundary Layer in Beijing, China
    Mingyu Zhou — 202

11. Influence of the Nocturnal Low-Level-Jet on the Vertical and Mesoscale Structure of the Stable Boundary Layer as Revealed from Doppler-Sodar-Observations
    Frank Beyrich, Dieter Kalass and *Ulrich Weisensee* — 236

12. Dynamics of the Continental Boundary Layer: The CRPE Sodar Results (1984–1993)
    Christine Amory-Mazaudier — 247

13. Sodar Investigations of Gravity Waves by Cross Spectral Analysis
    Günther Bull — 275

14. Sodar Monitoring of Nocturnal Boundary Layer During the Harmattan in Ile-Ife, Nigeria
    J.A. Adedokun — 293

15. An Overview of Similarity Methods to Estimate Turbulence Quantities from Sodar Measurements in the Convective Boundary Layer
    Dimitrios Melas — 307

16. Sodar: A Tool to Characterize Hazardous Situations in Air Pollution and Communication
    S.P. Singal, B.S. Gera and *Neeraj Saxena* — 325

17. Application of Sodar in Urban Air-Quality Monitoring Systems
    Jacek Walczewski — 385

18. Operational Use of Sodar Information in Nowcasting
    Th. Foken, H.-J. Albrecht, K. Sasz and *F. Vogt* — 395

PART THREE: Ocean Acoustics — 407-582

19. Acoustic Remote Sensing of Ocean Flows
    Antony Joseph K. and *Ehrlich Desa* — 409

20. Acoustic Remote Sensing of Ocean—Atmosphere
    Interactions
    A. Weill and *H. Dupuis* **449**

21. Range-Average Inversions in Ocean Acoustic
    Tomography
    R. Michael Jones **476**

22. Acoustic Measurements of Currents and Effluent Plume
    Dilutions in the Western Edge of the Florida Current
    John R. Proni and *Robert G. Williams* **537**

23. Acoustic Propagational Characteristics and Tomography
    Studies of the Northern Indian Ocean
    S. Prasanna Kumar, Y.K. Somayajulu and
    T.V. Ramana Murty **551**

INDEX **583**

Part ONE

Atmospheric Acoustics and Instrumentation

Acoustic Remote Sensing Applications
S.P. Singal (Ed)
Copyright © 1997 Narosa Publishing House, New Delhi, India

1. Physical Grounds for Acoustic Remote Sensing of the Atmospheric Boundary Layer

M.A. Kallistratova

A.M. Obukhov Institute of Atmospheric Physics, Russian Academy of Sciences,
3 Pyzhevsky, Moscow 109017, Russia

1. Introduction

Acoustic sensing of the atmosphere, in broad sense, refers to any of the methods for obtaining data on meteorological parameters from measurements of characteristics of sound wave propagation. These methods include, estimation of wind velocity and temperature (in both tropo- and stratospheres from the sound propagation time), acoustic tomography of the atmosphere, determination of turbulence intensity from fluctuations of amplitudes and phases of sound wave. Different methods of sounding use different sound sources of both natural (volcanic eruptions, thunderstorms, sea storms) and artificial (explosions, sirens, detonation sources) origin. However, in the last ten years acoustic sensing most often implies sounding of the atmosphere by short pulses (filled with a carrier frequency) from a narrow-beamed sound transmitter. It is in this sense this term will be here.

Acoustic sensing is based on the sound wave scattering by turbulent air inhomogeneities that are always present in the atmosphere. This technique does not differ basically from clear air radiolocation. However, the former has some specific properties:

- Speed of sound is more sensitive to temperature variations than that of electromagnetic waves: for the same temperature variations the changes in the values of the refractive index for sound waves are 1000 times greater than those in the values of the refractive index for electromagnetic waves. Therefore a cross-section of acoustic scattering from the atmospheric temperature inhomogeneities is about a million times greater than that of electromagnetic waves. Thanks to this fact sodar[1] is more simple in design than clear air radar and its prime cost is much lower.

---

[1]The abbreviation SODAR was first used in [1]: SOnic Detection and Ranging.

- Speed of sound is responsive to changes in wind velocity. Therefore sodar provides a more detailed information on dynamic properties of turbulence.
- The velocity of sound propagation is a million times lower than that of electromagnetic waves. Due to this fact the processing of sodar information is considerably simplified, moreover, a better spatial resolution and a short dead zone can be reached.
- Acoustic waves of both centimetre and decimetre bands used usually in sodars are readily absorbed in the air. Therefore the altitude range for acoustic sounding is rather low and bounded to the heights of the order of 1 km.

The above-enumerated peculiarities of sound propagation in the atmosphere make acoustic waves especially suitable for sounding the atmospheric boundary layer (ABL). The ABL is the most time- and space-varying part of the atmosphere. The data on the ABL parameters and the processes occurring in this layer are very important for the solution of both fundamental and practical problems of atmospheric physics such as:

- The processes of heat, moisture and momentum exchange between the underlying surface and the upper layers occur within the ABL. Therefore its parameterization and development of its models are required for improving weather forecast and studying climatology.
- Both accumulation and short-range transfer of anthropogenic pollutants occur within the ABL. Hence, to solve the problems of environmental protection, it is necessary to study the mechanism of turbulent mixing within this layer and to take regularly its parameters.
- The ABL strongly affects the propagation of optical- and microwaves. Therefore the data on the ABL refractive index variations are essential for comprehension of the atmospheric interference nature in location and navigation systems and in line-of-sight communication links.
- The wind shears in the lower part of this layer can be dangerous for landing and taking-off heavy aircrafts. So the information on wind conditions in the ABL are of great importance to air transport safety.

At the same time the data on meteorological parameters routinely obtained by meteorological services are still lacking to solve the problems mentioned above. The meteorological standards were developed many years ago to be used for synoptic forecast and aeronavigation safety and they are not suitable enough to determine the ABL parameters. Radiosonde data have a poor vertical resolution in the lower 1 km layer. Moreover, such measurements are taken only twice a day and they cannot supply the data on daily variations in the ABL structure. Until recently, basic experimental data were obtained with contact sensors mounted on meteorological masts, free or kite balloons and aircrafts. Obviously, a complete volume of information required now for a comprehensive study of the ABL cannot be obtained from such a kind of measurements.

Therefore, remote sensing, (especially acoustic sounding), the most simple and cost effective method, becomes very important for the ABL study. However, it should be borne in mind that the results of acoustic sounding, as well as those of any indirect method, require a thorough comparison with the results of conventional measurements and cannot supersede completely these latter tools.

2. Acoustic Sounding Method and Its Development

2.1 Physical Principle of the Method

Adiabatic speed of sound c in motionless air is determined from the Laplace formula:

$$c = (\gamma P/\rho)^{1/2} \tag{2.1}$$

where P is the air pressure, ρ the air density and $\gamma = 1.4$ is the ratio of heat capacities for constant pressure and constant volume. Using the equation for ideal gas condition $P/\rho = RT$, equation (2.1) can be rewritten in the form:

$$c = (\gamma RT/\mu)^{1/2} \tag{2.2}$$

where T is the mean temperature (Kelvin), R the universal gas constant and μ the molecular weight of the mixture of gases that are the constituents of air. In the real atmosphere, water vapour is always present and the values of ρ, μ, and γ depend on vapour concentration. Fluctuations of sound speed in wet air are due to both adiabatic changes in density when air temperature varies, and changes in density owing to turbulent fluctuations in water vapour concentration q.

In the atmosphere the phase sound velocity c_{ph}, that governs the process of scattering, depends also on the projection of wind velocity vector $\vec{v}$ on a normal to the wave front:

$$c_{ph} = c + \vec{v}\vec{k}/|\vec{k}| \tag{2.3}$$

where $\vec{k}$ is the wave vector. In what follows just a phase sound speed will be implied, the subscript ph being omitted. Atmospheric turbulence results in random fluctuations of T, q and $\vec{v}$ that are responsible for fluctuations of the sound refractive index:

$$n = c_0/c \tag{2.4}$$

where c_0 is the mean speed of sound. The intensity of a scattered wave carries information of the intensity of turbulent fluctuations $n' = n - 1$.

Characteristic frequencies of turbulent fluctuations are below audio frequencies; therefore not temporal fluctuations but a spatial field of random inhomogeneities n' (i.e. as if a momentary photo of turbulence) is essential for audio sound scattering. Only those inhomogeneities, whose characteristic scales, l_t are comparable with the lengths of sound waves λ, are responsible for scattering; irregularities of considerably larger scales result in wave refraction.

Inhomogeneities of the refractive index for the atmosphere are too slight and n' values usually do not exceed $n' \approx 10^{-2}$. That is why the intensity of scattering from chaotically occurring inhomogeneities is low. Scattering is known to increase in certain directions due to a constructive interference, when inhomogeneities are periodic and the Bragg condition holds (Fig. 1):

$$l_t = \lambda/(2 \sin \Theta_B) \qquad (2.5)$$

where l_t is the period (scale) of inhomogeneities and Θ_B the angle of wave incidence (the Bragg angle) which is half the scattering angle θ. As the spatial power spectrum of atmospheric turbulent inhomogeneities is continuous, the spectral component $\kappa = 2\pi/l_t$ satisfying (2.5) will always be found, which will determine the intensity of scattering at the angle θ.

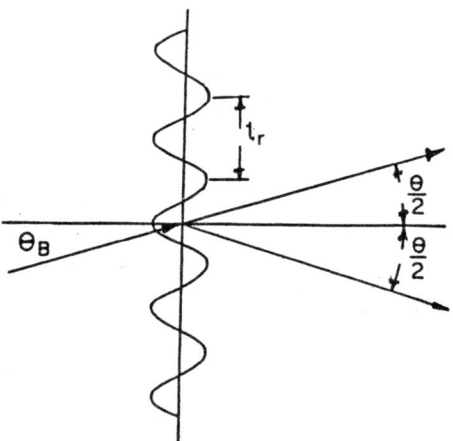

Fig. 1. Wave scattering by periodical structure of inhomogeneities; l_t is the period, $\Theta_B = \theta/2$ is the Bragg angle.

Since spectral density of the small-scale atmospheric turbulence rapidly increases with increase in the scale l_t, scattering indicatrix is extended toward small angles. Variations in temperature and wind velocity enter in different ways in the expression for sound speed and refractive index. It is important that in this case temperature is the scalar value and velocity is the vector; due to this fact an angle dependence of scattering intensity is different for temperature and velocity fluctuations. In particular, backscattering (at $\theta = 180°$) occurs only from temperature and humidity inhomogeneities. This fact offers a possibility to determine separately the fluctuations of scalar parameters and wind velocity from the measurement of sound scattering at different angles.

Mean wind flows that carry small-scale inhomogeneities are always present in the atmosphere. The time of transfer of such inhomogeneities by wind flows through the sodar beam usually is quite less than the characteristic time of their evolution. Therefore it is possible to consider that the spatial field of scatterers moves as a "frozen" one. In this case frequency of the received scattered signal f_s is shifted from the radiation frequency f_0 due to the Doppler effect. At backscattering:

$$f_s = f_0(1 - 2v_r/c) \qquad (2.6)$$

where v_r is the projection of wind velocity in the direction of sounder beam. Thus, using a Doppler sodar it is possible to determine a beam component of the mean wind velocity and using a three-component sodar it is possible to determine the whole vector of wind velocity $\bar{v}$.

2.2 Experimental Schemes of Sounding

Figure 2 shows two basic variants of a geometrical configuration of antennae for acoustic sounding: monostatic and bistatic. In a monostatic sodar the same reversible electroacoustic transducer serves to radiate a sound pulse and to receive the echo-signal scattered at $\theta = 180°$. The scattering volume V, bounded by both the beam width $\gamma_{0.5}$ and the pulse duration τ, moves away from a source at a sound speed. Time changes in the intensity of echo-signal give information on the changes in the intensity of temperature and humidity fluctuations as a function of distance. The minimal height resolution is $\delta H = c\tau/2$ at a vertical sounding. Transmitter and receiver in a bistatic sodar are separated by some distance. Scattering radiation comes from a stationary region of space cut by the intersection of the beams.

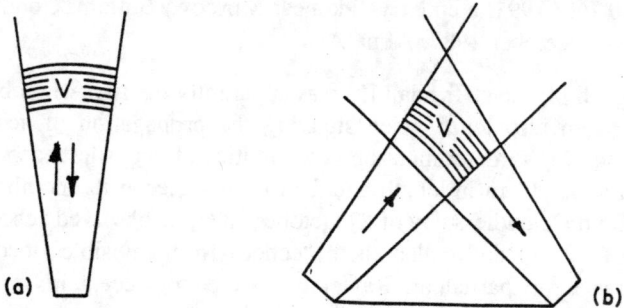

Fig. 2. Two basic variants of a geometrical configuration of sodar: (a)—monostatic, (b)—bistatic.

To visualize the scatterers' position and reflectivity, a continuous record of the level of echo-signal on the height-time co-ordinates is used. Fig. 3 gives two examples of the echograms of a vertical monostatic sodar signal. Such echograms are very useful in studying the morphology of turbulent formations and their evolution.

2.3 Brief History of Acoustic Sensing

Development of the method for acoustic sounding the lower troposphere was preceded by a long period of both theoretical and experimental studies of acoustic wave scattering from turbulent inhomogeneities.

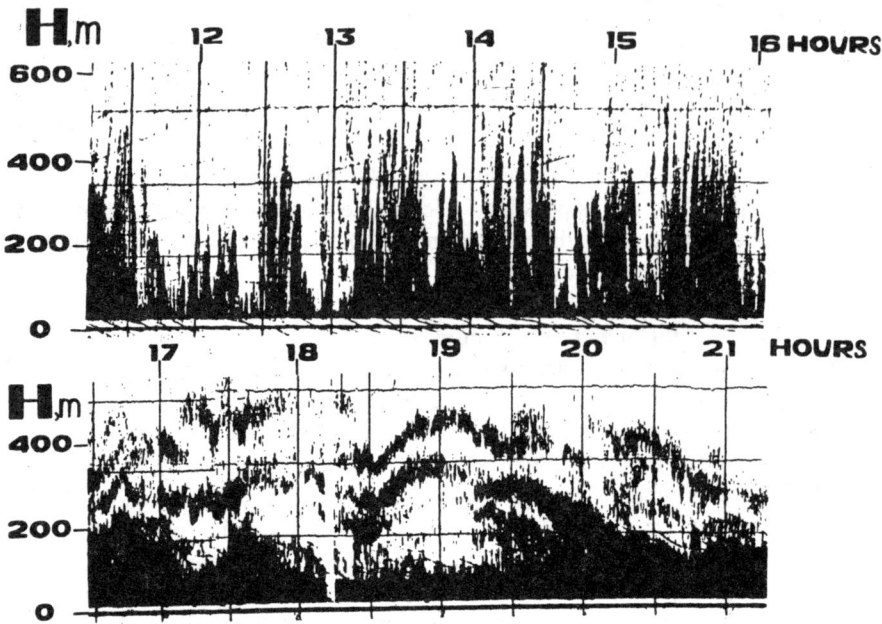

Fig. 3. Examples of sodar echograms (facsimile records). Top: hot summer day, 10 July 1991, a country-side near Moscow; Bottom: cloudy winter day, 10 December 1987, Alma-Ata.

An English physicist Tyndall [2] was apparently the first who observed sound scattering from turbulence when studying the propagation of acoustic signals through a sea fog to determine the potentialities of acoustic beacons. He used a huge horn about 10 m in length and 2 m in diameter at its mouth for the beam forming. On the cloudless day of 27 October 1873, he observed echo-signals from a height of 200 m and called them "echoes from invisible flocculence". His contemporaries, in particular, Rayleigh [3] did not accept his hypothesis and considered refraction to be responsible for this effect.

Fresh interest in this phenomenon was awakened only half a century later in studying an anomalous propagation of sound from powerful explosions and detecting of sound level within the area of acoustic shadow. However, only during World War II a purposeful study of sound scattering began owing to the development of sound ranging of artillery. To explain the deterioration of audibility by a strong wind, the hypothesis that decay of sound is due to its scattering from wind velocity fluctuations was used. In 1940 Sieg [4] conducted experimental studies in the range of frequencies from 2500 to 4000 Hz. He observed an additional (as compared to molecular absorption) decay of sound with an increase of wind velocity and a weak dependence of this decay on frequency.

A theoretical problem of sound scattering by turbulence was first formulated and generally solved by Obukhov [5] in 1941, who applied the theory of locally isotropic turbulence that he had developed together with Kolmogorov at that time. Even the statement of this problem contained potentialities of acoustic remote sensing because the objective of Obukhov's work was to calculate not sound decay but the intensity of scattering by "elementary volume" of a turbulent medium.

During the following two decades the fundamental theory of sound wave scattering was independently developed in Russia by Obukhov [6], Blokhintsev [7, 8], Tatarskii [9, 10], Monin [11], and in the USA by Pekeris [12], Kraichnan [13], Lighthill [14], Batchelor [15].

In 1946, in connection with the study of microwave propagation, the lower atmosphere was acoustically sounded by a special instrument that was named sodar (see Gilman et al, [1]). The objective of the experiment was to obtain echo-signals reflected from the inversion layers to study the sources of fading in the line-of-sight communication links. Using the carrier frequency of 4 kHz the authors observed strongly pulsating echo-signals from heights of 100–200 m that occurred even without inversion layers. Note that both the articles [1] and [7] were published in the same issue of the Journal of Acoustical Society of America. However, the relation between these theoretical and experimental results became evident only 25 years later in the paper by Little [16].

A purposeful experimental verification of the theory was performed in the late 1950s in Russia by Kallistratova [17, 18, 19]. The results of her measurements of scattering cross-section in the atmospheric surface layer reliably confirmed the physical scattering mechanism predicted by Obukhov. At small scattering angles these experiments gave good quantitative agreement with the calculations according to the formulae of Tatarskii [9]. The discrepancy at large angles demanded revision of the theoretical calculations. Improved expressions for the effective scattering crosssection were obtained in [10, 11], which are widely used now all over the world. Meanwhile, two other important experiments were also conducted:

- In 1964, Kelton and Bricout [20] measured the Doppler shift in the frequency of scattered signals (at a carrier frequency of 11 kHz). Being quite unfamiliar with the American [13–15] and Russian [11] studies and basing on Pekeris' incorrect formula that does not describe the gap in scattering indicatrix at $\theta = 90°$, they used a bistatic configuration, scattering angle being close to 90°. Nevertheles, they succeeded in recording scattered signals and conducting Doppler measurements of wind velocity at a height of 18 m; the results obtained proved to be in good agreement with the anemometric data.
- In 1968, in Australia McAllister [21] repeated Gilman's experiment to find a way for studying temperature inversions that were important for propagation of radio microwave. He succeeded in obtaining a qualitative pattern of echo-signal from heights of the order of 1 km using sound frequency (950 Hz) and increasing radiation intensity (a matrix of 196 loudspeakers was used as a transmitter). His main achievement was the use of the facsimile recorder of the type which was long employed in sonars (sea

ultrasonic locators) to record continuously a vertical distribution of echo-signals (as in Fig. 3). McAllister attributed the observed signals to the reflections from sharp gradients of mean temperature and employed a model for multi-layered inversions to explain a rather high level of the signal.

The theoretical and experimental papers of the 60's contained everything necessary for performing acoustic sounding as a method to study the lower troposphere. The decisive step in this direction was made in 1969 by Little [16] who generalized the experience gained by the Russian, American and Australian researchers and analyzed systematically the potentialities of the new method. Since that time the method of acoustic sensing has mainly been developed in the following directions:

- developing and improving the instrumentation;
- checking the reliability of acoustic sensing via comparison with the results of conventional *in situ* measurements;
- theoretical and experimental evaluation of the effect of different atmospheric factors on the accuracy and range of acoustic sensing;
- widening the scope of application of remote acoustic sensing in the atmospheric investigations;
- combining sodars with some other means for remote sensing (e.g. RASS or wind profilers).

The works related to the above have been analysed in the perfect review by Brown and Hall [22], and in the papers by Neff and Coulter [23], Singal [24, 25], Weill and Lehmann [26], Asimakopoulos and Helmis [27], and Kallistratova [19].

It should also be noted that the physical mechanism of wave scattering in the atmosphere is still being refined within the scope of the theory of propagation of acoustic and electromagnetic waves though random media. In the last two decades, a few monographs [28–33] as well as number of theoretical papers on this topic have been published that are impossible to be reviewed here. These papers contain some results different from Tatarskii's result [10] that has become classical. To make orientation in this ocean of literature easier, it is reasonable to formulate the main principles of theoretical treatment of the problem of scattering and to indicate the main differences in the solutions obtained by some authors.

3. Calculation of Effective Scattering Cross-Section

3.1 Way of Solving the Problem

The theoretical problem of sound wave scattering in a randomly inhomogeneous medium is solved by approximate methods. In case of the atmosphere, where relative fluctuations in the meteorological parameters are small and a scattered acoustic field is weak in comparison with an initial one, the method of small perturbations is usually used and the single-scattering approximation is applied. Just this approximation, called the Born's approximation, was used in all the papers mentioned in section 2.2. A general procedure to solve the problem is the following:

1. A complete system of hydrodynamic equations for the atmosphere is considered. These contain the law of conservation of matter, the law of conservation of pulse, the equation of gas condition, and the condition of adiabatic motion. Irreversible processes of energy loss due to viscosity and heat conduction are neglected (these losses are taken into account only at the final stage of solving the problem by introducing a molecular sound absorption). Moreover, both the Coriolis and gravity forces are also neglected.

2. Each of the medium parameters (pressure P, velocity $\vec{v}$, density ρ, temperature T, entropy E) entering the equations is presented as a sum of three values: a mean value which is constant for given conditions, a fluctuation addition which characterizes turbulence, and a small acoustic addition caused by sound wave propagation: $P = P_0 + p' + p$; $\vec{v} = \vec{v}_0 + \vec{v}' + \xi$; etc. In this case a number of simplifying assumptions are made:

- turbulent motions are assumed to be incompressible;
- oscillation velocity of air particles in wave is assumed to be noneddy (in contrast to the velocity of turbulent motions);
- density changes in sound wave are considered to be an adiabatically reversible process, so that the entropy of every small portion of air remains constant in time.[2]
- the inherent frequencies of turbulent fluctuations are assumed to be small as compared to frequency of oscillations of sound wave, i.e. "frozen turbulence" is considered;
- mean wind velocity that does not cause scattering is assumed to be equal to zero;
- relative fluctuations of atmospheric pressure are assumed to be small as compared to those of other parameters.

A number of authors make some other simplifying assumptions. However, even after the introduction of all the simplifications, the system of equations remains rather complex. Therefore the system is linearized by omitting the small terms containing the product of two or more acoustic additions, or the product of an acoustic addition and the square or a higher power of fluctuation additions.

3. The linearized system is reduced to one equation for any parameter of sound wave, e.g., for the acoustic pressure p, or for the potential $\Pi = p/\rho_0$ (where ρ_0

---

[2] However, in this case, the entropy in a given point of space changes due to turbulent fluctuations. Note here that an unambiguous classification of medium motions into turbulent and wave is a complex problem and requires, in particular, an accurate treatment of spatial velocity derivatives. The excellent descriptions of the physical nature of the problem are contained in [8, 22].

12 Atmospheric Acoustics and Instrumentation

is the mean air density), or for the oscillation velocity ξ. This equation can be written in the form:

$$\Delta p + k^2 p = G(p, n') \tag{3.1}$$

where $k = \omega/c$ is the wave number, ω is the angular sound frequency, c is the mean phase sound velocity, G is some function of p, and n' are the fluctuations in the index of acoustic refraction (see 2.4), that depend on turbulent fluctuations in temperature, density, and wind velocity.

4. To solve an equation of the (3.1) type, is assumed that:

$$p = p_0 + p_1 \tag{3.2}$$

where p_0 satisfies the homogeneous wave equation: $\Delta p_0 + k^2 p_0 = 0$, and p_1 is the singly scattered field. To calculate the first approximation of p_1, it is necessary to replace the value of p_1 by the value of p_0 in the right-hand part of the equation which will be the result of substitution of (3.2) into (3.1).

The solution of p_1 corresponding to divergent waves at a great distance $\mathbf{r}$ from the centre of the volume, as compared to its linear dimensions, is found in the form of the retarded potential:

$$p_1(\mathbf{r}) = -\frac{1}{4\pi} \cdot \int_V G(p_0, n', \mathbf{r}_1) \cdot \frac{e^{ik|\mathbf{r}-\mathbf{r}_1|}}{|\mathbf{r}-\mathbf{r}_1|} d^3\mathbf{r}_1 \tag{3.3}$$

where $\mathbf{r}_1$ is the alternating vector that runs through the scattering volume V (Fig. 4). Then the mean flux S of scattered energy is calculated as:

$$S = \langle p_1 p_1^* \rangle / 2\rho_0 c_0 \tag{3.4}$$

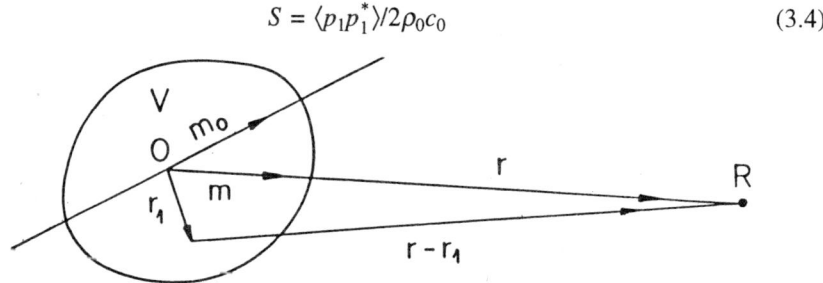

Fig. 4. Co-ordinates for the determination of scattering.

When calculating the mean product $\langle p_1 p_1^* \rangle$ from (3.4), in the right-hand part of the obtained equation under the integral we have the product $\langle G(\mathbf{r}_1) G^*(\mathbf{r}_2) \rangle$ that is just a linear combination of correlation functions for the fluctuations of the atmospheric parameters $B_f(\mathbf{r}_1, \mathbf{r}_2)$. In the case of uniform isotropic fields, B_f depends on $|\mathbf{r}_1 - \mathbf{r}_2|$ only.

5. To apply the solution to atmospheric turbulence that is (according to the Kolmogorov model) a locally isotropic and locally homogeneous medium where correlation functions B_f are not determined, one usually turns to spectral functions.

The three-dimensional functions $\Phi_T(\kappa)$ and $\Phi_v(\kappa)$ for temperature and wind velocity fluctuations can respectively be presented in the form:[3]

$$\Phi_T(\kappa) = \frac{1}{8\pi^3} \int\!\!\!\int\!\!\!\int_{-\infty}^{\infty} B_T(\mathbf{r}) \cos \kappa \mathbf{r} d^3\mathbf{r} \qquad (3.5)$$

$$\Phi_v(\kappa) = \frac{1}{8\pi^3} \int\!\!\!\int\!\!\!\int_{-\infty}^{\infty} B_v(\mathbf{r}) \cos \kappa \mathbf{r} d^3\mathbf{r} \qquad (3.6)$$

In the case of the problem of scattering, the argument of spectral functions κ is the so-called scattering vector **K** that is the difference between the wave vectors of incident and scattered waves. As shown in Fig. 5:

$$|\mathbf{K}| = k|\mathbf{m}_0 - \mathbf{m}| = 2k \sin \frac{\theta}{2} \qquad (3.7)$$

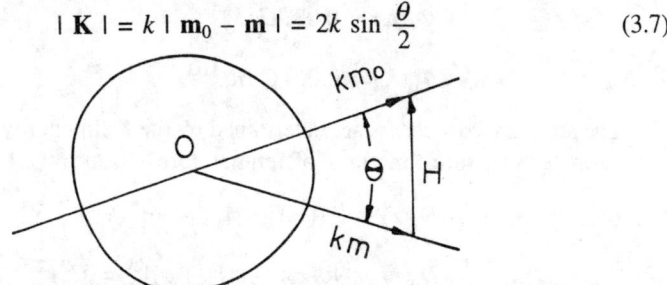

Fig. 5. On calculation of scattering vector **K**. $\mathbf{m}_0$ and $\mathbf{m}$ are the unit vectors for both incident and scattered wave directions; θ is the scattering angle.

6. The effective scattering cross-section σ from a unit volume into a unit solid angle in the θ direction is the ratio between the mean flux density of scattered energy S and the incident energy flux density S_0 in the volume:

$$\sigma(\theta) = S(\theta)r^2/S_0 V \qquad (3.8)$$

3.2 Effective Scattering Cross-Section

The presented way of solving the problem is simple enough. However, when deducing the wave equation for nonhomogeneous medium and solving it, one has to make a number of simplifying assumptions whose allowances cannot be always estimated in advance. Moreover, there are different understandings of statistical characteristic of turbulence. Therefore the results of different authors noticeably differ and the theory of sound scattering is still being refined. Let us dwell upon the best known theoretical results.

---

[3] A very simplified presentation is given here. In fact, the wind field is vector and it is described by the correlation tensor B_{ik} whose Fourier transform is the spectral tensor Φ_{ik}. For the case of incompressible (for turbulent motions) medium, $\nabla v = 0$, the tensor Φ_{ik} can be expressed through the scalar three-dimensional spectral density $\Phi_v(\kappa)$ (see [34] for details).

14 Atmospheric Acoustics and Instrumentation

The solutions of the wave equation for scattering from wind velocity inhomogeneities only were obtained in [6, 8, 13, 14]; temperature inhomogeneities only were under consideration in [12]; Batchelor [15] obtained the solution for both the types of scatterers (wind and temperature). These solutions were expressed through spectral densities of fluctuations. Blokhintsev [8] showed that, for an audible sound, the scales of turbulent inhomogeneities $l_t = 2\pi/K$ for a wide range of θ lie within the inertial interval of turbulence $l_0 < l_t < L_0$ and that, to calculate a scattering crosssection, one can use the Kolmogorov spectrum of locally isotropic and locally homogeneous turbulence.

Tatarskii [9] was the first who took this opportunity. He expressed the spectral densities (3.5), (3.6) through structure parameters of temperature C_T^2 and wind velocity C_v^2:

$$\Phi_T(K) = 0.033\, C_T^2 K^{-11/3} \qquad (3.9)$$

$$\Phi_v(K) = 0.061\, C_v^2 K^{-11/3} \qquad (3.10)$$

The structure parameters are determined by the Kolmogorov-Obukhov 2/3 power law for the structure functions of temperature D_T and wind velocity D_v:

$$D_T(r) \equiv \langle [T(r_1) - T(r_1 + r)]^2 \rangle = C_T^2 r^{2/3} \qquad (3.11)$$

$$D_v(r) \equiv \langle [v(r_1) - v(r_1 + r)]^2 \rangle = C_v^2 r^{2/3} \qquad (3.12)$$

The structure parameter values are accessible to *in situ* measurements.

Note that in this book Tatarskii, following Obukhov, proceeded from the acoustics equation by Andreev and Rusakov [35] which was deduced when wind velocity was assumed to be noneddy (rot $\mathbf{v} = 0$) and the entropy of the medium was assumed to be constant ($E(\mathbf{r}) = 0$). As a result, in the formula for scattering cross-section the factor $\cos^2\theta$ was omitted, however the latter has been presented earlier in [13, 15]. After revealing the discrepancies between the theory and the experimental results for the large scattering angles [17, 18], this formula was refined in [36, 11, 10] and took on the form that is used in most of the modern papers on acoustic sounding:

$$\sigma(\theta) = 2\pi \cdot k^4 \cos^2\theta \left[\frac{\Phi_T\left(2k \sin\frac{\theta}{2}\right)}{4T_0^2} + \frac{\Phi_v\left(2k \sin\frac{\theta}{2}\right)}{c_0^2} \cos^2\frac{\theta}{2} \right] \qquad (3.13)$$

or after substitution of both (3.9) and (3.10):

$$\sigma(\theta) = 0.03 k^{1/3} \left(\sin\frac{\theta}{2}\right)^{-11/3} \cos^2\theta \left[0.54 \frac{C_T^2}{4T_0^2} + \frac{C_v^2}{c_0^2} \cos^2\frac{\theta}{2} \right] \qquad (3.14)$$

3.3 Consideration of Humidity Fluctuations

Note that all the authors mentioned above, did not take into account scattering from humidity fluctuations. The density of humid air is the sum of densities of dry air ρ_d and water vapour ρ_w:

$$\rho = \rho_d + \rho_w = \frac{P\mu_d}{RT}(1 - 0.61q) \qquad (3.15)$$

where μ_d is the molecular weight of dry air, q is the specific humidity that is approximately equal to the concentration of water vapour in the air, ρ_w/ρ_d:

$$q \equiv \rho_w/\rho \approx \rho_w/\rho_d \qquad (3.16)$$

For sound speed in a motionless air, instead of (2.2), the following formula with account of (3.15) is valid:

$$c = \left(\gamma \frac{RT}{\mu_d}\right)^{1/2}(1 + 0.225q) \qquad (3.17)$$

Fluctuations in the refraction index and density in a humid air are equal:

$$n' = \frac{T'}{2T_0} + \frac{0.45}{2}q' \qquad (3.18)$$

$$\frac{\rho'}{\rho} = -\frac{T'}{T_0} + 1.66q' \qquad (3.19)$$

An attempt to take into consideration the effect of humidity in a final formula was made by Wesely [37] who replaced the term $C_T^2/(4T_0^2)$ in square brackets of (3.14) by the structure parameter of the refraction index C_n^2 following from (3.18):

$$C_n^2 = \frac{C_T^2}{4T_0^2} + \frac{0.45}{4T_0}C_{Tq} + \frac{(0.45)^2}{4}C_q^2 \qquad (3.20)$$

where C_q^2 is the structure functions of humidity determined similarly to (3.11) and C_{Tq} is the mutual structure function of temperature and humidity:

$$C_{Tq} = r^{-2/3}\langle[T(r_1) - T(r_1 + r)] \cdot [q(r_1) - q(r_1 + r)]\rangle \qquad (3.21)$$

The value of C_{Tq} can be presented in the form $C_{Tq} = R_{Tq}(C_T^2 \cdot C_q^2)^{1/2}$ where R_{Tq} is the mutual correlation coefficient of temperature and humidity differences at two points, that can be either positive or negative according to the atmospheric stratification (see [38, 39]). In this case it is assumed (based on the results of measurements [40, 41]), that the spectral density is related to the structure parameter C_q^2 through the formula analogous to (3.9).

However, it should be noted that such a method of taking into account humidity is incorrect. Referring to [8], in case of a medium with a complex composition

16 Atmospheric Acoustics and Instrumentation

(e.g., sea water with a varying salinity or humid air), an additional term, which depends on the derivative of density fluctuations, appears in the right-hand side of the wave equation (3.1). The equation for a medium of a complex composition was solved by Chernov [28]. However, he expressed the scattered energy through the correlation functions of turbulent fluctuations (both exponential and Gaussian) that unsatisfactorily described the atmospheric turbulence. The accurate calculations of a contribution of humidity fluctuations into the scattering cross-section were made by Ostashev [33, 42] who, following [8] and [28] obtained the wave equation for a turbulent medium of complex composition. Ostashev's solution for scattering cross-section has the form:

$$\sigma(\theta) = 0.03 k^{1/3} \left(\sin \frac{\theta}{2}\right)^{-11/3} \left\{ 0.136 \left[\frac{C_T^2}{T_0^2} \cos^2\theta \right. \right.$$

$$\left. + 2\left(0.45 - 1.22 \sin^2 \frac{\theta}{2}\right) \frac{C_{qT}}{T_0} \cos\theta + \left(0.5 - 1.22 \sin^2 \frac{\theta}{2}\right)^2 C_q^2 \right]$$

$$\left. + \frac{C_v^2}{c_0^2} \cos^2 \frac{\theta}{2} \cos^2\theta \right\} \tag{3.22}$$

The peculiarities of angular dependence of scattering for different scatterers, noted in section 2.1, follow exactly from this formula. Fig. 6 gives the scattering indicatrixes calculated from (3.22). It is seen that only humidity fluctuations scatter sound at an angle of 90°. Proceeding from this, a possibility of remote measuring C_q^2 was suggested in [33].[4]

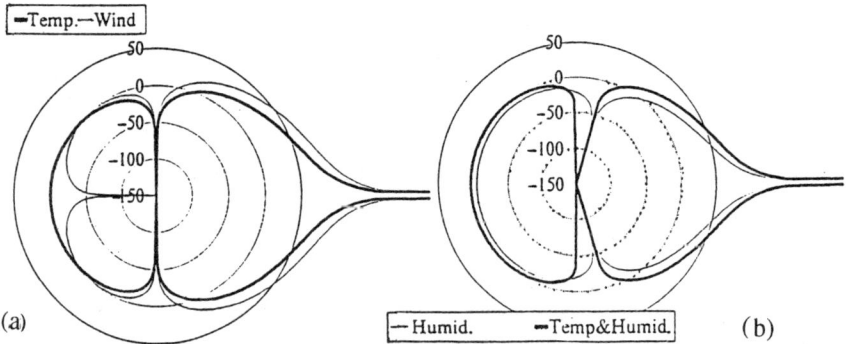

Fig. 6. Indicatrtixes of scattering (in the polar co-ordinates) by inhomogeneities of temperature and wind velocity (a), and humidity (b); in (b) the contribution of the term depending on the correlation between the fluctuation of temperature and humidity in (3.22) is also shown. The number near circles are in dB.

---

[4]Note that pressure fluctuations which were neglected in deducing the wave equation can also contribute into scattering at an angle of 90° (see [43]).

Let us present here the estimates of structure characteristics entering (3.22). The values $0.136\, C_T^2/T_0^2$ and C_v^2/c_0^2 for a height of 100–200 m above the underlying surface in the middle latitudes on a summer day are within $3 \cdot 10^{-11}$–$9 \cdot 10^{-8}$ m$^{-2/3}$ and 10^{-9}–$2 \cdot 10^{-6}$ m$^{-2/3}$, respectively. Under other meteorological conditions the value C_v^2/c_0^2 is also, as a rule, greater than C_T^2/T_0^2. Therefore at all the scattering angles, except for the region near $\theta = 180°$, the contribution of wind velocity fluctuations is usually greater than that of temperature fluctuations.

The terms for humidity in (3.22) are equal to $0.8 R_{qT} C_q C_T$ and $0.6 C_q^2$ at $\theta = 180°$. Over land they usually are lower by 1 and 2 orders respectively than the term for temperature at back scattering [30]. However, over warm ocean and also over lakes in sultry summer the terms of humidity can be of the same order as that of temperature; the sign of the correlation coefficient R_{qT} for these two types of the underlying surface is different. Humidity inhomogeneities can also essentially contribute to scattering over wet soil after rain on a hot day.

3.4 Consideration of Anisotropic Turbulence
In recent papers [44, 45] scattering cross-section is calculated for anisotropic turbulence by using the spectrum of turbulence in the form:

$$\Phi_T(\kappa) = C(\kappa_v^2 + \eta^2 \kappa_h^2)^{-\mu} \tag{3.23}$$

where the parameter C determines the turbulence intensity, η characterizes anisotropy or flatness of inhomogeneities, μ is the power that may differ from the Kolmogorov one, the wave number κ_v refers to the vertical direction and κ_h to the horizontal one. The anisotropy coefficient $\eta > 1$ corresponds to inhomogeneities stretched in the horizontal direction. In this case intensity of back scattering decreases when a sound beam is out of the vertical, and, when the symmetric bistatic scheme of sounding is used, the intensity of a received signal can perceptibly exceed the value calculated from (3.22).

4. Experimental Verification of the Theory
The theoretical results obtained by the late 1950's required an experimental verification. First, the very mechanism of generating an acoustic echo due to scattering by turbulence needed corroboration. It was apparent that this mechanism, rather than reflection from the gradients of mean temperature, gave every reason to take the remote measurements of wind velocity by the Doppler effect. Second, the formulas of scattering cross-sections obtained with approximated methods using a great number of simplifying assumptions needed a quantitative verification.

4.1 Field Experiment in the Atmospheric Surface Layer
The first purposeful experiments [17–19, 46] were carried out over a flat uniform steppe area on a summer day and night at the Tsimlyansk Field Station of the Institute of Atmospheric Physics. They were accompanied by independent

determination of the structure parameters C_T^2 and C_v^2 from the measurements of gradients of the corresponding parameters from a 16-m mast.

To measure scattering indicatrix, the special reversible electrostatic transducers (MKI) placed on a turning mounting were designed (Fig. 7). The MKI had a beam width of $2\gamma_{0.5} = 3°$, and gave a pressure $p = 20$ Pa at the basic operating carrier frequency of 11 kHz. The symmetric bistatic scheme in pulse mode was used in the measurements. The distance between transmitter and receiver changed from 20 to 140 m. To measure the backscattering, the transmitter and the receiver were placed side by side and directed upwards.

Fig. 7. Photo of the MKI reversible electrostatic transducers.

The deflections from the symmetric scheme (Fig. 8) did not result in a decrease of echo-signal, which suggested that there were no wave reflections in the mirror direction.

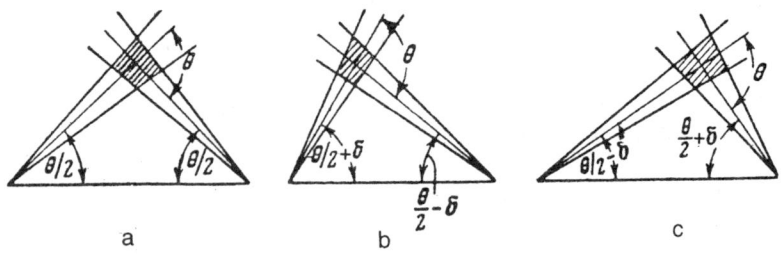

Fig. 8. Geometric scheme for revealing the mirror reflections.

Experimental angular dependence of scattering (Fig. 9) was in good agreement with the theoretical one for the wind term in formulas (3.14) and (3.22). Rather good agreement with the theory was obtained not only in the angular dependence but also in absolute values of scattering cross-section. The intensity of back scattering manifested good correlation with the values of C_T^2 and no correlation with the values of C_v^2. Thus it was confirmed that no contribution was made by wind field inhomogeneities into back scattering. The experiment also proved a weak frequency dependence of scattering cross-section in the frequency range 5 and 40 kHz (Fig. 10).

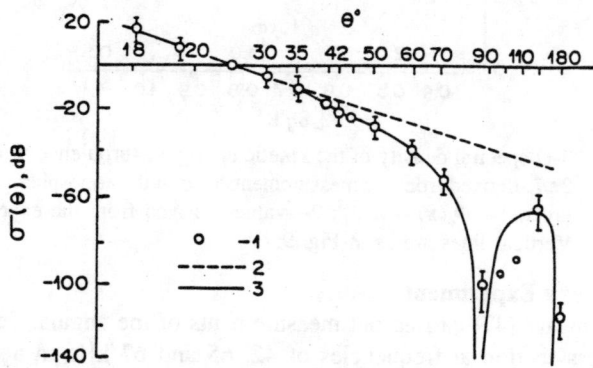

Fig. 9. Dependence of cross-section σ on the scattering angle θ. 1—experiment, vertical lines are the 5% confidence interval; 2—$\sigma \sim (\sin \theta/2)^{-11/3}$; 3—$\sigma \sim (\sin \theta/2)^{-11/3} \cdot (\cos \theta/2 \cdot \cos\theta)^2$.

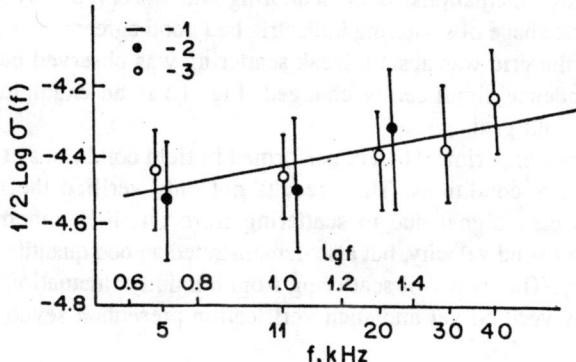

Fig. 10. Dependence of σ on the carrier frequency f. 1 – $\sigma \sim f^{1/3}$; 2—experiment 1960, 3—experiment, 1961. Vertical lines are as in Fig. 9.

The experimental data allowed the three-dimensional spectrum to be calculated using (3.14) and *in situ* measurements of C_T^2 and C_v^2. The slope of the spectrum in the scale range between 3.5 and 0.5 cm coincides with the theoretical slope $\Phi_v(\kappa) \sim \kappa^{-11/3}$ within the 5% confidence interval. (Fig. 11).

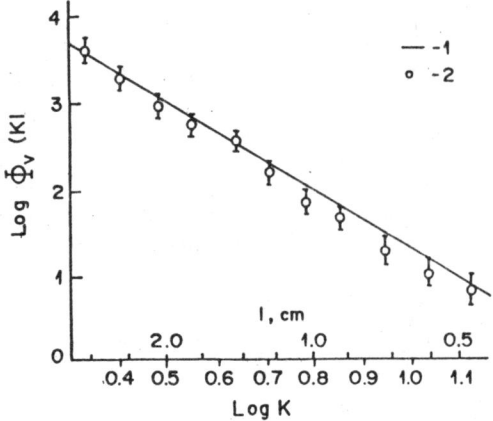

Fig. 11. 3-D spectral density of the kinetic energy of turbulence $\Phi_v(\kappa)$, $\kappa = 2\pi/l_t$, derived from the measurements of σ in the atmospheric surface layer. 1—$\Phi_v(\kappa) \sim \kappa^{-11/3}$; 2—values derived from the experiment. Vertical lines are as in Fig. 9.

4.2 Laboratory Experiment

Baerg and Schwarz [47] carried out measurements of the angular dependence of scattering cross-section at frequencies of 42, 65 and 67 kHz in a wind tunnel. Flow passed through a metal grid, the values of Reynolds number were Re = 4000–40000. When the grid was cold, scattering occurred from wind velocity fluctuations alone and when the grid was heated, scattering occurred from both wind velocity and temperature fluctuations. Back scattering was observed only when the grid was heated. The shape of scattering indicatrix had good agreement with the formula (3.13). When the grid was absent, weak scattering was observed but the shape of angular dependence significantly changed. Fig. 12 is an example of scattering indicatrix for cold grid.

The described experiments were performed in field conditions at low humidity and in laboratory conditions. Their results not only verified the mechanism of generating an echo-signal due to scattering from turbulent inhomogeneities of temperature and wind velocity, but also demonstrated a good quantitative agreement with the theory. The theory of scattering from humidity fluctuations has not been experimentally verified yet and such verification presents a severe problem.

5. Peculiarities and Limitations of Acoustic Sensing of the Atmospheric Boundary Layer

When sounding the atmospheric boundary layer up to the heights of several hundreds of meters, many additional factors, that were negligible in the special experiments, affect the echo-signal parameters. On its path to and back from a scattering volume, sound wave is absorbed by air molecules and additionally attenuated due to turbulence. Sound refraction related mainly with the mean wind velocity gradient

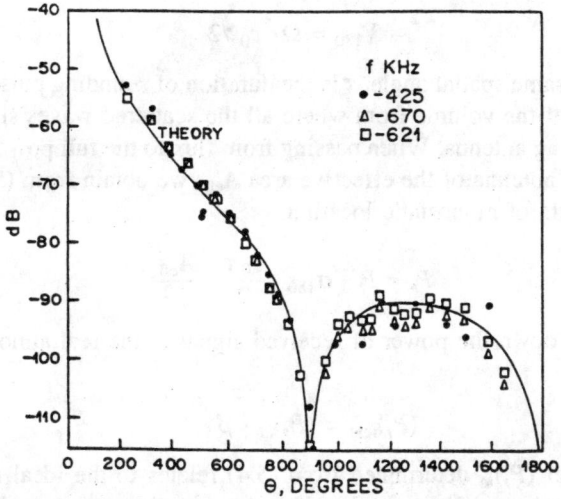

Fig. 12. Angle dependence of sound scattering in the wind tunnel [47].

provokes an inclination of the sound wave front and slight deviations of the value of scattering angle from that for a motionless medium. These deviations can be significant in the region of the scattering angles $\theta = 90°$ and $\theta = 180°$, where narrow gaps in the scattering indicatrix can be concealed. Both strong wind and wind refraction also result in errors in the Doppler measurements of the wind velocity components [33, 48]. Here we consider the influence of these factors on back scattering, a mode commonly used in monostatic sodars, and the limitations of the method.

5.1 Echo-Signal Intensity at Back Scattering in the ABL

It follows from (3.22) that the back scattering cross-section is equal to:

$$\sigma_{180} = 3.9 \cdot 10^{-3} k^{1/3} \frac{C_T^2}{T_0^2} \cdot (1 + b_q) \tag{5.1}$$

where
$$b_q = 1.54 T_0 C_{Tq}/C_T^2 + 0.59 \, T_0^2 C_q^2/C_T^2$$

Density of scattering energy flux on a receiving-transmitting antenna, as it follows from (3.8), is equal to:

$$S = \frac{\sigma_{180} V}{r^2} S_0 = \frac{\sigma_{180} V P_t}{r^2 \Omega r^2} \tag{5.2}$$

where r is the distance from the centre of scattering volume to the antenna and P_t the acoustic energy radiated by sodar to the spatial angle Ω, which is determined from the directional pattern of sodar.

The value of scattering volume V at back scattering is determined by the formula:

22 Atmospheric Acoustics and Instrumentation

$$V_{180} = \Omega r^2 c_0 \tau/2 \, , \tag{5.3}$$

where Ω is the same spatial angle, τ is the duration of sounding pulse, $c_0\tau/2$ is the vertical extent of the volume from where all the scattered waves simultaneously arrive to receiving antenna. When passing from flux to the full power P_r of signal received by the antenna of the effective area A_{eff}, we obtain from (5.2) and (5.3) the basic formula of monostatic location:

$$P_r = P_t \cdot \sigma_{180} \cdot \frac{c_0 \tau}{2} \cdot \frac{A_{\text{eff}}}{r^2} \tag{5.4}$$

Let us write down the power of received signal in the real atmosphere in the form:

$$(P_r)_{\text{real}} = (P_r)_{id} \cdot \beta \tag{5.5}$$

where the power $(P_r)_{id}$ determined from (5.4) relates to the ideal nonabsorbing motionless atmosphere turbulized only within the scattering volume and the multiplier β is the product of the coefficients describing the effect of different factors:

$$\beta = \beta_1 \beta_2 \beta_3 \beta_4 \tag{5.6}$$

The coefficient β_1 take into account the molecular absorption of sound on its paths form antenna to the scattering volume and back:

$$\beta_1 = e^{-4\alpha \cdot r} \tag{5.7}$$

where α is the absorption coefficient (related to the amplitude of sound pressure) in inverse meters.

Table 1 gives the distances up to the scattering volume, $r = 1/(4\alpha)$, where a received signal is attenuated due to molecular absorption by a factor of e (for standard meteorological conditions: $T = 293$ K, $P = 1013$ mb, $h_r = 70\%$ ($q = 10^{-2}$ g/g); note that the value α markedly varies according to temperature and humidity).

Table 1. The effect of molecular absorption on the range of sounding

Carrier frequency f, Hz	500	1000	1600	2000	4000	6300
$r = 1/(4\alpha)$, m	800	400	280	230	100	45

In practice, α is usually calculated from the surface values of pressure, temperature and humidity using the American National Standard [49].

The coefficient β_2 describes an additional signal attenuation caused by turbulence. Due to distortion of wave front, sound beam generated by sodar antenna is widened and a part of energy, scattered in the volume so widened, fails to fall within the receiving antenna beam. Moreover, if the radius of wave coherence is smaller than the diameter of receiving antenna $2a_s$, then the waves scattered by every element of the volume will be combined incoherently in the receiving antenna.

These two effects will result in an additional attenuation of received signal. According to [50], the coefficient β_2 can be written in the form:

$$\beta_2 = (1 + 4a_s^2/\rho_c^2)^{-2} \tag{5.8}$$

where the radius of coherence $\rho_c(r)$, with account of the wave path to and back from the scattering volume, is:

$$\rho_c = \{k^2 \int C_n^2(s)\,[(1-s/2r)^{5/3} + (s/2r)^{-5/3}]\,ds\}^{-3/5} \tag{5.9}$$

At mean values typical of the refractive index structure parameter C_n^2 (including wind velocity fluctuations) in the layer of sounding, $C_n^2 = C_T^2/4T_0^2 + C_v^2/c_0^2 > 10^{-7}\text{ m}^{-2/3}$, and at typical sodar antenna diameters $2a_s \leq 1.5$ m, the value $2a_s/\rho_c < 1$. This results in $\beta_2 \geq 0.5$.

The third cofactor β_3 takes into account the attenuation of received signal caused by a turn of sound wave front due to refraction. The fact is that the difference between the group and phase speeds of sound in a moving medium results in lack of coincidence between the direction of a tangent to the trajectory of sound beam and the normal to wave front (therefore sound beam in a moving medium is irreversible). According to [51], under the linear profile of the mean wind velocity $v(z) = A + Bz$,

$$\beta_3 = \exp\{-2(\langle M\rangle/\gamma_{0.5})^2\} \tag{5.10}$$

where $\langle M\rangle = \langle v\rangle/c_0$ is the mean value of the Mach number for a layer sounded, $\gamma_{0.5}$ is the half width of sodar beam at 0.5 level.

At narrow directional patterns $\gamma_{0.5} < 0.05$ ($2\gamma_{0.5} < 6°$) and at strong wind $\langle M\rangle \geq 0.03$, the cofactor β_3 can halve the power of received signal in comparison to that for calm conditions. However, at the values $2\gamma_{0.5} \approx 10\text{–}12°$ usually used in sodars, this cofactor gives no more than a 10% attenuation.

The fourth cofactor $\beta_4 > 1$ describes a possible contribution of wind field inhomogeneities into the back scattering due to deflection $\delta\theta$ of scattering angle from 180° in response to refraction:

$$\beta_4 = \left(1 + \frac{C_v^2 T_0^2}{0.136 C_T^2 c_0^2}\sin^2\frac{\delta\theta}{2}\right) \tag{5.11}$$

where the value $\delta\theta$, as shown in [50], has the order $\delta\theta \approx 2\langle M\rangle$. In a rather strong wind, $\langle M\rangle \geq 0.02$, and at large ratio between the structure parameters of temperature and wind velocity fluctuations, $(C_v^2/c_0^2)/(C_T^2/T_0^2) \geq 100$, which is typical of the near-neutral atmospheric stratification, the contribution of wind inhomogeneities can reach 30%.

At vertical monostatic sounding some increase of scattered signal may be due to turbulence anisotropy in the ABL if the anisotropy coefficient η in (3.23) is greater than 1. As shown in papers [52, 44, 45], in case of axisymmetric anisotropic inhomogeneities stretched in the horizontal direction with $\eta > 10$, scattering intensity can exceed that calculated from the Kolmogorov spectrum by a factor of several tens. This effect is observed when radiolocating both the troposphere and the lower stratosphere where the value of η can reach a few tens. However, in the atmospheric boundary layer, the experiments [53–55] on the study of angular sensitivity of back scattering have demonstrated, $\eta \cong 1.5$ as the maximal observed value even in the presence of stratified turbulized structures. Such values on η may result merely in a 30% increase of scattering cross-section.

5.2 Limitations of the Method of Acoustic Sensing and Sodar Sensitivity

Acoustic sounding can be conducted during day- and night-time in a wide range of air temperatures and weather conditions. However, it cannot be employed under extreme conditions, such as storms and heavy shower. Monostatic sodar gives echo-signal only if the ABL is thermally stratified. Fortunately, the cases of the neutral stratification, when the temperature gradient is adiabatic through the whole ABL, are very rear.

More severe limitations are, however, present in sounding range and sodar sensitivity. This is due to ambient acoustic noises. Even in quiet rural locations the intensity of acoustic noise P_{Na} caused by wind blowing over the sodar noise-proof cuff is usually considerably greater than the intensity of electronic noises P_{Ne} of sodar receiving amplifier $P_{Ne} = K_B T_0 \Delta f$ (K_B is the Boltzmann constant, Δf is the receiver frequency pass band). So the sodar potential $\Pi_s = 10 \lg P_t/P_N$ is not a constant value for a given sodar and can change significantly during 24 hours according to wind velocity, traffic intensity, birds' songs, insects shirring, etc. Under given Π_s the sensitivity of sodar decreases with height and depends on the mean air temperature and relative humidity h_r. Fig. 13 gives an

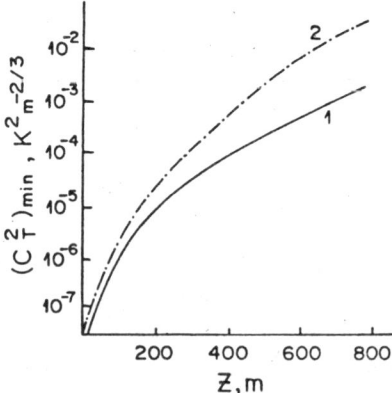

Fig. 13. The sensitivity $(C_T^2)_{min}$ of the vertical monostatic sodar against height at $\Pi_s = 165$ dB. $f = 2$ kHz. $1 - \alpha = 2.5 \cdot 10^{-3}$ m^{-1} (either for $T_0 = 20°C$, $h_r = 20\%$ or for $T_0 = 0°C$, $h_r = 50\%$); $2 - \alpha = 1.3 \cdot 10^{-3}$ m^{-1} ($T_0 = 20°C$, $h_r = 50\%$).

example of such dependence under two different values of the molecular absorption index α.

The usual maximal range of acoustic sounding is about one kilometre, although most of the time it is lower. There are few possibilities to increase the upper range of this remote method. First, as it follows from Table 1, a decrease in sounding frequency reduces the effect of molecular absorption. However, the spectrum of intensity of acoustic noise F_{Na} has, as a rule, the form of power function [31]:

$$F_{Na}(f)/F_{Na}(f_1) = (f/f_1)^{-a} \qquad (5.12)$$

where $f_1 = 1000$ Hz is the reference frequency and the value of 'a' can range between 0.6 and 4.

Figure 14(a) presents examples of noise spectra F_{Na} from [55–59]. Fig. 14(b) gives the values for $F_{Na}(f_1)$ and a according to the data by different authors summarized in the book [31]. Due to the relationship (5.12), a decrease in sounding frequency below 1 kHz results in a increase in sounding height only at a very low level of acoustic noise. At high-level noise, e.g. in cities, increase of the carrier frequency up to 2–3 kHz proves to be more cost-effective. Note also that lower

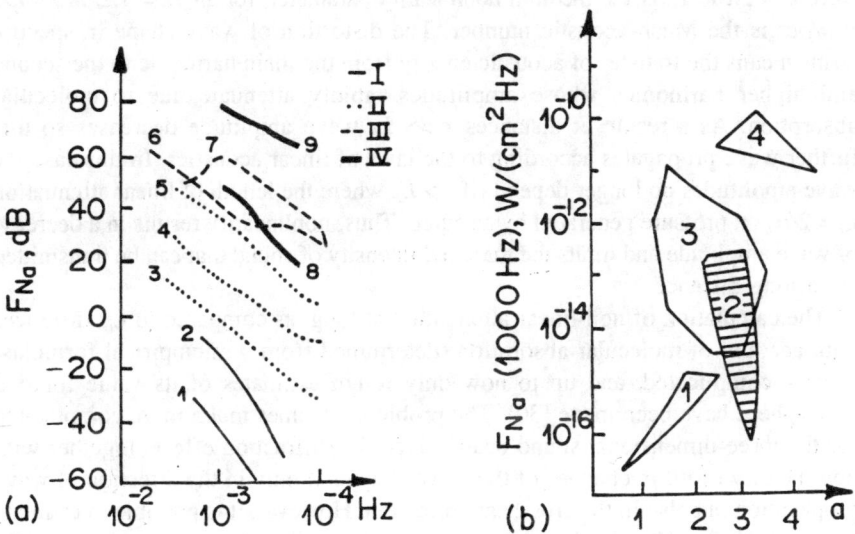

Fig. 14. (a) Spectra of the ambient acoustic noise. I—noise of wind on the sodar cuff [55]: 1—$v = 1$ m/s, 2—$v = 2$ m/s; II-urban noise [56, 57]: 3, 4—residential districts at night and at day, 5—downtown 30 m apart from a route; III-transport noise [58, 59]: 6—15 m apart from highway, 7—heavy lorries; IV-take-off of a jet plane [59]: 8—from its side at a distance of 1800 m, 9—60 m apart from its front. (b) The parameters of the noise model (5.12). 1—rural area; 2—urban condition; 3—near highways; 4—near airport (after [31]).

frequencies require bigger antenna apertures and result in deterioration of sodar spatial resolution and in increase of its "dead zone". Therefore the attempts to use low frequencies (see e.g., [60, 61]) have found no frequent applications.

The sounding range can be somewhat extended if longer sounding pulses are used. However, this way gives a small effect and impairs resolution.

Increasing the sodar acoustic power P_t seems to be the most attractive way to make the range of sounding greater. Thus in [63], a matrix of loudspeakers generating pressure up to 700 Pa at a frequency of 1000 Hz was used in the radioacoustic sounding system (RASS). However, changing P_t from 350 to 1700 W, no increase in the range of sounding was observed. The fact is that such an increase in the pressure amplitude triggers the mechanism of non-linear attenuation.

The physical mechanism of non-linear attenuation consists in the following [64]. As a wave propagates, its originally sinusoidal shape is distorted and it becomes saw-shaped at some distance L_b. The distance of discontinuity formation L_b for plane wave is given as:

$$L_b = (k\varepsilon_0 M_a)^{-1} \tag{5.13}$$

were $k = 2\pi/\lambda$, ε_0 is the medium nonlinearity parameter, for air $\varepsilon_0 = 1.2$, $M_a = \xi/c = p/\rho c^2$ is the Mach acoustic number. The distortion of wave shape in spectral terms means the transfer of acoustic energy from the main harmonic to the second and higher harmonics whose amplitudes rapidly attenuate due to molecular absorption. As a result, at distances $r \gg L_b$ wave amplitude decreases so that further wave propagates according to the laws of linear acoustics. In this case the wave amplitudes no longer depend (if $r > L_l$, where the length of linear attenuation $L_l = 2/\alpha$) on pressure generated by a source. Thus, nonlinearity results in a decrease of wave amplitude and limits the maximal intensity of sound that can be transmitted for a long distance.

The calculation of non-linear attenuation at long, as compared to L_b, distances with account of molecular absorption (determined from semiempirical formulas) is very complicated, and up to now only rough estimates of its value for real atmosphere have been made [30]. The problem becomes much more complicated for the three-dimensional sound beam, since the diffraction effects together with nonlinearity result in changes of the wave shape not only in the direction of wave propagation but also in the crossbeam direction. However, it is possible to estimate the border value $(P_t)_{\max}$ whose excess results in a marked non-linear attenuation. Since this latter in a diverging spherical wave is considerably weaker than in a plane wave, the criterion for the degree of non-linear attenuation is the parameter $N = L_d/L_b$, where $L_d = ka_s^2$ is the diffraction length. At $N > 1$ the break in acoustic pressure occurs in the near-field zone of a transmitter antenna and the effect of nonlinearity is very strong, and at $N < 1$ wave becomes spherical before the break. As shown in [65], an increase in power of a transmitter is effective only up to the value $N \approx 1$. In this case, sodar operates in the mode of non-linear saturation, which does not prevent the Doppler measurements of wind velocity and the obtaining of a qualitative pattern of the space-time distribution of strongly turbulized regions.

For quantitative measurements of the turbulence structure parameters, according to [65], it is necessary to have $N \leq 0.4$. If the value of N is higher than 0.4, the non-linear attenuation causes significant errors in scattering cross-section measurements in the upper part of the sounding range.

For sodar with the carrier frequency $f_0 = 2000$ Hz and the antenna diameter $2a_s = 1.3$ m, the value $N = 1$ corresponds to the acoustic power $P_t = p^2/(2\rho c) \cong 30$ W of a transmitter in its main lobe or to the electric power of about 500 W (at the transfer coefficient of the order of 6%). The value $N \cong 0.4$ for the same sodar parameters corresponds to the acoustic power of the order of 6 W (in the main lobe) or to the equivalent electric power of about 100 W.

6. Verification of the Method for Acoustic Sensing of the ABL

Acoustic sounding is, in principle, an absolute method, i.e. any reference to other contact measurements is unnecessary. The Doppler measurements of wind velocity need no special calibration. To determine the structure parameters of turbulence electroacoustic calibration of sodar in both transmission and reception modes is required, which can be performed in an anechoic chamber [66].

However, the effects of a number of geophysical factors on the results of measurements, corrections for which are most often difficult to introduce, naturally call for comparison between the results of synchronous remote and *in situ* measurements.

It is impossible to take simultaneously remote and contact measurements in the same air volume. Besides, the data of remote measurements are averaged over a rather big volume, while the *in situ* data are obtained practically at one point. So a perfect coincidence should not be expected for these two kinds of measurements of the ABL parameters that are characterized by a great natural variability. Therefore such comparisons are usually made with considerable averaging time (20–40 min) and for great bodies of data.

6.1 Wind Measurements

The best agreement was obtained for the measurements of the mean horizontal wind velocity and wind direction and for the vertical velocity variances. Figs. 15 and 16 give some examples of such comparisons made at the A.M. Obukhov Institute of Atmospheric Physics, Russian Academy of Sciences [76, 68].

Thorough comparisons of wind data by four commercial sodars with *in situ* measurements by anemometers mounted on the 300 m mast were made in the USA [69]. The measurements were taken within a wide range of wind velocities, from 0.5 to 18 m/s. The correlation between the sodar and *in situ* values of the wind velocity ranges from 0.90 to 0.97 and that of the vertical velocity variances ranges from 0.85 to 0.89 for different sodar units. However, these experiments

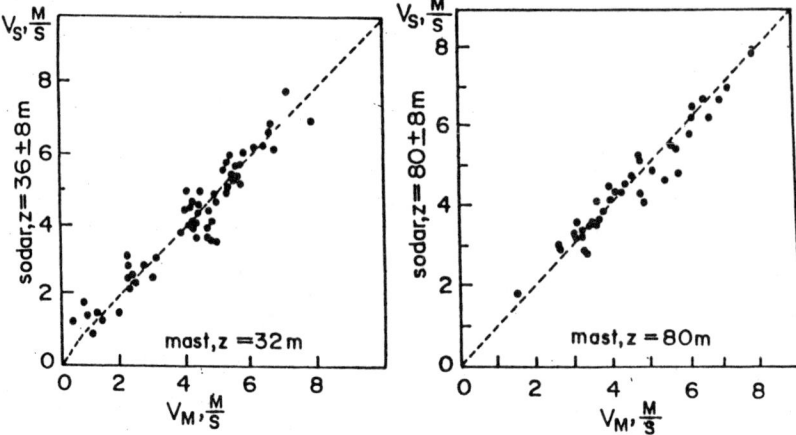

Fig. 15. Comparison of sodar and *in situ* measurements of mean wind [67, 68].

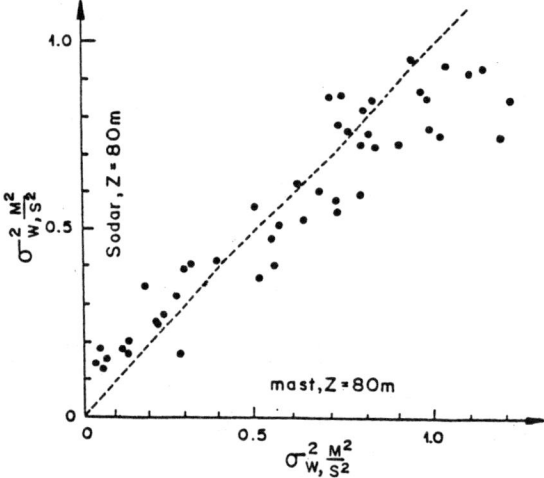

Fig. 16. Comparison of sodar and *in situ* data on standard deviations of the vertical velocity σ_w^2 [68].

demonstrated that the values of root-mean-square deviation of the horizontal velocity components σ_v and σ_u at heights of 100–200 m obtained by different sodar units poorly correlate between themselves as well as with the values obtained from the mast measurements [70]. The reasons for these discrepancies are yet to be explained; the inherent peculiarity of such parameters as σ_v and σ_u could be one of the sources of poor correlation. Therefore before judging the potentialities of sodar to measure these parameters it would be useful to compare these values measured by anemometers at the points spaced at the same distances (of the order of 250–300 m).

6.2 Measurements of Temperature Structure Parameter

The comparisons of the C_T^2 values measured by vertical sodars and microthermometers, that were performed in different places, yielded conflicting results.

Figure 17 presents the scatter plot of sodar and aircraft measurements over a uniform steppe surface [71]. The correlation coefficient is close to 0.9 and the root-mean-square deviation of the value $lg[(C_t^2)_{sod}/(C_t^2)_{in\ situ}]$ is about 0.1. Good agreement was also obtained in the papers [72, 73]. A number of other authors [74–76] found systematic discrepancies: $M_s \equiv (C_t^2)_{sod}/(C_T^2)_{in\ situ} \leq 1$ under developed convection conditions, but $M_s > 2$ under inversions. The reason for such a contradiction is, apparently, the difference in conditions of the measurements, that results either in a difference of the values of factors $\beta_2 - \beta_4$, or in appreciable contribution of humidity fluctuations into back scattering (see section 5.1). However, taking into account a very wide range of variability in the C_T^2 values within the ABL, that can reach a few orders of magnitude, sodar measurements of C_T^2 with an accuracy of factor 2 could be very useful, e.g. in studying the disturbance of optical wave propagation in the atmosphere [77].

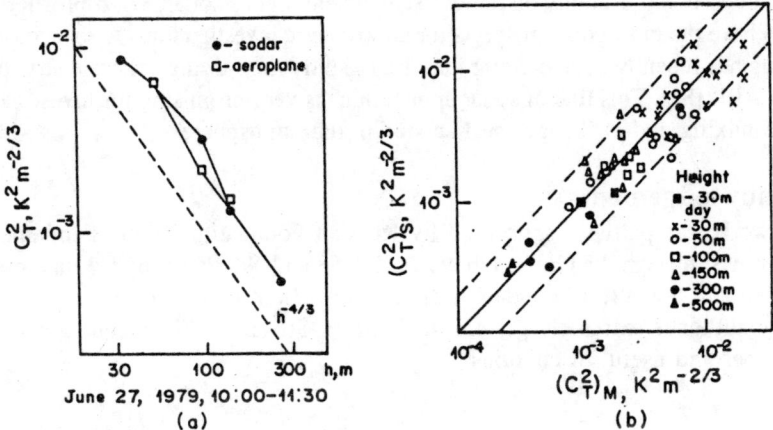

Fig. 17. Comparison of sodar and *in situ* (mast, aeroplane) C_T^2 measurements [71].

It should be noted that very little published information on the comparisons of *in situ* and bistatic sodar measurements of C_v^2 is available.

7. Conclusion

By now acoustic sensing is considered as a well evaluated advantageous technique for a number of the ABL investigations. The chief value of acoustic location is in an easy interpretation of the data obtained, which is based on the physically clear and well evaluated theory of sound scattering in a turbulent medium.

The sodars are a very useful tool in studying the dynamics of the ABL, and especially in the regions difficult of access. Sodars are also very important for providing a continuous monitoring of the ABL parameters for ecological purposes. However, as shown in the review [78], much work is to be done on adapting (or developing new) dispersion models of the ABL for incorporating sodar data.

At present acoustic sensing is used widely enough. However, the development of the method has not yet been completed. There are prospects for technical perfection of sodar design as well as for obtaining with its aid more vast information than we have now. Further development of the methods is required for obtaining the profiles of kinetic energy dissipation rate $\varepsilon = 0.4(C_v^2)^{3/2}$ as well as the development of the procedures for determining humidity profiles, precipitation intensity, turbulent fluxes, etc. The methods of automated identification of the stratification type and the mixing layer height from a space-time pattern of echo-signal intensity [79] also need an improving. Further, purposeful field experiment has been carried out in 1995 [80] to check the formula (3.22). The analyses of the data obtained revealed no evidence of impact of humidity fluctuations on the parameters of sound scattering in the vicinity of $\theta = 90°$.

The combined systems of remote sensing involving sodar, wind-profiler and RASS have the most great prospects for atmospheric investigations. Also, acoustical sensing has recently been coming into use to study long-living coherent structures in the ABL [81]. This line of sodar applications is very promising for investigation of the mixing and exchange mechanisms in the atmosphere.

Acknowledgements

This work was partially sponsored by Russian Foundation of the Fundamental Researches through the Grants No. 93-05-08753 and No. 96-05-65741, and by the International Scientific Foundation through the Grant MAC300.

The author is extremely grateful to Dr. S.P. Signal for his encouragement of this paper and useful discussions.

References

1. G.W. Gilman, H.B. Coxhead and F.H. Willis, "Reflection of sound signals in the troposphere", J. Acoust. Soc. Amers., 18, 274–283, 1946.
2. J. Tyndall, "On the atmosphere as a vehicle of sound", Phil. Trans. Roy. Soc. London, Ser. A, 164, 183–244, 1874.
3. Rayleigh (J.W. Strutt), "The Theory of Sound", Macmillan, New York, 1877. (Amer. ed., Dover, New York, 1945).
4. H. Sieg, "Uber die Schallaus breitung im Freien und ihre Abhangigkeit von den Wetterbedingungen", Elektr. Nachr. Techn., 17, No. 5, 193–199, 1940.
5. A.M. Obukhov, "Scattering of sound in turbulent flow", Doklady Akademii Nuak SSSR., 30, 611–614, 1941 (In Russian).
6. A.M. Obukhov, "On sound propagation in vortical flow", Doklady Akademii Nauk SSSR., 39, 46–49, 1943 (In Russian).

7. D. Blokhintzev, "The propagation of sound in an inhomogeneous and moving medium" I, II, J. Acoust. Soc. Amer., 18, 322–334, 1946.
8. D.I. Blokhintzev, "Acoustics of inhomogeneous moving medium", Moscow 1946, 2nd ed. Moscow "Nauka", 206 p., 1981' (In Russian).
9. V.I. Tatarskii, " Wave Propagation in a Turbulent Medium", Moscow, 1959 (translated from Russian by R.A. Silverman, Dover, New York, 1961).
10. V.I. Tatarskii, "The Effects of the Turbulent Atmosphere on Wave propagation", Moscow, 1967 (translated from Russian by J.W. Strohbehn, Israel Program of Scientific Translations, 1971. (Avilable from National Technical Information Service, Springfield, Va).
11. A.S. Monin, "Characteristics of the scattering of sound in a turbulent atmosphere", Sov. Phys. Acoust., Engl. Transl., 7, 370–373, 1962.
12. C.L. Pekeris, "Note on scattering in an inhomogeneous medium", Phys., Rev., 71, No. 4, 268, 1947.
13. R.H. Kraichnan, "The scattering of sound in a turbulent medium", J. Acoust. Soc. Amer., 25, 1096–1104, 1953.
14. M.J. Lighthill, "On the energy scattered from the interaction of turbulence with sound or shock waves", Proc. Combridge Phil. Soc., 49, 513–555, 1953.
15. G.K. Batchelor, "Wave scattering due to turbulence", in: Symposium on Naval Hydraulics, edited by F.S. Sherman, 409–423, National Academy of Sciences, Washington, D. C., 1957.
16 C.G. Little, "Acoustic methods for the remote probing of the lower atmosphere", Proc. IEEE, 57, 571–578, 1969.
17. M.A. Kallistratova, "An experimental investigation in the scattering of sound in turbulent atmoshere, "Doklady Akademii Nauk SSSR, 125, 69–72, 1959a (In Russian).
18. M.A. Kallistratova, "Experimental investigation of sound wave scattering in the atmosphere", Tr. Inst. Fiz. Atmos., "Atmos. Turbulentnost", 4, 203–256, 1961. (In Russian, English translation, U.S. Air Force FTD TT–63–441).
19. M.A. Kallistratova, "Acoustic and radio-acoustic remote sensing studies in C.I.S. (Former U.S.S.R.)-current status", Int. J. Remote Sensing, 15, No. 2, 251–266, 1994.
20. G. Kelton and P. Bricout, "Wind velocity measurements using sonic techniques", Bull. Amer. Meteorol. Soc., 45, 571–580, 1964.
21. L.G. McAllister, "Acoustic sounding of the lower troposphere", J. Atmos. Terr. Phys., 30, 1439–1440, 1968.
22. E.H. Brown and F.F. Hall, Jr., "Advances in atmospheric acoustics", Review of Geophysics and Space Physics, 16, No 2, 47–110, 1978.
23. W.D. Neff and R.L. Coulter, "Acoustic remote sensing", in: Probing the Atmospheric Boundary Layer, D.N. Lenschow editor, p. 201–266, Amer. Meteor. Soc., Boston, M.A., 1986.
24. S.P. Singal, "Acoustic sounding stability studies", In: P.N. Cheremisinoff (Ed.), Air Pollution Control, Encycl. of Environ. Control. Technology, vol. 2, Gulf Publ. Comp., 1003–1061, 1989.
25. S.P. Singal, "Current status of air quality related boundary layer meteorology studies using sodar", Proc. 5th Intern. Symposium on Acoustic Remote Sensing of the Atmosphere and Oceans, 453–476, 1990.
26. A. Weill and H.-R. Lehmann, "Twenty years of acoustic sounding—a review and some application", Z. Meteorol., 40, 241–250, 1990.

27. D.N. Asimakopoulos and C.G. Helmis, "Recent advances on atmospheric acoustic sounding", Int. J. Remotes Sensing, vol. 15, No. 2, 223–233, 1994.
28. L.A. Chernov, "Waves in randomly inhomogeneous media", Moscow, Nauka, 173 p., 1975 (in Russian).
29. A. Ishimaru, "Wave Propagation and Scattering in Random Media II", 317 p., 1978.
30. M.A. Kallistratova and A.I. Kon, "Radio-acoustical sounding of the atmosphere", Moscow, Nauka, 197 p., 1985 (In Russian).
31. N.P. Krasnenko, "Acoustical sounding of the atmosphere", Novosibirsk, Nauka, 167 p., 1986 (In Russian).
32. S.M. Rytov, Yu.A. Kravtzov and V.I. Tatarskii, "Principles of statistical radio physics, Part 4, Wave propagation through random media", Berlin, Springer, 463 p., 1989.
33. V.E. Ostashev, "Sound propagation in moving media", Moscow, Nauka, 208 p., 1992 (In Russian).
34. A.S. Monin and A.M. Yaglom, "Statistical Fluid Mechanics" I, MIT Press, Cambridge, Mass., 1971.
35. N.N. Andreev and I.G. Rusakov, "Acoustics of moving media", Moscow, GTTI, 150 p., 1934 (In Russian).
36. M.A. Kallistratova and V.I. Tatarskii, "Accounting for wind turbulence in the calculation of sound scattering in the atmosphere", Sov. Phys. Acoust., Engl. Transl., 6, 503–505, 1960.
37. M.L. Wesely, "The combind effect of temperature and humitity fluctuations on refractive index", J. Appl. Meteorol., 15, 43–49, 1976.
38. W. Kohsiek, "Measuring C_T^2, C_q^2, C_{Tq}^2 in the instable surface layer and relations to the vertical fluxes of heat and moisture", Boundary Layer Meteorol., 24, No. 1, 89–107, 1982.
39. S.W. Fairal, D.E. Schacher, K.L. Davidson, "Measurements of the humidity structure function parameters C_{tq}^2, C_q^2 over the ocean", Boundary Layer Meteorol., 19, No. 1, 81–92, 1980.
40. E.E. Gossard, "Power spectra of temperature humidity and refractive index from aircraft and tethered balloon measurements", IEEE Trans. Antennas Propagat., AP-3, 186–201, 1960.
41. L.G. Yelagina, B.M. Koprov and D.F. Timanovskiy, "Certain characteristics of the atmospheric surface layer above show", Izvestiya, Atmosph. and Ocean Physics, 14, No 9, 652–655, 1978.
42. V.E. Ostashev, "Sound propagation and scattering in media with random inhimogeneities of sound speed, density and medium velocity", Waves in Random Media, 4, Sept., 1–26, 1994.
43. A.S Monin and A.M. Yaglom, "Statistical Fluid Mechanics" II, MIT Press, Cambridge, Mass., 1974.
44. A.S. Gurvich, and A.I. Kon, "Aspect sensitivity of radar returns from anisotropic turbulent irregularities", "J. Electromagnetic Waves and Applications, 7, No 10, 1343–1353, 1993.
45. A.S. Gurvich, "On the scattering of sound and radio waves by turbulent structures in the stratosphere", Izvestiya, Atmosph. and Ocean Physics, 30, No. 1, 3–12, 1994.

46. M.A. Kallistratova, "Procedure for investigating sound scattering in the atmosphere", Sov. Phys. Acoust., 5, 512–514, 1959 (In Russian).
47. W. Baerg and W.H. Schwarz, "Measurements of the scattering of sound from turbulence", J. Acoust. Soc. Amer., 39, No. 6, 1125–1132, 1965.
48. A. Spizzichino, "The refraction of acoustic waves in the atmosphere and its effect on sodar winds measurement", Ann. Telecommun., vol. 29, 301–310, 1974.
49. American National Standard Method for the Calculation of the Absorption of Sound by the Atmosphere. — ANSI S1; 26-1978/ASA-27, 1979.
50. S.F. Clifford and E.H. Brown, "Excess attenuation in echosonde signals", J. Acoust. Soc. Am., 67(6), 1967–1973, 1980.
51. V.M. Bovsheverov and G.A. Karyukin, "Influence of wind on the accuracy of determination of the temperature structure parameter by acoustic sounding", Izvestiya, Atmosph. and Ocean Physics, 17, No. 2, 151–156, 1981.
52. R.J. Doviak and D.S. Zrnic, "Reflection and scatter formula for anisotropic air", Radio Sci., 19, No. 1, 325–236, 1984.
53. W.D. Neff, "Quantitative evaluation of acoustic echoes from the planetary boundary layer", NOAA Tech. Rep., ERL 322-WPL 38 Boulder CO, 1975.
54. M.A. Kallistratova and I.V. Petenko, "Aspect sensitivity of the sound backscattering in the atmospheric boundary layer", J. Appl. Phys. B:, 57, 41–48, 1993.
55. S.P. Singal, B.S. Gera, M.A. Kallistratova, I.V. Petenko, "Sodar aspect sensitivity studies in the convective boundary layer", Proc. 7th ISARS, editor: W.D. Neff, 3–7 October 1994, Boulder, Colorado, 1.19–1.24, 1994.
56. H. Ottersten, M. Hurting, G. Stilke, et al, "Shipborne sodar measurements during Jonswap 2", J. Geoph. Research, 79, No. 36, 5573–5584, 1974.
57. G.L. Bonvallet, "Levels and spectra of traffic, industrial and residential area noise", J. Acoust. Soc. America, 23, No 3, 435–439, 1951.
58. P.B. Ostergaard and R. Donley, "Background-noise levels in suburban communities", J. Acoust. Soc. America, vol. 36, No, 3, 409–413, 1964.
59. R.H. Macmillan, "The control of noise from surface transport", J. Sound and Vibration, 43, No 2, 173–87, 1975.
60. N. Olson, "Survey of motor vehicle noise", J. Acoust. Soc. America, 52, No. 5, Part 1, 1291–1306, 1972.
61. M. Fukushima, K. Akita and H. Tanaka, "Night time profiles of temperature fluctuations deduced from two-year sodar observations", J. Meteorol. Soc. Jap., 53, 487–491, 1975.
62 I.A. Bourne and T.D. Keenan, "High power acoustic radar", Nature, 251, 206, 1974.
63. M.S. Frankel, N.S.F. Chang, M.J. Sanders, "A high frequency radio acoustic sounder for remote measurement of atmospheric winds and temperature", Bull. Amer. Meteorol. Soc., 58, No. 9, 928–934, 1977.
64. O.V. Rudenko and S.I. Solujan, "Theoretical backgrounds the nonlinear acoustics" Moscow, Nauka, 288 p, 1975, (In Russian).
65. S.D. Danilov, M.A Kallistratova, "Nonlinear sound attenuation in acoustic and radioacoustic sounding of the atmosphere", Izvestiya, Atmosph. and Ocean Physics, 23, No. 9, 740–742, 1987
66. S.D. Danilov, A.E. Gur'yanov, M.A. Kallistratova, I.V. Petenko, S.P. Singal, D.R. Pahwa, B.S. Gera, "Acoustic calibration of sodar", Meas. Sci. Technol., 3, 1001–1007, 1992.

67. M.A. Kallistratova, I.V. Petenko, E.A. Shurygin, "Sodar studies of the wind velocity field in the lower troposphere", Izvestiya, Atmosph. and Ocean Physics, 23, No. 5, 339–345, 1987.
68. I.V. Petenko, E.A. Shurygin, "Comparison of sodar and turbulent measurement", In Proc. of the 'Field Experiment KOPEX-86, Prague 1988, 37–54, 1988.
69. P.L. Finkelstein, J.C. Kaimal, J.E. Gaynor, M.E. Graves and T.J. Lockhart, "Comparison of wind monitoring systems, Part II: Doppler sodars", J. Atmos. and Oceanic Tech., 3, 594–604, 1986.
70. J.E. Gaynor, "Accuracy of sodar wind variance measurements", Int. J. Remote Sensing, 15, No. 2, 313–324, 1994.
71. A.E. Gur'yanov, M.A. Kallistratova, G.A. Karjukin, V.P. Kukharetz, I.V. Petenko, S.L. Zubkovskii, "Reliability of determinations on the vertical profile of the temperature structure parameter in the atmosphere by acoustic sounding", Izvestiya, Atmosph. and Ocean Physics, 17, No. 1, 107–111, 1981.
72. D.N. Asimakopoulos, R.S. Cole, S.J. Caughey, B.A. Crease, "A quantitative comparison between acoustic sounder returns and the direct measurement of atmospheric temperature fluctuations", Bound. Layer Meteor., 10, 137–147, 1976.
73. T.J. Moulsley, D.N. Asimakopoulos, R.S. Cole, B.A. Crease, S.J. Caughey, "Measurement of boundary layer structure parameter profiles by acoustic sounding and comparison with direct measurements", Guart. J.R. Met. Soc., 107, 203–230, 1981.
74. D.A. Haugen and J.C. Kaimal, "Measuring temperature structure parameter profiles with an acoustic sounder", J. of Appl. Met., 17, 895–899, 1978.
75. W.D. Neff, "Beamwidth effects on acoustic backscatter in the planetary boundary layer", J. Applied Meteorology, 17, 1978.
76. Y. Chen, S. Li, N. Lu, "The C_T^2 by sodar and in situ measurement", Sci. Atmos. Sin., 8, No. 2, 153–160, 1982.
77. A.E. Gur'yanov, M.A. Kallistratova, A.S. Kutyrev, I.V. Petenko, P.V. Shcheglov, A.A. Tokovinin, "The contribution of the lower atmospheric layers to the seeing at some mountain observatories", Astronomy and Astrophysics, 262, 373–381, 1992.
78. M.A. Kallistratova, "Sodar data as short-range dispersion models input", Proc. 7th Intern. Symposium on Acoustic Remote Sensing of the Atmosphere and Oceans (Ed. W.D. Neff), Boulder, 3–7 October, 1994.
79. E.A. Shurygin, "On automatic recognition of echo-sounders pattern using Doppler sodar", in Proc. 8th Intern. Symposium on Acoustic Remote Sensing of the Atmosphere and Oceans (Ed. M.A. Kallistratova), Moscow, pp. 3.61–3.66, 1996.
80. M.A. Kallistratova, I.V. Petenko, M.S. Pekour, B.S. Agrovskii, "Experimental investigation of sound scattering at an angle of 90°", Ibid., pp. 1.23–1.27, 1996.
81. V.I. Petenko, "Coherent structure properties derived from sodar data", Ibid., pp. G.51–G62, 1996.

Acoustic Remote Sensing Applications
S.P. Singal (Ed)
Copyright © 1997 Narosa Publishing House, New Delhi, India

2. An Overview of the Technological Development of Atmospheric Echosounders (SODARS)

John A. Kleppe
Electrical Engineering Department, University of Nevada, Reno,
Reno, Nevada, USA 89557-0153

1. Introduction and Historical Overview

Atmospheric acoustics can be considered to have two major fields of study. The first is concerned with the effects on sound due to the variations and inhomogeneities of the earth's atmosphere; for example, the random deformation of acoustic waves as they propagate through atmospheric fluctuations. The second general area of study has to do with investigations into the application of sound as a measurement tool in atmospheric research. An example of this would be the use of sound waves to measure the height and intensity of temperature inversions for air quality studies. These two general areas are interrelated, because knowledge of one is necessary to gain an effective understanding of the other.

The history of atmospheric acoustics is interesting, since it does span several centuries of investigations by a wide range of researchers. The early interest appears to have been centered around the outdoor measurement of the speed of sound. Details on sound speed calculations and measurements have been reported over a period of some 300 years. An excellent paper by Wong [1] provides historical information and proposes a calculation that results in a new value for the speed of sound in "standard dry air" (C_0 = 331.29 m/s) at 0°C and a barometric pressure of 101325 Pa. Wong also points out that most past investigations into sound speed provided no information on the gas composition of the air supply used. In most of these studies, investigators obtained their sound speed through direct methods; that is, the actual measurement of sound over relatively long distances in open air, short distance measurements in laboratories, and measurements in tubes. It was observed by Foley [2] that, from 1738 to 1919, there seemed to be a progressive decrease in the numerical value of sound speed recorded by the cannon flash open air method. He suggested that this might be due to a slight change in the constitution of the atmosphere. Wong, however, concluded that the decrease in the measured sound speed was not likely due to a change in the major constituents of the atmosphere but may have been caused by an increase in the CO_2 content or a change in the correction methods adopted by investigators to reduce their measured data to standard conditions. Wong further concluded that contrary to the beliefs of some investigators, the outdoor environment can offer a relatively constant standard air composition. On a calm day over long distances, the averaging effects on the environmental conditions are some of the advantages of "free" sound speed measurements.

An excellent paper by Brown and Hall [3] presents a brief historical overview of atmospheric acoustics and a detailed review of developments in atmospheric acoustics during the period 1968–1978. The authors noted that much of the early acoustical research done by Galileo, Mersenene, Sauveur, and others during the rebirth of science in the seventeenth century has become relatiely inaccessible in its original form. However, descriptions of most of the results appear in Rayleigh's (1896) *The Theory of Sound* [4] or A. Wood's (1946) *Acoustics* [5].

The field of atmospheric acoustics seems to have been initiated by an English clergyman named Derham, who received a letter in 1704 that claimed that there was a greater audibility of sound in England than in Italy. Derham and an Italian acquaintance, Averroni, conducted experiments in England and Italy that showed, after proper accounting for the effects of wind, that sound propagation did not differ in the two countries.

The field of atmospheric acoustics again went dormant until the last half of the nineteenth century. During this period, Stokes (1857) [6], Tyndall (1874, 1875) [7, 8], Henry (1874) [9], Reynolds (1876) [10], and Rayleigh (1877) [4] investigated such topics as atmospheric temprature profiles and their effects on the refraction of sound, refractive effects of wind, and the scattering of sound by an apparently still atmosphere. Of particular interest is the work by Tyndall (1875) [8], who studied sound propagation in the atmosphere over the sea using the apparatus shown in (Fig. 1). Tyndall detected acoustic backscatter from temperature and wind structure in the atmosphere, which he attributed to "acoustic clouds". He was also able to demonstrate that air heated by a flame could attenuate the direct propagation of sound. It is noteworthy, however, that even after this work, Tyndall's technique and results lay dormant for over 70 years, before being studied again as a method to probe the atmosphere.

Shortly before the end of World War II, Gilman, Coxhead, and Willis [11] used acoustic backscatter to study the structure of low-level temperature inversions as they affected propagation in microwave communication links. During the late 1950s, acoustic scattering from the atmosphere was investigated experimentally in the Soviet Union by Kallistratova [12–14] and theoretically by Tatarskii [15], but it was the investigations by McAllister [16–18] in Australia that showed that echoes could be reliably obtained to heights of several hundred meters.

The most serious efforts to bring atmospheric acoustics into a modern engineering science was accomplished by a group of scientists and engineers at the Wave Propagation Laboratories of the National Oceanic and Atmospheric Administration

Fig. 1. Tyndall's apparatus for studying sound propagation in the atmosphere (from [8]).

(NOAA) in Boulder, Colorado. The effort, beginning in the late 1960s and early 1970s [19–22] demonstrated that acoustic sounders could also be used to measure winds in the atmosphere by means of the doppler shift in the scattered sound [23–26] and to show the practical feasibility of utilizing acoustic sounders to monitor temperature inversion structures [27]. Little [22] discussed the detectability of fog, clouds, rain, and snow using acoustic echosounding methods.

The engineering design of acoustic sounders began to gain serious interest during the early 1970's. Parry and Sanders [28] discussed the design and operation of an acoustic "radar" and focused attention on a particular sounder similar to that used by McAllister. This paper brought focus on the various subsystems that make up an acoustic sounder, Wescott et al, [29]. Cronenwett, Walker, and Inman [30] reported results of their acoustic "radar" investigations using a sounder design similar to that described by Wescott and Parry. Chadwick and Little [31] compared the sensitivities of various atmospheric sounders.

One of the earliest commercial acoustic sounder was developed by AeroVironment, Inc., in Pasadena, California. This system, as reported by Tombach, MacCready, and Baboolal [32], was designed as a monostatic system for use in low-level air quality diffusion studies. The system, called the Model 300, was a monostatic (collocated sound source and receiver) acoustic radar. It used a 70 W (electrical), 2000 Hz sound burst, which was emitted from a speaker located at the focal point of a 4-ft diameter parabolic reflector (antenna). Sound, scattered back from small-scale atmospheric temperature variations, was received at the same reflector-pickup transducer and recorded on a time-height facsimile recorder. The system used a relay as the duplexer switch that changed the transducer from the transmit to the receive mode. Since the acoustic receiver must be very sensitive to receive the low-level backscattered signals, it was necessary to also use an enclosure around the transducer-parabolic reflector assembly. This enclosure (cuff) provided a shield from wind noise and general background noise that typically propagates near the earth's surface.

In 1974, a portable acoustic echosounder was reported by Owens [33]. This system, designated the NOAA Mark VII, consisted of a central unit, power amplifier, facsimile recorder, and a 4-ft parabolic reflector. Owens was able to use his acoustic sounder for many of the early experiments conducted by NOAA, including shipboard experiments and the first operation of an acoustic sounder in Antarctica (1975) by Hall and Owens [34].

Early acoustic sounders used facsimile recorders to record backscatter data. The facsimile recorder provided, in general, an inexpensive, reliable method for displaying time-height records. The "fish finder" technology of SONAR was easily adapted to atmospheric echo sounding. Facsimile recording, however, had some major drawbacks including the production of records that were poorly suited to quantitative data logging, lack of easily changed pulse repetition rates, and a "messy" record to handle and interpret. Also, because the dynamic range of received signals from atmospheric scattering could cover some 40 dB, the use of a facsimile recorder with a maximum range of 12–15 dB (or in some cases 6–8 dB) required significant signal compression [35].

In 1975 with these limitations in mind, research efforts at the University of Nevada (UNR) and Scientific Engineering System, Inc. (SES), both of Reno turned to the problem of developing a digital-based acoustic doppler sounder system. Owens [36, 37], at UNR, was the first to report on an acoustic sounder display using a digital dot matrix printer and a special character matrix. The concept of using a microcomputer as an integral part of an acoustic sounder stimulated many ideas impossible to consider using just a facsimile recorder. Following the advance work by Little, Brown, Hall, Beran, Neff, and others at NOAA, Kleppe and Dunsmore [38, 39] of SES proposed using advanced digital signal processing techniques to integrate doppler signal processing into Owen's backscatter system. Comparisons were made of several doppler processing techniques including the *Fast Fourier Transform* (FFT), adaptive linear prediction filtering (ADAP), *real correlation* (RC), *and complex covariant* (CC) methods. This work was undertaken mainly to determine which method would offer the most effective analysis, keeping in mind a desire to accomplish real-time three-axis acoustic doppler processing while displaying it with the backscatter data. The main emphasis was to bring all of this together in a form realizable by microcomputer technology.

This work was successful and contributed to the development of a commercially available microcomputer-based acoustic doppler system having an advanced display. Called the *Echosonde*®, the system was produced by SES in Reno, Nevada. It used the complex covariant method to process the acoustic returns.

Several commercial firms continued the further development of acoustic sounders through the early 1980s. Radian Corporation used the SES *Echosonde*® as a building block to develolp a sophisticated multiaxis doppler sounder that used colors to display the doppler wind information.

There were also parallel developments of acoustic sounders by other United States companies, such as Xonics, Inc. The Xondar system was designed for a broad range of applications, from mean wind measurements to turbulence and diffusion studies (Balser, McNary, and Nagy [40]).

AeroVironment, Inc., developed the AVIT (AeroVironment Invisible Tower) which was a pulsed doppler acoustic system. The AVIT was a flexible, modular system. The basic three-axis system used three adjacent pencil-beam reflectors. One was tilted north (or south) 30° from the vertical to observe the N-S wind; one was tilted east (or west) similarly to observe the E-W wind; and one pointed vertically to observe the vertical component. The reflectors (antennas) were operated sequentially, in a monostatic mode.

Several foreign entities have also developed acoustic sounders. The University of Uppsala in Sweden, through its Department of Meteorology, joined forces with Sensitron AB of Stockholm to create a doppler sounder. The French also developed an acoustic doppler sounder, marketed by Bertin and Cie. It was capable of measuring wind speed, wind direction, turbulence, and vertical wind motion up to a maximum height of 1500 m above the ground. The French company, REMTECH, continues to offer several products including a doppler SODAR, and a Radio Acoustic Sounding System (RASS), (Fig. 2).

Fig. 2. A bistatic radar can be added to a phased array SODAR which creates a wind profiler that also provides remote sensing of air temperature.

The development of acoustic sounders (SODARS) continues as new applications arise and the general technology used in these systems improves. One of the most notable advancements has been the commercialization of acoustic phased array antennas for sonar use, (Fig. 3). For example the REMTECH PA1 is a sophisticated three dimensional phased array SODAR capable of measuring up to 1000 m. The

40 *Atmospheric Acoustics and Instrumentation*

Fig. 3. The REMTECH PA1 phased array SODAR.

signal processing uses a 386/387-33 MHz personal computer and uses multiple frequency coding and Fast Fourier Transform (FFT) digital signal processing methods.

REMTECH also manufactures a longer range phased array acoustic wind profiler, (Fig. 4). This system has a maximum range (in a quiet environment) of 2.5 km.

Fig. 4. The REMTECH HPPA1 long-range acoustic wind profiler.

It measures doppler as well as turbulence parameters without ground clutter in the first few hundred meters thus giving this system some distinct advantages over 915 MHz radar type wind profilers.

Similar advancements in SODAR technology have been made by the Radian Corporation of Austin, Texas and AeroVironment Inc. located in Pasadena, California.

The fundamentals of operation and the general design of SODAR systems have remained fairly static over the past few years, making it possible to gather together some of the more important engineering design principles involved. It should be noted that the design procedures and technical approaches covered in the rest of this chapter may be useful in other areas of acoustic instrumentation design and analysis beyond atmospheric echosounders.

2. The Atmosphere

It is useful to review briefly that which a SODAR measures; that is, a portion of the atmosphere. The atmosphere is the gaseous envelope of the earth. It extends from the surface of the earth to several earth radii where it merges with the interplanetary medium. Most of the atmosphere's mass is below a height of 10 km. In comparison to the radius of the earth (6371 km), the bulk of the atmosphere is contained in a relatively thin layer. However, when this collection of viscous gasses, water vapor, and floating particles is acted upon by an equator-to-pole air temperature gradient, the evaporation and condensation of water, the rotation of the earth, and surface friction, it becomes a very complex physical system. Temperature and wind gradients affect the propagation of sound in a nonturbulent atmosphere. For many applications at lower frequencies, the concept of a smooth mean atmosphere has cetrtain agreed upon average properties that can be useful in system design. The atmospheric pressure and air temperature, and consequently the air density, vary with their elevation above the surface of the earth.

It is important to understand in some detail the absorption and speed of sound in air. The design of a SODAR system requires that the design engineer take into account absorption effects.

Air consists of molecular oxygen and nitrogen with traces of other gases. The addition of water vapor to air has a profound effect on the absorption especially for frequencies below approximately 5 kHz. An excellent paper by Bass and Shields [41] provides a comparison of theoretical expressions for sound absorption and measured values for the frequency range between 4 kHz and 100 kHz, temperature between 18°C and 38°C and relative humidity values between 0 and 100%. The expressions are somewhat complex to use. Because of this, a computer program was developed using ANSI.S1.26 (1978) [42, 48]. A copy of this program has been included as Appendix A for easy use by the design engineer.

This program requires that the user enter frequency (Hz), ambient temperature (°C), atmospheric pressure (10^6 Pa)) and relative humidity (per cent). The output of the program is the absorption coefficient a (dB/100 m), the molar concentration of water vapor h (percent), the relaxation frequency for oxygen f_{ro} (Hz), the relaxation frequency for nitrogen, f_{rn} (Hz), the speed of sound for dry air (m/s), the speed of sound for the humid air (m/s), the density of the humid air (kg/m^3), the vapor pressure of the water vapor (Pa), the molar mass for the humid air (g/mole) and the specific acoustic impedance (rayls).

EXAMPLE

Determine the total absorption coeficient a(dB/m) in a standard atmosphere at 20°C, 1 atmosphere pressure, and a relative humidity of 60%, for a soundwave of 2000 Hz. Use the computer program ATMOS1 provided in Appendix A.

Solution

The use of this program is quite straightforward since we simply have to enter the frequency,

$$f = 2000 \text{ (Hz)}$$
$$T = 20(°C)$$
$$P = 0.101325(10^6 \text{ Pa})$$
$$\text{RH} = 60(\%)$$

The input-output for the computer run using the program of Appendix A follows for easy reference:

RUN
ENTER IN FREQUENCY (HZ), TEMP (C), PRESSURE (10E6PA), RH (%)
? 2000,20,.101325,60

THE TOTAL ABSORPTION (dB/100M)	= 0.9642441
THE MOLAR CONCENTRATION OF H20 (%)	= 1.384712
THE OXYGEN RELAXATION FREQUENCY (HZ)	= 49358.39
THE NITROGEN RELAXATION FREQUENCY (HZ)	= 493.6493
THE SOUND SPEED FOR HUMID AIR (M/S)	= 344.2977
THE DENSITY (KG/M3)	= 1.198385
THE VAPOR PRESSURE E (PA)	= 1403.06
THE MOLAR MASS	= 28.82834
THE SPECIFIC ACOUSTIC IMPEDANCE (RAYLS)	= 412.6012

Limited ranges of atmospheric and acoustic variables exist where the agreement between laboratory measurements of atmospheric absorption from a number of investigators is clearly at a maximum and sufficient to form a purely empirical basis for accuracy. Where $2 \leq f \leq 15$ kHz, $T = 20°C$, $P = 0.101325$, and $10\% \leq \text{RH} \leq 100\%$, the accuracy of the absorption calculation is approximately ± 5% [42].

For values 40 Hz $\leq f \leq 1 \times 10^6$ Hz, $0 \leq T \leq 40°C$, $P \leq 0.2 \times 10^6$ (less than 2 atm), and $10\% \leq \text{RH} \leq 100\%$ the program provides an accuracy of approximately ± 10%. For more accurate results or for use over an extended range of these variables, the reader should refer to ANSI S1.26–1978 (ASA 23-1978) directly.

Provided $\alpha/k \ll 1$, the effects of separate mechanisms for the absorption of sound combine, so that the total absorption coefficient is nearly the sum of the absorption coefficients of the individual loss mechanisms calculated as if they were operating alone:

$$\alpha = \sum_i \alpha_i \text{ nepers}, \quad k = 2\pi/\lambda$$

The method we have used for calculating the attenuation coefficient for air has assumed that the sound wave is passing through a standard, still, homogeneous atmosphere of humid air. However, the method does provide a practical basis for the dominant sources of atmospheric attenuation in most cases for which ground effects, scattering by atmospheric turbulence, and nonlinear effects associated with very high-intensity sounds are not of concern. Also, effects due to atmospheric pollutants such as dust, fog, or precipitation have not been considered.

These effects will be discussed later in this paper as they arise during specific applications.

2.1 The Turbulent Atmosphere

There is a vast body of literature concerning atmospheric turbulence; the review by Brown and Hall [3] is an excellent reference for this subject. The distinction between a smooth, refracting atmosphere and a turbulent, scattering atmosphere, in part, is dependent on the application. Because of the frequency dependence of attenuation and the path length over which SODARS operate, some basic assumptions can be made, however. The practical frequency range for SODAR operation is between 1 kHz to 4 kHz with the usual selection being 2 kHz, as it is the best trade-off between maximum range, attenuation, and scattering wavelength.

Atmospheric turbulence, whether produced dynamically by the flow of air over a rough surface or convectively by solar heating of the ground, results in localized fluctuations in atmospheric parameters, such as temperature, humidity, and velocity. These localized variations in atmospheric conditions correspond to fluctuations in the local refractive index for acoustic waves. The resultant three-dimensional field of refractive index is most conveniently expressed in terms of acoustic refractivity that is, the intensity (power) of the refractive index fluctuates as a function of spatial wavelength and direction.

2.2 Structure Functions [43–47]

At the acoustic wavelengths normally used in SODARS, the turbulence is usually thought to be in the Kolmogorov inertial subrange, which therefore may be considered locally homogeneous and isotropic. Under such conditions, it is convenient to describe the refractivity fluctuations in terms of the structure constant of the refractivity C_n defined by

$$D_n(r) = \langle \, | \, n(x) - n(x + r) \, |^2 \, \rangle_{AV} = C_n^2 r^{2/3} \qquad (1)$$

where $D_n(r)$ is the structure function of the refractivity and describes the way in which the mean square difference in refractivity between two test points, x and $x + r$, varies as a function of their spacing r. The indicated average is over possible realizations of refractive index. Thus, for a Kolmogorov spectrum of turbulence, C_n is the rms difference in refractive index at two points a unit distance apart. The two structure constants of most interest for acoustic sounding are those for temperature C_T and velocity C_V with

$$C_T^2 = \overline{\left[\frac{T(x) - T(x+r)}{r^{1/3}}\right]^2}, \quad C_V^2 = \overline{\left[\frac{V(x) - V(x+r)}{r^{1/3}}\right]^2} \tag{2}$$

2.3 Scattering Cross Sections

We can see that the scattering cross section, σ, can be written in the following form [14]:

$$\sigma(\theta) = 0.03 k^{1/3} \cos^2\theta \, [\sin(\theta/2)]^{-11/3} \left[\frac{C_V^2}{c^2} \cos^2\left(\frac{\theta}{2}\right) + \frac{0.13 C_T^2}{T^2}\right] \tag{3}$$

where θ = scattering angle (relative to the incident wave at the scatterer), $k = 2\pi/\lambda$, c = speed of sound and T = absolute temperature (K).

Various interesting observations can be made from these equations depending upon the scattering angle, θ. For example, when a monostatic system is used, $\theta = \pi$ and the scattering is due to C_T only, whereas for $\theta \neq \pi$, the scattering is due to both C_T and C_V. It is also observed that for $\theta = \pi/2$, the total scattering is zero.

It is also possible to obtain doppler signals from an acoustic sounder. Acoustic energy transmitted into a moving atmosphere is frequency shifted as it is scattered. The amount of frequency shift of the received signal is proportional to the wind speed at the height from which the received signal was scattered. The well-known doppler equation can be applied:

$$V = \frac{\lambda \Delta f}{2} \tag{4}$$

where V = doppler velocity, Δf = the doppler frequency shift and λ = transmitted wavelength.

Depending on the system configuration chosen, then, there is a wealth of information to be obtained from the acoustic returns. In principle it is also possible to obtain values of C_V^2 from intensity measuremenst of bistatic echoes after the contribution from C_T^2 is subtracted; however, this requires an accounting for attenuation effects in the measurement of C_T^2.

The preceding discussion dealt with the scattering cross sections of a gaseous atmosphere resulting from random small-scale fluctuations in the refractive index produced by localized fluctuations in temperature and velocity. We can also examine the continuous per-unit *volume* reflectivity of the atmosphere due to hydrometeors; that is the case for rain, snow, hail, cloud, or fog particles.

The reflectivity of hydrometeors for acoustic waves has been studied and reported by Little [22] and Chadwick and Little [47]. It is the normal case, in which the operating wavelength is large compared with the hydrometeor diameter and the SODAR reflectivity, η (the backward-scattering cross section per unit volume of the region), may then be written in the following form:

$$\eta = \frac{\pi^5}{\lambda^4 V} \sum_V D^6 \frac{1}{9}\left[\frac{m'-m}{m'} + \frac{3(\rho'-\rho)}{\rho+2\rho'}\right]^2 = 4\pi\sigma_a (m^2/m^3) \qquad (5)$$

where σ_a = backward scattering cross section per unit volume per unit solid angle in which a plane wave of wavelength λ propagating through a medium of bulk modulus m and density ρ is incident upon spheres of bulk modulus m' and density ρ'; $\sum_V D^6$ = sum over the pulse volume of the diameter of each drop raised to the sixth power; λ = operating wavelength; and V = pulse volume = volume that includes the scatterers intercepted by the beam of sound.

For a SODAR with a parabolic antenna (reflector), the pulse volume at a range r from the SODAR would be

$$V = \frac{\pi r^2 \theta_{BW}^2 h}{8} \qquad (6)$$

where θ_{BW} = beamwidth (– 3 dB), $h = c\tau$, c = speed of sound and τ = pulse length.

Equation (5) is applied appropriately to small spherical raindrops or hailstones. For snowflakes, assuming that the snowflake dimensions are small compared to the wavelength, we can select appropriate values for the diameter, density, and bulk modulus. We may select the values of m' and ρ' corresponding to solid ice, provided we use the corresponding value of D, that is, the diameter of the (solid) ice sphere of mass equal to the snowflake. In Equation (5), if we assume that $m' \gg m$ and $\rho' \gg \rho$:

$$\frac{1}{9}\left[\frac{m'-m}{m'} + \frac{3(\rho'-\rho)}{\rho+2\rho'}\right]^2 \approx \frac{25}{36}$$

which when substituted back into Equation (5) yields

$$\sigma_a = \frac{\eta}{4\pi} \approx \frac{25\pi^5}{144\pi\lambda^4}\left(\frac{1}{V}\right)\sum_V D^6 = \frac{16.91}{\lambda^4}\left(\frac{1}{V}\right)\sum_V D^6 \; (m^2/m^3 \; \text{SRADIAN}) \qquad (7)$$

For a given SODAR, λ is usually constant and the reflectivity of a region of precipitation is proportional to $\sum D^6$. In meteorological radar, it is conventional to write

$$Z = \left(\frac{1}{V}\right)\sum_V D^6 \qquad (8)$$

where Z, termed the *reflectivity factor*, is a function only of the number and size of the particles per unit volume. It is important to note that the acoustic reflectivity is proportional to Z in the same way that electromagnetic radar reflectivity is proportional to Z, and hence, the numerous published measurement of Z obtained with meteological radars can be used in estimating the acoustic reflectivities of fog and precipitation.

3. Acoustic Sounder (SODARS) [48]

It is now instructive to discuss the SODAR equations for various system configurations and applications. Since SODARS can be used to measure atmospheric structure functions as well as wind velocity, SODAR equations are essential to the design, analysis, and use of a system. SODAR equations in general follow a development parallel to the system equations used in electromagnetic radar. It is useful to consider first a monostatic system and then extend the development to other possible system configurations.

3.1 Monostatic SODAR

The simplest system configuration is a vertically pointed monostatic system, as is shown in (Fig. 5 a). For wind measurements, this system can also be operated by tilting the parabolic disk at some prescribed angle, as is shown in (Fig. 5 b).

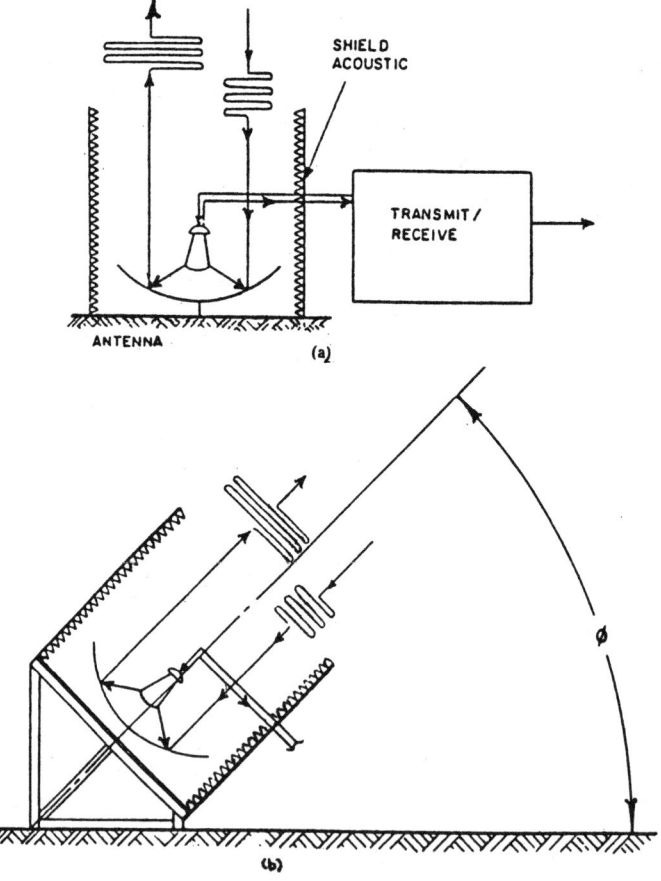

Fig. 5. Monostatic SODAR. (a) Vertical pointing monostatic SODAR and (b) Tilted nonostatic SODAR.

Because the vertical pointing system is a special case ($\phi = \pi/2$) for the tilted system, it is useful to develop the SODAR equation for it and then simply substitute ($\phi = \pi/2$) for the vertical pointing configuration.

A number of investigators have studied this problem. Brown [49] reported theoretical work on the effects of a nonzero mean wind on the propagation and scattering of sound by turbulent refractivity fluctuations. Spizzichino [50, 51] developed a description of the errors to be expected in measured wind velocity and height definition by taking into account the parabolic dish (antenna) characteristics, pulse and receiving gate durations, spectral broadening of the received echo, and atmospheric refraction effects. He showed that these measurement errors can be minimized by an adequate adjustment of the SODAR operating parametric optimization technique that predicts the relative accuracy of horizontal wind measurement for both monostatic and bistatic SODAR configurations.

Referring to (Fig. 5a) the monostatic SODAR equation can be written in the following form (Neff [52]):

$$P_R = [P_T \eta_T \eta_R] \cdot [e^{-4\alpha R}] \left[\sigma_o(R, f) \cdot \left(\frac{c\tau}{2}\right) \cdot \left(\frac{A}{R^2} \cdot G\right) \right] \quad (9)$$

where P_R = received power, i.e. measured electrical power; η_R = efficiency of conversion from received acoustic power to electrical power; $P_T \eta_T$ = radiated power (P_T = electrical power applied to the transducer, η_T = efficiency of conversion to radiated acoustic power); $e^{-4\alpha R}$ = round-trip loss of power resulting from attenuation by air where α = average attenuation (m^{-1}) to the scattering volume at range $R(m)$; $\sigma_0(R, f)$ = scattering cross section per unit volume; that is, fraction of incident power backscattered per unit distance into unit solid angle at frequency f; $c\tau/2$ = maximum effective-scattering volume thickness, c = local speed of sound (ms^{-1}), τ = pulse length (s); $A/R^2 \cdot G$ = solid angle subtended by the antenna aperture $A(m^2)$ at range $R(m)$ from the scattering volume, modified by an effective-aperture factor G, arising from the antenna's directivity; where

$$G = \frac{G(0)\lambda^4 L^2(\theta_0)}{4\pi^2 \theta_0^2 A^2} = \text{beam shape compensation factor}$$

$$L(\theta_0) = \left[1 - f_0^2\left(\frac{2\pi a\theta_0}{\lambda}\right) - f_1^2\left(\frac{2\pi a\theta_0}{\lambda}\right) \right] \text{ (piston source)}$$

a = antenna radius at the aperature A

θ_0 = one-half the total beam width $\frac{\theta_{BW}}{2} = 0.67 \frac{\lambda}{D} = 0.34 \frac{\lambda}{a}$

$G(0)$ = on-axis antenna gain = $\frac{4\pi A}{\lambda^2}$

48 Atmospheric Acoustics and Instrumentation

The scattering cross section $\sigma(R, f)$ is given by Equation (3), which is repeated for easy reference:

$$\sigma(\theta) = 0.03k^{1/3} \cos^2\theta \left[\sin\left(\frac{\theta}{2}\right)\right]^{-11/3} \left[\frac{C_V^2}{c^2} \cos^2\left(\frac{\theta}{2}\right) + 0.13 \frac{C_T^2}{T^2}\right] \quad (9.1)$$

where θ = scattering angle, k = wavenumber = $2\pi/\lambda$, c = speed of sound and T = absolute temperature.

For the monostatic case, only backscatter is involved, that is $\theta = \pi$, which yields

$$\sigma(\pi) = (0.03k^{1/3})\left(0.13\frac{C_T^2}{T^2}\right) = 0.0039k^{1/3}\frac{C_T^2}{T^2} \quad (9.2)$$

Using this and the SODAR equation, we obtain a volume-averaged measure of C_T^2 from each SODAR range gate. Application of the SODAR equation and the expression for the SODAR cross section yields a discrete time series of volume-averaged C_T^2 values for any selected SODAR range gate. The errors in the values of C_T^2 so determined depend to a large extent of our ability to measure or estimate the correction terms in the SODAR equation. Once the total electronic gain of the system is accurately measured (within a few percent), the major error left in the evaluation for C_T^2 is that arising from estimates of the parabolic dish (antenna)-transducer efficiencies and, to a lesser extent from the calculated attenuation based on meteorological conditions. Neff [52] described in some detail the calibration procedure that can be used and made comparisons with tower measurements to obtain quantitative measurements of C_T^2. Neff also noted that examination of spatial distributions of C_T^2 have revealed the presence of quasi-horizontal patchy lamina in the stably stratified boundary layer. The high degree of anisotropy and inhomogeneity in these spatial distributions can lead to larger-than expected acoustic scatterers. The SODAR assumes that the thickness of the layer L' scattering the sound is much less than the SODAR pulse length, that is:

$$L' \ll c\tau$$

This assumption can be made to avoid the problem of including the spatial distribution of C_T^2. For cases where this assumption does not hold, we must use the shortest pulse length possible while maintaining sufficient power to obtain the SODAR range necessary to reach the heights being measured. When it is not practical to reduce the pulse length, detailed calculations can be made to determine what errors to expect. Neff provides examples of these calculations for the vertical and tilted SODARS used during his experiments.

A monostatic SODAR can also be used to measure radial components of the wind velocity in the direction of the sound pulse (Fig. 6). An examination of the scattering reveals theoretically that the backscatter at $\theta = \pi$ is due to thermal gradients

An Overview of Technological Development of Atmospheric Echosounders

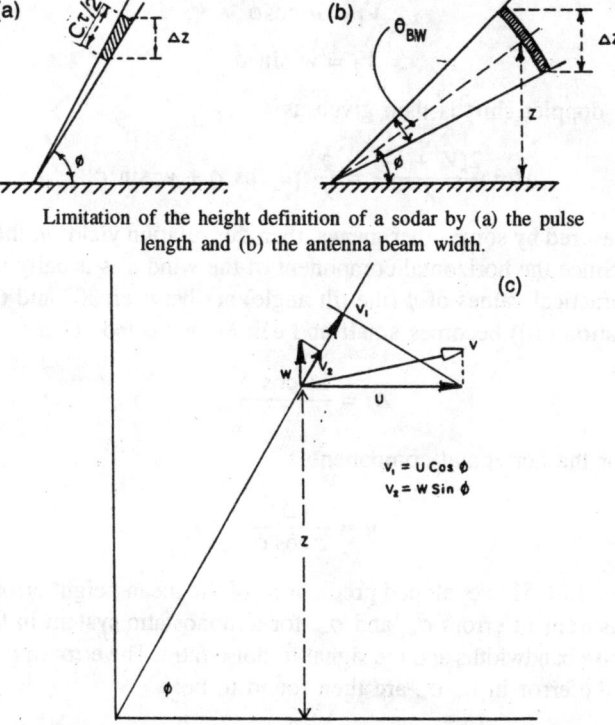

Limitation of the height definition of a sodar by (a) the pulse length and (b) the antenna beam width.

(c) Representation of the wind and its components within the pulse volume at height, Z.

Fig. 6. Geometry for a tilted monostatic SODAR used to measure horizontal winds.

only; however, in practice the velocity structure function is not exactly equal to zero, and hence some scattering may be picked up and processed for doppler content. Repeating Equation (4) the *radial* velocity component of the wind can be measured:

$$V = \frac{\lambda \Delta f}{2}$$

where V = doppler velocity, Δf = the doppler frequency shift and λ = transmitted wavelength.

A monostatic SODAR may be operated in a vertical mode to measure the vertical component of the wind or tilted at an appropriate angle to measure horizontal winds. Consider the problem of measuring a single horizontal wind component u, using a monostatic doppler SODAR. The geometry, in Fig. 6 includes the true wind vector V, which also has a vertical component w. The doppler shift in frequency as seen by the tilted monostatic SODAR will have two radial components; that is V_1 due to horizontal component of the true wind u and V_2 due to the vertical component of the true wind w. From geometrical considerations

50 Atmospheric Acoustics and Instrumentation

$$V_1 = u \cos\phi$$
$$V_2 = w \sin\phi$$

and the total doppler shift is then given as

$$\Delta f = \frac{2(V_1 + V_2)}{\lambda} = \frac{2}{\lambda} [u \cos\phi + w \sin\phi] \qquad (10)$$

If w is measured by some other means, then this relation yields u, the horizontal component. Since the horizontal component of the wind u is usually much larger than w and practical values of ϕ (the tilt angle) are between 30° and 60°, the last term of Equation (10) becomes small and can be neglected. Then

$$\Delta f = \frac{2u \cos\phi}{\lambda}$$

or solving for the horizontal component

$$u \approx \frac{\lambda \Delta f}{2 \cos\phi}$$

Spizzichino [50, 51] developed predictions of the mean height error σ_z and the velocity measurement errors σ_u and σ_w, for a monostatic system in terms of the signal and noise bandwidths and the signal-to-noise ratio. The error in u (neglecting w), σ_u and the error in w, σ_w are then found to be:

$$\sigma_u = \left(\frac{\lambda}{2 \cos\phi}\right)\sigma_f \text{ and } \sigma_w = \left(\frac{\lambda}{2}\right)\sigma_f$$

where σ_f = error in measuring Δf.

Spizzichino [50] also shows that for practical signal-to-noise ratios of greater than 5 dB

$$\sigma_u^2 \approx \frac{\sqrt{3}\lambda^2 B}{4n\tau \cos^2\phi}$$

and

$$\sigma_w^2 = \frac{\sqrt{3}\lambda^2 B}{4n\tau}$$

The mean height error σ_z, is also shown to be:

$$\sigma_z^2 = \frac{1}{6}\left(\frac{c\tau}{2} \sin\phi\right)^2 + (z\theta'_{BW} \cot\phi)^2$$

where λ = operating wavelength, B = receiver bandwidth, τ = pulse length, n = number of pulses to be averaged, ϕ = tilt angle of SODAR, z = mean height of the pulse volume v, θ_{BW} = antenna beamwidth (3 dB) and $\theta'_{BW} = \theta_{BW}/\sqrt{10}$ = variance of beamwidth.

Spizzichino also studied the problem when atmospheric refraction becomes important. the error on height determination due to refraction becomes important for $\phi < 25°$, a risk of the echoes not being received within the main antenna beam appears for $\phi < 40°$ (at a frequency of 2500 Hz). The error on horizontal wind determination due to refraction is on the order of 0.5 m/s and increases with ϕ. However, because this error is systematic and depends only on the wind itself, a correction can be applied in practice.

When these errors are sizable, one may have to choose a longer pulse at the expense of range resolution or a higher frequency at the expense of range due to attenuation. One also may select a narrower receiver bandwidth; however, the trade-off here is the need to maintain the required doppler dynamic range. The only choice may be to use more samples in the average. However, here, one must keep in mind the wind dynamics to ensure that a proper statistical average is obtained.

3.2 Bistatic and Monostatic SODARs:

Monostatic SODARs have major applications in thermal gradient detection and single-axis doppler measurements. The monostatic systems, however, have a major drawback in that their use may be limited for continuous wind profiling. Monostatic echoes (in theory) are entirely the result of temperature-produced fluctuations in the sampling volume. The returns, therfore, can be marginal, discontinuous, and provide gaps in the wind data. In a "neutral" atmosphere, no information would be obtained. Referring again to Equation (3), we can see that for bistatic SODARs having a scattering angle $0 < \theta < \pi/2$, there will be echoes from both temperature and velocity structure functions. Brown [49] analyzed in detail Equation (3) and proposed a modified version that includes the effects of a nonzero mean wind on the propagation scattering of sound by turbulent refractivity fluctuations. Following the development by Brown, suppose an initial acoustic beam launched at an elevation angle ϕ_0 and a scattered wave received at an elevation angle ϕ_1, with the apparent scattering angle $\theta = \phi_0 + \phi_1$ (Fig. 7) defines the magnitude of the components of the mean wind in the initial and scattered direction by $m_0 = e_0 \cdot m$ and $m_1 = e_1 \cdot m$, where m is the vector with components $m_i = (w_i/c_0)$. Then, for angles not too close to grazing, such that $\sin^2 \phi_0 \gg 2m_0$ and $\sin^2 \phi_1 \gg 2m_1$, use of Snell's law implies that the acattering cross section in coordinates fixed at the ground becomes

$$\sigma_m(\theta) = 0.033 k_0^{1/3}(1 - m_0)^{1/3} \cdot (M \cos \theta + m_0 \Sigma + m_1 \Sigma^{-1})^2 X^{-11/3} \cdot Y \quad (11)$$

where

$$Y = \left\{ 0.13 \frac{C_T^2}{T_0^2} + \frac{C_V^2}{c^2} \left[M \cos^2 \frac{\theta}{2} - \frac{1}{2} m_0 (1 - \Sigma) - \frac{1}{2} m_1 (1 - \Sigma^{-1}) \right] \right\}$$

with $M = 1 + m_0 + m_1$, $\Sigma = \sin \phi_0 / \sin \phi_1$, and the square of the log amplitude

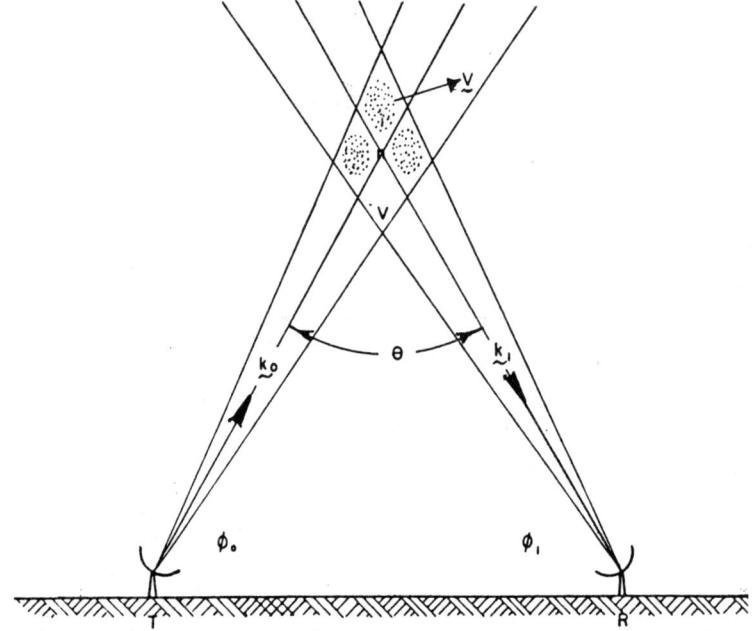

Fig. 7. A schematic drawing of the acoustic scattering for a bistatic SODAR.

$$X^2 = M \sin^2\left(\frac{\theta}{2}\right) - \left(\frac{1}{2}\right) m_0 (1 + \Sigma) - \frac{1}{2} m_1 (1 + \Sigma^{-1})$$

When the mean wind vanishes, $m_0 = m_1 = 0$, $M = 1$ and Equation (11) reduces to Equation (3).

For apparent backscatter $\theta = \pi$, both Σ and X reduce to one. Then, except for the factor $(1 - m_0)^{1/3}$, which differs only negligibly from 1 when $m_0 \ll 1$, the backscatter cross section becomes

$$\sigma_m(\pi) = 0.004 k_0^{1/3} (1 - m_0)^{1/3} \frac{C_T^2}{T_0^2} \approx \sigma_0(\pi) \tag{12}$$

approximately the same as the backscatter cross section $\sigma(\pi)$ shown in Equation (9.2). However, Equation (11) gives nonnegligible wind-induced distortion of the scattering cross section at apparent scattering angles equal to or smaller than 90°. For example, for $\theta = 90°$ with a launch angle of $\phi_0 = 45°$, the wind-distorted cross section becomes

$$\sigma_m(\pi/2) \approx 275 m^2 \, \sigma(\pi)$$

Typical values of m, ranging from 0.01 to 0.05, give the result that $\sigma_m(\pi/2)$ varies between 3 and 70% of the backscatter signal strength. Even larger percentages for $\sigma_m(\theta)$ occur as θ approaches the critical angle.

Fig. 8 shows curves of the normalized scattering cross section $\sigma_m(\theta)/\sigma_0(\pi)$ as a function of apparent scattering angle θ seen from the ground, with the various values of the parameter $\gamma = C_V^2 T_0^2 / C_T^2 c_0^2$ for a 45° launch angle and a mean horizontal wind of magnitude of $0.05 c_0$. Fig. 8a shows that large fluctuations may occur in measurements near $\theta = \pi/2$. For backscattering and near backscattering, however, the presence of a mean wind makes little difference in $\sigma(\pi)$. Fig. 8b shows the angular dependence of the scattering cross section σ for small hydrometers ($D \ll \lambda$) and the velocity and temperature fluctuations.

Bistatic SODARs are systems is which all of the transmitting and receiving antennas are separated in space but are aimed at a common volume. This configuration will yield a higher signal-to-noise ratio since returns from both temperature and wind fluctuations contribute to the scattering. Wind profiles can be obtained by moving the common volume up and down or by steering one of the beams mechanically or electronically. Alternative systems with a combination of fan-beam antennas have also been used.

Kaimal and Haugen [53] reported a simplified doppler SODAR that could be built around an existing vertically pointed receiver system. This instrument consisted of two fan-beam transmitters and a pencil-beam receiver in an orthogonal configuration. A signal-to-noise ratio considerably better than other techniques was realized from the bistatic operation and a vertically pointing receiver that could be shielded from background noise. The SES-Echosonde reported by Kleppe and Dunsmore [54–56] was an improvement over this system, because monostatic doppler SODAR could be used to provide simultaneous measurements of the vertical wind and temperature structure. This system configuration is shown in Fig. 9.

The derivation of vertical wind speed from the measured doppler frequency for the vertical pointing monostatic system has already been discussed. For the combined configuration, we again refer to Fig. 9. The system consists of a vertical monostatic doppler and two orthogonal bistatic transmitters. The vertical monostatic antenna is used to obtain values for C_T^2 and the vertical component of wind, V_z. The two bistatic transmitters each consist of a fan-beam horn of sufficient width to illuminate the scattering volume over the desired altitude range. The angle to the scattering volume from the bistatic transmit antennas is given by θ.

The total wind vector, u, v, and w can then be determined to be

$$w = \frac{c \Delta f z}{2f} \text{ (vertical component)} = V_z$$

$$u = \left[\frac{c}{f \cos \theta}\right] \Delta f_{xz} + V_Z \left[\frac{1 + \sin \theta}{\cos \theta}\right] \text{ (x component)}$$

$$v = \left[\frac{c}{f \cos \theta}\right] \Delta f_{yz} + V_Z \left[\frac{1 + \sin \theta}{\cos \theta}\right] \text{ (y component)} \quad (13)$$

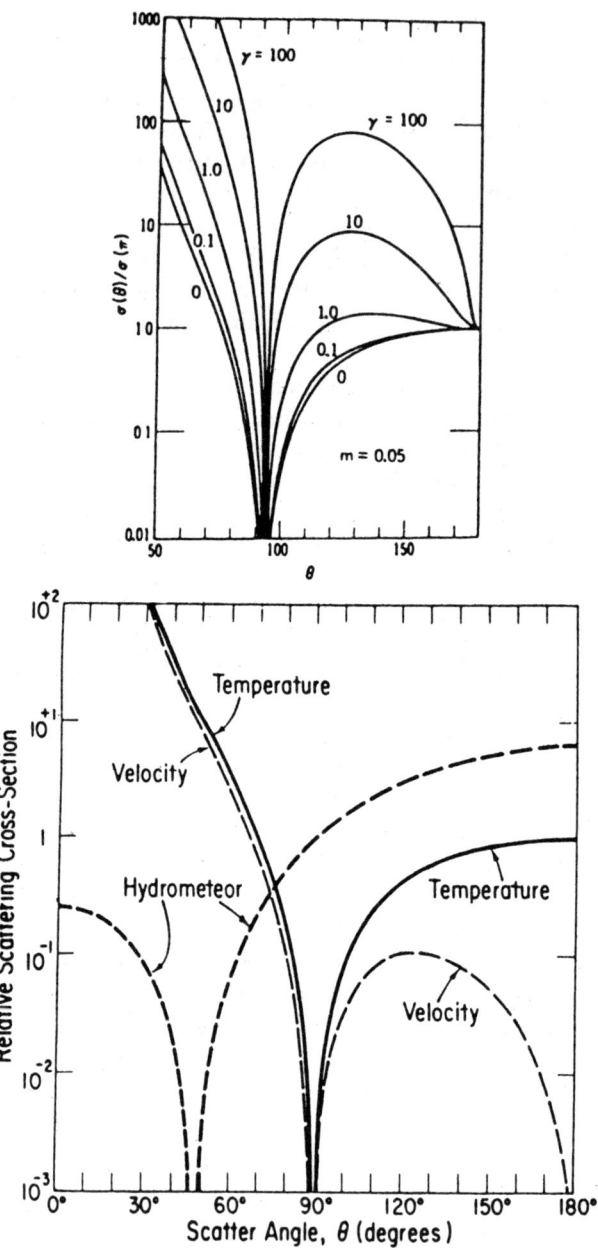

Fig. 8. Curves of the normalized scattering cross section. (a) Typical deformation due to a 15 m/s mean wind, where $\gamma = C_V^2 T_0^2 / C_T^2 c_0^2$.
(b) Angular dependence for small hydrometers ($D \ll \lambda$) and for the Kolmogorov spectrum of velocity and temperature variations.

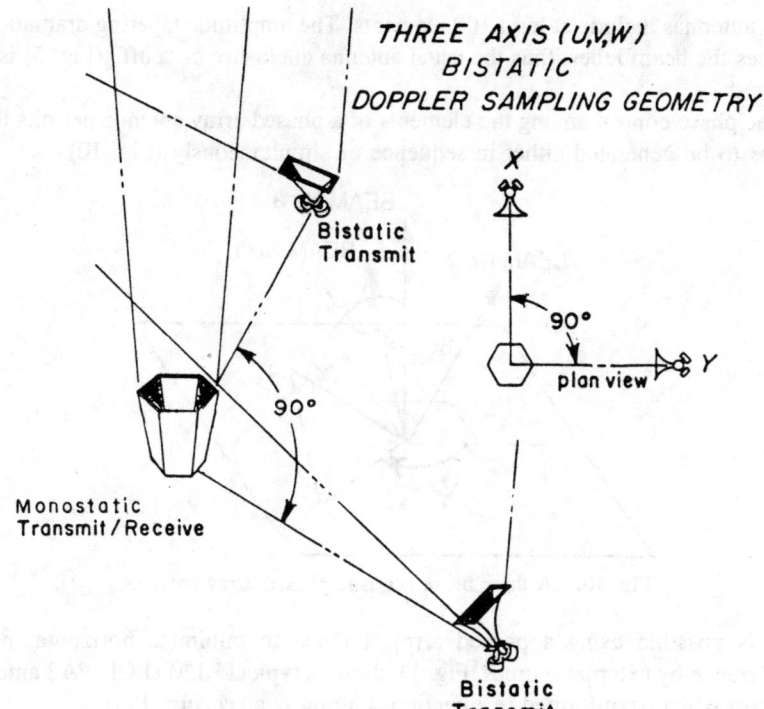

Fig. 9. The SES-Echosonde three-axis monostatic-bistatic doppler SODAR.

where f = transmitted frequency, c = speed of sound, Δf_{xz} = doppler frequency shift in xz plane and Δf_{yz} = doppler frequency shift in yz plane.

It is possible to use a single frequency for all three antennas by time multiplexing the transmit tone bursts; that is, $T_1, T_2, T_1, T_3, ...,$ *et cetera*. A microcomputer can then compute the various doppler frequency shifts via time of arrival to the desired scattering volume.

There have been a number of investigations into the various possible SODAR system configurations. It can be seen that there is no one system that can be used for all possible applications. The paper by Balser et al, [40] presents an analysis of several configurations including a *continuous wave* (CW) system. A vertical pointed transmitter sending to tilted receivers has an inherent problem in its increased susceptibility to background and wind noise over a vertically pointed receiver. Davey [57] developed an optimization technique that provides a systematic method for selecting the antenna geometry of a doppler SODAR system.

3.3 Phased Array Antennas for SODAR Applications

The development of the acoustic phased array antenna has greatly advanced SODAR applications. There are available on the market today a complete family of electronically steered beam acoustic phased array antenna for sodars. Most of these antennas include at least 100 elements. The amplitude tapering dramatically reduces the beam lobes, thus the usual antenna enclosure or "cuff" (Fig. 5) is not required.

The phase control among the elements of a phased array antenna permits three beams to be generated either in sequence or simultaneously (Fig. 10)

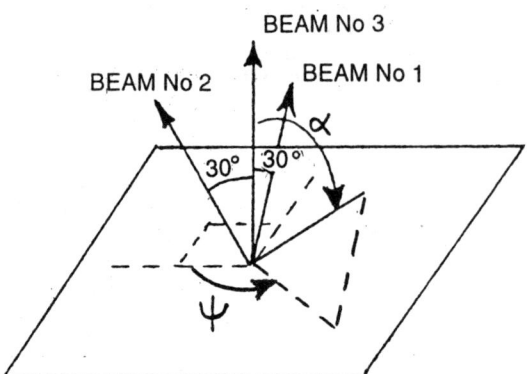

Fig. 10. A three beam acoustic phased array antenna.

It is possible using a phased array antenna to minimize horizontal noise interference by external sources. Fig. 11 shows a typical REMTECH PA2 antenna diagram when beamformed in direction 2 along α at ψ (Fig. 11a).

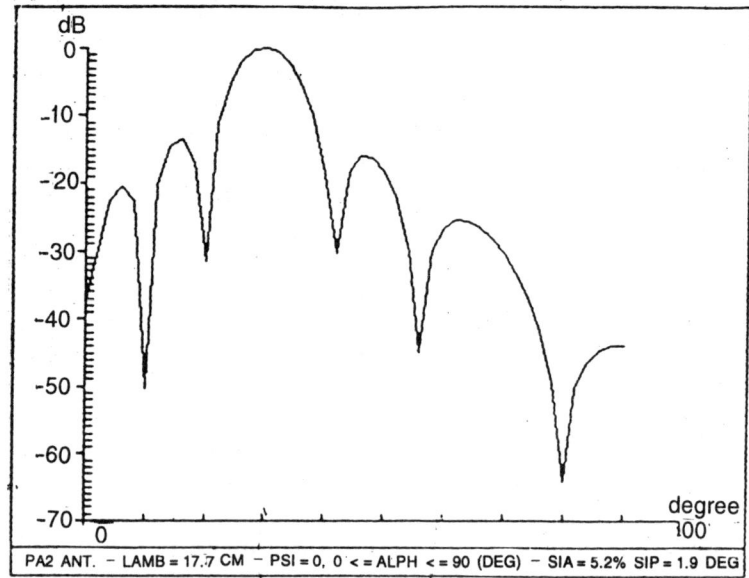

Fig. 11(a). REMTECH PA2 antenna diagram when beamformed in direction 2.

Fig. 11(b) shows the diagram for this same antenna when α is fixed at 90° and ψ varies from 0 to 360°. This configuration corresponds to a tilted beam and shows the secondary lobes parallel to the ground which is usually the prevailing direction for incoming noise.

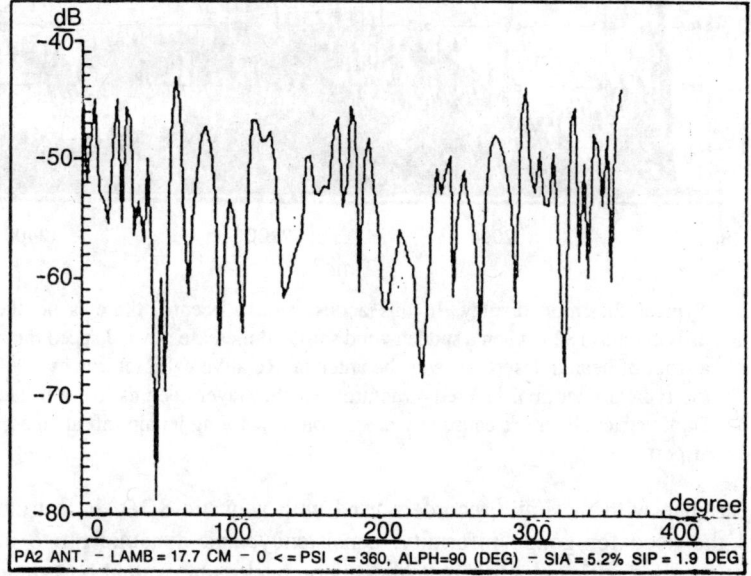

Fig. 11(b). Same antenna as of Fig. 11a when α is fixed at 90° and ψ varies from 0 to 360°.

The advantages of the phased array antenna are numerous and include:

- Only one antenna is needed, the steering of the beam is performed by phase control among the antenna elements.
- The power increase is significant thus leading to increase in range compared to previous systems.
- No antenna enclosure is needed due to tapering of the amplitude and the antenna is a simple flat plate.

A single phase array antenna such as that shown in Fig. (3) is small, light weight and portable.

4. Signal Processing and Display Methods

4.1 Amplitude Information

The first method for displaying SODAR intensity returns was through the use of analog facsimile recorders, a typical output of which is shown in Fig. 12. The display was usually generated by electrical engraving on special paper, using a current writing amplifier. Such facsimile records were of high quality but were also quite cumbersome and difficult to analyze.

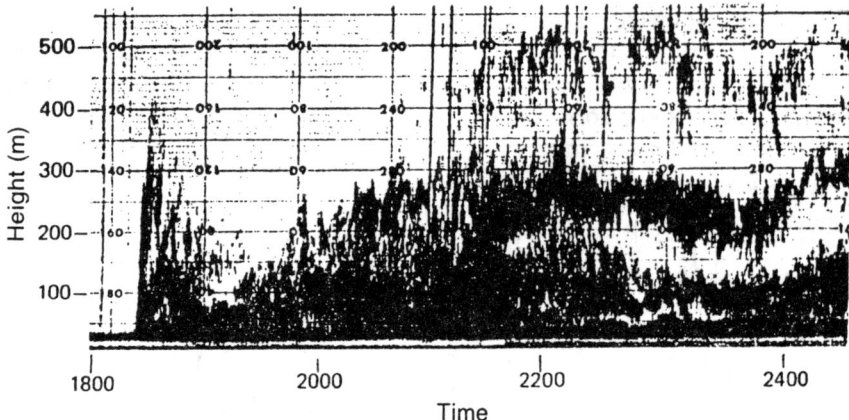

Fig. 12. Typical facsimile display. In this acoustic radar record, there is no thermal activity until 1820, when a sudden wind shift and increase in wind speed transports a front of heated desert air over the antenna. Relative calm returns by 2000 and the typical nocturnal layered structure of stable layers begins to form later on. Dark vertical lines are caused by noise from low flying jet aircraft at an adjacent airport.

Experience with facsimile recorders used with acoustic SODAR instruments proved the need for good uniformity, consistency, and dynamic range in the display. To achieve this in a SODAR, the analog backscatter signal must be gain-ramped for range correction and converted to a representative power measurement by a precision rms-to-dc voltage converter. This signal is then sampled by an *analog-to digital converter* (ADC), which sends these digitized values to a computer.

The sampling rate, range-gate resolution, and shading constants are then selected according to the specific application being addressed. For example, as in the case of a meteorological radar, the contributing volume depth (spatial) resolution is equal to $c\tau/2$, where c is the speed of sound and τ the pulse width of the SODAR.

Minimum range depends on the pulse width as well as on the duration of antenna "ringing" for a monostatic system or the direct path time for a bistatic system. This could practically be on the order of 2 pulse lengths.

Many different displays for SODAR amplitude data have been devised. One early system using a dot matrix printer was reported by Owens [37]. The forward-thinking work of Owens showed that a dot matrix printer could be used for a SODAR display. And, according to Owens [37] ten levels of gray scale were found to be adequate. A benefit of this dot matrix approach was the simultaneous display of wind characteristics determined from doppler analysis at the appropriate height on the display. When interpreting the charts, it was especially useful to have wind speed and direction, for example, at the appropriate location on the inversion height display. A sample SODAR display using this dot matrix approach is shown in Fig. 13, which shows that the printer output could provide meaningful

An Overview of Technological Development of Atmospheric Echosounders 59

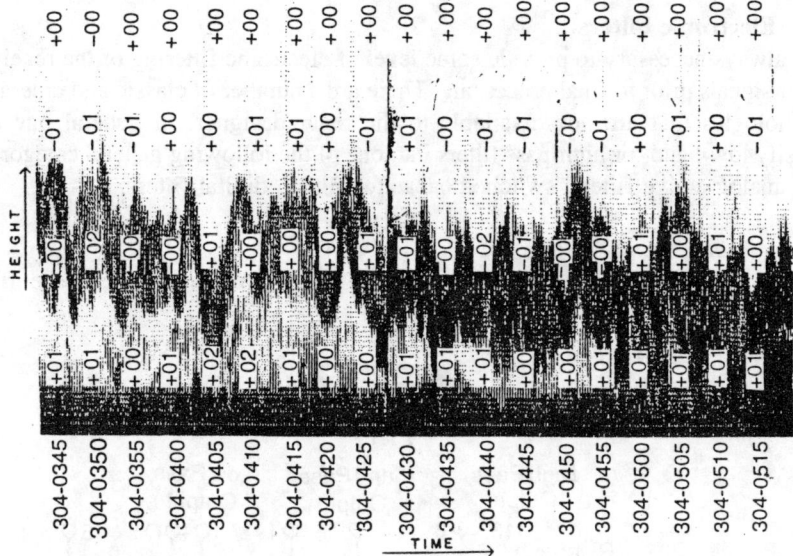

Fig. 13. Digital computer dot matrix printer generated acoustic radar display. Intensity information was averaged in time and space. Doppler information is shown averaged at five elevations over 5-minute intervals. This record shows an inversion being eroded from the top by wind and weakened by thermal plumes from within.

interpretation and recognition of atmospheric processes. In addition, its versatility in producing different types and combinations of displays provide the user with a permanent, nondeteriorating record of computer commands and operations and program output.

Many other display methods that have been developed since that include the use of modern computer graphics and color monitors. One example is shown in Fig. 14.

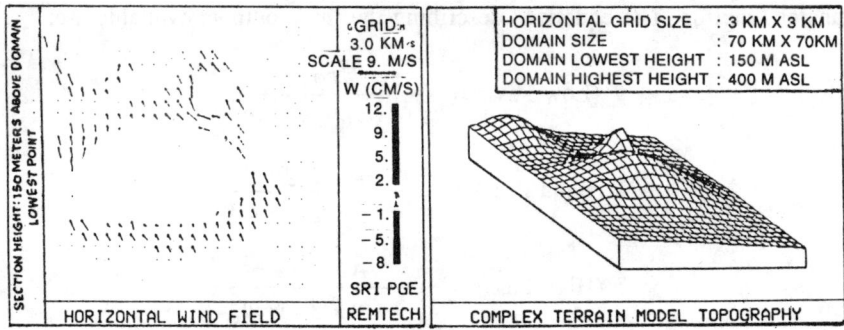

Fig. 14. Real time SODAR generated wind field analysis model.

4.2 Electronic Filters

It is always necessary to provide some level of electronic filtering of the received sodar signals prior to final processing. There are a number of classic and emerging technologies that are now available to the radar designer. In general one can classify low-band- or high-pass filters into one of the following general categories, vig. analog active filters, switched capacitor filters, digital filters.

4.2.1 Analog Active Filters

One of the most widely used active filters is the so called state space filter (Fig. 15).

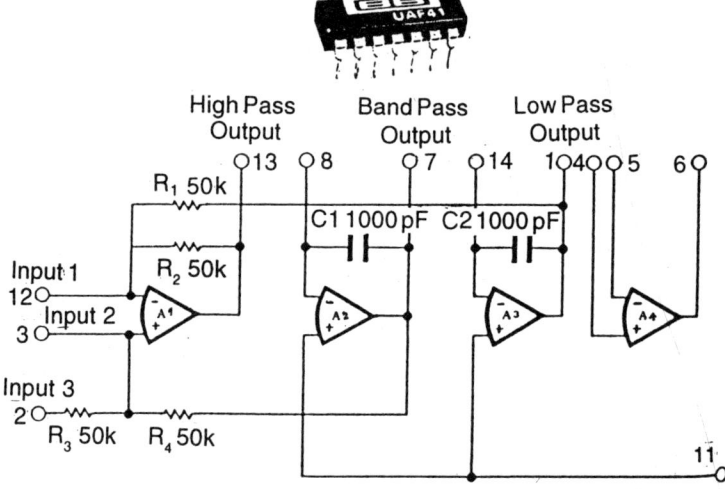

Fig. 15. Universal active filter UAF41 [58].

The UAF41 uses the state variable technique to produce a basic second order transfer function. The equations describing the three outputs available are:

$$T \text{ (Low Pass)} = \frac{A_{LP}\omega_0^2}{s^2 + (\omega_0/Q)s + \omega_0^2}$$

$$T \text{ (Band Pass)} = \frac{A_{BP}(\omega_0/Q)s}{s^2 + (\omega_0/Q)s + \omega_0^2}$$

$$T \text{ (High Pass)} = \frac{A_{HP}s^2}{s^2 + (\omega_0/Q)s + \omega_0^2}$$

To obtain band reject characteristics the low-pass and high-pass outputs are summed to form a pair of $j\omega$ axis zeros:

$$T \text{ (Band Reject)} = \frac{A(s^2 + \omega_0^2)}{s^2 + (\omega_0/Q)s + \omega_0^2}$$

where $A_{LP} = A_{HP} = A$

The state variable approach uses two op amp integrators (A2 and A3 in the simplified schematic of Fig. 15) and a summing amplifier (A1) to provide simultaneous low-pass, band-pass and high-pass responses. One UAF41 is required for each two poles of low-pass or high-pass filters and for each pole-pair of band-pass or band reject filters.

The Universal Analog Filter is quite straightforward to use. For example, consider the following design problem.

Example 1: It is desired to design a 4 pole Butterworth, Band-Pass Filter with $Q = 25$ and a center frequency $f_c = 19$ kHz. The band-pass gain is to be U $A_{BP}=1$.

Use of a simple program provided by the UAF manufacture [58] will show the resulting hardware design shown in Fig. 16.

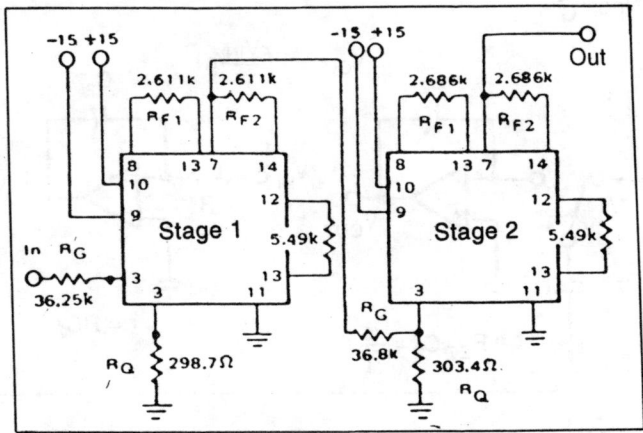

Fig. 16. Final design for the band-pass filter of Example 1.

It can be seen that complex filter design has been reduced to the selection of a few resistors. The main drawback to the UAL concept is the limited bandwidth and the fact that the filter constants cannot be changed in real time under computer control.

4.2.2 Switched Capacitor Filter (SCF) [59]

Switched-capacitor filters (SCF) are not new; they first appeared in IC form in the late 1970s. During their early years, however, SCFs acquired the reputation of being noisy and having limited frequency response—less than 30 kHz. As a result, they were used on only a limited number of applications. That reputation is no longer deserved. Technological changes over the past few years have resulted in a fourfold reduction in noise levels and the operating frequency has been multiplied. Some SCF can now operate to 150 kHz.

SCFs are based on the principle that a capacitor, if switched in the circuit, can be made to function as a resistor (Fig. 17(a)). A switched-capacitor's effective resistance is proportional to the switching rate. Because the frequency response of an active filter is controlled by its resistor and capacitor values (Fig. 17(b)), substituting a swithced-capacitor for the resistor in a filter allows you to tune the filter simply by changing clock rates. This tuning ability is one of the primary benefits of SCFs.

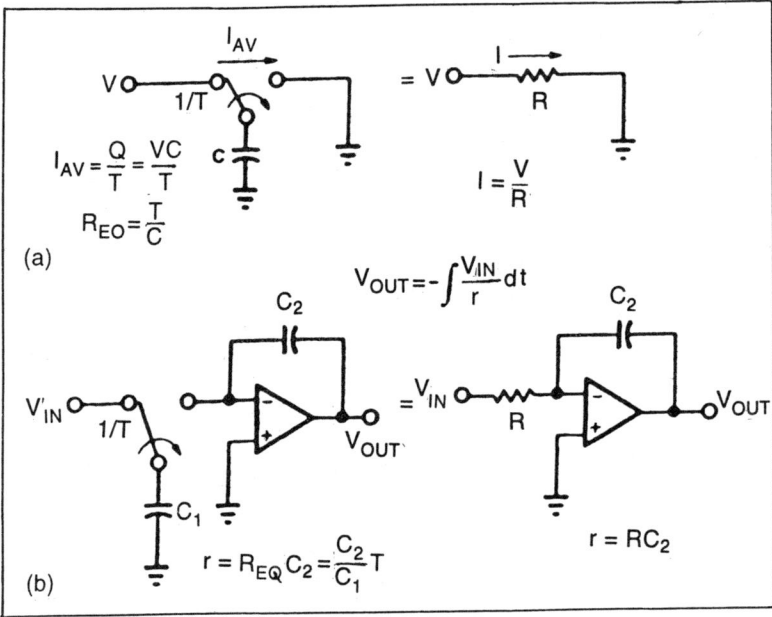

Fig. 17. A capacitor (a) can be made to behave as a resistor when switched at a rate of $1/T$. Its effective resistance is T/C; a switched-capacitor integrator (b) has its frequency response controlled by a capacitor ratio and the clock frequency.

There are digitally programmable universal active filters based on switched capacitor principles. For example the EG&G RETICON RU5620A (Fig. 18) [60].

The RU5620 is a digitally programmable Universal Active Filter (UAF). It is implemented using Reticon's proven doublypoly NMOS switched capacitor filter technology. It consists of a single second-order section with Q and f_0 programmable by separate 5-bit digital control words. Basic filter types are selected by hard wiring or switch selecting the filter inputs. Higher order filters can be obtained by cascading the RU5620. Thus cascadability, microprocessor control, and universality make the RU5620 an ideal, cost effective filter building block.

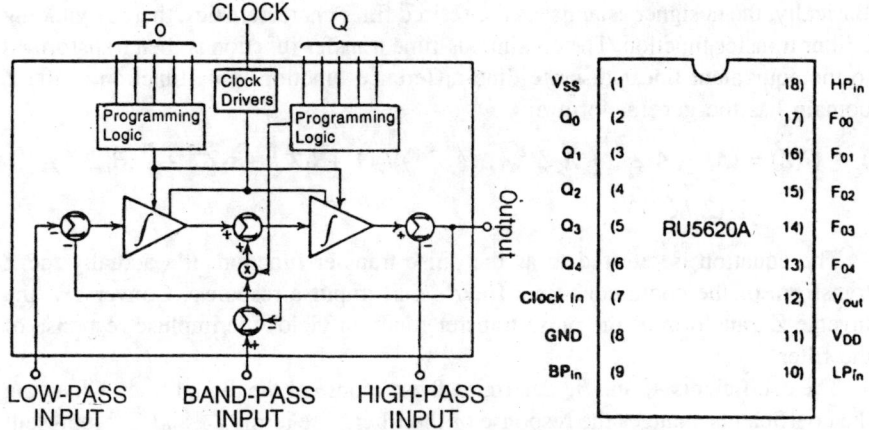

Fig. 18. (a) Functional block diagram and (b) Pinout.

Key Features
- Universal: Implement any classical filter
- Easy to use: No external components required
- Look up table responses: No calculations required
- Q and F_0 independently programmed
- Clock and control lines are TTL and CMOS compatible
- Small size: 18 pin DIP
- Low current consumption: Typically 4.5 mA
- Wide supply Voltage Range: ± 5 V to ± 10 V

Filter Types Available
- Low-pass
- Band-pass
- High-pass
- Low-pass elliptical
- High-pass elliptical
- Notch
- All-pass

4.2.3 Digital Filters [61]

With the advances made in digital-signal processing, digital filters are becoming a more attractive design alternative to traditional analog techniques. Because digital-system information is in digital form, filtering can be accomplished relatively easily by passing the data through a filter algorithm. In addition, digital filters have the advantages of no filter-characteristic drift over time, temperature, or voltage. And they can easily be designed to filter low-frequency signals. Moreover, the filter response can be made to closely approximate the ideal response, and linear phase characteristics are possible.

There are many well established methods of determining the filtering algorithm.

Basically, the designer establishes the sesired filter characteristics, thereby yielding a filter transfer function. The continuous-time transfer function is then transformed to the equivalent linear discrete-time-difference function. This function in the Z domain has the general form of:

$$G(Z) = (A_0 + A_1 Z^{-1} + A_2 Z^{-2} + \ldots A_n Z^{-n}) / (1 + B_1 Z^{-1} + B_2 Z^{-2} + \ldots B_m Z^{-m})$$

$$= Y(Z)/X(Z)$$

The equation is referred to as the pulse transfer function. It's actually the Z transform of the continuous-time filter's unit impulse response. Conversely, the inverse Z transform of the pulse transfer function yields the impluse response of the filter.

The coefficients A_n and B_m determine the response of the digital filter. Changing the coefficients changes the response of the filter. The terms Z^{-n} and Z^{-m} represent sampling delays or taps. The $G(Z)$ equation represents the algorithm of sampling the input, multiplying it by A_0, and adding it to the previous sample that's been multiplied by A_1, then adding that value to the next previous sample which has been multiplied by A_2, and so on. An output value occurs when all N values have been multiplied and accumulated.

In parallel, each output value is stored, multiplied by B_1, then added to the previous output value which has been multiplied by B_1, and so on. The equation can be rearranged so that the result of the output multiply accumulate is added to the result of the input multiply accumulate to produce an output. This procedure is referred to as convolution. An output sample is produced for every input sample (Fig. 19).

The key to digital-filter design is to determine the filter coefficients that will produce the desired frequency response. Recursive digital filters, or infinite-impulse-responsive (IIR) filters, are a type of digital filter in which the design methodology closely follows that of an analog filter. One method for determining the coefficients is to define a realizable continuous-time domain Chebyshev, Butterworth, or equal-ripple filter then use Z transforms to transform the continuous-time-domain transfer function to the equivalent discrete-time transfer function that yields the filter coefficients.

A second popular method is the bilinear transform. In this method, engineers first design an analog filter so that after it's transformed to a digital filter, the resulting filter meets a set of desired digital-filter specifications. This analog filter is then transformed to a digital filter via the bilinear transform from the S variable of the Laplace transform to the Z variable of the Z transform.

In a non-recursive digital filter or finite-impulse-response (FIR) filter, the output is computed using the present input X_n and the previous inputs X_{n-1}, X_{n-2}, $\ldots X_{n-N}$. This implies that the coefficients B_m, are all 0, and there's no feedback from the output. Designing non-recursive digital filter (FIR) involves defining an ideal desired frequency response from which the ideal impulse response is computed.

An Overview of Technological Development of Atmospheric Echosounders   65

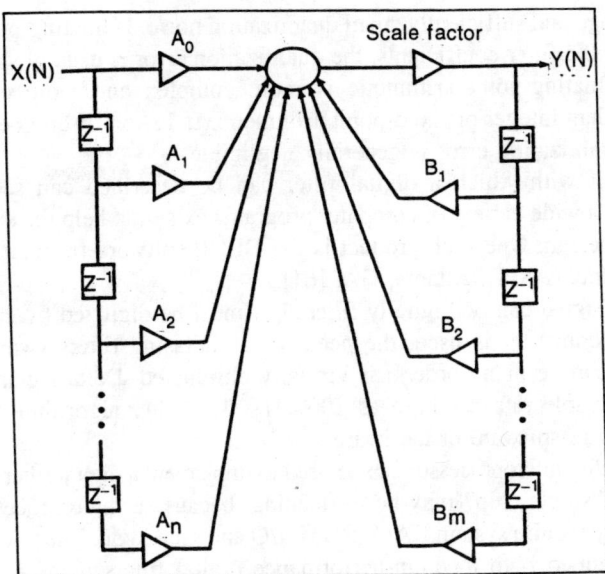

Fig. 19. In functional structure of a digital filter, A and B coefficients determine the response of the filter and Z terms represent sampling delays called taps (From [61]).

The ideal impulse response is truncated to a finite number of non-zero samples using a windowing function, which is judiciously chosen. A common windowing function is the Kaiser window function.

There are advantages and disadvantages to each type of digital filter (IIR and FIR). An IIR filter is always stable because there is no feedback from the output and the impulse response is finite. In addition, the amplitude and phase can be arbitrarily specified. On the other hand, an FIR filter will generally require more taps, and consequently more math, to compute the output value. The design methodology does not resemble the familiar analog design techniques.

An IIR will generally have fewer coefficients, but the required output feedback can make circuit implementation more complex. A stable IIR filter can become unstable if the coefficients arent' chosen properly to account for digital math errors.

There are four main types of errors that can arise in the design of digital filters. These are referred to as quantization errors. They are:

Quantization errors of the input analog-to-digital conversion.
Quantization errors of the coefficients.
Quantization errors due to arithmetic computations, including overflow.
Limit cycles.

In most cases, a 12-bit analog-to-digital converter (ADC) provides enough dynamic range and sufficiently small quantization noise. If floating-point numbers are used for the filter coefficients, the quantization error is usually small enough. However, floating-point arithmetic is more complex and more expensive to implement than integer or fixed-point arithmetic. If 12- or 16-bit coefficients are used, the quantization error is generally negligible.

The detail with which a digital filter can be described can seem endless. Fortunately, a wide variety of computer programs exist that help the engineer with the filter's design. One such product is the DEDP software from Atlanta Signal Processing Inc. (ASPI), Atlanta, GA, [61].

Before a signal can be digitally filtered, it must be digitized by an ADC. If a delta-sigma converter is used, the need for antialiasing filters (which must be analog and can be many orders) is virtually eliminated. Delta-sigma converters may have sample rates as high as 100 kHz. The filter algorithm can then be implemented in software or hardware.

A single-chip microprocessor can be used to implement a digital filter in software. However, a "single chip" may be misleading, because a microprocessor system will generally require system RAM, ROM, I/O and glue logic. The microprocessor can implement low- to medium-performance digital filters if the only function they are performing is the digital filtering. As the work load of the microprocessor increases, its capability to digitally filter a signal in real time decreases. Once the system is designed, changing the filter's characteristics is as easy as changing variables in software and downloading the code to the system.

For higher performance and moderate flexibility, the filter can be implemented in dedicated hardware using programmable logic for design flexibility. The limiting parameter will be the time to do a multiply-accumulate function and the amount of physical space required for the hardware implementation of the taps. Consider a circuit that uses a single-port 16-bit multiplier-accumulator capable of an 85-ns clock speed (Fig. 20.)

The device can work in twos-compliment numbers and has output saturation capabilities. As stated before, these two features are desirable when implementing digital filters. In addition, the device can be easily controlled with a programmable logic device (PLD) because it's microcoded based.

It will be seen later that the type of filter to select, i.e., analog, switched capacitor or digital will depend to a great extent on the final signal processing method used. For example if Fast Fourier Transform (FFT) methods are used then simple analog anti-aliasing filters are all that may be required with the remaining filter functions being provided by an FFT Digital Signal Processing (DSP) hardware/ firmware arrangement.

4.3 Spectral Mean and Variance Estimation Using Digital Signal Processing

Interest in the application of digital methods to signal processing has rapidly increased, and numerous approaches to digital signal processing can be applied to

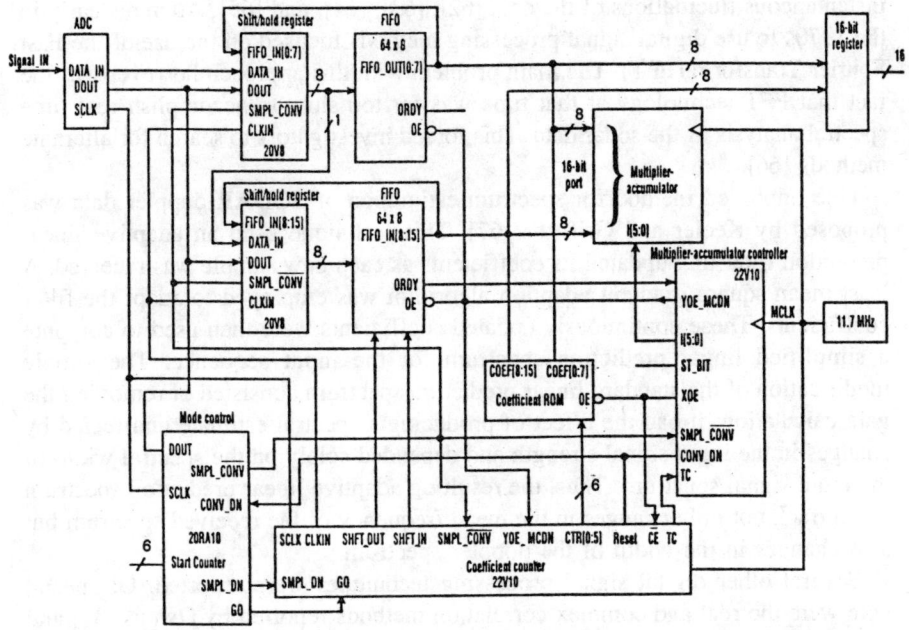

Fig. 20. An FIR filter is implemented in a circuit that uses a single-port 16-bit multiplier accumulator capable of a 85-ns clock speed. Because it's based on microcode, the multiplier accumulator can be controlled with a PLD, from [61].

SODARs. The major concern is to select a method that minimizes the number of computations and amount of storage required. The method for obtaining spectra for SODAR data differs from radar primarily in its *pulse repetition frequency* (PRF). In the acoustic case, a spectrum can be generated for each pulse. For a given signal-to-noise and duty cycle, an increased PRF does not improve the accuracy of the estimate; however, some improvement results in averaging acoustic spectra from succeeding pulses with a resulting loss of horizontal resolution.

The input data to a SODAR signal processor can be assumed to consist of analog variations representing signal plus noise (or noise alone) corresponding to a given time or range. For many applications, the complete doppler spectrum is not needed, so only the mean velocity (or some suitable central value) and its velocity spread are recorded. This greatly simplifies data handling by avoiding the problems attendant to the measurement and storage of complete spectral information. Moreover, greater accuracy can be attained in measurements of the mean doppler velocity and velocity spread than in complete spectral measurements, although the same general constraints apply to the required length of record. The duration must be long compared with the decorrelation time but short compared with the time scale or target variability. The time constants or integration times for measurements of the mean velocity and velocity spread can be determined for SODARs. The signal can generally be modeled as "locally" stationary, because in such cases the range gates can be chosen so that the time trends are very slow relative to the instantaneous fluctuations of the data [62], [63], [64], and [65]. Attempts early in the 1970s to use digital signal processing methods focused on the use of the Fast Fourier Transform (FFT). The main problem with this approach however was the fact that FFT technology at that time was far too slow to accomplish real time spectral analysis of the sodar data. This forced investigators to search for alternate methods [66].

One improved method for spectrum estimation of SODAR doppler data was proposed by Keeler and Griffiths [67]. This technique used an adaptive linear prediction filter that updated its coefficients as each new sample was received. A least mean square gradient adaption algorithm was employed to adapt the filter coefficients. These continuously updated coefficients were then used to compute a simplified linear prediction spectrum for the input sequence. The simple modification of the standard linear prediction spectrum consisted of removing the gain calculation; it had the effect of producing a spectral estimate unaffected by changes in the input signal strength and depended solely on the spectral width of the input signal spectrum. Thus, the resulting adaptive linear prediction spectrum could track not only changes in the mean frequency of the received spectrum but also changes in the width of the doppler spectrum.

Several other digital signal processing techniques were reported. Of special note were the real and complex correlation methods reported by Owens [37] and Kleppe [38, 39, and 68].

4.3.1 Real Correlation Approach to Spectral Moment Estimation
[37, 62, 63, 64]

Mathematically, the problem is to determine the statistics of certain functions of the estimators of the spectral density of a random process. The autocorrelation function $R(\tau)$ for a real stationary random process $x(t)$ is defined as

$$R(\tau) = \lim_{T \to \infty} \frac{1}{T} \int_0^T x(t)x(t + \tau)\, dt \tag{14}$$

The quantity $R(\tau)$ is always a real valued even function with a maximum at $\tau = 0$. It may be either positive or negative.

For a sampled signal of N data values $x(n)$, $n = 1, 2, ..., N$ from a record $x(t)$ that is stationary with zero mean, the estimated autocorrelation function at the displacement $\tau = rT$ can be written in the following form:

$$\hat{R}_x(rT) = \frac{1}{N - r} \sum_{n=1}^{N-r} x(n)x(n + r) \quad r = 0, 1, 2, ..., m \tag{15}$$

where T = sample period, r = lag number, m = maximum lag number and N = sample record length.

For cases involving doppler SODAR data, the function $R(\tau)$ persists periodically with the same period as the underlying sine wave. Because of turbulence, the signal $x(t)$ is not a pure sinusoid but is rather composed of a set of sine waves whose mean frequency changes with the motion of the signal scatterers. The signal therefore is a nonstationary, narrowband, random, process. The SODAR data however may be considered as locally stationary, because in such a case the time trends are time-varying. Power spectrum then computed from an ensemble of sample records will be closely approximated by a short time-averaged power spectrum computed from a single sample record, (Kleppe and Dunsmore [38] and Owens [37]).

A typical plot of $R(\tau)$ for a SODAR record is shown in Fig. 21 for zero doppler shift, that is, $\omega = \omega_0$.

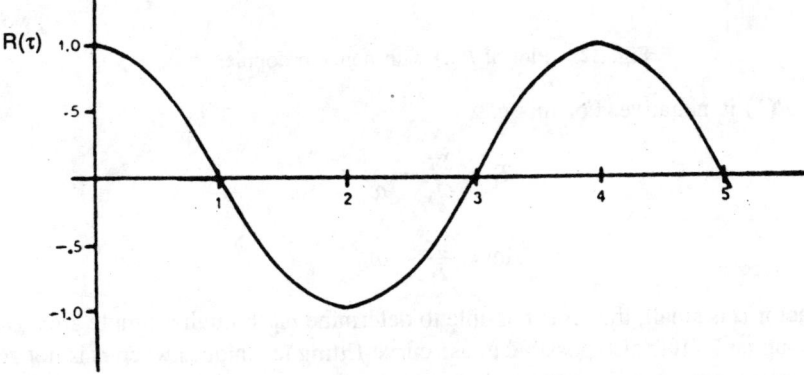

Fig. 21. Plot of $R(\tau)$ with zero doppler shift.

Function $R(\tau)$ has the following general form:

$$R(\tau) = e^{-\alpha|\tau|} \cos(\omega\tau)$$

For SODAR data, the value of α appears to be quite small. This has been confirmed by calculating $R(\tau)$ for many data runs and observing the results.

The model of $R(\tau)$ over the first few cycles is given with $\alpha = 0$:

$$R(\tau) \approx \cos(\omega\tau) \qquad (16)$$

for values $\omega = (2n + 1)\,\pi/2$

$$R(\tau) = \cos(\omega\tau) = 0$$

and for

$$\omega_0 = \frac{\pi}{2} \text{ and } \tau = 1$$

$$R(1) = 0$$

When a doppler shift occurs, it is possible to calculate $R(\tau)$ for each of the values $\tau = 0, 1, 3, 5....$ and compare the sign of $R(\tau)$ with that calculated for the zero doppler shift case. Fig. 22 shows the case where the doppler shift is $\omega_1 > \omega_0$.

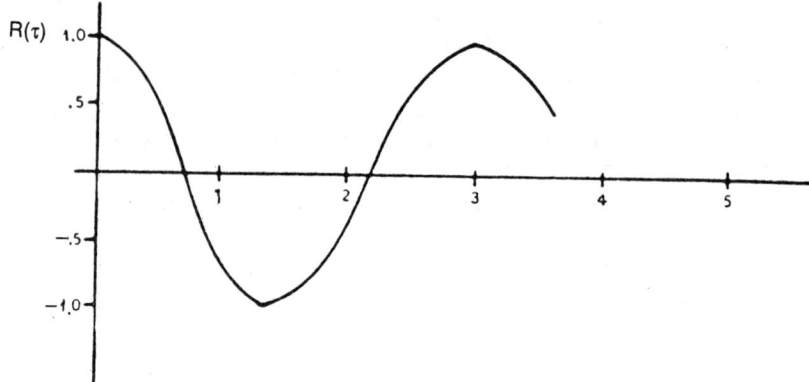

Fig. 22. Plot of $R(\tau)$ with nonzero doppler shift.

and $R(1)$ is negative. For this case

$$T_1 = \frac{2\pi}{\omega_1} = \frac{2\pi}{\omega_0 + \Delta\omega}$$

$$\Delta\omega = \frac{2\pi}{T_1} - \omega_0$$

If factor α is small, then it is possible to determine ω_1 through a simple arc cosine look-up table. It is also possible to use curve-fitting techniques when α is not zero or when more lag values are calculated.

4.3.2 Complex Correlation Approach to Spectral Moment Estimation

Mathematically, the problem is to determine the statistics of certain functions of the estimators of the spectral density of a random process.

Let $x(t)$ be a stationary random process having an estimated spectral density function denoted by $S_x(f)$. The estimated mean frequency $\dot{f}$ of $x(t)$ is defined as

$$\dot{f} = \frac{\int_{-\infty}^{\infty} f S_x(f)\, df}{\int_{-\infty}^{\infty} S_x(f)\, df} \tag{17}$$

Also let $R(\tau)$ represent the complex autocorrelation function of the process $x(t)$. $R(\tau)$ and $S(f)$ are a Fourier transform pair defined by

$$R(\tau) = \int_{-\infty}^{\infty} S_x(f) e^{j2\pi f \tau}\, df \tag{18}$$

Moments of $S(f)$ correspond to the derivatives of $R(\tau)$ evaluated at $\tau = 0$. In equation form for the bandlimited case

$$R^{(n)}(\tau) = \int_{-\infty}^{\infty} (j2\pi f)^n S_x(f) e^{j2\pi f\tau} \, df \qquad (19)$$

Then it follows that

$$R(0) = \int_{-\infty}^{\infty} S_x(f) \, df \qquad (20)$$

and, using the notation of $\cdot$ to denote derivatives

$$\dot{R}(0) = \int_{-\infty}^{\infty} (j2\pi f) S_x(f) \, df = j2\pi \int_{-\infty}^{\infty} f S_x(f) \, df \qquad (21)$$

Therefore, using Equation (17)

$$\bar{f} = \frac{\dot{R}(0)}{j2\pi R(0)} = -j\frac{\dot{R}(0)}{2\pi R(0)} \qquad (22)$$

The squared spectral width or variance σ^2 can be determined from

$$\sigma^2 = \mu_2 - \mu_1^2$$

where μ_1 = first moment = $\bar{f}$ and

$$\mu_2 = \frac{\int_{-\infty}^{\infty} f^2 S_x(f) \, df}{\int_{-\infty}^{\infty} S_x(f) \, df} \qquad (23)$$

Applying Equation (19) once more yields

$$\ddot{R}(0) = \int_{-\infty}^{\infty} (j2\pi f)^2 S_x(f) \, df = -4\pi^2 \int_{-\infty}^{\infty} f^2 S_x(f) \, df$$

and using Equation (20)

$$\mu_2 = \frac{-\ddot{R}(0)}{4\pi^2 R(0)}$$

Therefore

$$\sigma^2 = \mu_2 - \mu_1^2 = -\frac{1}{4\pi^2}\left[\frac{\ddot{R}(0)}{R(0)} - \left(\frac{\dot{R}(0)}{R(0)}\right)^2\right] \qquad (24)$$

The complex autocorrelation function $R(\tau)$ may be defined as the complex function

$$R(\tau) = A(\tau) e^{j2\pi\phi(\tau)}$$

where $A(\tau)$ = a real *even* function of τ and $\phi(\tau)$ = a real *odd* function of τ. Using the properties that

$$\dot{A}(0) = \phi(0) = 0$$

we see that

$$\dot{f} = -\frac{j\dot{R}(0)}{2\pi R(0)} = \dot{\phi}(0) \qquad (25)$$

The derivative can be approximated using an appropriate small nonzero value for τ. Thus

$$\dot{f} = \dot{\phi}(0) \approx \frac{\phi(\tau) - \phi(0)}{\tau} = \frac{\phi(\tau)}{\tau} \qquad \tau \neq 0$$

noting that $2\pi\phi(\tau)$ is the arg of $R(\tau)$ we can write:

$$\dot{f} = \frac{1}{2\pi\tau} \arctan\left[\frac{\mathrm{Im}R(\tau)}{\mathrm{Re}R(\tau)}\right]$$

applying Equation (24) with an approximation for the derivatives yields:

$$\sigma^2 = \frac{-1}{4\pi^2}\frac{\ddot{A}(0)}{A(0)} = \frac{1}{2\pi^2\tau^2}\left[1 - \frac{A(\tau)}{A(0)}\right]$$

and using the fact that

$$|R(\tau)| = |A(\tau)e^{j2\pi\phi(\tau)}| = |A(\tau)| = A(\tau)$$

and

$$\phi(0) = 0$$

then

$$\sigma^2 = \frac{1}{2\pi^2\tau^2}\left(1 - \frac{R(\tau)}{R(0)}\right) \qquad (26)$$

A processing algorithm may now be proposed so that complex samples may be obtained from $x(t)$ and used to form $z(n)$ where

$$z(n) = x_r(n) + Jx_i(n)$$

The sample at time n is multiplied by the complex conjugate of samples at time $n + 1$. The products of the samples taken pairwise are averaged and the mean of the spectrum is directly proportional to the argument of the average of the complex products. This average, by definition, is the complex autocorrelation function of the signal $R(\tau)$ evaluated for $\tau = 1$. It follows then that $R(1)$ may be estimated by Equation (15), where

$$\hat{R}(1) = \frac{1}{N-1}\sum_{n=1}^{N-1} z(n)z^*(n+1) \qquad (27)$$

and

$$\hat{R}(0) = \frac{1}{2N} \sum_{n=1}^{N} [|z(n)|^2 + |z(n+1)|^2]$$

For the radar case, where τ is the radar interpulse period, the technique would be termed *pulse pair estimation*. The various aspects of the complex correlation method have been compared to *Fast Fourier Transform* (FFT) and *vector phase change* (VPC) techniques by Sirmans and Bumgarner [65] and Kleppe and Dunsmore [38].

We can also show that the variance term can be corrected for uncorrelated noise, as in Sirmans and Bumgarner [65].

$$\sigma^2 = \frac{1}{2\pi^2\tau} \left[1 - \frac{|\hat{R}(\tau)|}{\hat{R}(0)} \left(\frac{10^{SNR/10} + 1}{10^{SNR/10}} \right) \right] \quad (28)$$

where SNR = signal-to-noise ratio (dB).

Now assume that $z(n)$ is a quadrature sampled signal of N data pairs $x_r(n)$ and $x_i(n)$ for $n = 1, 2, ..., N-1$. The estimated complex autocorrelation $R(1)$, represented by Equation (27), may be written in the form:

$$\hat{R}(1) = \frac{1}{N-1} \sum_{n=1}^{N-1} z(n)z^*(n+1) = \hat{R}_r + j\hat{R}_i \quad (29)$$

A possible method for obntaining $x_r(n)$ and $x_i(n)$ from $x(t)$ is shown in (fig. 23). The resulting complex signal $z(n)$ can be written in the following form:

$$z(n) = x_r(n) + jx_i(n)$$

The mean frequency estimate, $\hat{f}_0$, can then be determined from

$$\hat{f}_0 = \frac{1}{2\pi\tau} \arctan \left[\frac{Im\hat{R}}{Re\hat{R}} \right] \quad (30)$$

and the variance σ^2 as

$$\sigma^2 = \frac{2}{(2\pi\tau)^2} \left[1 - \frac{|\hat{R}|}{R(0)} \right] \quad (31)$$

The mean velocity, $\bar{v}$ of the process measured is then given as

$$\bar{V} = \frac{\lambda}{2} \hat{f}_0 \quad (32)$$

4.3.3 Quadrature Sampling [55], [56], [68]

The quadrature components of $x(t)$ can be determined using a hardware process such as that shown in (Fig. 23). However, it will now be shown that the entire

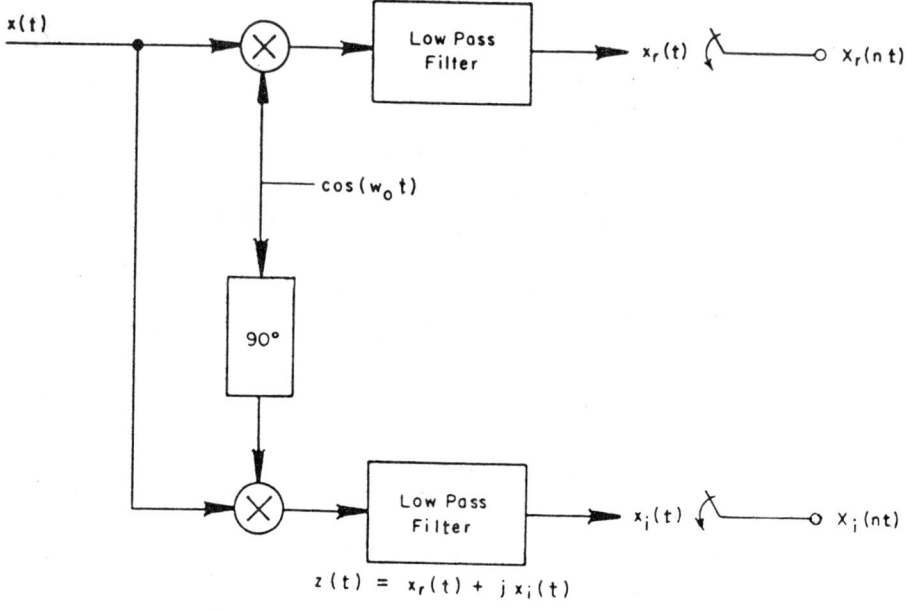

Fig. 23. Analog method for generating complex signals.

quadrature process can be replaced by a sampler running at a frequency f_s that is four times the transmitted frequency, f_0. This reduces the analog hardware needed to implement the complex covariant method. Assume that we have a received signal $x(t)$,

$$x(t) = a(t) \cos(\omega_0 t + \phi(t))$$

where $\phi(t) = 2\pi f_d t$, f_d = doppler frequency shift
We select the sampling interval, $T = 1/f_s$ such that

$$T = \frac{T_0}{4}$$

then
$$\omega_0 = \frac{\pi}{2T}$$

The representation of the sampled signal can then be written in the following form:

$$x(nT) = a(nT) \cos(\omega_0 nT + \phi(nT))$$
$$= a(nT) \cos\left(\frac{n\pi}{2} + \phi(nT)\right) \quad (33)$$

For n = even; that is, 0, 2, 4,

$$x_r(nT) = (-1)^{n/2} a(nT) \cos[\phi(nT)] \quad (34)$$

and for n = odd
$$x_i(nT) = (-1)^{(n+1)/2} a(nT) \sin[\phi(nT)] \qquad (35)$$

The complex signal $\hat{z}(m)$ then can be written as

$$\hat{z}(m) = x_r(m) + jx_i(m), \quad m = 1, 2, 3, \ldots, N' = \frac{N}{2}$$

The estimate of the complex autocorrelation function for one lag is then calculated easily using

$$R(1) \approx \frac{1}{N'-1} \sum_{n=1}^{N'-1} \hat{z}(n)\hat{z}^*(n+1) = R_r + jR_i \qquad (36)$$

and the mean frequency, $\hat{f}$, using

$$\hat{f} = \frac{1}{2\pi\tau} \arctan\left[\frac{R_i}{R_r}\right] \qquad (37)$$

with the variance σ^2 being

$$\sigma^2 = \frac{1}{2\pi^2\tau^2}\left[1 - \frac{|R|}{R(0)}\right] \qquad (38)$$

The hardware required to obtain the doppler information, therefore, has been greatly reduced, requiring only an A/D converter directly. A microcomputer can quite efficiently provide the sign changes, averaging, and table look-up values for calculating the final desired output.

The quadrature sampling process can be implemented using a microcomputer. It is possible to acquire and display in real time three-axis doppler information as well as the backscatter intensity data. The information may then be displayed simultaneously using an appropriate display system.

The quadrature sampling and complex correlation method described is an excellent one to use for real time SODAR data signal processing. However because of recent developments in hardware Fast Fourier Transform methods (FFT) are currently gaining renewed interest.

4.3.4 Fast Fourier Transform (FFT) Methods:

The theoretical basis for the Fast Fourier Transform is well covered in the literature [63], [64] and hence it will not be repeated here. It is instructive however to present a brief review.

An infinite-range Fourier transform of a real-valued or a complex-valued record $x(t)$ is defined by the complex-valued quantity

$$X(f) = \int_{-\infty}^{\infty} x(t) e^{-j2\pi ft} \, dt \qquad (39)$$

Theoretically, as noted previously, this transform $X(f)$ will not exist for an $x(t)$ that is a representative member of a stationary random process when the infinite limits are used. However, by restricting the limits to a finite time interval of $x(t)$, say in the range $(0, T)$, then the finite-range Fourier transform will exist, as defined by

$$X(f, T) = \int_0^T x(t) \, e^{-j2\pi ft} \, dt \tag{40}$$

Assume now that this $x(t)$, is sampled at N equally spaced points a distance Δt apart, where Δt has been selected to produce a sufficiently high cutoff frequency. As before, the sampling times are $t_n = n\Delta t$. However, it is convenient here to start with $n = 0$.

$$x_n = x(n\Delta t) \qquad n = 0, 1, 2, \ldots, N - 1$$

Then, for arbitarary f, the discrete version of Equation (40) is

$$X(f, T) = \Delta t \sum_{n=0}^{N-1} x_n \exp[-j2\pi fn\Delta t] \tag{41}$$

The usual selection of discrete frequency values for the computation of $X(f, T)$ is

$$f_k = \frac{k}{T} = \frac{k}{N\Delta t} \qquad k = 0, 1, 2, \ldots, N - 1 \tag{42}$$

At these frequencies, the transformed values give the Fourier components defined by

$$X_k = \frac{X(f_k)}{\Delta t} = \sum_{n=0}^{N-1} x_n \exp\left[-j\frac{2\pi kn}{N}\right] \qquad k = 0, 1, 2, \ldots, N - 1 \tag{43}$$

where Δ has been included with $x(f_k)$ to have a scale factor of unity before the summation. Note that results are unique only out to $k = N/2$ since the Nyquist frequency occurs at this point. Fast Fourier transform methods are designed to compute these quantities, X_k.

Referring to Equation (39), the inverse Fourier transform of $X(f)$ is

$$x(t) = \int_{-\infty}^{\infty} X(f) e^{j2\pi ft} \, df \tag{44}$$

This leads to the discrete inverse Fourier transform formula

$$x_n = \frac{1}{N} \sum_{k=0}^{N-1} X_k \exp\left[j\frac{2\pi kn}{N}\right] \qquad n = 0, 1, 2, \ldots, N - 1 \tag{45}$$

where the Fourier components X_k of Equation (43) are computed by the FFT procedures. The constant $(1/N)$ in Equation (45) is a scale factor only and is not otherwise important. This inverse Fourier transform can be computed by the same FFT procedures previously described. The various cross correlation and complex spectrum relationships are shown in (Fig. 24).

An Overview of Technological Development of Atmospheric Echosounders

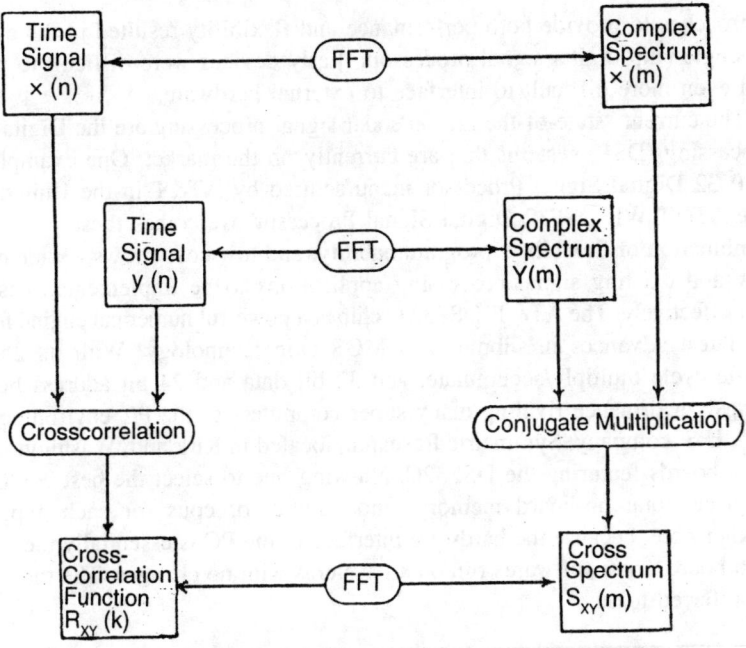

Fig. 24. Various cross correlation relationships.

For example the autocorrelation function $R_x(rT)$ described earlier in section (4.3.1) as equation (15) can be calculated using FFT methods as follows with $T = 1$:

$$R_{xx}(r) = F^{-1}[S_{xx}(m)] = F^{-1}[\overset{*}{X}(n)X(n)]$$

In words this means that the autocorrelation function of the signal $x(t)$ is equal to the inverse Fourier Transform of the autospectrum of $x(t)$. In the same manner the cross correlation function $R_{xy}(r)$ between the signals $x(t)$ and $y(t)$ can be calculated as the inverse Fourier Transform of the cross spectrum of $x(t)$ $y(t)$, i.e.,

$$R_{xy}(r) = F^{-1}[S_{xy}(m)] = F^{-1}[\overset{*}{X}(n)Y(n)]$$

The dramatic improvement in FFT hardware and firmware has led to a renewed interest in FFT based signal processing for sodar and a wide range of other fields. The rapid advances in component technology are having a profound effect on all fields of digital systems design. High-performance systems for digital signal processing that were impossible to realize only several years ago can now be implemented easily and very cost-effectively. Until recently, there have been two general approaches to the implementation of digital signal processing algorithms. Depending on performance requirements, one can chose either a dedicated hardware approach or a software implementation on a general-purpose computer. The first approach is very inflexible, expensive, and typically requires long design cycles. The second approach is tolerable only when low performance is acceptable. Initial approaches to provide both performance and flexibility resulted in the evolution of single-chip digital signal processors. Early devices were difficult to program and even more difficult to interface to external hardware.

The current "state-of-the-art" in sodar signal processing are the Digital Signal Processing (DSP) systems that are currently on the market. One example is the DSP 32 Digital Signal Processor manufactured by AT&T in the United States. The AT&T WE DSP32 Digital Signal Processor overcomes these problems. Its combination of flexibility, programmability, and low cost allows a wide range of new and existing signal processing applications to be implemented easily and cost-effectively. The AT&T DSP32 C chip is a powerful numerical engine featuring the latest advances in submicron CMOS chip technology. With its 25 Mflop single cycle multiply/accumulate, and 32 bit data and 24 bit address busses, it brings performance rivaling many super computers to the PC environment.

A U.S. company, Symmetric Research, located in Kirkland, Washington offers three boards featuring the DSP32C, allowing one to select the best combination of price, total on-board memory, and number of cpus for each application. Furthermore, because the hardware interface to the PC is essentially the same for each board, most softwares run on all systems with no changes. The three boards are referred to as:

Board	Number of cpus	On board memory	Parallel port	Serial ports
DSP 400	1	3 Mbyte	1 32 bit 12.5 MHz	1 DSP32C
DSP MOD	1	8 Mbyte	1 32 bit 12.5 MHz	1 DSP32C
DSP MUL	up to 4	1 Mbyte per cpu	none	1 DSP32C per cpu

where the DSP 400 board uses 400 mil wide 128K × 8 SRAM memory chips, and the DSP MOD and DSP MUL boards use 256K × 32 SRAM memory modules. Each board is equipped with a 16 bit PC/AT interface, and can have all of its on-board memory transparently accessed by the PC while the DSP32C is executing.

As the potential for DSP grew another U.S. company, Burr Brown, located in Tucson, Arizona, began to introduce component products that made the application of DSP more practical. Burr Brown's first single chip A/D unit was introduced in 1986. Burr Brown has since gone the next step in making real time DSP affordable with the introduction of hardware and software tailored to the specific needs of DSP. One of these products designed for use with the IBM®XT/AT and compatibles in DSPLAYXL. This software package allows the users to easily execute their real-time application by simply drawing a block diagram. This simple approach to running real-time DSP hardware will speed the user's development cycle.

5. Concluding Statement

It was beyond the scope of this chapter to provide a review of possible applications for SODARs. These systems have found widespread use for qualitative and quantitative measurements of the microstructure of temperature and velocity fluctuations in the lower planetary boundray layer. SODARs can be used as stand-alone systems or in conjunction with other devices such as radiosondes, meteorological towers, captive balloons, high frequency radars, and lidars.

The type of SODAR system to use depends entirely on the proposed application; that is, monostatic, bistatic, amplitude, doppler, or a combination of these configurations. Investigators continue to develop and apply SODARs in a wide range of new applications while gaining a better understanding of the results obtained. Development work for the near future will involve the design of more advanced steerable narrow-beam antenna systems (or arrays), multifrequency systems for attenuation measurements, combined instruments involving high-frequency radars and lidars, advanced transducers, and improved signal processing and display techniques. A great deal of work is yet to be done in this most exciting field of atmospheric probing by acoustic devices.

References

1. G.S. Wong, "Speed of Sound in Standard Air", *J. Acoust. Soc. Amer.* 79, May 1986, pp. 1359–1366.
2. A.L. Foley, "Velocity of Sound in Free Air", *Proc. Indiana Acad. Sci.,* 1927, pp. 205–207.
3. E.H. Brown and F.F. Hall, Jr., (1978), "Advances in Atmospheric Acoustics", *Rev. Geophys. and Space Phys.,* 16, 1978, pp. 47–110.
4. Lord Rayleigh [1896] *"Theory of Sound",* 1 and 2, 2d ed., Dover, New York, 1945.
5. A. Wood, *Acoustics,* Interscience, 1946.
6. G.G. Stokes, "On the Effect of Wind on the Intensity of Sound", Report, Brit. Assoc., Dublin, 1857, p. 22. Reprinted in *Mathematical and Physical Papers of G.G. Stokes,* 4, Cambridge University Press, New York, 1904, pp. 110–111.
7. J. Tyndall, "On the Atmosphere as a Vehicle of Sound", *Phil. Trans. Roy. Soc.,* London, Ser. A. 164, 1874, pp. 183–244.
8. J. Tyndall, *Sound,* 3d ed., Appleton, New York, 1875.
9. J. Henry, *Report of the Lighthouse Board of the United States,* US Government Printing Office, Washington, D.C., 1874.
10. O. Reynolds, "On the Refraction of Sound in the Atmosphere, " *Phil. Trans Roy. Soc.,* London, 166, 1876, p. 315.
11. G.W. Gilman, H.B. Coxhead, and F.H. Willis, "Reflection of Sound Signals in the Troposphere", *J. Acoustical Soc. America,* 19, 1946. pp. 274–283.
12. M.A. Kallistratova, "An Experimental Investigation into the Scattering of Sound in Turbulent Atmosphere," [in Russian], *Dakl. Aka. Nauk. SSSR,* 125, 1959, pp. 69–72.
13. M.A. Kallistratova, "Procedure for Investigating Sound Scattering in the Atmosphere", *Sov. Phys. Acoust. Eng. Trans.,* 5, 1959, pp. 512–514.
14. M.A. Kallistratova, "Experimental Investigation of Sound Wave Scattering in the Atmosphere," [in Russian], *Tr. Inst. Fiz. Atmos., Atmos. Turbulentnost,* 4, 1961, pp. 203–256; English Trans., U.S. Air Force FTD TT-63-441.

15. M.A. Kallistratova and V.I. Tatarskii, "Accounting for Wind Turbulence in the Calculation of Sound Scattering in the Atmosphere," *Sov. Phys. Acoust., Eng. Trans.,* 1960, pp. 503–505.
16. L.G. McAllister, "Acoustic Sounding of the Lower Troposphere," *J. Atmos. Terr. Phys.,* 30, 1968, pp. 1439–1440.
17. L.G. McAllister, J.R. Pollard, A.R. Mahoney, and P.J.R. Shaw, "Acoustic Sounding— A New Approach to the Study of Atmospheric Structure," *Proc. IEEE,* 57, 1969, pp. 579–587.
18. L.G. McAllister, "Wind Velocity Measurements in the Lower Atmosphere Using Acoustic Sounding Techniques," Tech. Note-A204 (AP), Weapons Res. Estab., Salisbury, Australia, 1971.
19. C.G. Little, "Acoustic Methods for the Remote Probing of the Lower Atmosphere," *Proc. IEEE,* 57, 1969, pp. 571–578.
20. C.G. Little, "Status of Remote Sensing of the Troposphere," in *Remote Sensing of the Troposphere,* ed. V.E. Derr, U.S. Government Printing Office, Washington, D.C., 1972, pp. 30-1-30-16.
21. C.G. Little, "Status of Remote Sensing of the Troposphere, *Bull. American Meteorological Soc.,* 53, 1972, pp. 936–949.
22. C.G. Little, "On the Detectability of Fog, Cloud, Rain and Snow by Acoustic Echo Sounding Methods," *J. Acoust. Soc. Amer.,* 29, 1972, pp. 748–755.
23. D.W. Beran, "Acoustics: A New Approach for Monitoring the Environment near Airports," *J. Airer.,* 8, 1971, pp. 934–936.
24. D.W. Beran, C.G. Little, and B.C. Willmarth, "Acoustic Doppler Measurements of Vertical Velocities in the Atmosphere," *Nature,* 230, 1971, pp. 160–162.
25. D.W. Beran, B.C. Willmarth, F.C. Carsey, and F.F. Hall, Jr., "An Acoustic Doppler Wind Measuring System," *J. Acoustical Soc. America,* 55, 1947, pp. 334–338.
26. D.W. Beran and S.F. Clifford, "Acoustic Doppler Measurements of Total Wind Vector," *Proc. 2nd Symp. Meteorological Observatories and Institutes,* American Meteorological Society, 1972.
27. W.R. Simmons, J.W. Wescott, and F.F. Hall, Jr., "Acoustic Echo Sounding as Related to Air Polluton in Urban Environments," *Tech. Rep. ERL 216-WPL 17,* Nat. Oceanic and Atmos. Admin., Boulder, CO, 1971, 77 pp.
28. H.D. Parry and M.J. Sanders, Jr., "The Design and Operation of an Acoustic Radar," *IEEE Trans. Geosci. Electron.,* GE-10, 1972, pp. 58–64.
29. J.W. Wescott, W.R. Simmons, and C.G. Little, "Acoustic Echo Sounding Measurements of Temperature and Wind Fluctuations," Tech. Memo. ERLTM-WPL5, Environ. Sci. Serv. Admin., Boulder, CO, 1970.
30. W.T. Cronenwett, G.B. Walker, and R.L. Inman, "Acoustic Sounding of Meteorological Phenomena in the Planetary Boundary Layer," *J. Appl. Meteorol.,* 11, 1972, pp. 1351–1358.
31. R.B. Chadwick and C.G. Little, "The Comparison of Sensitivities of Atmospheric Echo-Sounders," *Remote Sens. Environ.,* 2, 1973, pp. 223–234.
32. I. Tombach, P.B. MacCready, and L. Baboolal, "Use of a Monostatic Acoustic Sounder in Air Pollution Diffusion Estimates," *2d Joint Conf. on Sensing of Environmental Pollutants,* Washington, D.C., December 10–12, 1973.
33. E.J. Owens, "Development of a Portable Acoustic Echo Sounder," Tech. Rep. ERL 298-WPL 31, Nat. Oceanic and Atmos. Admin., Boulder, CO, 1974.
34. F.F. Hall, Jr., and E.J. Owens, "Atmospheric Acoustic Echo Sounding Investigations at the South Pole," *Antarct. J. U.S.,* 10, 1975, pp. 191–192.

35. D.W. Thomson, and J.P. Scheib, "Improved Display Techniques for SODAR Measurements," *Bul. American Meteorological Soc.,* 59, No. 2, pp. 147–152.
36. E.J. Owens, "NOAA Mark VII Acoustic Echo Sounder," Tech. Memo ERL WPL-12, Nat. Oceanic and Atmos. Admin., Boulder, CO, 1975.
37. E.J. Owens, "Microcomputer-Controlled Acoustic Echo Sounder," Tech. Memo. ERL WPL-21, Nat. Oceanic and Atmos. Admin., Boulder, CO, 1977.
38. J.A. Kleppe and H.L. Dunsmore, "A Complex Correlation Approach to Spectral Moment Estimation for Doppler Echosonde Data," Report submitted to NOAA, ERL, Boulder, CO, December 1976.
39. J.A. Kleppe, H. Dunsmore, and E. Owens, "New Techniques for Estimation of the Mean and Variance of Acoustic Frequency Doppler Spectra," *Proc. of Fall Meeting, American Geophysical Union,* San Francisco, December, 1976, p. 36.
40. M. Balser, C.A. McNary, and A.E. Nagy, "Remote Wind Sensing by Acoustic Radar," *J. Appl. Met.*, 15, 1976, pp. 50–58.
41. H.E. Bass and F. Douglas Shields, "Absorption of Sound in Air: High Frequency Measurements," *J. Acoustical Soc. America,* 62, September 1977, pp. 571–576.
42. ANSI, 26-1978, *American National Standard Method for the Calculation of the Absorption of Sound by the Atmosphere,* Acoustical Soc. America, New York, 1978, pp. 1–29.
43. E.H. Brown and R.J. Keeler, "Application of Propagation Parameters to Atmospheric Echosondes," *Proc. of 16th Radar Meteorological Soc.,* American Meteorological Society, Boston, MA, 1975, pp. 272–277.
44. F.F. Hall, Jr., J.W. Wescott, and W.R. Simmons, "Acoustic Echo Sounding of Atmospheric Thermal and Wind Structure," *Proc. of 5th Int. Symp. on Remote Sensing of the Environment,* University of Michigan Press, Ann Arbor, 1971, pp. 1715–1732.
45. F.F. Hall, Jr., "Temperature and Wind Structure Studies by Acoustic Echo-Sounding," in *Remote Sensing of the Troposphere,* ed. V.E. Derr, U.S. Government Printing Office, Washington, D.C., 1972, Chapter 18.
46. F.F. Hall, Jr., "Report on the July 1972 Workshop in Atmospheric Acoustics," Tech. Memo, ERL WPL 8, Nat. Oceanic and Atmos. Admin., Boulder, CO, 1973.
47. R.B. Chadwick and C.G. Little, "The Comparison of Sensitivities of Atmospheric Echo-Sounders," *Remote Sens. Environ.,* 2, 1973, pp. 223–234.
48. J.A. Kleppe, "Engineering Applications of Acoustics," Artech Press, Boston, 1989, ISBN-0-89006-260-9.
49. E.H. Brown, "Some Recent NOAA Theoretical Work on Echo Sounding in the Atmosphere," *J. Geophys. Res.,* 79, December 1974.
50. A. Spizzichino, "Discussion of the Operating Conditions of a Doppler Sodar," *J. Geophys. Res.,* 79, 1974, pp. 5585–5591.
51. A. Spizzichino, "The Refraction of Acoustic Waves in the Atmosphere and Its Effect on Sodar Wind Measurement," *Ann. Telecommun.,* 29, 1974, pp. 301–310.
52. W.D. Neff, "Qualitative Evaluation of Acoustic Echoes from the Planetary Boundary Layer," NOAA Tech. Ref. ERL 3226 WPL-38, Boulder, CO, 1973.
53. J.C. Kaimal and D.A. Haugen, "An Acoustic Doppler Sounder for Measuring Wind Profiles in the Lower Boundary Layer," *J. Appl. Meteor.,* 16, December 1977.
54. J.A. Kleppe and H.L. Dunsmore, "Digital Signal Processing of Acoustic Radar Amplitude and Doppler Data," *Proc. of 4th Symp. of Turb. Dif. and Air Pollution,* American Meteorological Soc., Reno, 1979, pp. 549–554.

55. J.A. Kleppe and H.L. Dunsmore, "The Application of Microcomputers to Acoustic Radar Systems," *Proc. 1st Int. Symp., Mini and Microcomputers in Control,* San Diego, 1979, pp. 178–180.
56. J.A. Kleppe and H.L. Dunsmore, "The Applicationof Microcomputers to 'Real Time' Acoustic Radar Signal Processing", *Proc. 7th Int. Symp., Mini and Microcomputers,* Anaheim, C, A, 1979, pp. 137–140.
57. R.F. Davey, "A Comparison of Doppler SODAR Antenna Configurations Used for Horizontal Wind Measurement," *J. Acoustical Soc. America,* 63, May 1978, pp. 1335–1341.
58. Universal Active Filter UAF41–Burr Brown, Tucson, Arizona, U.S.A.
59. R.A. Quinnell, "Switched-Capacitor Filters," EDN Special Report, EDN January 4, 1990, pp. 86–98.
60. EG&G RETICON–Sunnyvale, California, U.S.A.
61. M. Trapp, "Learn the Fundamentals of Digital Filter Design," *Electronic Design,* July 25, 1991, pp. 83–93.
62. J.S. Bendat and A.G. Piersol, *Measurement and Analysis of Random Data,* John Wiley and Sons, New York, 1967.
63. J.S. Bendat and A.G. Piersol, *Engineering Application of Correlation and Spectral Analysis,* John Wiley and Sons, New York, 1980.
64. J.S. Bendat and A.G. Piersol, *Random Data-Analysis and Measurement Procedures,* 2nd ed. John Wiley and Sons, New York, 1986.
65. D. Sirmans and R.J. Doviak, "Meteorological Radar Signal Intensity Estimation," NOAA TM ERL-NSSL-64, 1973.
66. D. Sirmans and B. Bumgarner, "Estimation of Spectral Density Mean and Variance by Covariance Argument Techniques," *Proc. 16th Radar Meteorological Conf.,* American Meteorol. Soc., 1975.
67. R.J. Keeler and L.J. Griffith, "Acoustic Doppler Extraction by Adaptice Inverse Filtering," *J. Acoustical Soc. America,* 61, May 1977.
68. J.A. Kleppe, "An Overview of Acoustic Sounder Developments: Past, Present and Future," *Proc. Int. Symp. on Acoustic Remote Sensing of the Atmosphere and Oceans,* University of Calgary, Canada, 1981.

Appendix A

ATMOS1

ATMOS1 is a program for computing the total absorption of sound by the atmosphere using ANSI S1.26–1978 (ASA 23–1978). Also included is the computation of the velocity of sound as a function of temperature, pressure, and relative humidity. The reader is referred to [48] for the equations used to write this program.

```
10  REM ATMOS1
20  PRINT "Enter in frequency (Hz), temp (c), pressure (10E6PA), RH(%)
30  Input F, T, P, RH
32  LET T = T + 273.16
34  LET TO1 = 273.16
36  LET X1 = (T/TO1)
38  LET X2 = (TO1/T)
40  LET X3 = .4343*LOG (X1)
42  LET E1 = - 8.29692* (X1 - 1)
44  LET E2 = 4.76955*(1 - X2)
46  LET PR1 = 10.79586*(1 - X2)–(5.02808*X3)
48  LET PR2 = 1.50474E-04*(1 - (10^E1)) + 4.2873E - 04* (- 1 + (10^E2))
    - 2.219598
50  LET BB = PR1 + PR2
52  LET PT = (10^BB)
54  LET X4 = P/.101325
55  LET H = RH*PT/X4
56  LET E = 1013.25*PT*RH
58  LET X5 = (T/293.16)
60  LET X6 = - 6.142*((X5^ - .3333333) - 1)
62  LET FRO = X4*((24 + 44100!*H)* ((.05 + H)/(.391 + H)))
64  LET FRN = X4*(1/SQR(X5))*(9 + 350*H)*EXP(X6)
66  LET A1 = (1.84E - 11*(1/X4)*SQR(X5))
68  LET A2 = (.01278*EXP(- 2239.1/T))
70  LET A3 = (FRO + (F^2/FRO))
72  LET A4 = .1068*EXP(-3352/T))
74  LET A5 = (FRN + (F^2/FRN))
76  LET A6 = (A2/A3) + (A4/A5)
78  LET AB = (F^2)*(A1 + (X5^ - 2.5)*A6)
79  LET AB = 869*AB
85  LET D = 1.2929*(273.16/T)*(X4 - ((.3783*E)/101325!))
86  LET VH = SQR((1.402*P*1000000!)/D)
87  LET Z = VH*D
```

```
 88 LET M = (D*T*8314.481)/(P*1000000!)
 95 PRINT "THE TOTAL ABSORPTION (DB/100M) =", AB
 96 PRINT "THE MOLAR CONCENTRATION OF H20 (%) =", H
 97 PRINT "THE OXYGEN RELAXATION FREQUENCY (HZ) =", FRO
 98 PRINT "THE NITROGEN RELAXATION FREQUENCY (HZ) =", FRN
100 PRINT "THE SOUND SPEED OF HUMID AIR (M/S) =", VH
103 PRINT "THE DENSITY(KG/M3) =", D
104 PRINT "THE VAPOR PRESSURE E (PA) =", E
105 PRINT "THE MOLAR MASS =", E
106 PRINT "THE SPECIFIC ACOUSTIC IMPEDANCE (RAYLS) =", Z
107 END
```

Acoustic Remote Sensing Applications
S.P. Singal (Ed)
Copyright © 1997 Narosa Publishing House, New Delhi, India

3. Design of a Tri-Monostatic Doppler Sodar System

Yoshiki Ito
Kaijo Corporation, Hamura, Tokyo, Japan

Introduction

Today a monostatic Doppler sodar is widely used not only for the meteorological research works but also for the routine works such as the investigation of atmospheric environments and the diffusion monitoring of a power plant. The fundamental design of acoustic components doesn't change significantly as compared with the system developed in 1970s, but the electronic circuit has made rapid progress. For example, an integrated circuit by CMOS technology achieved low power consumption of the system and brought out the development of a battery operated equipment. A high-speed microprocessor made it possible to carry out real time processing of digitalized Fast Fouriel Transform (FFT) at the Doppler frequency analysis. Moreover, a microprocessor reduced the hardware such as a conventional analog filter bank, and contributed to improve the reliability and capability of the equipment. This paper outlines the system hardware and the signal processing of general purpose Doppler sodar, Model AR410 made by Kaijo corporation as an example.

Basic Design of Doppler sodar

Principle of Sodar Observation:
The principle of acoustic sounding is based on the detection of the backscattered signal from acoustic refractive index discontinuities or disturbances in the atmosphere. A usual acoustic sounder transmits the tone burst signal of several acoustic watts and receives the scattered signal from scattering region with the reflectivity larger than 10^{-12}. A wind finding acoustic sounder detects the Doppler frequency shifts of received signals. Fig. 1(a) shows the monostatic method transmitting the tone burst to three directions in the atmosphere by time sharing from one point on a ground. On the other hand, Fig. 1(b) shows a method radiating signal upward and receiving it by three receivers on ground. This method is a bistatic Doppler sodar, and has advantages that there are no time lag and no beam separation in the

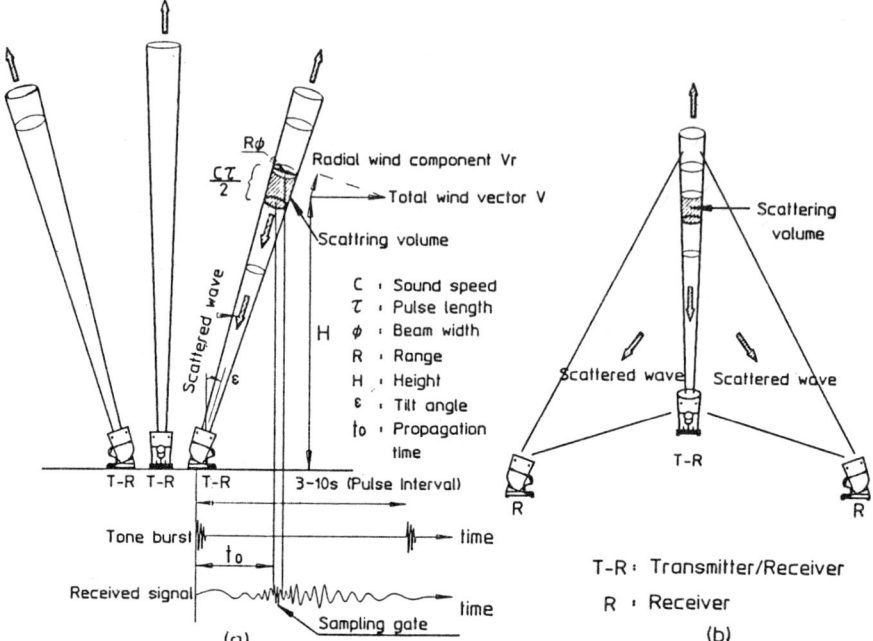

Fig. 1. Schematic diagram of Doppler wind measurement by acoustic means
(a) Tri-monostatic mode sodar, (b) Bistatic mode sodar

sounding of three directions of wind component with a shorter cycle than the monostatic type, but a long baseline is necessary in this case. The instrument described here is the monostatic type.

The received signal power, P_r from scattering volume for the monostatic sodar can be written as follows by the radar equation [1, 2]

$$P_r = P_t \cdot \eta_t \cdot \eta_r \cdot \sigma(\pi) \cdot C \cdot \tau \cdot A_r \cdot G \cdot e^{-2\alpha R}/2R^2 \tag{1}$$

where $\sigma(\pi)$ is the acoustic backscatter cross section per unit volume, P_t the transmitted power, η_t and η_r the efficiencies of transmitter and receiver respectively, C the speed of sound, τ the pulse length, A_r the antenna effective aperture, G the directivity compensation factor, α the acoustic attenuation coefficient and R being the range to scattering region. $\sigma(\pi)$ is related to the turbulent state of air temperature represented by the structure constant for air temperature C_T^2 as follows

$$\sigma(\pi) = 0.0039 \ k^{1/3} \ C_T^2/T^2$$

where k is the wave number, T being the air temperature in the scattering volume. The parameters except $\sigma(\pi)$, C, α and R in Eq. (1) are the characteristic values inherent to the system hardware. For further discussion, we introduce the received acoustic signal intensity, I_s defined by

$$I_s = P_r/\eta_r \cdot A_r = P_t \cdot \eta_t \cdot \sigma(\pi) \cdot C \cdot \tau \cdot G \cdot e^{-2\alpha R}/2R^2 \qquad (3)$$

On the other hand, the total received noise intensity, I_n is assumed to be the received acoustic noise level, N_a, and the equivalent electric noise level, N_e of the system. N_a is the value of received acoustic noise power P_{Na} divided by A_r, and N_e is the value of electric noise power P_{Ne} divided by $\eta_r A_r$. N_a is regarded as an equivalent of the acoustic ambient noise level, N_{am} multiplied by the directivity compensation factor for noises or sidelobes, G_N and the diffraction attenuation effect, L_D. Thus we get

$$I_n = P_{Na}/A_r + P_{Ne}/\eta_r \cdot A_r = N_a + N_e = G_N \cdot L_D \cdot N_{am} + N_e \qquad (4)$$

As usually $N_a >> N_e$, the signal to noise (S/N) ratio SN is expressed as

$$SN = I_s/I_n = I_s/G_N \cdot L_D \cdot N_{am} = P_t \cdot \eta_t \cdot \sigma(\pi) \cdot C \cdot \tau \cdot \left(\frac{G}{G_N}\right) \cdot e^{-2\alpha R}/2N_{am}L_D R^2 \quad (5)$$

Choice of Sodar Parameters and Antenna Design

In order to increase the detectability of the sodar system, it is necessary to enhance the S/N ratio. There may be following methods to achieve it: (1) the increase of output Power, (2) the improvement of the transmitting and receiving efficiencies, (3) the increase of pulse length, (4) the improvement of antenna gain or sidelobe suppression, (5) the decrease of atmospheric attenuation, (6) the installation in a low ambient noise area or the reduction of the noise level received by an acoustic antenna.

The electro dynamic speaker employed as an electro-acoustic transducer has the maximum electric input power about a hundred watts and the efficiency of 20 to 30% in audio frequency range, although the system efficiencies η_r, η_t in Eq. (1) are usually assumed to be about 0.1 [2]. Some efforts were done such as the improvement of system efficiency by employing offset antenna or the increase of output power by assembling many transducers. A phased array Doppler sodar developed in these days [3], which can easily synthesize multidirectional beams with phase shifted signals, makes it possible simultaneously to increase the output power and the antenna area by adding the transducer element.

It should be carefully considered to increase the pulse length, since it causes decreasing of range resolution. As the range resolution, Δh is expressed by $\Delta h = C\tau/2$, the pulse length is required to be shorter than $\tau = 0.12$ sec in order to keep the range resolution of $\Delta h = 20$ m that is needed for most of the purposes. The choice of low operating frequency offers increasing of the measurement altitude due to less acoustic attenuation, but it is necessary to use a large aperture antenna in order to make a narrow beam. Fig. 2 shows the radiated beam width as a function of the ratio of antenna diameter D to acoustic wave length λ. Even though the narrow acoustic beam gives better spatial resolution, the beam width ψ less than about 4 or 5 degrees is inadequate since turbulent refractive effect in the

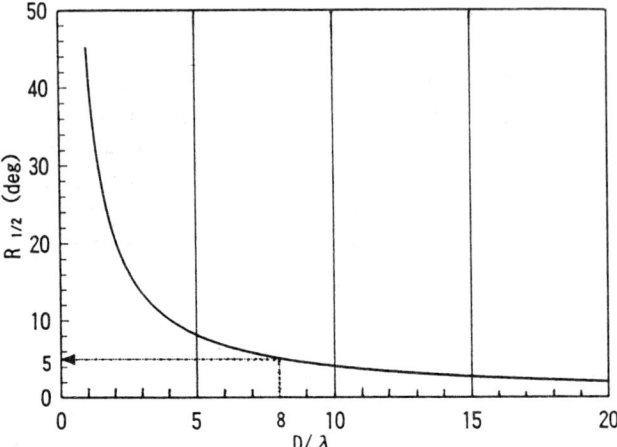

Fig. 2. Half-angle beam width $R_{1/2}(=\psi/2)$ as a function of D/λ defined by the angle that the directivity function $R(\theta)$ in Eq. (6) decreases to 1/2 compared with the center of the beam.

atmosphere may cause the backscattered signal to deviate randomly from the beam axis [4], thus it is necessary to design it to be as broad as $\psi \leftrightarrows 10°$. The refractive or bending effect for the uniform wind U in the atmosphere of sound speed C is approximated by $\tan^{-1}(U/C)$. The effect of power loss for $\psi = 10°$ gets to 1/2 in acoustic pressure at a wind of 30 m/s. Fig. 2. shows that it is necessary to use an antenna diameter of $D \leftrightarrows 1.4$ m when the beam width is $\psi = 10°$ at the operating frequency of $f_0 = 2$ kHz ($\lambda \leftrightarrows 0.17$ m) that is adequate for the measurement up to 500 m in general use [5].

By the way it is not always advantageous to operate at a low frequency for the improvement of detectability because of not only the increase of antenna diameter but also the larger background noise level. Fig. 3 shows the sound pressure level (SPL) of the typical ambient noises measured through one octave band filter Noct, and the converted values to the spectral noise level or noise level per unit hertz, Nspc. As shown in Fig. 3, the background noise level generally decreases with increasing frequency. If we design the electric circuit carefully, the electric noise power, P_{Ne} will be suppressed as low as the thermal noise ($P_{Ne} \leftrightarrows kBT$), where k is Boltzmann's constant ... $k = 1.38 \times 10^{-23}$ Joule/deg ..., B the band width, T the absolute temperature. For example, the thermal noise power, P_{Ne} per unit band width in the case of $B = 1$ and $T = 290$ is about 4×10^{-21} W(or Ne = 4×10^{-20} W/m² for $\eta_r A_r \leftrightarrows 1$ in Eq. (4)), which is smaller than the received acoustic noise level by 20 dB and more as discussed later.

As the S/N ratio is usually limited by ambient acoustic noise, it is necessary to reduce the received acoustic noise by means of the improvement of antenna directivity and by the use of the acoustic shield around an antenna in order to suppress the sidelobe. The basic beam pattern of the circular aperture antenna is approximated by introducing the directivity function $R(\theta)$ [6] as follows

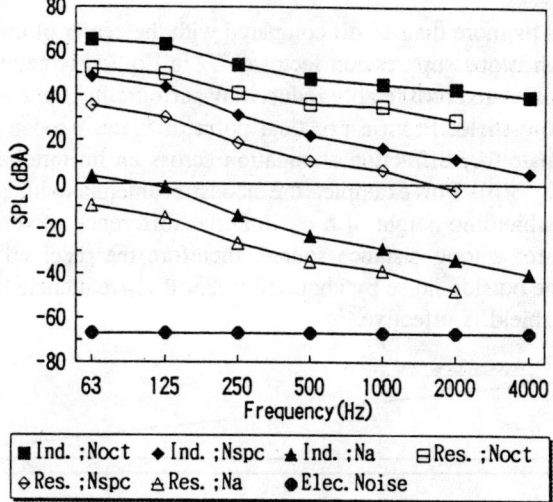

Fig. 3. Examples of the typical ambient noise level in the industrial area (Ind.) and the residential area (Res.) measured through one octave band filter, Noct, expressed with the converted values to spectral noise level, Nspc. 0 dBA = 10^{-12} W/m².

$$R(\theta) = |\, 2J_1(x)/x \,| \qquad (6)$$

where

$$x = \pi D \sin \theta / \lambda$$

and J_1 is a first-order Bessel function, D the antenna diameter, θ the angle from the center of the beam, λ the wavelength. We should use $R^2(\theta)$ for the round trip signal and $R(\theta)$ for the acoustic noise.

Figure 4 is the radiation pattern in the case of $D/\lambda = 8$ that gives the beam width of $\psi = 10°$. As shown in Fig. 4, the radiation pattern $R(\theta)$ at angles of 30°

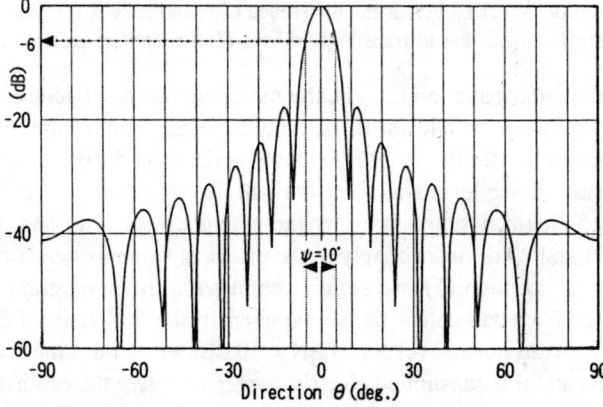

Fig. 4. Radiation pattern of the circular aperture antenna in the case of $D/\lambda = 8$

to 90° decreases by more than 25 dB compared with the center of the beam, which means that the sidelobe suppression factor G/G_N in Eq. (5) is better than 25 dB. The ambient noises are considerably reduced by surrounding the antenna aperture with the acoustic shield. From a practical point of view, we can introduce the method to estimate the diffraction attenuation across an infinite barrier [7] with the help of Fig. 5(b). For example, the acoustic antenna with a diameter of $D = 1.4$ m and shielding height of $h = 1.6\,D$ has difference of sound pass $\delta \leftrightharpoons a - r = 1.6$ m for a long distance source, therefore the received noise can be reduced from the outside noise by about 22 to 25dB at frequencies of 1 to 2 kHz if the acoustic shield is effective.

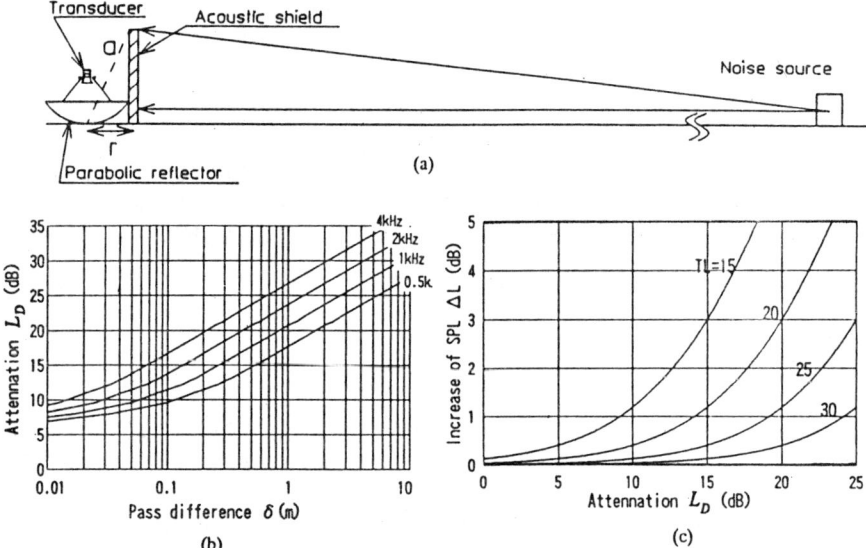

Fig. 5. Diffraction attenuation across an acoustic barrier; (a) Schematic model figure, (b) Attenuation effect L_D versus the difference of sound pass δ, (c) Increase of sound pressure level ΔL due to transmission loss TL through an acoustic barrier.

Based on the radar equation Eq. (1), the maximum range of acoustic sounding, R_{max} can be estimated as follows. If the ambient noise level at the observational area is assumed to be 30 dBA in 2 kHz octave band or -1.5 dBA in spectral noise level, the reduced noise level Na in Eq. (4) will be -46.5 dBA ($= 2.2 \times 10^{-17}$ W/m^2) when each sidelobe suppression by the antenna directivity and the acoustic shield is 25 dB and 20 dB respectively as the practical value. Instead of calculating with the Eqs. (1) through (5), we suggest employing the nomogram to estimate R_{max} in Fig. 6. If we can detect the signal even in the S/N ratio of SN = 0 dB, the received acoustic noise level Na (if SN = 10 dB, we can use the scale position of Na + 10 dB) and the transmitted electric energy Pt τ give the estimation of R_{max} under the various conditions of the acoustic attenuation coefficient α and the temperature structure constant C_T^2. C_T^2 varies usually from 10^{-3} to 10^{-5} deg^2 m$^{-2/3}$ with increasing of the height from ground up to 2 km under convective conditions in the summer [8]. When the tone burst signal of the transmitted electric energy of Pt τ = 100 J propagates through the atmosphere with the mean acoustic attenuation of α = 0.005 m^{-1} and is scattered from the temperature fluctuating region of $C_T^2 = 10^{-4}$ deg^{2}m$^{-2/3}$, for example, Fig. 6 indicates the maximum range of acoustic sounding, Rmax to be 670 m in the acoustic noise level of Na = $-$ 46.5 dBA. As described later in Eq. (11), the signal processing requires a margin of S/N ratio by 10 dB, thus the practical estimation by using the scale position of Na + 10 dB in Fig. 6 gives 500 m to Rmax. Fig. 6 can also be utilized to know the required transmission power or the required acoustic noise level to get the data of

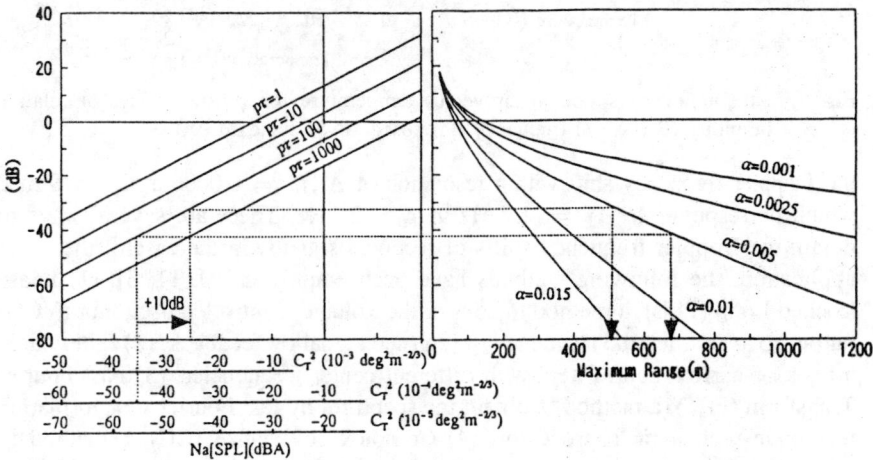

Fig. 6. Nomogram to estimate the maximum range of acoustic sounding, Rmax related with the received acoustic noise level N_a, Transmitted electric energy $Pt\ \tau$ and atmospheric attenuation α

the desired altitude when the values of α and C_T^2 are given. The characteristic curves of the acoustic attenuation coefficient α for various frequencies are shown in Fig. 7.

Signal Processing

Doppler frequency shift, Δf of the backscattered signal from the scattering volume moving with velocity component, V_r in the radial direction is given as follows

$$V_r = C\Delta f/2f_0 \qquad (7)$$

As the operating frequency f_0 is known and the speed of sound C is also known and can be treated as a constant, the Doppler sodar can work as a remote sensing instrument to detect the Doppler frequency shift caused by the moving scatterer. If we need to detect V_r with the resolution of 0.1 m/s, it is necessary to estimate

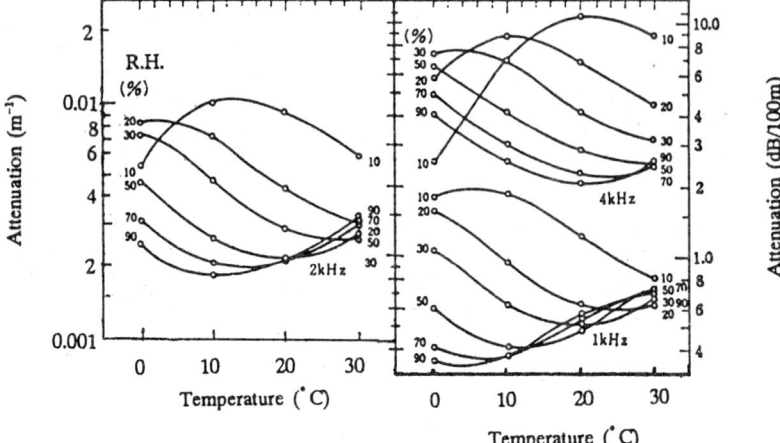

Fig. 7. Attenuation of sound in air versus temperature for various values of relative humidity (R.H.) and frequency (Proposed ANSI standard [9]).

the Doppler frequency shift with a resolution of $\Delta f/f_0 \backsimeq 0.0006$ at $C = 340$ m/s, which corresponds to $\Delta f \backsimeq 1.2$ Hz at $f_0 = 2$ kHz. There are several ways to obtain the Doppler frequency shifts of received signals. In the case of the sodar application, the following methods have been employed [10, 11, 3]; (1) Phase Locked Loop (PLL), a method to control the voltage controlled oscillator (VCO) output so as to track the frequency of returned signal by feedback, (2) Filter bank, a bank of narrow band filters with different center frequencies, (3) Fast Fourier Transform (FFT), a method to obtain the spectrum by fast Fourier transformation technique with a microprocessor, (4) Complex covariance (CC), a method to calculate the complex covariance function from in-phase and quadrature components of received signal to obtain the first and second moments of a spectral density function. Among these methods, (3) and (4) are regarded as excellent ones from the standpoint of noise immunities and easy construction with the microprocessor and peripherals. The FFT method is characterized by flexibility in Doppler frequency estimation since full spectrum is calculated, so that a better estimation can be expected even when the S/N ratio is low. On the other hand, the CC method has the advantages that the computation of spectral moment is easy and the frequency resolution is better than the FFT method when the S/N ratio is high. Considering these features, the CC method is usually used in the submerged Doppler current meter in the water where the S/N ratio is high and a better resolution of velocity and range is required, while the FFT method is useful to remove the inevitable noises in atmospheric acoustics. The actual processing of the FFT method that is used in AR410 is described in the following.

The sampling theory suggests that the sampling frequency, f_s must be more than twice of received signal frequency, f_r. The frequency resolution δf of the FFT method is written as

$$\delta f = f_s/N \tag{8}$$

where N is the number of sampling data. The frequency resolution δf is related to the velocity resolution δVr by $\delta Vr = \delta f\, C/2f_0$. On the other hand, the time resolution of sampling gate δt is written as

$$\delta t = N/f_s \tag{9}$$

The time resolution δt is related to the range resolution δR by $\delta R = \delta t\, C/2$. It should be noticed that the total range resolution is the sum of the resolution $C\tau/2$ caused by tone burst and the resolution $CN/2f_s$ caused by sampling gate. As the Doppler frequency shift induced by wind is less than 10% of the operating frequency f_0 i.e. $\Delta f/f_0 < 0.1$, the sampling frequency must be $f_s > 2(f_0 + \Delta f\text{max}) \leftrightarrows 2.2 f_0$ and also satisfy the resolutions of velocity and range shown in Eq. (8) and (9), where these Eqs. are related in the tradeoff. Model AR410 sodar works with the FFT processing of $f_s = 7040$ Hz and $N = 1024$ at $f_0 = 2.4$ kHz. Therefore the range resolution due to sampling gate is about 25 m and the radial velocity resolution is about 0.5 m/s. In order to improve the velocity resolution and to get a better estimation of the first moment without bias caused by the spectral broadening noise, Model AR410 calculates the first moment using several data around the spectral peak

$$f_r = \int f \cdot S(f) df \Big/ \int S(f) df \leftrightarrows \sum_{i=p-2}^{p+2} f_i \cdot S(f_i) \Big/ \sum_{i=p-2}^{p+2} S(f_i) \tag{10}$$

where five points around the spectral peak are averaged to calculate the first moment. This averaging number should be chosen from the spectral spreads of actual received signals.

Scattered acoustic waves attenuate proportionally to the altitude, therefore it is necessary to discriminate the reliability of received signals. This can be carried out by monitoring the S/N ratio of received signals. The ratio between the signal passed through band pass filter and the noise passed through band eliminate filter can be used in analog method, while in the digital method, the ratio between the signal represented by the spectral peak and the noise given by the spectral density off the peak is available. Model AR410 defines the S/N ratio γ of received signal in the spectral region as shown in Fig. 8 and Eq. (11)

$$\gamma = S(f_p) \Big/ \sum_{i=1}^{n} S(f_i) \cdot n^{-1} \tag{11}$$

where n is the number of spectral lines including noises in the frequency band of $\pm 10\%$ around f_0. A number of field tests have shown that $\gamma \leftrightarrows 10$ is good enough for the threshold level to discriminate the reliable signals.

Description of the Total System

The specifications and the functional block diagram of AR410 sodar system are

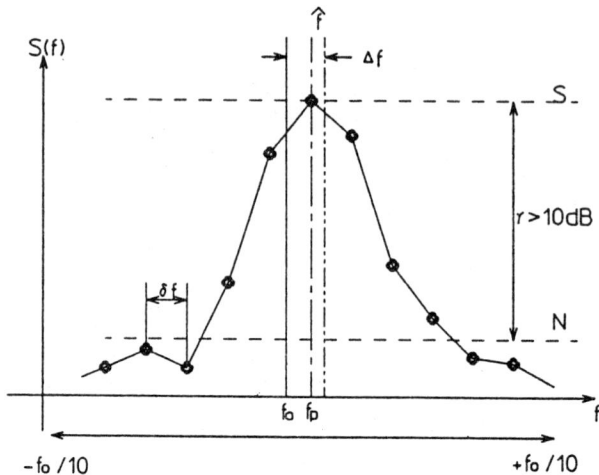

Fig. 8. Schematic diagram of signal processing of received power spectrum

shown in Table 1 and Fig. 9 respectively. This system is characterized by the mixed pulse transmission and the offset parabolic reflector. The mixed pulse transmission, which means the combination of short and long pulses τ_1, τ_2 as shown

Table 1. Specifications of Model AR410 Doppler sodar

Measurement system	Tri-monostatic mode
Operating frequency*	2.4 kHz (std.), 1.6k, 3.2 kHz
Output power*	900 W (max)
Antenna diameter	1.1 × 1.0 m; elliptical offset
Pulse length*	10 ~ 900 msec
Pulse repetition*	3, 5, 10 sec
Beam width	10 degrees at 2.4 kHz
Receiver gain	130 dB (max)
Receiver bandwidth	± 10% around f_0

*user programmable

in Fig. 10, performs better altitude resolution at lower layers and better detectability at higher altitudes. τ_1 is usually selected from a range of 50 to 100 msec for lower than 100 or 200 m (programmable) and τ_2 from 150 to 300 msec for higher altitudes. A better efficiency of the transmitting and receiving is performed by the offset arrangement of the transducer. The system has also many changeable parameters. Operating frequency, pulse width, repetition period, output power, measuring height, averaging time and operating interval can be easily changed by the user programmable switch. The function of each part is described as follows.

As shown in Fig. 9, the carrier signal generated by Oscillator is provided to Pulse Gate and modulated to give the tone burst carrier, which is then led into

Design of a Tri-Monostatic Doppler Sodar System   95

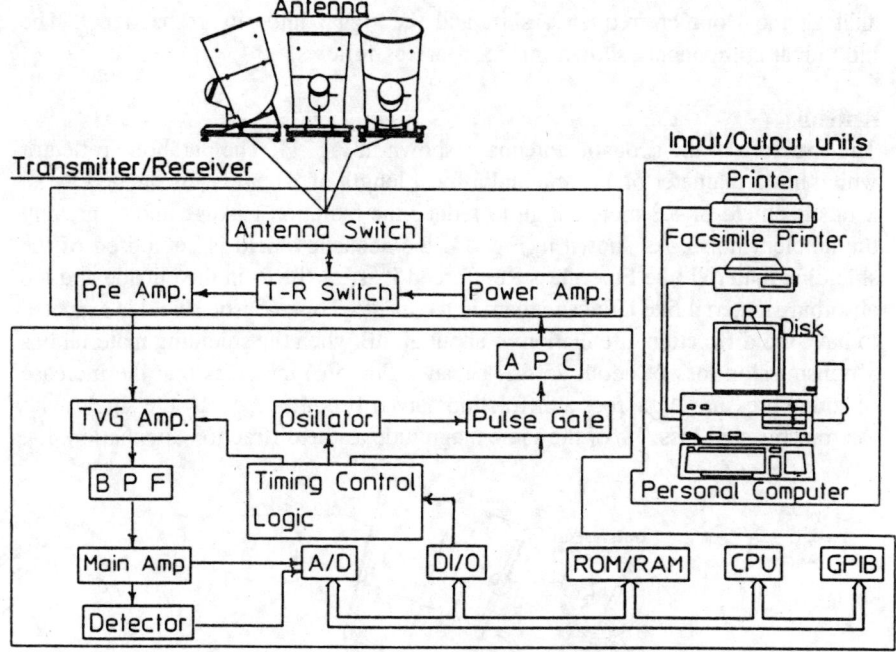

Fig. 9. Block diagram of Model AR410 Doppler sodar.

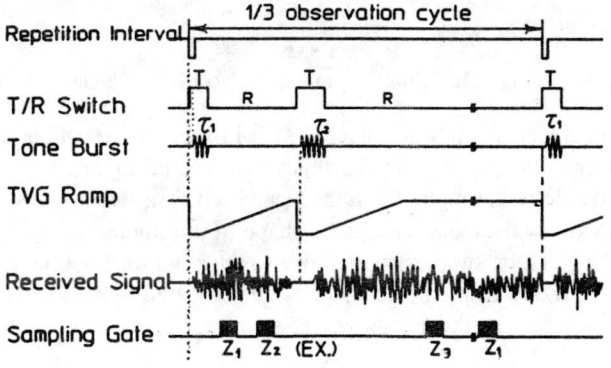

Fig. 10. Timing chart of control logic.

Power Amplifier through a long wire in order to feed electric power to the transducer assembled in Antenna. Acoustic pulses are transmitted from three transducers in turn and scattered signals are received by the same transducers. Received signals amplified in the Pre-Amplifier are delivered through a long shielded wire to the Main unit. After passing through the Time Variable Gain (TVG) amplifier to compensate for the spherical expansion of acoustic waves, received signals are led into the Main Amplifier and then divided into A/D converter and Detector to analyze the Doppler frequency shift and the signal intensity respectively. The individual components shown in Fig. 9 are as follows.

Antenna
The structure of the acoustic antenna is shown in Fig. 11. The parabolic reflector which has a diameter of 110 cm and a focal length of 50 cm is surrounded by an acoustic shield of 1.8 meters high to reduce the radiation leakage and to prevent the ambient noise. As shown in Fig. 11, the acoustic shield is composed of the shielding material like Fiberglass Reinforced Plastics (FRP) in the outside and the absorbing material like Urethane foam in the inside. This acoustic shield is expected to have the diffraction attenuation of about 20 dB when the shielding material has a transmission loss of about 25 dB, because Fig. 5(c) indicates that the increase of sound pressure ΔL across a diffraction barrier becomes 3 dB by the shield with the transmission loss, TL of the same magnitude as the diffraction attenuation, L_D.

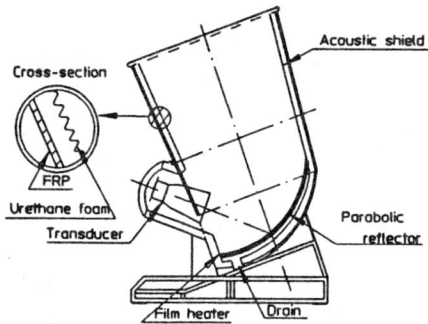

Fig. 11. Cross section of the acoustic antenna.

The arrangement of the transducer shown in Fig. 11, so called an offset feed horn, is designed to improve the efficiency of transmitting and receiving without blocking acoustic radiation by the transducer itself. This techniques improves the efficiency by a few decibels. The flared shape of the shielding barrier contributes slightly to the sidelobe suppression relative to the main lobe, while the round edge of the shield top decreases the acoustic noise generated by wind. The film heater can be sandwitched in the parabolic reflector and the transducer horn to melt snow.

Transmitter/Receiver
Transmitter/Receiver consists of Power Amplifier, T-R switch, Antenna Switch and Pre-Amplifier. The tone burst signals delivered from the Main unit are amplified by Power amplifier and transmitted to the atmosphere through the transducer. Power amplifier is a single ended push-pull audio frequency amplifier with a maximum output of 300 watts. The acoustic combination of 12 transducers increases the maximum output up to 900 watts with the three channel power amplifier. T-R switch protects the receiver from high voltage transmission by switching the

transducers in every repetition interval with mercury relays or protective circuits using diode base. Pre-amplifier circuit consists of a low-noise operational amplifier, wide band BPF (band pass filter) and line driver with total gain of 70 to 80 dB. Amplified signals in this band are transmitted to the Main unit through the shielded wires that connects the outdoor sensing section to the indoor processing section.

Main Unit

Main unit plays the main part of signal processing to control the transmitted and received signal. Timing control logic regulated by a stabilized crystal oscillator delivers the timing clock shown in Fig. 10 to the peripheral circuits. The operating hardware parameters such as pulse repetition period, pulse length, operating frequency, receiver gain and output power are transmitted as binary coded digits from DI/O to Timing control logic is composed of programmable counters and AND/OR gate logic integrated circuits. TVG amplifier compensates for a round trip loss of the acoustic signal caused by spherical expansion of the wave front. The system gain increases with the time after transmission of the tone burst in order to compensate for $1/R$ attenuation in the received output voltage, which is proportional to the scattered acoustic pressure rather than the received power corresponding to $1/R^2$ attenuation in the radar equation. As shown in Fig. 12,

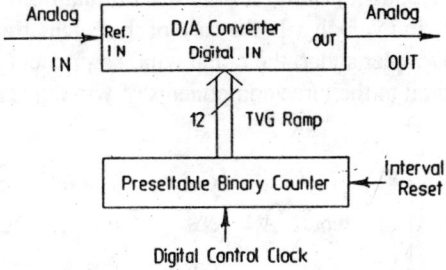

Fig. 12. Circuit components of TVG amplifier.

TVG amplifier employs D/A converter for an analog/digital multiplier to make the gain ramp signal, therefore the compensation curve can be adjusted by the digital control clock from Timing control logic. BPF consists of the two stage switched capacitor filter with the band of quality factor $Q = 5$ and the sharp cut-off characteristics. This circuit design to employ the switched capacitor filter whose band characteristics are easily controlled by an external clock makes it possible to switch the operating frequency through a command from I/O unit. Total gain of receiver circuits including Pre-amplifier and Main amplifier is about 120 dB. Digitized signals through A/D converter are processed to detect the Doppler frequency shift and the backscattered signal intensity. The calculation of FFT and power spectrum from 1024 data points is carried out within 80 msec by 32 bits CPU (INTEL 80486DX) with a clock speed of 33 MHz.

Input/Output (I/O) Unit

I/O unit that consists of a personal computer is able to communicate with the Main unit through GPIB interface. I/O unit works as (1) the man-machine interface to set the operation parameters such as averaging time, measuring height and other operating hardware parameters, (2) the data display and storage of wind velocity, received signal spectra and signal intensity. Automatic Power Control (APC) and the automatic change of operation parameters are available by the programmable function. Measured data are usually stored in floppy or magneto-optical disks. The echo intensity is also sent to the color display and the dotmatrix printer as visual information. The graphic display of received signal spectra is useful for signal diagnosis at the time of the noise check in the installation or the investigation of unusual data.

Arithmetic Operation in Main Unit

The setting of Timing control logic according to the command information from I/O unit is carried out at the beginning of each observation cycle, therefore the dominant job of Main unit is to process the received signals. After Timing control logic opens the sampling gate in the specified altitudes, A/D conversion is made. The signal operation to get the wind velocity in Main unit is as follows; (1) the detection of Doppler frequency shift, (2) the discrimination of signal reliability, (3) the calculation of wind velocity. In the case of the calculation of wind velocity in the tri-monostatic Doppler sodar, the radial wind component V_1, V_2, V_3 towards each beam can be related to the three components of wind in Cartesian coordinate V_x, V_y, V_z as follows.

$$-\begin{pmatrix} V_1 \\ V_2 \\ V_3 \end{pmatrix} = \tilde{A} \begin{pmatrix} V_x \\ V_y \\ V_x \end{pmatrix} \quad \text{where} \quad \tilde{A} = \begin{pmatrix} \cos x_1 & \cos y_1 & \cos z_1 \\ \cos x_2 & \cos y_2 & \cos z_2 \\ \cos x_3 & \cos y_3 & \cos z_3 \end{pmatrix} \quad (12)$$

$\cos x_j$, $\cos y_j$ and $\cos z_j$ ($j = 1, 2, 3$) being direction cosines of j-th beam axis relative to x-, y-, z-axes. Solving Eq. (12)

$$\begin{pmatrix} V_x \\ V_y \\ V_z \end{pmatrix} = -\tilde{A}^{-1} \begin{pmatrix} V_1 \\ V_2 \\ V_3 \end{pmatrix} \quad (13)$$

Therefore V_x, V_y and V_z can be obtained from V_1, V_2, and V_3. And V_1, V_2, and V_3 in turn are calculated from the Doppler frequency shift of the received signal in each direction by Eq. (7). The antenna arrangement of the tri-monostatic Doppler sodar is usually set up so that two antennas are in orthogonal direction at a zenith angle of ε and the remaining antenna in zenith direction to measure the vertical wind directly. The wind velocity U and the wind direction θ are calculated as

$$U = \sqrt{V_x^2 + V_y^2} \quad (14)$$

$$\theta = \tan^{-1} \frac{V_x}{V_y} \qquad (15)$$

And the standard deviation of vertical wind, σw is obtained from the vertical velocity W_i of the total available data number of N_w

$$\sigma w = \sqrt{\frac{1}{N_w} \sum_{i=1}^{N_w} (W_i - \overline{W}_i)^2} \qquad (16)$$

As the observation cycle is three times as long as the pulse repetition interval in the tri-monostatic sodar, i.e. usually longer than 10 sec, therefore σw estimated by Eq. (16) lacks high frequency fluctuations. Combined with the special averaging caused by the pulse length, it may cause under-estimation of more than 10% in convective conditions [12]. The actual estimation of σw, however, often shows slight over-estimation and large scatter due to random errors in the measurement. As there is the sampling volume separation and the sampling time lag among each beam together with the beam wandering [13], the accurate estimation of the standard deviation of horizontal wind σu, σv is not made.

Observational Examples

A lot of research or routine works have been done with AR410 Doppler sodar in Japan. Some examples of the comparison with in situ measurements are shown below.

Fig. 13 shows a comparison of 10 min averaged wind speed from sodar with the sonic anemometer mounted on a meteorological tower at a height of 100 m. The distance between the sodar and the meteorological tower is about 500 m and the location of these sensors is on the plateau near the sea coastline. The data observed every one hour during a whole year are plotted on the scatter diagram. The comparison of more than 8000 data appears highly correlative with a correlation coefficient of 0.96. Fig. 14 shows the examples of received signal spectra which varied within half a day. Fig. 14(a) is observed in the moderate wind, Fig. 14(b) in the strong wind and Fig. 14(c) is observed under precipitation.

With the increase of wind speed, Fig. 14(a) and (b) show that the received spectrum becomes broad, i.e. the increase of wind speed from 2 m/s to 10m/s brings out the spectral broadening of 25 Hz to 70 Hz, which corresponds to 1.8 m/s to 5.0 m/s at an operating frequency of 2.4 kHz. On the other hand, Fig. 14(c) shows the twin peak in the received spectrum. The broad spectral peak shifted by about 120 Hz from 2.4 kHz, which corresponds to 8.5 m/s in velocity, seems to be caused by the downward motion of rain drops. Such unusual received spectra seem to cause the large scatter under precipitation in Fig. 13. The analysis in the case of Fig. 13 showed that the data loss rate during a whole year was about 1.3% at the height of 100 m.

Table 2 indicates the monthly variation of the data acquisition rate at each altitude. This table suggests that the maximum altitudes of acoustic sounding

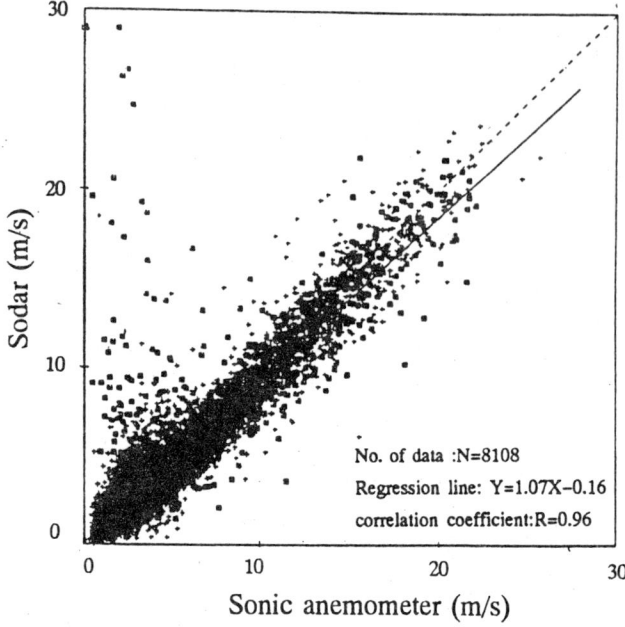

Fig. 13. Scatter diagram of 10 minutes averaged wind speed from sodar versus sonic anemometer (after Akai et al. [14]). ■ denotes under precipitation and + without precipitation.

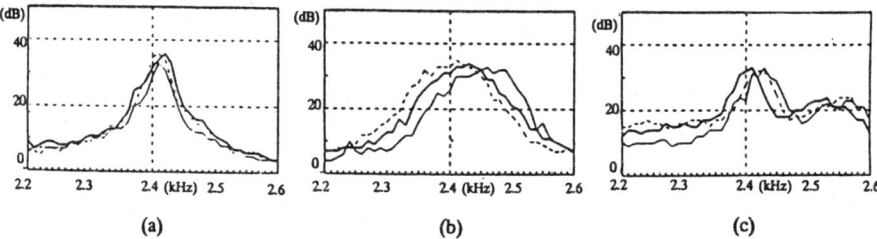

Fig. 14. Received power spectrum; (a) in the moderate wind, (b) in the strong wind, (c) under precipitation

become higher in summer and lower in winter, and that the major factor of the data loss is deduced to the small temperature fluctuations and the low humidity. The mean values of temperature and relative humidity at the height of 1000 m above the site were about 20°C, 85% during August respectively, which may give the corresponding acoustic attenuation of $\alpha = 0.0025$ at the operating frequency. When we assume that C_T^2 is about 10^{-4} deg^{2}m$^{-2/3}$ at the maximum range of acoustic sounding, Rmax, referring to the measurements of C_T^2 in the summer by Koprov and Tsvang [8], Fig. 6 gives the Rmax to 800 m. This estimation of Rmax from Fig. 6 almost agrees with the observational result of Table 2.

Table 3 shows the comparisons of wind speed and direction with the

Table 2. Monthly variation of the data acquisition rate (%) operated at 1.6 kHz
(after Ito et al, [15])

Height (m)	Month							
	8	9	10	11	12	1	2	3
50	100	100	100	100	100	100	100	100
100	100	100	100	100	100	99	100	99	95
150	100	100	99	100	98	91	93	89
200	100	99	98	100	92	82	85	88
300	99	99	96	92	82	71	80	79
400	87	93	88	81	58	41	61	73
600	59	75	53	42	17	11	27	46
800	26	45	23	16	5	3	7	16
1000	6	12	9	7	1	1	0	2
No. Data	458	462	584	309	456	207	347	357

meteorological tower instruments of the Meteorological Research Institute of Japan, which is located in a suburban area surrounded by trees and low buildings. The mean difference of wind speed V and the standard deviations of the difference of wind speed and direction S and D respectively are compared in each wind class. The mean difference V is less than 0.7 m/s regardless of wind speed, while the standard deviation S is in the range of 0.7 to 1.2 m/s for wind speeds below 5 m/s and 1.2 to 2.3 m/s over more than 5 m/s. The standard deviation D decreases with the wind speed from 30 to 10 deg over more than 2 m/s.

Table 4 shows the data comparison of wind speed and direction among the sodar, tower and pilot balloon. Broadly speaking, the standard deviations of the difference of wind speed and direction are in the range of 0.7 to 1.3 m/s and 20 to 35 deg respectively between one another. The correlation coefficients are better than 0.9 at higher than 100 m altitude. Thus there seems to be no systematic difference in the wind data among the sodar, tower and pilot balloon.

Concerning turbulent variables σw, σu, σv and C_T^2, we have some comparisons mainly about σw [15]. However as mentioned already, there are the problems of beam separation and beam wandering, therefore we are examining the method to separate the random errors to obtain better estimation of the turbulent variables [16].

Summary

This paper outlined the general purpose Doppler sodar AR410. Together with the acoustic antenna design and the circuit function, the present author suggested the signal processing technique to estimate the Doppler frequency shift based on the FFT method and the way to reject unavailable signals. It may be helpful to refer to the nomogram to estimate the maximum altitude of acoustic sounding.

Table 3. Comparison of wind speed and direction with the tower instrument. N is the number of data, V the mean difference of wind peed, S the standard deviation of the difference of wind speed, D being the standard deviation of wind direction. (after Ito et al, [15])

W.S. (m/s)	Height (m)			
	50	100	150	200
1 ~ 2	N = 503 V = − 0.7 S = 0.8 D = 53	N = 425 V = − 0.7 S = 0.7 D = 49	N = 371 V = − 0.6 S = 0.8 D = 45	N = 371 V = − 0.6 S = 0.8 D = 45
2 ~ 3	N = 656 V = − 0.3 S = 0.8 D = 34	N = 533 V = − 0.3 S = 1.0 D = 31	N = 480 V = − 0.2 S = 0.9 D = 33	N = 429 V = − 0.2 S = 1.0 D = 34
3 ~ 4	N = 536 V = 0.1 S = 0.9 D = 24	N = 450 V = 0.1 S = 2.2 D = 25	N = 411 V = 0.1 S = 1.0 D = 26	N = 409 V = 0.1 S = 1.2 D = 22
4 ~ 5	N = 397 V = 0.3 S = 1.1 D = 20	N = 405 V = 0.3 S = 1.0 D = 19	N = 348 V = 0.3 S = 1.2 D = 0	N = 331 V = 0.3 S = 1.4 D = 23
5 ~ 6	N = 216 V = 0.7 S = 1.3 D = 23	N = 329 V = 0.6 S = 1.2 D = 23	N = 318 V = 0.5 S = 1.5 D = 19	N = 297 V = 0.5 S = 1.7 D = 19
6 ~ 8	N = 160 V = 0.6 S = 1.7 D = 20	N = 320 V = 0.7 S = 1.4 D = 14	N = 348 V = 0.6 S = 1.5 D = 16	N = 355 V = 0.7 S = 1.9 D = 18
8 ~ 10	N = 78 V = 0.7 S = 2.1 D = 19	N = 86 V = 0.8 S = 2.1 D = 10	N = 109 V = 0.6 S = 1.9 D = 12	N = 125 V = 0.4 S = 2.3 D = 12
10 ~	N = 66 V = 0.8 S = 1.8 D = 8	N = 134 V = 0.7 S = 2.5 D = 6	N = 126 V = 0.6 S = 2.4 D = 8	N = 117 V = 0.5 S = 2.1 D = 10

Some results on the wind data compared with in situ measurements show the reliability of the Doppler sodar. The correlation coefficient of wind speed is better than 0.9, and the standard deviation of the difference of wind speed and wind

Table 4. Comparisons of wind speed and wind direction between; (a) sodar and tower instrument; (b) sodar and pilot balloon; (c) tower instrument and pilot balloon (after Ito et al [15])

(a)

Height (m)	No. Data	Corr.	rms difference (m/s)	(deg.)
50	102	0.85	0.7	34
100	102	0.94	0.7	33
150	99	0.94	0.9	33
200	97	0.93	1.2	31

(b)

Height (m)	No. Data	Corr.	rms difference (m/s)	(deg.)
50	24	0.78	1.0	23
100	25	0.94	0.9	25
150	25	0.93	1.1	19
200	24	0.92	1.4	26
300	27	0.96	1.0	28
400	16	0.96	1.3	20
600	10	0.87	0.8	30

(c)

Height (m)	No. Data	Corr.	rms difference (m/s)	(deg.)
50	24	0.86	0.8	35
100	28	0.93	0.9	30
150	27	0.93	1.1	36
200	30	0.94	1.2	20

direction is 0.7 to 2.3 m/s and 10 to 35 deg respectively dependent on wind speed. The sodar system helps the meteorological observation of the lower atmospheric boundary layer. It gives wind profile up to several hundred meters with the condition in which the annual data loss rate is less than 2% at the height of 100 m and is less than 10% at 200 m. Saying about turbulent parameters, σ_w seems to be the available function, whereas σ_u and σ_v are under development to make a better estimation. The detection of temperature inversion and qualitative monitoring of thermal structure from echo intensity is reliably carried out, but the quantitative analysis with sufficient accuracy such as the temperature fluctuation or the temperature lapse remains as future works.

Acknowledgments

The author is grateful to Prof. Y. Mitsuta of Kyoto University for his helpful advice and encouragement.

References

1. Neff, W.D. "Quantitative evaluation of acoustic echoes from the planetary boundary layer", NOAA Technical Report ERL 322-WPL 38, 1975.
2. Hall, F.F. Jr and J.W. Wescott, "Acoustic antennas for atmospheric echo sounding", J. Acoust. Soc. Am., vol. 56, 1376–1382, 1974.
3. Ito, Y., Y. Kobori, M. Horiguchi, M. Takehisa and Y. Mitsuta "Development of wind profiling sodar", J. Atmos. Ocean. Technol., 6(5), 779–784, 1989.
4. Simmons, W.R., J.W. Wescott and F.F. Hall, Jr "Acoustic echo sounding as related to air pollution in urban environments", NOAA Technical Report ERL 216–WPL 17, 1971.
5. Krasnenko N.P. and S.L. Odintsov "Optimal frequencies for meteorological sodar", J. Acoust. Soc. Am., 75(2), 390–394, 1984.
6. Olson, H.F. "Acoustical engineering", D. Van Nostrand Co., Inc, 1957.
7. Maekawa, Z., "Noise reduction by screens", Applied Acoustics vol. 1, 157–173, 1968.
8. Koprov, B.M. and L.R. Tsvang, "Characteristics of small-scale turbulence in a stratified boundary layer", Bull. Acad. Sci. SSR, Atmospheric and Oceanic Physics, 2, No. 11, 1142–1150, 1966.
9. Harris, C.M. (Ed.) "Handbook of noise control 2nd Ed.", 3–10, McGraw-Hill, 1979.
10. WMO, "Instruments and observing methods, Report No. 3, Lower tropospheric data compatibility, Low level intercomparison experiment", 1980.
11. Sirmans, D. and B. Bumgarner, "Numerical comparison of five mean frequency estimators", J. Appl. Meteorol., vol. 14, 991–1003, 1975.
12. Finkelstein, P.L., J.C. Kaimal, J.E. Gaynor, M.E. Graves and T.J. Lockhart; "Comparison of wind monitoring systems. Part II: Doppler; sodars", J. Atmos. Ocean. Technol., vol. 3, 594–604, 1986.
13. Neff, W.D., J.E. Gaynor and Ye Jing-Ping, "The effect of sodar beamwander and tilt on Doppler-derived turbulence measurements", Preprints of the 7th Symp. on Meteorol. Obs. and Instru., 422–427, 1991.
14. Akai, Y., K. Asakura and N. Katayose, "An evaluation of wind measurements in the lower atmosphere by a Doppler sodar", Tenki, 40(7), 21–33, 1993 (in Japanese).
15. Ito Y., Y. Watanabe, T. Mizukoshi, T. Hanafusa, T. Yoshikawa, K. Naito and N. Kodaira, "Development of a Doppler sodar and some applications to the observational study of atmospheric boundary layer" Tenki, 33(8), 375–385, 1986 (in Japanese).
16. Takehisa, M., Y. Ito, T. Kataoka and Y. Mitsuta, "Precision and relative accuracy of a phased array Doppler sodar", Bull. Disas. Prev. Res. Inst., Kyoto Univ., 42, 65–70,1992.

Acoustic Remote Sensing Applications
S.P. Singal (Ed)
Copyright © 1997 Narosa Publishing House, New Delhi, India

4. A Modular PC-Based Multiband Sodar System

G. Mastrantonio and S. Argentini

Istituto di Fisica dell' Atmosfera-CNR, Via G. Galilei,
cp. 27-00044 Frascati (Roma), Italy

Introduction

The acoustic remote sensing technique is a powerful tool for investigating the phenomena that take place in the Planetary Boundary Layer (PBL). It provides a "snapshot" of the thermal structure of the PBL by displaying the echoes in a facsimile format. Through the harmonic analysis of the received signal, it is possible to derive the wind profile and to visualize the dynamic evolution of the airflow crossing the antenna beams. In the past, in order to analyse the sodar echoes and to retrieve the Doppler information, powerful computers were needed, unless simple algorithms, such as, for example, the zero-crossing or the real covariance methods, were used [1]. In 1978 our Laben 70 minicomputer took more than one day to retrieve the Doppler shifts of one hour of digitized sodar echoes, using the FFT (Fast Fourier Transform). Moreover low cost mass storage peripherals were not available to record instantaneous measurements of echo intensity and radial wind components so that only average quantities were recorded during long period campaigns. More recently the explosive development of the personal computer (PC) technology and of the digital signal processor (DSP) cards which plug into the PC-bus, provide a much simpler way to analyse sodar echoes in real time. It is now very common to find sodar systems on the market, that use PC-like computers for data processing. These systems give measurements that are usually integrated over periods lasting from 5 minutes to half an hour. This is satisfactory for most routine applications, but for research purposes it is preferable to have the possibility of recording all the data the system is able to provide. In this case it is important to have the software tools that permit an easy handling of large amounts of data (20 Mbytes/day).

In what follows, after a brief overview of the hardware operating features, a new PC-based acquisition and data processing system will be described. The capability of the developed software in testing the system, handling and processing the instantaneous data, and printing the results in a compact form, will be highlighted.

It will also be shown how the new system allows the tuning of some of the system parameters and the double band running mode. In this way, by alternately using high and low frequencies, the higher resolution of the minisodars in the lower layers matches the usual sodar range in the same system.

Electronics and Data Acquisition System

The fundamentals of the electronics and of the signal processing have been already discussed in the past [2, 3]. The system is basically a three-axis monostatic Doppler sodar, able to simultaneously radiate tone bursts at different frequencies. The operational parameters, i.e. carrier frequencies, tone length, reception delay, pulse repetition rate, etc., can be set manually by local keyboard or remotely by serial line. A microprocessor synthesizes the three carrier frequencies and schedules the several functions in the electronics. In the most recent realization, switched capacitor filters are automatically tuned to predetermined frequencies to filter the received signal. Since each channel signal lies in a different frequency band, the three signals are added; the combined signal is then digitized and analyzed. As part of this process, a sampling pulse train at a suitable frequency is produced by multiplying one of the carrier frequencies by a rational number in the range 1–1/16. In order to satisfy the requirements of Shannon's theorem, the sampling frequency has to be at least twice the bandwidth of the signal to be sampled (and not necessarily twice its maximum frequency). Carrier frequencies and sampling pulses can be chosen so that the sample's number is minimized [3]. This under sampling process, used with a proper choice of the band and sampling frequency, has the same result as heterodyning the signal before digitalization.

The digitized signal goes to the input of a DSP (Digital Signal Processor) card plugged into the bus of a compatible PC where two FFT analyses are carried out. The basis for the card's computational power (a SKY 321 card) is the TMS32010 signal processing chip produced by Texas Instrument. In the card, program (8 Kbytes) and data (128 Kbytes) memory are separate and mappable in the memory addressable by DOS (see Fig. 1). The presence of two uni-directional 16 bits input and output ports permits I/O operation without engaging the PC-Bus. The I/O flow can be performed in DMA (Direct Memory Access) and driven by an external device. It is then possible to retrieve the digitized signal and at the same time carry out the harmonic analysis of previously stored data. In our case size buffers have been configured in the 128 Kbytes SKY data memory. While data are stored in the buffer A1, a first analysis is carried out on the data previously stored on buffer B1 and the results of this analysis are stored in buffers B2-B3. Then the tasks of the buffers A and B are reversed. It should be noticed that up to this point the PC has not been engaged and all the I/O operations as well as the first analysis have been carried out by the SKY321; the PC is only informed about the presence of the results of the first analysis in the part of its memory corresponding to the buffers A2-A3-B2-B3. The PC is then used to perform the successive calculations to get Doppler shifts and to visualize results.

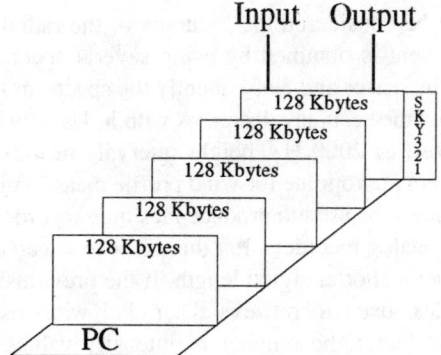

Fig. 1. Schematic view of the SKY321 card plugged into the PC-Bus.

Spectral Analysis

The spectral analysis used to retrieve the Doppler shifts and the intensity values for each channel is carried out through a two-step procedure [2]: the spectrum of the echo is first localized and the first moment calculation is then limited to the expected signal bandwidth rather than to the full channel spectrum. Especially for low S/N (Signal to Noise) values, this procedure minimizes the bias toward zero due to the transfer function of the receiver chain and the statistical error [4]; moreover the S/N level in the intensity values is improved. The intensity and radial components of the wind are thus obtained for the three channels at the different range gates. A parameter that is related to the S/N ratio is used to reject unreliable data [2, 3]. The range gate width is a function of the time length of the analyzed signal stretch and consequently of the number of points over which the FFT (Fast Fourier Transform) is carried out. For example suppose that the three carrier frequencies are 1750-2000-2250 Hz, and the bandwidth of the combined signal is 800 Hz. If a digitalization frequency of 1600 Hz is used, and 256 successive samples are taken to perform the FFT analysis, the vertical resolution Δz will be:

$$\Delta z = \frac{c}{2} * \frac{256}{1600} = 27.2 \text{ m}$$

if a value of 340 m/s is assumed for the sound speed.

The FFT analysis performed on these samples produces a spectrum having a resolution of

$$\Delta f = \frac{1}{T} = \frac{1600}{256} = 6.25 \text{ Hz}$$

where T is the time length of the signal analyzed. At a frequency of 2000 Hz, a Doppler shift of 6 Hz is related to a radial wind

$$v_r \cong \frac{c}{2} * \frac{6}{2000} = 0.51 \text{ m/s}$$

Naturally this cannot be considered the accuracy of the radial wind measurement because the first moment is obtained by using several spectral bins. Anyway, if the height resolution increases and consequently the spectrum frequency resolution decreases, undoubtedly the accuracy decreases with it. Usually in the sodar systems using frequencies close to 2000 Hz, height intervals in the range 25–50 m are used. These intervals are appropriate for wind profile measurements. The associated intensity, however, cannot be used to produce facsimile records that are comparable in resolution with the analog recorders. For this purpose a second harmonic analysis may be performed over a shorter signal length. If the previously considered signal length of 256 samples, used to retrieve the radial wind, is divided by 4, we multiply by the same factor the number of intensity values associated to each layer so that in the intensity profile a value every 6.8 m is available. In this case the frequency spectrum resolution is not so critical since only the zero moment is needed. In Fig. 2 the improvement in the facsimile records gained in this way is shown. Further progress may be obtained in the case of a noisy environment if the two-steps procedure is also used for the determination of intensity. Since the lower frequency spectrum resolution does not permit an accurate location of the signal, the Doppler shift found for the original 256 sample signal length may be used to locate the signal in the four derived spectra. The knowledge of the spectral position of the signal permits the carrying out of the intensity calculation in a narrower spectral region and decreases the influence of noise. In Fig. 3 the effect of the procedure on a noisy facsimile record is shown.

Examples of Accuracy in Radial Wind Measurements

The advantages of using the two-steps procedure on the errors associated with radial wind measurements has been reported elsewhere [2]. Some of the results obtained in the recent past, during a campaign in which a network of sodars has been used, confirm the accuracy with which the radial component of the wind may be measured using the two step procedure previously mentioned. In particular [5] the daily cycle of the vertical component of the wind associated with a sea breeze has been measured close to the coast (Fig. 4a) and 20 km inland (Fig. 4b). The vertical component of the wind due to the Rome heat island circulation is seen by comparing the daily averages of w in the city and in the rural area surrounding Rome (Fig. 5), [6]. We emphasize the care needed in these measurements in order to minimize errors due to antenna pointing and to terrain slope. Moreover the bias that may result from an unwanted elimination of data due to ambient noise must be taken into account. For example, in a noisy environment, when convective activity is present, positive values of w associated with plumes are more likely above the acceptance threshold than negative values.

The Sodar Real-Time Analysis Program: Diagnostic and Data Handling Tools

The diagram in Fig. 6 illustrates the capability of the real time analysis program.

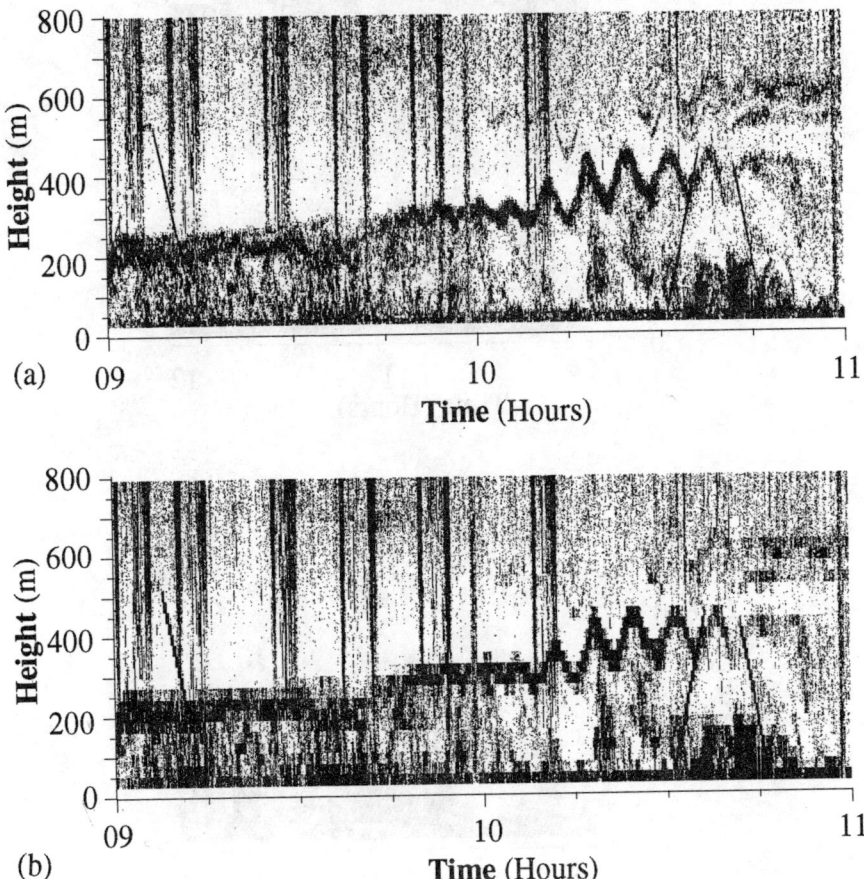

Fig. 2. Improvement in the facsimile digital recording obtained using the procedures mentioned in the text. The pixel vertical dimension is 6.8 m in (a) and 27.2 m in (b). The gray scale has been obtained by varying the number of dots, randomly positioned in each pixel.

Input signals can come directly from the sodar system, from played-back recorded echoes [3], or from files containing digitized echoes. Also it is possible to simulate the input to test the algorithms. The output of the program consists of the average wind profile which can be recorded on disk files, as well as drawn on the PC display and on the printer on request. If enough room is available on the disk, an option is present that permits the storage of the instantaneous values on files hereafter referred to as MT files. In our system, for this purpose we use a Sony optical disk driver that supports 5.25 inch rewritable diskettes whose capacity (600 Mbytes) allows us to record almost a month of sodar data. In the MT files for each antenna the radial wind profile, the intensity profile (low and high resolution), and the parameter giving information on the S/N ratio at each range

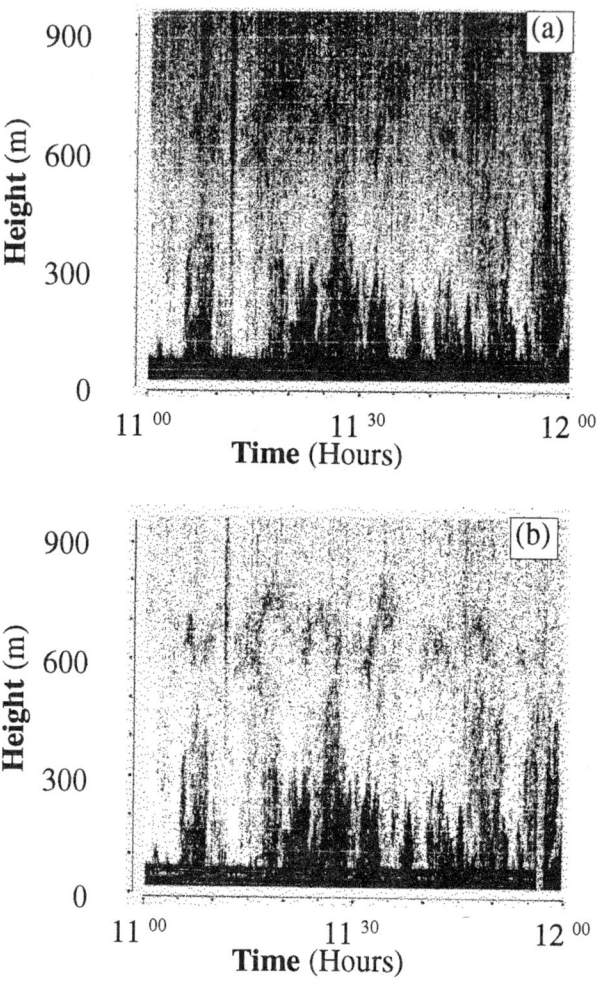

Fig. 3. A facsimile record obtained in a noisy environment processed (a) without and (b) with the two-steps procedure.

gate are recorded. Several programs are available that, using MT files as input, carry out successive analyses and generate several type of graphs, including daily facsimile records to be printed on a laser printer. All the graphs may also be produced in standard format such as PIC and TIFF files. This allows us to manipulate these graphs using commercial software and/or to include them in reports, using word processors or publishers as it was done for most figures in this paper.

In addition to this software, some programs have been developed that help to determine some of the adjustable parameters such as frequencies, relative amplification of the three channels, etc. One of the programs displays the digitised

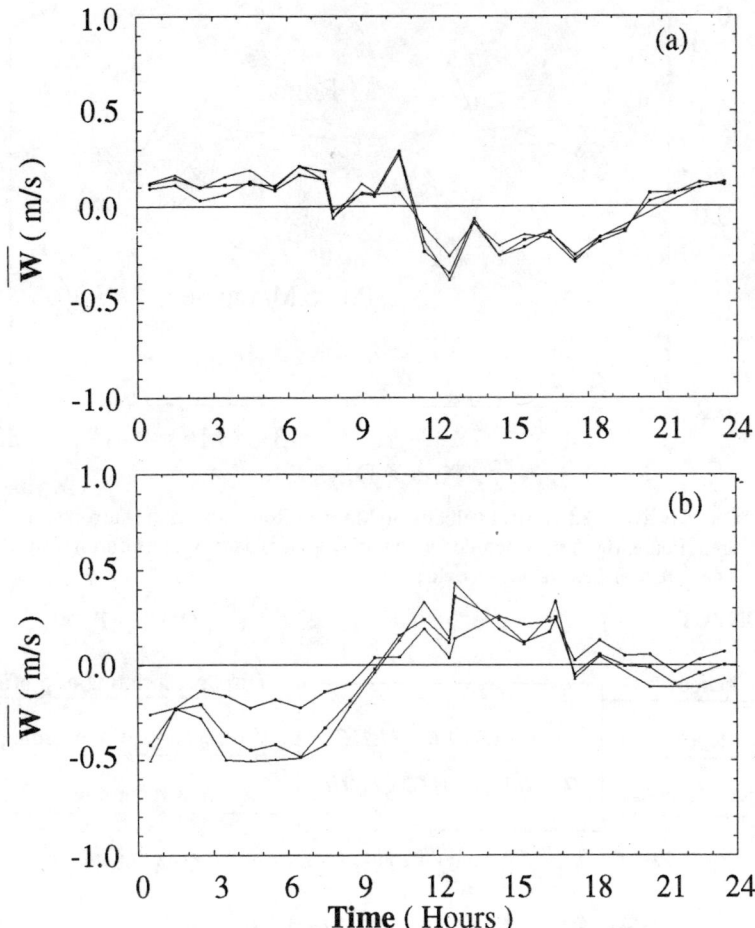

Fig. 4. One hour averaged vertical velocity at three range gates below 200 m during a sunny day, (a) close to the coast and (b) 20 km inland. The phases of the sea breeze circulation at the two places are opposite.

echo and its harmonic analysis. It aids in setting the signal dynamics to the proper values and in getting the transfer function of the signal receiving chain. In Fig. 7, a typical output of this program is shown. In Fig. 8 the 2 kHz channel transfer function is represented. The graph shows the average of 35 spectra carried out while the system was passively recording ambient noise.

The second program draws a pseudo-three-dimensional representation of spectra from successive scans. It is particularly useful for determining the dumping time of the system reverberation after the burst emission; it also allows us to optimize the choice of the carrier frequency for a given transducer in order to minimize the reverberation and decrease the first range gate height. In Figs. 9a-c typical outputs

112 *Atmospheric Acoustics and Instrumentation*

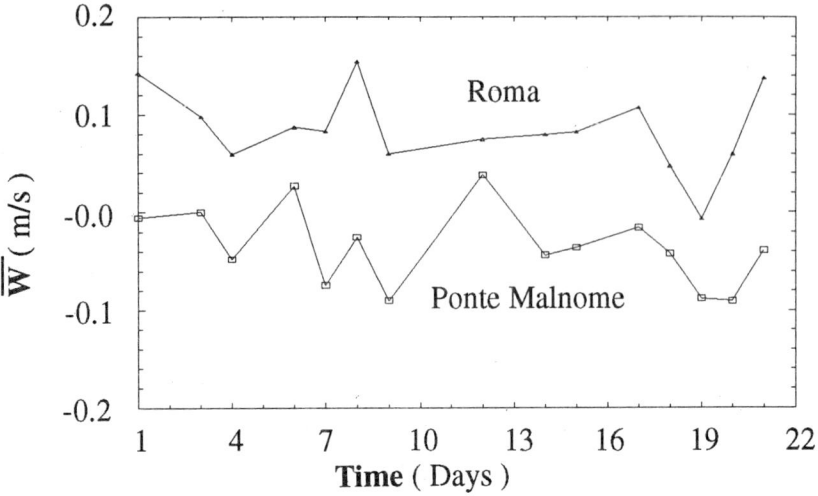

Fig. 5. One day averaged vertical velocity at 200 m in Rome and in the surrounding rural area (Ponte Malnome) recorded with two Doppler sodar systems during July 1992. The effect of heat island is evident.

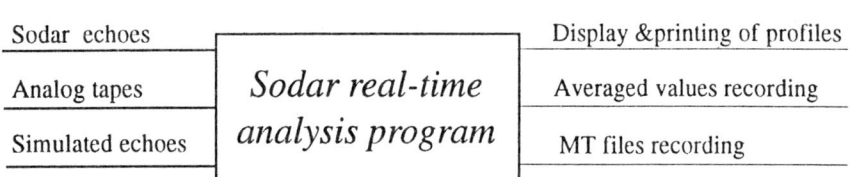

Fig. 6. I/O schematic diagram of the real-time program.

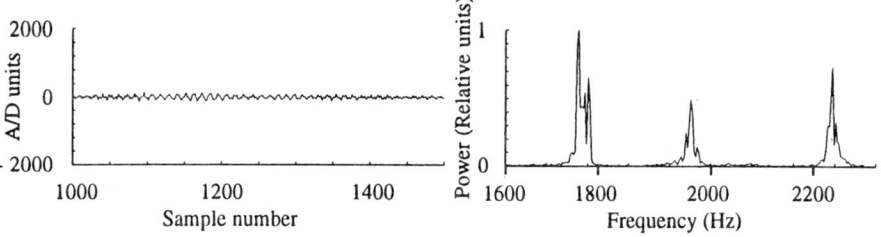

Fig. 7. Output of one of the diagnostic programs; (left) digitized echo, (right) related spectrum.

of this program are shown. The spectra of 20 successive scans are displayed: while the first one is obtained using a sample starting 0.1 seconds after the emission, each successive sample is delayed 6 ms with respect to the previous one. The reverberation dumping is evident, as well as the minimization of the problem by moving the carrier frequency from the critical value.

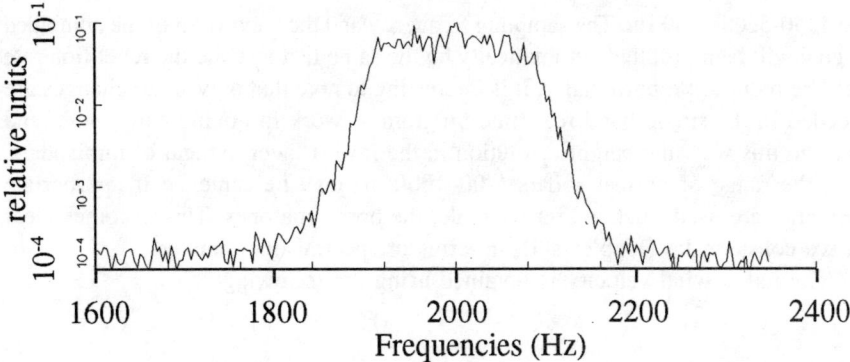

Fig. 8. Channel 2 logarithmic transfer function obtained by averaging 35 spectra while the sodar was passively receiving ambient noise.

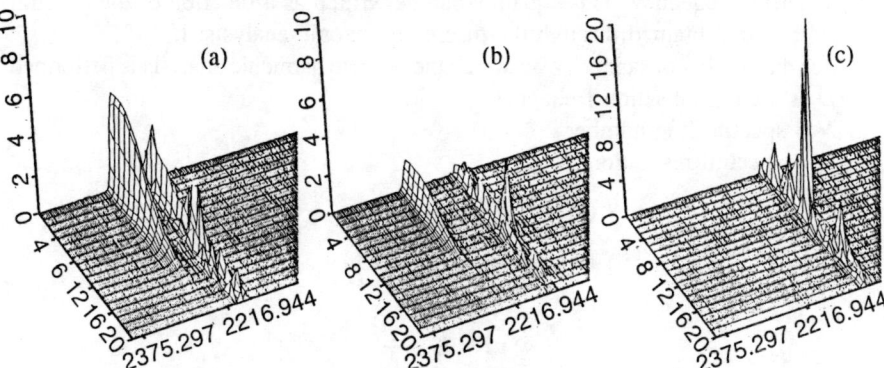

Fig. 9. Display of the reverberation effect on the spectra (see text). When the carrier frequency is moved far from the critical value (from (a) to (c)), the reverberation disappear.

Single and Double Band Doppler Sodar Operation

When working in real time analysis, only 20% of the PC (a 80386-20 MHz) and SKY321 time is kept busy. When needed, time is available to carry out more calculations or to handle more complex sodar system configurations [6].

In the past we have used only a single band sodar so that, when running, each antenna was able to send only one carrier frequency: usually for the three-axis sodar the frequencies 1750-2000-2250 Hz have been used. Considering the availability of computing time and the automatic tuning of the switched capacitor filters to the chosen carrier frequencies, in co-operation with the Multimicro S.a.S. that make the sensors, the hardware and the software have been modified so that the system is able, if requested, to switch alternately between two frequency bands. For this purpose the "base band" and a multiplying factor in the range 2–3 have to be chosen. For example if the base carrier frequencies are 1750-2000-2250 Hz and a factor of 2.6 are chosen, the second set of carrier frequencies will be 4550-5200-5850 Hz. The sampling frequency and the bandwidth of the combined signal will be multiplied automatically by the same factor while the repetition rate will be reduced proportionally. It is interesting to note that only small changes are needed in the single band real time program to work in double band mode and that, in this way, the height resolution in the lowest layer, typical of minisodars, and the range of normal sodars (700–1000 m) may be achieved if appropriate antennas are used, such as, for example, the horn-type ones. This becomes clear if we consider the Doppler shift in terms of spectral line numbers.

The radial wind velocity is obtained using the following:

$$v_r = \frac{c}{2} \cdot \frac{\Delta f}{f} \qquad (1)$$

where v_r is the radial velocity, c is the sound velocity, Δf is the Doppler shift and f the carrier frequency. This equation can be written as a function of the spectral lines that are obtained through the discrete harmonic analysis. If

n is the number of samples over which the discrete harmonic analysis is performed
D is the digitalisation frequency
N = spectral line number
δf = spectral resolution

then

$$N(f) = \frac{f}{\delta f} = n \frac{f}{D} \qquad (2)$$

$$v_r = \frac{c}{2} \cdot \frac{\Delta N}{N} \qquad (3)$$

If the same n for the discrete Fourier Transform and the same factor for multiplying the carrier and the digitalisation frequencies are used, we see in the Eq. (2) that the spectral line number corresponding to the carrier frequency is the same. In this condition the radial wind will produce the same spectral line shift independent of the frequencies used. The band change will produce a change, by the same factor, of the height resolution. If, as in our case, the hardware of the sodar system is able to alternately radiate two frequencies bands, and the same number of samples are digitised after each emission, very few changes need to be made in the software: only the recording of a flag may be necessary for the selection of the height scale associated with the data. At this time we have tested the double band system using a dish type antenna: in Figs. 10a-d an example of data obtained while the system was using the double band feature is shown. In the future we are planning to use more appropriate antennas that minimize the problem of matching the emitting transducer to the reflecting part of the antenna.

Conclusions

A description of the sodar system used at the present time in our laboratory has

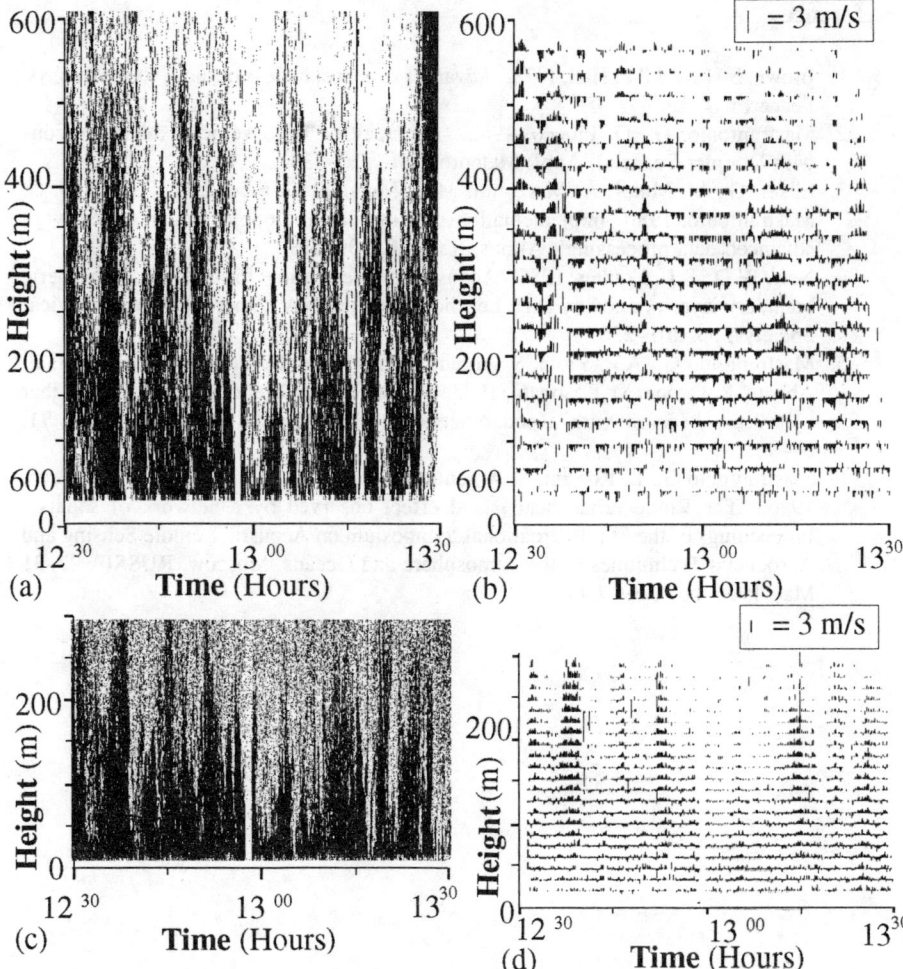

Fig. 10. Facsimile and w records obtained while the system was operating in the double-band mode. Lower band (a-b), higher band (c-d).

been presented. Its modularity and the computing power allow for the addition of more real time processing. Several software tools allow for easy handling of the instantaneously recorded data and the printing of the graphs in a compact form: for example a daily facsimile record may be obtained on a A4 sheet if a 300 dots/inch laser printer is used. The two-step procedure reduces the influence of the noise and increases the accuracy of the radial wind measurement. The double band operation allows for greater resolution of the minisodar in the lower layers and, simultaneously, the range of a "standard" sodar.

Acknowledgements

The authors are grateful to Dr. G. Greenhut for his helpful suggestions.

References

1. Brown, E.H., and F.F. Hall, 1978, 'Advances in atmospheric acoustics', Rev. Geophys. Space Phys. **16**, 47–109.
2. Mastrantonio, G., G. Fiocco, 1982, 'Accuracy of wind velocity determinations with Doppler Sodar' J. Appl. Meteorol., **21**, 820–830.
3. Elisei, G., M. Maini, A. Marzorati, M.G. Morselli, G. Fiocco, G. Cantarano, G. Mastrantonio, 1986 'Implementation of a multiaxial Doppler Sodar system with advanced data processing', Atmos. Res., **20**, 109–118.
4. Neff, W.D., R.L. Coulter, 1986, 'Acoustic remote sensing, Probing the atmospheric boundary layer', edited by D.H. Lenshow (Boston, MA: American Meteorological Society), pp. 201–236.
5. Mastrantonio G., A.P. Viola, S. Argentini, G. Fiocco, L. Giannini, L. Rossini, G. Abbate, R. Ocone, M. Casonato, 1993, 'Aspect of low level circulation in the Tiber valley as observed with a sodar network', Boundary Layer Meteorology, **71**, 67–80.
6. Mastrantonio G., L. Rossini, S. Argentini, A.P. Viola, L. Giannini and G. Fiocco, 1996, "The Rome urban heat island effect observed by a network of sodars". Proceedings of the 8th International Symposium on Acoustic Remote Sensing and Associated Techniques of the Atmosphere and Oceans. Moscow, RUSSIA 27–31 May 1996, pp. 7.39–7.44.

Acoustic Remote Sensing Applications
S.P. Singal (Ed)
Copyright © 1997 Narosa Publishing House, New Delhi, India

5. Mini Acoustic Sounding, A Powerful Tool for ABL Applications: Recent Advances

D.N. Asimakopoulos[1,2], C.G. Helmis[1] and M. Petrakis[2]

[1]Department of Applied Physics, University of Athens
Panepistimioupolis, Build. Phys.-5, 15784, Athens, Greece

[2]Institute of Meteorology and Physics of the Atmospheric Environment,
National Observatory of Athens, PO Box 20048 Athens 11810, Greece

1. Introduction

The development of acoustic mini sounding of the lower atmosphere proceeded rapidly after the early work of Moulsley and Cole (1979) and Asimakopoulos et al (1987). In principle the so called acoustic mini sodar, (henceforth AMS) or high resolution acoustic Radar is a straight-forward electronic design system which is not significantly different from the conventional ones. In fact it could easily be described as a system that combines all improvements made through the years towards reducing the antenna size and the operating capability in adverse environment conditions. The high resolution time and space characteristics are not necessarily associated with the system dimensions but in many cases can be encountered as operating parameter options of these modern design systems.

To construct a conventional AMS it is essential to design a reasonably small acoustic antenna which will maintain the necessary directional and efficiency characteristics. In this respect it is worth mentioning the work of Moulsley et al (1978) and Stephanou and Mavrakis (1986), who dealt with the detailed study of acoustic antenna array design with active or passive phase and altitude element adjustments. Their work covered different operating frequencies and antenna configurations indicating also some small size and light weight acoustic antenna design operating at relatively high frequencies, (normally higher than 2.5 kHz), with very good directional capabilities. Following this work several miniature systems appeared in the field during different experimental campaigns and on some occasions unusual applications of the technique were employed, (e.g. Horizontal acoustic sounding, Moulsley et al, 1985).

Following this technical progress which was associated as expected with a good and interesting number of applications, brief mention of which will be made in this paper, other research groups progressed even further producing miniature stand-alone flexible systems. Mursch et al (1994) and Coulter et al (1994) presented recently highly portable and low power consuming systems which could operate almost unattended in remote locations incorporating novel tranceiving antennae and fully computerized basic electronic circuitry.

With systems like this it is also possible to significantly improve the data processing and the display techniques to match the experimental needs of a number of applications some times in adverse ambient environmental conditions, (e.g. see Hogstrom et al, 1988; Helmis et al, 1995a).

It is the purpose of this paper to present the recent advances on atmospheric AMS considering both the technological aspect of the instrumentation and the main experimental field achievements.

2. Technical Progress

The technical description of a conventional acoustic sodar is well documented in the literature, (e.g. Asimakopoulos and Cole, 1977, Asimakopoulos and Helmis, 1994). The same applies for the basic applications of the technique for qualitative and quantitative studies, (Brown and Hall, 1978). However it is worth mentioning some recent advances which have been employed to imporve the sodar performance.

The mini acoustic antenna, the most important component of the system, comprises an array of Philips or Motorola piezoelectric horns or similar other elements. These elements have efficiencies, of up to 45% in the range of operating frequencies, are of small size and weight and are widely available. Because of their small size (5–8 cm) optimal geometric arrangements with more elements per area are possible, which result in enhanced directivity at frequencies around 3.5 kHz (Stephanou and Mavrakis 1986). A picture of a recently designed tweeter acoustic antenna used by the University of Athens research group is shown in Fig. 1.

It should be noted that amplitude and phase measurements made on the tweeter elements of the array showed maximum deviations of 100% and ± 60° respectively. In this particular antenna these differences were not corrected by means of active or pasive electronic circuits. However, care was taken to minimize the effect of these differences on the pattern of the antenna by placing the elements in such a way to produce symmetrical current and phase distribution patterns with respect to the center of the antenna.

The electronic circuits that drive the AMS must also be improved. In particular, a preamplifier utilizing a screened miniature audio-transformer and a high-impedance low-noise differential amplifier must be installed directly at the antenna array. In addition, improvements are necessary on the receiver analog, or in some cases in the digital filters by utilizing digital dedicated hardware which can be tuned at any frequency improving at the same time the circuit system response time. These modifications on the University of Athens AMS resulted in a recovery time of only 0.8 ms from the end of transmission, at 3.0 kHz, and a 4 ms pulse duration. This recovery time corresponds to a minimum discernible height of less than 1 m,

Mini Acoustic Sounding, A Powerful Tool For ABL Applications

Fig. 1. Picture of tweeter array acoustic antenna, which measures 45 × 45 cm.

measured from the middle of the transmitting pulse. Further increase of the operating frequency does not necessarily improve the aforementioned characteristics of the system, significantly. On the contrary, due to higher acoustic attenuation and sidelobes level, it will reduce drastically the effective range of the system. It was also stated earlier that now a days a dedicated computer is used to store the required data. This allows for the automatic selection of a number of different operating frequencies which improves the system performance, especially when operation in a noisy environment is compulsory. Furthermore, the dedicated computer is also used for processing on-line and storing the desired data.

Nowadays the above mentioned electronic improvements and technological availability allows for the construction of Sodars and acoustic arrary which operate satisfactorily without any shielding, and other complicated electronics. Furthermore, in some cases electronic steering of the antenna is possible which permits automatic selection of the antenna beam orientation, thus reducing unnecessary reflections from nearby obstacles while more sensing directions can be monitored.

3. Data Processing and Display Techniques

The data processing development has been focused on the techniques which allow weak acoustic echo detection especially in high ambient noise level environment. In this respect dual or triple frequency operation of the sodar and rather sophisticated spectral method analysis have been employed recently. These are normally accompanied by selected routines for signal qualification and validation of the block results. Along those lines it is worth mentioning the following techniques:

The first moment technique, (May and Strauch, 1989). This method involves the following steps:

Several Doppler spectra from each radar resolution cell window are averaged and the noise level is estimated. Then a guess for the position of the signal peak is obtained by finding the maximum spectral power density when the spectrum has been smoothed. The moments are calculated from the unsmoothed spectrum over the interval surrounding this guess out to the first point where the density falls below the noise level on either side. Of course before calculating the moments the noise level is subtracted from the values.

The Simple Homodyne Complex Covariance, (Ito and Kobori, 1989). In this method a phased array alongside a fully computerized Sodar is employed. In this technique as the received signal is obtained in a digital form, the sampling frequency is adjusted to synchronize with the transmitting wave 4 times the transmitting frequency. Then the reference signals for the homodyne detection is simulated following a well determined pattern.

The FFT method. This is the rather classical way of analyzing the Doppler acoustic returns. In this respect a single or dual FFT approach is employed through which the Doppler shift can be intensified. Most interesting is the dual method where the first FFT is only used to indicate the interesting part of the spectrum and the second one to accurately read the true Doppler shift signal.

Naturally the different research papers introduce certain small improvements or innovations in the process of reading the shifted signal which improves in one way or the other the accuracy and the resolution.

The above methods present different advantages and disadvantages which are very much related with the specific application. In this respect is is worth mentioning the case of the noisy environment operation under which a combination of a dual frequency operation with a double spectrum analysis can give best results. It is worth mentioning here that the multi frequency operation does not seem to provide good results in cases where an acoustic array of speakers is used as an antenna.

Finally in order to produce a high quality data set a continuity method could be applied which relies mainly upon the consistency of data over time and height. The algorithm does this by looking for large changes in the neighboring measurements over the sodar sample intervals (in space and time), which later on are rejected. Also winds are computed only at those heights and times for which all three antenna beams reported a radial velocity.

The display techniques used recently draw their origin from the well-known electromagnetic Radar field. The work published by Papageorgas et al (1993) is a typical example of the recent progress. In this paper the authors used the so-called "on line false-color display technique". This allows the use of different systems such as the ink-jet printer or the color VGA monitor. In the first case an ink-jet printer is used to obtain a color hardcopy of the AMS echoes with different intensity regions. For the application described below, the AMS signal intensities were categorized into eight 3 dB spaced successive levels. Each level was presented by a discrete color, produced by a combination of yellow, magenta and cyan, which are provided by the printer. This arrangement gave to the system a total dynamic range of 24 dB. The lower and the higher amplitude signals were presented by the darker and the brighter colors, respectively, while the noise level was presented by white. The size of the ink drops was so small, that a resolution of 300 drops per inches (dpi) was achieved. Considering the effective range of the AMS used in this case the signal intensity was only displayed up to 250 m, incorporating 480 dots, (i.e. 0.5 m per dot). This performance is almost ideal for high-frequency AMS system that operate on short pulse lengths to achieve high-resolution records. The time resolution of the system was set to 26 cm per hour.

The color VGA monitor approach can be described as follows:

Most personal computers, are equipped with low-cost Video Graphics Array (VGA) color monitors. The resolution of such a display system is 640 (horizontally) by 480 (vertically) pixels, while colors are produced by programming the intensity of three beams (red, green and blue) in 64 steps. Although it is possible to display 256 K colors, only 4 bits correspond to a pixel, leaving 16 colors to be displayed on screen simultaneously. The colors used here can be chosen in a way that their brightness was analogous to the signal strength, thus the higher intensity areas can be distinguished by their white or red color. The 16 colors used correspond to a dynamic range of 48 dB.

The echoes were again displayed up to 250 m using 480 pixels i.e. a spatial resolution of 0.5 m per pixel, with a time resolution of 640 pulses per screen. The display used was an Acer 7015 VGA monitor with maximum resolution of 640 (horizontally) by 600 (vertically) and an active display area of 245 mm by 180 mm. Figure 2 is a photograph (black and white copy only) of the color VGA monitor display of a sodar recording during the 1st of November 1990 Athens experiment between 0400 to 0430 LST. This figure shows clearly the potential capability of the color display to visualize such a fine scale atmospheric structure that is not available by other conventional displays. Both time and space resolution of the VGA record are better than the ink-jet ones, allowing for a more detailed examination when low spatial resolution is used. The VGA method, however is less usable, since lengthy records can only be shown on the screen when low spatial resolution is used.

Recently because of the very many applications especially of the high resolution acoustic sounding, it became of prime importance the automatic and accurate classification of the acoustic records.

Thus the development of a layer detection and classification algorithm was necessitated for the different research groups, (see Kalogiros et al, 1995). The algorithm consists of the image processing and the layer classification modules. The image processing module is carried out at two levels: a regional level and a local one, in much the same way as humans interpret an image. At the regional level the algorithm detects the number and the approximate height of the layers in the fascimile window. At the local level the algorithm traces the boundaries of the layers and applies rules of consistency to them. This method of image processing was preferred instead of known clustering techniques, which are more time

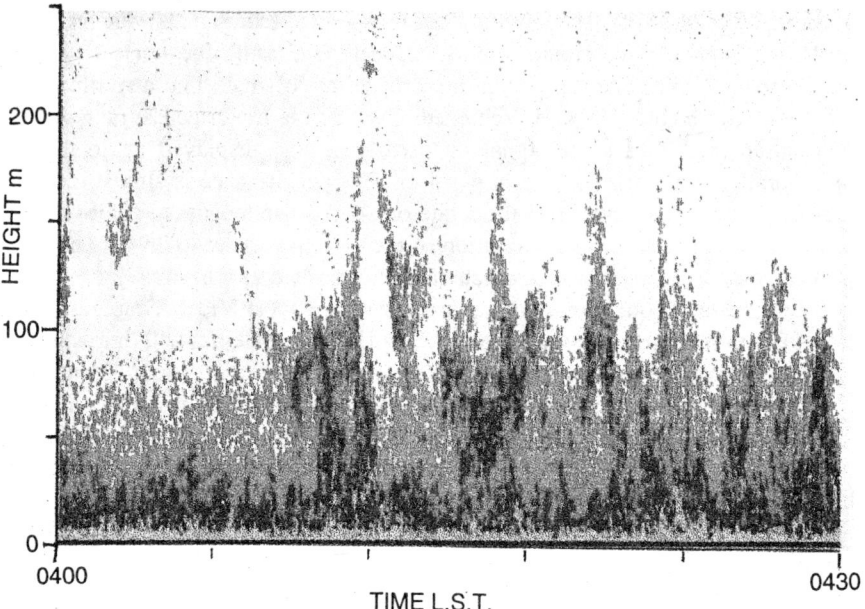

Fig. 2. High frequency AMS echo record displayed by color VGA monitor.

consuming since they work at pixel level. Also, they are more difficult to apply because only distance measures of similarity can be used into the clustering process in our case. The classification module extracts the features from the boundaries of the layers.

The time length of each window is selected in a way to contain a sufficient number of shots. This secures a reliable representation of the atmospheric thermal structure and eventually a reliable classification. On the other hand, the time length of the window should be limited to avoid the inclusion of different structures within a window.

Initially, the acoustic noise is detected and removed from the record to avoid false detection of layers. Following this stage, the image is smoothed using cell of e.g. 4×4 pixels. In this way, the detection becomes insensitive to temporal loss of signal. Also, the components of the image are reduced significantly by a factor of 16.

The image is divided in a number of sub-windows with equal number of shots to detect rapid changes of the height of the layers. The image typically is divided into ten sub-windows. The layer detection in each sub-window is accomplished using a histogram of the number of echo-pixels per height zone with a 4 pixels width. A layer is identified from the presence of a local maximum in the histogram. In order to trace a layer in areas where there is low density of echo pixels, the no echo pixels are replaced by echo ones inversely to the local modulation of the density of echo pixels.

This procedure is applied only in areas of low density of echo pixels (between the limits 25% and 75% of the cell fill: 16 pixels). This procedure made possible the tracing of the intermittent layers at the left side of the image.

The remaining noise at the right side of the image was exaggerated but the layer detection step was applied prior to this step. In this way, no layer was detected and traced in this area. The estimated height of the boundaries of each layer detected with the histogram method is used to initialize a following modified contour. This gives with time single-valued upper and lower boundaries of each layer in every sub-window of the window. A typical example of this procedure is shown in Fig. 3 which was recorded during an experimental campaign in 1989 over a flat inland terrain in central Greece. From this figure the capability of the procedure becomes clear which allows the description even of weak layers of short duration and depth.

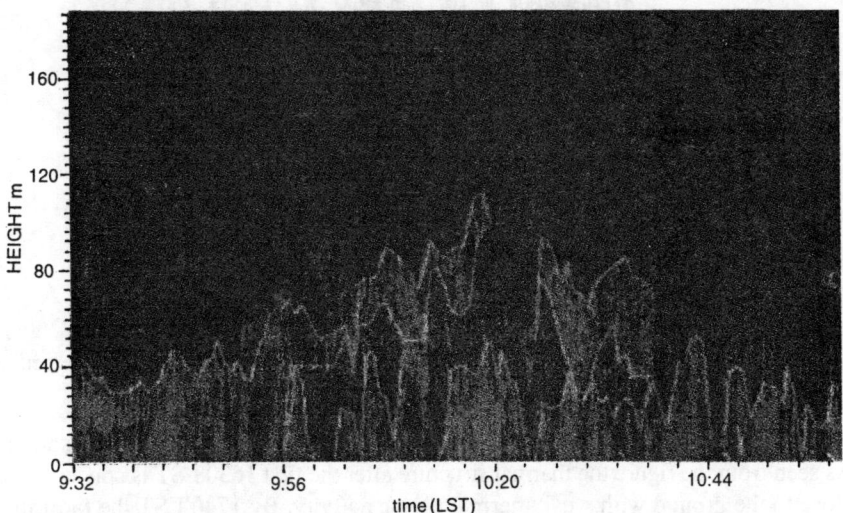

Fig. 3. An example of image processing technique and layer classification procedure on AMS record.

4. Selected Applications

Because of the system flexibility there are numerous applications reported in the literature. In this chapter however the most common and recent and representative ones will be presented.

4.1 Atmospheric Boundary Layer High Resolution Monitoring

Among other common ABL applications one can refer to the recently published work of Helmis et al (1994a) where monostatic AMS observations were used as a key system to study the sea-breeze flow at the top of a close to the coast 1000 m height mountain. Figure 4 shows the detailed picture of the thermal structure of the first 300 m as given by the 4.7 kHz AMS.

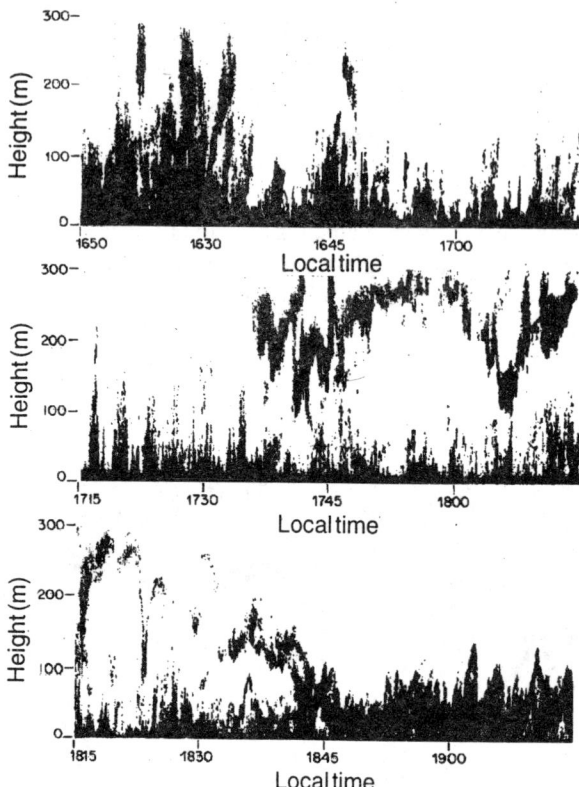

Fig. 4. Facsimile record of the 4.7 kHz monostatic acoustic mini sounder for the period 1615 to 1915 LST.

In this figure the fascimile record of a typical late arrival sea-berrze is shown. As seen from the figure the thermal structure after the first 1630 LST is concentrated close to the ground with weak thermal plume activity. By 1740 LST the facsimile record shows an elevated inversion at a height of 240 m which oscillates in height until 1930 LST when it finally sinks and joins with the surface layer activity. This layer is a part of the interface of the upper level synoptic flow and the sea-breeze circulation pattern.

Figure 5(a) shows the vertical velocity variance profiles, σ_w^2, estimated also by the 4.7 kHz AMS for two time periods, 1636-1706 LST and 1736-1806 LST before and after the sea breeze arrival correspondingly. Both profiles seem to follow an increase in height at a rate close to $z^{2/3}$ which is predicted by free convection theory. Also there is a strong decrease of the σ_w^2 values after the sea breeze arrival. Same behaviour is shown by the temperature structure parameter C_T^2 profile (Fig. 5(b)) which were determined by the 4.7 kHz AMS. In this case the observed slope follows a steeper than the −4/3 power law, predicted by free convection theory. The major difference between the two profiles is the decrease of C_T^2 values after

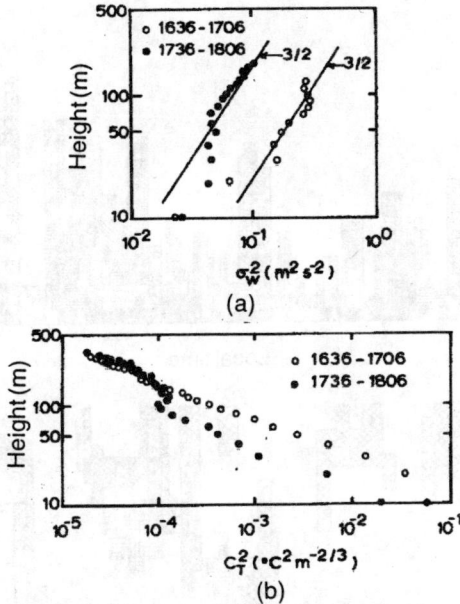

Fig. 5. Mean profiles of the vertical velocity variance σ_w^2 (a) and temperature structure parameter C_T^2 (b) for the period 1636-1706 snd 1736-1806 LST.

the sea-breeze arrival and for heights smaller than 200 m. At higher levels the C_T^2 values remain unchanged.

In Fig. 6 the isopleths of C_T^2 estimated by the 4.7 kHz AMS for two periods 1636–1706 and 1736–1806 LST are shown. This presentation gives more quantitative details of the internal structure of the first 300 m. Before the sea-breeze arrival the atmospheric boundary layer exhibited a well organized plume activity from the ground up to 200 m. After the sea-breeze arrival a weak plume activity is present close to the ground and a well defined inversion layer is evident at heights between 200 and 300 m. This pattern coincides with the oscillatory activity observed by the AMS and presented in Fig. 4. It is worth noting that the bottom part of the pattern is quite uniform while the layering on top is variable with values ranging between 10^{-4} to 10^{-5} °C m$^{-2/3}$.

Another example of recent ABL application is the estimation of the surface momentum and the heat fluxes which can also be monitored by a monostatic AMS.

In this respect vertical velocity variance (σ_w^2) profiles have been recently used to estimate these parameters. A comparison was made by Helmis et al (1994b, 1995a) using in-situ measurements and a high resolution AMS, operating in a noisy environment (at the coast line). Figure 7 is a typical example of this comparison giving Q_0 and U^* values computed from the AMS data using the σ_w^2 profile method and the eddy correlation method using in-situ measurements. The comparison

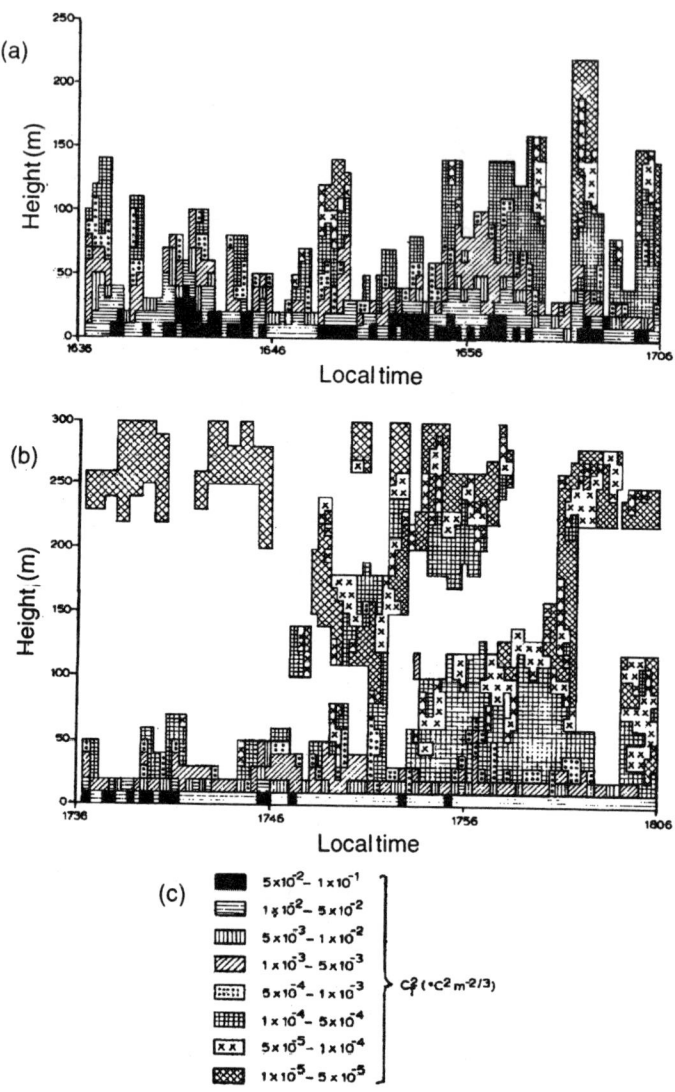

Fig. 6. Echosonde-derived isopleths of C_T^2 for two periods 1636–1706 and 1736–1806 LST.

confirms the ability of the σ_w^2 profile method to provide estimates of these parameters with reasonable accuracy on a continuous basis. The above clearly demonstrates the multiple use of the AMS for the study of the structure of the ABL and its dynamics.

4.2 Wind Energy Applications

Because of the very short "dead time" after transmission and the system configuration

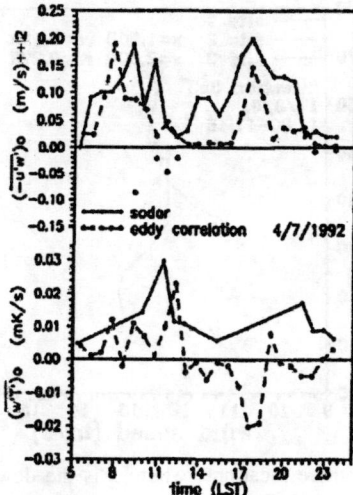

Fig. 7. Time series of momentum and heat flux estimates using the sodar data and in-situ measurements on 4th July 1992

flexibility the AMS provides a unique remote sensing platform for wind energy applications. Thus both wake wind turbine studies as well as citing and wind energy profiling can be performed over almost any terrain.

4.2.1 Wake studies

A very interesting application of high resolution AMS taken in a very noisy environment (close to a medium size wind turbine) in Samos island wind Park (Greece) is presented by Helmis et al (1995b). In this case the horizontal wind component was monitored by three synchronized tilted sodars placed in three different locations, alongside the wind direction producing the corresponding wind profile upstream (site 1) and downstream (sites 2 and 3) of the turbine. Figure 8 gives the corresponding sodar profiles where the observed downstream reduction of the velocity is due to the wake effect, with a minimum in wind speed at hub height (25 m). The maximum of the wind speed at the upstream location is due to the speed-up effect over this complex terrain. It is worth mentioning here that the minimum detecting height is about 10 m while the strongest wind speed was about 15 m/s.

Similar performance was also reported earlier by Hongstrom et al (1988) during an extended experiment in Sweden where a large and noisy wind turbine was studied and a combination of various meteorological instrumentation operating alongside the two AMS systems was used. For this particular application the AMS system was split and operated in novel configuration which was named by the authors as umbrella, serial and conventional tristatic.

4.2.2 Wind energy studies

A lot of work was reported on wind profiling and wind energy studies using

128 *Atmospheric Acoustics and Instrumentation*

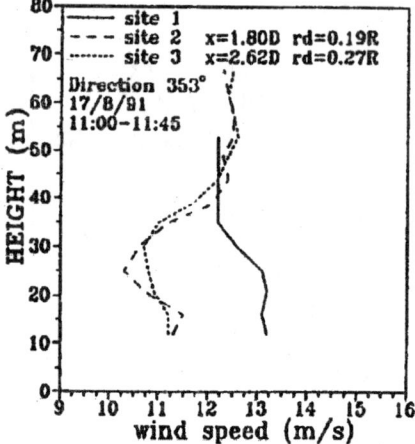

Fig. 8. Sodar profiles at three locations, where x is the downstream distance at 25 m height and rd the radial distance from the wake center.

tristatic AMS systems. A typical example along these lines was recently reported by Helmis et al (1995b) which describes an extended experiment under high wind conditions and over complex terrain, (see Fig. 9). Observing the figure, detailed information on the wind characteristics are revealed. In fact despite the figure scatter, a wind maximum in the range 12 to 21 m is observed, while a continuous decline is noticed higher up. Obviously this is a local characteristic related with the terrain structure, and the wind strength and the inner layer depth.

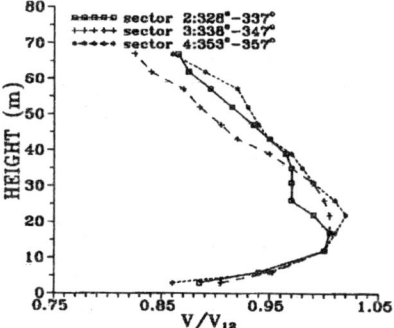

Fig. 9. Normalized AMS and meteorological mast wind profiles for three different directional sectors.

4.3 Air Pollution Microscale and Local Climatological Studies

Since a high resolution AMS can resolve small scale local characteristics such as those related to the wind and temperature structure, it can easily be used to study the plume of an operating chimney or the local climatological map of the city center or elsewhere.

4.3.1 Study of a methane plume

This is an early but quite representative study which shows the use of monostatic AMS system for microscale applications (Moulsley and Cole, 1980; Cole and Moulsley, 1979). In the set of experiments that the authors reported, their high resolution AMS managed not only to identify the methane plume track but to measure its diameter with a very small error of the order of 10%. To define the experiment scale it is worth mentioning that the methane plume was released at 30 m and the AMS was placed 70 m away from the release stack. The authors following a novel theoretical approach claimed that their AMS can estimate not only the size of the methane plume but also the mean gas concentration. The latter can be achieved through a new parameter which is named "concentration stucture parameter", C_m^2, which links to the gas concentration in any part of the plume.

4.3.2 Local climatological studies

It is now well documented that the acoustic sounders with their modern automatic chart classification facility can be used to climatologically categorize the wind and the temperature structure of the local ABL. This was further improved by the AMS system capabilities which can provide information for short distances from the antenna at high space and time resolution configurations. This performance makes the AMS system a powerful tool for Air Pollution and Climatological studies, (e.g. see Asimakopoulos et al, 1994). Following the work presented in this paper, a typical example of AMS classification work is shown in Fig. 10. This figure gives the frequency distribution of the AMS chart categories by hour of day for one year operation at the Athens city center. As seen from the figure local effects, such as the heat island effect, the urban boundary layer etc are shown through different than the rural area categories percentages, which are in depth analyzed by the authors. Thus longer duration convection category and stronger and shallow surface based radiation inversions are present in Fig. 10.

It is obvious that similar approaches can be used for environmental impact assesment studies or industrial planning where detailed temperature structure information is necessary.

5. Concluding Remarks and Suggestion for Further Work

From the presentation and the results presented above, some interesting conclusions can be drawn and suggestions for further work can be made.

- The acoustic mini sounding technique provides a new practical low cost platform for the remote sensing of the first few hundred meters of the ABL.
- The use of the instrument can be extended in high ambient noise environment while the high resolution performance introduces new areas of applications.

Summarizing, a well calibrated AMS system can perform as follows:

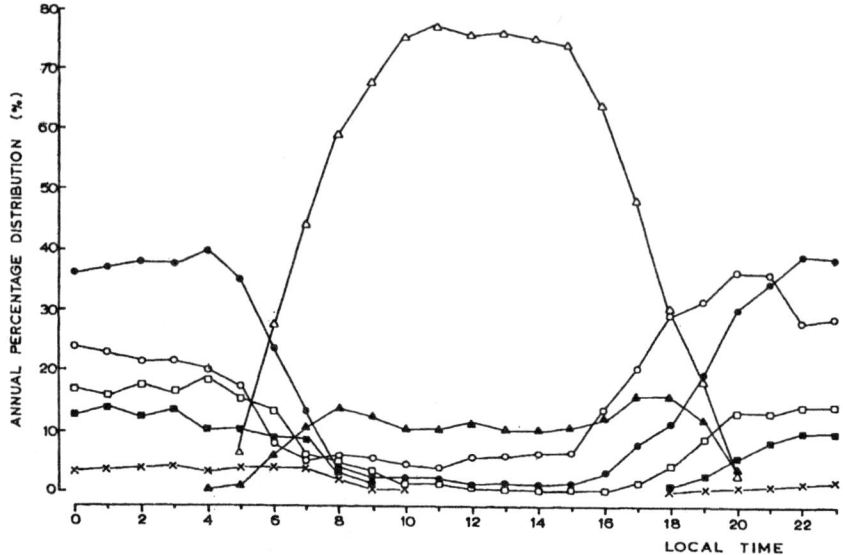

Fig. 10. Annual percentage distribution of Sodar categories by hour of day for the Athens city center.

The code of Acoustic radar records is:
- △ convection activity
- ■ ground-based inversion with single layer above
- ▲ convection activity with elevated layers
- × ground-based inversion with multiple layering above
- ○ no echo; neutral conditions
- □ ground-based inversion
- ● single elevated layer or multiple-layered above

1. The AMS are highly portable low cost systems with a very short dead time after transmission. Their effective maximum range extends from 1 to 250 m for both Doppler and intensity measurements.
2. The AMS can operate in noisy areas such as airports, city centers, wind turbines and coastal and marine environments. From these studies very interesting results have been recently reported in the literature.
3. In addition to the standard ABL parameters that a conventional sodar can monitor, the AMS provide very good performance estimating turbulence components such as turbulence intensity, variances, heat fluxes etc. However to establish a better understanding more research is needed in these fields.
4. The AMS systems allow new mode of operations which can be chosen in such a way to fit the experimental needs. This can be achieved through the high portability and capability to split the conventional tristatic system into 3 monostatic serial or umbrella synchronized monostatics or double bistatics with a large number of unusual back scattered modes.
5. The combination of an AMS and a conventional low frequency sodar can

make an almost ideal profilometer filling all heights from the surface up to 1 or 2 kms of the ABL.

References

1. Asimakopoulos D.N. and Cole R.S., 1977: "An acoustic sounder for the remote probing of the lower atmosphere" Journal of Scientific Instruments, 9, 47–50.
2. Asimakopoulos D.N., Stefanou G. and Helmis C.G., 1987: "Atmospheric acoustic mini-sounder" Journal of Atmospheric and Oceanic Technology, 4, 345–347.
3. Asimakopoulos D.N. and Helmis C.G., 1994: "Recent advances on atmospheric acoustic sounding" Int. J. of Remote Sensing, 15, 223–233.
4. Asimakopoulos D.N., Helmis C.G. and Deligiorgi D.G., 1994: "Climatological evaluation of sodar recordings over complex terrain", Int. Journal of Remote sensing, 15, 383–392.
5. Brown E.H. and Hall F.F., 1978: "Advances in atmospheric acoustics" Review of Geophysics and Space Physics, 16, 47–110.
6. Cole R.S. and Moulsley T.J., 1979: "The use of the atmospheric acoustic sounder to track methane plumes", Atmospheric Environment, 13 1437–1441.
7. Coulter R.L., Martin T.J. and Hart R.L., 1994: "A battery–powered minisodar for rapid, short term deployment or long-term operations" Int. J. of Remote Sensing, 15, 245–250.
8. Helmis C.G. Deligiorgi D.G., Asimakopoulos D.N. and Lalas D.P., 1994a: "Evidence of sea-breeze flow over the top of Hymettos mountain" Int. J. of Remote Sensing, 15, 479–187.
9. Helmis C.G., Kalogiros J.A., Papadopoulos K.H., Soilemes A. and Asimakopoulos D.N., 1994b: "Estimation of the Atmospheric Surface Momentum and Heat Fluxes using a High Resolution Acoustic Radar" Journal de Physique, IV, 287–290.
10. Helmis C.G. Kalogiros J.A. Asimakopoulos D.N., Papadopoulos K.H. and Soilemes A., 1995a: "Acoustic Sounder measurements of atmospheric turbulent fluxes on the shoreline" The Global Atmosphere-Ocean System, 2, 351–362.
11. Helmis C.G., Papadopoulos K.H., Asimakopoulos D.N., Papageorgas P.G. and Soilemes A.T., 1995b: "An experimental study of the near wake structure of a wind turbine operating over complex terrain" Solar Energy, 54, 413–428.
12. Hogstrom U., Asimakopoulos D.N., Kambezidis H., Helmis C.G. and Smedman A., 1988: "A field study of the wake behind a 2 MW wind turbine" Atmospheric Environment, 22, 803–820.
13. Ito Yoshiki and Kobori Yasuhiro, 1989: "Development of wind profoling Sodar", J. of Atmospheric and Oceanic Technology, 6, 779–784.
14. Kalogiros J.A., Helmis C.G., Asimakopoulos D.N., Papageorgas P. and Soilemes A., 1995: "A layer detection and classification algorithm for Sodar Facsimile records" Int. Journal of Remote Sensing, 16, 2939–2954.
15. May P.T. and R.G. Strauch, 1989: "An examination of wind profiler Signal Processing algorithms" J. of Atmospheric and Oceanic Technology, 6, 731–735.
16. Moulsley T.J., Asimakopoulos D.N., Cole R.S. and Crease B.A., 1978: "Design of arrays for acoustic sounder antennas" Journal of Physical Environment (Scientific Instruments), 11, 675–662.

17. Moulsley T.J. and Cole R.S., 1979: "High Frequency atmospheric acoustic sounders" Atmospheric Environment, 13, 347–350.
18. Moulsley T.J. and Cole R.S., 1980: "The evaluation of acoustic sounder returns from a methane plume" Atmospheric Environment, 14, 1063–1066.
19. Moulsley T.J., Asimiakopoulos D.N., Helmis C.G., Lalas D.P. and Gaynor J., 1985: "A quantitative comparison of horizontal and vertical Acoustic Sounding with in-situ measurements" Boundary layer Meteorology, 33, 85–100.
20. Mursch-Radlgruber E., Wolfe D.E., Gregg D.W., King C.W., Neff W.D., Sharp K.A.H. and Ruffieux D., 1994: "NOAA's portable high-frequency minisodar design and first results" Int. J. of Remote Sensing, 15, 325–332.
21. Papageorgas P.G. Helmis C.G., Soilemes A.T., Asimakopoulso D.N. and Metaxaki-Kosionides C., 1993: "Real-time color display techniques for High Resolution Acoustic Sounder Echoes", Applied Physics, B57, 37–39.
22. Stephanou G.J. and Mavrakis D.H., 1986: "Switched beam planar arrays suitable for high resolution acoustic sounders" IEEE Trans. Geosci. Rem. Sensing, GE-24, 5, 745–750.

Acoustic Remote Sensing Applications
S.P. Singal (Ed)
Copyright © 1997 Narosa Publishing House, New Delhi, India

6. Radio-Acoustic Temperature Profiling in the Troposphere

G. Bonino
Istituto di Fisica Generale dell'Universitá and Istituto di
Cosmogeofisica del CNR, Torino- Italy

Introduction

The vertical thermal profile of the lower atmosphere can be remotely measured by the Radio Acoustic Sounding System (RASS) technique. The basic principle consists in the determination of the upward sound speed by means of a Doppler radar. Since this velocity is proportional at every height to the square root of the local temperature the vertical thermal profile can be derived. The backscattered radioecho is due to the periodic variations of the refractive index of the atmosphere produced by the acoustic waves.

The first RASS experiment was performed in the early 1960's and designed for wind measurement [1]. The RASS technique for temperature measurement was first developed in the 1970's and adopted by several groups [2–10]. These systems were essentially designed for temperature profile measurements and allow a maximum sounding range between few hundred meters and few kilometers [11]. Recently, wind profiler radars have been utilized for the RASS technique, see e.g. [12–15]. With this new generation of RASS a sounding range of several km can be obtained. This is a consequence of greater potentiality, large radar antennas and sophisticated signal processing. However these systems start the temperature measurements at altitudes of few hundred meters and in some cases of few kilometers, therefore they are not suitable, in general, for Planetary Boundary Layer (PBL) sounding.

A review of the sounding schemes adapted, the theory of the radio-acoustic sounding of the atmosphere and results obtained for PBL sounding by RASS combined with wind profiler radars are given.

Methods of Sounding

The adopted schemes for the radio-acoustic soundings are: (a) pulsed acoustic waves and continuous electromagnetic waves (AC-PM, EM-CW); (b) continuous acoustic waves and pulsed radio waves (AC-CW, EM-PM); (c) continuous acoustic waves and frequency modulated radio waves (AC-CW, EM-FM); (d) both acoustic and radio pulsed waves (AC-PM, EM–PM).

The method AC-PM, EM-CW consists essentially of a powerful acoustic source beaming a short pulse of sinusoidal waves toward the zenith. The upward speed of this pulse is proportional at every height to the square root of the local absolute virtual temperature. The pulse speed is continuously measured from the ground by means of a Doppler radar. The radar echo is due to the change in the refractive index of air compressed by the acoustic waves. The faint echo is maximized by choosing an acoustic wave in Bragg resonance with the radio wave

$$\lambda_a = \lambda_r/2$$

where λ_a and λ_r are the acoustic and radar wavelength, respectively. The record of the measured sound speed as a function of the delay leads to the acquisition of the temperature vertical profile [e.g. 2, 3, 6, 16]. The accuracy of temperature measurement is of few tenth of K degree if the Bragg condition is satisfied with a good approximation and short acoustic pulses are adopted.

In the method AC-CW, EM-PM the frequency shift of the scattered radio signal is not equal to the Doppler shift, but it is exactly equal to the sound frequency. Since the radioecho is maximized when the Bragg condition holds, by tuning the acoustic frequency v_a to the maximum scattered signal from a certain altitude. From the tuned v_a and the value of $\lambda_a = \lambda_r/2$, the sound speed and subsequently the temperature at the considered altitude are derived, since the temperature as also the acoustic frequency which statisfies the Bragg condition are unknown. In practice a wide enough band of acoustic waves is transmitted to cover the Bragg condition in the whole sounding range. Peters et al, [17] used a comb of sinusoids for the acoustic source. A frequency-modulated continuous-wave source is often used for the RASS associated to wind profilers [e.g. 13].

The method AC-CW, EM-FM adopts an acoustic source as in the scheme (b) and uses a FM CW radar for the measurement of atmospheric sound velocity profiles [18].

Finally the method (d) is applied in systems with a high power radar (10^3–10^6 W) for very high sounding range [12]. The choice of the parameters of a RASS (wavelengths, mode of sounding, size of antennas, and so on) is a complex task and depends on the applications of the measurements.

Theory of the Radio-Acoustic Sounding

The basic theory of the radio-acoustic sounding in the ideal conditions of no wind and no atmospheric turbulence has been developed by Marshall [2]. The minimum power $P_{r\,min}$ (W) received by the Doppler radar, yielding a reliable temperature measurement is given by:

$$P_{r\,min} = 1.38 \times 10^{-16} \frac{P_t P_a G_t G_r G_a (1 - \cos\theta_a/2)^2 N^2 10^{-(L z_{max}/10)}}{z_{max}^2} \quad (1)$$

where P_t and P_a are the radiated power of the radar and acoustic sources (W), G_t and G_r the radar transmitting and receiving antenna gain (ratio), G_a the acoustic source gain (ratio), θ_a the beamwidth of the acoustic source, L the absorption loss of acoustic waves (dB/m), z_{max} the sounding range (m), N the number of wavelengths of the acoustic pulse.

The problem of the interaction of electromagnetic and acoustic waves in the real atmosphere is complicated by the presence of wind and atmospheric turbulence.

The principal causes limiting the sounding range of a RASS are: (a) attenuation of the acoustic waves, (b) strong horizontal winds and (c) atmospheric turbulence.

Sound attenuation in the atmosphere, in addition to the spherical divergence of the waves, is mainly due to atmospheric molecular absorption and to scattering of acoustic waves by atmospheric inhomogeneities. Such attenuations strongly increase with the decrease of acoustic wavelength [19].

The principal effect of horizontal wind is the displacement of the acoustic waves in the wind direction, which causes a shift of the focus of the radio echo with respect to the origin. It is possible to reduce or even to eliminate this effect by moving, for example, the radar receiving antenna downwind [20] or by utilizing an array of antennas [21]. The main effect of atmospheric turbulence on the sounding range, in addition to the scattering effect, is due to perturbation of coherence of the radiowaves reflected by the different periods of the acoustic pulse and by the different portions of the acoustic wavefront [22]. This effect cannot be eliminated. It depends on atmospheric turbulence structure, on the acoustic wavelength and on the effective sizes of the antennas, and becomes less effective as the acoustic wavelength increases. Since an increase of the antenna sizes and a decrease of the vertical resolution of temperature measurements is associated with wavelength increases, it is necessary to choose the RASS parameters in respect of any specific requirements.

The predominant effect of atmospheric turbulence on a RASS is that of a reduction of the sounding range with respect to the ideal case of a nonturbulent atmosphere. The mechanisms determining such a reduction are: (a) attenuation of the acoustic intensity due to scattering of sound by turbulent temperature and velocity fluctuations; (b) pertubation of coherence of the radioecho coming from the different layers of the acoustic wavetrain (pertubation of the longitudinal coherence); (c) pertubation of coherence of the radioecho coming from the different portions of the wavefront crosswise the beam direction (perturbation of the transversal coherence). It has been demonstrated that the mechanism (c) is predominant.

The perturbation of the transversal coherence is due to the fact that the acoustic wavefront propagation in a turbulent atmosphere begins to be affected and gives rise to phase fluctuations. Cross-size radius of the wave coherence ρ_c is deduced by making equal to unit the structure function of the phase fluctuations on base ρ. For a wave reaching an altitude z in a turbulent environment, it is [22].

$$\rho_c = (0.73\, C_n^2 k_a^2 z)^{-3/5} \qquad (2)$$

where C_n^2 is the structure parameter for refractivity fluctuations depending on atmospheric conditions and $k_a = 2\pi/\lambda_a$ is the acoustic wavenumber. As is shown by (2), ρ_c decreases with altitude. Until ρ_c is greater than the radius of the acoustic wavefront intersected by the radiobeam, the influence of turbulence on the phase of the wavefront is negligible. Since the crosswise size of the acoustic wavefront is $\sim z\,\theta_a/2$, where θ_a is the acoustic beamwidth and $\rho_c \propto z^{-3/5}$, there is an altitude z_c at which $\rho_c = z_c\,\theta_a/2$. Then ρ_c becomes smaller than the radius of the wavefront; the surface of the wavefront $S \sim z^2(\theta_a/2)^2$ becomes divided in zones of coherence each having an area $S_c \sim \rho_c^2$. While in the absence of turbulence the diffused radiowaves result from the coherent addition of all parts of the intersection between acoustic and radar beams, in a turbulent atmosphere the composition occurs only for the zones of coherence.

Assuming that the Bragg condition holds and disregarding the effect of horizontal wind in reducing the radioecho, the ratio between the intensity of the signal I in a turbulent atmosphere and the signal intensity I_o in the absence of turbulence is given with sufficient approximation by the formula [22].

$$I/I_o = (1 + \rho_0^2/\rho_c^2)^{-1} \qquad (3)$$

where $\rho_c^2 \approx 4z^2/k_a^2\,(2r_a^2 + r_e^2)$ is the effective radius of the zone of interaction of the electromagnetic and acoustic beams at the altitude z; $2r_a$, and $2r_e$, are the diameters of acoustic and radar emitters, respectively. In Eq. (3) the strong dependence of I as a function of z is prominent. One derives $I \propto z^{-26/5}$; while in the case of absence of turbulence $I_o \propto z^{-2}$.

Assuming that the Bragg condition holds and that the effect of bistaticity of the system is negligible at altitudes far from the acoustic and radio antennas, the maximum sounding range z_{max} in turbulent atmosphere is given by [23]:

$$z_{max} = \left[\frac{P_t P_a}{CP_n(C_n^2)^{6/5}}\right]^{5/26} \left[\frac{G_t G_r G_a(G_t + G_a)}{(G_t + G_r + G_a)(G_r + G_t/2 + G_a/2)}\right]^{5/26} N^{10/26} \lambda_a^{6/13} \qquad (4)$$

where C_n^2 is the structure parameter for refractivity fluctuations depending on atmospheric conditions, P_n the noise power of the receiver, $C \sim$ constant. In (4) it should be noted that the maximum sounding range increases with the acoustic wavelength, but more weakly than in the ideal case; even though z_{max} has a weak dependence on P_t, some researchers have opted for very large values of P_t in order to increase range. For example, in the case of monostatic pulsed Doppler radar, Matuura et al, [12] used 1 MW of peak power. Considering the performance of several RASS systems of comparable characteristics, the proportionality to $\sim \lambda_a^{1/2}$ holds for both the maximum sounding range and the typical sounding range.

The effect of horizontal wind on the sounding range is not contained in (4), which yields the maximum sounding range for a given system and for given values of atmospheric turbulence. This range corresponds to the actual sounding range for a negligible average wind speed within the sounding altitude or for RASS equipped with arrays, for example of radar receiving antennas, capable of collecting the radio echo shifted by displacement of the acoustic pulse dragged by horizontal wind.

An exhaustive experimental validation of equation (4) is a complex task. The influence of the acoustic antenna size and of the number of acoustic wavelengths forming the acoustic pulse has been investigated [24] obtaining a good agreement between theoretical and experimental values.

Recently, Lataitis [25] has developed a theoretical model in the form of a radar equation for the echo power of a RASS. The effects of horizontal wind, turbulence, vertical temperature gradient, and absorption of sound have been included. This model is consistent with the earlier models of Marshall [2], Kon [22] and Masuda et al [26].

RASS for PBL Investigations

The first radio-acoustic sounding system was developed in early 70's and successfully tested at Stanford University [2–4]. The maximum range varied from about 400 m to 3 km for strong wind or calm conditions, respectively. Since 1972 different types of radio acoustic sounders designed for temperature profile measurements in the PBL have been developed in the USA, Europe and Japan. The list of teams involved, together with the acoustic frequency adopted and the range of operations, are reported in Table I of Bonino [11]. The metric RASS (acoustic wavelengh λ_a = 1 m) developed by our group is considered here in more detail.

This system was successfully tested at the remote-sensing station of Turbigo (Italy) [6, 7], where the RASS thermal profile, together with measurements of a Doppler SODAR and a fluxmeter is being utilized for PBL studies and for air quality management of the area [27–30]. Comparison among vertical thermal profiles measured by the metric RASS and in situ thermosonde measurements has demonstrated that the system can measure thermal profiles in the PBL with accuracy and vertical resolution of traditional apparatus with advantages of remote sensing techniques.

We adopted the AC-PM and EM-CW sounding scheme reported in the previous paragraph. At the beginning of the measurements, the acoustic frequency is established on the basis of the absolute temperature and relative humidity measured by means of sensor at 2 m above ground level. This frequency is approximately equal to the expected beat frequency output from the Doppler radar receiver. A series of soundings are then performed for successively increasing and decreasing values of acoustic frequency until the best radio-echo signal is established. Once this condition is attained, a preset number of soundings is performed. Measurements are accepted if the signal to noise ratio S/N > K, with K presettable. For each series of soundings, the average temperature vertical profile is provided together with the temperature standard deviation profile obtained from measurement in individual soundings. The mean thermal profile can be obtained at a preset time by averaging a series of profiles, each one obtained with n soundings [31].

An important application of the metric RASS concerns the monitoring of fog episodes. During late Autumn and Winter, fog occurs very often in the Po Valley (the most industrialized and populated area in Italy) where three metric RASS are operating. These episodes occur mainly as a consequence of radiative cooling at night. There are also occurrences of frontal and advective fogs. In these last cases we are generally in the presence of multiday fog episodes. As they are long lasting (as far as 10 days) with deep fog layer (from ground to 200–600 m), they give rise to critical air pollution conditions due to the fog-capping thermal inversion. Other important problems (like safety of surface and air transportation, insulation of high voltage electric lines, diseases and so on) are related to the fog. This explains the interest involved in studying fog formation and evolution.

The metric RASS is suitable for thermal profile measurement in fog conditions with good temperature accuracy and vertical resolution. It gives: the thickness of the fog layer; the intensity of thermal inversion at the top of the fog; the atmospheric stability parameter above the fog capping inversion [27–28].

RASS-Wind Profilers

Wind profilers are clear-air Doppler radars that measure the vertical profile of the three wind components in almost all meteorological conditions. They operate at wavelengths of 0.3–6 m. Their advanced data processing allow the detection and analysis of the very weak scattered signal from the atmosphere. Wind profiler radars together with an acoustic source are very well suited for RASS [15]. At the Radio Atmospheric Science Center in Kyoto, Japan, it was demonstrated that a RASS added to wind profiler radar (MU radar) can measure the temperature profile throughout the troposphere [12]. This radar uses a peak power of 1 MW and has a very large antenna with the possibility of beam steering that allows tracking of the acoustic wavefronts.

Several types of wind profilers have been developed for different meteorological applications. All wind profiler radars can be used for RASS.

These systems, compared to the ones considered in the previous paragraph have in general larger radar antennas and transmitted power and much more sophisticated data processing which allows the temperature measurements when the signal-to noise ratio is as low as –35 to –40 dB [15]. Very important results with RASS added to wind profiler radars have been obtained in Boulder at NOAA/ERL/Wave Propagation Laboratory, Aeronomy Laboratory and University of Colorado. A 50 MHz radar was capable to measure temperature profile from 2.1 km to 5.9 km above ground level (AGL). A 404 MHz radar was able to measure RASS profiles from 400 m AGL to 1.5–2.5 km AGL. A 915 MHz wind profiler obtained RASS data in the altitude range from 0.2 km to 0.7–1.5 km.

Conclusions

The RASS can measure thermal vertical profiles of the lower atmosphere with an accuracy generally within a few tenths K. Several systems have been designed and built with different purposes. The RASS of the first generation, produced up to mid-1980's are suitable to measure temperature profile in the PBL. The RASS of second generation, obtained by the addition of an acoustic source to wind profiler radars can measure the temperature profile throughout the troposphere starting the measurement at altitude of few hundred meters and in some cases of few kilometers.

The results obtained demonstrate that the RASS is a useful tool for meteorological studies. A RASS suitable for PBL sounding can be utilized also for air pollution dispersion studies and for monitoring particular situations such as for episodes.

Acknowledgments

I am indebted to Prof. C. Castagnoli for his support and to Profs P.P. Lombardini, A. Longhetto, P. Trivero and Drs. G. Elisei and A. Marzorati co-authors in papers reported in the references.

References

1. R.W. Fetter, "Remote measurement of wind velocity by the electromagnetic-acoustic probe", II Experimental system Rep. 420, 50 pp., Midwest Res. Inst., Kansas, City, Mo, 1961.
2. J.M. Marshall, "A radio acoustic sounding system for the remote measurement of atmospheric parameters", Sci. Rep No. 39, US-EL-72-003, Stanford Electron Lab., Stanford, CA, 152 pp., 1972.
3. E.M. North, A.M. Peterson and H.D. Pang, "A remote sensing system for measuring low-level temperature profiles", Bull. Am. Meteor. Soc., 54, 912–919, 1973.
4. E.M. North, "A radio acoustic sounding system for remote measurement of atmospheric temperature", Sci., Rep. No. SU-SEL-73-021, 108 pp., Stanford Electron Lab., Stanford, Calif., 1974.
5. M.S. Frankel and A.M. Peterson, "Remote temperature profiling in the lower troposphere", Radio Sci., 11, 157–166, 1976.
6. G. Bonino, P.P. Lombardini and P. Trivero, "A metric wave radio-acoustic tropospheric sounder", IEEE Trans., Geosci., Electron, GE-17, 179–181, 1979.
7. G. Bonino, P.P. Lombardini and P. Trivero, "Comparison of RASS temperature profiles with other tropospheric soundings", Nuovo Cimento, 1C, 207–214, 1980.
8. M. Fukushima, K. Akita and Y. Masuda, "Development of radio acoustic sounding system (RASS)", Rev. Radio Res. Lab. (Japan), 26, 555–567, 1980.
9. V.M. Bovshererov, M.A. Kallistratova, L.V. Knysev, A.G. Gorelik and M.Y. Egorov, "A radio device for thermal probing of the atmosphere by radioacoustic sounding", Meteorologiya i Gidrologia, No. 3, 120–123, 1981.
10. G. Peters, H. Hinzpeter and H. Timmermann, "Design criteria of a radio acoustic sounding system (RASS) for the atmospheric boundary layer", Proc. 1st Intern.

Symp. Acoustic Remote Sensing Atmos. and Ocean (Calgary, Canada, (1981), printed by University of Calgary, Cat. No. ISBN: 88953-031-9-1982, II-62-67.
11. G. Bonino, "Remote sensing of thermal profile in the lower atmosphere by radio acoustic system", Proc. 2nd Intern. Symp. Acoustic Remote Sensing Atmos. and Oceans (Rome, Italy, 1983), printed by IFA-CNR, Rome, XXXV, 1–32.
12. N. Matuura, Y. Masuda, H. Inuki, S. Kato, S. Fukao, T. Sato and T. Tsuda, "Radio acoustic measurement of temperature profile in the troposphere and stratosphere", Nature, 323, 426–428, 1986.
13. P.T. May, G. Strauch and K.P. Moran, "The altitude coverage of temperature measurements using RASS with wind profiler radars", Geophys. Res. Lett., 15, 1381–1384, 1988.
14. P.T. May, R.G. Strauch, K.P. Moran and W.L. Ecklund, "Temperature sounding by RASS with wind profiler radars: A preliminary study", IEEE Trans. Geosci. Remote Sens., 28, 19–28, 1990.
15. R.G. Strauch, P.T. May, K.P. Moran, D.A. Merritt and W.L. Ecklund, "Advances in RASS", in Acoustic Remote Sensing, Proc. 5th Intern. Symp. Acoustic Remote Sensing Atmos. and Oceans (New Delhi, India, 1990), Editor S.P. Singal, Tata McGraw-Hill Publ. Co. 97–108, 1990.
16. G.V. Azizyan, V.M. Bovsheverov, G. Gorelik, M. Yu, M.A. Yegorov, M.A. Kallistratova, G.A. Karyukin and L.V. Knyazev, "An experiment in temperature profiling in the lower troposphere by radioacousting sounding", Izv Acad. Sci, USSR Atmos. Oceanic Phys. Engl. Transl 17, 112–116, 1981.
17. G. Peter, H. Timmermann and H. Hinzpeter, "Temperature sounding in the planetary boundary layer by RASS system analysis and results", Int. J. Remote Sens., 4, 49–63, 1983.
18. G. Peters, "RASS-temperature sounding using a FM-CW radar", Proc. Alfred-Wegener-Konf. on Ground Based Remote Sensing Techniques for the Troposphere (Hamburg, Germany, 1986), 20–28, 1986.
19. E.H. Brown and F.F. Hall, "Advances in atmospheric acoustic", Rev. Geophys. Space Phys., 16, 47–109, 1978.
20. M.S. Frankel, N.J.F. Chang and M.J. Sanders, "A high-frequency radio acoustic sounder for remote measurement of atmospheric winds and temperature", Bull. Amer. Meteor. Soc., 58, 928–934, 1977.
21. N. Matuura, Y. Masuda, M. Fukushima and K. Akita, "Development of a radio acoustic sounding system (RASS) for wind and temperature", Proc. Second Int. Symp. on Acoustic Remote Sensing of the Atmosphere and Oceans, Rome, Ist. di Fisica dell'Atmosfera, CNR, No. 41, 1–10, 1983.
22. A.I. Kon, "Power of the signal in radioacoustic sounding of the turbulent atmosphere", Izv. Akad Nauk SSSR Ser. Fiz Atmos. Okeana, 20, 178–184, 1984.
23. M.A. Kallistratova and A.I. Kon, "On maximal ranges of radioacoustic sounding of the atmosphere, Izv Akad. Nauk SSSR Ser Fiz. Atmos. Okeana, 19, 1267–1271, 1983.
24. G. Bonino, P. Trivero, G. Elisei and A. Marzorati, "On the sounding range of 1-m wavelength radio acoustic sounder", J. Atmos. Oceanic Tech., 6, 851–855, 1989.
25. R.J. Lataitis, "Signal power for radio acoustic sounding of temperature: the effects of horizontal winds, turbulence, and vertical temperature gradients", Radio Sci., 27, 369–385, 1992.
26. Y. Masuda, J. Awaka, K. Okamoto, T. Tsuda, S. Fukao and S. Kato, "Echo power

loss with RASS (radio acoustic sounding system) due to defocusing effects by distorted acoustic wave front", Radio Sci., 25, 979–981, 1990.
27. G. Bonino, P.P. Lombardini, A. Longhetto and P. Trivero, "Radio Acoustic measurement of fog-capping thermal inversions", Nature, 290, 121–123, 1981.
28. G. Bonino, D. Anfossi, P. Bacci and A. Longhetto, "Remote sensing of stability conditions during severe fog episodes", in Air Pollution Modelling and Its Applications, Vol. IV, C. De Wispelaere Ed., Plenum Press, 601–619, 1985.
29. G. Bonino, G. Elisei, A. Marzorati and P. Trivero, "Results on planetary boundary layer sounding by automatic RASS", Atmos. Res., 20, 309–316, 1986.
30. G. Bonino, G. Elisei, A. Longhetto, A. Marzorati and P. Trivero, "Evolution of the atmospheric convective boundary layer monitored by the metric RASS", Nuovo Cimento, 12C, 163–171, 1989.
31. G. Bonino and P. Trivero, "Automatic tuning of Bragg condition in a radioacoustic system for PBL temperature profile measurement", Atmos. Environ., 19, 973–978, 1985.

Acoustic Remote Sensing Applications
S.P. Singal (Ed)
Copyright © 1997 Narosa Publishing House, New Delhi, India

7. Radio Acoustic Sounding System (RASS) for Studying the Lower Atmosphere

S.P. Singal and Malti Goel*

National Physical Laboratory, New Delhi, India
*Department of Science & Technology, Ministry of Science and Technology, Technology Bhawan, New Delhi-110016, India.

1. Introduction

The Radio Acoustic Sounding system (RASS) is a novel remote sensing technique to measure temperature and wind profiles in the lower atmosphere. This technique can be understood as an extention of electromagnetic—acoustic probe (EMAC) concept used for measuring winds [1, 2]. The technology development of RASS is primarily associated with the rapid advancements made in Doppler radar and acoustic technologies, as well as hardware and software developments in computers and signal processing techniques. It is indeed an example of development of technology resulting from the advancements made in related science and technology fields.

A first demonstration of RASS concept was made in early 1970s by researchers in USA [3–7] who succeeded in retrieving temperature profiles upto altitudes that varied from 1 to 3 Km. The interest in the RASS as a tool for atmospheric monitoring was, however, revived in 1987 when its operation in field was tested in combination with the existing wind profile radars at three frequencies viz. 50, 400 and 915 MHz, [8–16] at NOAA Aeronomy Laboratory and Wave Propagation Laboratory in the United States. It was seen that different altitudes of temperature scounding could be reached from these systems varying from 2 Kms to 5–9 Kms, 400 m to 1.5–2.5 Kms, and 200 m to 0.7–1.5 Kms, respectively [8].

RASS has gained further importance for its ability to measure a number of parameters in the lower atmosphere viz. wind and temperature, their temporal variance of less than few hours as well as heat flux with a greater resolution compared to a radiometer. Looking at the operational use of RASS in weather service, the availability of measured informations from different heights is very important. The altitude coverage of a RASS is determined by structural parameters of the system and by the atmospheric conditions. The accuracy of RASS temperature retrievals has been documented by May et al, [10] to about ±1.0 degree centigrade for 5 minutes integration time, which with further refinements can approach ± 0.5 degree centigrade [17] or less.

Various tests for implementing RASS technique have been conducted by researchers not only in USA but also in the former Soviet Union [18–23], Italy [24–29], West Germany [30–33] and Japan [34–40]. The greatest altitude coverage to date has been achieved by Japanese researchers who used high power acoustic sources and the high-power meso and upper atmosphere (MU) radar facility [40]. They could attain maximum altitudes that varied from 8 to 22 Km, (depending on wind conditions) by using the beam steering capabilities of the MU radar to compensate for the various deleterious atmospheric effects. In India it is planned to develop a high frequency RASS as well, in conjuction with MST (Mesosphere, Stratosphere and Troposphere) Radar at Gadaanki [41].

In the light of the large body of work that exists on RASS, in this study we present a review and analyze the impact of a realistic atmosphere on the performance of a RASS of selected configuration and evaluate its potential in retrieving accurate time series of temperature and vertical wind in comparison to radiosonde in different atmospheric conditions. The ability of the RASS to measure second-order quantities such as vertical wind and temperature variances, and the corresponding heat flux, is also presented by comparing the RASS data with measurements obtained from in-situ sensors onboard air-craft.

2. Principle of Operation

The idea of RASS originated from the application of Bragg's law to radiowave propagation in the atmosphere. An acoustic wave causes permittivity or refractive index variations in the atmosphere. The radar backscatter from such inhomogeneties can be measured, only when the Bragg condition i.e. $K_a = 2K_r$ is met, K_a and K_r being wavenumber vectors of acoustic and radar waves, respectively. This Bragg reflection condition is equivalent to two conditions. The wavelength condition i.e. $2\lambda_a = \lambda_r$ and the wavefront condition i.e. $n_a = n_r$ where n_a and n_r are unit vectors along the acoustic and radar wavelengths λ_a and λ_r respectively. These two conditions ensure specular reflection so that the incident electromagnetic radiation is reflected directly on to the receiver by the acoustically induced spherical density fluctuations, they being the necessary conditions for efficient backscattering and a strong radar return.

The observed speed of sound C_a at a particular height deduced from the Doppler frequency shift is directly related to the local atmospheric virtual temperature T_v,

$$C_a = 20.047\sqrt{T_v}$$

where C_a is in ms^{-1} and T_v in Kelvins. The configuration of a pulsed Doppler radar and an acoustic source in the centre offers such a capability (Fig. 1). The acoustic source is used to induce a vertically propagating density wave travelling at the speed of sound. The resulting density perturbations serve as a target for the Doppler radar, which is used to measure the speed of the disturbance. Gating the

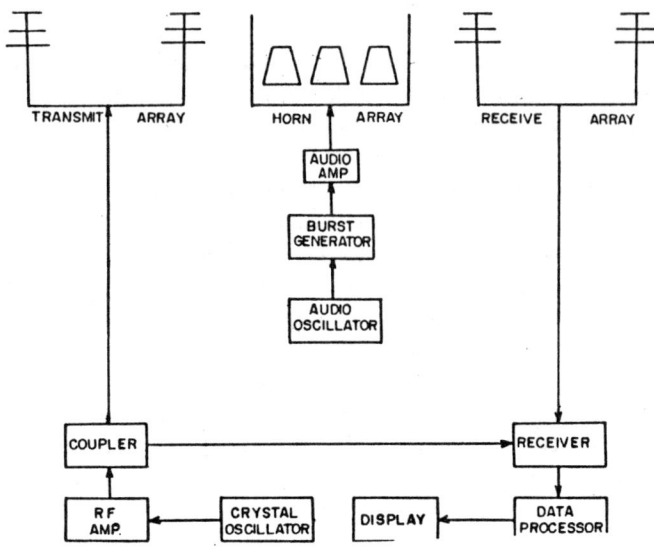

Fig. 1. Block diagram of RASS at National Weather Service (Stanford University, USA) [3].

returns provides the required range resolution. The corresponding angular Doppler shift ω_B in the radar signal is $2K_r C_a$.

Application of this well known concept of Bragg's reflection to the constantly changing atmosphere however, imposes certain complexities. The backscatter radar echo of RASS is affected by Doppler frequency shift arising from

(a) Speed at which longitudinal acoustic wave propagates, and
(b) Bulk velocity of the atmosphere

To achieve the Bragg condition, there are many potential RASS configurations. The system can be either monostatic or bistatic. Both the radar and acoustic sources can be either pulsed or continuous wave [42–44]. In addition, various types of frequency modulations can be applied to either radar or acoustic signals. One can either vary λ_r (radar wavelength) to track the acoustic pulse or keep λ_r constant and vary the emitted acoustic frequencies. This can be optimized for different altitude ranges. Several investigations have shown that changing acoustic frequency is more convenient. If one uses a single frequency acoustic signal, the observed frequency shift is equal to the frequency ω_a of the tone and no information about sound speed or temperature can be retrieved. Only when a spectrum of acoustic frequencies is emitted bracketing Bragg frequency $\omega_B = 2K_r C_a$, it gives complete information about the temperature profile. The criticality of Bragg matching therefore depends on the frequency spectrum of the acoustic waves in a pulse.

Various forms of acoustic excitation not only determine the nature of the radar backscatter signal but also affect overall RASS performance. A RASS utilizing an FM-CW (frequency modulated continuous wave) radar and continuous acoustic excitation was studied by Peters et al, [32]. The combination of a pulsed Doppler Radar and a continuous broadband acoustic source (Fig. 2) was suggested and tested by Strauch et al, [12]. The frequency of the acoustic excitation can be modulated in a deterministic manner using for example, a linear or sawtooth modulation function or it can be random modulation consisting of band limited noise [11]. In this case at Bragg frequency, a sharp peak is produced at each altitude. The other possible alternative is to transmit a comb of frequencies for which the Bragg matched condition will be satisfied over the range of altitudes of interest [31]. The choice of acoustic/radar frequencies on the other hand, determines the range. One main criterion that limits the choice is the acoustic attenuation in the atmosphere. The use of higher acoustic frequency leads to higher atmospheric absorption, which can become as high as 45 dB/km at 2000 Hz.

There are two categories of RASS [45]: Doppler RASS, in which frequency shift of reflected electromagnetic waves gives a measure of sound velocity. The sharpness of the radar signal depends on how close the acoustic frequency is to the Bragg frequency, with the sound propogation time, however, limiting the

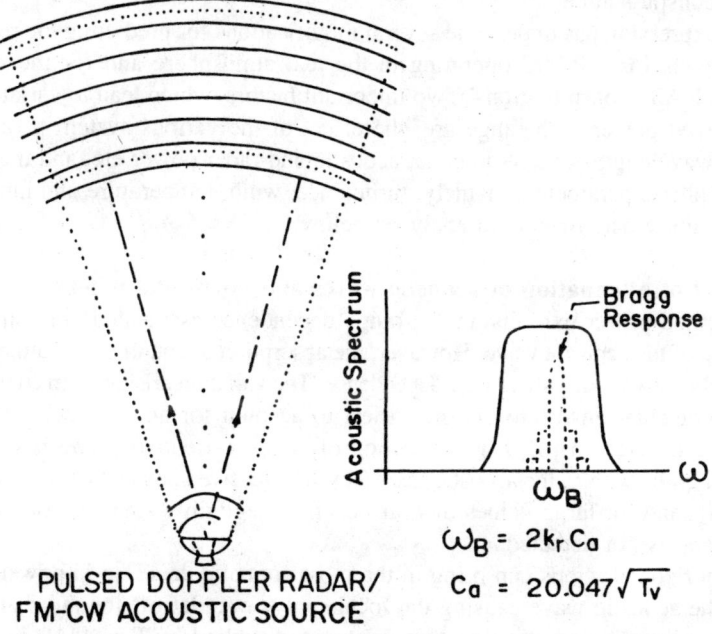

Fig. 2. Schematic of a RASS comprising a monostatic pulsed Doppler radar, a colocated FM-CW acoustic source and observed Doppler spectrum; Acoustic excitation and the spectrum (solid line); the radar resolution volume and Bragg response function (dashed line). (From [54])

sampling rate. The Bragg RASS, on the other hand is one where sound velocity is determined from that acoustic frequency which yields the maximum reflectivity. In this case height information is derived from the travel time of electromagnetic waves, resulting in a higher sampling rate.

3. RASS Echo Power

RASS, physically conceptualized by Marshall [3] and Marshall et al, [4] has the received echo power P_R in which an ideally static and homogenous atmosphere is given as:

$$P_R = \frac{N^2 G_r G_t G_a P_r P_a \left[1 - \cos\left(\frac{\theta_a}{2}\right)\right]^2 (1.38*10^{-16}) 10^{-LR/10}}{R^2}$$

where N is the number of wave lengths of the acoustic pulse in the radar resolution range, G_r is the radar receiving antenna gain, G_t is the radar transmitting antenna gain, G_a is the acoustic source gain, P_r is the radar radiated power, P_a is the acoustic source radiated power, R is the distance traversed from source by the acoustic pulse, L is the absorption loss of the acoustic waves (dB) and θ_a is the beam width of the acoustic source.

This expression has undergone several modifications for predicting the received power levels for a RASS operating in the real atmosphere and for the various potential RASS configurations. Two important factors, which lead to variations in the received power with range are; the effect of the various system parameters such as wavelength, size of antenna, acoustic and radar power etc. and the effect of atmospheric parameters, namely, turbulence, wind, temperature and humidity. Some of these parameters are analysed below.

(a) Effect of attenuation of acoustic waves and horizontal winds.

The returned power exhibits an R^{-2} range dependence associated with spherical spreading of the acoustic wave. However, the absorption of sound by the atmosphere can significantly modify this range dependence. The variation arising from attenuation due to molecular absorption is the easiest to account for and has been dealt by Frankel and Peterson [6]. L is a function of acoustic frequency and is found to increase rapidly with f_a, the acoustic frequency. The received power P_R is proportional to $10^{-LR/10}$ and for large values of L, it becomes negligibly small beyond certain heights and is not detected.

The horizontal winds can bring in the most dramatic decrease in power. They advect the acoustic wave causing the focal spot to wander off the radar antenna. The influence of winds on the echo power was studied by Kon [44], Karyukin [46] and Peters et al, [30]. They showed that the received power decreases according to the relation:

$$I = \exp[-4\rho_v^2/r_s^2]$$

where I is the intensity factor of the received power, and $r_s = D/\sqrt{2}$ defines the diffraction limited RASS focal spot radius, with D as the effective diameter of the radar antenna; ρ_v is the virtual horizontal displacement in the focal spot at range R due to advection caused by horizontal wind and is given by the relation:

$$\rho_v = C_a^{-1} \int_0^R dz\, V_H(z)$$

with $V_H(z)$ as the horizontal wind velocity at a height z. The effective focus of RASS is displaced by $2\rho_v$. The effect of horizontal winds on received power can be compensated by increasing radar antenna diameter.

(b) Effect of turbulence and vertical temperature gradient.
The turbulence and vertical temperature gradient modify the range dependence by distorting the acoustic wave fronts. The distortion due to turbulence is quantified in terms of a transverse coherence length ρ_0 which is a measure of the maximum transverse separation of two points on the acoustic wavefront such that the turbulence induced relative phase fluctuation at these two points is less than 1 radian, higher is the acoustic frequency, smaller is the coherence length. ρ_0 depends on the weighted path average value $\langle C_n^2 \rangle$ of the acoustic refraction structure parameter which in turn depends on the distribution of the turbulence strength C_n^2 with height. This broadens the RASS focal spot and reduces the back scattered power received by the radar antenna. Quantitatively the influence of turbulence has been considered by Nalbandyan [47], Clifford and Wang [48], Kallistratova and Kon [49]. However, their efforts were confined to a uniform distribution of the acoustic refractive index structure parameter, C_n^2, with height, while the current belief is that there is a exponential decay in C_n^2 above the boundary layer height. C_n^2 profiles can be inferred from profiles of the temperature structure parameter C_T^2 and velocity structure parameter C_v^2 from the following relation,

$$C_n^2 = \frac{C_T^2}{4T} + \frac{C_v^2}{C_a^2}$$

The contribution of C_T^2 to C_n^2 is generally small in a convective boundary layer, except close to top and bottom.

The ratio of radar beam width θ_e to the coherence length ρ_0 determines the extent to which turbulence affects the received signal power. The decreasing intensity can be expressed as $I \propto \left[1 + \left(\frac{D_t}{2r_s}\right)^2\right]^{-1}$ where D_t is the turbulent induced broadening of focal spot. It can be shown that there is a $R^{-16/5}$ range dependence for turbulence alone assuming that the effects of horizontal wind and vertical temperature gradient were negligible.

A vertical temperature gradient t_1, on the other hand, changes only the curvature of the acoustic wave front and is described by Z_v, the virtual vertical displacement of the acoustic source,

$$Z_v = t_1 R^2 / 4 T_0$$

with T_0 as the ground temperature measured in Kelvins. It can cause the power to decrease as R^{-6} for altitudes exceeding 8 km.

The combined effect of horizontal winds and turbulence has been considered by Nalbandyan [50], Bhatnagar and Peterson [51] and Kon [52]. The influence of vertical temperature gradient and winds on the acoustic wavefront distortion has been examined by Takahashi et al, [53] and Masuda [38] who numerically integrated the impact of this distortion on the RASS echo power. However, Lataitis [54] has considered the combined effect of all these environmental parameters on the RASS echo power. In the following, we describe his approach to derive such a radar equation.

4. The RASS Equation

Lataitis [54] has developed a simple equation for the average signal power of a RASS comprising a monostatic pulsed Doppler Radar surrounded by continous wave broad beam acoustic sources. Every acoustic source is characterized by the spectrum of its excitation, which is assumed to be relatively flat with a centre frequency $f_a = 2\pi/\omega_a$ near the Bragg frequency $f_B = 2\pi/\omega_B$ and a bandwidth 'b_a' broad enough to bracket the so called Bragg onset i.e. $b_a \gg C_a/\Delta$ where Δ is the radar range resolution. A Gaussian distribution of transmitted intensity is assumed and the acoustic beam width is broader than radar beam width.

The derivation takes into the account the effects of horizontal winds, turbulence and temperature gradient as described above. A physically displaced acoustic source is included in the development via an additional transverse displacement ρ_a. However, a few factors are excluded in this process, namely the lateral misalignment of the acoustic wavefronts (including higher order distortion) on the radar resolution volume caused by horizontal wind. This effect has been considered by Kon [51] but was shown to the negligible provided $V_H \Delta/(C_a D) < 1$. The second order effects associated with the finite width of the acoustic beam and the excess attenuation of sound, have also been found to be small for sufficiently broad beams [55] and are thus excluded. The other factors are changes in the acoustic intensity associated with any divergence or convergence of the acoustic beam by a background sound speed gradient, the increase in the relative refractive index perturbations and therefore in the echo power and any attenuation of the RASS signal due to presence of hydrometeors.

Starting with the governing equation given by Kon and Tatarskii [42] for the singly scattered field that results when an electromagnetic wave irradiates the refractive index perturbations, for a vertically pointed monostatic RASS, the average echo power, P_R is given by,

$$P_R = 1.2 \times 10^{-14} \{\Delta/\lambda_r R\}^2 \, P_r G_a I \cdot 10^{-L/10} \int_0^\infty dK' \, P_a(K') \cdot \text{sinc}^2 \, \{[2K_r - K'$$
$$(1 - \rho_t^2/2R^2)] \, \Delta/2\}$$

Here $K' = \omega'/C_a$ is the wave number corresponding to the centre frequency of the acoustic modulation, $P_a(K')$ is the acoustic power spectrum associated with the refractive index perturbations, ρ_t is the transverse displacement of the virtual acoustic source. Other symbols have the same meaning as defined earlier.

This equation describes the expected RASS signal power in terms of various length scales [56]. In practice, the ratio of total displacement of the acoustic source ρ_t (sum of its actual displacement ρ_a and the wind induced virtual displacement ρ_v) to the radius r_s of the RASS focal spot determines the extent to which the actual position of the acoustic source relative to the radar and horizontal winds, affects the RASS signal power.

In the above expression sinc^2 term (sinc^2 is Fourier transform of the range weighted function) describes the Bragg response function which is centered about the modified Bragg wave number $K_B' = 2K_r/\{1 - \rho_t^2/(2R^2)\}$, where $\rho_t = \rho_a + \rho_v$ $= \rho_a + C_a^{-1} \int_0^R dz \, V_H(z)$. The factor $\rho_t^2/(2R^2)$ in the argument of the sinc^2 term describes the modification in the Bragg condition due to a displaced acoustic source.

This expression is further evaluated in the limits of either a broad band or a narrow band excitation as follows.

(i) For a case when the acoustic spectrum is sufficiently broad, much broader than the radar beam width, that it brackets the Bragg response, we can set $P_a(K') = P_a/2B$ where B is the wave number bandwidth of the excitation, the echo power expression becomes:

$$P_R = 3.7 \times 10^{-14} \frac{\Delta}{(\lambda_r R)^2} \frac{P_r \rho_a G_a I \, 10^{-L/10}}{B}$$

under the condition that $B = 2\pi b_a/C_a \gg \Delta^{-1}$ or $b_a \gg C_a/\Delta$. The term I is expressed by setting the Bragg condition i.e. $\lambda_r = 2\lambda_a$ to be true (where $\lambda_a = 2\pi/K_a$ is the centre wavelength of the acoustic excitation)

$$I = \left\{ \frac{1}{1 + \frac{\theta_e^2}{2\rho_0^2} + (Z_v/Z_0)^2} \right\} \exp \left\{ \frac{-4(\rho_t^2/r_s^2)}{1 + \frac{\theta_e^2}{2\rho_0^2} + (Z_v/Z_0)^2} \right\}$$

Z_0 the virtual vertical displacement for which gradient induced spot spread equals diffraction limited spot sizes and is given by $Z_0 = \pi D^2/(2\lambda_r)$, θ_e, the width of radar beam at range R is given by $\theta_e = 4R/K_r D$ and the coherence length is given by

$$\rho_0 = (0.546 K_a^2 R C_n^2)^{-3/5}$$

(ii) In the limit of narrowband acoustic excitation (or for a single tone) where acoustic power spectrum $P_a(K')$ becomes $P_a(K') = (P_a/2)\delta(K' - K_a)$ with δ representing the Dirac delta function, δ is given by $\delta^2 = \rho_0^2[2/\theta_e^2 + \{\omega'z_v/(\sqrt{2} K_r D\ C_a R)\}^2]$. The echo power can be expressed as:

$$P_R = 6.0 \times 10^{-15}\ (\Delta/\lambda_r R)^2\ P_r P_a G_a\ I \cdot 10^{-L/10} \cdot \text{sinc}^2\ \{[2K_r - K_a\ (1 - \rho_t^2/2R^2)]\Delta/2\}$$

Here $b_a \ll C_a/D$. In the absence of winds, turbulence and gradients, the factor $\rho_t^2/(2R^2)$ in the argument of the sinc^2 term is zero and $I = 1$. In this limit the expression becomes consistent with the development of Marshall (3) and Marshall et al, [4], wherein a narrow band acoustic beam was used for excitation and the effect of environmental factors was ignored.

Results of analysis of the dependence of echo power intensity of a RASS on various atmospheric parameters can be summarised into the following:

(1) An acoustic source displacement of distance ρ_a from the radar results in a corresponding decrease in RASS echo power described by $I = \exp[-4\rho_a^2/r_s^2]$ and is constant with height.

(2) Horizontal wind V_H is responsible for a displacement in the effective focus of the RASS (Fig. 3) which produces a decrease in the received power according to the relation $I = \exp[-4\rho_v^2/r_s^2]$ and depends on height. The displacement ρ_v is related to $V_H(z)$.

(3) Assuming a uniform distribution of C_n^2 within the boundary layer (i.e. $R \leq 1$ km) and an exponential decay in C_n^2 above the boundary layer, it has been seen that the echo power with range exhibits three distinct regions. For $R < R_c$, where R_c is the lowest altitude for which the coherence length $\rho_0 < \theta_e$, the width for the radar resolution cell, the loss corresponds to an R^{-2} dependence of the echo power; For $R_c < R < R_b$, where R_b is the boundary layer height, there is an $R^{-16/5}$ range dependence which corresponds to the $R^{-26/5}$ dependence of the echo power and for $R > R_b$, the echo power once again decreases as R^{-2}.

(4) The effect of temperature gradient is to shift the vertical position of the virtual acoustic source as shown in Fig. 4. However for a realistic temperature gradient of 1–10 K per km and assuming horizontal wind to be negligible, there is little effect on the back scattered power below altitudes of roughly 8 km. For greater ranges the temperature gradient can cause the power to decrease as R^{-6}.

(5) The turbulence has a greater impact on the received RASS power at lower altitudes and temperature gradients have a greater impact at higher altitudes.

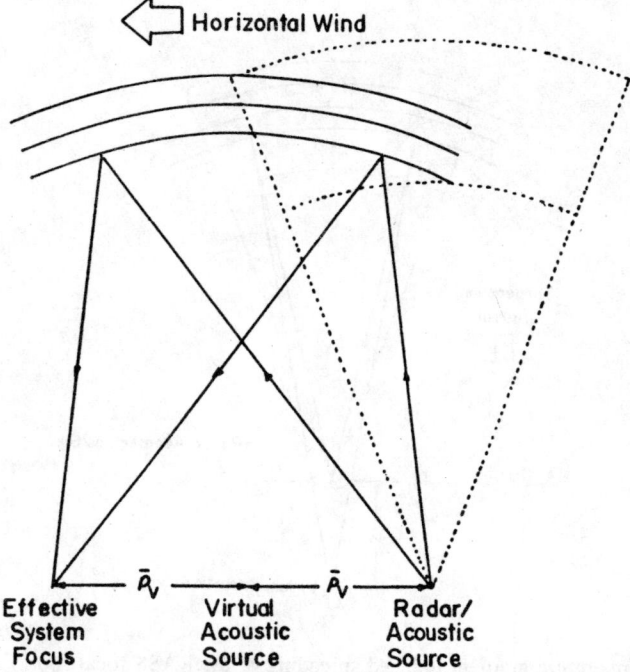

Fig. 3. Displacement of acoustic wavefronts in the radar resolution cell due to horizontal wind. The system focus is displaced by double of the virtual acoustic source displacement ρ_v (From [54]).

For light enough winds, the broadening of the RASS focal spot due to turbulence and temperature gradient can increase the height coverage by as much as a factor of 2.

5. RASS Doppler Spectrum

A theoretical explanation [54] for the Doppler spectrum broadening arising from real environment conditions is given below. The RASS echo power can be expressed in terms of RASS Doppler Spectrum W_R as

$$P_R = \int_{-\infty}^{\infty} dw \, W_R(\omega)$$

The first moment of Doppler spectrum approaches the Bragg frequency ω_B [42]. However the broadening of RASS Doppler spectrum from atmospheric conditions causes error or bias in the temperature retrieval.

To determine the error due to the effects of horizontal wind, turbulence and vertical temperature gradient etc., the expression for the RASS Doppler Spectrum can be written as,

$$W_R(\omega) \propto P_a(\omega) \, \text{sinc}^2 \{2\pi(\omega - \omega'_B)/\Delta_1\}$$

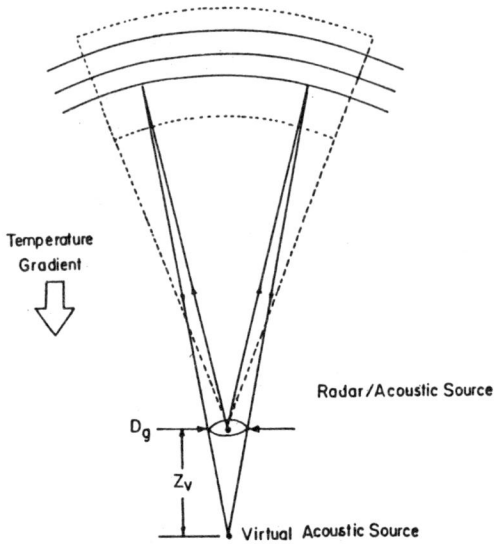

Fig. 4. Temperature gradient induced spreading of the RASS focal spot. For a negative temperature gradient the virtual acoustic source lies a distance Z_v below the actual source [54].

In this equation $W_R(\omega)$ is mainly a product of two terms, the power spectrum $P_a(\omega)$ of the acoustic excitation and the Bragg response function described by sinc^2 term. The terms, modified Bragg frequency ω'_B and modified spectral width Δ_1, are expressed as

$$\omega'_B = \omega_B (1 + \varepsilon)/(1 + \alpha) \cong \omega_B(1 + \varepsilon - \alpha)$$

and

$$\Delta_1 = 4\pi C_a (1 + \varepsilon)/\Delta_e(1 + \alpha) \cong 4\pi C_a (1 + \varepsilon - \alpha)/\Delta_e$$

ε and α are factors determining the impact of turbulence, and horizontal wind & presence of temperature gradient respectively and Δ_e is the effective range resolution.

For $|\varepsilon|, |\alpha| \ll 1$, i.e. in the absence of winds, the spectral width becomes

$$\Delta_1 = \frac{4\pi C_a}{\Delta} = \frac{4\pi f_a}{N}$$

where $C_a = \lambda_a f_a$ and $N = \Delta/\lambda_a$, the number of acoustic wavelengths in the radar range resolution. It can be seen that when N increases, Δ_1 decreases.

For a narrow band excitation of the acoustic frequency ω_a, the power spectrum P_a is a very narrow function centered about $\omega = \omega_a$. In this case, the RASS Doppler spectrum can be approximated by

$$W_R(\omega) \propto \text{sinc}^2\,[2\pi(\omega_a - \omega'_B)/\Delta_1]\,P_a(\omega)$$

for $b_a \ll C_a/\Delta_e$. The spectrum is narrow and is centered about ω_a with an amplitude proportional to the deviation of the acoustic frequency from the Bragg condition as described by the sinc^2 term, a result consistent with the work of Marshall et al, [4]. As discussed earlier this limit is not useful for temperature sensing because the Doppler shift depends only on the acoustic frequency and therefore does not contain any temperature information. However, in case the turbulence is strong, the observed frequency is shifted away from ω_a by an amount $K_a\,\rho_t \cdot V/R$ and the spectrum is broadened in proportion to the speed V of the acoustic wavefront irregularities which is not, in general, equivalent to the wind speed.

For a broadband acoustic signal with a centre frequency $\omega_a = \omega'_B$ and a flat response about ω'_B, the RASS spectral width can be written as:

$$W_R(\omega) \propto \text{sinc}^2\,[2\pi(\omega_a - \omega'_B)/\Delta_1]$$

For $b_a \gg C_a/\Delta_e$, the spectrum has a peak at ω'_B which is proportional to the local sound speed and therefore temperature.

Spectral broadening Δ_1 in the presence of uniform horizontal wind can be written as:

$$\Delta_1 = \frac{4\pi C_a[1 + (V_H\Delta/DC_a)^2]^{1/2}}{\Delta}$$

The second term within the square root is the ratio of the horizontal range $V_H\Delta/C_a$, (over which the virtual acoustic source displacement varies between the bottom and top of the radar range resolution) to the radar antenna diameter D. If the displacement is large compared with D, then Δ_1 approaches to $\Delta_1 = 4\pi V_H/D$. This describes the broadening of the RASS Doppler spectrum due to the variation of the Bragg condition with zenith angle within the radar range resolution, which does not exist for a simple displacement of the acoustic source.

In the absence of temperature gradient and in the limit of strong turbulence (i.e. $\rho_o \to 0$) the factor $\varepsilon = \rho_t V/(C_a R)$ as a first approximation contributes to spectral broadening, and in the presence of uniform horizontal wind and weak turbulence the factor α contributes and is given by:

$$\alpha = V_H^2/\{2C_a^2 - \rho_a^2/(2R^2) - 2Z_v(R)/R\}$$

The spectral width under these conditions becomes,

$$\Delta_1 \cong 4\pi C_a/\Delta \cdot \{1 + (\rho_t V/C_a R) + (\rho_a^2/2R^2) - (V_H^2/2C_a^2) + (2Z_v(R)/R\}$$

This expression can be analyzed as follows:

(1) For turbulence alone, $\Delta_1 \cong 4\pi C'_a/\Delta$, where $C'_a = C_a + \rho_t V/R$

Here the second term which represents the projection of the speed V of the acoustic wavefront purturbations onto the radar axis, contributes directly to the apparent sound speed C'_a observed by the radar and hence to the spectral width. This contribution can be either positive or negative depending on whether the acoustic source is upwind or downwind of the radar, respectively.

(2) If we consider the contribution due to the displacement ρ_a of the acoustic source alone, then also the spectral width Δ_1 can be expressed in the same way but with

$$C'_a = C_a [1 + \rho_a^2/(2R^2)] \cong C_a[1 + \rho_a^2/(R^2)]^{1/2}$$

The effect of the horizontal displacement is, therefore an increase in the sound speed along the radar axis and a consequent increase in the spectral width.

(3) If we consider only the horizontal wind contribution to spectral width, then $C'_a \cong C_a\sqrt{(1 - V_H^2/C_a^2)}$. This result indicates that the effect of horizontal wind is to decrease the apparent sound speed along the radar axis resulting in a decrease in the spectral width.

(4) Similarly the contribution of the temperature gradient alone can be studied. In this case the spectral width is given by,

$$\Delta_1 \cong 8\pi C_a Z_v(R)/(R\Delta) = 2\pi C_a t_1 \, R/(\Delta T_0)$$

6. Sources of Temperature Error and Corrections

The temperature error ΔT due to various environmental effects is proportional to,

$$\Delta T \propto (\omega'_B - \omega_B)$$

which means that the fractional temperature error $\Delta T/T$ at the actual temperature T can be found from:

$$\omega'_B \cong \omega_B \left(\frac{1 + \Delta T}{2T} \right)$$

or

$$\frac{\Delta T}{T} = 2 \left[\frac{\omega'_B}{\omega_B} - 1 \right]$$

(a) Approximations in Temperature Retrieval: Correction for humidity

For a gaseous medium the relationship between sound welocity, molecular weight M and temperature T is given by

$$C_a = \left(\frac{\gamma R_o T}{M} \right)^{1/2} = K_D \sqrt{T}$$

R_o is the universal gas constant, γ the ratio of specific heats, M is molecular weight and K_D a constant having the value $\left(\dfrac{\gamma R_o}{M}\right)^{1/2}$. In a dry atmosphere for ideal gas conditions, K_D is computed as 20.047.

The temperature T as measured from RASS is given by

$$T = \dfrac{C_a^2 M}{\gamma R_o}$$

where C_a is the measured velocity of sound wave. The effect of humidity on it can lead to some error. Harris [57] has computed variation in the sound velocity in the presence of moist air. The correction is to modify the value of constant K_D, which is replaced by K_1, where $K_1 = R_c K_D$. Here, R_c is Harris ratio corresponding to a given humidity. However, if humidity does not vary with height more than 10%, the correction will be the same at all heights. Only when humidity changes abruptly i.e. when a fog layer is capped by very dry air, or in a similar other situation, the accuracy of measurements is adversely affected, if the correction is not applied. It was estimated that a change of approximately 25% in relative humidity would cause an error of 0.5°C in the virtual temperature measured by RASS.

Analysing its impact on each parameter, in a real atmosphere [58],

$$M_d = M \dfrac{1 + q(\eta)}{1 + q}$$

Here, M_d is value of M for dry air = 28.964, q is specific humidity and η is ratio of molecular weight of water vapour to air.

Hence T_v can be written as [72].

$$T_v = T \dfrac{1 + q(\eta)}{1 + q}$$

For RASS measured virtual temperature, the equation becomes,

$$T_v = \dfrac{v_m^2 M_d'}{\gamma' R_o}$$

Here, v_m is the measured sound speed, M_d' the value of M_d used and γ' is the value of γ used.

Also

$$v_m = \dfrac{C_o f_d}{2 f_r}$$

C_o is the velocity of light in vacuum = 2.998×10^8 m/Sec
f_d is the Doppler frequency
f_r is the radar carrier frequency

The correction for actual virtual temperaure w.r.t that calculated from RASS measurements is given by

$$T_{vm} = \frac{C^2 \gamma' M_d}{C_o^2 \gamma M_d'}$$

where, $C = \frac{C_o}{n}$, n is the refractive index of the medium, which changes with moisture. In the above expression the ratio of specific heats γ is a function of temperature and humidity and its determination with sufficient accuracy is still an active research topic. Cramer [59] has given the following expression for the accurate values of γ in terms of C_a

$$C_a = \gamma \frac{R_0 T}{M} \left(1 + \frac{\rho B(T)}{R_0 T} \right)$$

where $B(T)$, the second virial coefficient in cm^3/mol, is given by $B(T) = a - b e^{c/T}$ with a, b, c defined as constants for the standard atmosphere,

Using this formula for ROSE II experiment data set [60] conducted in Alabama, USA and assuming that CO_2 mole fraction is constant at 314 ppm, yielded a value of γ between 1.3993 and 1.4015. When this correction is applied, a substantial reduction in the difference between RASS measurements and radiosonde measurements was seen (reducing from 0.7 to 0.3°C).

(b) Corrections for the Maximum Height Range
The maximum height range of RASS is given by

$$R_{max} = DC_a/u$$

where u is mean wind speed between ground and measuring volume. Maximum altitude coverage of RASS is largely affected by

(i) Specific system parameters such as wavelengths, size of antennas, acoustic and radar power etc.
(ii) Effect of atmospheric parameters such as turbulence, wind, temperature and humidity.
(iii) Distance between centres of acoustic and radar antenna,
(iv) Acoustic attenuation which is a strong function of frequency.

Bauer and Peters [61] have observed RASS echoes outside the focal spot and have given a theoretical discription of this "tail field" of the focal spot, which can extend the actual height range of a RASS. The outside of the focal spot is explained by diffuse scattering of sound wave distortions, which are caused by turbulent eddies, smaller in size compared to scattering volume.

The distortions are described by complex phase shift $\varphi(\Omega)$. In its simplified form spot aperture can be written as $\langle \rho \rangle = A \cdot F(\rho_c)$ where A describes the attenuation due to sound propagation and is proportional to $\dfrac{10^{-LR/10}}{R^2}$, L is sound absorption, $F(\rho_c)$ describes the turbulence induced horizontal intensity distribution.

For a radar with collocated electromagnetic antennas and an acoustic source which is shifted by ρ_a to compensate for the wind drift of the sound waves, the receiving antenna is shifted by ρ_c, given by

$$\rho_c = 2\left[\rho_a + \int_0^R \frac{u(z)}{C_a}\right]$$

The acoustic signal near 2 kHz (for a RASS with 915 MHz of radar singal) is strongly attenuated in the atmosphere. Here the absorption of sound which is more than 2 dB/100 m [57] becomes the limiting factor for height coverage. This attenuation can vary with changes in temperature and humidity.

In practice the virtual temperature measured by the profiler is a weighted average of the virtual temperatures at each height in a cell. Measured temperature in a cell m is given by Angevine et al. [58], taking into account non uniformity due to advection of acoustic wave by the wind

$$T_m = T_o + \xi \frac{\int_{r-\Delta r/2}^{r+\Delta r/2} z\, 10^{az}\, dz}{\int 10^{az}\, dz}$$

where T_o is virtual temperature at $z = 0$
    $a = a(z)$ is the signal strength at range z in dB
and ξ is virtual temperature lapse rate

Signal to Noise Ratio (SNR) varies strongly with altitude and also depends on the wind drift of acoustic beam, which acts to displace or distort the focussing spot. At some heights spot becomes no longer detectable. With the advancement in technology, possibility of operating with low SNR has contributed to increasing effective range of RASS.

Theoretical estimates of the effect of wind on range of RASS and the efficiency of possible methods of compensation were also analysed by Belyavaskaya et al, [62]. Steinhagen and Neisser [63] have recently examined various possibilities of increasing the altitude coverage of temperature measurement using RASS. An acoustic excitation in frequency sweep mode with automatically adopted acoustic frequency window gives maximum improvement in altitude coverage. It is seen that better SNR is obtained for small acoustic window width (~ 100 Hz.). The benefit of changing distance between antennas is however, limited.

Multiple acoustic sources in various configurations have been suggested to enhance the height coverage of a RASS in operational use. When the radar is surrounded with a number of acoustic sources, it is hoped that in the presence of harizontal wind, the acoustic wave from one of the sources is eventually advected directly over the radar, which maximizes the echo power. Another way to describe the effect of multiple acoustic sources is in terms of the multiple focal spots produced by an acoustic array. This distributed focal spot has an area larger than the spot associated with a single acoustic source and, therefore, has a greater probability of overlapping the radar antenna for a given range. From theoretical studies, Lataitis [56] has computed the expected enhancement in height coverage for a RASS using multiple acoustic sources and shown that the temperature error for acoustic sources located at equal distances from the radar is essentially the same as for a RASS comprising a single source even in the presence of turbulence.

(c) Correction for turbulence
The turbulence can affect the signal in two ways

(i) Sound wave velocity, derived from propagation time between two reference points, gets lower than in the homogenous medium.
(ii) The acoustic wavefront gets a random tilt and the apparent sound velocity (derived from λ as measured parallel to undisturbed wave propagation in this case) is higher than in the homogenous case.

Presence of turbulence degrades or smears the signal, as it decreases transverse coherence of the acoustic wavefront [59]. The distortion can be described by random complex phase shifts $\Psi(\Omega)$. In case of spherical waves the perturbed phase surface can be written as.

$$D_{\Psi(\Omega)} = \left(\sigma_\Psi \frac{K_a}{2} \Omega\right)^{5/3}$$

$D_{\Psi(\Omega)}$ is structure function of complex phase shift $\Psi(\Omega)$, with Ω as solid angle. The parameter σ_Ψ contains effect of turbulence in terms of structure parameter C_n^2, and is expressed as,

$$\sigma_\Psi = 2K_a^{1/5} \left(2.91 \int_0^R dz\, C_n^2(z)\, Z^{5/3}\right)^{3/5}$$

The weighted path average of the acoustic refractive index structure parameter is given by

$$\langle C_n^2 \rangle = \frac{8}{3} R^{-8/3} \int_0^R dz\, C_n^2(z) Z^{5/3}$$

Peter and Angevine [60] related the weighting fractions to transverse coherence length ρ_o of the acoustic field and derived the fractional temperature error due to turbulence as

$$\frac{\Delta T}{T} = \frac{1}{K_a^2 \rho_o^2}$$

For the purpose of correction, using the expression for the local coherence length yields,

$$\frac{\Delta T}{T} = 0.483 \, R^{6/5} \, K_a^{2/5} (C_n^2)^{6/5}$$

Alternatively, the effect of turbulence on the temperature error is described by the factor ε which is related to the delta dirac function. The error can be expressed as [54].

$$\Delta T/T = 2\rho_t V / \{C_a R[1 + 2\rho_0^2/\theta_e^2 + 2\rho_0^2 Z_v^2/(DR)^2]\}$$

When the turbulence is weak, the perturbations in the acoustic wavefront are small and are not sensed by the radar. In this limit $\rho_0 \to \infty$ and $\Delta T \to 0$. On the other hand, when the turbulence is strong, the wavefront perturbations are very prominent and are easily observed by the radar. In this limit $\rho_0 \to 0$ and $\Delta T/T = 2\rho_t V/(C_a R)$. The contribution induced by turbulence of varying intensities gets smaller as the distance increases and approaches the limit $2V_H^2/C_a^2$ at longer ranges.

(d) Correction due to horizontal wind

Horizontal wind causes a bias in temperature [54], as follows.

$$\frac{\Delta T}{T} = \frac{2\rho_v V_H}{C_a R_1 \left(1 + 2\dfrac{\rho_0^2}{\theta_e^2}\right)}$$

for $\dfrac{\rho_0^2}{\theta_e^2} \ll 1$

$$\frac{\Delta T}{T} = \frac{\rho_v V_H}{C_a R}$$

The analysis as a function of the range has shown that temperature error at short ranges is determined by physical displacement of the acoustic source, the bias is positive and decreases with range. At long ranges ΔT is dominated by horizontal wind. The error due to strong horizontal winds is possible to be eliminated by moving the radar receiving antenna downwind or by utilizing an array of antennas.

Steinhagen et al, [64] constructed a RASS with fixed electromagnetic antenna and a movable acoustic antenna mounted at a turntable to optimise the distance and direction between both antennas to compensate for horizontal wind. The optimum acoustic beam width is given by

$$\theta_a = 2 \arctan\left(\frac{V_H}{2C_a} + \tan\frac{1}{2\theta_e}\right)$$

For $\theta_e = 6°$ and $V_H = 20$ m/s, θ_a is computed as 9.5°.

Temperature gradients do not appear to directly influence the temperature retrieval. The influence of temperature gradient when it is sufficiently strong is to decrease the temperature bias associated with turbulence effects. In the presence of horizontal wind, the temperature error ΔT approaches TV_H^2/C_a^2 in weak turbulence and $3TV_H^2/C_a^2$ in strong turbulence using an acoustic source with a beam width narrower than the radar beam width.

The combined effect of these parameters is shown by Lataitis [54] as depicted in Fig. 5.

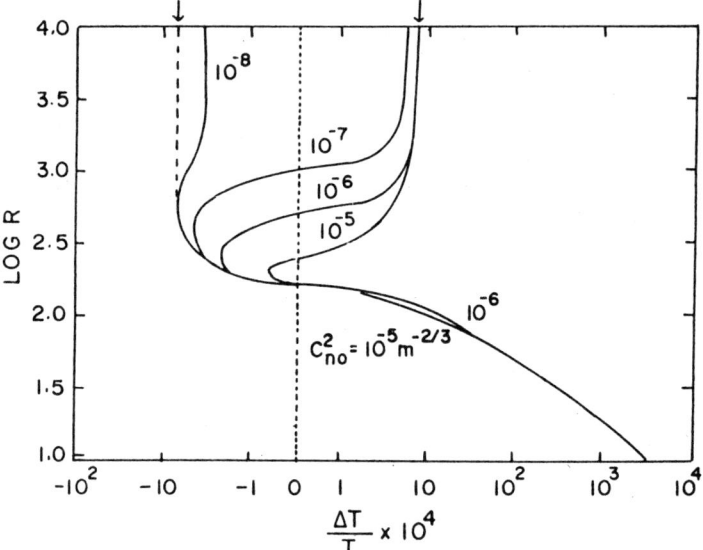

Fig. 5. The total relative temperature error $\frac{\Delta T}{T}$ plotted as a function of range R for an acoustic source displacement $\rho_a = 5$ m, a uniform horizontal wind $V_H = 10$ m/sec and various values of turbulence strength C_n^2 [54].

(e) Correction due to vertical motion

The other most important correction in RASS temperature measurements is due to the presence of vertical motion of the atmosphere, modifying the expression for sound velocity as follows.

$$C_a = w + K_D \sqrt{T_V} \quad \text{or} \quad T_v = \left[\frac{C_a - w}{K_D}\right]^2$$

where w is the vertical wind velocity.

Each 1 m/s of vertical wind, can produce a change in temperature by 1.7°C. North et al, [5] categorised vertical motion in atmosphere in four categories. To determine its effect on RASS profile the following was proposed,

(i) Random, high frequency motions associated with small scale turbulence. This would average out over measurements carried out for longer durations.
(ii) In thunderstorm, substantial high winds in convective cells would occur. This must be measured so that appropriate correction to temperature can be applied.
(iii) A downward flow compensating the upward motion in convective cells. It is normally distributed over a large area and reduced by an order of magnitude.
(iv) Large scale motion associated with topography and synoptic scale phenomena. They normally produce low winds resulting into minor corrections.

Moran and Strauch [17] demonstrated that a vertical wind of 0.3 m/sec resulted in temperature error of about ± 0.5°C and sugested that the wind should be measured simultaneously along with the acoustic velocity, for applying the correction.

In the absence of direct measurement of wind velocity, the strong temperature anomalies observed in the RASS profiles due to presence of events like active sea breeze or rapidly moving cold fronts can provide an estimate of vertical motion of the atmosphere [65].

Jordan and Lataitis [66] have found that significant errors in the measured wind speed profiles can occur due to non-uniform acoustic illumination across the radar beam. Wind speed can be under or overestimated depending on the relative radar and acoustic beam pointing angles. One of the methods, to remove the beam overlap error is to calculate it approximately and take care of it in the measurements if the radar and acoustic beam widths and pointing angles are known. However, a better solution to the beam overlap problem is suggested to use a steerable acoustic source that changes pointing angles in unison with the radar beams behaving like multiple acoustic sources which point along the different radar beam pointing angles.

7. State-of-the-Art in RASS Measurements

Different frequency RASS have been deployed in field measurements and at long term sites as simple additions to wind profiling radars. Several workers [9, 16, 22, 36, 67, 68] have experimentally explored the RASS capabilities in different conditions. Neiman et al, [67] observed an arctic front and arctic airmass over Denver during 1–5 Feb 89. The RASS measurements extended to approximately 15 km and were taken at 15 min intervals during frontal passage and at 1 hour intervals in normal conditions. Simultaneous RASS and radiosonde observations showed good agreement with regard to key thermal features. Bonino et al [68] successfully tracked thick fog episodes from 12th to 20th Jan. 1992 in the Po valley, a vide and flat region of north Italy. A meteric RASS demonstrated the capability to follow evolution of the thermal profile which is a fundamental parameter characterising the fog conditions. It was concluded that measuremnt of fog capping thermal inversions which reduce thickness of dispersion layer also help in investigating critical air pollution conditions.

162 Atmospheric Acoustics and Instrumentation

In the following paragraphs we compare the RASS field measurements with and without corrections with those of well known radiosonde observations and also with other means of measuring these parameters.

(a) Measurement of Temperature profiles

The performance test for RASS conducted during ARM-91 (Feb. and March 1991) in Colorado revealed excellent agreement with rawinsonde data. Three different RASS configurations with frequencies 50 MHz, 404 MHz and 915 MHz indicated standard deviation of about 1.0°C for virtual temperature [69]. In this large scale winter time experiment remote wind measurements at different heights were also made, thus maximising the temporal and vertical resolution of upper air measurements and exhibiting complementarity with microwave radiometer measurements.

Complementarity of RASS with satellite based microwave radiometer measurements throughout the troposphere and more accurate temperature retrieval from a combined system was demonstrated by Schroeder et al [70] who devised an integration method for two profiles. Measurements from a 915 MHz RASS ($R_{max} < 2$ km) were extrapolated to cover the tropospheric height using linear statistical inversion techniques, having a prior knowledge of local climatology. This facilitated blending it with satellite measurements in the troposphere. Sixteen sets of measurements between 1–9 Feb. 1989, revealed RASS ability to resolve lower level inversions in the composite temperature profile in the combined system [Fig. 6]. This combination technique also gave data prospect of continuous unattended accurate ground based measurements of certain parameters for modelling studies.

Peters and Angnevine [60] have compared RASS field measurements with 31 radiosonde profiles during Rural Oxidants in the Southern Environment II (Rose II) experiment conducted in June 1992. RASS profiler with 4 acoustic sources confined around the 915 MHz radar having a range resolution of 103 m was located in a clearing in a pine plantation in west-central Alabama. Radiosonde flights were also made. The mean wind speed in the boundary layer averaged to 4 m/sec and the air was hot and humid. It was seen that uncorrected virtual temperature measured by RASS is higher than that measured by radiosonde at all but the lowest heights. The median temperature difference over the height range 450–980 m was 0.69°C. However after the turbulence correction, RASS median virtual temperature became lower than from the sonde mean value in nearly all range gates. In particular the median difference over height 450–980 m was – 0.16°C. These results were very encouraging for temperature profiling using RASS in the lower atmosphere.

A boundary layer RASS of 1290 MHz was tested in Berlin. It was located in a relatively flat farm land south-east of Berlin at Lindeberg Meteorological Laboratory of German Weather Service [71]. A 2.7 m × 2.7 m antenna and single acoustic source were mounted on an apparatus that rotates to keep acoustic source always upwind of the antenna. The radar pulse length was 90 m. During a rain free period from 22 July–6 Aug. 1994, 36 soundings were made 4 times a day and a comparison

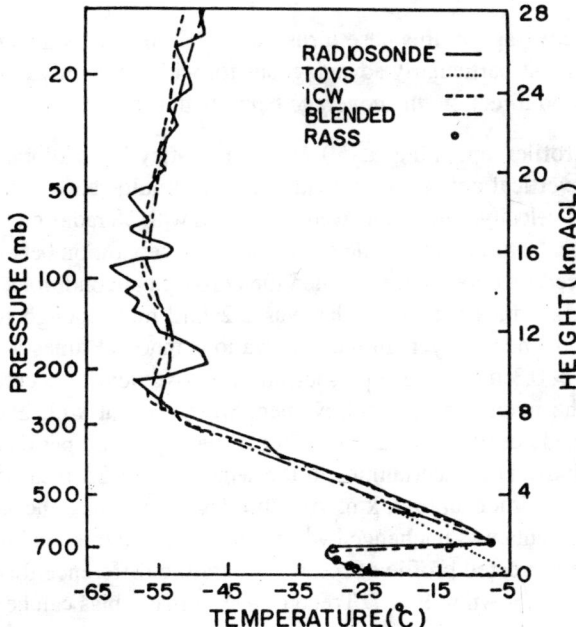

Fig. 6. Temperature profile measured by RASS, TOVS sounding, combined RASS—TOVS profile using inverse co-variance weighting (ICW) and direct blending [70].

was made with RASS data. The median difference was found to be 2.1°C at heights 360–650 m. The scatter was also large. After applying corrections for turbulence and wind, a median difference of 1.3°C still remained except that for 0600 and 1800 UTC, the median difference reduced to 0.55°C. This could not be explained. These results required further study especially of the phenomenon affecting lowest range gates in RASS measurements for correction to be applied.

(b) Measurement of vertical wind velocity

One of the largest source of error identified in RASS temperature profiling is neglect of wind velocity along the beam. Certain advancements in RASS technology have made it possible to measure the wind velocity which is a few m/sec and the acoustic velocity which is of the order of a few hundred m/sec using the same instruments simultaneously. Bauer and Peters [61] suggested following options.

(i) By using greater speed Digital Signal Processors (DSP) to carry out a 2048 point FFT over a frequency range that covers both speed of sound (330 m/sec and the wind speed (0.3 m/sec) [57].
(ii) By developing a system using DSP to produce the acoustic signal controlled by radar control programme i.e. the use of a acoustic source that is controlled by the same computer that controls the radar.
(iii) By using oblique beams in the operation of the wind profiler. Though the

height coverage in this case turns out to be poor as compared to vertical beam, it was particularly advantageous for wide angle measurements and in downwind direction, the coverage being better than upwind direction [72].

A RASS profiler operating at 50 MHz at Platteville, Colorado, facilitated correction for vertical atmospheric conditions by simultaneous measurements of sound and wind velocity. The results were compared with 58 radiosonde observations made at Denver, Colorado [17]. The effect of 40 km separation between these two sites was reduced by chosing times when meteorological conditions were similar. The lowest height measured by radar was 2.2 km, high enough to neglect the effect of local boundary layer. In the observations made at times when moderate vertical winds > 0.3 m/sec were present, the improvement in accuracy was seen upto 0.7°C. The magnitude of improvement was constant with height from the observations made during several weak frontal passages and periods of moderate upper level winds. The uncertainties in the wind and RASS measurements were of the same magnitude upto 5 km. At altitudes above this, the uncertainty in RASS measurements was unchanged while uncertainty in the wind measurements decreased. The corrected profile of mean temperature difference for the entire set of observations is shown in Fig. 7. Trend of a systematic bias can be seen in these measurements.

At higher altitudes, other factors such as strong horizontal wind can influence the measured value. For conditions with moderate turbulence, perturbations in the acoustic wavefronts due to turbulence are advected by the wind and when tracked by the radar, cause a slight shift in the measured velocity. Using a surface wind of 6 m/sec and a gradient of 1 m/sec/km, the theoretical model underestimated the error by nearly 0.5°C. Larger wind gradients would produce theoretical results that were closer to observed errors.

It has also been seen that under conditions of high clutter or low radar reflectivity, clear air radars (wind profilers) perform poorly in determining wind directions. High quality harizontal wind measurements under such conditions can be obtained by measuring the difference between the speed of sound for the two pairs of antennas pointing in orthogonal directions [33, 73], while the vertical wind component can be measured with the addition of a fifth vertically pointing beam, allowing two separate estimates of the vertical velocity. Initial testing of the acoustically enhanced profilers [74] have, however, shown that this increases the height coverage and estimates the wind direction accurately but underestimates the wind speed.

(c) Measurements of heat flux
The heat flux has been calculated by Peters et al, [31] as,

$$\theta_v = \rho \frac{C_p}{2} [\text{Cov}(T_v' w')(-1) + \text{Cov}(\dot{T}_v' w')(1)]$$

As a first approximation

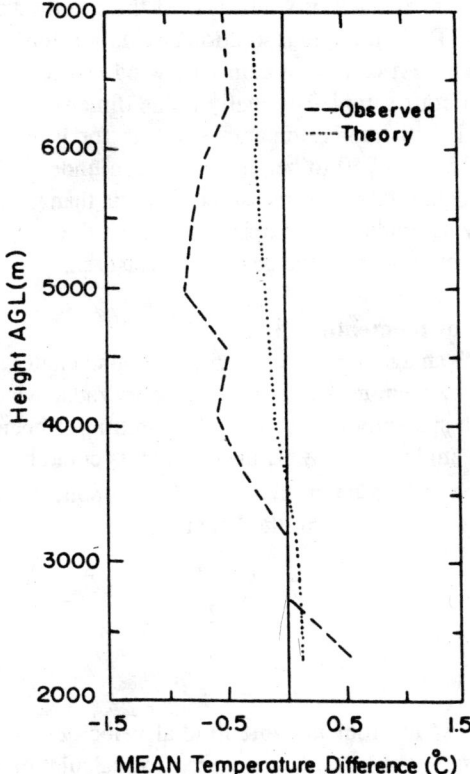

Fig. 7. The mean temperature difference btween RASS measurements corrected for vertical air motion and radiosonde observations [17].

$$\theta_v = \rho C_p \langle w' \, T_v' \rangle$$

Here ρ is air density

C_p is heat capacity of air at constant pressure

w' represents perturbations of vertical velocity

T_v' represents perturbations of virtual temperature.

Cov is covariance and the argument of covariance is the lag between temperature and vertical velocity. The covariance at zero lag is not used. Simultaneous measurement of acoustic and wind velocity enables the calculation of heat flux by eddy correlation.

A comparison of vertical heat flux measurements to aircraft measurements during ROSE II experiment [75] has shown that RASS technique measurements agree well (within the observation limits) with aircraft measurements. Virtual heat flux drives the boundary layer mixing under light wind conditions and is therefore of considerable interest in any boundary layer observations or modelling programme.

RASS sampled a vertical stack of volumes over the site for a time period varying from 2 to 14 hours. Typical range resolution cell dimensions above 150 m height were taken as 100 m vertical, 60 m along the wind (30 sec × 2 m/sec) and 16 m cross wind. The aircraft sampled virtually a one dimensional path covering 50–100 km in about 10–20 min. Comparable results for heat flux of 2 h average RASS profiles at 150 and 250 m heights were seen under light wind conditions. Both of these techniques, however, measured less flux than the surface instruments, with a discrepancy as much as 27% of the surface value. It is presumed that the discrepancy arises mainly from the scales of measurements.

(d) Measurement of momentum flux

In view of RASS advantage in its insensitivity to ground clutter, direct measurement of vertical flux of momentum has been made from radar wind profiler in RASS mode [76]. From a specially designed RASS emitting 4 beams with orthogonal azimuths and a tilt angle of $\Psi = 8.5°$, measurements could be made over a height range of 800 m. The orthogonal horizontal wind components, $\bar{u}, \bar{v}$ are determined from measured radial velocities in the beams

$$\bar{u} = \frac{(r_+ - r_-)}{2 \sin \Psi}$$

$$\bar{v} = \frac{(s_+ - s_-)}{2 \sin \Psi}$$

Here, r_+, r_- and s_+, s_- are four measured radial velocities in the 4 beams. Undisturbed time series of $r_\pm$ and $s_\pm$ are used for calculation of the momentum

$$\overline{u'w'} = \frac{\overline{r_+'^2} - \overline{r_-'^2}}{4 \sin \psi \cos \psi} - \frac{C_a' u'}{\cos \psi}$$

$$\overline{v'w'} = \frac{\overline{s_+'^2} - \overline{s_-'^2}}{4 \sin \psi \cos \psi} - \frac{C_a' u'}{\cos \psi}$$

The fractional statistical error e' in the momentum flux is estimated from

$$e'^2 = \frac{2 \, \text{Fuw}}{n_i}$$

where $\text{Fuw} = \dfrac{\overline{u'^2}\,\overline{w'^2}}{\overline{u'\omega'}} + \dfrac{\overline{u'w'}\,\overline{v'w'}}{\overline{v'\omega'}}$ and n_i is no. of independent samples

RASS measured momentum flux when compared with those from surface based in-situ measurements showed an agreement in the gross structure of diurnal variation. The discrepancies were attributed to statistical properties of the turbulent field and low sampling rate of 75 sec. It was concluded that more experiments are needed for providing utility of RASS in direct measurement of momentum flux in the boundary layer. It was also suggested to use a combination of RASS and spaced antenna drift technique, which can make better resolution feasible for tracking the momentum, because their operation is conditioned by turbulence generally associated with a discontinuity in the thermal gradient.

Vertical flux of momentum is an important parameter in the equation of motion for the atmospheric boundary layer (ABL). With certain modifications, RASS can become a valuable tool because in surface measurements, fast response sensors are deployed on meteorological towers upto a certain height only, and using the climatogical informations, model calculations are made to cover the entire region of ABL. Hence, these measurements are not true representative of turbulent ABL.

(e) Measurement of water vapour flux

Preliminary tests of 1235 MHz RASS and lidar combination for retrieving vertical profiles of water vapour flux were conducted by Senff et al, [77] at a flat site north west of Hamburg. Under predominently convective conditions water vapour profiles at a time resolution of 60 sec. and height coverage upto 700 m were recorded. No other measurements are reported on this parameter.

8. Future Outlook

Over 50 Radio Acoustic Sounding systems have been developed world wide and tested so far. Many prototypes systems have also been successfully developed, fabricated and evaluated for their performance. The detailed specifications of different RASS along with the range covered are given in Table 1. An attempt is also being made to set up a RASS system in India [78, 79]. India has a MST radar and implementation of RASS can accrue benefits for measuring remote temperature profiles upto heights > 10 km. As seen from the NOAA data the upwind motion is almost negligible in the atmosphere over the Indian region except during monsoon season, thus favouring RASS application.

Generally, tropical regions are characterised by dynamic instability of the boundary layer mainly influenced by conditions at the earth surface. In most regions there have been scanty observations on temperature profiling and almost no observation of wind profiling in the lower atmosphere. Operation of RASS for large scale experiments will provide much needed data on basic atmospheric parameters in the boundary layer. Hopefully, a selective network of boundary layer RASS complemented by satellite based microwave radiometer system would become a future choice for temperature profiling to fulfil the data needs of the forecasting models [80, 81]. The concept of integrated sounding system that combines state-of-the-art remote and in-situ sensors including RASS into a single transportable facility has been introduced during TOGA-COARE [82]. The RASS is proving to be a multifunctional tool not only for meteorological studies but also for applications in the boundary layer studies and in air pollution dispersion studies. The perameters

Table 1. System details alongwith range covered for the various Radio Acoustic Sounding Systems (RASS) developed over the world

SITE (Location and Year)	RADAR Characteristics							Acoustic characteristics				
	Freq. MHz	Wave Length m	Antenna Size/ Separation	Beam width	Vertical Range Resolution	Mean Power	Radar Type	Frequency Hz / Wavelength m	Beam-Width / Pules Width	Electrical/ Acoustic Power	Mode/Antenna	Temp. Accuracy	Min./Max. Altitude	Ref
Stanford Univ. 1969–70	36.8	8.15	2 Yagi / 73.2 m	45°	—	320 w	Pulsed	85 Hz / 4.07 m	16.8° / 100	85 w (El) / 6.5 w (AC)	Pulsed / 3 × 3 Array (15 m)	0.5°C Max. Error	600 m / 1500 m	[3]
Stanford Univ. 1972–75	36.8	8.15	2 × 2 Yagi / 50 m	30°	—	8 w	CW	85 Hz / 4.07 m	24° / Variable	70 w (Ac)	Pulsed / 3 × 3 Array (15 m)	0.45°C	400 m / 3000 m	[5]
Stanford Research Institute 1976	440	0.68	2 Helix / 60 m	45°	—	1.0 w	CW	1000 Hz / 0.34 m	10°	340 w (Ac)	Pulsed / 10 × 10 Array	—	200 m – 1100 m	[7]
TURBIGO Metric RASS	159	1.88	40 m² / 60 m	30°	20–30 m	20 w	CW	360 Hz / 0.94 m	25° upto 0.5 sec	1500 W (El) / 150 w(Ac) to 200 w (Ac)	Pulsed / 3 × 3 Array of folded horns each having 6 loudspeakers	0.2°C	80 m / 1.0 km	[24]
Russian RASS	1025	0.3	2.5 × 3.5 = 7 m / 8 m	—	—	5 w–10.0 w	CW	2300 Hz / 0.15 m	0.1 sec	6 w (Ac)	Pulsed Parabolic and spiral	0.5°C	50 m / 800 m	[20]
Hamburg (German) RASS	593.5	0.5	60 m² / Parabolic dish 9 m	8°	150 m	30 w	Pulsed	1250–1400 Hz / .272 – .242 m	6.5°	20 w (Ac)	CW	—	100 m / 1 Km	[31]
RRL RASS Japan 1977–79	445	0.67	Parabolic dish 3 m / 40 m	15°	20–30 m	100 w	CW	998 – 140 Hz / 0.34 m	15° / 50 – 200 ms	300 w (El) / 4.6 w (Ac)	Pulsed / 1.5 m φ with 50 w × 4 driver units	0.6°C	Ground to 2.0 km	[34]
RRL RASS Japan 1983	890.5	0.335	—	—	—	10 w	CW 36 receivers	2000 Hz / 0.17 m	—	300 w (El)	Pulsed / 1.5 m φ	—	Ground to 1200 m	[35]

System	Freq (MHz)		Area			Power	Type	Freq / Wavelength	Beam	Acoustic Power	Acoustic Source		Range / Ref
JRC RASS Japan 1987	915	0.32	1.2 m–10.7 m	20°/29°	—	10 w	CW 36 receivers	$\dfrac{2100 \text{ Hz}}{0.162 \text{ m}}$	$\dfrac{12°}{120 - 200 \text{ ms}}$	—	$\dfrac{\text{Pulsed}}{1 \text{ m}\phi}$	—	$\dfrac{\text{Ground to}}{1000 \text{ m}}$ [37]
Stugaraki RASS Japan 1986	46.5	6.4	8330 m²	3.6°	300 m	1 MW	MU Radar Pulsed Doppler Radar	$\dfrac{88 - 102 \text{ Hz}}{3.2 \text{ m}}$	$\dfrac{45°}{15}$	100 w (Ac)	Pulsed, pneumatic acoustic generator	—	$\dfrac{2.0 \text{ km}}{20 \text{ Km}}$ [36]
Platteville USA	49.8	6.0	10,000 m²	3°	300 m	200 w	Pulsed	$\dfrac{110 \text{ Hz}}{3.09 \text{ m}}$	60°	50 w (Ac)	$\dfrac{\text{FM-CW}}{\text{Parabolic dish 3 m }\phi}$ fed from an array of 4 transducers	0.2°C	$\dfrac{2.1 \text{ km}}{5 - 9 \text{ km}}$ [11]
Eric USA	404.37	0.7414	25 m²	7.8°	250 m	30 w	Pulsed	$\dfrac{900 \text{ Hz}}{0.378 \text{ m}}$	18°	5 w (Ac)	$\dfrac{\text{FMCW}}{\text{Parabolic dish 1.3 m }\phi}$	—	$\dfrac{400 \text{ m}}{3 - 5 \text{ km}}$ [11]
Denver USA	915	0.32	100 m²	2°	150 m	100 w	Pulsed	$\dfrac{2000 \text{ Hz}}{0.17 \text{ m}}$	8°	5 w (Ac)	$\dfrac{\text{FMCW}}{\text{Parabolic dish 1.3 m }\phi}$	—	$\dfrac{200 \text{ m}}{0.7 - 1.5 \text{ km}}$ [11]
Boulder USA	915	0.32	4 m²	9°	150 m	3 w	Pulsed	$\dfrac{2000 \text{ Hz}}{0.17 \text{ m}}$	9°	5 w (Ac)	$\dfrac{\text{FMCW}}{\text{Parabolic dish 1.3 m }\phi}$	—	$\dfrac{120 \text{ m}}{0.6 - 1 \text{ km}}$ [11]

which can be measured include u, v, w, T, their variances, heat flux and momentum flux. The data for RASS combined with a wind profiler allows one to probe the vertical structure and the accompanied termporal changes in the boundary layer that occur with the passage of an outflow, and to trace the boundary layer recovery process. It can fill up several gaps in the existing in-situ and remote sensing meteorological probes, namely,

(i) Traditionally used meterological tower measurements with in-situ sensors used for boundary layer parameterisation can measure upto a certain height only. Data need to be extrapolated with model assumptions to cover ABL.
(ii) Radiosonde which does provide altitude coverage gives two or at the most four observations in 24 hrs. More number of radiosonde operations become expensive and it also does not provide finer data in the lowest regions needed for ABL paramerisation.
(iii) Sodar and FM radar used for profiling do not yield complete picture because their operation is conditioned by turbulence generally associated with a discontinuity in the thermal gradient. The radar also has a limitation in the lower region due to presence of ground clutter.
(iv) A RASS has an advantage as it can provide continuous monitoring of ABL and measure atmospheric parameters at 15 min intervals for obtaining comparable height range as that of a Doppler sodar, and also it uses lower acoustic power.

9. Conclusions

In the recent years the RASS method for measuring vertical profiles of the virtual temperature and wind speed has been tested world wide in meteorological studies as well as in weather prediction and has been found acceptable. It is thus becoming an important tool in the development of next generation upper-air observing systems.

A large body of literature exists describing theoretical and experimental verification studies of the technique. However, so far, the capabilities are best demonstrated for measuring temperature profiles upto 3 to 5 km height, with a range resolution of 100 to 250 m. For the application in routine weather observing systems, a combination of RASS with in-situ sensors and/or satellite radiometery has been proposed for improving retrieved temperature soundings, particularly in the lower troposphere.

The accuracy of RASS temperature measurements is limited by the radar ability to measure vertical wind, by the effect of horizontal wind and turbulence as well as other environmental parameters and approximations. One can apply various corrections due to structure peculiarities of the system and atmospheric parameters. In practice, several factors can simultaneously influence the echo power and it is virtually impossible to isolate the contribution from a single parameter. The major factors are vertical wind, that can affect the accuracy to a large extent, followed by turbulence. In rainy season, further limitations are introduced by precipitation echoes which can prevent measurements of vertical air motion, so the corrections are limited to non-precipitating conditions. Studies conducted during TOGA COARE [83] revealed the nature of contamination. The 915 MHz RASS data appeared to be greatly affected by precipitation and represented typical difficulty in finding algorithms and in obtaining a representative mean wind in the presence of convective precipitation with sharp gradients in vertical motion and horizontal winds.

As far as second order quantities are concerned, such as temperature and wind variances and heat flux, relatively few studies have been made. Only preliminary results of measurement of heat flux, momentum flux and water vapour flux using RASS [54, 75–77] compared with in-situ sensors mounted on towers upto 250 m, and other remote sensors as well as onboard aircraft are available. From these it is evident that more studies need to be made to identify potential sources of errors especially in flux measurement studies and to verify the theoretical analysis.

Acknowledgements

Authors thank Dr. R.G. Strauch and Dr. R.J. Lataitis for providing, the NOAA Technical Memorandum ERL WPL-230 and some of the other recently published papers on the subject as also helping us to review this chapter through their critical comments. The second author expresses her grateful thanks to Prof. V.S. Ramamurthy, Secretary, Department of Science and Technology, Govt. of India for the encouragement.

References

1. Smith, P.L. Jr., "Remote measurement of wind velocity by the electromagnetic acoustic probe, I: System analysis," Rept. No. 419, 50 pp, Midwest Research Institute, Kansas City, MO, USA, 1961.
2. Fetter, R.W., "Remote measurement of wind velocity by the electromagnetic acoustic probe, II: Experimental system," Rept. No. 420, 50 pp., Midwest Research Institute, Kansas City, MO, USA, 1961.
3. Marshall, J.M., "A radio acoustic sounding system for the remote measurement of atmospheric parameters," SU-SEL-70-050, 139 pp, Radio Science Laboratory, Stanford Electronics Laboratory, Stanford university, USA, 1970.
4. Marshall, J.M., Peterson, A.M. and Barnes, A.A. Jr., "Combined Radar-Acoustic sounding System," Appl. Opt., 11, 108–112, 1972.
5. North, E.M., Peterson, A.M. and Pang, H.D., "RASS, a remote sensing system for measuring low level temperature profiles," Bull. Amer. Meteorol. Soc., 54, 912–919, 1973.
6. Frankel, M.S. and Peterson, A.M., "Remote temperature profiling in the lower atmosphere," Rad. Sci., 11, 157–166, 1976.
7. Frankel, M.S., Chang N.J.F. and Sanders, M.I. Jr., "A high frequency radio acoustic sounder for remote measurement of atmospheric winds and temperature," Bull. Amer. Meteorol. Soc., 58, 928–934, 1977.

8. May P.T., Strauch, R.G. and Moran, K.P., "The altitude coverage of temperature measurements using RASS with wind profiler radars," Geophys. Res. Lett., 15, 1381–1384, 1988.
9. May, P.T., Strauch, R.G., Moran, K.P. and Neff, W.D., "High resolution weather observations with combined RASS and wind profilers," Proc. 24th Conference on Radar Meteorology, pp. 746–749, Amer, Meteorol, Soc., Boston, MA, USA, 1989.
10. May, P.T., Moran, K.P. and Strauch, R.G., "The accuracy of RASS temperature measurements," J. Appl. Meteorol, 28, 1329–1335, 1989.
11. May, P.T, Strauch, R.G., Moran, K.P. and Ecklund W.L., "Temperature sounding by RASS with wind profiler radars: A preliminary study," IEEE Trans. Remote Sens., 28, 19–28, 1990.
12. Strauch, R.G., Moran, K.P., May, P.T., Bedard, A.J., Jr. and Ecklund, W.L., "RASS temperature sounding techniques," NO. AA Technical Memorandum ERL WPL-158, 22 pp. NOAA, Environmental Research Labs., Boulder, Colo. USA, 1988.
13. Strauch, R.G., Moran, K.P., May, P.T., Bedard, A.J., Jr. and Ecklund, W.L., "RASS temperature soundings with wind profiler radars," Proc. 24th Conference on Radar Meteorology, pp 741–745. Amer. Meteorol. Soc. Boston, AM. USA, 1989.
14. Schroeder, J.A., "A comparison of temperature soundings obtained from simultaneous radio-metric, radio-acoustic and rawin sonde measurements," J. Atmos. Oceanic Technol., 7, 495–503, 1990.
15. Moran, K.P., Wuertz, D.B., Strauch, R.G., Abshire, N.L. and Law, D.C., "Rass demonstration on a NOAA network wind profiler". NOAA Tech. Memo. ERL WPL-184, 24pp. NOAA, Environ. Res. Labs., Boulder, Colo., USA, 1990.
16. Moran K.P., Strauch, R.G. and May, P.T., "Lower tropospheric temperature profiling," Preprints 25th Conference on Radar Meteorology, pp. 237–24, Amer. Meteorol. Soc., Boston, M.A., USA., 1991.
17. Moran, K.P. and Strauch, R.G, "The accuracy of RASS temperature measurements corrected for vertical air motions," J. Atmos. Oceanic Technol. 11, 995–1001, 1994.
18. Makarova, T.I., "Non linear absorption of sound in problems of acoustic and radio acoustic sounding of the atmosphere," Izv. Acad. Sci. USSR Atmos. Oceanic Phys., 16, 123–125, 1980.
19. Makarova T.I., "Measurement of temperature profiles in the surface layer of the atmosphere by radar-acoustic sounding," Izv. Acad. Sci., USSR Atmos. Oceanic Phys., 16, 453–455, 1980.
20. Azizyan, G.V., Belyavskaya, V.M., Gorelik, A: G., Yegorov, M. Yu., Kallistratova, M.A., Karyukin, G.A. and Knyazev, L.V., "An experiment in temperature profiling in the lower troposphere by radio-acoustic sounding," Izv. Acad. Sci. USSR Atmos. Oceanic Phys., 17, 112–116, 1981.
21. Azizyan, G.V., "The frequency spectrum of the scattered signal in radar acoustic atmospheric sounding systems," Izv. Acad. Sci. USSR Atmos. Oceanic Phys., 17, 657–659, 1981.
22. Babkin, S.I., Prashkin, E.G. and Ul' yanov, Yu. N., "Experimental results of temperature-wind atmospheric sounding by the radioacoustic method," Izv. Acad. Sci. USSR Atmos. Oceanic Phys., 20, 470–474, 1984.
23. Gurvich A.S., Kon, A.I., and Tatarskii, V.I., "Scattering of electromagnetic waves on sound in connection with problems of atmospheric sounding (review)," Radio Phys. Quant. Electron., 30, 347–366, 1987.

24. Bonino, G., Lombardini, P. and Trivero, P., "A metric wave radio-acoustic troposphere sounder," IEEE Trans. Geosci. Electron., GE-17, 179–181, 1979.
25. Bonino, G., Lombardini, P. and Trivero, P., "Comparison of RASS temperature profiles with other tropospheric soundings," Nuovo Cimento, 1 c(3), 207–214, 1980.
26. Bonino, G., and Trivero, P., "Automatic tuning of Bragg condition in a radio-acoustic system for PBL temperature measurement," Atmos. Environ., 19, 973–978, 1985.
27. Bonino, G. Elisei, G., Marzorati, A. and Trivero, P., "Results on planetary boundary layer sounding by automatic RASS," Atmos. Res., 20, 309–316, 1986.
28. Bonino, G., Trivero, P., Elisei, G. and Marzorati, A., "On the sounding range of a 1m wavelength radio-acoustic sounding," J. Atmos. Oceanic Technol., 6, 851–855, 1989.
29. Bonino, G., Lombardini, P., Longhetto, A. and Trivero, P.L., "Radio-acoustic measurement of fog-capping thermal inversions." Nature, 290, No. 5802, 121–123, 1981.
30. Peters, G., Timmerman, H. and Hinzpeter, H., "Temperature sounding in the planetary boundary layer by RASS-System analysis and results," Int. J. Remote Sensing, 4, 49–63, 1983.
31. Peters, G., Hinzpeter, H., and Baumann, G., "Measurements of heat flux in the atmospheric boundary layer by SODAR and RASS: A first attempt," Radio Sci., 20, 1555–1564, 1985.
32. Peters, G., Hasselmann, D. and Pang, S., "Radio-acoustic sounding of the atmosphere using a FM-CW radar," Radio Sci., 23, 640–646, 1988.
33. Peterman, K.R., Riese, C.E., Batson, D.T., Eaton F.D. and Chadwick, R.B. "FM-CW Radar-Acoustic Sounding system", Proc. Acous. Remote. Sensing, Tata McGraw Hill, pp. 104–114, 1990.
34. Fukushima, M., Akita, K. and Masuda, Y., "Development of radio acoustic sounding system (RASS)," Collected Research Results on Environmental Preservation, pp. 2, Radio Research Labs. Japan pp. 102-1 to 102–12, 1978.
35. Matuura, N., Msuda, Y., Fukushima, M. and Akita, K., "Development of a radio-acoustic sounding system (RASS) for wind and temperature, "Proc. Intl. Symp. on Acoustic Remote Sensing Rome, Italy, pp. XL I–10, 1983.
36. Matuura, N., Masuda, Y., Inuki, H., Kato, S., Fukao, S. and Tsuda, T., "Radio-acoustic measurement of temperature profiles in the troposphere and stratosphere," Nature, 323, 426–428, 1986.
37. Fukushima, M., "Received signal characteristics of a radio-acoustic sounding system (RASS). Influence of horizontal winds for temperature measurements," Tran. of the IEICE, E70, 476–483, 1987.
38. Masuda, Y., "Influence of wind and temperature on the height limit of a radio-acoustic sounding system," Radio Sci., 23, 647–655, 1988.
39. Tsuda, T., Masuda, Y., Inuki, H., Takahashi, K., Takami, T., Sato, T., Fukao, S. and Kato, S., "High time resolution monitoring of the tropospheric temperature with a radioacoustic sounding system (RASS), "PAGEOPH, 130, 497–507, 1989.
40. Kato, S. Ogawa, T., Tsuda, T., Sato, T., Kimura, I. and Fukao, S., "The middle and upper atmospheric radar: First results using a partial system," Radio Sci., 19, 1475–1484, 1984.
41. Reddi, C.R., DST Core Project Report on Indian RASS Programme, DST Report, 1992.

42. Kon, A.I. and Tatarskii, V.I, "The scattered signal frequency spectrum for radio acoustical atmospheric soundings," Izv. Acad. Sci., USSR Atmos. Oceanic Phys., 16, 142–148, 1980.
43. Kon, A.I. and Tatarskii, V.I. "The power of a signal under conditions of radioacoustic sounding of the atmosphere," Radioteknika Electronika, 10, 1903–1908, 1986.
44. Kon, A.I., "A bistatic radar acoustic atmospheric sounding system," Izv. Acad. Sci. USSR, Atmos Oceanic Phys., 17, 481–484, 1981.
45. Vogt, S., "Advances in RASS since 1990 and Practical Application of RASS to Air Pollution and ABL studios", Proc. of 8th Intl. Symp on Acoustic remote Sensing and Associated Techniques of the Atmosphere & Oceanic (ISARS, 96, 27–31 May 1996, Mocow, Russia.
46. Karyukin, G.A., "Influence of wind on operation of radar acoustic atmospheric sounding systems," Izv. Acad. Sci. USSR, Atmos. Oceanic Phys. 18, 26–30, 1982.
47. Nalbandyan, O.G., "Scattering of electomagnetic waves from a sound wave propagating in a turbulent atmosphere," Izv. Acad. Sci., USSR Atmos Oceanic Phys. 12, 877–880, 1976.
48. Clifford, S.F., Ting-i Wang, and Priestley, J.T., "Spot size of the radar return from a radar-acoustic sounding system (RASS) due to atmospheric refractive turbulence", Radio Science Vol. 13, No. 6, pp. 985–989, 1978.
49. Kallistratova, M.A. and Kon, A.I., "Distance limits in radio-acoustic sounding of the atmophere," Izv. Acad. Sci., USSR Atmos Oceanic Phys., 19, 956–959, 1983.
50. Nalbandyan, O.G., "The theory of radioacoustic sounding of the atmosphere," Izv. Acad. Sci. USSR Atmos. Oceanic Phys., 13, 172–177, 1977.
51. Bhatnagar, N. and Peterson, A.M., "Interaction of electomagnetic and acoustic waves in a stochastic atmosphere," IEEE Trans. Antenna Propar., AP-27, 385–393, 1979.
52. Kon, A.I., "Combined effect of turbulence and wind on the signal intensity in radio-acoustic sounding of the atmosphere," Izv. Acad. Sci., USSR Atmos. Oceanic Phys., 21, 942–947, 1948.
53. Takahashi, K., Masuda, Y., Matuura, N., Kato, S., Fukao, S., Tsuda, T. and Sato, T., "Analysis of acoustic wavefronts in the atmosphere to profile the temperature and wind with a radioacoustic sounding system." J. Acoust. Soc. Amer., 84, 1061–1066, 1988.
54. Lataitis, R.J. "Theory and application of a radio-acoustic sounding system (RASS)," NOAA Tech. Memo ERL WPL-230, 207 pp, Wave Propagation Laboratory, Boulder, Colo., USA, 1993.
55. Brown, E.H. and Clifford, S.F., "On the attenuation of sound by turbulence," J. Acoust Soc. Amer., 60, 788–794, 1976.
56. Lataitis, R.J., "Signal Power for radio acoustic sounding of temperature: The effects of horizontal wind, turbulence, and vertical temperature gradients", Radio Science, Vol. 27, pp. 369–385, 1992.
57. Harris, C.M., "Effects of Humidity on the Velocity of sound in Air". The Acoustic Society of America, Vol. 49, No. 3 (part 2), pp. 890–893, 1971.
58. Angevine, W.M., Ecklund, W.L., "Errors in Radio Acoustic Sounding of Temperature", Atmospheric and Oceanic Technology, Vol. 11, pp. 837–842, 1994.
59. Cramer, O., "The variation of the specific heat ratio and the speed of sound in air with temperature, pressure, humidity and CO_2 concentration", J. Acoust. Soc. Am., Vol. 93, No. 5, pp 2510–2516, 1993.

60. Peters, G., Angevine, W.M., "On the Correction of RASS Temperature errors due to turbulence" 27th Conf. on Radar, Meteorology, Colorado, pp. 317–319, Oct. 9–13, 1995.
61. Bauer, M., and Peters, G., "On the Altitude Coverage of Temperature Profiling by RASS", Appl. Meteor., pp. 487–489, 1993.
62. Belyavskaya, V.D., Kallistratova, M.A., Karyukin, G.A., and Petenko, I.V., "Real Heights of Radio Acoustic Sounding in the Decimeter Wavelength Range", Izvestiya, Atmospheric and Oceanic Physics, Vol. 20, No. 4, pp. 277–279, 1984.
63. Steinhagen H. and Neisser J., "Improvements of the Altitude Measurements using RASS, ISARS'96, May 27–31, 1996. Moscow, Russia.
64. Steinhagen, H., Christoph, A., Ritoph, W.S., Gordorf, U., Lippmann, J., Neisser, J., "A 1290 MHz. Wind Profiler with RASS for the boundary layer", 3 Intern. Symp. Trop. Profiling Vol. 2 (2), 343–346, 1996.
65. Goel, Malti, "Potential of RAS system in India," National Symposium on Engineering Application of Acoustics, National Physical Laboratory, New Delhi, Oct. 30–Nov. 1, 1991.
66. Jordan, J.R. and Lataitis, R.J., "Wind estimation using an acoustically enhanced five beam RASS system," Proc. ISARS'94, Boulder Colorado, USA, pp. 367–371, 1994.
67. Neiman, P.J., May, P.T., Stankov, B.B., and Shapiro, M.A., "Radio Acoustic Sounding System Observations of an Arctic Front", Appl. Meteor. Vol. 30, pp. 881–892, 1991.
68. Bonino, G. Longhetto, A., Majorati, A., and Trivero, P.L. "A case study of winter fog developments monitored by RASS and Soder," Proc. of 8th Intl. Symp on Acoustic Remote Sensing and Associated Techmiques of the Atmosphere and Oceans, ISARS'96, 27–31 May, 1996, Moscow, Russia.
69. Martner, B.E., Wuertz, D.B., Stankov, B.B., Strauch, R.G., Westwater, E.R., Gage, K.S., Ecklund, W.L., Martin, C.L., and Dabberdt, W.F., "An evaluation of Wind Profiler, RASS, and Microwave Radiometer Performance", Bull. Amer. Meteor. Soc. Vol. 74, pp. 599-613, 1993.
70. Schroeder, J.A., Westwater, E.R., May, P.T. and McMillan, L.M., "Prospectus for temperature sounding with satellite and ground based RASS measurements," J. Atmos. Oceanic Tech., 8, pp. 506–513, 1991.
71. Peters G., and Angevine, W.M., "On the Correction of RASS temperature errors due to turbulence", Beitr. Phys. Atmos., 69, No. 1, p 81–96, (1996).
72. Angevine, W.M., Ecklund, W.L., Carter, D.A., Gage, K.S., and Moran, K.P., "Improved Radio Acoustic Sounding Techniques", Atmospheric and Oceanic Technology, Vol. 11, pp. 42–49, 1994.
73. Makarova, T.I., "Wind velocity measurements with double beam radio-acoustic sounding",: Izv. Acad. Sci., USSR Atmos. Oceanic Phys., 16, 366–368, 1980.
74. Neff, W.D., "Studies of variability in the tropospheric and atmospheric boundary layer over the South Pole: 1993. Experimental design and preliminary results," Antarctic Journal of the United States, 1994–95.
75. Angevine, W.M., Avery, S.K., and Kok, G.L., "Virtual Heat Flux Measurements from a Boundary-Layer Profiler-RASS Compared to Aircraft Measurements", Applied Meteorology, Vol. 32, pp. 1901–1907, 1993.
76. Hirch, L., Peters, G., "Momentum flux profiles in the PBL measured with a Radar RASS wind profiler", Feb. 1996.

77. Senff, C., Bosenberg, J. and Peters, G., "Measurements of water vapour flux profiles in the convective boundary layer with lidar and radar—RASS", J. Atom. Oceanic Tech. **11**, 85–93, 1994.
78. Goel, Malti, "Potential of Radio Acoustic Sounding System in India," J. Acous. Soc. of India, Vol. 19, No. 1, p 1–15, 1991.
79. "A Feasibility study on utilisation of Radio Acoustic Sounding System in WAT (Wind and Temperatures) Profiling", DST Technical Publication, 1991.
80. Goel, M. and Ramnath, S., "Radio acoustic remote sensing of temperature: a possibility under MONTBLEX," Proc. of International Symposium on Remote Sensing of the Atmospheric Environment, National Physical Laboratory, New Delhi, pp. 293–294, Oct. 24–26, 1990.
81. Goel, Malti, "Feasibility of radio acoustic sounding system for measuring crucial atmospheric parameters," Brainstorming Session on Atmospheric Technology and Instrumentation, Calcutta University, Calcutta, Sept. 21–22, 1991.
82. Riddle, A.C., Angivine, W.M., Ecklund, W.L., Miller, E.R., Parsons, P.D.B., Carter, D.A. and Cage K.S. "In-itu remotely sensed horizontal winds and temperature intercomparisons obtained using integrated sounding system during TOGA COARE", Contri. to Atmos Phys. in Procs. 1996.
83. Parsons, D., Dabberdt, W., Cole, H., Hock, T., Martin, C., Barrett, A.L., Miller, E., Spowart, M., Howard, M., Ecklund, W., Carter, D., Gage, K., and Wilson, J., "The Integrated Sounding System Description and Preliminary Observations from TOGA COARE", Bull. Amer. Meteor. Soc., Vol. 75, No. 4 pp. 553–567, 1994.

Part TWO
Applications in the Atmosphere

8. Determination of the Turbulent Structure Parameters

Pan Naixian
Department of Geophysics, Peking University, Beijing, China

Introduction

Through the evolution of sodar technique, today sodars have been developed into sophisticated equipments capable of not only displaying echo intensity on facsimile charts to show the temperature structure in boundary layer, but measuring Doppler wind velocities and some turbulence parameters as well. The advancement in measuring the turbulence parameters attributed to individual researchers' efforts started from the mid 70s, not long after the first sodar prototype was produced [1–11].

The coefficient of temperature structure is a significant parameter in boundary layer physics. Neff [1] carried on the measurement of C_t^2 in 1975. In his experiment, a comparison between sodar measurements and tower instruments measurements was given at one level. In his case study, the ratio of C_{ta}^2/C_{td}^2* varied from 2 to 0.5, while the atmospheric stratification changed from stable to unstable. Statistically the ratios of C_{ta}^2/C_{td}^2 are about 0.7 under unstable conditions and 1.16 under stable conditions. Asimakopoulos et al [2] got more or less a similar result: the ratios of C_{ta}^2/C_{td}^2 were about 0.86 under unstable conditions and 1.43 under stable conditions. Later Haugen et al, [4] made a comparison at three levels, but the differences between sodar and tower instrumentation measurements were large. The largest error may reach to 400% or more. He attributed the large differences to attenuation of turbulence and beam bending. He suggested using the bistatic sodar to measure C_t^2 and C_v^2 for estimation of the excess attenuation. For verifying the deductions of Haugen et al and Brown et al [12], Neff [13] carried out an experiment in which the sodar worked at two frequencies 1250 Hz and 2500 Hz. The experiment showed that the return signals get more rapid fading with narrower beams (i.e., that corresponding to the higher frequency). Also much reduced average

---

*C_{ta}^2 are measured by sodar and C_{td}^2 are measured by tower instruments.

amplitudes were received with narrower beams when winds got strong. Neff attributed these phenomena to the excess attenuation and wind effects.

The coefficient of velocity structure is another important parameter. Some methods for determining C_v^2 were sugested and relative experiments for comparison of the sodar values with the direct measurements have been done [3, 5, 6]. The results show the same general trend but the differences between the two values are often large.

Among the echo-sounders, sodar is more sensitive than lidar and radar [14], but it follows that the attenuation caused by turbulence in sodar is much more serious than in the other echo-sounders. The transmission of acoustic wave as a mechanical wave is strongly influenced by wind while optic and radio waves are not. An error factor of two in C_t^2 measurements was considered to be not very serious as the values of C_t^2 and C_v^2 vary by several orders in the atmosphere. However, if the accuracy can not be enhanced, their applicable value will be reduced. It is necessary to treat the errors caused by wind and turbulence in more detail, then we will have more accurate parameters of turbulence.

Scattering Theory of Acoustic Wave and Some Problems

The scattering cross section of acoustic waves was deducted under the condition of homogeneous and isotropic random fields as follows [15]:

$$\sigma(\theta) = \frac{\pi}{2} k^4 \cos^2 \theta \left[\frac{\Phi_T\left(2k \sin \frac{\theta}{2}\right)}{T^2} + \cos^2 \frac{\theta}{2} \frac{\Phi_E\left(2k \sin \frac{\theta}{2}\right)}{\pi C^2 \left(2k \sin \frac{\theta}{2}\right)^2} \right] \quad (1)$$

where θ is the scattering angle, k is the wave number, T is the air temperature, C is the sound speed, Φ_T and Φ_E are spectral functions of temperature and turbulence respectively. Equation (1) shows that the acoustic wave scattering is produced by the inhomogeneities in temperature field and velocity field and only the eddies whose scales satisfy to

$$l(\theta) = \frac{\lambda}{2 \sin \frac{\theta}{2}} \quad (2)$$

are effective for scattering at θ direction. That is the so called Bragg's condition. Here we need mention that in deduction of Eq. (1), the acoustic wave is assumed to be a plane wave.

In local homogeneous and isotropic random fields, Eq. (1) was reduced to a convenient expression for practical use by using the Kolmogorov's spectrum:

$$\sigma(\theta) = 0.033 k^{1/2} \cos^2 \theta \left[\frac{C_v^2}{C^2} \cos^2 \frac{\theta}{2} + 0.136 \frac{C_t^2}{T^2} \right] \left(\sin \frac{\theta}{2} \right)^{-11/3} \quad (3)$$

Eq. (3) holds true only under the condition of $l_0 \ll l(\theta) \ll L_0$, where l_0 is called the inner scale of turbulence and L_0 is the outer scale of turbulence. There are some weak points in Eq. (1) and Eq. (3) for practical use. One is $\sigma(\theta) \to \infty$ while $\theta \to 0$. It is unreasonable in physics. The other is that not all the scattering angles, θ, fit Eq. (3). The $l(\theta)$ varies inversely as θ. When θ become small enough then $l(\theta)$ is not much smaller than the outer scale of turbulence L_0. Therefore Eq. (3) will not hold true. This limitation will not affect the use of Eq. (3) in conventional monostatic and bistatic sodar but will cause trouble in estimation of excess attenuation with the following expression:

$$\alpha_e = \int_0^{2\pi} \int_0^{\pi} \sigma(\theta) \sin\theta \, d\theta \, d\varphi \qquad (4)$$

The integral in Eq. (4) is divergent.

The Estimation of Excess Attenuation

In fact, the outer scale of turbulence is limited, then the minimum scattering angle is determined by

$$\theta_{min} = 2 arc \sin\left(\frac{\lambda}{2L_0}\right) \qquad (5)$$

So Eq. (4) becomes

$$\alpha_e = \int_0^{2\pi} \int_{\theta_{min}}^{\pi} \sigma(\theta) \sin\theta \, d\theta \, d\varphi \qquad (6)$$

Eq. (6) avoids the divergence of integral at $\theta = 0$. Two questions remain to be solved for Eq. (6). First, how to determine the outer scale of turbulence, L_0? Weber [16] suggested

$$L_0 = 2\pi U \int_0^{\infty} R(\tau) \, d\tau \qquad (7)$$

where U is the average wind speed and $R(\tau)$ is the autocorrelation coefficient of wind speed. Theoretically we have $R(\tau) \to 0$, when $\tau \to \infty$. But, in reality, when τ is large enough $R(\tau)$ will not be zero and will take different values. Whether we take the average of L_0 or not, L_0 may not have clear sense for Eq. (6) in real situations. Quintarelli [17] suggested using the sodar data to determine L_0. Using only the section of autocorrelation coefficient curve at small lags he got more consistent values of L_0 than by the spectral method. But he did not show the variation of L_0 with time. We must choose the value of L_0 carefully, because forward scatter is very strong and Eq. (6) is sensitive to L_0, that is $\alpha_e \propto L_0^{5/3}$ [18].

Second, when $l(\theta)$ becomes large enough and goes out of the inertial subrange, Eq. (3) will not hold true. These two problems are difficult to solve in practice.

Generally, C_T^2 will be overestimated if a realistic L_0 of atmosphere is used in Eq. (5) and (6).

The beam width of commercial sodar is about 10 degree and the detectable level is about 1 to 2 kilometers high. Treating such a finite width beam as a plane wave, which has an infinitive wave front, is unreasonable. In a finite width beam, the Bragg's condition, Eq. (2), is limited by the diameter of the beam. In other words, the scales of eddies, which scatter the acoustic waves, are limited by cross section of the beam. Those eddies whose scales are larger than or close to the diameter of the beam will not scatter the acoustic wave. In optics, the diffractive interfering spectrum can not be produced if the light passes through a "grating" whose grate is less than two. Here the situation is similar. Intead of L_0 in Eq. (5), Pan and Liu [18] defined an effective turbulent outer scale, L_e, as follows:

$$L_e = R \tan \frac{\gamma}{2} \qquad (8)$$

Here R is the distance from the antenna to scattering volume and γ is the beam angle. Then the minimum scattering angle θ_{min} in Eq. (6) is limited by L_e, not by the L_0. Substituting Eq. (3) into Eq. (6) and calculating Eq. (6), we have

$$\alpha_e = 0.00206 k^{1/3} \frac{C_v^2}{T} (15 m^{1/3} + 0.6 m^{-5/3} - 13.09)$$

$$+ 0.113 k^{1/3} \frac{C_t^2}{T^2} (12 m^{1/3} + 0.6 m^{-5/3} - 10.89) \qquad (9)$$

where $m = \lambda/2L_e$ and the smallest terms are neglected.

Now, the largest eddies which can scatter the acoustic wave in the beam satisfy the condition of $l_0 \ll l(\theta_{min}) \ll L_0$.

The results of calculation with Eq. (9) show that for the same C_t^2 and C_v^2, the larger the distance from the acoustic source, the larger is the coefficient of excess attenuation; the smaller the beam angle, the smaller is the coefficiant of the excess attenuation. For sodar case the back scattering waves suffer more excess attenuation during its way to the antenna than the transmitted wave from the antenna running the same distance. For the possible values of C_t^2 and C_v^2 in the atmosphere, the first item in Eq. (9) is larger than the second item and is about 4.5 to 5.5 times when C_t^2 equals to C_v^2. Considering that C_v^2 is larger than C_t^2 about one or two orders in most cases, we can generally say, turbulence of velocity is the main factor in excess attenuation but turbulence of temperature is not. Additionally, calculation of Eq. (9) also shows that the excess attenuation has a relation with wave length as

$$\alpha_e \propto \lambda^{-2} \qquad (10)$$

The Determination of C_v^2

There are some approaches to determine C_v^2 or equivalently the rate of dissipation of turbulent kinetic energy ε.

The main approaches are as follows. The first one is based on Eq. (3). When scattering angle θ equals π, C_v^2 has no contribution to scattering cross section. The echo intensity depends on C_t^2 only, that is the monostatic mode of sodar system. The contribution of C_v^2 in Eq. (3) increases as θ departs from π. Chosing a proper θ we can measure the echo intensity in bistatic mode. Joining these two modes we can seperate C_v^2 and C_t^2 from each other. Thomson et al, [5] measured the profiles of both C_v^2 and C_t^2 with a combined monostatic-bistatic system. They pointed that values of C_v^2 exceed C_t^2 by one or two orders, therefore we can possibly determine C_v^2 by using bistatic system only. Caughey et al, [6] did in the same way and using the relation of

$$\varepsilon = (C_v^2/1.97)^{2/3} \tag{11}$$

got the time series of ε.

The second approach is determining C_v^2 directly from the vertical velocity spectrum. Under the condition of Kolmogorov's local homogeneous isotropic random field, the one dimension spectrum of energy can be deduced from structure function as follows [19]:

$$E(\kappa) = \frac{1}{2\pi} \Gamma(5/3) \sin(\pi/3) C_v^2 \kappa^{-5/3} \tag{12}$$

But this requires a fast sampling rate to cover enough length of inertial subrange in the velocity spectrum. For most sodars it is impossible to do so. Desbraux and Weill [20] and Underwood and Coulter [21] used a relatively high sampling rate in their equipments and measured C_v^2 with the help of spectral models [22, 23].

The third approach is determining C_v^2 from the variance of echo spectrum. The variance of vertical velocity spectrum of sodar echo, σ_{tu}^2, is related with the turbulent energy in the scattering volume as follows [24]:

$$\sigma_{tu}^2 = \langle V'^2 \rangle \tag{13}$$

The one dimension spectrum can also be expressed by von Kármán spectrum as

$$E(\kappa) = \Gamma(5/6) L_0 \langle V'^2 \rangle / (3\sqrt{\pi}\Gamma(1/3)(1 + \kappa^2 L_0^2)^{5/6}) \tag{14}$$

Combination of Eq. (12) with Eq. (14) and let $\kappa = 1$, yields

$$C_v^2 = 0.636 L_0 \langle V'^2 \rangle / (1 + L_0^2)^{5/6} \tag{15}$$

Equation (15) relates the coefficient of velocity structure in the local homogenous isotropic random field with the variance of velocity fluctuation and the outer scale of turbulence. In probing the atmosphere with sodar, the value of $\langle V'^2 \rangle$ in Eq. (13) is smaller than that in Eq. (14), because the value of $\langle V'^2 \rangle$ measured by sodar is limited by scattering volume. For fitting the sodar measurements, instead of L_0 we take the effective outer scale L_e in Eq. (15) and yield an approximation:

$$C_v^2 = 0.636 \, L_e \langle V'^2 \rangle / (1 + L_e^2)^{5/6} \tag{16}$$

where L_e is expressed by Eq. (8). For Doppler sodar C_v^2 can be thus obtained from Eq. (13) and (16).

It is necessary to mention the spectral broadening by the cross wind [25]. The variance of vertical velocity spectrum measured by sodar can be expressed as

$$\sigma_s^2 = \sigma_{cw}^2 + \sigma_{tu}^2 \tag{17}$$

where σ_{cw}^2 is caused by cross wind effect and σ_{tu}^2 is the real variance of turbulence spectrum. Pan [26] gave an estimation of σ_{cw} under the assumption of conoidical acoustic beam and homogeneous distribution of scatterers:

$$\sigma_{cw} = 0.004363 \, U\gamma \tag{18}$$

where U is the speed of cross wind in meters per second and γ is the beam angle in degree. Eq. (18) is close to an empirical expression deducted from sodar data [24]. Withdrawing σ_{cw}^2 from σ_s^2 is necessary for the third approach.

The Wind Effect on Received Echo Power

The echo signal at antenna is found very weak when wind is strong [13]. In fact the acoustic ray will be bent by wind. For a vertical emitting beam, wind will cause the axis of beam to incline at an angle [26]:

$$\delta = \arctan \left(\frac{\sqrt{\langle u(Z) \rangle^2 + \langle v(Z) \rangle^2}}{C} \right)$$

$$= \arctan \left(\frac{1}{20.05} \sqrt{\frac{\langle u(Z) \rangle^2 + \langle v(Z) \rangle^2}{T}} \right) \tag{19}$$

where $\langle u(Z) \rangle$ and $\langle v(Z) \rangle$ are average wind components in the layer from surface to Z level and T is average temperature of the layer. The angle δ is also the scattering volume incline angle from the axis of antenna. The arrival angle of echo from the scattering volume at Z is [26]

$$\psi = 2\delta$$

$$= 2\arctan \left(\frac{1}{20.05} \sqrt{\frac{\langle u(Z) \rangle^2 + \langle v(Z) \rangle^2}{T}} \right) \tag{20}$$

Eq. (20) shows that the displacement of scattering volume at Z level depends on the composition of the wind vectors below. Using Eq. (20) and the wind profiles from Doppler sodar data we can get the arrival angles.

For convenience, we assume that the intensity of sound on a section of the beam is uniform. Beam bending will cause a loss of a part of the echo energy at the antenna. The volume whose scattering wave can reach the antenna is the overlap part of the original section with the moved section. So the attenuation factor caused by wind will be

$$\alpha_w = \frac{P_{rw}}{P_{r0}} = \frac{A_w}{A_0}$$

$$= \frac{2}{\pi} \left[\frac{\arccos\left(\frac{\psi}{\gamma}\right)}{180} \cdot \pi - \sqrt{1 - \left(\frac{\psi}{\gamma}\right)^2} \cdot \frac{\psi}{\gamma} \right] \quad (21)$$

where γ is the beam angle, P_{r0} is the received echo power with no wind, P_{rw} is the received echo power with wind, A_w is a part section of the beam which contributes to the received echo signal at the antenna and A_0 is the beam section.

Figure 1 shows the dependence of the arrival angle of echo, which comes from Z level, on the average wind speed in the layers below Z level. Figure 2 shows the relation between the wind attenuative factor, α_w, and the ratio of arrival angle to the beam angle, ψ/γ. Apparently the narrow beam is influenced more by wind. Using Eq. (8, 9, 20, 21) and the data from Neff [1], we estimated the variation of attenuation at two different frequencies used in his experiment [13]. The results are qualitatively consistent with each other.

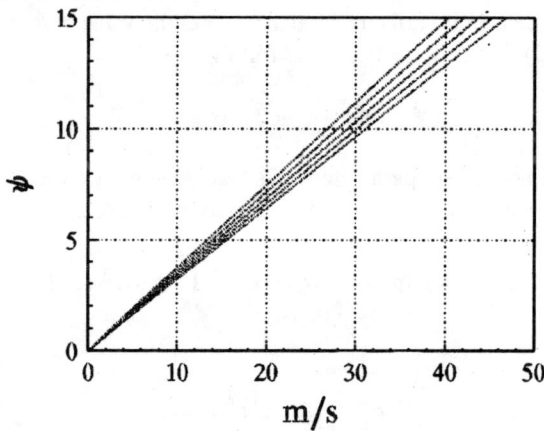

Fig. 1. Dependence of arrival angles with wind speed, from the top to bottom $t = -40°C$, $-20°C$, $0°C$, $20°C$, $40°C$

The Modified Sodar Equation and The Determination of C_v^2

Accepting the effective outer scale of turbulence, equation (8) and (9) can be used

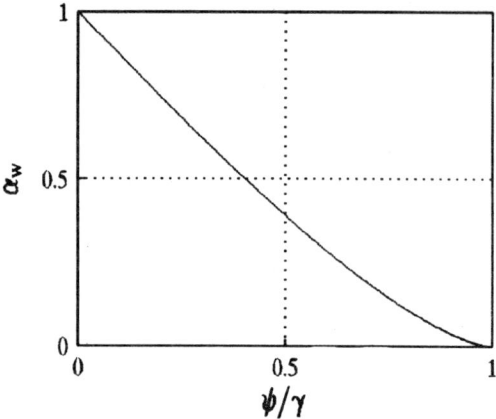

Fig. 2. Dependence of wind attenuation factor α_w with the ratio of arrival angle to beam angle ψ/γ.

to estimate excess attenuation. Adding the wind attenuation, the sodar equation will be modified as follows:

$$P_r = P_t \eta_r \eta_t \frac{c\tau A G}{2R^2} \sigma(\pi) \alpha_w \exp[-2\int_0^R \alpha_m \, dr$$
$$- \int_0^R \alpha_e(r) \, dr - \int_0^R \alpha_e(r, R) \, dr] \qquad (22)$$

Here α_w is the wind attenuative factor, α_m is the attenuation coefficient of molecular absorption, $\alpha_e(r)$ is the coefficient of excess attenuation for the path from antenna to the point R and $\alpha_e(r, R)$ is the coefficient of excess attenuation for path from R to antenna. The R in $\alpha_e(r, R)$ means that L_e takes the value at R and is not changed with r. Eq. (22) could be rewritten as follows:

$$P_r = C_s \frac{1}{R^2} \cdot C_t^2 \cdot \alpha_w \cdot T_m \cdot T_e \qquad (23)$$

where C_s is composed by parameters of sodar system, T_m is the transmissivity of molecular absorption and T_e is the transmissivity of excess attenuation

$$T_e = \exp[-\int_0^R \alpha_e(r) \, dr - \int_0^R \alpha_e(r, R) \, dr] \qquad (24)$$

Then we have

$$C_t^2 = \frac{P_r R^2}{C_s \cdot \alpha_w \cdot T_m \cdot T_e} \qquad (25)$$

There are two ways for estimating the C_t^2. Steps of the first way are: (1) to calibrate all of the parameters of sodar system; (2) to calculate the first guess of C_t^2 as follows:

$$C_t^2 = \frac{P_r R^2}{C_s} \tag{26}$$

(3) to measure the C_v^2 with bistatic sodar; (4) to calculate the coefficient of excess attenuation by substitutiong C_t^2 and C_v^2 into Eq. (9); (5) Estimating α_w and T_m based on meteorological data; (6) to calculate the corrected value of C_t^2 from Eq. (24) and (25); and (7) Iteration may be used if high accuracy of C_t^2 is needed.

The second is based on the combination of Doppler sodar with a platinum wire thermometer. The thermometer is set at the lowest level where the sodar is able to measure. The C_{td}^2 measured with thermometer is used to be the reliable C_t^2 value of sodar at the lowest level. As the height is low we assume $T_m \approx T_e \approx 1$. Choosing the weak wind day, the α_w will be equal to 1. Then the coefficient C_s can be determined. The tasks remained are the determination of T_e and α_w. For Doppler sodar C_v^2 can be obtained from Eq. (13) and (16). We have obtained the first guess of C_t^2 from Eq. (26) then T_e can be determined through Eq. (24) and (9). Wind attenuation α_w is determined by Doppler wind data. Finally the value of C_t^2 is determined by Eq. (25). If iteration is taken the accuracy will be better, but it is not necessary.

The primary experiments have been done with the second method at the Peking University [27]. Table 1 and Fig. 3 show the results. In Fig. 3a, the convection was developing during this period and the mixing height was changing from 100 m to 340 m. It may be the reason of deviation of curve from the $-4/3$ line

Table 1. The correction of the excess attenuation of C_t^2

		$t = 15°C$, 9:52–10:26, Oct. 29, 1989					$t = 7.4°C$, 11:52–12:50, Nov. 30, 1989			
Height	L_e	C_{t0}^2	C_v^2	α_e	T_e	C_t^2	C_{t0}^2	C_v^2	α_e	T_e	C_t^2
50	2.78	720	93	27	0.997	721	1090	2061	630	0.939	1160
70	5.84	520	156	162	0.987	526	560	1281	1401	0.818	684
90	7.59	350	74	120	0.978	357	330	1281	2118	0.694	475
110	9.34	300	94	215	0.963	311	120	908	2191	0.569	210
130	11.09	260	76	231	0.947	274	71	1054	3390	0.433	164
150	12.84	200	100	389	0.922	216	30	738	3035	0.322	93
170	14.59	150	109	528	0.890	168	39	477	2430	0.238	163
190	16.34	119	76	447	0.860	127	12	256	1575	0.177	67
210	18.09	66	144	993	0.810	81	10	239	1745	0.128	77
230	19.84	37	62	503	0.775	47					
250	21.59	42	87	804	0.730	57					
270	23.34	39	82	866	0.683	57					

Note: The real values of C_{t0}^2, C_t^2 and α_e in the table need to be multiplied with 10^{-6} and C_v^2 need to be multiplied with 10^{-4}.

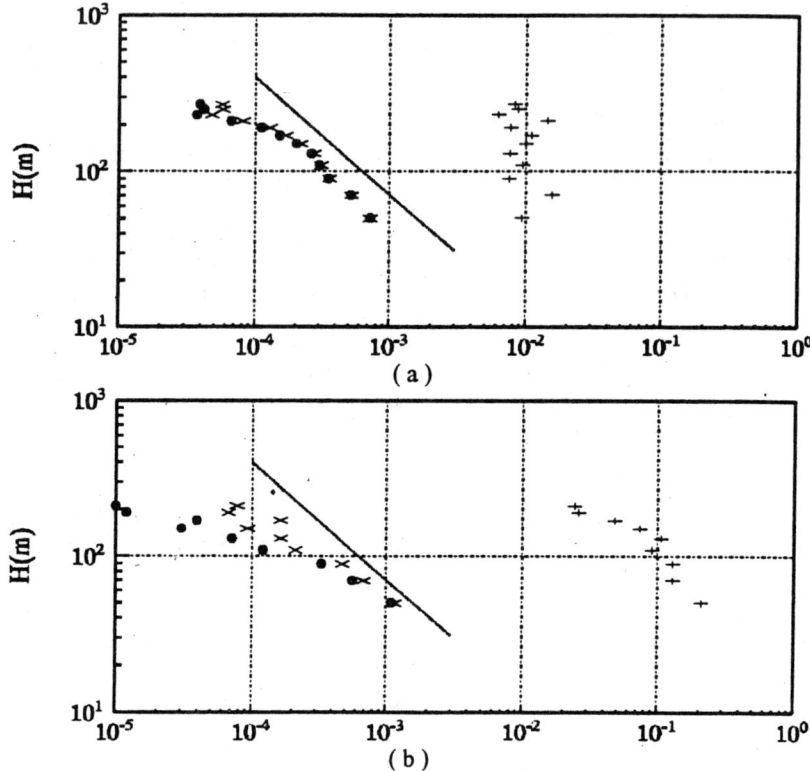

Fig. 3. The comparison of the corrected profiles of C_T^2 ($°C^2\ m^{-2/3}$) and C_v^2 ($m^{4/3}s^{-2}$) with their uncorrected values. • are the uncorrected values of C_T^2, × are the corrected values; + are the C_v^2. $\gamma = 10°$. The slope of the straight lines is $-4/3$. ((a) Oct. 26 1989, local time 9:52–10:26, $t = 15°C$. (b) Nov. 30 1989, local time 11:52–12:50, $t = 7.4°C$)

between 100 m to 200 m. In Fig. 3b the convection was well developed and the curve below 100 meter fitted the $-4/3$ power very good. Above 100 meter the data were not reliable as the signal was too weak. The correction of excess attenuation has been applied to the data from Moulsley et al, [7]. The results show that in convective condition the profiles of C_t^2 become closer to $-4/3$ power after the correction [18].

Conclusion

The coefficient of temperature structure, C_t^2, has been used in estimating the atmospheric attenuation in microwave communication and the heat flux in the boundary layer at unstable condition, etc. The coefficients of velocity structure, C_v^2, reflect the distribution and variation of kinetic energy of turbulence in space and in time. The applications will lose their improtance if the errors in parameters are too big. The errors in C_t^2 and C_v^2 are caused by various factors. Large variation of atmospheric dynamics and thermodynamic characteristics in space and time makes the interaction between acoustic waves and atmosphere very complicated. Analysis shows that the attenuation caused by absorption, turbulence and beam bending should be involved in sodar equation, as neglect of these factors may cause an underestimation of C_t^2 by 5% to 500% and even more. The group of Eq. (22) and Eq. (8, 9, 20, 21) should be more proper for quantitative measurement of C_t^2 than the original sodar equation. An improvement on estimation of C_v^2 is necessary. The modified sodar equation implies that the non-Doppler sodar has a limited potential for the measurement of C_t^2. If we use a Doppler sodar and the Eq. (22, 8, 9, 20, 21), the error of C_t^2 may be controlled with in 100%.

References

1. W.D. Neff, "Quantitative evaluation of acoustic echoes from the planetary boundary layer", NOAA *Technical Report* ERL 322–WPL 38, 1975.
2. D.N. Asimakopoulos, R.S. Cole, S.J. Caughey and B.A. Crease, "A quantitative comparison between acoustic sounder returns and the direct measurement of atmospheric temperature fluctuations", *Boundary Layer Meteor.*, 10, 137–147, 1976.
3. J.E. Gaynor, "Acoustic Doppler measurement of atmospheric boundary layer velocity structure functions and energy dissipation rates", *J. Appl. Meteor.*, 16, 148–155, 1977.
4. D.A. Haugen and J.C. Kaimal, "Measuring temperature structure parameter profiles with an acoustic sounder", *J. Appl. Meteorol.*, 17, 895–899, 1978.
5. D.W. Thomson, R.L. Coulter and Z. Warhaft, "Simultaneous measurements of turbulence in the lower atmosphere using sodar and aircraft", *J. Appl. Meteor.*, 17, 732–734, 1978.
6. S.J. Caughey, B.A. Crease, D.N. Asimakopoulos and R.S. Cole, "Quantitative bistatic acoustic sounding of the atmospheric boundary layer", *Quart. J. Roy. Meteor. Soc.*, 104, 147–161, 1978.
7. T.J. Moulsley, D.N. Asimakopoulos, R.S. Cole, B.A. Crease and S.J. Caughey, "Measurement of boundary layer structure parameter profiles by acoustic sounding and comparison with direct measurements", *Quart. J. Roy. Meteor. Soc.*, 107, 203–230, 1981.
8. M.Y. Zhou, Y.J. Chen, N.P. Lu and S.M. Li, "A comparative study between the temperature structure coefficients detected by sodar and direct methods", *Acta Geophysica Sinica*, Vol. 25, No. 6, 492–499, 1982 (in Chinese).
9. D.N. Asimakopoulos, T.J. Moulsley, C.G. Helmis, D.P. Lalas and J. Gaynor, "Quantitative low-level acoustic sounding and comparison with direct measurements", *Boundary Layer Meteor.*, 27, 1–26, 1983.
10. Y.J. Chen, S.M. Li and N.P. Lu, "The C_n^2 by sodar and in-situ measurements", *Scientia Atmospherica Sinica*, Vol. 8, No. 2, 153–160, 1984 (in Chinese).
11. T.J. Moulsley, D.N. Asimakopoulos, C.G. Helmis, D.P. Lalas and J. Gaynor, "A quantitative comparison of horizontal and vertical acoustic sounding with in-situ measurements", *Boundary Layer Meteor.*, 33, 85–100, 1985.

12. E.H. Brown and S.F. Cliford, "On the attenuation of sound by turbulence", *J. Acoust. Soc. Amer.*, 60, 788–794, 1976.
13. W.D. Neff, "Beamwidth effects on acoustic backscatter in the planetary boundary layer", *J. Appl. Meteorol.*, 17, 1514–1520, 1978.
14. R.B. Chadwick and C.G. Little, "The comparison of sensitivities of atmospheric echo-sounders", *Remote Sensing of The Enuironment*, 2, 223–234, 1973.
15. V.I. Tatarskii, *"The effects of the turbulent atmosphere on wave propagation"*, translated from Russian, Israel Program for Scientific Translations, Chap. 2, § 35, 157–162, 1971.
16. A.H. Weber, J.S. Irwin, J.J. Mathis and J.R. Kahler, "Spectral scales in the atmospheric boundary layer", *J. Appl. Meteor.*, 21, 1622–1632, 1982.
17. F. Quintarelli, "Some atmospheric turbulence parameters which can be measured using acoustic radar", *Proceedings of Fourth International Symposium on Acoustic Remote Sensing of the Atmosphere and Oceans*, Vol. 1, 26, 1–4, Australian Defence Force Academy, Canberra, Australia, 1988.
18. N.X. Pan and J.Q. Liu, "The estimation of Excess attenuation of acoustic waves", *Scientia Atmospherica Sinica*, 15, 4, 116–122, 1991. (in Chinese with English abstract)
19. V.I. Tatarskii, *"Wave Propagation in a Turbulent Medium"*, translated from Russian by R.A. Silverman, Dover, New York, Chap. 2, 1961.
20. K.H. Underwood and R.L. Coulter, "Vertical velocity spectra from Doppler radar", *Proceedings of the Second International Symposium on Acoustic Remote Sensing of the Atmosphere and Oceans*, Rome, Italy, X(1–11), 1983.
21. G. Desbraux and A. Weill, "Mean turbulent properties of the stable boundary layer observed during 'COAST' experiment", *Proceedings of the Third International Symposium on Acoustic Remote Sensing of the Atmosphere and Oceans*, France, 195–221, 1985.
22. J. Høstrup, "Velocity spectra in the unstable planetary boundary layer", *J. of Atmos. Sci.*, 39, 2239–2248, 1982.
23. J.C. Kaimal, "Turbulence spectra, length scales and structure parameters in the stable surface layer", *Boundary Layer Meteor.*, 4, 289–309, 1973.
24. N.X. Pan, "Digital Characteristics of spectra of echo signal of Doppler sodar and their pickup", *Kexue Tongbao (Science Bulletin), Academia Sinica*, 30, 9, 1196–1201, 1985.
25. E.H. Brown, "Turbulent spectral broadening of backscattered acoustic pulses", *J. Acoust. Soc. Am*, 56, 1398–1460, 1974.
26. Pan Naixian and Zheng Yi, "Wind effect on measurements of wind velocity with Doppler sodar", *ACTA Scientiarum Naturalium Universitatis Pekinensis*, 1, 98–105, 1986. (in Chinese with English abstract)
27. N.X. Pan, "Some problems of measurement of C_t^2 by sodar", *Chinese Journal of Atmospheric Sciences*, Vol. 17, No. 1, 41–48, 1993.

Acoustic Remote Sensing Applications
S.P. Singal (Ed)
Copyright © 1997 Narosa Publishing House, New Delhi, India

9. Turbulence Variables Derived from Sodar Data

R.L. Coulter
Argonne National Laboratory, Argonne, IL 60439, USA

Introduction

Sodar can provide valuable information about the turbulent nature of the lower atmosphere over a range of scales. The phase speed of sound (c) is a function of virtual temperature (T_v), and the relative wind speed; thus the strength of the scattering of acoustic energy is determined by inhomogeneities in the temperature and wind structure of the atmosphere. It follows that atmospheric turbulence on the scale of the wavelength of the acoustic energy (5–25 cm) can be interrogated by measurements of the relative amount of scattered acoustic energy (Kallistratova, 1961). In addition, Doppler extraction techniques can be applied to the returned signal to measure atmospheric wind components and mechanical turbulence on the scale of the acoustic pulse length (5–50 m). Analysis of velocities and signal intensities over successive range gates for several minutes with sophisticated methods such as Fourier or wavelet transforms can define coherent atmospheric structures such as thermal plumes on the scale of the daytime mixed layer (500–2000 m) and Kelvin-Helmholtz and gravity waves in the nocturnal boundary layer (NBL) (100–400 m).

Derivation of the relationship between acoustic scattering and atmospheric turbulence has been shown elsewhere (Tatarski, 1961, 1971; Brown, 1972). Here we will discuss the signals obtained with sodars and their relation to atmospheric turbulence, and illustrate methods, examples, and problems in their calculation.

Theory

Signal Amplitude
The intensity of the scattered acoustic signal is proportional to the temperature and velocity structure parameters (C_T^2 and C_V^2, respectively) (see, e.g., Tatarski, 1961, 1971).

$$\sigma(\theta) = 0.05\lambda^{-1/3} \cos^2(\theta) \left[\frac{C_V^2}{c^2} \cos^2(\theta/2) + 0.13 \frac{C_T^2}{T^2} \right] \sin^{-11/3}(\theta/2) \quad (1)$$

where σ is the scattering cross section, θ is the scattering angle between incident and scattered energy, T is the absolute temperature, and λ is the acoustic wavelength. It is immediately apparent that C_T^2 can, in principal, be determined uniquely by measurement of the back scattered energy ($\theta = \pi$). The structure parameters are defined as

$$C_T^2 = \frac{D_T(r)}{r^{2/3}} = \frac{\langle [T(x+r) - T(x)]^2 \rangle}{r^{2/3}} \quad (2a)$$

$$C_V^2 = \frac{D_V(r)}{r^{2/3}} = \frac{\langle [V(x+r) - V(x)]^2 \rangle}{r^{2/3}} \quad (2b)$$

where $D_T(r)$ and $D_V(r)$ are the temperature and velocity structure functions, respectively, r is the separation distance between simultaneous measurements of temperature or velocity and < > indicates an ensemble average. In the case of backscattered acoustic energy the effective separation distance is $\lambda/2$, which is almost always within the inertial subrange of atmospheric turbulence. Simultaneous measurements of scattered energy at two angles can lead to estimates of both C_T^2 and C_V^2. This is most efficiently accomplished with a collocated transmitter and receiver (often pointed vertically) and a second, separated receiver (or transmitter). The collocated, or monostatic, receiver is used to obtain C_T^2 and the two signals are combined to evaluate C_V^2 (Thomson et al, 1978).

The accuracy of this method is usually limited to about a factor of 2 because of problems such as excess attenuation (Neff, 1978), beam pattern definition, and the inherent variability of the acoustic transducers used. The success of such measurements is highly dependent on the accuracy of the calibration procedure. One method for calibration involves measuring the response of the system to a known signal (determined with a calibrated microphone) and measuring the output power of the transmitter (see Sisterson and Coulter, 1979; Coulter, 1979). This method requires care and attention to detail including the beam pattern and the exact geometry of the transmitter, receiver and microphone. A second, somewhat less satisfying method for calibration is to make simultaneous measurements of returned signal strength and direct measurements of C_T^2 (with separated temperature measurements, e.g.).

There are several useful turbulence parameters that can be derived using C_T^2, as long as the turbulence is in the inertial subrange and is isotropic. For example, the dissipation rate of turbulence (ε) and the temperature destruction rate (N) can be written as

$$\varepsilon = (C_V^2/1.97)^{3/2} \quad (3)$$

$$N = \frac{C_T^2}{3.2} \varepsilon^{1/2} = 0.22 \, C_T^2 C_V \tag{4}$$

A most useful relationship is the observation (Wyngaard, 1971) that the sensible heat flux (H) in the surface layer can be written as

$$H = 0.48 \, \rho c_p \left(\frac{g}{T}\right)^{1/2} z(C_T^2)^{3/4} \tag{5}$$

where ρ and c_p are the density and heat capacity of air, respectively, g is acceleration due to gravity, and z is the height above the surface. This relationship also holds for measurements in the mixed layer (Neff, 1957; Tsvang, 1969) but should be corrected for moisture effects (Coulter and Wesely, 1980). That is, H is modified to become

$$H = 0.48 \, \rho c_p \left(\frac{g}{T}\right)^{1/2} (2TC_a)^{3/2} z\gamma \tag{6}$$

where

$$C_a^2 = C_T^2 \, [1 + (0.06/\beta)^2]/(4T^2), \tag{7}$$

$$\gamma = \{[1 + (0.07/\beta)^2]/[1 + (0.03/\beta)^2]\}^{3/4} \, (1 + 0.07/\beta^{-1}, \tag{8}$$

and β (Bowen ratio) is the ratio of sensible to latent heat flux. The height of the measurement implies a volume average defined by the pulse length (about 3–30 m) and the area subtended by the beam width of the antenna (5–10 deg).

Doppler Shift

The signal from the scattered energy will, in general, be shifted in frequency by an amount (f_D) proportional to the component of the mean speed (v_r) of the scatterers within the scattering volume along the radial direction between the transmitter-receiver and the scattering volume. That is,

$$f_D = -2 \frac{v_r}{c} f \tag{9}$$

where f is the transmitted frequency and c is the phase speed of sound in air. If the transmitter-receiver is pointing vertically the vertical velocity (w) will be measured. This is a particularly useful component of the wind to measure because it is related to a number of useful turbulence variables, in particular the standard deviation of w (σ_w). In the case of a separated transmitter and receiver the component is perpendicular to an ellipsoid of revolution about the transmitter and receiver that passes through the scattering volume.

Mean Frequency Shift

Methods for the determination of the Doppler frequency shift include, among others, phase locked loop detection, complex covariance estimators, and spectral methods (see for example, Neff and Coulter, 1986). At present spectral methods are generally recognized to have significant advantages:

(1) the Fast Fourier Transform (FFT) is computationally efficient and can be implemented in hardware or software;
(2) definition of the total spectrum rather than an estimation of its mean permits better definition of the signal region (Mastrontonio, 1982); and
(3) extraneous signals, including that from precipitation and/or extraneous noise can be either removed or isolated for study.

Problems with this technique include artificial broadening of the spectrum caused by the finite width of the frequency "bins", and multiple peaks in the signal spectrum.

Standard Deviation of Vertical Velocity

This parameter may be related directly to the spread of pollutants, for example. In principal, the standard deviations of the horizontal components of motion can also be derived from combinations of the velocity components along tilted (from vertical) radials and the vertical component; however, the spatial and temporal separation imposed by these measurements creates significant overestimates of the standard deviations of horizontal motions (Kristensen and Gaynor, 1986). Values for σ_w are generally obtained in two ways.

(1) The variance is calculated directly from a series of estimates of w, i.e.,

$$\sigma_w = \sqrt{\frac{\sum w_i^2 - n\overline{w}^2}{n - 1}} \tag{10}$$

where n is the number of samples. The estimate of w_i can be determined from a single average of a number of spectra. If, however spectra are averaged, some information about the variance will be lost because of the averaging process. On the other hand, averaging over several spectra can potentially provide a better estimate for each value, if the noise component of the signal is random. This type of tradeoff must be faced continually when determining the appropriate operating parameters of a sodar.

(2) σ_w is calculated from the broadening of the average spectrum (Quintarelli, 1993). That is, the spectra from each of the samples at a given range gate are averaged in frequency space. The broadening of the spectrum over and above that of the spectrum from a single estimate is proportional to σ_w. Here potential problems are presented by the false broadening of the spectrum due to the digital spectrum that cannot be fully predicted because of the variable location of the "true" frequency from each pulse within a frequency bin (Harris, 1978).

There are several problems with any estimate of σ_w that must be addressed.

(1) Any sodar system has a number of sources that contribute to the signal variance in addition to atmospheric motions (Spizzichino, 1974; Neff and Coulter, 1986; Coulter, 1990). It is critical that this value be determined and accounted for. This is particularly important for nighttime values of σ_w, which can be quite small, and the contribution due to "system variance" becomes relatively more important (Gaynor, 1977; Coulter, 1990).

(2) The signal to noise ratio (snr) must be large throughout the measurement period; if values from only a portion of the sample period are included due to small values of the snr, the variance is likely to be overestimated because the more turbulent portions of the signal that are associated with larger values of snr are preferentially sampled.

(3) Because of the finite size of the sample volume, signals associated with larger turbulence within the volume will dominate the received signal; this can also lead to overestimates. For example, C_T^2 is generally larger in actively rising air as it mixes with the cooler air above; when the sample volume encompasses both rising and descending air, the more turbulent, rising portions may dominate the signal.

Figure 1 illustrates that values of σ_w can be calculated with sodars (in this case a minisodar was used, which has some advantages over conventional sodar, as discussed below) if appropriate care is taken to account for system variance.

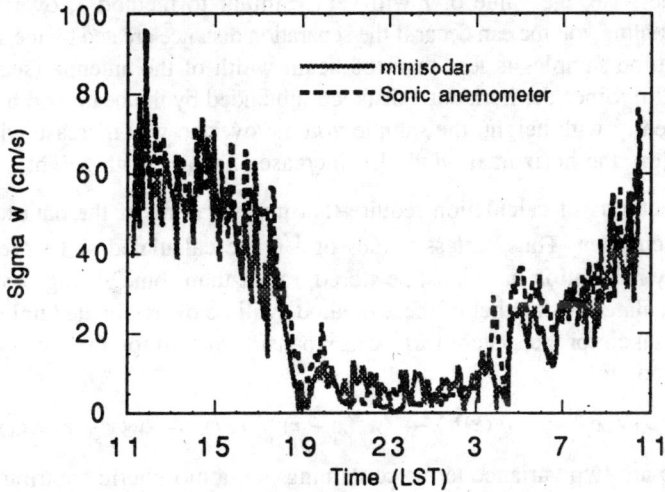

Fig. 1. Variation of σ_w over a 24-hr period calculated with a minisodar and a sonic anemometer at a height of 24 m above the ground, separated by 20 m from the minisodar (adapted from Coulter and Martin, 1986).

Profiles of σ_w and related parameters are useful, for example in estimates of local heat flux in convective conditions through calculation of the profile of σ_w^3/z (Weill, 1980).

Velocity Structure Parameter

The vertical velocity can be used to determine C_V^2 from (2b) in 2 ways:

(1) Computation of the difference in vertical velocity between two range gates. The separation between range gates is then equivalent to r in (2b). This is appealing because the estimates are almost simultaneous (separated by the time for sound to traverse the distance between range gates). The dependence of the numerator in (2b) should be proportional to $r^{2/3}$ if the separation is within the inertial subrange of turbulence; this can be tested simply by calculating C_V^2 over several different range gates. Care must be taken to account for the overlap of portions of the signals from neighboring range gates caused by the finite length of the acoustic pulse and the number of samples required to calculate the spectrum (see Coulter, 1990; Kristensen, 1978). This effect is most important for neighboring range gates and can lead to significant underestimates.

(2) Computation of the difference in vertical velocity between values at the same range gate but separated in time. The value for r in (2b) is now determined by the product of the horizontal wind speed and the time separation between transmit pulses. Various length acales can be investigated by calculating differences between samples separated by more than a single time step. The implicit assumption in this type of calculation is that the horizontal wind speed is constant over the averaging time necessary; otherwise, the value of r will vary. Similar to method 1, overlap of the sampling volume can occur if the separation distance created by the separation in time samples is less than the beam width of the antenna (see Coulter, 1990). Since the horizontal distance subtended by the beam width increases linearly with height, the sample volume overlap will increase with height unless the horizontal wind also increases linearly with height.

Both methods of calculation require that no averaging of the data take place before calculation. Thus, unless values of C_V^2 are calculated and stored in real time, every realization of w must be stored, rather than some average value. Also values calculated with either of these methods will be overestimated unless proper account is taken for the system variance. Expansion of (2b) for a vertically pointing antenna leads to

$$D_V(r) = \langle [V(x + r) - V(x)]^2 \rangle \rightarrow \langle w^2(x + r) + \omega^2(x) - 2w(x + r) w(x) \rangle \quad (11)$$

Here there are two variance terms containing non-atmospheric contributions and thus the system variance is doubled in importance compared to the calculation for σ_w.

Minisodars

The advent of the minisodar (Moulsley and Cole, 1979; Asimakopoulos et al, 1983; Weill, 1986; Coulter, 1986) provided several advantages pertinent to the measurement of turbulence variables. These devices are of small physical stature, use relatively high frequencies (4.6 kHz), and often use phased arrays of piezo ceramic tweeters for antennae because of their high efficiency. Generally it is not necessary to use heavy shielding with minisodars because the background noise is smaller at larger frequencies and the high frequency transmissions are more readily absorbed by the atmosphere. The high frequency, piezo ceramic tweeters and small size of the minisodar each provide advantages for turbulence measurement.

High Frequency: High transmit frequencies provide better Doppler velocity resolution because the Doppler shift of the scatterers is proportional to frequency. The velocity resolution is generally 2–3 times better with minisodars than with conventional sodars. Higher frequencies also provide narrower beam widths. In general, the beam width is proportional to λ/d where d is the size of the antenna. Smaller beam widths result in less extraneous noise and facilitate comparisons with in-situ instrumentation. Finally, the pulse widths may be decreased because of the increased velocity resolution; this results in finer profiling capabilities.

Piezoceramic tweeter arrays: The use of arrays eliminates the need for parabolic dish antenna and the attendant "ringing" of the dish after signal transmission. Thus the first usable height above the surface can be as small as 5 m above the surface or even less if pulse lengths are short enough (Asimakopoulos, 1983). In addition, the frequency response of these may often be peaked; this provides additional rejection of noise outside the signal band, hence larger snr. These piezoceramic transducers often have efficiencies near 50%, compared to approximately 10% for their electromagnetic counterparts.

Small size: The small size allows the instrument to be used in locations difficult to reach such as in mountainous terrain. Thus profiles of turbulence can be determined near obstructions, in valleys, and within forests where the profile is not expected to be easily predicted by classical theories of turbulence that require homogeneity of flow. The light weight allows one to orient the antenna at radical angles to capture unique aspects of atmospheric flow.

The principal drawback of the minisodar is that it is limited in its normal operating range to approximately 200 m because atmospheric absorption at these higher frequencies is very significant. A compensation for this limitation is that the maximum rate of sampling, which is inversely proportional to maximum signal height, is much greater.

Large-Scale Structures
The sodar (or minisodar) is capable of sampling the mixed layer of the atmosphere at range gates as small as 5 m apart and as often as once every second (if the height is limited to 170 m and only the vertical axis is used). This time-height cross section can be used to investigate the organized structures that evolve in the lower atmosphere and often dominate the transfer of energy between the surface and upper levels of the atmosphere.

198 Applications in the Atmosphere

During daytime convective conditions, these "coherent" structures are often called thermal plumes: large regions (.5–5 km diameter) of organized, rising air that travel with the mean wind and penetrate into the inversion that caps the mixed layer, transferring energy from the surface layer to above the mixed layer and, in fact, mixing the mixed layer. One drawback of the sodar is that it is an Eulerian device, sampling only that part of the plume that passes within its beam width. Thus only a section of the plume is sampled. This results in a non-regular set of data that is easily visualized but difficult to analyze objectively. Conventional frequency analysis of the velocity and/or intensity field provides some information but cannot localize the structures in time. New methods such as wavelet transforms can be used to determine the relative strength of different scales of motion as a function of time (Gao and Li, 1993). Figure 2 compares the vertical velocity field

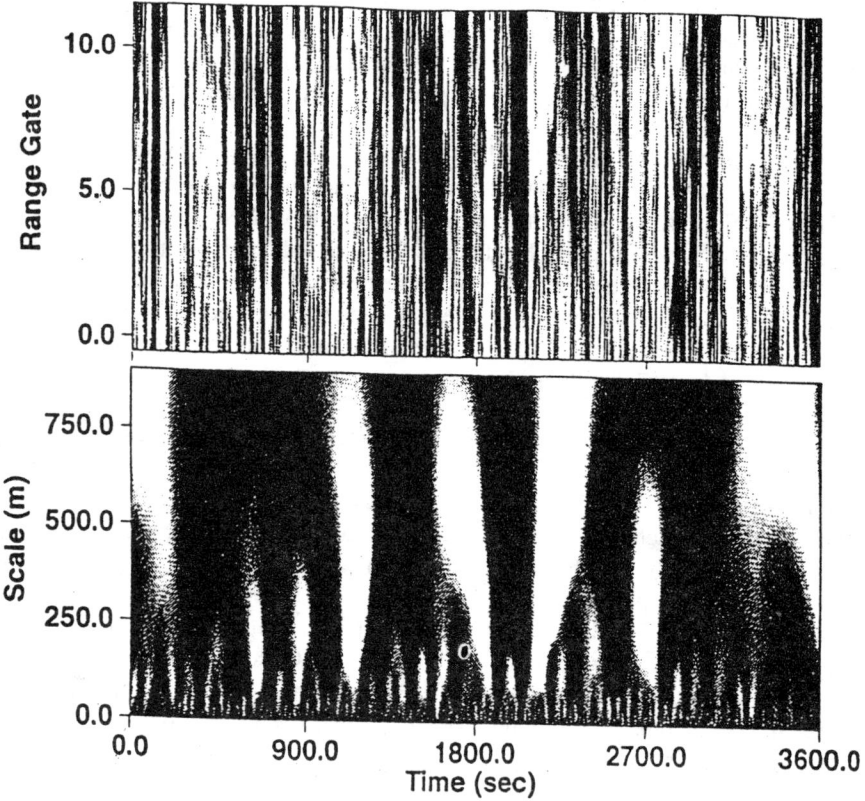

Fig. 2. (Top) Vertical time section of vertical velocities for one-hr period in convective conditions. Vertical velocities are sampled approximately 1/sec and range gate thickness is 8 m. Dark and light areas are rising and descending air respectively; maximum dark/light occurs for 200 cm/s magnitude.
(Bottom) Wavelet transform field derived from the vertical velocities at 84 m (range gate 11) over the same time period. Dark and light areas are positive and negative coefficients respectively. Scales are defined by mean horizontal wind speed and number of elements included in wavelet computation.

in the lowest 100 m of the convective boundary layer (sampled at a rate of about 1/sec) with a wavelet transform of the velocity at 84 m above the surface over the same time period. Of particular interest is the similarity between the vertical time section of Fig. 2a and the time variation of the horizontal scales shown in Fig. 2b. Note also how, at large scales, the transform isolates the coherent structures from the combination of small and large scale motions in the vertical time section. A method to use the variance of the wavelet transform field to isolate plumes of a selected intensity from within a data field in an objective manner is suggested by Coulter and Li (1994). This selected data field can then be analyzed to estimate, for example, a convective velocity scale, w^*, that is associated with these structures. This can then be related to the surface heat flux (H) and mixed layer height (z_i) via

$$w^* = \left[\frac{gz_i}{\rho c_p \theta_v} H \right]^{1/3} \tag{12}$$

where θ_v is potential temperature. Since the thermal plume moves with the mean wind while maintaining its organized structure, it represents of a relatively large surface area compared to in-situ measurements of H.

During nighttime, the structures are much different, consisting predominantly of gravity waves and Kelvin-Helmholtz instabilities. In either case a rapid sampling rate (4 min^{-1} or less) and fine vertical resolution are necessary in order to resolve the structures and determine their strength and influence. Because the boundary layer is normally much shallower during nighttime, the minisodar is often the appropriate instrument.

The sodar is a mature, well-founded instrument for atmospheric investigations. Recent computer technological advances provide the option of performing many turbulence computations in real-time. These parameters can then be used, for example in emergency response models (Coulter, et al, 1988; Coulter et al, 1992) as well as for detailed velocity and turbulence input to complex models of the boundary layer such as those that use large eddy simulation techniques.

References

1. Asimakopolous, T.J. Mousley, C.G. Helmis, D.P. Lalas, and J. Gaynor, 1983: Quantitative low-level Acoustic Sounding and comparison with direct measurements. *Boundary Layer Meteorol.* **27**, pp 1–26.
2. Brown, E.H., 1972: Acoustic-Doppler-radar scattering equation and general solution. *J. Acoust. Soc. Am.* **52**, 1391–1396.
3. Coulter, R.L. 1979: Sodar Calibration Results. Argonne National Laboratory Radiological and Environmental Division Annual Report ANL-78-65 Pt. IV pp 14–18.
4. Coulter, R.L. and M.L. Wesely, 1980: Estimates of surface heat flux from sodar

and laser scintillation measurements in the unstable boundary layer. *J. Appl. Meteorol.* **19**, 1209–1222.

5. Coulter, R.L. and T.J. Martin, 1986: Results from a high power, high frequency sodar, *Atmospheric Research* **20** # 2–4, pp 257–270.
6. Coulter, R.L., T.J. Martin, and R.E. Meyers, 1988: A Minisodar for Emergency Response and Military Applications. Extended Abstracts, Lower Tropospheric Profiling: Needs and Technologies, May 31–June 3 Boulder, Colorado. pp 270–228.
7. Coulter, R.L. 1990: Minisodars—Applications and Potential. *Proceedings. 5th International Symp. Acoustic Remote Sensing of the Atmosphere and Oceans.* 88–96., Tata McGraw-Hill, Pub. New Delhi, India.
8. Coulter, R.L., J.D. Shannon, T.J. Martin, and D.R. Cook, 1992: A Portable system for Prediction of a Transient Plume *J. Air & Waste Management Assoc.* **42**, 433–436
9. Coulter, R.L., and B.L. Li, 1994: A technique using the wavelet transform to identify and isolate coherent structures in the planetary boundary layer; Proceedings, 11th Symposium on Boundary Layers and Turbulence, American Meteorol. Soc. March, 1995 Charlotte, N.C. (to be published)
10. Gaynor, J.E., 1977: Acoustic doppler measurements of atmospheric boundary layer velocity structure functions and energy dissipation rates *J. Appl. Meteorol.* **16**, pp 148–155.
11. Gao, W. and B.L. Li, 1993: Wavelet analysis of coherent structures at the atmosphere forest interface, *J. Appl. Meteorol.* **32**, 1717–1725.
12. Harris, F.J., 1978: On the use of windows for harmonic analysis with the discrete Fourier Transform. *Proc. of IEEE* **66**, 51–83.
13. Kallistratova, M.A. 1961: Experimenal investigation of sound wave scattering in the atmosphere. Tr. Inst. Fiz. Atmos., *Atmos. Turbulentmost,* **4**, pp 203–256 (English translation, U.S. Air Force FTD TT-63-441)
14. Kristensen, L., 1978: On sodar techniques. Riso National Laboratory, Riso Report # 381, August.
15. Kristenson, L. and Gaynor, J., 1986: Errors in second moments estimated from monostatic Doppler sodar winds, Part I: Theoretical Description. *J. Atmos. Oceanic Technol.* **3**, pp 523–528.
16. Mastrantonio, G. and G. Fiocco, 1982: Accuracy of wind velocity determination with Doppler sodar. *J. Appl. Meteorol.*, **21**(6): 823–830.
17. Mousley, T.J. and R.S. Cole, 1979: High frequency atmospheric acoustic sounders. *Atmos. Envir.* **11**, pp 347–350.
18. Neff, W.D., 1975: Quantitative evaluation of acoustic echoes from the planetary boundary layer. *Tech. Rep.* ERL 322-WPL 3834 pp., NOAA, Boulder, COLO.
19. Neff, W.D., 1978: Beamwidth effects on acoustic backscatter in the planetary boundary layer. *J. Appl. Meteor.* **17**, 1514–1520.
20. Neff, W.D., and R.L. Coulter, 1986: Chapter 13, "Acoustic Remote Sensing", in *Probing the Atmospheric Boundary Layer.* American Meteorol. Society, Boston MA. pp 201–235.
21. Quintarelli, F., 1993: Acoustic sounder observations of atmospheric turbulence parameters in a convective boundary layer. *Journ. Appl. Meteorol.* **32** # 7, pp 1433–1440.

22. Sisterson, D.L. and R.L. Coulter, 1979: Sodar Calibration Method. Argonne National Laboratory Radiological and Environmental Division Annual Report ANL-78-65 Pt. IV pp 7–18.
23. Spizzichino, A. 1974: Discussion of the operating conditions of a doppler sodar. *J. Geophys. Res.* **79**, pp 5585–5591.
24. Tatarskii, V.I., 1961: Wave Propagation in a Turbulent Medium. Translated from Russian by R.A. Silverman, Dover, New York.
24. Tsvang, L.R., 1969: Microstructure of temperature fields in the free atmosphere. *Radio Sci.*, **4**, 1175–1178.
25. Weill, A., C. Klapisz, B. Strauss, F. Baudin, C. Jaupart, P. VanGrundebeeck and J.P. Goutorbe, 1980: Measuring heat flux and structure function of temperature fluctuations with an acoustic Doppler sodar. *J. Appl. Meteorol.* **19**, 199–205.
26. Weill, A., C. Klapisz, and F. Baudin, 1986: The CRPE minisodar: applications in micrometeorology and in physics of precipitation. *Atmos. Res.* **20**, 317–335.
27. Wyngaard, J.C., J. Izumi, and S.A. Collins, Jr, 1971: Behavior of the refractive index structure parameter near the ground. *J. Opt. Soc. Am.* **61**, 1646–1650.

Acoustic Remote Sensing Applications
S.P. Singal (Ed)
Copyright © 1997 Narosa Publishing House, New Delhi, India

10. Development of Sodar Detection and Its Application for Studies of Atmospheric Boundary Layer in Beijing, China

Mingyu Zhou

National Research Center for Marine Environmental Forcosts, Beijing, China

Introduction

Since the acoustic sounder system was first developed in Australia [1], many research works on sodar have been carried out in the world [2–8]. In China a monostatic acoustic radar was developed in the seventies [9]. After that a Doppler sodar system was made by Lu et al, [10], and a ship mounted Doppler sodar has been designed by Chen et al [11].

In the recent ten years, by using the investigative potential of these acoustic radars, many studies on vertical distribution of wind and atmospheric stratification, and the structure of atmospheric boundary layer in Beijing (China) area have been completed [12–15]. In addition, the calculation of heat flux with ship mounted sodar data in the western equatorial Pacific has been carried out under the China-U.S. joint TOGA Project [16].

This paper is a comprehensive report of investigations on sodar detection and its application for studies of atmospheric boundary layer completed in Beijing, China. It contains Doppler sodar system, some characteristics in atmospheric boundary layer, calculation of temperature structure coefficient and heat flux, and application of sodar sounding to atmospheric dispersion.

Doppler Sodar System and a Comparison with in-situ Measurements

Principles of Doppler Sodar Wind Measurement

In order to measure total wind vector, a three-antenna system is used. Three antennas are set in two planes which are perpendicular to each other as shown in Fig. 1a. Two antennas T_A and T_B are inclined with an angle α to horizontal direction. Antenna T_C is vertically pointed. These three antennas when used for transmitting and receiving, with a transmitted frequency f_0, the radial wind velocities are

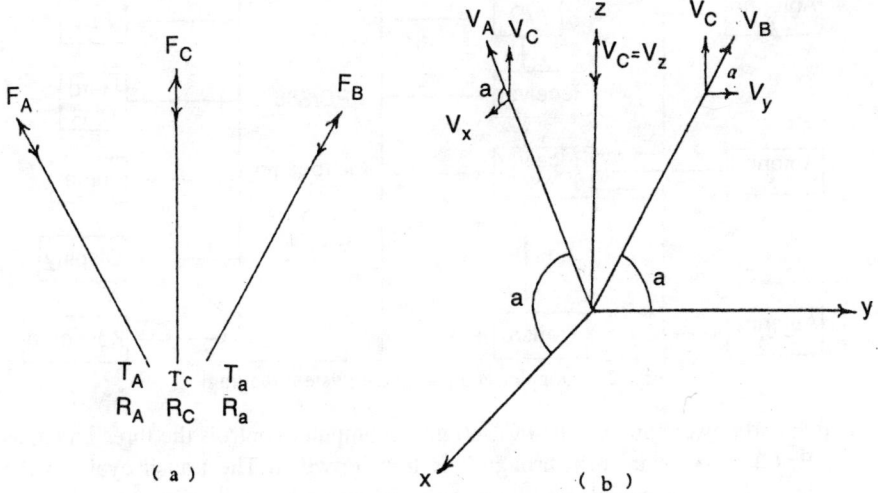

Fig. 1. Antenna system configuration and principle of measuring wind.

$$V_a = -\frac{C}{2}\left(\frac{\Delta f_a}{f_0}\right), V_b = -\frac{C}{2}\left(\frac{\Delta f_b}{f_0}\right), V_c = -\frac{C}{2}\left(\frac{\Delta f_c}{f_0}\right) \quad (1)$$

where C is velocity of sound, minus sign "–" represents that wind direction is opposite to the sign of Δf, i.e. when $\Delta f > 0$, the air flow moves down along the axis, while $\Delta f < 0$, represents air flow moving upward. From Fig. 1b we can easily obtain

$$V_x = (V_a - V_c \sin \alpha)/\cos \alpha,$$
$$V_y = (V_b - V_c \sin \alpha)/\cos \alpha, \quad (2)$$
$$V_z = V_c.$$

The total wind vector can be written as

$$V = V_x i + V_y j + V_z k \quad (3)$$

According to the ratio of V_x to V_y and its sign the wind direction can be defined.

Configuration of Doppler Sodar System

The Doppler sodar system consists of antenna, transmitter, receiver, antenna switch, PC/386 microcomputer, FFT board and other external devices. The system diagram is shown in Fig. 2.

Sodar antenna consists of a flat trailer, parabolic reflector, absorbing foam and transducer. The parabolic reflector with a diameter of 1.2 m and a screen around it for attenuating the ambient noise are made by fiber glass. The working frequency is at 1600 Hz.

The main parts of Doppler sodar consist of transmitter, receiver, T-R transit

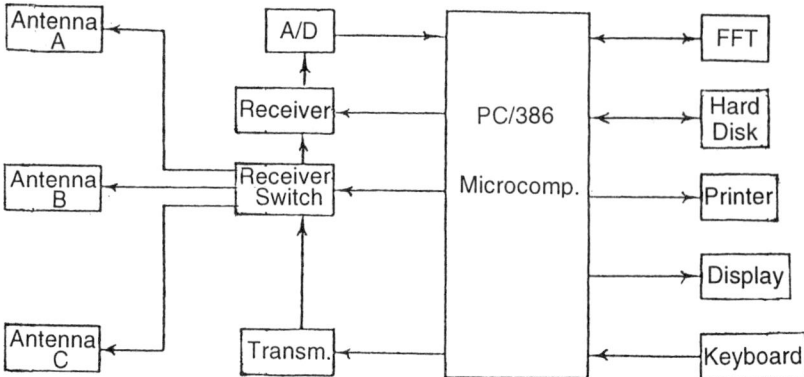

Fig. 2. Doppler acoustic radar system diagram.

switch and power supply. The PC/386 microcomputer controls the three antennas (A, B, C) working in turn through T-R transit switch. The transit cycle of the whole system can be selected between 6–90 seconds, while the corresponding transit cycle for each antenna is 2–30 seconds. The transmitter and receiver systems are used for transmitting the sound pulse and extracting echo signal respectively.

PC/386 microcomputer has 640KB RAM, Great-Wall EGA card and color display terminal, and 80MB hard disk. DOS 3.3 operating system can support operating assembly, Fortran, and Basic languages.

A/D conversion board is a multifunction A/D board with 32-channel analog input and 2-channel analog output. It has a resolution of 12-bit, 32-channel DI/O and 4-channel 16-bit timer/counter.

FFT board is one of the main parts of this system. By use of TMS 32020 VLSI signal processing chip, it takes only 42 ms for 1024 point complex FFT operation. The spectrum analysis of Doppler signals and the frequency shift detection are conducted with the board.

This system is equipped with an M1570 color printer. As output equipment used for copying the data of wind speed and echo diagram, it can clearly show the diagram of echo intensity which varies with height and time, and the profiles for wind speed and direction.

Comparison of Wind Detected by Doppler sodar and In-Situ Measurements

The accuracy of the Doppler sodar measurements is 0.3 m/s in horizontal wind speed, 0.05 m/s in vertical, and $\pm 3°$ in wind direction. The measured wind speed ranges from 0 to 28 m/s, and detection height from 30 to 1000 m. A comparative experiment has been done at Nanyuan (south of Beijing) airport by using Doppler sodar, three-component anemometer and ADAS tethersonde. Shown in Fig. 3 is a correlation diagram for the comparison. The variance of wind velocity is 0.5, which is calculated from the wind velocity deviation between sodar and three-

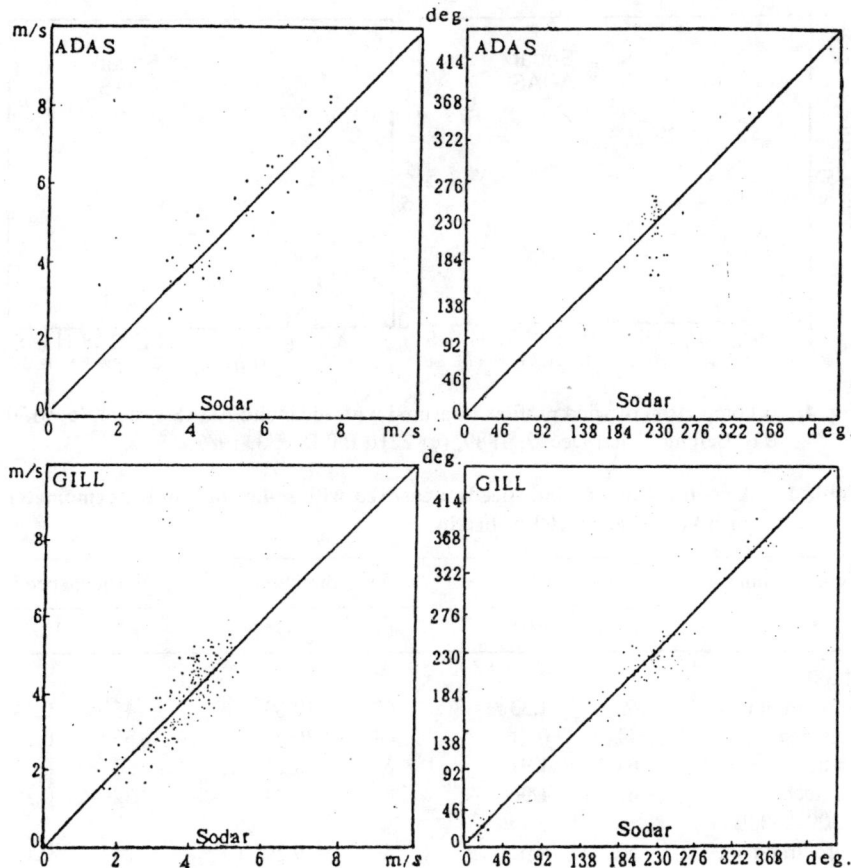

Fig. 3. The correlation of wind detected with sodar and direct measuring system.

component anemometer basd on 177 samples, the variance of wind direction deviation is 14.13, and the variance of vertical velocity deviation is 0.15. On the other hand, the variance of wind velocity is 0.54 which is calculated from the deviation between Doppler sodar and tethersonde based on 46 samples, and the variance of the wind direction deviation is 25.92. Two examples of profiles for wind velocity and direction measured by sodar and ADAS tethersonde shown in Fig. 4 indicate that these results are very close to each other.

In order to make a comparison of the experimental results mentioned above with that in foreign countries, the results of comparative experiments carried out at Boulder, CO U.S.A. are shown in Table 1. The Doppler sodars made in four companies were compared at BAO station in Boulder, and the comparative experiment was conducted in 1982 supported by Wave propogation Laboratory, NOAA, U.S.A. The results listed in Table 1 show that the results of our comparative experiment are in agreement with that in Boulder experiment.

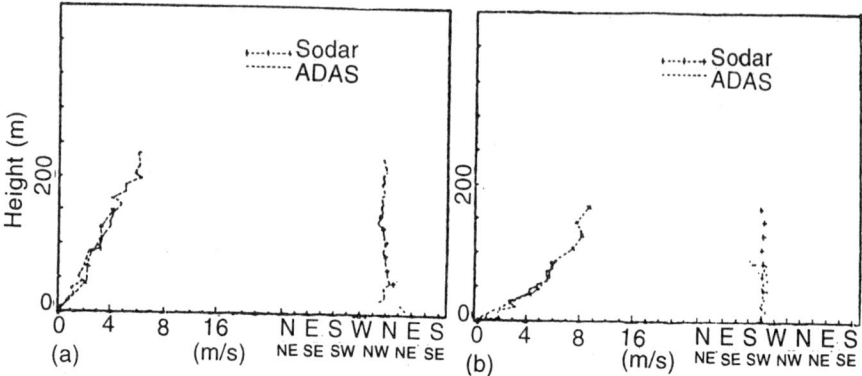

Fig. 4. A comparison of wind profiles measured with sodar and ADAS system. (a) 0150 BT (Beijing Time) Dec. 3, 1989, (b) 2210 BT Dec. 3, 1989.

Table 1. A comparison of wind speeds measured with sodar and sonic anemometer on BAO tower at 200 m height

Type of sodar	Wind speed		Wind direction		Vertical speed	
	n^*	VD^{**}	n^*	VD^{**}	n^*	VD^{**}
(1200–1700BT)						
Aerovironment	69	1.33	69	19.69	45	0.34
Remtech	44	0.91	44	30.69	35	0.15
Radian	64	1.91	64	18.25	37	0.32
Xontech	67	1.96	67	29.24	32	0.38
(0000–0500BT)						
Aerovironment	59	0.62	59	19.29	25	0.16
Remtech	41	0.42	41	11.65	20	0.10
Radian	53	0.88	53	24.31	27	0.22
Xontech	63	0.87	63	34.45	27	0.22

n^* sample number VD^{**} the variance of deviation

Some Characteristics of Atmospheric Boundary Layer

Lump Structure Echo

In recent years, in connection with some weather processes, meteorologists have obtained some special echoes by acoustic radar. Wyckoff [4] has detected synoptic scale subsidence inversion layer in atmospheric boundary layer. In thunderstorm weather condition, Hall et al, [7] have succeeded in obtaining obvious density current structure echo. Under the lower layer cumulus condition, Gaynor et al, [5] have found echo of a cylindrical shape with a vertical scale of about several hundred meters. These echoes help us much in understanding the inner characters of some synoptic processes and boundary layer structure caused by these processes.

In a continuous observation in three periods from 1976 to 1979 in Beijing area, we found some smallscale lump echoes that sometimes exist in boundary layer atmosphere [13]. The appearance of these echoes are usually connected with some synoptic processes.

The experiment was made in the western suburb of Beijing by a monostatic acoustic radar [8]. We found sometimes some strong echoes, appearing suddenly at a certain height, lasting generally for several seconds and then disappearing abruptly. The outline of these strong echoes was very obvious. After a moment analogous strong echoes appeared at another height. Fig. 5 shows the appearance of these strong echoes. From the figure it is seen that during the period from 18:30 to 00:20, 12 September, 1979, upto a height of 300 m these strong echoes appear very frequently. Morphologically, these echoes are different from the usual inversion or thermal ones. These strong echoes sometimes presented themselves as rhombic, and sometimes as rectangular. In facsimile photograph, these echoes show configurations of irregular behavior, different scales and clear outline in boundary. This may be called "lump echo". These lump echoes sometimes appeared very frequently, sometimes very rarely. As seen from Fig. 5, in the period from 18:30 of 11th to 00:20 of 12th, September, 1979, there are 875 lump echoes of different sizes. The data observed on the tower show that in the same period the inversion layer is formed and the stratification is stable between 300 and 600 m. From Fig. 5, it can also be seen that, in the lower part of the atmospheric boundary layer there exist echoes from inversion or stable stratification sometimes with lump echoes added to these echoes. Before this period, there was almost no indication of these lump echoes. It started raining at 00:20, the dark stripes from 00:20 to 01:20 in Fig. 5 are its interference signals. After the rain the lump echoes usually become rare or even disappear.

This means that this lump echo has a certain space scale and "living" time. According to the assumption of "frozen turbulence", the horizontal scale of the lump echo can be calculated in terms of its dwelling time and mean wind speed at a height at which the lump echo appears. By means of the wind speed observed at the tower we have calculated the scale of lump echo as shown in Fig. 5, and given the frequency number distribution chart of lump scale (see Fig. 6). From Fig. 6, it is seen that most of the horizontal scales are less than 30 m, while the horizontal scales larger than 50 m are rare. The maximum horizontal scale of lumps is 190 m.

From Fig. 5 we can see that the vertical scale of most lump echoes is not greater than an acoustic pulse width (equivalent to 17 m). The lump echoes with vertical scale larger than 17 m is only 25 in number.

Another type of lump echoes with different morphological characters are sometimes found by sodar detection. Lump echoes of this kind often are concentrated in a short time range. Their vertical range is large (the maximum range is up to 200 m). They usually form a cluster of lump echoes. As seen from Fig. 7, in the period of 06:20–06:30, 06:40–06:50, 07:08–07:12, etc. some such typical clusters were observed. In general, they appear at a height from 100 to 300 m. In the inner

208 *Applications in the Atmosphere*

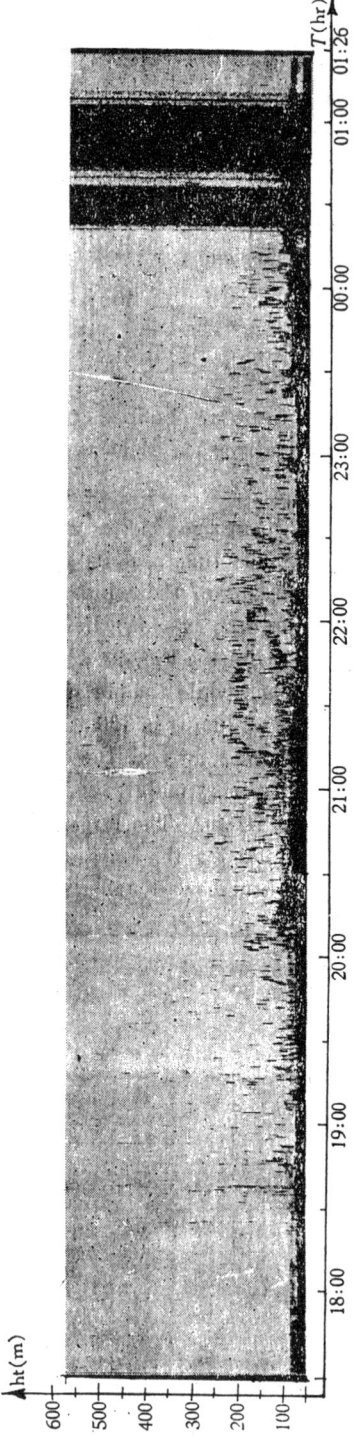

Fig. 5. The facsimile photograph of acoustic sounder from 17:29 of 11th to 01:26 of 12th September, 1979.

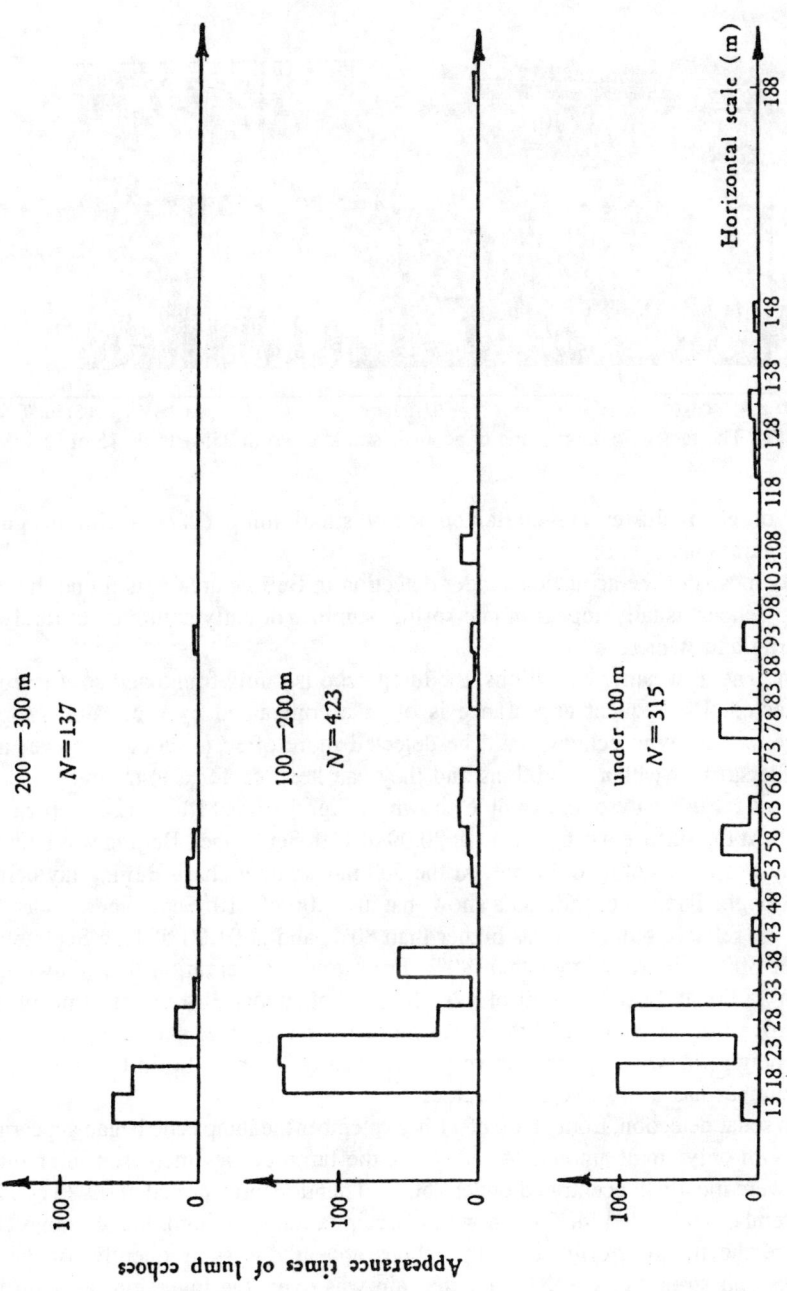

Fig. 6. The frequency number distribution of horizontal scale in different heights from 18:30 of 11th to 05:00 of 12th September, 1979.

210 Applications in the Atmosphere

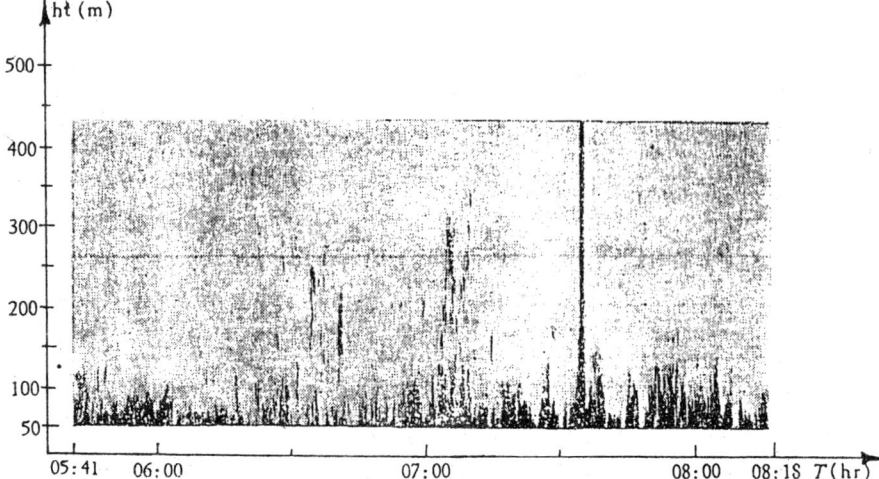

Fig. 7. The facsimile photograph of acoustic sounder. From 05:41 to 08:15 of 25th May 1977.

part of each cluster are contained many small lump echoes with irregular configurations.

By means of the acoustic sounder detection in Beijing area, it is found that the lump echoes usually appear in late spring, summer or early autumn, but rarely in autumn and winter.

In general weather conditions, the lump echo is rarely found and so it is often neglected. Its frequent appearance is often accompanied by a certain synoptic process. The lump echoes could be detected more often in cloudy, shower and thunderstorm weather conditions and they can keep on for a long time.

Now let us analyse an example shown in Fig. 5 from 11th to 12th September 1979. At the surface weather chart at 20:00 of 11th September, Beijing was situated in the warm area of a cold front. At the 500 mb weather chart, Beijing lay before the trough. The radiosonde data show that at 19:00 of 11th September, under 1.5 km, the relative humidity was higher than 80%, and at 01:00 of 12th September, under 300 m it was higher than 90%. From tower observation it is shown that from 18:00 of 11th to 00:00 of 12th September, under 300 m the atmospheric stratification was stable and there was an inversion layer on surface layer. Under this height the wind speed was very low, usually 1-2 m/s. So in this period the Ri number has a positive large value.

In sodar detection, from 18:00 of 11th September the lump echo began appearing, but with only small amount. After 20:00, the lump echoes increased in amount, and were mostly concentrated under 250 m. Thunderstorm started at 00:20 of 12th September and lasted for 50 min with a precipitation of 39 mm. Fig. 5 shows that before the thunderstorm the lump echoes appeared very frequently, on larger scales and strength as well. After the rain was over, the inversion layer on the surface layer disappeared, the echo of acoustic radar became weak, and the lump echo also vanished.

From the examples as described above, we can see that frequent appearance of lump echo in boundary layer atmosphere is often under such weather conditions as light wind, stable stratification and large relative humidity.

Characteristics of Vertical Distribution for Wind Velocity in Mountain Area

A Doppler sodar is set in the courtyard of a storehouse which is situated in northwestern suburb of Beijing and attached to State Oceanic Adminstration. The storehouse is 70 m above sea level. About 5 km north of the storehouse rises a mountain of Jiansanzui, 700 m above sea level, to west of the storehouse is Shuanglongling, further west there is a mountain, 1000 m above sea level. To the east of the storehouse the plain stretches out and the terrain in the south of the storhouse slopes gently and 8 to 9 km south of the storehouse there are some hills. Continuous detection of the vertical distribution of wind and temperature stratification in the atmospheric boundary layer was carried out by using the Doppler sodar.

Because the observational site (the storehouse) is near mountain area, obviously local wind characteristics are present under weak wind weather. When the weather is controlled by severe weather systems, they are not so obvious. Below we will analyze the vertical characteristics of local wind under weak wind weather. On the 700 hpa weather map of 20:00 BT (Beijing Time) 24 May, Beijing is situated in the rare of trough. On the simultaneous surface weather map, eastern China is controlled by a high pressure, and the surface wind is very weak, Beijing is on the left side of the high pressure and southwest wind prevails. On the surface weather map of 20:00 BT 25 May, it is calm and still controlled by a high pressure, and at 08:00 BT southwest wind prevails with speed 2 to 4 m/s. In the conditions as described above, it is clear that in Beijing area, in the condition of weak wind, the local wind and inversion formed due to radiation cooling play an important role. On a calm night and weak wind condition, the pollutants are difficult to diffuse, resulting in serious air pollution. Notice that in this case the local wind induced by the terrain will appear as a good atmospheric condition which may be benificial to diffusion of pollutants. Therefore, study on the effect of local wind and inversion is important.

Fig. 8a shows the wind profiles detected with Doppler sodar in the afternoon (14:30–15:00) of 24 May. Every wind profile in the figure is obtained with 10 minutes average. It is seen that in the lower atmosphere below 400 m the wind blows from the south with velocity 4-6 m/s between 100-400 m. Below 100 m there exists a wind shear layer.

The observational results also showed that after 17:00 BT an inversion layer appeared at 200 m height and the wind velocity gradually increased. A measured result of Doppler sodar from 17:30 to 18:00 BT is given in Fig. 8b, showing that a maximum wind velocity larger than 8 m/s appears at level 170 m, and it is close to the velocity of low-level jet. This indicates that near the mountain area the maximum wind velocity of low-level jet in the lower atmosphere may appear before sunset, which is different from the situation in flat area.

212 *Applications in the Atmosphere*

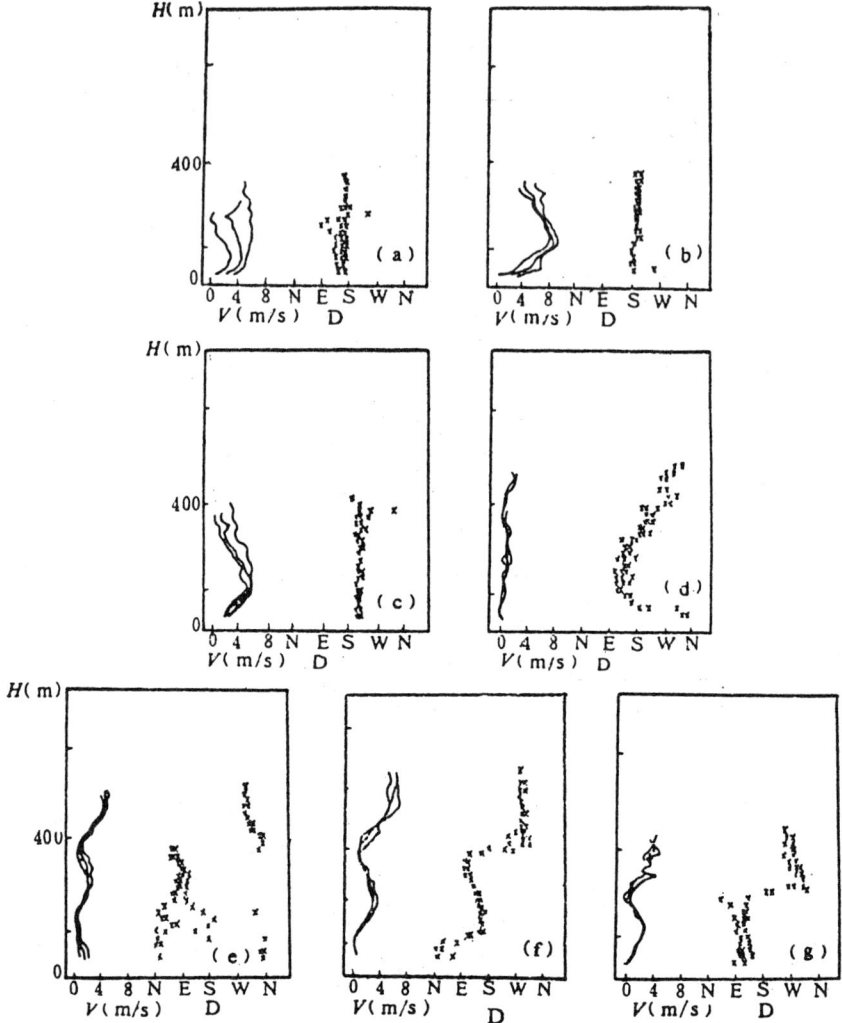

Fig. 8. The vertical distribution of the wind detected with the Doppler Sodar (May 24–25, 1989).

Fig. 9 is the vertical distribution of wind measured at Beijing Meteorlogical Observatory at 19:00 BT 24 May. It can be seen from Fig. 9 that in the atmospheric boundary layer the southerly wind prevails below 1 km. Near the height of 350 m there is a low level jet with a maximum velocity of 12 m/s. The geostrophic wind is 8.2 m/s estimated from surface pressure gradient at 20:00 BT 24 May. The 19:00 BT radiosonde data measured at Beijing Meteorological Observatory showed that the lower atmosphere is under stable condition. Therefore it is easy to understand the above fact that the lowlevel jet is larger than the geostrophic wind.

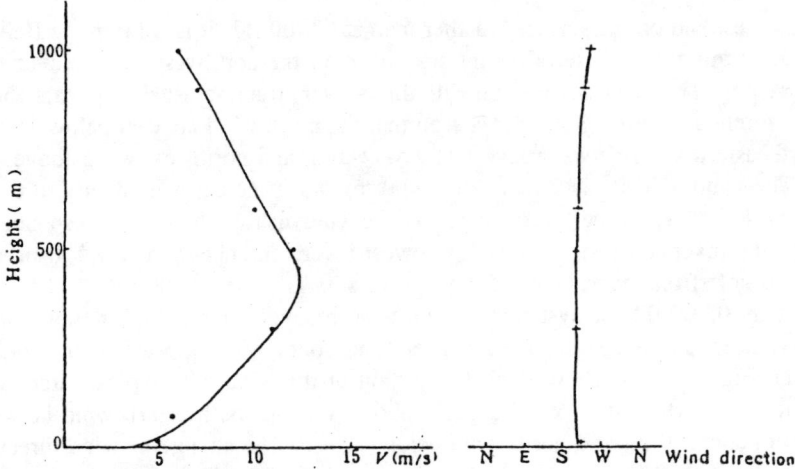

Fig. 9. The vertical distribution of wind detected at Beijing Meteorological Observatory. (1900 BT 24 May 1989).

The observational results detected with Dopplar Sodar at the storehouse shown in Fig. 8c illustrate a maximum southern wind velocity of 6 m/s at level 160 m in the wind profile at 19:00 BT. Comparing Fig. 9 with Fig. 8c, we can see that the southwesterly wind prevails above the storehouse and Beijing Meteorological Observatory. But the wind velocity measured at the storehouse is obviously lower than that measured at Beijing Meteorological Observatory and the height of maximum wind speed at the storehouse is about 180 m lower than that at Beijing Meteorological Observatory. It may be explained that the sourthern wind produced by the weather system decreases due to the increasing terrain resistance effect on entering into the mountain area. In addition, in the north and west of the storehouse, the drainage flow induced by radiation cooling of the mountain declivity can balance a part of the south wind. Therefore it will decrease the wind which is controlled by weather system. the The height of maximum wind velocity at the storehouse is different from that at Beijing Meteorological Observatory. The elevation of Beijing Meteorological Observatory is 55 m, and it is only 15 m lower than the elevation of storehouse. Seemingly, the vertical inhomogeneity of the drainage flow may be the main cause for the height differences of the maximum wind velocity at the two stations. In fact, the wind velocity is weak at the level of 300 m above the storehouse. It means that the drainage flow is strong above 300 m height. The corresponding detection of temperature stratification with sodar also illustrates that the strong inversion layer exists near 300 m.

In the mountain area the drainage flow is gradually increasing with time. A vertical wind profile detected with sodar from 21:30 to 22:00 is shown in Fig. 8d Below 400 m the wind velocity is weak, only 2 m/s, the wind is from south in the layer between 100 to 150 m. The wind direction becomes west and northwest near the surface layer and above 300 m. After midnight, the system wind gradually decreases, and on the surface weather map at 02: 00 BT, it is calm in the Beijing area. At this time, the terrain wind prevails over the northwestern mountain area of Beijing. The wind distribution with three-layer structure usually appears above the storehouse. From Fig. 8e it is seen that the north wind appears below 150 m, northeastern wind between levels 150 to 400 m, and northwest wind above 400 m The wind velocity is 2 m/s below 400 m, and it increases gradually to 6 m/s above 400 m. The three-layer structure of the wind distribution is in correspondence with the inversion distribution. The lowest inversion appears below 150 m, and the top of strong echo layer of upper inversion is close to 400 m.

After 03:00 BT the systematic south wind recovers gradually. The north wind prevails in the upper and lowest layers and south wind appears in the middle layer. Fig. 8f gives the vertical distribution of the wind with typical three-layer structure. Northwestern wind appears above 400 m, southeastern wind between 100 m and 350 m, and north wind below 100 m. The change of wind direction is very quick between layers, i.e. the transition layer of wind direction is very thin, and the wind shear is strong. The multiple layer structure also can be seen in the vertical distribution of wind speed. The wind is weak below 100 m and increases gradually with height above 100 m, and reaches a maximum value at 250 m. Then the speed decreases gradually with height up to 400 m. This type of wind distribution is maintained until 06:00 BT. This vertical structure of wind velocity may be formed by the overlapping of systematic southern wind on down slope wind induced by the mountain slopes at different heights in the complex mountain area. The northwestern wind above 400 m may be correlated with the systematic wind at the upper boundary layer. The down slope wind from northwest induced by the terrain is consistent with the systematic wind, and contributes to downward extension of northwestern wind. It may intensify the wind shear between the layer with northwest wind and the layer with south wind below 400 m. After 06:00 BT, because the slope surface is heated by the solar radiation, the northern wind in the surface layer below 100 m vanishes gradually and changes to southern wind. The observational data of Beijing Meteorological Observatory at 07:00 BT show that it is calm on the ground and the northern wind prevails in the upper layer. The detection of Doppler acoustic radar at 07:00 BT at the storehouse shows (see Fig. 8g) that there is eastern/southeastern wind below 250 m and northwestern wind above 300 m, and an obvious wind shear layer appears between 250 and 300 m. Correspondingly the data of the vertical distributions of wind velocity show a clear two-layer structure. There is a minimum value of wind speed near 300 m. The southern wind is maintained until 09:00 BT in the lower layer below 200 m. After that, as the systematic northwestern wind increases, the whole boundary layer is controlled by the northwestern wind.

The nocturnal multi-layer inversion is also a notable problem. A result detected with sodar on 21 December, 1988 presented a typical process of multi-layer inversion. In the night of 21 December, Beijing was controlled by a high pressure, the wind with northern direction was very weak. After sunset, a surface radiation inversion layer was formed, with a multi-layer inversion structure gradually formed. Fig. 10 is an example of observation at the building top of the National Research Center for Marine Environment Forecasts, in the western suburb of Beijing. Although the nocturnal multi-layer inversion has been discussed by many scientists, but in this example, we have to point out that there is a thin stable echo layer between 100 m and 200 m. This layer is very thin, and is almost maintained throughout whole night. Especially, it is clearly seen with a color terminal. As shown in Fig. 10, there is a stable layer at level 150 m, the depth of this echo layer is about 10 m. Sometimes the depth of the echo layer may be larger than 10 m, about 20 to 30 m. This layer has a clear edge. In some cases an obvious wave with billow structure as shown in Fig. 10 exists on this stable layer. The amplitude of this wave could reach 200 m. Even so, the strong wave activity could not break the thin stable layer. In a stable layer, although the turbulent intensity is very weak, but the turbulence with certain intensity still exists. Even in the lower atmosphere with very stable stratification the intermittent turbulence could be formed by interaction between the radiation cooling and wind shear in the surface layer (Zhou et al, [17, 18]). Why the turbulent movement even the wave with large amplitude could not break the thin stable layer? What physical process could produce and maintain this thin stable layer for a long time? It is certainly a valuable subject for future research.

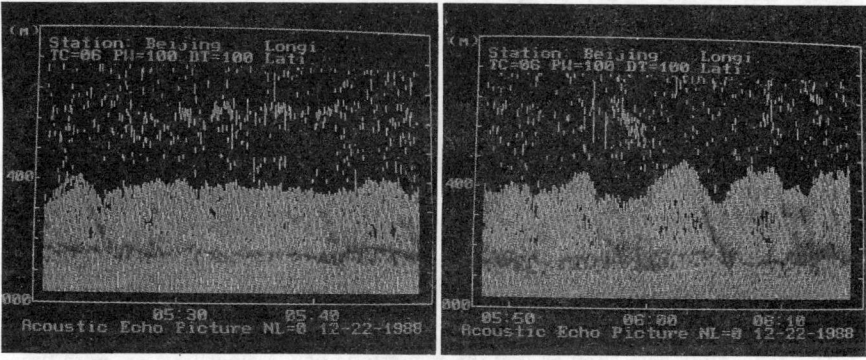

Fig. 10. The temperature structure detected with sodar (22 Dec., 1988).

Calculation of Temperature Structure Coefficient and Heat Flux

Calculation of Temperature Structure Coefficients
The echosonde equation for the monostatic backscatter case may be written as follows:

$$P_r = P_t \eta_r \eta_t G R^{-2} A\ \sigma(\theta)\ (c\tau/2)\ e^{-2\alpha R} \qquad (4)$$

where P_r is the received electric power, P_t is the transmitted electric power, η_t is the transmitting electro-acoustic transducer coefficient, η_r is the receiving acoustoelectric transducer ocoefficient, σ is the scattering cross section, c is the velocity of sound in the scattering region, τ is the pulse length, R is the height to the scattering region, A is the effective area of antenna, G is antenna direction compensation factor, and α is the attenuation coefficient, which depends on the transmitted frequency, the atmospheric temperature, and the relative humidity.

According to the theory of locally isotropic turbulence, in dry air the scattering of an acoustic wave in turbulent medium is given by the formula

$$\sigma(\theta) = 0.033 K^{1/3} \cos^2 \theta \left[\frac{C_V^2}{C^2} \cos^2 \frac{\theta}{2} + 0.13 \frac{C_T^2}{T^2} \right] \times \left(\sin \frac{\theta}{2} \right)^{-11/3} \qquad (5)$$

where $\sigma(\theta)$ is the scattered power per unit volume, unit incident flux and unit solid angle from the initial direction of propagation, θ is the scattering angle measured from the original incident direction, C_V^2 is the velocity structure coefficient, C_T^2 is the temperature structure coefficient, $K = 2\pi/\lambda$ is the acoustic wavenumber, λ is the acoustic wavelength, and T the average absolute temperature in the scattering volume. When $\theta = 180°$ (backscatter), from similarity theory, Eq. (5) becomes

$$\sigma(180°) = 0.0288 \, (C_T^2/T^2) \, \lambda^{-13} \qquad (6)$$

Substituting Eq. (6) in Eq. (4) gives

$$Pr = P_t \eta_r \eta_t GAR^{-2} \, (c\tau/2) e^{-2\alpha R} \, 0.0288 \, (C_T^2/T^2) \, \lambda^{-1/3},$$

or,

$$C_T^2 = P_r R^2 T^2 \lambda^{1/3} / [0.0288 \, \eta_t \eta_r GP_t (c\tau/2) \, A e^{-2\alpha R}] \qquad (7)$$

If the values of the parameters of an acoustic sounder system are given, we can compute the magnitude of C_T^2 from the echo sounder power.

The sodar data and meteorological tower data obtained in the countryside of Beijing from 1975 to 1977 were used to study C_T^2 and atmospheric stratification.

The echosonde data showed that the magnitude of C_T^2 clearly varies with time, and on condition whether in stable or unstable stratification, that it is stochastic in character, and that sometimes it is strongly intermittent. We made a statistical analysis of the probability distribution of the value of C_T^2 at different heights. From Fig. 11 we can see the probability distribution of C_T^2 at 50 m. Both in stable and unstable stratification the magnitude of C_T^2 has an obvious positive bias. Based on the sodar data at different heights below 300 m, we discovered that the probability distribution of C_T^2 at these heights shall have the anomalous character described above. The skewness is usually between 1.0–3.0, and in unstable stratification it increases with height, but in stable stratification its distribution with height follows no obvious rule.

Development of Sodar Detection and its Application 217

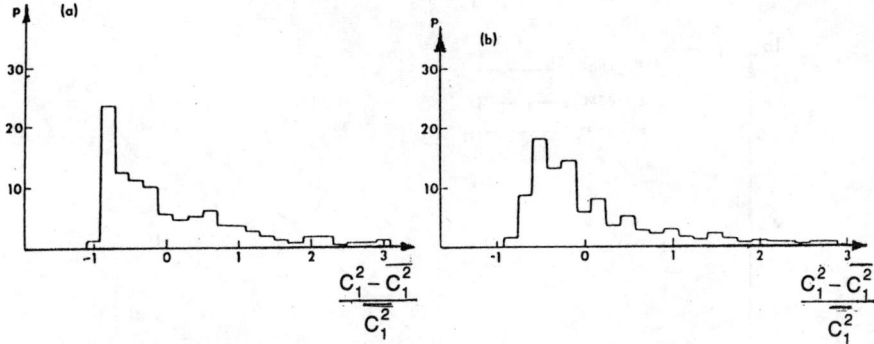

Fig. 11. (a) The probability distribution of C_T^2 at 50 m altitude. (Unstable stratification, 13:00–13:35, 19 October 1977.) (b) The probability distribution of C_T^2 at 50 m altitude. (Stable stratification, 22:35–23:10, 19 October 1977.)

Cumulative probability distributions of C_T^2 are shown in Fig. 12 and Fig. 13 (In Fig. 12, 13, and 14 the values of C_T^2 on the ordinate are relative.) In these figures, the ordinate system is logarithmic versus Gaussian, the ordinate is the magnitude of C_T^2 and the abscissa is cumulative percentage of values less than the given ordinate value of C_T^2. Figures 12 and 13 show that in both stable and unstable stratification the curves of cumulative porbability distribution are straight lines, i.e., they fit the lognormal distribution.

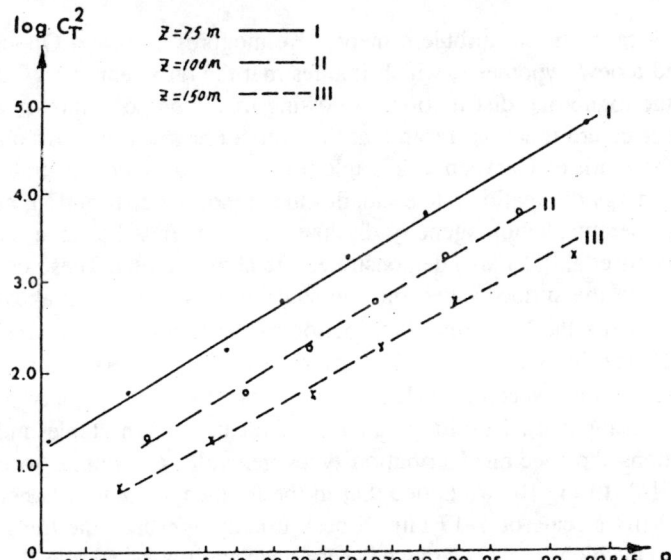

Fig. 12. Cumulative probability distribution of C_T^2 in stable stratification: Abscissa gives percentage of values less than given ordinate value of C_T^2 (05:23–05:53, 20 October 1977.)

218 Applications in the Atmosphere

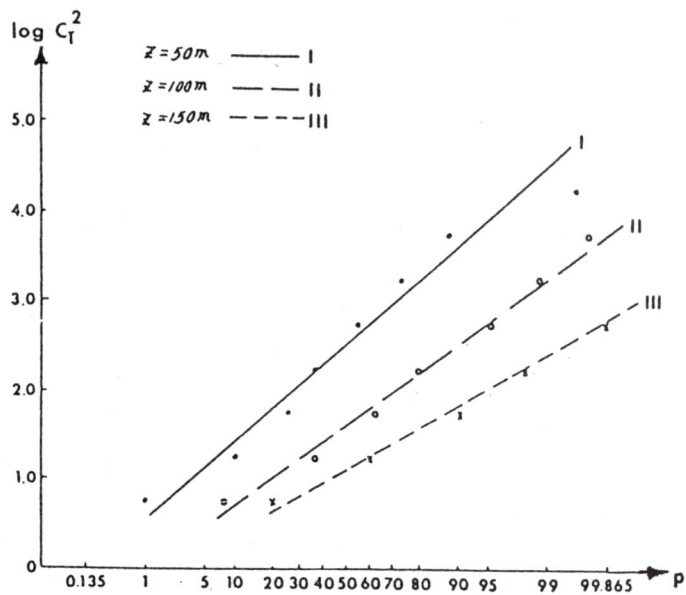

Fig. 13. Cumulatlve probability dibtribution of C_T^2 in unstable stratification: Abscisa gives percentage of values less then given ordinate value of (13:00-13:35, 19 October 1977.)

In the local isotropic turbulent theory, Kolmogorov [19] and Oboukhov [20] introduced a new hypothesis, which implies that the turbulent energy dissipation rate fits the lognormal distribution. Following this hypothesis they modified the turbulent structure function. From that time on, a great deal of field observation, such as the work of Gurvech and Yaglom [21], not only demonstrated that the turbulent energy dissipation rate ε satisfies the lognormal distribution, but also the rate of temperature inhomogeneity dissipation N satisfies the same distribution. Later Stewart et al, [22] also demonstrated the above results. These observations were made in the surface layer, but the sodar results mentioned above indicate that C_T^2 also fits the lognormal distribution in the atmospheric boundary layer. This illustrates that the lognormal distribution is more or less a general rule.

We then made a spectral analysis of the magnitude of C_T^2 (only the relative values are computed). Results of analysis indicate that in stable and unstable stratifications the spectral distribution types generally are similar to each other (see Fig. 14). In Fig. 14 we notice that in the frequency range between $10^{-3} - 3 \times 10^{-3}$ s^{-1} (time scale of 5–17 min) a peak usually appears. The transition from the peak to the high-frequency range is very steep and fits the f^{-3} rule. These results clearly indicate that in the atmospheric boundary layer periodic activity usually exists, with a period of about several minutes to more than ten minutes. This periodic activity reflects the thermal plumes during the day and may reflect gravity waves at night. When the same stability prevails at different heights, the location

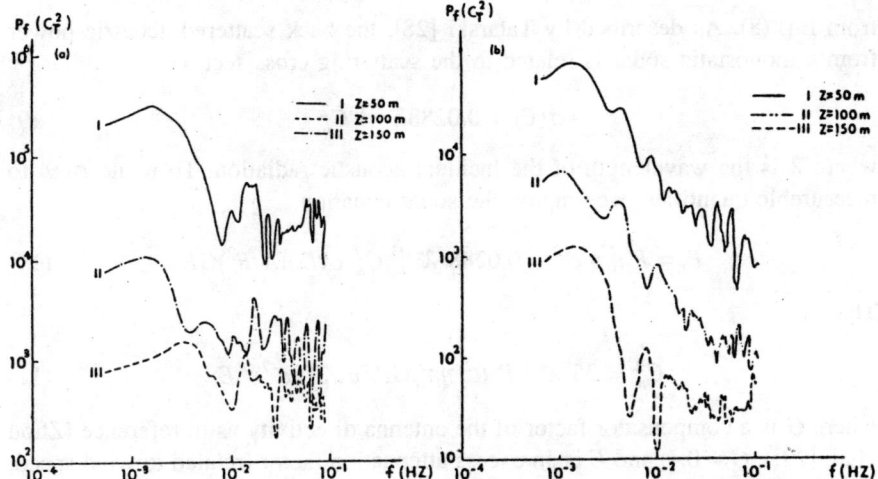

Fig. 14. (a) The spectra of C_T^2 in stable stratification (22:35–23:10, 19 October 1977).
(b) The spectra of C_T^2 in unstable stratification (14:00–14:37, 19 October 1977).

of the peaks of C_T^2 spectra are almost in the same frequency range. When the stability increases, the peaks of the spectra move towards low frequencies.

Sethuraman [23] has analyzed the rule of wind variability in the atmosphere in the presence of gravity waves and computed the energy spectra of the three wind components (u, v, w). The spectra obtained by him are similar to those observed by us. In the spectra of the three wind components there exists a peak range of energy near a frequency of 2×10^{-3} Hz. Between the peak range and high frequency range there is a clear steep transitional range. These results are similar to our observational results.

Calculation of Sensible Heat Flux
Since 1970, the acoustic remote sensing technique has been applied to observe the marine boundary layer. For example, in the GATE experiment, the monostatic sodar and Doppler sodar were installed on a NOAA ship (Mandics et al, [2], Gaynor et al, [24, 25]). Ottersten [26] also set up a monostatic sodar on a ship in Jonswaps 2 experiment over the North Sea. These observations provided a good understanding of the characteristics of the marine boundary layer.

According to the similarity theory, Wyngaard [27] obtained an expression for C_T^2,

$$C_T^2 = 2.68(g/T)^{-2/3}\left(\frac{Q}{C_P\rho}\right)^{4/3} Z^{-4/3} \tag{8}$$

where C_T^2 is the temperature structure coefficient, Q is the surface sensible heat flux, z is the height, g/T is a buoyancy factor. If we can quantitatively measure the magnitude of C_T^2 at different heights, the sensible heat flux can be estimated from Eq. (8). As described by Tatarski [28], the back scattered acoustic power from a monostatic sodar is related to the scattering cross section.

$$\sigma(\theta) = 0.0288\lambda^{-1/3} C_n^2 \qquad (9)$$

where λ is the wavelength of the incident acoustic radiation. To relate $\sigma(\theta)$ to measureble quantities, we employ the sodar equation,

$$P_r = P_t \eta_t \eta_r e^{-2\alpha R} \, 0.0288 \, \lambda^{-1/3} \, C_n^2 (c\tau/2)(A/R^2) GE \qquad (10)$$

That is

$$C_n^2 = 35 \, \lambda^{1/3} \, P_r R^2 / \eta_t \eta_r G P_t (c\tau/2) A e^{-2\alpha R} E \qquad (11)$$

where G is a compensator factor of the antenna directivity as in reference (Zhou et al, [29]), $G = 0.4$, and E is an excess attenuation factor (related to wind speed, turbulence levels, width and wavelength of incident radiation). Assuming $E \approx 1$, the underestimation of the surface heat flux due to excess attenuation is believed to < 5%. The factor $\exp(-2\alpha R)$ is an assumed constant. For our sodar system, we adopt the parameters as $\eta_t = 0.40$, $\eta_r = 0.34$, and the effective receiver aperture $A = 1.13 \, m^2$. The structure coefficient of refractive index C_n^2 in Eq. (11) was given by Wesely [30]

$$C_n^2 = [C_T^2 / 4T^2] \, \alpha_a^2$$

$$\alpha_a^2 = 1 + \gamma_{eT}(2DC_e T/C_T P) + (DC_e T/C_T P)^2$$

in which $D = 0.307$, $\gamma_{eT} = C_{eT}^2 / C_e C_T$, C_{eT} is a structure coefficient of temperature fluctuation and moisture fluctuation, C_e^2 is the structure coefficient of moisture fluctuation, $C_e T/C_T P = (5.03 \mid \beta \mid)^{-1}$, $\beta = Q/L_W E$ is the Bowen ratio.

Coulter and Weseley [31] calculated the sensible heat flux over land surface using sodar and laser scintillation measurements. Because the air over land surface is drier than that over marine surface, so γ_{eT} is small. If we neglect γ_{eT} in Eq. (12), there is no significant error appearing in the estimation of sensible heat flux over land surface. But we can not neglect γ_{eT} in the marine boundary layer, As analyzed by Gaynor [32] in GATE experiment, here γ_{eT} is large. In fact, it is very difficult to obtain the variation of γ_{eT} with height; for simplicity, the parameterization method should be adopted. Wyngaard [33] analyzed the AMTEX data and suggested a relation between γ_{eT} and Z/Z_i, where Z_i is the depth of mixed layer, i.e.

$$\gamma_{eT} = 0.9 - 1.8 \, Z/Z_i \qquad (13)$$

The α_a^2 in Eq. (12) can be written as

$$\alpha_a^2 = 1 + (0.9 - 1.8 \, Z/Z_i)(0.122/\mid \beta \mid) + (0.06/\mid \beta \mid)^2 \qquad (14)$$

From here, we can get the virtual temperature structure coefficient

$$C_{TV}^2 = C_T^2 \{1 + \gamma_{eT}(0.756\, C_e T/C_T P) + (0.378 C_e T/C_T P)]$$
$$= C_T^2 \{1 + (0.9 - 1.8 Z/Z_i)(0.15/|\beta|) + (0.075/|\beta|)^2\} \quad (15)$$

Notice that an expression of the virtual heat flux is given as

$$Q_V = Q(1 + 0.07/\beta_s) \quad (16)$$

Now if we use C_{TV}^2 and Q_V to replace C_T^2 and Q in Eq. (8), then the complete expression of C_T^2 describing the effect of humidity fluctuation on C_n^2 can be given as

$$C_T^2 = 2.68\, (g/T)^{2/3}\, (Q/C_p \rho)^{4/3}\, Z^{-4/3}\, (1 + 0.07/|\beta_s|)$$
$$\{1 + (0.9 - 1.8\, Z/Z_i)(0.15/|\beta|) + (0.075/|\beta|)^2\}^{-1} \quad (17)$$

The surface sensible heat flux in the convective marine boundary layer can be then estimated as

$$Q = 0.48 \rho C_P (g/T)^{1/2}\, (C_T^2)^{3/4}\, Z \times \gamma \quad (18)$$

where
$$\gamma = \gamma_0 \gamma_1 \quad (19)$$
$$\gamma_0 = (1 + 0.07/|\beta_s|)$$

$$\gamma_1 = \left\{ \frac{1 + (0.9 - 1.8 Z/Z_i)(0.15/|\beta|) + (0.075/|\beta|)^2}{1 + (0.9 - 1.8 Z/Z_i)(0.122/|\beta|) + (0.06/|\beta|)^2} \right\}^{3/4}$$

Here in the Bowen ratio β is related to β_s (sea surface Bowen ratio), which can be calculated from the meteorological data observed routinely on board of the ship by the bulk transfer method. The relation between β and β_s is given by Wyngaard [33]

$$\overline{w'\theta'} = Q(1 - 1.5 z/z_i) \quad (20)$$

$$\overline{w'q'} = M(1 - 1.5\, z/z_i) \quad (21)$$

$$\beta = \overline{w'q'}/\overline{w'\theta'} = Q(1 - 1.5 z/z_i)/M(1 - 1.5\, z/z_i)$$
$$= \beta_s(1 - 1.5 z/z_i)/(1 - 0.5\, z/z_i) \quad (22)$$

Thus we have obtained a formula which can estimate the sensible heat flux in the convective marine boundary layer using the profiles of C_T^2 measured by monostatic sodar.

Several examples of the distribution of C_T^2 with height are shown in Fig. 15. Under the convective condition, the variation of C_T^2 with height fits the −4/3 power

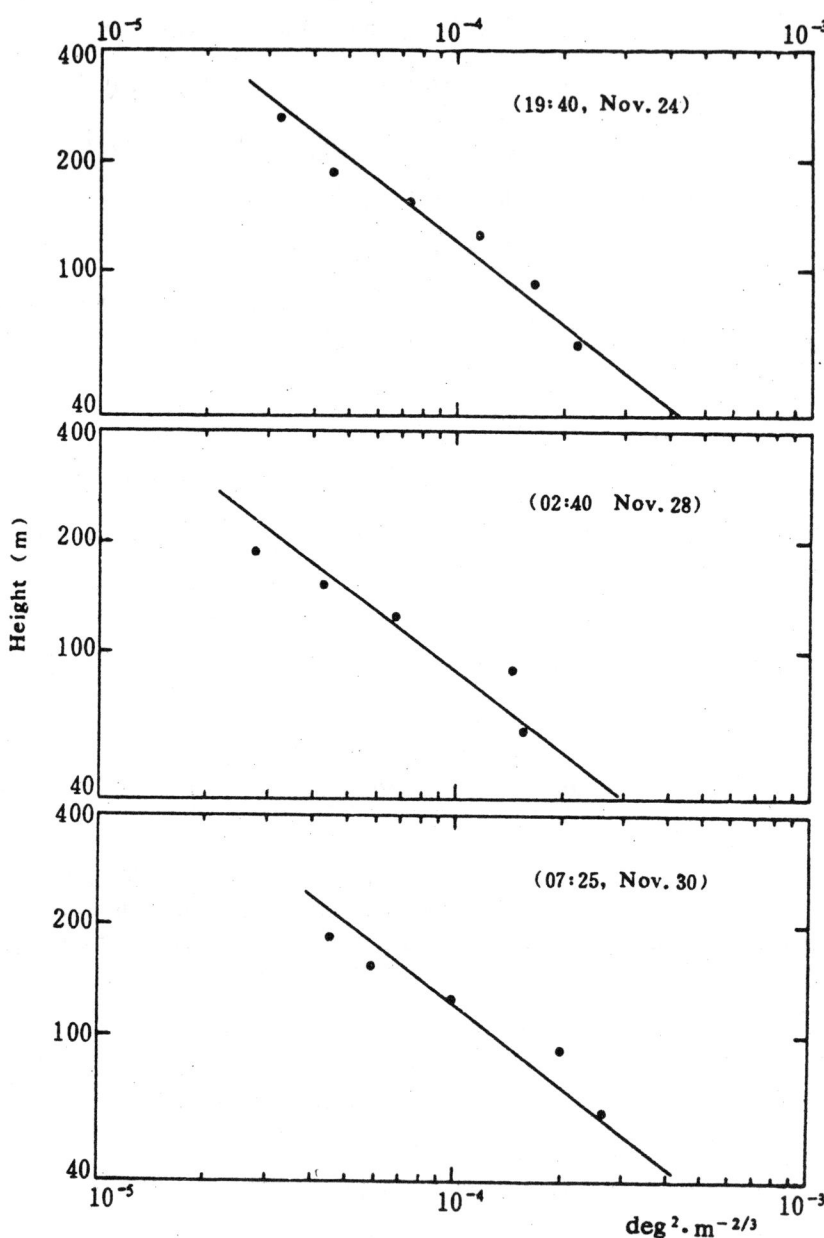

Fig. 15. The distribution of C_T^2 with height

law when the heights are lower than 150–200 m. If the slope of C_T^2 deviates from $-4/3 \pm 1/5$, we will omit these data for the estimation of sensible heat flux. We take out 5 sets of data from a total of 39 sets of data. At heights of 50–500 m, the values of C_T^2 are in the range of 10^{-3}–10^{-5} $K^2 m^{-2/3}$.

The Bowen ratio β_s is large over land surface. If we take $\gamma \approx 1$, the errors for the calculation of heat flux will be less than 10%. But over marine surface, the value of β_s is small. Then, it is necessary to consider the influence of γ on the estimation of sensible heat flux. The values of β_s, for three sections observed in the West Pacific (1986–1987) are 0.088, 0.107, 0.071 respectively, and the average value is 0.088. Therefore, when we estimate the sensible heat flux over the equatorial western Pacific Ocean, it is necessary to make a correction to the function γ. There are 34 sets of C_T^2 profiles observed during the same time period as with routine meteorological data. We estimated the sensible heat flux by sodar data and by the bulk transfer method. A comparison of these two methods is shown in Fig. 16 where the correlation coefficient is 0.947.

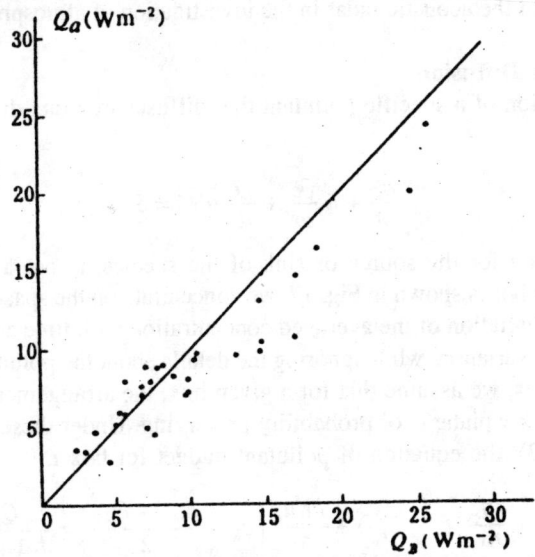

Fig. 16. A comparison of sensible heat flux calculated by sodar measurements (Q_a) and by bulk transfer meathod (Q_b).

Strictly speaking, the sensible heat flux calculated by sodar measurements should be compared with the direct flux measurements, as we know, direct measurements of sensible heat flux over marine surface are difficult. Some scientists (Large et al) [34], have made a comparison of sensible heat flux by both the direct flux measurement and the bulk transfer methods, and the results are close to each other. Thus, it has a representative character to compare the sensible heat flux calculated by both the sodar measurements and the bulk transfer method. The scattering of the points in Fig. 16 may be related to the accuracy of measurements and to the different spatial volumes of these two methods.

Application of Sodar Sounding to Atmospheric Dispersion

Acoustic sounder can be used for measuring, with real-time advantage, such quantities in the atmospheric boundary layer as velocities, mixing height, capping inversion thickness and strength, which are necessary for investigating the transport and diffusion of atmospheric pollutants. In particular, it can provide fairly reliable results in a case where an ordinary model calculation would not work well, because certain synoptic processes bring about a singular variation of the structure of boundary layer.

In the following, the theoretical formula for the modified box model and its operation method are described. The local parameters in the formula are obtained by using sodar records. The modified box model is useful for the quality assessment of atmospheric environment. Besides, a special example of mixed layer development and its effect on the surface layer concentration are studied. These reveal the important role of the acoustic radar in the investigation of atmospheric dispersion.

Box Model for Diffusion

The concentration of a specific pollutant that diffuses in a one-dimensional flow is given by

$$\frac{\partial \bar{c}}{\partial t} + \bar{u}\frac{\partial \bar{c}}{\partial x} + \frac{\partial}{\partial x}\overline{u'c'} = S \tag{23}$$

where S standing for the source or sink of the species, is not a function of $\bar{c}$. Considering the boxes shown in Fig. 17, we concentrate on the space-time averaged concentration, variation of the averaged concentration with time and asymptotical behavior of this variation, while ignoring the details about the pollutant distribution in a box. Besides, we assume that for a given box, the arrangement of the source strength on the x-y plane is of probability preserving. Under these conditions, we obtain from (23), the equation of pollutant budget for box i,

$$\frac{\partial \bar{c}_i}{\partial t} + \frac{\bar{u}}{\Delta x_i}(\bar{c}_i - \bar{c}_{i-1}) + \frac{\partial K u_c}{\Delta x_i}\left(\bar{c}_i - \frac{\bar{c}_{i+1} + \bar{c}_{i-1}}{2}\right) = \frac{Q_i}{h \Delta x_i \Delta y_i} \tag{24}$$

where h is the mixing height, K is the eddy diffusivity, Δx is the length of box along the wind, Q_i is the total area effluent strength in box i nd u_i is the turbulent

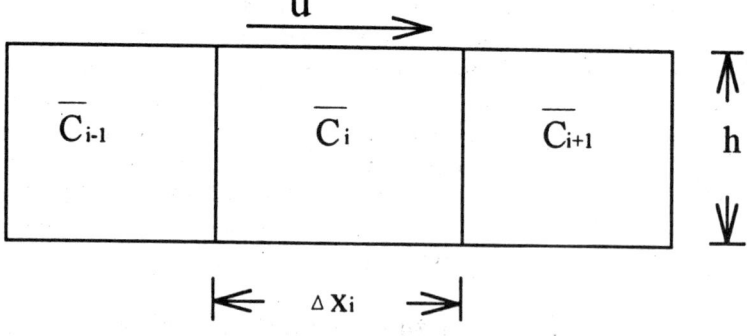

Fig. 17. Conceptual representation of the boxes.

characteristic velocity. The third term on the left hand side in Eq. (24) has been obtained by assuming an exchange of air parcels with identical volume in the adjacent two boxes. If we ignore the diffusion term and assume that h, $\bar{u}$ and $\bar{c}_{i-1}$ are constants ($\bar{c}_{i-1}$ can be considered approximately as constant for the case, for example, when box $i-1$ is located at a rural site upwind of the city concerned), the solution of Eq. (24) is

$$\bar{c}_i = \frac{Q_i}{h\bar{u}\Delta y_i} + \bar{c}_{i-1} + \left[\bar{c}_i(0) - \left(\frac{Q_i}{h\bar{u}\Delta y_i} + \bar{c}_{i-1}\right)\right]\exp - \frac{\bar{u}}{\Delta x_i} t \quad (25)$$

After some simplifications, a semiempirical formula for the box model is obtained as follws,

$$\bar{c} = a\bar{u}^b \quad (26)$$

where a and b are constants. The same expression was first proposed by Benarie [35].

It should be noted that a and b can be considered as constants only for specific time period. Thus if the daily average is taken as a unit, it would deviate from the physical conditions of the model. Therefore we use the box model following the procedure given below.

(1) Divide the time of a day into four periods, $\Delta\tau_j$ (j = 1, 2, 3, 4) according to the sodar records. Each of these periods corresponds to a specified step in daily evolution of the atmospheric boundary layer. For $\Delta\tau_1$ (Beijing Time 05:00 to 08:00 for autumn), the radiation inversion formed at night is being dissipated, and the concentrations of pollutants at the ground usually have their maximum values of the day since the pollutant storaged in the stable layer would be transported down to the ground. During the period $\Delta\tau_2$ (08:00 to 11:00), the radiation inversion has been dissipated to disappear, the concentrations of pollutants begin to decrease but this decrease is limited by the mixing height and the strength of capping inversion. As to the period $\Delta\tau_3$ (11:00–17:00), the capping inversion will either disappear or become very weak so that the ascending air parcels generated by convective activities could penetrate across it, and the mixed layer reaches its maximum height of the day. The concentrations of pollutants at the ground will therefore significantly decrease. In the periods $\Delta\tau_4$ (17:00–05:00), the elevated sources have no effects on the concentrations of pollutants in the weakly mixed layer formed by disturbances of the urban dynamic roughness and heat island. h_i, the mixing height with respect to $\Delta\tau_j$ can be determined conveniently according to the sodar records.

(2) Calculate the space-time average of wind speeds obtained by both acoustic Doppler radar and meteorological tower for each of $\Delta\tau_j$.

(3) Carry out regressive analysis of Eq. (26) by using the concentrations of pollutants obtained from continuous observations to determine the values of a_j and b_j.

Figs. 18–20 show the comparisons of SO$_2$ concentration between the results calculated by Eq (26) and those measured by the instrument KZL-SO$_2$. The data used for calculation are from November to December, 1981 where the values of a_j and b_j are specified by using the data of October 1981. Of course, an appropriate adjustment has been given to the source strength term. In drawing these results, the profiles of wind from 320 m meteorological tower and an assumption that the shape of the profiles is not affected by ground roughness are utilized. Besides if no inversion can be found from the sodar records, the mixing height is considered as 1 km, and if the mixing layer develops to higher than 320 m, the mean wind speed within it is considered to be equal approximately to that within 320 m.

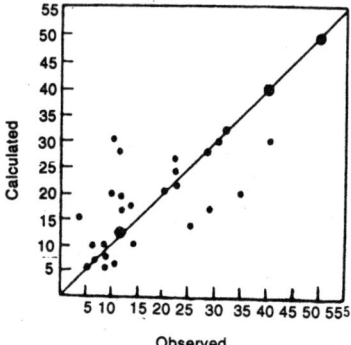

Fig. 18. A comparison of calculated and observed values for SO$_2$ concentrations in the urban district, Beijing, 24 November–3 December, 1981.

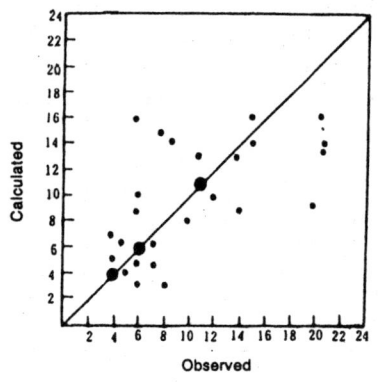

Fig. 19. As in Fig. 18 except for the northwest suburb.

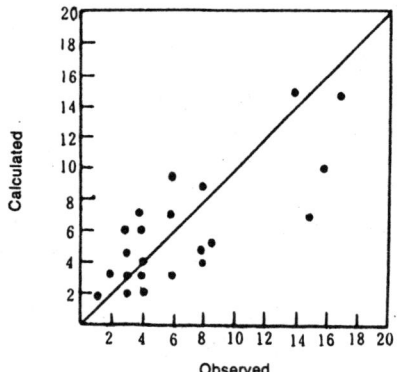

Fig. 20. As in Fig. 2, except for the southeast suburb.

The values of a_j and b_j are presented in Table 2. Inspecting these values we can find out that, on an average, the correlation of concentrations at the ground in the suburbs to wind speeds is higher than that in the urban section, and further that the concentrations in the suburban areas for nighttime have higher correlation to

Table 2. The Values of a_j and b_j

Locations	Periods	a_j	b_j
The Northwest Suburb	$\Delta\tau_1$	0.20	−0.39
	$\Delta\tau_2$	0.13	−0.42
	$\Delta\tau_3$	0.16	−0.70
	$\Delta\tau_4$	0.30	−1.00
The Urban	$\Delta\tau_1$	0.50	−0.31
	$\Delta\tau_2$	0.20	−0.50
	$\Delta\tau_3$	0.30	−0.84
	$\Delta\tau_4$	0.30	−0.30
The Southeast Suburb	$\Delta\tau_1$	0.13	−0.06
	$\Delta\tau_2$	0.18	−0.72
	$\Delta\tau_3$	0.12	−0.73
	$\Delta\tau_1$	0.50	−1.44

wind than for daytime. $|b_j|$ can be greater than 1, and this may be possible from the strong interactions between turbulence and advection during nighttime.

The dots in Figs. 18–20 are somewhat sporadic, and this is partly ascribed to scattered limited data we havel had. Furthermore, the limited data implicate a possibility of non-constant a_j and b_j. In spite of these deficiencies, the results shown in Figs 18–20 still support the modified box model. The r.m.s. errors, by taking Fig. 19 as an example, are 0.06, 0.028, 0.05 and 0.049 respectively for periods $\Delta\tau_1$ to $\Delta\tau_4$.

A Case Study of Mixed Layer Development and its Effect on the Concentrations of Pollutants at the Ground

In the field of atmospheric dispersion studies, we are concerned not only with the air quality under normal meteorological conditions but also with that in special synoptic processes. In this section we will give an example to discuss the ability of the acoustic radar in studies of special pollutant events.

Fig. 21 shows the facsimile photograph of the variation of mixing height measured by sodar on November 19, 1981 in Beijing. At local time 16:00 the mixing height is 520 m, and the strength of the capping inversion is rather weak. After that, the mixing height should continuously increase and the capping inversion should tend to be broken down. But, on the contrary the sodar facsimile photogrph shows a decreasing mixing height and a rising capping inversion strength, and the decreasing capping inversion goes to superpose upon the radiation inversion. Apparently, under the influence of this process, the concentrations of pollutants at the ground will increase and the usual model calculations will not be able to predict it. For being specific, let us consider the Gaussian plume model for $x = y = 0$ [36]:

$$\bar{c} = \frac{Q}{2\pi u \sigma_y \sigma_z} \left\{ \sum_{-\infty}^{+\infty} \exp\left[-\frac{(2Nh - h_Q)^2}{2\sigma_z^2}\right] + \sum_{N=-\infty}^{+\infty} \exp\left[-\frac{(2Nh + h_Q)^2}{2\sigma_z^2}\right] \right\} \quad (27)$$

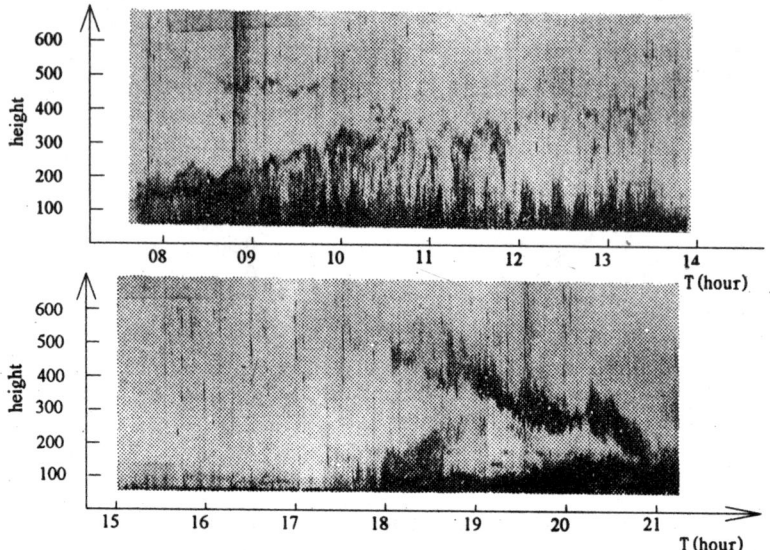

Fig. 21. A special development of mixing height, November 19, 1981, Beijing.

where h is the mixing height, h_Q the source's effective height and N an integral. If $\sigma_z \ll h$,

$$\frac{\partial \bar{c}}{\partial h} = 0 \qquad (28)$$

It indicates that the capping inversion does not affect the concentration distribution in the mixed layer. If $\sigma_z \sim h$, $h \gg h_Q$ and $N = 1$ (the condition $N = 1$ is usually employed for $\sigma_z \sim h$), it follows from Eq. (27) that

$$\frac{\partial \ln \bar{c}}{\partial h} = -\frac{\Delta h}{\sigma_z^2} \approx -\frac{4}{h} \qquad (29)$$

If $\sigma_z \geq h$, the plume will uniformly distribute between the ground and the base of the capping inversion, i.e. the mean concentrations in the air column can be recognized as the surface concentrations. Having considered the concentration distribution for the plume in the infinite atmosphere and putting the total pollutant mass into the layer with a thickness h, we obtain the mean concentrations between the ground and the base of the inversion,

$$\langle \bar{c} \rangle = \frac{1}{h} \int_{-\infty}^{+\infty} \bar{c}_0 \, dz = \frac{Q}{2\pi u \sigma_y \sigma_z h} \int_{-\infty}^{+\infty} \exp\left[\frac{-(Z-h_Q)^2}{2\sigma_z^2}\right] dz = \frac{Q}{\sqrt{2\pi}\,\sigma_y u h} \qquad (30)$$

$$\frac{\partial \langle \bar{c} \rangle}{\partial h} \propto -\frac{1}{h^2}, \text{ for } \sigma_z \geq h \qquad (31)$$

From Eqs. (28), (29) and (31) it is readily seen:

(1) For a specified pollutant source, the effect of the mixing height on the concentration distribution is significant only for downstream regions of wind at some distance from the source. This effective distance would decrease with the growth in strength of turbulent activities, since σ_z is controlled by turbulence.

(2) The smaller the h, the greater the effect on the concentration, in particular, when $\sigma_z > h$. In general, the mixed layer begins to grow within 1–2 hr after sunrise. At this time, the structure of atmospheric boundary layer is dynamically non-stationary, which makes it very difficult to determine theoretically the initial value of h. Acoustic radar then can be used to provide visual photograph to cope with the difficulty.

Table 3 shows the effect of the circumstances described in Fig. 21 on the concentration at the ground after local time 16:00. In the scheme we assume that wind velocity $u = 1$ ms^{-1}, the source strength $Q = 30$ g/sec, the effective source height $h = 80$ m and the downstream distance $x = 5$ km.

Table 3. The Effects of the Special Development of the Mixed Layer on the Concentrations (in mgm^{-3})

Local Time	Concentrations			
	$x = 1$ km		$x = 5$ km	
	B	A	B	A
16:00	1.86×10^{-1}	1.86×10^{-1}	2.50×10^{-2}	2.50×10^{-2}
17:00	1.87×10^{-1}	1.86×10^{-1}	3.00×10^{-2}	1.20×10^{-2}
18:00	1.87×10^{-1}	1.86×10^{-1}	3.20×10^{-2}	1.20×10^{-2}
19:00	2.34×10^{-1}	1.86×10^{-1}	4.20×10^{-2}	1.20×10^{-2}

The results in column A are obtained by usual model computation, and that in column B by model computation adjusted according to the sodar records. It is clear that at $x = 5$ km, larger errors would have occurred if there had been no real time observation.

Calculation of Vertical Diffusion Coefficients in the Convective Boundary Layer

Following Hay and Pasquil [37], Sheih [38] analyzed the physical process of a plume dispersion for a continuous source in more detail. This includes three different diffusion processes. The r.m.s. of the fluctuation of the plume centerline about its mean $\sigma_{z'c}(T, t_a)$, the r.m.s. of the mean plume centerline $\sigma_{\overline{zc}}(T, t_a)$ and the r.m.s of the instantaneous plume width due to relative diffusion $\sigma_r(T, t_a)$. The formulas are similar to those suggested by Gifford [39], when the relative diffusion is neglected. The vertical diffusion coefficients can be written as:

$$\sigma_{z'c}(T, t_a) = T \left\{ \int_0^\infty S_w(n) \frac{\sin^2 \frac{\pi nT}{\beta}}{(\pi nT/\beta)^2} \left[1 - \frac{\sin^2 \pi nt_a}{(\pi nt_a)^2} \right] dn \right\}^{1/2} \quad (32)$$

$$\sigma_{\overline{zc}}(T, t_a) = T \left\{ \int_0^\infty S_w(n) \frac{\sin^2 \frac{\pi nT}{\beta}}{(\pi nT/\beta)^2} \left[\frac{\sin^2 \pi nt_a}{(\pi nt_a)^2} \right] dn \right\}^{1/2} \quad (33)$$

in which $S_w(n)$ is the power spectral density of velocity: β is a ratio of integral time scale for Lagrangian to Eulerian, usually $\beta = .4$; t_a is the average time, T is the particle travel time. From (32) and (33), it can be seen that the plume dispersion coefficients are related to the spectra of fluctuation velocity and filter function. The different frequency comonents in turbulent energy spectra play different roles in the various diffusion processes. In (32) and (33), $(\sin^2 \pi nT/\beta)/(\pi nT/\beta)^2$ is a low pass filter. For larger travel time T the effect of turbulent eddy on the dispersion process will decrease with the increase of the high frequency component. For example the spectral density will rapidly decrease to less than 40% of the frequency component $n > 5 \times 10^{-2}$ Hz for $T = 60$ sec, and the value of spectral density can be neglected with $n > 10^{-2}$ Hz for $T = 300$ sec. The function $(1 - \sin^2 \pi nt_a/(\pi nt_a)^2)$ in (32) and (33) is a high pass filter. The value of spectral density will decrease to 10% with $n < 10^{-4}$ Hz for $t_a = 1800$ sec and with $n < 4 \times 10^{-4}$ Hz for $t_a = 3600$ sec. This means that the r.m.s of the fluctuation of the plume centerline about its mean $\sigma_{z'c}(T, t_a)$ is contributed mainly by turbulent eddies in the range $4 \times 10^{-5} < n < 10^{-2}$ Hz. The r.m.s of mean plume centerline $\sigma_{\overline{zc}}(T, t_a)$ is mainly contributed by the low frequency turbulent eddy; its influence on $\sigma_{\overline{zc}}(T, t_a)$ for $n > 10^{-3}$ Hz can be neglected.

Spectra of Vertical Velocity

The vertical diffusion coefficient can be calculated in terms of spectral density of vertical velocity. The spectra of vertical velocity at a height of 85 m measured by sodar and sonic anemometer were analyzed. An example, observed at 09:30–10:45 on October 19, 1983, is shown in Fig. 22. It can be seen that the spectra measured by Sodar correspond to those measured by sonic anemometer for energy peak and variation trend in the low frequency range, but some departures appear in the high frequency range. The observational results show that the magnitude of spectral power measured by sodar is larger than that measured by sonic anemometer, which is in accord with the results given by Lu et al, [40].

Vertical Diffusion Coefficient

$\sigma_{z'c}(T, t_a)$: According to formula (32), the vertical diffusion coefficient $\sigma_{z'c}(T, t_a)$

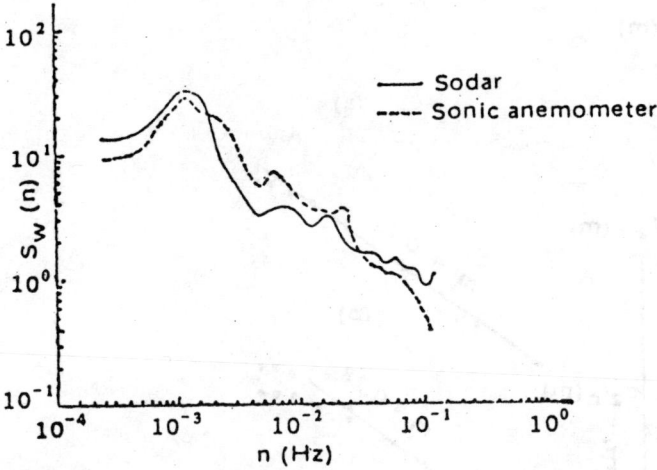

Fig. 22. The spectral of vertical velocity at 85 m measured by sodar and sonic anemometer (09:30–10:45, 19 Oct., 1983).

is calulated with t_a = 1800 sec for 9 sets of data for vertical velocity. The results for three sets of data are shown in Fig. 23. In general, in comparing $\sigma_{z'c}(T, t_a)$ for sodar and sonic anemometer the relative errors are about 8–20%; in some individual cases, especially for $T > 900$ sec, the error appears to be some-what large. The variability in the diffusion coefficient is produced mainly by the measuring error. The three curves in Fig. 23 satisfy the relationship $\sigma_{z'c}(T, t_a) \propto T^m$; The values of m corresponding to (a), (b) and (c) for $T < 900$ sec are 0.71, 0.78, and 0.74, respectively. For $T > 900$ sec, the $\sigma_{z'c}(T, t_a)$ on curve (a), which is calculated with 12-hour observation data, fit the relation $\sigma_{z'c}(T, t_a) \propto T^{0.71}$ until $T = 1800$ sec. These results are consistent with the variation of vertical diffusion coefficients as a function of the downwind distance or travel time T as given by Pasquill and Gifford and others [36–39].

Vertical Diffusion Coefficient $\sigma_{\overline{zc}}(T, t_a)$:

The fluctuation of a plume center, which is mainly influenced by low-frequency turbulent eddies, can be expressed by the vertical diffusion coefficient $\sigma_{\overline{zc}}(T, t_a)$. The $\sigma_{\overline{zc}}(T, t_a)$ was calculated using sodar data with formula (33). An example observed at 07:00–19:00 on 21 October, 1983 is shown in Fig. 24, which shows the approximate relationship $\sigma_{\overline{zc}}(T, t_a) \propto T^m$, where $m = 1$ for the cases of both $t_a = 3600$ sec and $t_a = 1800$ sec. The value of $\sigma_{\overline{zc}}(T, t_a)$ for $t_a = 1800$ sec is larger than that for $t_a = 3600$ sec. It is clear that because of the filter function, the low-frequency components will influence the average value of the fluctuation of the plume center $\sigma_{\overline{zc}}(T, t_a)$. Using the 12 hour continuous data on vertical velocity to calculate $\sigma_{\overline{zc}}(T, t_a)$ for $T < 1800$ sec, it was found that its magnitudes varied from several meters to more than one hundred meters. The magnitudes were less than

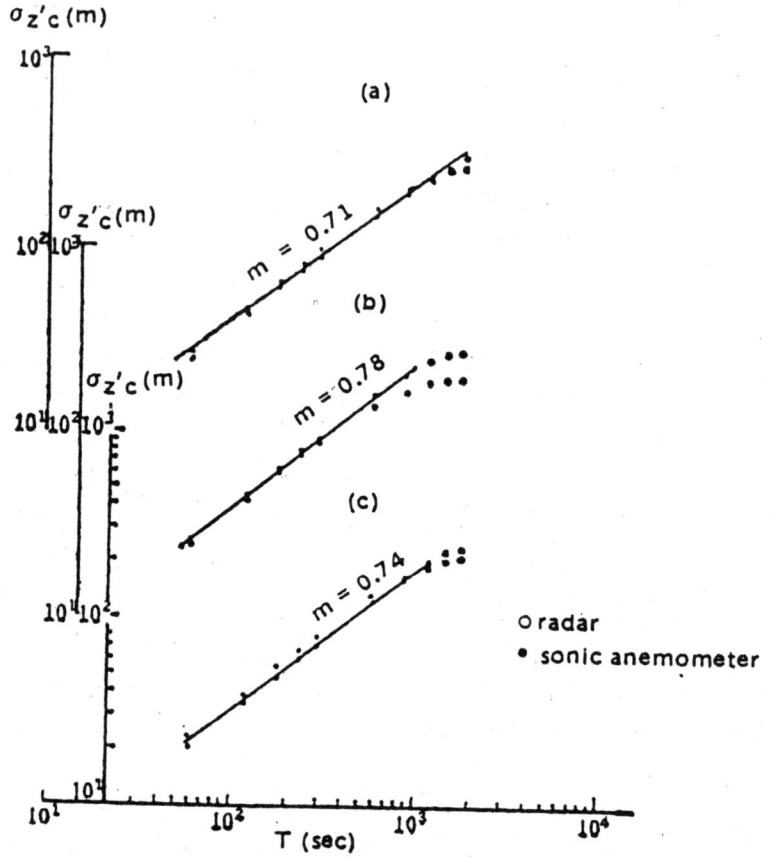

Fig. 23. The variation of $\sigma_{z'c}(T, t_a)$ versus travel time T.
(a) 07:00–19:00 October 21, 1983, (b) 11:00–12:15 October 19, 1983, (c) 09:30–10:45 October 19, 1983.

several to less than 100 m when 75-min observation data were used. It can be seen from the comparison of $\sigma_{z'c}(T, t_a)$ for curve (a) in Fig. 23 with $\sigma_{\overline{zc}}(T, t_a)$ in Fig. 24 that for $t_a = 1800$ sec, the magnitude of $\sigma_{z'c}(T, t_a)$ is about 1.9–3.7 times larger than $\sigma_{\overline{zc}}(T, t_a)$ for $T < 900$ sec, the slopes for $\sigma_{z'c}$ and $\sigma_{\overline{zc}}$ are different, and the difference in slopes becomes larger for small T. When the travel time T increases, the influence of turbulent eddy with a low-frequency component becomes more important. Hence, in this case the relative effect of $\sigma_{\overline{zc}}(T, t_a)$ on the dispersion of the plume will be increased. As mentioned above, for short-range diffusion, $\sigma_{\overline{zc}}(T, t_a)$ can be neglected, and $\sigma_{z'c}(T, t_a)$ becomes a key factor. But, for long-range diffusion it is necessary to calculate the w spectra and $\sigma_{\overline{zc}}(T, t_a)$ using data observed over a long time period. In this case, the diffusion process will obviously be influenced by the larger-scale turbulent eddies.

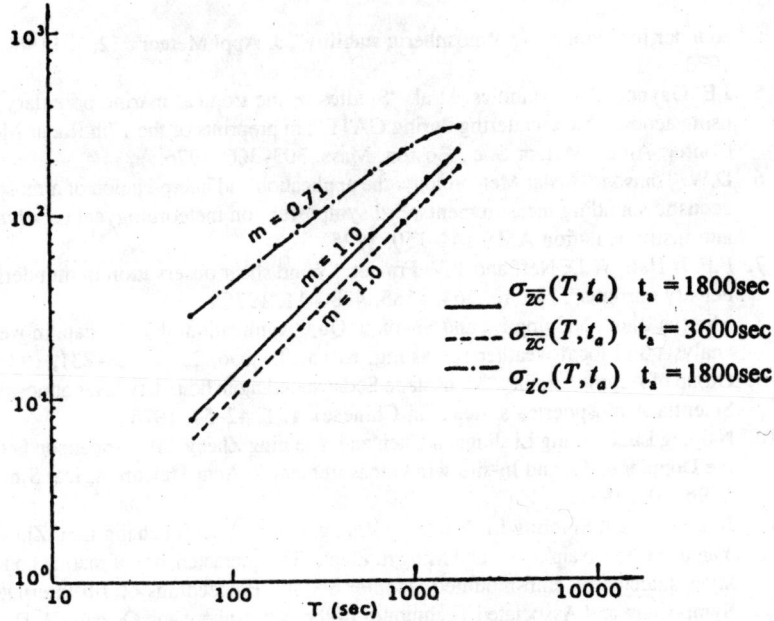

Fig. 24. The variation of $\sigma_{\overline{zc}}(T, t_a)$ versus travel time T (07:00–19:00, 21 October, 1983).

Summary

In China the dovelopment of acoustic detection and the studies on the structure of atmospheric boundary layer over land and ocean and the assessment of atmospheric environment with the sodar data have been carried out for more than twenty years. The main achievements of these investigations have been comprehended in this paper. Some phenomena in the atmospheric boundary layer, such as multiple-layer inversions, strong wind shear, lump structure echo and etc., have been discovered with sodar, which are not easily discovered with direct measuring systems. The knowledge of formation, existence and maintenance for these phenomena is still meager at present. The acoustic sounder could play an important role in the further studies on these problems and other problems of atmopheric boundary layer structure.

References

1. L.G. McAllister, "Acoustic sounding of the lower atmosphere", J. Atmos. Terr Phys, 30, 1439–1440, 1968.
2. P.A. Mandics and E.J. Owens, "Observations of marine atmosphere using a ship-mounted acoustic echo sounder", J. Appl. Meteor., 14, 6, 1110–1117, 1975.
3. A. Sppizichino, "Discussion of the operating conditions of a Doppler sodar", J. Geophys. Res., 76 36, 5585–5591, 1974.
4. R.J. Wyckoff, D.W. Beran and F.F., Jr Hall, "Radiosonde and the Acoustic echo

sounder for monitoring atmospheric stability", J. Appl Meteor., 12, 7, 1196–1204. 1973.
5. J.E. Gaynor, P.A. Mandics, et al, "Studies of the tropical marine boundary layer using acoustic backscattering during GATE", in preprints of the 17th Radar Meteor, Confer. Amer. Meteor Soc., Boston, Mass, 303–306, 1976.
6. D.W. Tomson, "Acdar Meteorology: the application and interpretation of atmospheric acoustic sounding measurements, 3rd symposium on meteorologycal observations and instrumentation AMS 144–150, 1975.
7. F.F. Jr Hall, W.D. Neff and T.V. Frazier, "Wind shear observation in thunderstorm density currents", Nature, 264, 5585, 408–411, 1976.
8. Mingyu Zhou, Naiping Lu and Shaohou Qu, "Application of sodar data in weather analysis and local weather forcasting, Koxue Tonbao, 25, 4, 328–331, 1985.
9. Group of Acoustic Radar, "Monostatic Sodar sounding in boundary layer atmosphere", Scientia Atmospherica sinica., (in Chinese), 1, 1, 42–54, 1976.
10. Naiping Lu, Shiming Li, Jingnan Chen and Yueming Zheng, "A comparison between the Doppler sodar and In-situ wind measurements", Acta Meteorological Sinica, 1, 1, 96–104, 1987.
11. Jingnan Chen, Shiming Li, Yueming Zheng, Lirong Su, Youchang Lan, Zhi Chen, Yanjuan Chen, Naiping Lu and Mingyu Zhou, "The characteristics of marine boundary layer detected by ship-mounted Doppler Sodar", Proceedings of 7th International Symposium and Associated Techniques of the Atmosphere and Oceans, W.D. Neff, Boulder Colorado U.S.A. 4-5-27, 1994.
12. Mingyu Zhou, Naiping Lu and Yanjuan Chen, "The detection of temperature coefficient of atmospheric boundary layer by acoustic radar", J. Acoustic Soc. Am. 68, 1, 303–308, 1980.
13. Mingyu Zhou, Naiping Lu, Yanjuan Chen and Shiming Li, "The lump structure of trurbulence field in atmospheric boundary layer", Scientia Sinica, 24, 12, 1705–1716, 1981.
14. Jingwei Xiao, Naiping Lu and Mingyu Zhou, "Application of Sodar sounding to atmospheric dispersion—mixing depth and concentration at the ground", Advances in Atmospheric Sciences, 2, 1, 63–71, 1985.
15. Naiping Lu, Yanjuan Chen and Shiming Li, "Application of sodar sounding to atmospheric Disperion (3)—Calculation of vertical diffusion coefficients in the convective boundary layer", Chinese Journal of Atmospheric Sciences, 11, 1, 95–104, 1987.
16. Naiping Lu, Shiming Li, Jingnan Chen and Mingyu Zhou, "A calculation method of the sensible heat flux by sodar", Acta Oceanologica Sinica, 8, 2, 191–198, 1989.
17. Mingyu Zhou and Yi Zhang, "The wave properties in process of the nocturnal radiative inversion", Kexue Tongbao 27, 3, 156–159 (in Chinese), 1982.
18. Mingyu Chou(Zhou), Hsian Fan, Naiping Lu, Shaohou Qu and Yenchuan Chen, "Acoustic radar and Remote sensing in the Boundary layer of atmosphere", Proceeding of the Twelfth International Symposium on Remote Sensing of Environmental Research, Institute of Michigan, Arbor, Michigan, 535–546, 1978.
19. A.N. Kolmogorov, "A refinement of previous hypotheses concerning the local structure of turbulence in a viscous incompressible fluid at high Renolds number", J. Fluid Mech. 13, 1, 82–85, 1962.
20. A.M. Oboukhov, "Some specific features of atmospheric turbulence", J. Fluid Mech. 13, 1, 77–81, 1962.

21. A.S. Gurvich and A.M. Yaglom, "Breakdown of eddies and probability distribution for small scale turbulence", Phys. Fluids, 10, 59–65 1967.
22. J.R. Stewart, J.R. Wilson, and R.W. Burling, "Some statistical properties of small scale turbulence", J. Fluid Mech., 41, 1, 141–152, 1970.
23. S. Sethuraman, "The observed generation and breaking of atmospheric internal gravity wave over the ocean", Boundary Layer Meteorology, 12, 3, 331–350, 1977.
24. J.E. Gaynor, and P.A. Mandics, "Analysis of the tropical marine boundary layer during GATE using acoustic sounder data", Monthly Weather Review, 106, 2, 223–232, 1978.
25. J.E. Gaynor, "Acoustic sounding in moist tropical marine atmospheric boundary layer during GATE". Proceeding of Fourth Symposium on Meteorological Observation and Instrumentation, 410–414, 1978.
26. H. Ottersten, M. Hurtig, G. Stilke and B. Brumm, "Ship borne sodar measurements during Jonswaps 2", J. Geophys. Res. 79, 36, 5573–5584, 1974.
27. J.C. Wyngaard, Y. Izumi and S.A. Collin, "Behavior of the refractive index structure parameter near the ground", J. of the Optical Society of America, 61 1646–1650, 1971.
28. V.I. Tatarskii, The effects of the turbulent Atmosphere on Wave Propagation, 472 pp, 1971.
29. Mingyu Zhou, Yanjuan Chen, Naiping Lu and Shming Li, "A comparative study between the temperature structure coefficients detected by sodar and direct methods", Acta Geophysica Sinica (in Chinese), 25 5, 492–499, 1982.
30. M.L. Wesely, "Combined effect of temperature and humidity fluctuations on refractive index", J. Appl. Meteor. 35, 1, 43–49, 1976.
31. R.L. Coulter and M.L. Wesley, "Estimates of surface heat flux from sodar and laser scintillation measurements in the unstable boundary layer", J. Appl. Mteor, 19, 10, 1209–1211, 1980.
32. J.E. Gaynor, "The importance of moisture fluctuations and wind shear in acoustic backscatter in GATE", J. Appl. Meteor. 18, 11, 1472–1480, 1979.
33. J.C. Wyngaard, W.T. Pannell, D.H. Lenschow and M.A. Lemone, "The temperature-humidity covariance budget in the convective boundary layer", J. Atmos, Scien, 35, 1, 47–58, 1978.
34. W.G. Large and S. Pond, "Sensible and latent heat flux measurements over the ocean", J. Phys. Ocean, 12, 3, 468–482, 1982.
35. M.M. Benarie, Atomspheric pollution, Proceedings of the 14th international Colloquiam, Paris France, May 5–8. 49–53. 1980.
36. D.B. Turner, "Work of atmosphere dispersion estimates", SDHEW. PHS. Pub. No. 995–AP-26, 1976.
37. J.S. Hay and F. Pasquill, "Diffusion from a continuous source in relation to the spectrum and scale of turbulence", Advances in Geophysics, 6, Academic Press, 345–365, 1959.
38. C.M. Sheih, "On lateral dispersion coefficient as functions of average time, J. Appl. Meteor, 59, 6, 557–561, 1980.
39. F.A. Gifford, An outline of diffusion in the lower layers of the atmosphere", U.S. Atomic Energy Commission, 65–116, 1968.
40. Naiping Lu, Jingnan Chen, Yueming Zheng, Yenjuan Chen and Shiming Li, "Acoustic Doppler radar for measuring wind" (in Chinese), Scientia Atmospherica Sinica", 10, 5, 1982.

Acoustic Remote Sensing Applications
S.P. Singal (Ed)
Copyright © 1997 Narosa Publishing House, New Delhi, India

11. Influence of the Nocturnal Low-Level-Jet on the Vertical and Mesoscale Structure of the Stable Boundary Layer as Revealed from Doppler-Sodar-Observations

Frank Beyrich*, Dieter Kalass and Ulrich Weisensee

Forschungseinrichtung für Luftchemie, Fraunhofer-Institut für Atmosphärische Umweltforschung Rudower Chaussee 5, D-12484 Berlin, Germany

1. Introduction

The nocturnal low-level-jet (LLJ) is a common phenomenon to be observed in the stable boundary layer (SBL). It has been found to occur quite regularly over different regions of the world (Mix, 1981, Brook, 1984, Mizuma, 1989, Jury and Tosen, 1989, Sjostedt et al, 1990, Savijärvi, 1991). Over northern Germany, a nocturnal jet can be observed in about 10% of all nights (Kottmeier et al, 1983).

A theoretical explanation of the nocturnal LLJ was given by Blackadar (1957). According to him, the jet results from an inertial oscillation of the ageostrophic wind vector in those layers that are decoupled from the influence of surface friction due to the rapid turbulence decay during the evening stabilization period. Characteristic features of the nocturnal LLJ are the appearance of a supergeostrophic velocity maximum most distinctly marked 4–7 hours after sunset (in midlatitudes) and a steady turning of the wind vector (clockwise on the northern hemisphere). The intensity of the LLJ (i.e., the ratio between maximum and geostrophic wind speeds – V_{max}/V_g) and the time, at which it is strongest, depend on the magnitude and phase of the ageostrophic wind component around sunset. Due to its small values in the upper part of the daytime boundary layer and the large ageostrophic deviations near ground, the LLJ first occurs at high altitudes and descends with time thereby increasing in strength.

Direct observations of the LLJ are possible only by continuous wind profile measurements within the lowest 200–500 m of the atmosphere at tall masts (e.g., Roth et al, 1979) or using remote sensing systems. Although an acoustic sounder

---

*Present affiliation: Lehrstuhl für Umweltmeteorologie, BTU Cottbus, PF 101344, D-03013 Cottbus, Germany

with Doppler-capability should be well suited to investigate the structure and evolution of the nocturnal jet in very detail, there are comparably few studies in the literature reporting on this phenomenon based on sodar data (Coulter, 1981, Peters, 1991, Kataoka et al, 1991, Culf, 1993). It is the main goal of this paper to demonstrate, how sodar can be used to obtain a lot of valuable information on SBL structure and processes in situations, where a LLJ is present.

2. Observations and Data

Several types of acoustic sounders have been designed and built at the Boundary Layer Department of the former Heinrich-Hertz-Institute of the Academy of Sciences of the G.D.R between 1985 and 1990 (Lehmann and Neisser, 1990). Based on these activities, a network of single-antenna monostatic sodars became operational in the Meteorological Service of the former G.D.R. at the end of the eightees (Foken, 1997).

The three-antenna, monostatic Doppler-Sodar ECHO-1D has been operated since 1989 during several national and international field experiments to study boundary layer structure, phenomena, and winds. Its basic technical parameters are given in Table 1.

Table 1. Technical parameters of Doppler-Sodar ECHO-1D

Sound frequency	1707 Hz
Emitted power (electrical)	150 W
Antenna	Parabolic, d = 1.2 m
Minimum sounding range	30–50 m
Maximum sounding range	optional 200/400/800 m
Height resolution	optional 6, 12.5, 25 m
Pulse length	optional 37.5, 75, 150 ms
Pulse repetition rate	optional 5, 10, 20 s

In the years 1990–1993, measurements were carried out during six field campaigns within the SANA programme (Schaller and Seiler, 1993) at different sites in the eastern part of Germany (see Fig. 1). This programme was aimed to study changes in air pollution character and level, that were associated with the reconstruction of industry and traffic in the former G.D.R. after re-unification of Germany. The programme included long-time monitoring of temporal and spatial variations in the amount and composition of atmospheric pollution, monitoring and modelling of pollutant transport and deposition, studies of multi-phase chemistry in heavily polluted air masses, and, finally, monitoring and modelling of the impact of changing air pollution on highly sensible, natural ecosystems. In addition to monitoring networks of ground-level stations, field experiments were carried out to study important processes in more detail. During these experiments, fast-response turbulence sensors, remote sensing systems, research aircrafts, radiosondes, and

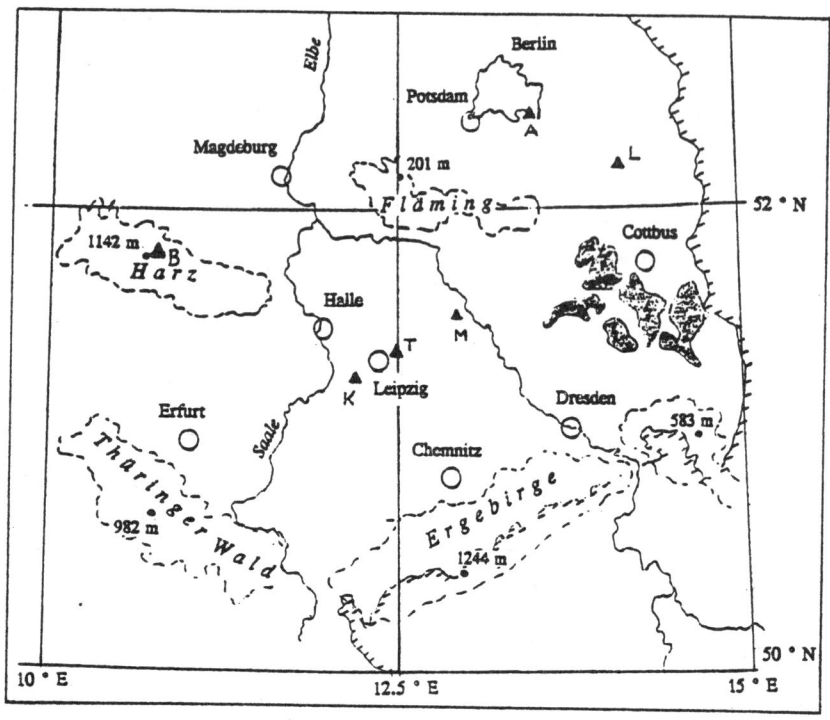

Fig. 1. Schematic map of south-east Germany with the sites at which the Doppler-Sodar ECHO-1D has been operated during the SANA experiments marked by letters A, L, K, and M (the shadowed area in the east are regions of lignite-burning industries, (f. Section 5)

chemical sensors for measuring gas and aerosol concentrations were operated in addition to the monitoring network.

Observations used in this study were mainly performed during the SANA field campaigns. In addition, data from the HAPEX-MOBILHY experiment (cf. Mazaudier and Weill, 1989) which took place in 1986 in the South-West of France are used in Section 6.

3. Jet-Like Wind Profiles and Sodar-Observed SBL-Structure

The single wind profile during a LLJ event is characterized by the existence of a pronounced velocity maximum normally at heights between 100 m and 300 m a.g.l. Wind speed at the jet axis, V_{max}, is supergeostrophic, and the intensity of the LLJ typically lies in the range 1.3–1.8. Below the jet axis, a nearly linear increase in wind speed with height can often be found over a range of 100–200 m and associated vertical wind shear may be as strong as 0.05–0.1 s^{-1} within this lower shear zone. A second shear zone normally exists above the jet axis, but with shear values often much weaker than in the lower layer. A series of wind speed profiles that are characteristic for the evolution of a nocturnal jet is shown in Fig. 2. from

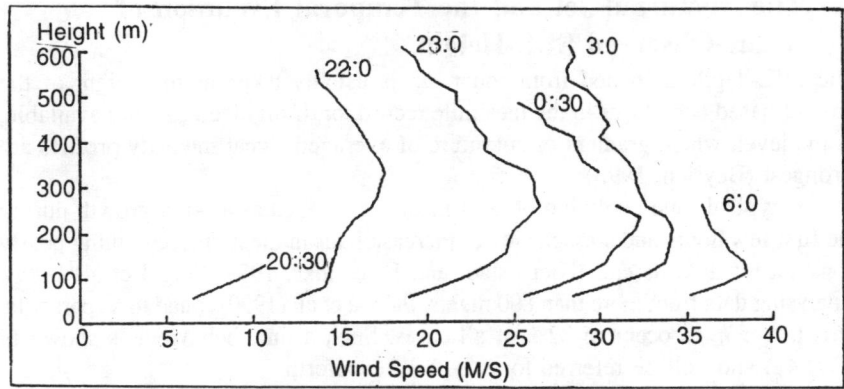

Fig. 2. Sequence of 30-minute averaged wind speed profiles measured in Melpitz (M), August, 31st, to September, 1st, 1991 (the single profiles are shifted along the abscissa in steps of 6 m/s)

which the features described above can be easily recognized. Additionally the increase in strength until approximately midnight and the decreasing intensity towards the end of the night as well as the decreasing height of the jet axis become obvious.

· According to observations in Antarctica discussed by Culf (1993), occurence of a LLJ can be inferred from the presence of a double-layer echo structure in the sodar facsimile record. Such a structure is attributed to the existence of the two shear layers below and above the jet axis. It can also be found for some of the LLJ cases during the SANA experiments as is illustrated by the example in Fig. 3. However the picture of a double layer structure seems to be not even typical for sodar measurements in midlatitudes. At least two reasons are seen for that discrepancy. On the one side, the antarctic LLJ is strengthened by terrain slope effects resulting in increased shear values. On the other side, the LLJ in Antarctica develops in the presence of background stable stratification, whereas in midlatitudes the upper shear zone often is situated within the residual layer, for which small potential temperature gradients (and hence C_T^2-values) are typical, thus only weak backscattered signals are to be expected from above the jet axis.

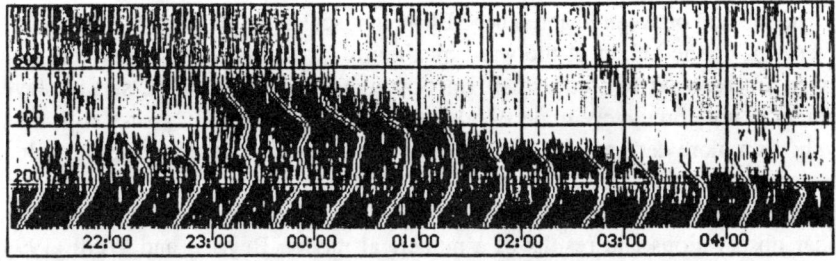

Fig. 3. Sodargram and wind speed profiles from measurements in Berlin, August, 14th, to August, 15th, 1993.

4. The Nocturnal Jet and the Temporal Evolution of Sodar-Observed SBL-Height

The SBL-depth estimated from sodar, h_s, is usually taken as the height of the ground-based echo layer in the facsimile record, or, if digitized data are available, as the level, where gradient or curvature of averaged signal intensity profiles are strongest (Beyrich, 1993).

The typical time evolution of h_s is mostly described as a rapid growth during the first few hours and a much slower increase later in the night, remaining nearly constant towards its end (Nieuwstadt and Driedonks, 1979, Singal et al. 1989). Analysing data from more than 800 nights, Pahwa et al, (1990) found this "parabolic growth" of h_s to occur in 82% of all cases. Such a time behaviour is shown in (Fig. 4a) and will be referred to as type I henceforth.

A different evolution, i.e., an increasing height during the first few hours followed by a marked decrease towards the end of the night (type II, cf. Fig. 4b), is described by van Dop et al, (1978) and is attributed to advective effects in this study. A type-II behaviour can also be inferred from observations presented by Arya (1981), Nieuwstadt (1984) or Beyrich and Weill (1993), but it has not received much attention in the literature. Especially, no attempt is known to explain the two different pictures of the time evolution of the depth of the ground based shear echo layer as observed with an acoustic sounder.

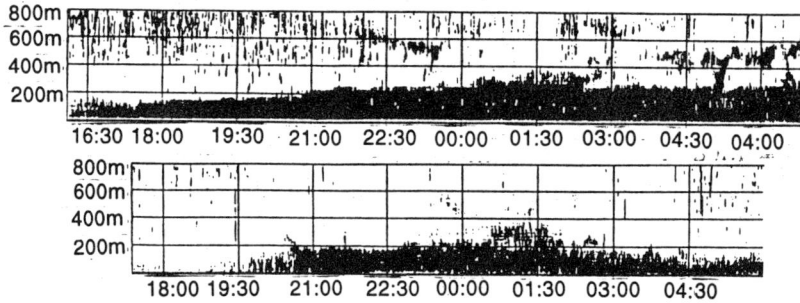

Fig. 4. Examples of the temporal evolution of the ground based shear echo layer-upper: October, 12th to October, 13th, 1990, Berlin (A)—type I; lower: August, 24th to August, 25th, 1992, Lindenberg (L)—type II

The relation between h_s and other SBL-height scales (as the height of the surface inversion, h_i, the height of the wind maximum, h_u, or the level, where turbulent fluxes are reduced to some small fraction of their corresponding surface layer values) is not completely understood yet. Whereas h_i normally grows with time, h_u shows a more decreasing tendency during the night with nearly constant values towards its end (Delage, 1974, Arya, 1981). Numerous investigators found good agreement between h_s and h_i, but others concluded just the opposite. Comparing sodar observations and results of a numerical model, Beyrich and Weill (1993) concluded that h_s may be closely related to different other height scales during different phases of even one night. Especially during the second part of clear nights they found the highest correlation between h_s and h_u. This raises the question, whether the type-II-behaviour of h_s might be associated with the presence of a nocturnal LLJ.

To identify such a possible relation, ten nights that clearly show the type-II temporal evolution of h_s have been selected from the SANA field campaigns. For comparison, a second set of ten nights showing a steady growth of h_s (type-I-behaviour) has been analysed. Hourly values of the wind speed at 100 m height have been normalized by their observed maxima during the night for all cases. In the same way, wind direction has been normalized by determining the difference between the hourly value and the initial one around sunset. Then the normalized values have been averaged over the ten cases for each of the two data sets of type-I- and type-II-nights. The resulting time behaviour is shown in Fig. 5a and 5b. Although the standard deviation is quite large, the differences are obvious. The mean "type-I night" does not show any pronounced nocturnal evolution in wind speed and wind direction. On the other hand, the mean "type-II-night" is characterized by a maximum in wind speed five hours after sunset and a steady clockwise turning of the wind direction, both in agreement with the inertial oscillation theory. It is thus concluded, that an observed type-II temporal evolution of h_s in the sodargram may be indicative of the nocturnal jet. This conclusion has been also verified for most of the single nights (Beyrich, 1994).

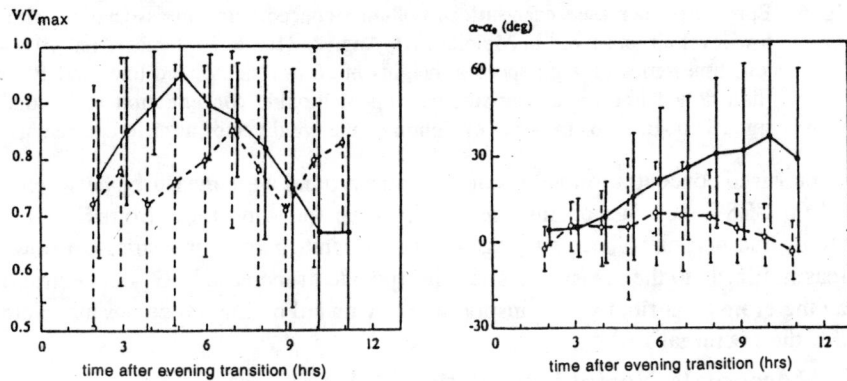

Fig. 5. Mean time evolution of normalized wind speed (left) and wind direction (right) for the nights of type I (dashed line) and type II (solid line) - thin bars indicate standard deviation

5. The LLJ and the Transport of Pollutants

Occurence of a LLJ can result in different trajectories at different levels within the SBL (Kottmeier, 1978) and thus significantly affect the transport of atmospheric pollutants (Watanabe, 1989). Enhanced vertical mixing in the shear zone below the jet axis in association with the changes in wind direction is postulated to cause episodes of increased pollutant concentration values near to the ground even in rural areas without local sources.

This is illustrated by the example given in Fig. 6. Fig. 6a shows time series of the wind speed at heights of 10 m and 100 m and of near-surface concentration of sulphur dioxide at a rural site in East Germany west of the Lausitz lignite basin (site "M" in Fig. 1). Whereas the wind speed at the surface is about 2.5 ms^{-1} during the whole night, a remarkable temporal evolution is observed at 100 m height. This is associated with the nocturnal inertial oscillation, as can be clearly seen from the hodographs of the temporal evolution of the wind vector at different heights (Fig. 6b).

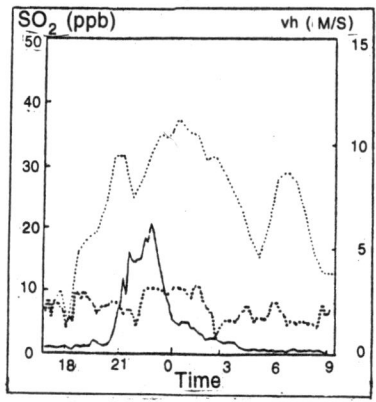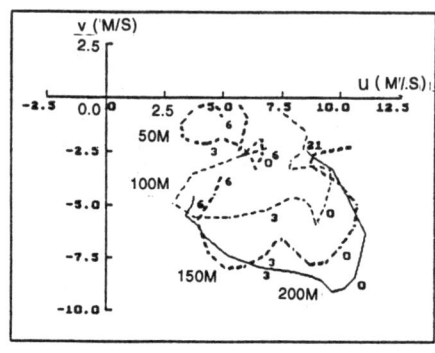

Fig. 6. Episode of increased near-surface pollutant concentration in association with a low-level jet, observed in Melpitz (M), August, 31st, to September, 1st, 1991 — left: time series of wind speed at heights of 10 m (thick dotted line) and 100 m (thin dotted line), and near-surface sulphur dioxide concentration (solid line); right: hodograph of the time evolution of the wind vector at different heights.

Increased concentration values of SO_2 near ground were measured between 21 and 00 CEST, i.e., just during the period when the wind blew from E to ESE thereby transporting polluted air masses from the lignite burning industries (shadowed areas in Fig. 1) to the measuring site. The episode of increased SO_2-concentration can therefore be attributed to transport and downward mixing processes associated with the nocturnal LLJ.

6. Mesoscale Variability of the LLJ

The inertial oscillation mechanism is typical for a horizontally homogeneous SBL, the nocturnal jet is therefore expected to be observed over extended areas, if the terrain is flat. However, horizontal homogeneity at scales of 50–100 km is not just typical for Central European conditions. Using simultaneous observations at different sites, the evolution of the LLJ is analyzed within different subscales of the mesoscale with horizontal separation ranging between ~ 5 km and ~ 150 km.

During the HAPEX-MOBILHY experiment, two Doppler Sodars of the Centre de Recherches en Physique de *l'*Environment (Issy-les-Moulineaux) were operated at two different sites in the Les Landes forest region in the SW of France (cf.

Mazaudier and Weill, 1989). One Sodar (A) was installed in the pine forest (tree height ~ 20 m), the second sodar (B) at a large clearing, approximately 5 km away. Figure 7a shows the hodograph of the wind vector at 100 m height at the two sites during the night June, 22, to June, 23, 1986. Although the overall behaviour of the wind is quite similar at both sites, the wind speed is significantly lower over the forest, especially during the first part of the night. This is attributed to the larger surface roughness.

During the SANA field campaign in August, 1992, two ECHO-1D Doppler-Sodars were in operation at Lindenberg Meteorological Observatory (German Weather Service) and at the Fraunhofer Institute in the South-East of Berlin (sites A and L in Fig. 1). Horizontal distance is ~ 45 km, and additionally, there is a height difference of approximately 60 m between both sites. However this difference in altitude is not associated with generally sloping terrain between both places, because the Lindenberg observatory is located on a small isolated hill. Due to this height difference and the greater roughness of the urban site, the temporal evolution of the wind at 200 m a.g.l. for Berlin is compared with that at 100 m a.g.l. for Lindenberg in Fig. 7b. Both curves are very close to each other.

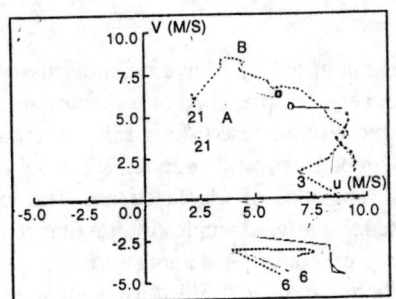
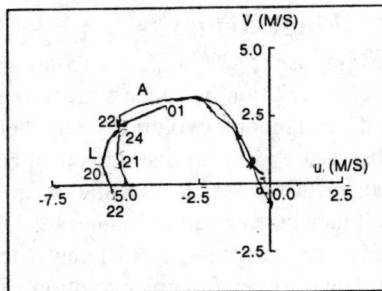
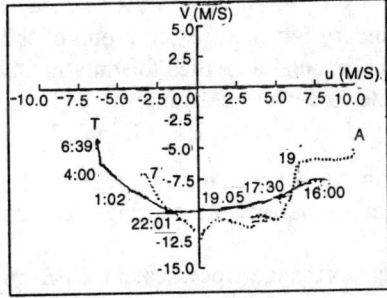

Fig. 7. Examples of horizontal quasi-homogeneity in the evolution of the nocturnal jet for different sub-scales of the meso-scale
(a) HAPEX-MOBILHY, June, 22nd, to June, 23rd, 1986, site separation is ~ 5 km
(b) Berlin and Lindenberg, August, 24th, to August, 25th, 1992, site separation is ~ 45 km
(c) Berlin and Taucha, October, 15th, to October, 16th, 1990, site separation is ~ 150 km

Doppler-Sodar measurements were carried out in Berlin (site A in Fig. 1) during the first SANA field experiment in October, 1990. Frequent radiosoundings were performed by a group of the Karlsruhe Nuclear Research Center at Taucha, east of Leipzig (site T), about 150 km to the South-West from Berlin. A strong nocturnal jet was observed at both sites within a southerly flow during the night October, 15th to October, 16th. Fig. 7c shows the behaviour of the wind vector at 200 m a.g.l. between early evening and next morning. Again, there is impressive similarity between the observations at both sites in the magnitude of wind speeds and the radius of the inertial oscillation as well as in the total turning angle of the wind vector during the night. The only remarkable deviation is a nearly 30 deg difference in wind direction values throughout the night, which should be attributed to a general slight curvature of the isobars over East Germany and corresponding differences in the geostrophic wind direction. However, this example clearly demonstrates that the nocturnal jet may develop over larger areas within the meso-α-scale in a similar way even in non-homogeneous, although not complex terrain. Note from Fig. 1, that sites A and T are separated by a chain of hills—Fläming-with maximum altitude of ~ 200 m a.m.s.l.

7. Final Remarks

Several aspects of sodar observations of the nocturnal LLJ have been discussed. A close relation has been shown to exist between the presence of a nocturnal jet and the temporal evolution of the height of the ground based shear echo layer, as observed even with a single-antenna, non-Doppler acoustic sounder. The role of the nocturnal inertial oscillation in causing episodes of increased near-surface pollutant concentration has been demonstrated. Finally, examples of simultaneous sodar measurements at different sites were presented that demonstrate a quasi-homogeneous temporal evolution of the nocturnal wind field during the night even over non-homogeneous terrain within the mesoscale.

At present, the Boundary Layer Research group of the Fraunhofer-Institute is mainly involved in experimental activities to study boundary layer structure and processes and their relevance for the transport and dispersion of atmospheric pollutants. Additionally, further activities will include

— the combination of sodar with in-situ meteorological and chemical measurements and other remote sensing techniques, as lidar, DOAS or FTIS,
— studies of vertical exchange processes in the PBL using a variable acoustic sounder configuration allowing for adapted operation according to the actual meteorological situation.

Acknowledgements

The SANA programme has supported by the BMFT (German Ministry for Research and Technology). The SO_2-concentration and near-surface wind values used in section 5 have been provided by the Meteorological Institute of Munich, thanks are to D. Müller, M. Weber and N. Beier. The radiosoundings in October, 1990, were performed by a group of the Nuclear Research Center of Karlsruhe and have been used in this study by kind permission of N. Kalthoff and E. Schaller. Data from the HAPEX-MOBILHY Experiment have been made available for use within a data base prepared by Ch. Amory-Mazaudier at the Centre des Recherches en Physique de l'Environnement in Issy-les-Moulineaux, France.

References

1. Arya, S.P.S., 1981: Parameterizing the height of the stable atmospheric boundary layer. *J. Appl. Meteorol.* **20**, 1192–1202.
2. Beyrich, F., 1993: On the use of sodar data to estimate mixing height. *Appl. Phys.* **B 57**, 27–35.
3. Beyrich, F., A. Weill, 1993: Some aspects of determining the stable boundary layer depth from sodar data *Boundary-Layer Meteorol.* **63**, 97–116.
4. Beyrich, F., 1994: Sodar Observations of the Stable Boundary Layer Height in relation to the Nocturnal Low-Level Jet. *Meteorol. Z.* (N.F.) **3**, 29–34.
5. Blackadar, A.K., 1957: Boundary layer wind maxima and their significance for the growth of nocturnal inversion. *Bull. Amer. Meteorol. Soc.* **38**, 283–290.
6. Brook, R.R., 1984: The Koorin Nocturnal Low-Level Jet. *Boundary-Layer Meteorol.* **32**, 133–154.
7. Coulter, R.L., 1981: Nocturnal Wind Profile Characteristics. *Proc. 1st Internat. Symp. Acoust. Remote Sensing, Calgary,* VI. 1-VI. 11.
8. Culf, A., 1993: Acoustic sounder observations of low-level-jets at Halley, Antarctica. in: *S.D. Mobbs and J.C. King (Eds.): Waves and Turbulence in Stably Stratified Flows.* Oxford: Clarendon Press, 121–138.
9. Delage, Y., 1974: A numerical study of the nocturnal boundary layer. *Quart. J. Royal Meteorol. Soc.* **100**, 351–364.
10. van Dop, H., R. Steenkist, D. Altena, R. Scholten, 1978: The use of acoustic methods for boundary layer studies near the coast of the Netherlands. *Proc. 4th Symp. Meteorol. Obs. & Instrum., Denver,* 326–329.
11. Foken, Th. et al, 1997: Operational Use of Sodar Information in Nowcasting *this issue.*
12. Jury, M.R., G.R. Tosen, 1989: Characteristics of the Winter Boundary Layer Over the African Plateau: 26°S. *Boundary-Layer Meteorol.* **49**, 53–76.
13. Kataoka, T., M. Takehisa, Y. Ito, Y. Mitsuta, 1991: A Low-Level Jet Observed by a Doppler-Sodar During the International Sodar Intercomparison Experiment (ISIE). *J. Meteorol. Soc. Jap.,* **69**, 171–177.
14. Kottmeier, Chr., 1978: Trajektorien unter Einfluß eines nächtlichen Grenzschichtstrahlstromes. *Meteorol. Rdsch.,* **31**, 129–133.
15. Kottmeier, Chr., D. Lege, R. Roth, 1983: Ein Beitrag zur Klimatologie und Synoptik der Grenzschichtstrahlströme über der Norddeutschen Tiefebene. *Ann. Meteorol. N.F.* **20**, 18–19.
16. Lehmann, H.-R., J. Neisser, 1990: Acoustic Sounding Work in the G.D.R.—in: *S.P. Singal (Ed.): Acoustic remote sensing.* New Delhi, Tata McGraw-Hill, 490–499.

17. Mazaudier, Chr., A. Weill, 1989: A Method of Determining the Dynamic Influence of the Forest on the Boundary Layer Using Two Doppler-Sodars. *J. Appl. Meteorol.*, **28**, 75–710.
18. Mix, W., 1981: Empirische Befunde über die vertikale Verteilung des horizontalen Windvektors an niedertroposphärischen Inversionen unter besonderer Berücksichtigung des Low-Level-Jet. *Z. Meteorol.* **31**, 220–242.
19. Mizuma, M., 1989: An Observational Study of the Nocturnal Low-Level Jet Over the Kyoto Reactor Site. *Kyoto Univ: Ann. Rep. Res. Reactor Inst.* **22**, 102–107.
20. Nieuwstadt, F.T.M., 1984: Some aspects of the turbulent stable boundary layer. *Boundary Layer Meteorol.* **30**, 31–55.
21. Nieuwstadb, F.T.M., A.G.M. Driedonks, 1979: The nocturnal boundary layer—a case study compared with model calculations. J. Appl. Meteorol. **18**, 1397–1405.
22. Pahwa, D.R., B.S. Gera, S.P. Singal, 1990: Study of the evolution of the nocturnal boundary layer at Delhi by acoustic sounding technique.—in: *S.P. Singal (Ed.): Acoustic remote sensing.* New Delhi, Tata McGraw-Hill, 363–371.
23. Peters, G., 1991: SODAR-ein akustisches Fernmeβverfahren für die untere Atmosphäre.—*Promet* **21**, 55–62.
24. Roth, R., Ch. Kottmeier, D. Lege, 1979: Die lokale Feinstruktur eines Grenzschichtstrahlstromes.—*Meteorol. Rdsch.* **32**, 65–72.
25. Savijärvi, H., 1991: The United States Great Plains Diurnal ABL-Evolution and the Nocturnal Low-Level Jet. *Mon. Wea. Rev.* **119**, 833–840.
26. Schaller, E., W. Seiler, 1993: Assessment of Photochemical Reactivity of the Atmosphere in Former East Germany: The SANA Project. *Proc. 86th Ann. Meeting Air & Waste Managm. Assoc., Denver/Co.*
27. Singal, S.P., V.K. Ohja, T. Pal, B.S. Gera, M. Sharma, M.N. Mohanan, 1989: Studies of atmospheric boundary layer of Delhi using acoustic sounders at two locations.—*Mausam* **40**, 193–196.
28. Sjostedt, D.W., J.T. Sigmon, S.J. Colucci, 1990: The Carolina Nocturnal Low-Level-Jet: Synoptic Climatology and a Case Study. *Wea. and Forecasting* **5**, 404–415.
29. Watanabe, A., 1989: Long-Range Transport of Polluted Urban Air Masses due to a Low-Level-Jet. *J. Meteorol. Soc. Jap.* **67**, 1015–1021.

Acoustic Remote Sensing Applications
S.P. Singal (Ed)
Copyright © 1997 Narosa Publishing House, New Delhi, India

12. Dynamics of the Continental Boundary Layer: The CRPE Sodar Results (1984–1993)

Christine Amory-Mazaudier
CETP, 4. Avenue de Neptune, 94107, Saint-Maur-Des-Fosses, France.

Introduction

Since the first acoustic sounding, in Australia by Mc Allister [1], and Little's work [2] on the theory of sound propagation and scattering, the doppler acoustic sounding technique has been intensively applied all over the world to study the atmospheric boundary layer.

In France, in 1971, Weill and Aubry [3] were the first to submit a proposal for developing acoustic sounding at **CNET** (**C**entre d'**E**tudes des **T**élécommunications) in the department **R.S.R** (**R**echerches **S**patiales **R**adioélectriques. An atmospheric acoustic program was set up in 1972 at CNET. This program involved the development of a Sodar with specially designed parabolic antennas; Doppler shift of the echoes was to be systematically measured. In 1973, the first acoustic sounding (Fig. 1 and 2) was performed [4]. After this first step, various theoretical, technical and experimental studies [5, 6, 7], led to the realization of the operational doppler sodar [8] used during more than one decade and a half (1976–1990) to study the dynamics of the continental boundary layer.

In 1975, CNET and **CNRS** (**C**entre **N**ational de al **R**echerche **S**cientifique) joined their efforts and created the **CRPE** (**C**entre de **R**echerche en **P**hysique de l'Environnement) laboratory which had in hand the development of continental atmospheric boundary layer studies.

In this paper we review the results obtained, during the last decade, with the CRPE doppler sodar operating during large international experimental campaigns. The first section recalls some characteristics of the doppler sodar, and the second one the parameters measured from acoustic soundings. In section 3, we briefly summarize the boundary layer studies made during the first decade (1974–1984) and refer to the previous review papers made on this subject [9, 10, 11]. The last section is devoted to the presentation of the results.

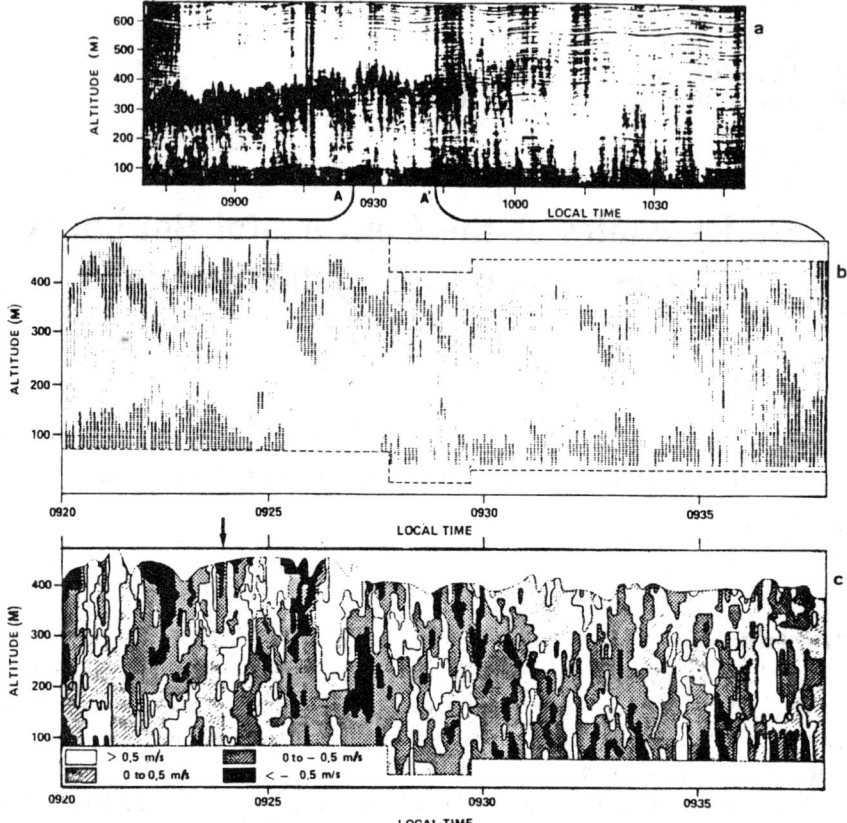

Fig. 1. Display of the Sodar data set obtained on 11.09.1973 (a) Facsimile record obtained at Issy-les-Moulineaux. Sounding frequency: 2000 Hz; pulse rate: 0.2 Hz; pulse length: 200 ms, filter bandwith: 100 Hz; the antenna was pointed vertically. With the operating conditions, the Sodar can isolate atmospheric regions of about 30 m in vertical and (altitude/10) in horizontal dimensions, (b) Numerical facsimile display of the AA' section in Fig. 1a. (c) Radial 'vertical' velocity associated with the reflectivity shown in Fig. 1b. Above about 400 m (dashed line), signal-to-noise ratio was too low and Doppler shifts could not be computed, (from Aubry et al [4].

Technical Characteristics of the CRPE Sodars

The acoustic and radar sounding techniques are quite similar. The acoustic wave emitted through the atmospheric medium is partially scattered by the temperature inhomogeneities (refractive homogeneities for radar). The received signal gives the return power and doppler shift (similar for radar). Parameters that can be derived from the received signal are namely: wind speed and direction, temperature structure parameter C_T^2, etc. (see section 3).

The CRPE doppler sodar [8] consists of a system of three monostatic antennas:

Dynamics of the Continental Boundary Layer: The CRPE Sodar Results 249

Fig. 2a. The CNET sodar doppler "Centre National d'Etudes des Télécommunications"

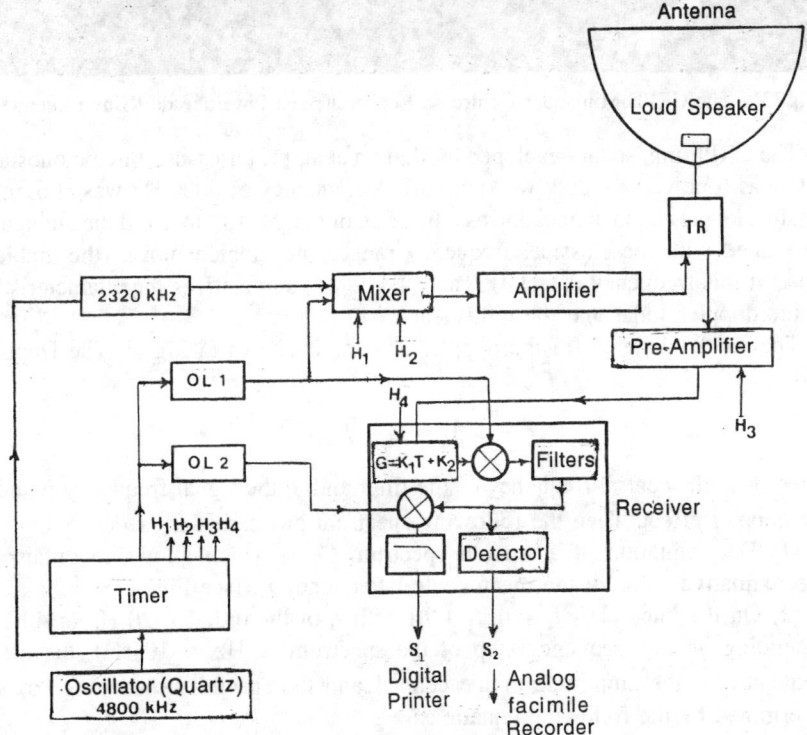

Fig. 2b. Block diagram of the CNET sodar

one vertical and two slanting at 30° from the vertical (Fig. 3). The angle of the horizontal projection is 90°. The acoustic pulse lasts 100 ms, the pulse repetition rate is 4 s, and the emission frequency is 2000 Hz. As pointed out by Spizzichino [6], the antenna axis elevation and the pulse duration are adequately chosen in order to minimize the wind measurement errors.

Fig. 3. The CRPE Minisodar Centre de Recherches en Physique de l'Environnement

The CRPE minisodar developed by Baudin et al, [12] uses also three monostatic antennas (one vertical and two slanting). A frequency of 6000 Hz was chosen to obtain a longer sound attenuation as a function of height [13], to avoid the ambiguity associated with the classical frequency range, and ambient noise (the ambient noise at this frequency is small). Table 1 [12, 14] summarizes the characteristics of the doppler sodar and the minisodar.

The received signal for the doppler antenna is shown in Fig. 4. The Doppler shift is:

$$\Delta f = f_0 - \overline{f}$$

where f_0 is the central frequency of the filter and $\overline{f}$ the signal frequency without the doppler effect. Then the following spectrum processing is made:

(1) Determination of a smooth spectrum $S'(f)$ which allows to obtain an approximative value of the mean central frequency f_1, (see Fig. 4).

(2) On the interval $J (f_1 - \eta, f_1 + \eta)$ with η of the order of 20 Hz or 40 Hz, depending on the frequency step of the spectrum (5 Hz or 10 Hz), the mean frequency $\overline{f}$, the amplitude of the echo A and the spectral broadening (σ) are determined by the following equations:

Table 1. Sodar and Minisodar characteristics

Characteristics	Minisodar	Doppler sodar
Frequency	6000 Hz	2000 Hz
Electric power	50 W	120 W
Radiated acoustic power	1.5 W	12 W
Beam width	1.3°	10°
Pulse duration period	50 ms	100 ms
Repetition period	1 s	4 s
Maximum range	170 m	680 m
Height of first gate	13.6 m	22.1 m
Resolution	8.5 m	17 m
Sampling	4.25 m	17 m
Receiver bandwith	560 Hz	200 Hz
Maximal radial velocity	8 m/s	8.5 m/s
Receiver gain	116–126 dB	100–110 dB
Output conversion	7 bits + sign	7 bits + sign

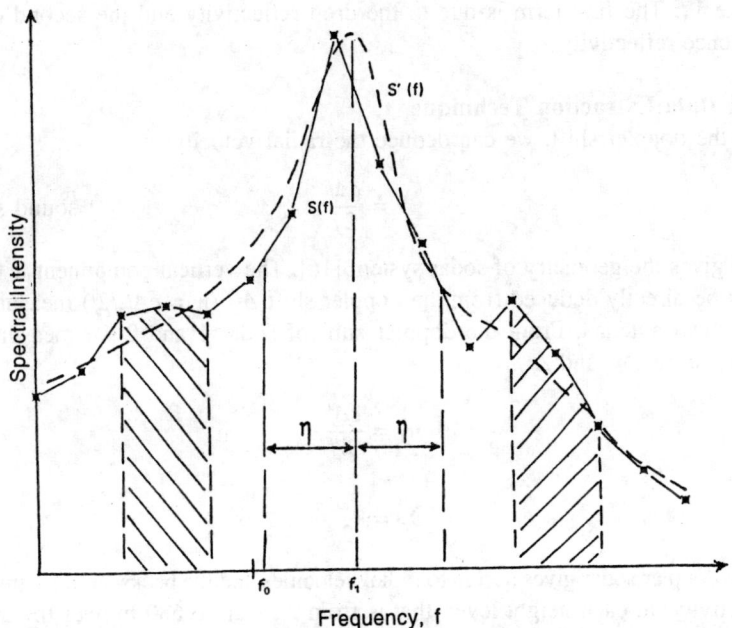

Fig. 4. Determination of the spectrum parameters

$$A = \int_J S(f) \, df$$

$$\overline{f} = (\int_J S(f) \, f \, df)/A$$

$$\sigma = (\int_J S(f) \, (f - \overline{f})^2 \, df)/A$$

(see Chong, [16] for more details)

Applications in the Atmosphere

For the minisodar, the spectrum is composed of two well separated peaks, as the emission frequency has been chosen in order to study precipitation following the suggestion of Little [15]. In fact, the acoustic backscattering intensity of sound by hydrometeors for drop diameters smaller than the acoustic wavelength (Rayleigh diffusion) is related to hydrometeor reflectivity factor Z by the relationship:

$$\eta = \frac{26\pi^5}{36\lambda^4} Z$$

and Z is proportional to the drop diameter to the power of six, as for radar.

For a monostatic sounder well isolated from the noise of the drops, the backcattering intensity can be expressed as:

$$\sigma = 2.110^2 \frac{1}{V_s} \sum_V \frac{D^6}{\lambda^4} + \frac{0.73 \, C_T^2}{T^2 \lambda^{1/3}} \quad \text{(Little [15])}$$

where T is the mean temperature in °K and $\sum_V$ is an integration inside the scattering volume V_s. The first term is due to the drop reflectivity and the second one to turbulence reflectivity.

Sodar Data Extraction Technique

From the doppler shift, we can deduce the radial velocity:

$$V = \frac{c\Delta f}{2f} \qquad \text{(c : sound speed)}$$

Fig. 5 gives the geometry of sodar system [16]. The vertical component w (sodar 3) can be directly deduced from the doppler shift Δf_3 ($w = c\Delta f_3/2f$) measured by the vertical antenna. From the doppler shift of sodar 1 and 2 we measure two doppler shifts Δf_1 and Δf_2.

$$u_r = \frac{c\Delta f_1}{2f}$$

$$v_r = \frac{c\Delta f_2}{2f}$$

The doppler sodar gives access to radial velocities and the backscattering intensity (reflectivity) at each height level, that is from 22.1 up to 680 m over the ground and with a height resolution of 17 m, and, the minisodar at each height level, from 13.6 up to 170 m over the gound and with a height resolution of 8.5 m

To deduce the three wind components u, v, w in an orthogonal coordinate system, we must take into account two facts:

(1) each radial component is the sum of projections of the horizontal and vertical components of the wind on the considered sodar axis

(2) when the vertical component is measured at an altitude z, at that time due

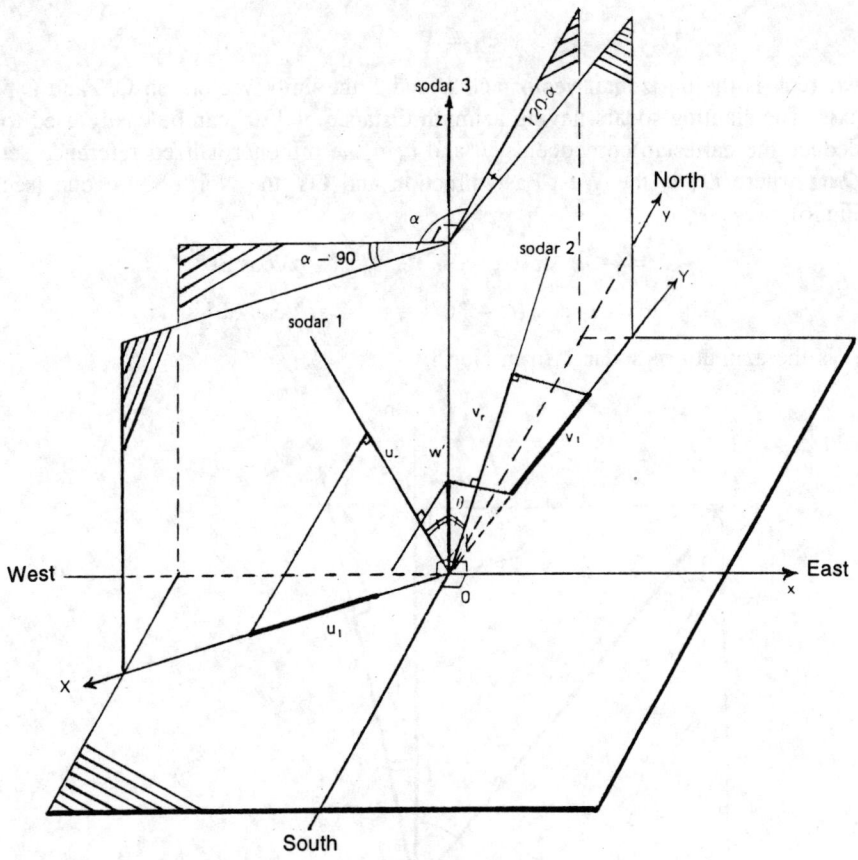

Fig. 5. Sodar geometry

α : azimuth of sodar
θ : Inclination of sodar 1 and 2 with the vertical
u_1, v_1 : Horizontal components of the wind in the direction of the slanting sodars 1 and 2
w' : Vertical component of the wind
u_r, v_r: radial component measured by the slanting sodars 1 and 2

to the sodar 1 and 2 at inclination (θ), the radial components u_r and v_r correspond to an altitude $z' = z \cos \theta$; therefore the horizontal wind components (u_1, v_1) in terms of the direction of the slanting sodars 1 and 2 are:

$$u_1 = (u_r - w' \cos \theta)/\sin \theta$$
$$v_1 = (v_r - w' \cos \theta)/\sin \theta$$

We have also:

$$u_1 = \bar{V} \cdot \bar{X}$$

$$v_1 = \bar{V} \cdot \bar{Y}$$

where V is the horizontal vector and X and Y the unity vectors on OX and OY axis. The slanting sodars having azimuth distance of 120° can be easily used to deduce the cartesian components, u and v in the orthonormalized reference set $Oxyz$ where Ox is the West-East direction and OY the North-South one (see Fig. 6)

$$u = (-u_1 \sin(\alpha - 30°) + v_1 \cos\alpha)/\cos 30°$$

$$v = (u_1 \cos(\alpha - 30°) + v_1 \sin\alpha)/\cos 30°$$

α is the azimuth of sodar 1 from North.

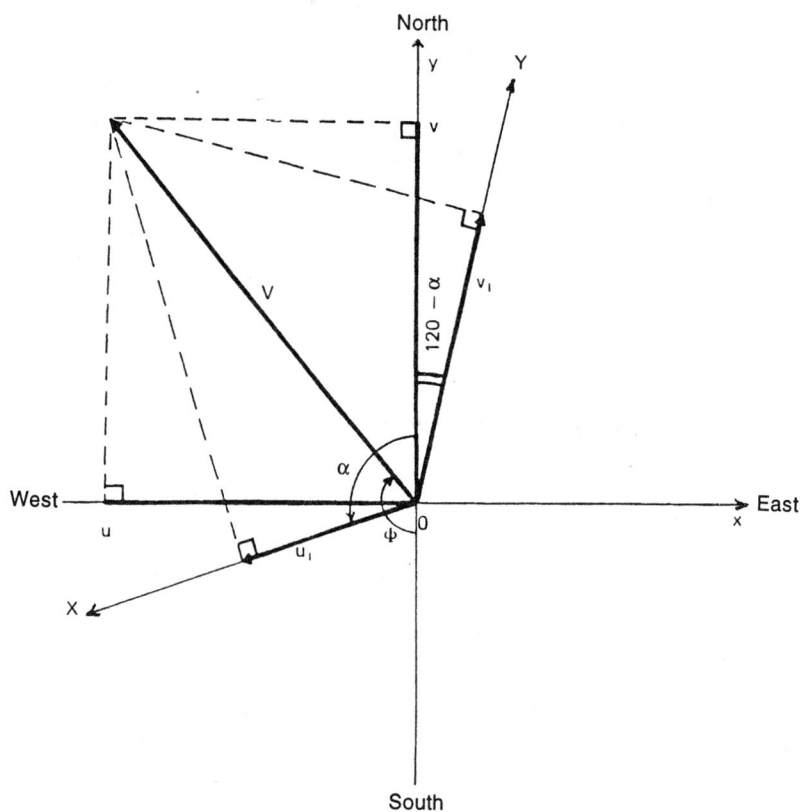

Fig. 6. The wind vector in the horizontal plane
u_1, v_1: wind components in the directions OX and OY
u, v: S-N and O-W components
V: wind amplitude
Φ: wind direction
α: azimuth of the slanting sodar 1

From the preceeding equations we can deduce the wind V and its direction Φ:

$$V = (u^2 + v^2)^{1/2}$$

$$\Phi = \text{Arctg}\,(u/v)$$

$\Phi = 0°$ corresponds to a northely wind

$\Phi = 90°$ corresponds to an easterly wind

Winds are in general averaged over 15 to 20 minutes to do mesoscale studies.

Brief Recalls on the First Decade of Sodar Measurements (1973–1983)

The first few years (1971–1976) of acoustic sounding development at CRPE were essentially devoted to the physics of the measurement ([3, 4, 5, 6, 7]), the development of acoustic technique [8] and the data processing [16].

Subsequently the scientific works were essentially devoted to document the homogeneous boundary layer by using one sodar. Various turbulent parameters were derived from sodar measurements; they are given by Table 2 [10]; During this period, some reviews were written by different authors [9, 10, 11, 17].

The scientific works, on the boundary layer, made at CRPE during the first decade (1973–1983) can be summarized as follows:

(1) documentation of the convective structure of the boundary layer ([27, 28, 29, 26]);

(2) characterization of the boundary layer flow, the mean horizontal component [30, 31], the vertical component [32, 33], and the gravity waves activity [34, 35];

(3) characterization of turbulent properties by computing turbulent parameters [36, 37, 38, 39, 25];

(4) modeling of the boundary layer [40];

(5) radio propagation studies [41, 42, 43];

(6) mean properties of the boundary layer and mesoscale studies [25, 43, 44];

Mesoscale Studies from 1984 to 1993

During the decade 1984–1993, international campaigns involving sodar network, radar network, meteorological stations, aeroplane etc. were conducted, in order to have:

(1) the continuity in wind altitude profile measurements by using complementary tools as ground anemometer, tower, minisodar, sodar and radar

(2) various horizontal distributions of wind measurements to connect the different scales (local, mesoscale and global)

(3) estimations of the boundary layer turbulent parameters by various tools and methods

(4) relations between the turbulent parameters and the mean properties of the boundary layer

(5) large data sets to document physical processes (frontal systems) or properties of the regional boundary layer.

Table 2 Determination of possible turbulent parameters [10] from sodar data

Parameter	Direct measurement	Not direct	References	Precision
Turbulent dissipation: ε; Struct. function of velocity: Cv	with bistatic sodar	with structure functions; with spatial density W	Gaynor (1977) [18] Weill (1978) [19]	40%
Struct. function of temperature C_T^2; Rate of Homogoneization: N	difficult with bistatic	with parametrization σ_w	Coulter and Weseley (1980) [20]; Weill et al. (1980) [21]; Neff (1975) [22]	30%
Friction velocity: U^*	yes, correlation between antenna	with parametrization dU/dZ	Strauss (1980) [23] Klapisz and Weill (1982) [24]	30%
Surface Heat flux: Qo	yes, need corrections (sodar equation)	with σ_w/Z and/or T.K.E budget	Coulter and Weseley (1980) [20]; Weill et al. (1980) [21]; Desbraux and Weill (1986) [25]	30%
σ_w	yes	with	Spizzichino (1975) [7]	function of sample duration
σ_u	no	σ_i = antennas		
σ_v	no	+ correlations		"
Skewness: σ_w^3	yes	with distribution function fitting	Taconnet and Weill (1983) [26]	
Height of the inversion: Z_i	yes if $Z_I < 10^3$ m	with $Q(Z)$ if $Z_i > 10^3$ m combination		20 m
W^*		Z_i, Qo		20%
Z/L		U^*, Qo		90%
Z_I/L		Z_i, Qo, U		100%

The CRPE sodars were involved in 4 campaigns: MESOGERS84, HAPEXMOBILHY86, FRONTS87, PYREX90. Table 3 gives the characteristics of these experiments and the references concerning the general paper of presentation of the experiment and the data base documentation. All the data are available and are still used at **CETP** (**C**entre d'études des **E**nvironnements **T**errestre et **P**lanétaire) formely **CRPE** (**C**entre de **R**echerches en **P**hysique de l'**E**nvironnement). Figure 7 shows the experimental areas of these four campaigns.

Mesogers 84

The Mesogers 84 campaign was conducted to study the boundary layer over complex terrain [45, 46] and took place in the south-west part of France from September 1st to October 4th, 1984. Five Doppler sodars were used routinely during the campaign and two other Doppler sodars during periods of intensive experiments.

- It was found, that the homogeneous boundary layer can sometimes develop over complex terrain [53] and in general during stable conditions (near 6.00 LT) the height of the inversion layer is not very different; the convection starts at the highest location.
- It was established that, if sodar are operating in very quiet locations, and if distances between sodars are typically larger than 15 km, we have really the means to validate satellite flux estimates and/or numerical meteorological medels either on complex terrain, or in complex meteorological situation, [54, 55].
- Estimation of virtual heat flux using 4 sodars and air craft measurements were done. The area is shown on Fig. 8 [45]. Sodar heat fluxes are estimated by extrapolation of virtual heat flux profiles down to the surface [37]. A priori sensible heat flux over complex terrain is supposed to be modified by typography, in the present case (Fig. 8), we observe a good homogeneity between all the data.
- An analysis of the mesoscale spectra and structure functions of the mean horizontal velocity fluctuations with 4 sodars have been done for three days of fair weather conditions [56]. It was found that the mean speed follows a $-5/3$ spectral law, this result, concerning our data sample can be justified by quasi-two-dimensional turbulence. The analysis of spatial and temporal structure functions with spatial increments limited to 38 km has shown the validity of Taylor's hypothesis at this scale. The turbulence advection speed has been found to be very close to mean spatial wind averaged over the four sodars and over the sample duration shown [56].
- The frontal slopes and friction-induced vertical momentum transfers were derived from sodar measurements [57, 58, 59]. Results show that the divergence of momentum transfer modifies the frontal slope and highlights the importance of estimating correctly the frontal friction in numerical modeling.

Table 3 International experimental campaigns (1984–1990)

Campaign	Scientific goals/country	Sodar measurements	Canopy/relief	Paper of presentation data base and catalog
MESOGERS 84	Dynamics of the boundary layer over complex terrain *Gers (France)*	from September 1 to October 5, 1984	Agricultural crops and trees small hills and valleys	Weill et al, 1988, [45] Mazaudier at al, 1986 [46]
HAPEX-MOBILHY 86	Hydric balance over various canopies (forest/agricultural crops) *Landes (France)*	from May 22 to June 30, 1986	agricultural crops and pine forest flat ground	André et al, 1988, [47] Mazaudier at al, 19 [48]
FRONTS 87	Frontal systems *Britain (France-England)*	from October 1 to January 14, 1987	bare soil, isolated trees flat ground	Clough 1987, [49] Bouvard et al, 1991, [50]
PYREX 90	Momentum fluxes near mountain *Pyrenees (France-Spain)*	from October 1 to November 30, 1990	bare soil mountain	Bougeault et al, 1993 [51] Mazaudier et al, [52]

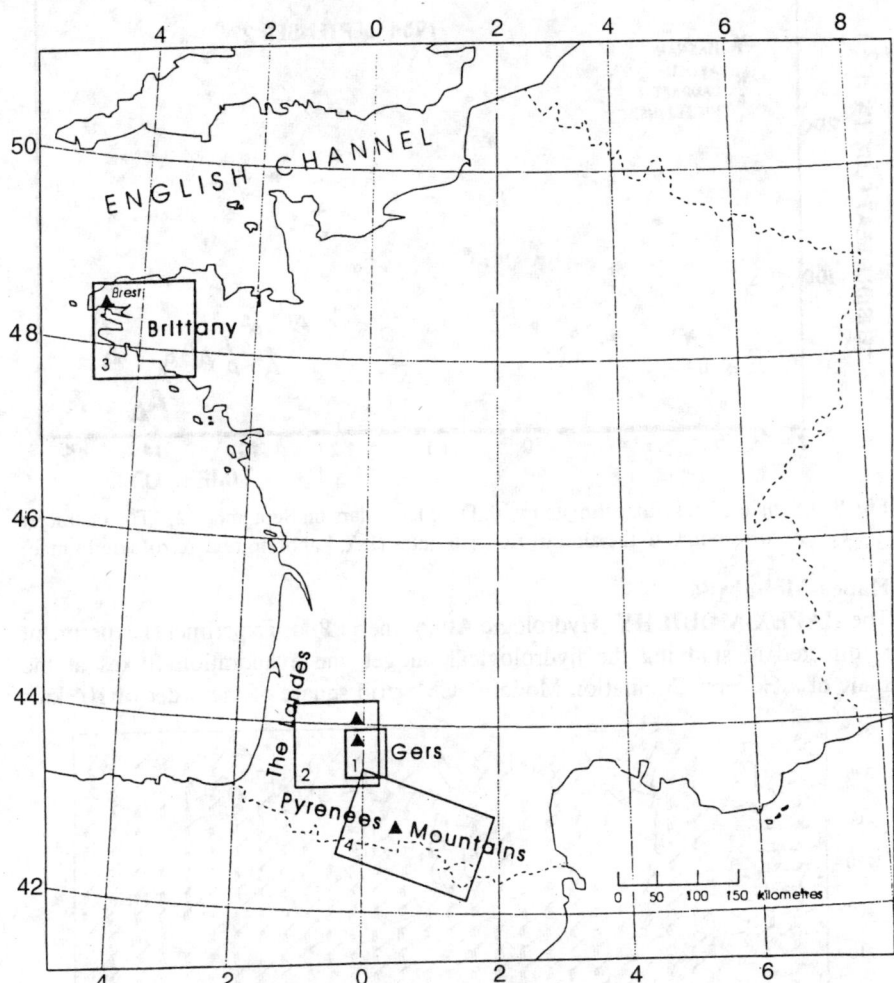

Fig. 7. Experimenal areas: 1: MESOGERS84 2: HAPEX-MOBILHY86 3: FRONTS87 4: PYREX90

— The sodar measurements were connected to radar ones, in order to study the height structure of gravity waves excited by a gravity current on September 28th, 1984 [60, 61]. Fig. 9a [61] illustrates the signature of the frontal passage on the sodar horizontal wind and figure 9b, the gravity wave activity observed on the sodar vertical drift. Figure 9c shows the signature of the fontal system on the radar vertical wind. These figures underline the importance of vertical continuity in the wind measurements. Wave periods were directly measured and it was possible to make reasonable inferences on the wave excitation mechanisms. Both Kelvin-Helmhotz waves generated in the regions of high wind shear found in association with the gravity current and Lee-type waves forced by the gravity current acting as an obstacle to opposing prefrontal flow were identified [61].

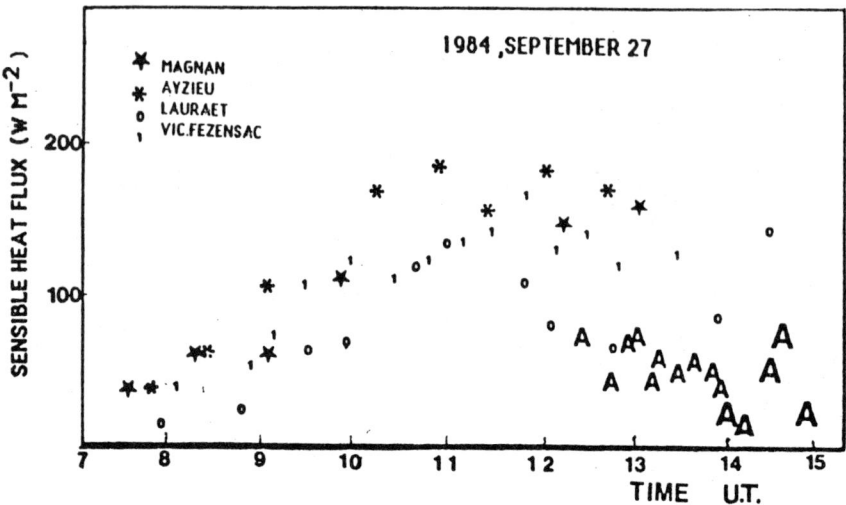

Fig. 8. Sensible heat flux estimated with Doppler sodars on September 27. The character A corresponds to the aircraft flux estimates (LA, Laboratoire d'Aérologie/France)

Hapex-Mobilhy86

The **HAPEX-MOBILHY** (**H**ydrologic **A**tmospheric **P**ilot **Ex**periment) experiment is directed at studying the hydrological budget and evaporation fluxes at the scale of a General Circulation Model (GCM) grid square of the order of 10^4 km^2

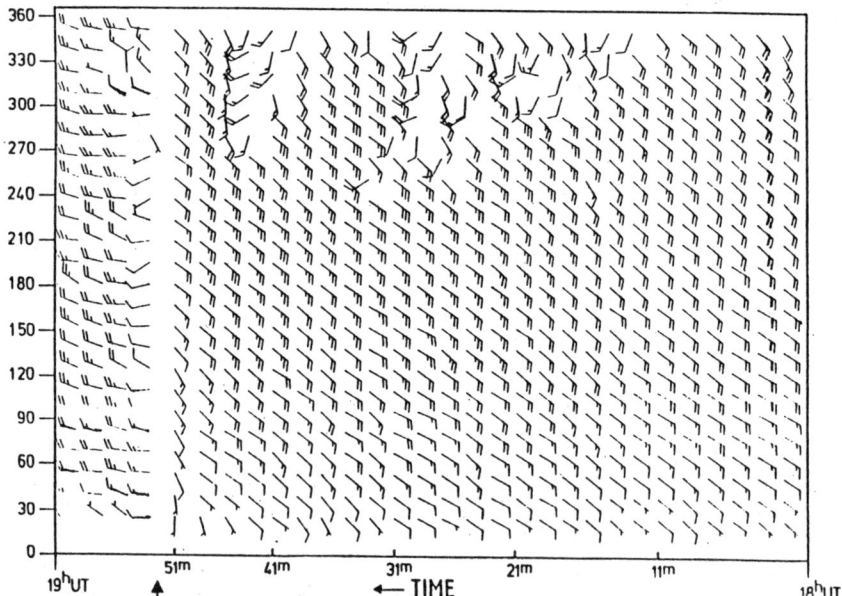

Fig. 9a. Two minutes averages of the horizontal wind measured by sodar (at Magnan Gers/France) from 1800 to 1900 UTC 28 September 1984. Full barb is 5 ms^{-1}. Time runs from right to left. The front is marked by an arrow.

Dynamics of the Continental Boundary Layer: The CRPE Sodar Results 261

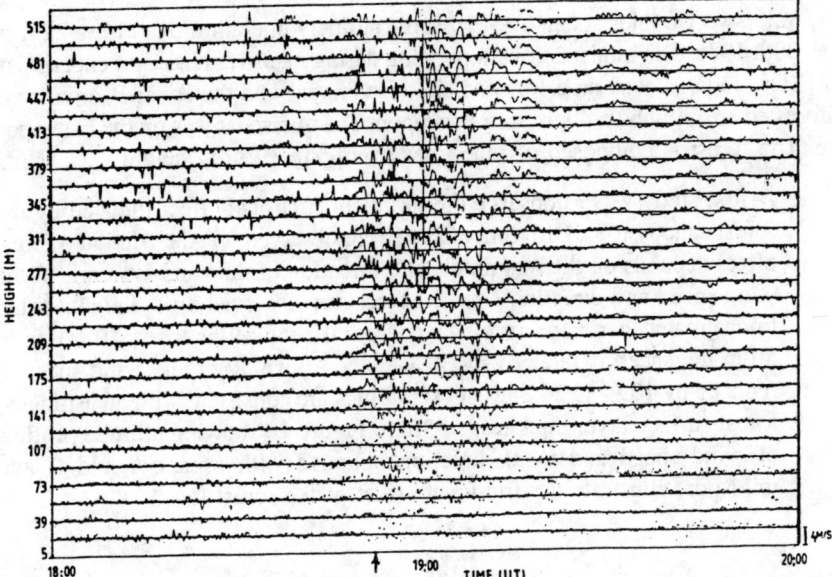

Fig. 9b. 4s vertical sodar data at Magnan (MESOGERS84), on September 28 1984. The front is marked by an arrow. Wave activity is observed.

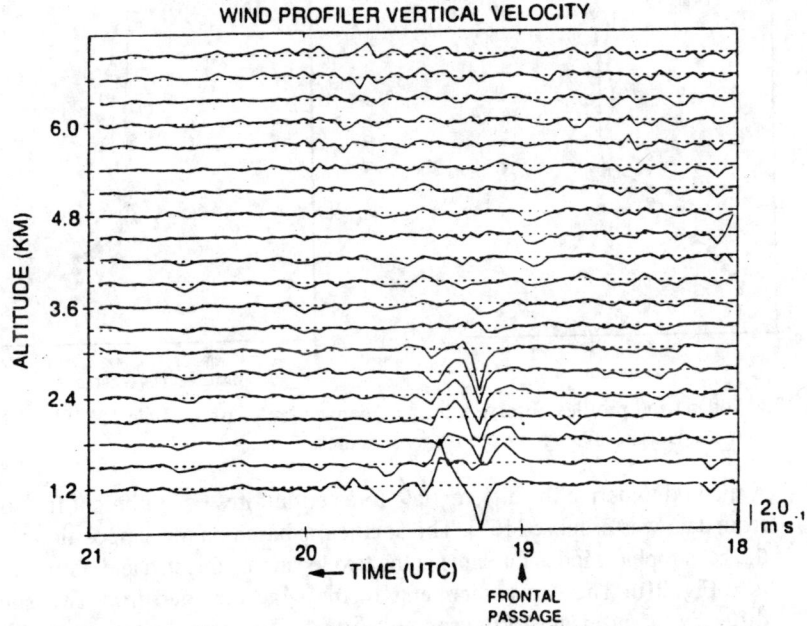

Fig. 9c. Time height cross section of vertical velocities observed by the 45 MHz radar wind profiler (LSEET: Laboratoire de Sondages Electromagnétiques de l'Environnement). Vertical velocities were measured continuously and represent roughly 2-min averages. The time of the frontal passage at the wind profiler site is marked by an arrow. The time runs from right to left.

[47, 62]. Two Doppler sodars (A, B) were routinely operating from May 22th to June 30th 1986. The minisodar was operating during intensive phases of experiment [48]. The sodar A was on a forested site (Estampon) and the sodar B on a large unforested area, Lubbon, 4 km from Estampon. The minisodar was on the Estampon site. The forest is composed of pine trees averaging 19 m in height.

— A first study was to observe the turbulent frictional effect induced by the forest canopy [63]. Different effects were observed: (1) the frictional forest effect depends on the magnitude and direction of the wind system; (2) the wind speed and direction over the forest are observed up to heights equal to or greater than 65m. In the layer from the top of the trees (19 m) up to 50 m the speed is reduced by 30%–60%. Above 50 m, the wind speed is reduced by 10%–15%. These observations are consistent with an estimated forest surface layer thickness of 30 m. Figure 10 showing altitude profiles of wind speed and wind direction obtained with sodars A and B and minisodar illustrate the frictional effect of the forest [64].

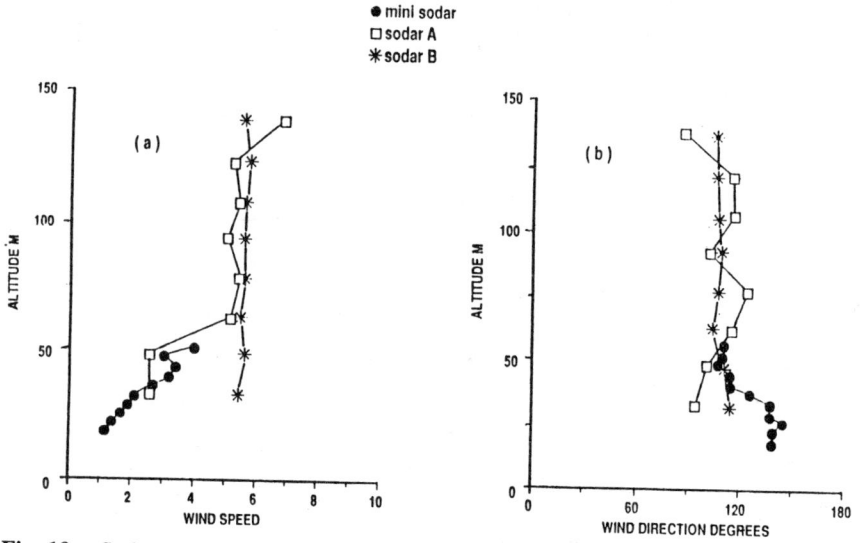

Fig. 10. Sodar wind profiles observed during a convective phase on June 15, 1986. The wind data are averaged over fifteen minutes.

— A method to derive the differential "forested-unforested" turbulent friction from data was proposed [63]. This method is based on the supposition that the geostrophic wind is the same at the two locations: forested and unforested (see Fig. 10). The two components of the wind obtained from two sites differ by the influence of the canopy ($\delta \tau/\delta z$). Therefore, by integrating the altitude wind profile difference, we can obtain an estimation of the differential "forested-unforested" turbulent friction.

— Systematic analysis of the minisodar wind profiles [65] shows: (1) that the minisodar wind profiles from the top of the canopy (19 m) up to 60 m can be reduced to a simple logarithmic law for eighty percent of the cases, whatever the stability conditions are; (2) the friction velocities derived from minisodar are in good agreement with estimations or measurement of this parameter made with sodar, radiosounding or hydra system; (3) the minisodar roughness lengths (5–6 m) are larger than the radiosounding ones (1–2 m) [66], this fact is explained by different data sets and parametrizations.

— A systematic study of the friction velocities obtained with the different tools (sodar A, sodar B, minisodar, radiosounding, hydra system) and by using different methods (aerodynamic or fluctuation) for neutral atmospheric conditions has been done. Figures 11a, 11b illustrate the very good correlation between all these estimations [67].

— The sodar data were used to parametrize the stable boundary layer height [68, 69, 70]. Vertical profiles of C_T^2 (3, 6, 9, 12 hours after neutrality) were derived from the amplitude of the sodar signal intensity [68] averaged over half an hour and compared to numerical simulation (Fig. 12). The shape of the profiles is well reproduced. Discrepancies between simulations and observations are observed for the height of the maximum. This can be explained by the fact that the model does not take into account explicitly the radiation effects.

— During the Hapex-Mobilhy campaign, momentum fluxes derived from sodar and satellite data [71] show a very good agreement.

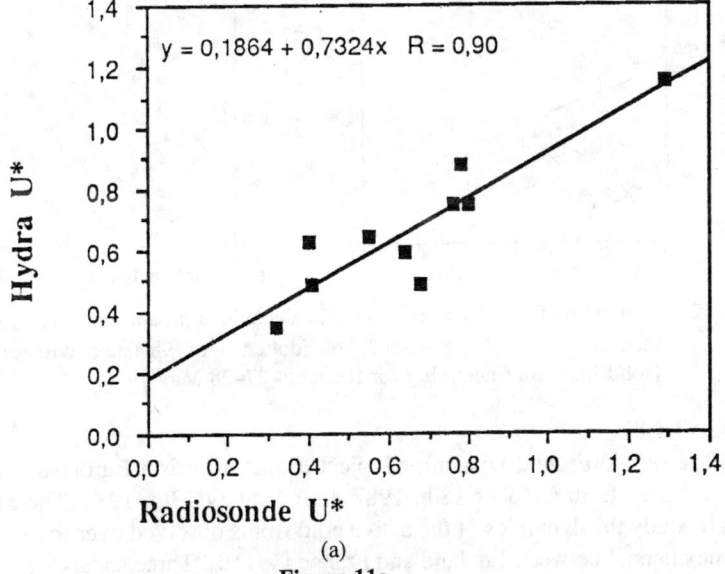

(a)
Figure 11a

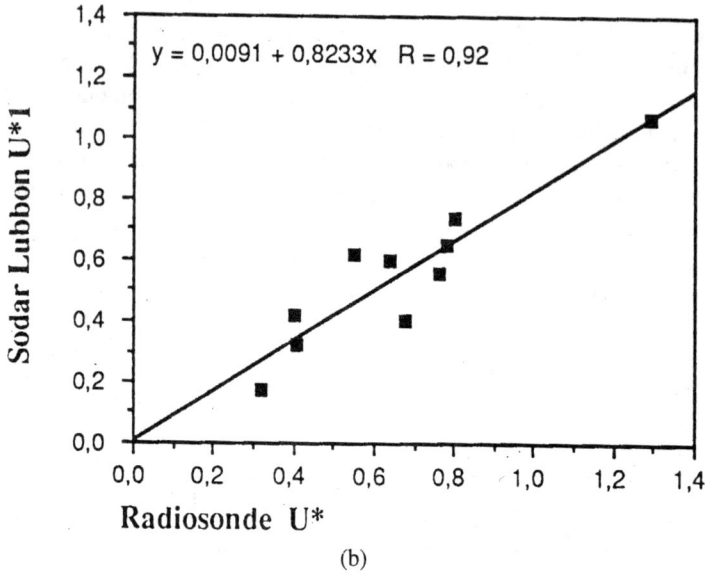

(b)

Fig. 11a, b Friction velocities measured by Hydra (Institute of Hydrology/England) and the sodar (clearing of Lubbon) as function of friction velocity deduced from radiosounding (**CNRM, C**entre **N**ational de **R**echerches en Météorologie/ France) for neutral atmospheric conditions

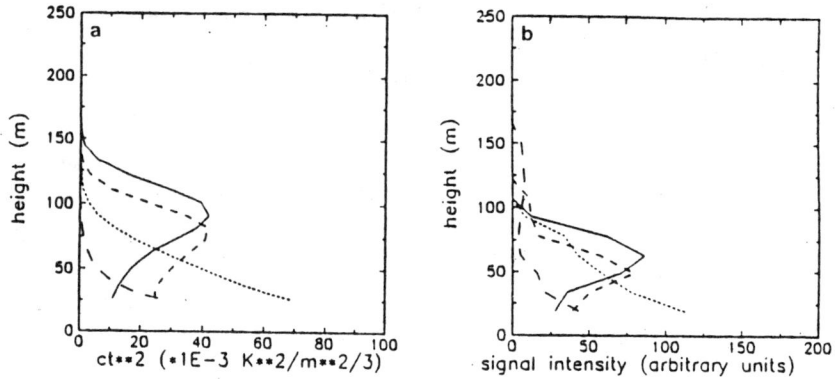

Fig. 12. Vertical profiles of C_T^2 from model calculations (a) and measured sodar signal intensity (b), 3h (long dashed), 6h (dotted line), 9h (short dashed line), 12h (solid line) after neutrality for the night 27–28 May 1986.

FRONTS 87

The Mesoscale Frontal Dynamics Project/Fronts 87 was a European experiment which lasted from October 18th, 1987 until January 13th, 1988. The main topic was to study the dynamics of the active cold fronts observed over the area centred on the channel between England and France [49, 50]. Three sodars were involved in this experiment.

— During this experiment, a systematic study of the sodar wind and friction velocity was made in relation to the radiosounding, radar and meteorological wind measurements.

The 4-seconds sodar data were used to compute the friction velocities [25, 72].

$$u^* = [\langle u'w' \rangle^2 + \langle v'w' \rangle^2]^{1/4}$$

u', v' w' are the fluctuations of the three wind components.

Figure 13 illustrates the relation between the Lower Level Jet intensity and the friction velocity [73, 74]. On the top panel, the sodar horizontal wind speed suddenly decreases when the frontal surface passes over the sodar. On the bottom panel, the wind deduced along the front from radiosoundings at the altitude of the LLJ, $z = 1100$ m (lower level jet) is superimposed on the sodar friction velocity. Friction velocity reaches its maximum value almost simultaneously with the along-front wind, and is more intense in the cold sector of the LLJ. Then u^* decreases behind the front. So frictional dissipation in the surface layer is increasing from the warm to the cold sector of the jet, as it was predicted by Mak's theory [75].

All the frontal systems observed during FRONTS87 have similar dynamical characteristics, and it was possible to draw a mean picture [76, 77]. Figure 14 illustrates the mean frontal system dynamical characteristics. From the top to the bottom the curves are respectively: (a) the maximum amplitude of the jet, (b) the horizontal sodar wind in the boundary layer, (c) the friction velocity derived from the sodar data, and (d) the horizontal wind at the ground surface. For all the frontal systems observed, friction velocity is more intense on the cold side of the Low Level Jet and its maximum represents 2% of the LLJ maximum magnitude. The amplitude of the wind in the low boundary layer is in general 50% of the maximum magnitude of the LLJ.

PYREX 90

The PYREX program is a major field study of the dynamical influence of the Pyrénées mountains (on the border between France and Spain) on the atmospheric circulation. Three sodars and 4 radars were operating continuously on sites along the main transect, on either side of the range and very close to the crest [51], [52].

Computations of the friction velocity have been performed from the 4s sodar data [78]; They show (Fig. 15) that the friction velocity oscillates between 0.4 and 0.6 ms^{-1}. The period of this oscillation, 2.5 h, is in fair agreement with the apparent period of the Lee waves observed on air velocity measurements. The Lee-wave system generates variations of the turbulent momentum transfer in the ABL.

Sodar data were also the study material for the modification of a cold frontal signature in the approach of a mountainous region [79] and modeling [80].

All Experiments

The data bases developed for each experiment were used together in order to

266 *Applications in the Atmosphere*

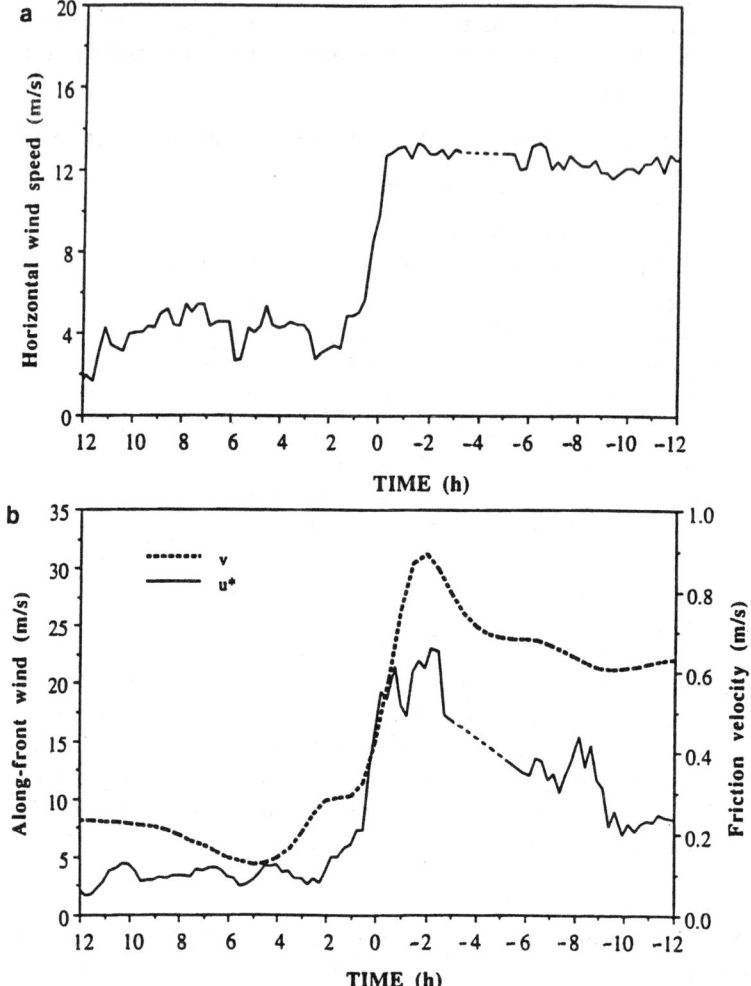

Fig. 13. Time variation of (a) horizontal wind speed at $z = 34$ m from the sodar, (b) friction velocity at $z = 34$ m, calculated from the sodar data and superimposed the along-front wind (line with dots) from the soundings at $z = 1100$ m, that is the height of the LLJ maximum. The hours in the time scale are relative to the passage of the front over Brest.

characterize the influence of different orographic and canopy conditions on the boundary layer dynamics during frontal system passage. The main characteristics observed for FRONTS87 (flat, bare ground) were found for all the types of canopies and orographic conditions (small hills and valleys with agricultural crops and trees, agricultural crops, forest, bare ground on the side of a mountain). We found that canopy and orographic conditions slightly modify the main signature. Mountain

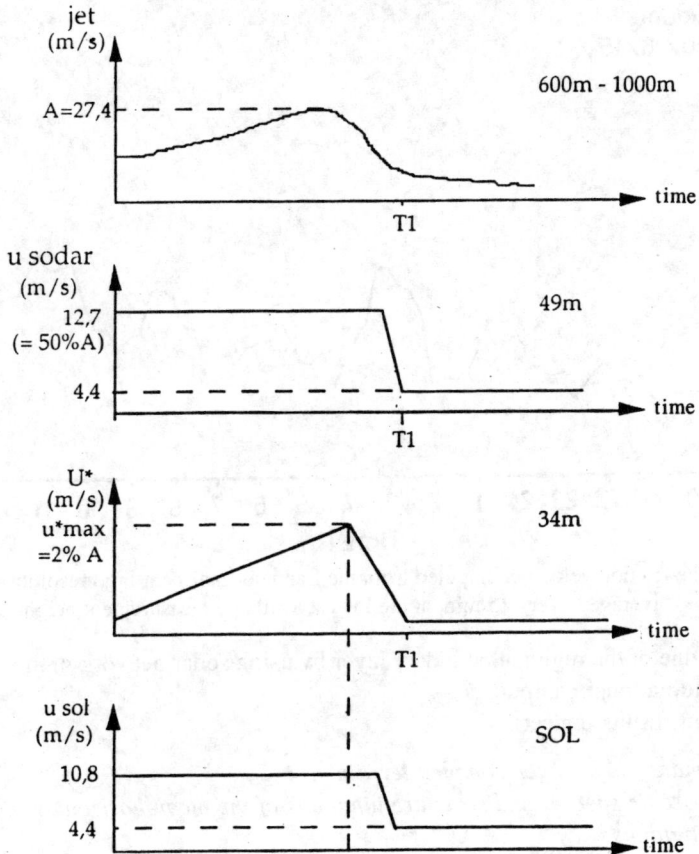

Fig. 14. Dynamical characteristics of "mean" frontal system over brittany, u^* is the friction velocity at $z = 34$ m; usol, the horizontal wind speed recorded by the meteorological ground station and u sodar the sodar horizontal wind.

effects yield to more intense friction-velocity values and to superpositon of an oscillating behaviour on the time variation of the friction velocity, while the forest effects induce a shift of the frontal signature on the time variation of friction velocity at higher height levels.

Reviews on the use of sodar for mesoscale studies [81, 82], and progress in the momentum flux computation were also made during this last decade [83].

Conclusion

In this paper we have briefly summarized the main results obtained with the CRPE sodars during the decade 1984–1993. Significant advances were made in

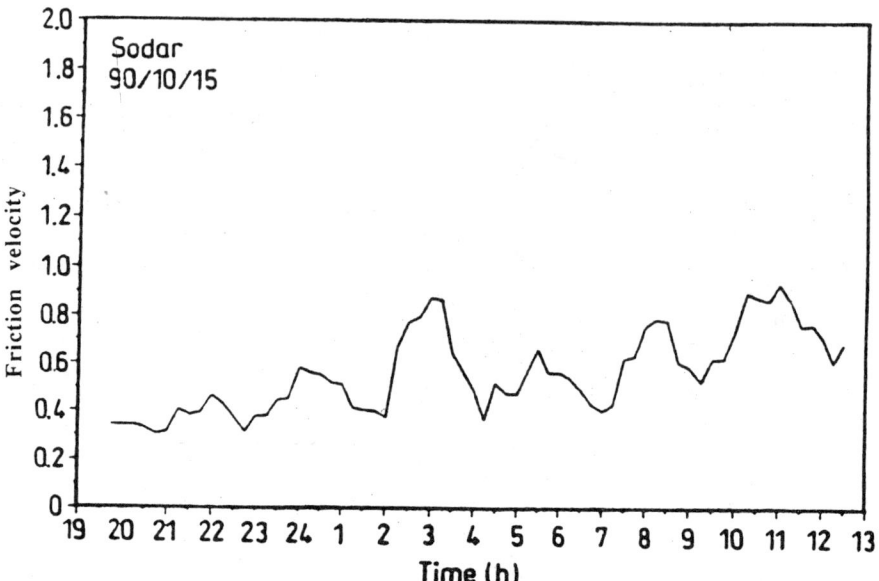

Fig. 15. The friction velocity computed from the Lannemezan sodar high resolution data (4s), averaged every 15 min, at the lowest available measurement height (34 m)

the knowledge of the regional boundary layer by using sodar network in the frame of large international campaigns.

The main results concern:

— *boundary layer over complex terrain*
— *mesoscale spectra and structure functions of the mean horizontal velocity fluctuations*
— *forested boundary layer*
— *frontal systems*
— *validation of satellite flux estimates*
— *influence of orographic and canopy conditions on the wind profile*
— *modeling of the boundary layer*

Sodar is still one of the most successfull remote sensing techniques to study the low level boundary layer. Sodar allows wind measurements at altitude levels where radar or tower measurements are not possible. In the future the use of sodar network jointly with radar network will provide a lot of new results in meteorology. These large data sets must be very useful for numerical modeling and meteorological forecasting.

Acknowledgement

The author thanks Dr. S.P. Singal for help in correcting the English. This work was financially supported by the **CNRS** (**C**entre **N**ational de al **R**echerche **S**cientifique).

References

1. McAllister, L.G., Acoustic sounding of the lower troposphere, *J. Atmos. Terr. Phys.*, 30, 1439–1440, 1969.
2. Little, C.G., Acoustic method for the remote probing of the lower atmosphere, *Proc. IEEE*, 57, 571–578, 1969.
3. Weill, A., and M. Aubry, Mesure de l'atténuation du son dans l'atmosphére—Projet d'émission, *Note Technique EST/RSR/71*, 1972.
4. Aubry, M., R. Chezlema, A. Spizzichino, Preliminary Results of the Atmospheric Acoustic Sounding Program at CNET, *Boundary Layer Meteorology*, 7, 513–319, 1974.
5. Aubry, M., F. Baudin, A. Weill, P. Rainteau, Measurement of the total attenuation of acoustic waves in the turbulent atmosphere, *J. Geophys. Res.* Vol 79, No. 36, 5589–5606, 1974.
6. Spizzichino, A., Discussion of the Operating Conditions of a Doppler Sodar, *J. Geophys. Res.*, 79, 36, 5585–5591, 1974.
7. Spizzichino, A., A spectral acoustic and radio waves scattered by atmospheric turbulence in the case of radar and sodar experiments, *Ann. Geophys.*, 31, 4, 433–445, 1975.
8. Baudin, F., G. Belbeoch and R. Chelzemas, Le sodar triple du CNET, *Note Technique CRPE/128, CRPE, Issy-les-Moulineaux*, France 1976.
9. Weill, A., Sodar micrometeorology: A review, *in Proceedings of the International Symposium on 'Acoustic Remote Sensing on the Atmosphere and Oceans'*, MATHEW T. Ed., pp. IV. 1–IV. 60, 1981.
10. Weill, A., Turbulent parameters measurements with sodar techniques: Momentum fluxes, virtual temperature, heat flux, turbulent dissipation rate: Methods, representativeness, *in proceedings of the Afred-Wegener-Konferenz on "Ground Based Remote Sensing Techniques for the Troposphere"*, pp 161–169, 1986.
11. Weill, A., and H.R. Lehmann, Twenty years of acoustic sounding—a review and some applications, *Z. Meteorol.*, 40(4), 241–450, 1990.
12. Baudin, F., A. Weill, J. Bilbille, S. Dubois, J.F. Fevre, and B. Piron, Le minisodar Doppler du C.R.P.E.: résultats préliminaires de sondage acoustique doppler de al couche de surface atmosphérique, *Météorologie (La)*, VII (4), 35–43, 1984.
13. Harris, C.M., Absorption of sound versus humidity and temperature, *Nasa Report*, CR-647, 1967.
14. Weill, A., C. Klapisz, and F. Baudin, The CRPE minisodar: Applications in micrometeorology and in physics of precipitations, *Atmos. Res.*, 20 (2–4), 317–333, 1986.
15. Little, C.G., On the detectability of fog, cloud, rain and snow by Acoustic echo sounding methods, *J. Atmos.*, 29, 748–755, 1972.
16. Chong, M. Mesure des profils de vent par sodar-doppler, *Note Technique CRPE/22* 1976.
17. Clarke, R.H., Observational studies in the atmospheric boundary layer, *Quart. J. Roy. Meteor.*, 105, 231–239, 1970.
18. Gaynord J.G., Acoustic Doppler measurements of atmospheric boundary layer velocity structure functions and energy dissipation rates, *J. Appl. Meteor*, 16, 127–137, 1977.
19. Weill, A., F. Baudin, J.P. Goutorbe, P. Van Grunderbeeck, and P. Leberre, Turbulence

structure in temperature inversions and in convection fields as observed by Doppler SODAR, *Bound. Layer Meteor.*, **15** (3), 375–390, 1978.
20. Coulter R.L., Weseley, estimate of surface heat flux from sodar and layer scintillations measurements, *J. Appl., Meteorol.* 19, 1209–1222, 1980.
21. Weill, A., C. Klapisz, B. Strauss, F. Baudin, C. Jaupart, P. Van Grunderbeeck, and J.P. Goutorbe, Measuring heat flux and structure functions of temperature fluctuation with an acoustic Doppler sodar, *J. Appl. Meteor.*, **19**(2), 199–205, 1980.
22. Neff, W., Qualitative evaluation of acoustic echoes from the planetary boundary layer, *NOAA, Tech. Ret. ERL 322, WPL, Boulder, Colorado*, 1975.
23. Strauss, Estimation du bilan d'énergie cinétique turbulente par sondage acoustique, *Document de Travail CRPR* No. 1063, 1980.
24. Klapisz, C., and A. Weill, Mean horizontal wind in inversion-capped convective boundary layer, *J. Appl. Meteor.*, **21** (5), 648–655, 1982.
25. Desbraux, G., A. Weill, Mean turbulent properties of the stable boundary Layer observed during the "COAST" experiment, *Atmospheric Res.*, 20, 151–164, 1986.
26. Taconet, O., and A. Weill, Convective plumes in the atmospheric boundary layer as observed with an acoustic Doppler sodar, *Bound. Layer Meteo.*, **25** (2), 143–158, 1983.
27. Weill, A., L. Eymard, M.E. Le Quere, C. Klapisz, F. Baudin, and P. Van Grunderbeeck, Investigation of the planetary boundary layer with an acoustic Doppler sounder, *in Proceedings of the IVth Symposium on 'Meteorological Observations and Instrumentation'*, pp. 415–421, A.M.S., 1978, (Denver, Co., U.S.A., juin 1978).
28. Van Grunderbeeck, P., P. Leberre, A. Weill, G. Dubosclard, and I. Itier, Mouvements convectifs de al couche limite atmphérique observés au moyen de sondeurs acoustiques Doppler et d'instruments portés par un ballon captif, *C.R. Acad. Sc. Paris*, **288B**, 209–212, 1979.
29. Eymard, L., Weill A., Investigation of Clear Air Convective Stuctures in the PLB Using Dual Doppler Radar and Doppler Sodar, *Journal of Applied Meteorology*, Vol. 21, No. 12, 1891–1906, 1982.
30. Weill, A., L. Eymard, O. Taconet, and C. Klapisz, La couche limite planétaire observée ponctuellement et son interprétation en terme de composante à moyenne échelle, *La Météorologie*, VI(9), 173–181, 1982.
31. Klapisz, C., A. Weill, Mean horizontal wind in inversion-capped convective boundary layer, *J. Appl. Meteor.*, 21(5), 648–655, 1982.
32. Weill, A., M.E. Le Quere, P. Van Grunderbreeck, J.P. Goutorbe, Measurements of vertical velocity by means of an acoustic sounder, *Proceedings of the IVth Symposium on "Meteorological Observations and Instrumentation"*, pp 415–421, A.M.S., (Denver CO., USA), 1978.
33. Taconet, O., and A. Weill, Vertical velocity field in the convective boundary layer as observed with an acoustic Doppler sodar, *Bound. Layer Meteo.*, 23, 133–151, 1982.
34. Eymard, L., and A. Weill, A study of gravity waves in the planetary boundary layer by acoustic sounding, *Bound. Layer Meteor.*, **17**, 231–245, 1979.
35. Weill, A., M. Blez, and F. Leca, Gravity waves and horizontal mixing in the atmospheric boundary layers, *Ann. Géophysicae*, **5** (B5), 413–420, 1987.
36. Weill, A., M. Aubry, F. Baudin, and J. Heissat, A study of temperature fluctuation in the atmospheric boundary layer, *Bound. Layer Meteo.*, **10** (3), 337–346, 1976.
37. Weill, A., C. Klapisz, B. Strauss, F. Baudin, C. Jaupart, P. Van Grunderbeeck and

J.P. Goutorbe, Measuring heat flux and structure functions of temperature fluctuation with an acoustic Doppler sodar, *J. Appl. Meteor.* 19(2), 199–205, 1980.

38. Klapisz, C., B. Strauss, and A. Weill, The budget of turbulent kinetic energy in convection mixed layers, in *Proceedings of the International Symposium on 'Acoustic Remote Sensing on the atmosphere and Oceans'*, MATHEW T. Ed., pp. IV. 84–IV. 97, 1981, (Calgary, Canada, 6 juin, 1981).

39. Louis, J.F., A. Weill, and D. Vidal-Madjar, Dissipation length in stable layers, Bound. Layer Meteo., 25(3), 229–243, 1983.

40. Klapisz, C., and A. Weill, Modéle semi-empirique d'évolution matinaledu profil de vent entre le sol et le sommet de l'inversion, *J. Rech. Atmos.*, 12 (2–3), 113–117, 1978.

41. Mon, J.P., A. Weill, and L. Martin, Effect of tropospheric disturbances on a 4.1 and 6.2 GHz line of sight path, *Ann. Télécomm.*, 35 (11–12), 468–471, 1980.

42. Klapisz C., A. Weill, Modélisation semi-empirique de la couche limite nocturne, Application au calcul du profil d'indice de réfraction, *Ann. Telecommun.* 40, No. 11–12, 1985.

43. Weill, A., F. Baudin, C. Mazaudier, G. Desbraux, C. Klapisz, J.P. Goutorbe, A.G.M. Driedonks, A. Druilhet, and M. Durand, A mesoscale 'shear convective' organization observed during COAST experiment: Acoustic sounder measurements, *Bound. Layer Meteo.*, 44(4), 359–371, 1988.

44. Desbraux, G., and A. Weill, Spectral density of velocity parameterization using the Doppler sounder of the C.R.P.E. during 'COAST' experiment, in *Proceedings of the Third International Symposium on 'Acoustic Remote Sensing of the Atmosphere and Oceans'*, pp. 195–221, 1986.

45. Weill, A., C. Mazaudier, F. Baudin, C. Klapisz, F. Leca, M. Masmoudi, D. Vidal-Madjar, R. Bernard, O. Taconet, B.S. Gera, A. Sauvaget. A. Druilhet, P. Durand, G. Dubosclard, J.Y. Caneil, P. Mery, A.G.M. Beljaars, W.A.A. Monna, J.G. Van Der Vliet, M. Crochet, D. Thomson, and T. Carlson, 'Mesogers 84, experiment: A report, *Bound. Layer Meteo.*, 42(3), 251–264, 1988.

46. Mazaudier, C., D. Landais, R. Raphalen, G. Desbraux, C. Davoust, La base de données "Mesogers 84", Utilisation du systéme de gestion de base de données relationnelle Multics Relational data Store pour l'agencement des observations (version juin 1986), *Document de Travail DT/CRPE/1143*, Novembre 1986.

47. André, J.C., J.P. Goutorbe, A. Perrier, F. Becker, P. Bessemoulin, P. Bougeault, Y. Brunet, W. Brutsaert, T. Carlson, R. Cuenca, J. Gash, J. Gelpe, P.H. Hildebrand, J.P. Lagouarde, C. Lloyd, L. Mahrt, P. Mascart, C. Mazaudier, J. Noilhan, C. Ottlé, M. Payen, T. Phulpin, R. Stull, J. Shuttleworth, T. Schmugge, O. Taconet, C. Tarrieu, R.M. Thepenier, C. Valencogne, D. Vidal-Madjar, and A. Weill, Evaporation over land-surfaces: First results from HAPEX-MOBILHY special observing period, *Ann. Geophysicae,* 6(5), 477–492, 1988.

48. Mazaudier, C., B. Tagett, J. Bouvet, M. Zhong, K. Han-Jeong, M. Lafeuille, C. Tiffon, Notice d'utilisation dela base de données Hapex-Mobilhy (réseaux sol durant la phase intensive d'observation, *DT/CRPE/1175*, mai 1989.

49. Clough, S.A., The mesoscale frontal dynamics project. Meteor. Mag., 116, 32–42, 1987.

50. Bouvard M., C. Mazaudier, R. Raphalen, La Base de données Fronts 87 sur le calculateur HP9000/70 du CNET (gestion sous Oracle), DT/CRPE/1208, 1991.

51. P. Bougeault, A. Jansa Clar, J.L. Attie, I. Beau, B. Benech, R. Benoit, P. Bessemoulin,

J.L. Caccia, B. Carissimo, J.L. Champeaux, M. Crochet, A. Druilhet, P. Durand, A. Elkhalfi, A. Genoves, M. Georgelin, K.P. Hoinka, V. Klaus, E. Koffi, V. Kotroni, C. Amory-Mazaudier, J. Pelon, M. Petitdidier, Y. Pointin, D. Puech, E. Richard, T. Satomura, J. Stein and D. Tannhauser, The Atmospheric Momentum Budget over a major mountain range: first results of the PYREX field program, *Annales Geophysicae,* 11, 395–418, 1993.

52. Mazaudier, C., J. Bouvet, S. Moreau, M. Bouvard, Catalogue des données sodar recueillies durant la campagne de mesure PYREX90 (1er octobre-30 novembre 1990), DT/CRPE/1225, MARS, 1993.

53. Dubosclard, G., C. Mazaudier, and A. Weill, Sodar network observations of the mixedlayer growth over complex terrain, *in Proceedings of the Third International Symposium on 'Acoustic Remote Sensing of the Atmosphere and Oceans',* pp. 171–184, 1986.

54. Weill, A., C. Mazaudier, F. Leca, and M. Masmoudi, Doppler sodar and fluxes measurement: Climates of fluxes at horizontal scales comparable to satellite pixels, *in Proceedings of the 4th Symposium of the International Society for 'Acoustic Remote Sensing', BOURNE I. Ed.,* pp. 155–160, Canberra, Australie, 16–24 février, 1988.

55. Weill, A., C. Mazaudier, M. Masmoudi, F. Leca, and G. Dubosclard, A complex terrain boundary layer experiment with a Doppler sodar network, the Mesogers 84 experiment, *in Proceedings of the 4th Symposium of the International Society for 'Acoustic Remote Sensing', BOURNE I. Ed.,* pp. 147–153, Canberra, Australie, 16–24 février, 1988.

56. Masmoudi, M., and A. Weill, Doppler sodar measurement of atmospheric mesoscale spectra: The Taylor hypothesis analysis, *J. Climate Appl. Meteor.* 27, (7), 864–873, 1988.

57. Gera, B.S., and A. Weill, Doppler sodar analysis of the frontal friction in relation to the frontal slope, *J. Climate Appl. Meteor.,* **26** (8), 885–891, 1987.

58. Gera, B.S., and A. Weill, Sodar observations of fronts in India and France, *in Proceedings of the 4th Symposium of the International Society for 'Acoustic Remote Sensing', BOURNE I. Ed.,* pp. 63–74, Canberra, Australie, 16–24 février, 1988.

59. Gera, B.S., and A. Weill, Doppler sodar observations of the boundary-layer parameters and a frontal system during the 'MESOGERS 84' experiment, *Bound. Layer Meteo.,* 54(1), 41–57, 1991.

60. Ralph M., C. Mazaudier, M. Crochet, S.V. Venkateswaran, Combined clear-air radar and sodar observations of mesoscale and microscale structures in a cold front, *Internal Geoscience and Remote sensing Symposium, 3–6 June, 1991, Espoo. Finland, Published by the Institute of Electrical and Electronics Engineers (IEEE),* New-York, NY, USA, 1991.

61. Ralph M., C. Mazaudier, M. Crochet, S.V. Venkateswaran, Doppler sodar and radar wind profiler observations of gravity wave associated with a gravity current, *Monthly Weather Review,* Vol 121, No. 2, 444–463, 1993.

62. Weill, A., Indirect measurements of fluxes using Doppler sodar, in 'Land Surface Evaporation', *SCHMUGGE T.J. & J.C. ANDRE Eds.,* pp. 301–311, Springer-Verlag, 1991, Banyuls, France, 10–21 octobre, 1988.

63. Mazaudier, C., and A. Weill, A method of determination of dynamic influence of the forest on the boundary layer using 2 Doppler sodars, *J. Appl. Meteo.,* 28(8), 705–710, 1989.

64. Mazaudier, C., J. Bouvet, and A. Weill, Sodar wind speed profiles over forested boundary layer during the Hapex-Mobilhy campaign, *Ann. Geophysicae,* 9(8), 501–509, 1991.
65. Mazaudier, C., B. Tagett, and A. Weill, Recent studies on the forested boundary layer, using two Doppler sodars, *in Proceedings of the Fifth International Symposium on 'Acoustic Remote Sensing of the Atmosphere and Oceans', SINGAL S.P. Ed.,* pp. 357–362, Tata McGraw-Hill Publ. Co. Ltd, New Delhi, Inde, 06–09 février, 1990.
66. Parlange, M.B., W. Brutsaert, regional roughness of the Landes forest and surface shear stress under neutral conditions, *Bound. Layer Meteorol.,* Vol. 48, No. 1–2, pp. 69–81, 1989.
67. Parlange, M.B., W. Brutsaert, Are radiosonde time scales appropriate to characterize boundary layer wind profiles? *J. of Applied Meteor.,* Vol. 29, No. 3–4, pp. 249–255, 1990.
68. F. Beyrich, "Some questions of mixing height, Determination from sodar observations", *in the proceeding of the sixth symposium on "acoustic remote sensing of the atmosphere and oceans",* pp 205–211, Greece, May, 1992.
69. Beyrich, F., and A. Weill, Some aspects of determining the stable boundary layer depth from sodar data, *Bound. Layer Meteo.,* Vol. 63, No. 1–2, pp 97–116, 1993.
70. F. Beyrich, V. Kotroni, Eststmation of Surface Stress over a Foret from sodar measurements and its use to parametrize the stable boundary layer Height, *Boundary Layer Meteorology, 66, No.* 1–2, pp 93–103, 1993.
71. Soares, J.V., R. Bernard, O. Taconet, D. Vidal-Madjar, and A. Weill, Estimation of a bare soil evaporation from airbone measurements, *J. Hydrol.,* 99(3), 281–296, 1988.
72. Kotroni, V., C. Mazaudier, Analysis of friction velocities during frontal events for various orographic conditions using a doppler sodar system, *in the proceeding of the sixth symposium on "acoustic remote sensing of the atmosphere and oceans",* pp. 271–280, Greece, May, 1992.
73. Kotroni, V., C. Mazaudier, Y. Lemaître and M. Petitdidier, "FRONTS87: European experiment for the study of active cold fronts. Part II: Combined observations from Sodar and ST radar of the circulation in the front of a cold front", *Proceedings of the First Hellenic Conference in Meteorology,* P. 426–468, 21–23 May 1992, Thessaloniki, Greece (in greek).
74. Kotroni, V., Y. Lemaitre, M. Petitdidier, Dynamics of a low-level Jet observed during the FRONTS 87 Experiment, Quaterly Journal of Royal Meteorological Society, 120, 277–304, 1994.
75. Mak, M-K., Steady, neutral planetary boundary layer forced by horizontally non-uniform flow, *J. Atmos. Sci.,* 29, 707–717, 1972.
76. Lefloch, C.C. Amory-Mazaudier, Analyse des caractéristiques dynamiques des fronts de la campagne Fronts 87, *Lettre No. 6 des radars ST, éditeur P. Bessemoulin, CNRM/GMEI, 42, Avenue G. Coriolis, 31057 Toulouse, septembre,* 1993.
77. Lefloch, C., Contribution à l'étude des fronts froids traversant la Bretagne, *DT/CRPE/ 1237,* 1993.
78. Kotroni V., C. Amory-Mazaudier, Influence of orography and canopy conditions on friction velocity observed during frontal events using doppler sodar observations, *Journal of Applied Meteorology,* Vol 32, No. 3, 506–521, 1993.
79. Kotroni V., C. Amory-Mazaudier, K. Lagouvardos, Doppler Sodar observations of

a cold front during the PYREX experiment: orographic effects, *Int. J. Remote Sensing*, Vol. 15, No. 2, 489–497, 1994.

80. Georgelin, M., E. Richard, Simulation numérique du contournement d'un relier: cas de la POI 6 de PYREX, *Atelier de Modélisation de l'Atmosphére*, pp 123–132, Toulouse, 1et 2 décembre, 1993.

81. Weill, A., Mesoscale studies, *in Proceedings of the Figth International Symposium on 'Acoustic Remote Sensing of the Atmosphere and Oceans', SINGAL S.P. Ed.*, pp. 259–271, Tata McGraw-Hill Publ. Co. Ltd. 1990, New Delhi, Inde, 06–09 février, 1990.

82. Weill, A., Perturbations de la couche limite atmosphérique au passage des fronts, *Météorologie (La)*, VII (12), 29–34, 1986.

83. Weill, A., and F. Goulam-Alli, Momentum fluxes measurements with SODAR and mini Sodar, *in Proceedings of the Fifth International Symposium on 'Acoustic Remote Sensing of the Atmosphere and Oceans', SINGAL S.P. Ed.*, pp. 161–166, Tata McGraw-Hill Publ. Co. Ltd, New Delhi, Inde, 06–09 février, 1990.

Acoustic Remote Sensing Applications
S.P. Singal (Ed)
Copyright © 1997 Narosa Publishing House, New Delhi, India

13. Sodar Investigations of Gravity Waves by Cross Spectral Analysis

Günther Bull*

Grimaustr 63, D-12439 Berlin, Germany formerly: Heinrich-Hertz-Institute for Atmospheric Research and Geomagnetism, Rudower Chaussee 5, D-12489 Berlin, Germany

1. Atmospheric Gravity Waves

In recent years the acoustic sounding technique has been widely applied for obtaining data of the atmospheric boundary layer (ABL). Using monostatic vertical sounding the backscattering cross section of a sound pulse is proportional to the temperature structure parameter C_T^2. Typical mesoscale phenomena of the ABL like inversion layers, low-level jets, convection cells or internal wave processes can be observed by sodar at heights between about 25 m and 1000 m. Gravity waves are readily detected in the ABL, where they perturb virtually all meteorological variables and thus appear in the records of instruments of every type, whether in situ or remote.

Various processes in the troposphere tend to generate gravity waves with periods between a few minutes and a few hours. Gravity waves play an essential role in the dynamic processes of the troposphere and upper atmosphere, see for instance Gossard and Hooke [1] and Einaudi et al, [2]. Gravity wave are important due to their ability to transport and redistribute energy, momentum and matter and to trigger various atmospheric phenomena. They are also important for many micro- and mesoscale processes in the ABL such as turbulence, diffusion, local flows, inversions, and act as trigger mechanism, for example, for convection and condensation and thunderstorm processes. Gravity waves provide a mechanism for the coupling of the ABL and the free atmosphere.

Contrary to direct wave measurements, e.g. with a microbarograph array, a remote sensing method like sodar requires for the measurement the presence of a sufficiently intense tracer. Sometimes the waves themselves produce the tracer C_T^2. For a review on sodar investigations see e.g. Singal, [3], Neff and Coulter [4] or Brown and Hall [5].

---

*Investigations were carried out at the former Heinrich-Hertz-Institute for Atmospheric Research and Geomagnetism, Berlin-Adlershof

2. Possible Measuring Configurations and Methods of Analysis of Wavelike Variations

In order to get the whole set of wave parameters we need a network of three sodar antennas. The present study aims to show that nevertheless it is also possible to get results on wavelike processes even with only one vertical-sodar using spectral and coherence analysis of digital sodargrams.

2.1 Sodar Investigations by Facsimile Records

Often wave analysis with vertical sodars were based on visual interpretation of the analog facsimile records. The sodar facsimile records reveal the wave motions rather clearly as undulations in the strata of strong returns. Often these strata are themselves produced by gravity wave encounters with the stable ABL. Facsimile recordings allow the evaluation of the frequency of occurrence of wavelike processes, the height of undulation layers, the periods of the waves and the vertical displacements of the undulating structures [6–10].

2.2. Investigations by Vertical Cross Section

Detailed analysis of wavelike processes in the ABL requires digitally recorded sodar backscatter measurements. For our investigations we have analysed height channels at height range gates of 25 or 50 m up to mostly 700 m height. For each height channel digital time series of the backscatter intensity were analysed. From these time series we can get the following information:

- the vertical profile of the backscatter intensity, which is proportional to C_T^2
- the corresponding vertical profile of the dispersion
- for each height channel the auto correlation function
- the cross correlation functions between neighbouring height levels and the cross correlation functions between different heights in relation to a fixed height
- auto spectra of the backscatter intensity at each height
- cross spectral and coherence spectra for neighbouring height levels.

Sodars with Doppler capabilities allow additionally the computation of the corresponding profiles and parameters for the vertical wind.

2.3. Computer Aided Trace Following Method

Instead of the usual visual analysis of analog facsimile records we used a computer aided trace following the height variations of absolute or relative maxima of the backscatter intensity. At the beginning the first height was given by the operator, then for each following single sodar profile a comparison with a given synthetic pattern was carried out by cross correlation within a predefined height range. So we get a digital time series of the height of an undulating layer:

- either the height variation of the upper boundary of a layer, mostly the upper level of a ground based inversion layer,
- or the height variation of an elevated wavy layer, mostly the boundary of an elevated inversion layer.

From these time series we can compute:

- the mean height of the wavy structure
- the mean vertical amplitude of the wavelike height variations
- the auto correlation function of the time series and the corresponding 1/e-value (correlation radius)
- the power spectrum of the vertical displacements.

If there are two or several wavelike layers present, we can further compute

- the cross correlation functions between the different layers and
- the cross spectra and coherence spectra of the different layers.

2.4 Simultaneous measurements with a Vertical Sodar and Other in situ or Remote Sensing Methods

Analysis of vertical cross section and trace following investigations allow the determination of several parameters of the behaviour of wavelike structures, but they do not allow the determination of the wave parameters, phase velocity, propagation direction and wavelength. But with a combination of analysis of vertical sodar backscatter investigations and some further suitable in situ or remote sensing methods we can add the missing data about the wave processes. Simultaneous measurements with a vertical sodar and e.g. a collocated microbarograph array can yield further these wave parameters, see section 6.

2.5 Measurements with a Sodar Network

In order to obtain the whole set of wave parameters we need a network of three sodar antennas. The passage of a wave across the three antennas can be evaluated by the recording of the time delays between the variations of the backscatter intensities at the corresponding three measuring points at different height levels. In addition to the forementioned data (article 2.2) for a series of height levels we obtain information on:

- propagation direction of the waves and phase velocity as mean values for a mean wave packet from cross correlation functions
- wave parameter propagation direction and phase velocity as functions of frequency from phase spectra
- horizontal wavelengths
- vertical transport of horizontal momentum and energy from wind measurements with a Doppler sodar network.

For wave parameters, derived from phase spectra, we can compute the significant levels for the corresponding coherence spectra.

The sodar network can be arranged in two ways. Firstly, using a three antenna Doppler system with one vertically pointing and two inclined antennas as for wind measurements, it is possible to compare the drift of the C_T^2 structure across the three antenna beams. A high pulse repetition rate is necessary. Wave investigations with this method were carried out successfully by Weill and colleagues [11, 12, 13]. This method has the advantage, that all three antennas are located at one point, but has the disadvantage of network triangles with different dimensions for different heights. Secondly, by a network with suitable baselengths of three vertically pointing antennas, the spatial coherence of the scattering inhomogenities in a series of height levels can be studied. The baselengths of a sodar network should be about some hundred meters to a few kilometers in dependence on the investigated frequency range. A sodar configuration with equal network dimensions for all height levels was used e.g. by Kjelaas et al [14], Fua and Mastrantonio [15], and Bull and Neisser [16]. The wave investigations with such a network can be done by the measurement of the backscatter intensities and (or) by the measurement of the drift of the spatial structure of the vertical wind field across the network.

3. Gravity Waves Investigations with Sodar Networks

The wave parameters can be evaluated with three sodar antennas, either by pointing the beams in different directions, or by using three vertically pointing sodar antennas arranged in a triangle. Mostly gravity wave investigations with sodar networks cover a measuring period of only a few hours or few days. The analyses were carried out using cross correlation or (and) cross spectral methods. Several studies show, that a large amount of the waves are excited in the ABL, mostly they have periods of some minutes (equal or larger than the Brunt Väisälä period) and propagate nearly in wind direction. On the other hand often large gravity waves with larger periods are detected with other propagation directions and with sources in higher levels of the atmosphere.

Gravity wave investigations using a three antennas Doppler system with one antenna pointing vertically and two inclined antennas as for wind measurements were carried out by Eymard and Weill [11], Blez and Weill [12] and Weill et al, [13]. The passage of a wave across the three-antennas field can be detected by the time delay between the features of the backscatter intensities recorded at different height levels of the triangles with increasing dimensions as a function of height. The results of Eymard and Weill indicate three types of oscillations:

— a "free" wave propagating in a different direction than the wind,
— several cases of waves near a critical level (where the phase velocity of the gravity waves equals the mean wind velocity) at the top of an echo layer and
— two cases of oscillating rising inversions interacting with convection.

For the herringbone waves phase velocity and direction are the same as that of the mean wind at the top of the layer with a very weak momentum transport. In contrast, a wave with a long period propagates in a different direction and with a large vertical momentum flux. Weill et al, described transitions from waves to turbulence and the reverse. They investigated the characteristic vertical backscatter profiles before and during the passage of waves and observed an increase of the turbulent dissipation rate when waves appeared.

Often a triangular array of sodar antennas was set up. For a sodar network of Fua and Mastrantonio [15] the legs of the triangle were 100 m in length. Kjelaas et al, [14] used a sodar network with baselengths of 300 m, Bull and Neisser [16] 350–450 m and 740–1190 m [7] and Nappo and Echman [17] 2.5–3 km. Fua and Mastrantonio pointed out that the accuracy of their analysis of a measuring period of 40 min was bad, because the lengths of their network were too small. Their analysis of the backscatter data revealed bad results, while the analysis of the vertical wind data showed better results. Kjelaas et al, investigated three events of waves, using time-lagged cross correlation analysis for the intensity patterns of the echo returns above the nocturnal inversion. The gravity waves analysed had horizontal wavelengths ranging from about 1 km to almost 6 km with horizontal phase velocities from 5 to 10 ms^{-1}. A comparison with a collocated array of microbarographs gave a good agreement. Nappo and Eckman reported on an episode of apparent wave instability and associated turbulence and described the breakdown of an unstable Kelvin-Helmholtz wave, collapsing into turbulence.

Our institute, the former Heinrich-Hertz-Institute for Atmospheric Research and Geomagnetism, performed two experiments with a sodar network. In the first experiment a network of three sodars with digital data acquisition was installed for a period of five days. The sodars were placed to form a triangular array having baselines between 740 m and 1190 m [16, 7]. During the international experiment JABEX in Slowakia in June/July 1989, a network using the Doppler-sodar of the Heinrich-Hertz-Institute-was erected, with three antennas arranged in a triangle with distances between 350 m and 450 m and transmitting vertically [16, 18]. The Doppler-sodar [19] was operated for about 10 days in this antenna configuration in order to investigate wave phenomena. For the primary processing of the three time series of the backscatter intensity and the three time series of the vertical wind velocity we used the synchronous time series of the measured noise level in order to correct the data, and cancel data with strong noise and extreme values. For the corrected time series correlation and spectral analyses were performed for 8 time intervals each of 2-hour duration for one night and 8 different heights. The following parameters were computed: variances of backscatter intensities and vertical wind velocities, auto- and cross-correlation functions, the power spectral densities, the phase spectra and the coherence spectra for different frequency channels from 0.55 mHz to 8.3 mHz (30 min to 2 min).

We found at most heights several frequency channels with significant coherence. Mostly the longer periods of 30–10 min or 30–6 min show significant wave parameter values, but in some cases a second period range with smaller periods (e.g. 4–2 min) was present. In addition meteorological information from a Sensitron-sodar (30 min mean of wind velocity and azimuth), meteorological tower (temperature at 1.5, 40, 100, and 200 m heights), aerological soundings (Station Wien) and weather charts was used.

Fig. 1 shows the wind velocity at 4 heights from Sensitron-sodar and the mean wave velocities for several significant frequency channels. Most wave velocities show values smaller than 8 m/s and are in rough agreement with the corresponding wind velocities. Therefore, it is probable that most of the waves have their sources in the ABL. Only two (or three) values at 406 m height, show larger values (upto 18 m/s). These waves are not generated in the ABL, but have their origin elsewhere

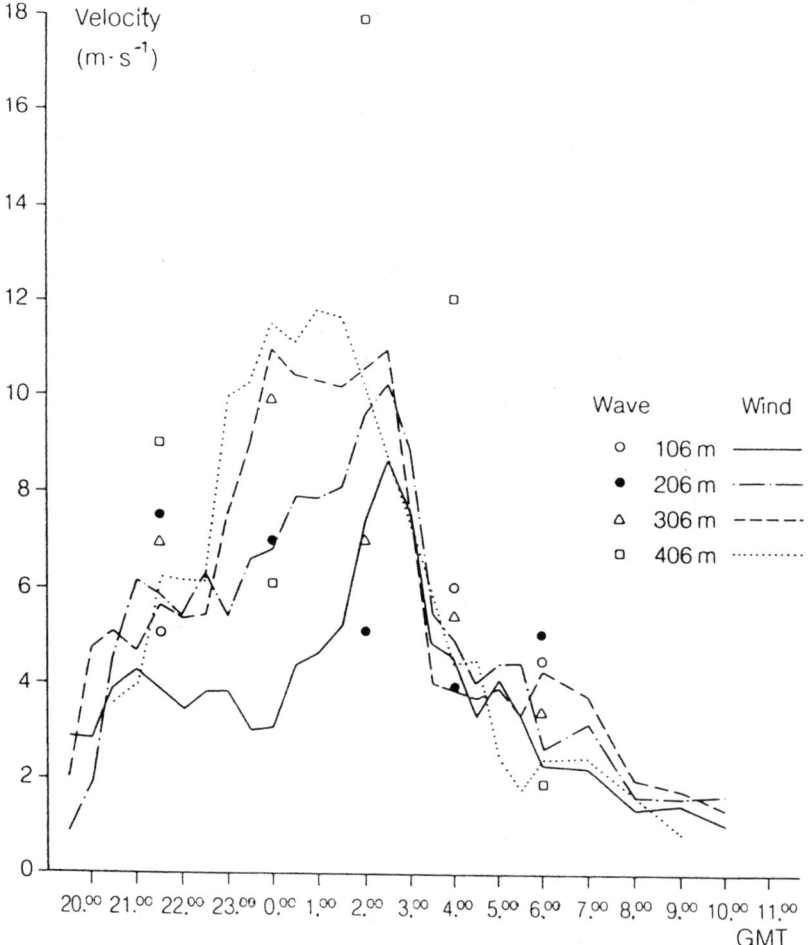

Fig. 1. Temporal variations of the wind velocity at different heights and the variations of the gravity wave velocities derived from sodar network 3./4.6.1989 [16]

at a higher level. The aerological wind profile of the Station Wien shows at 700 hPa such values, and we can assume that these waves originate from higher levels of the atmosphere. Investigations by Neisser [20] have shown that mostly waves are generated in higher atmospheric regions by jet streams, cold fronts and thunderstorms. Gravity waves possess mostly wave velocities of about 15–45 m/s in accordance with the properties of their sources, see Bull [21].

Fig. 2 shows the wind direction at the 4 heights. Most values of wave direction are again in rough agreement with the wind directions; but again some values around 200° show differences and agree with the Wien radiosonde data of 190–210°. A stationary coldfront was situated over a time of about three days in the region of the array. Probably this front crossed the network after midnight, according

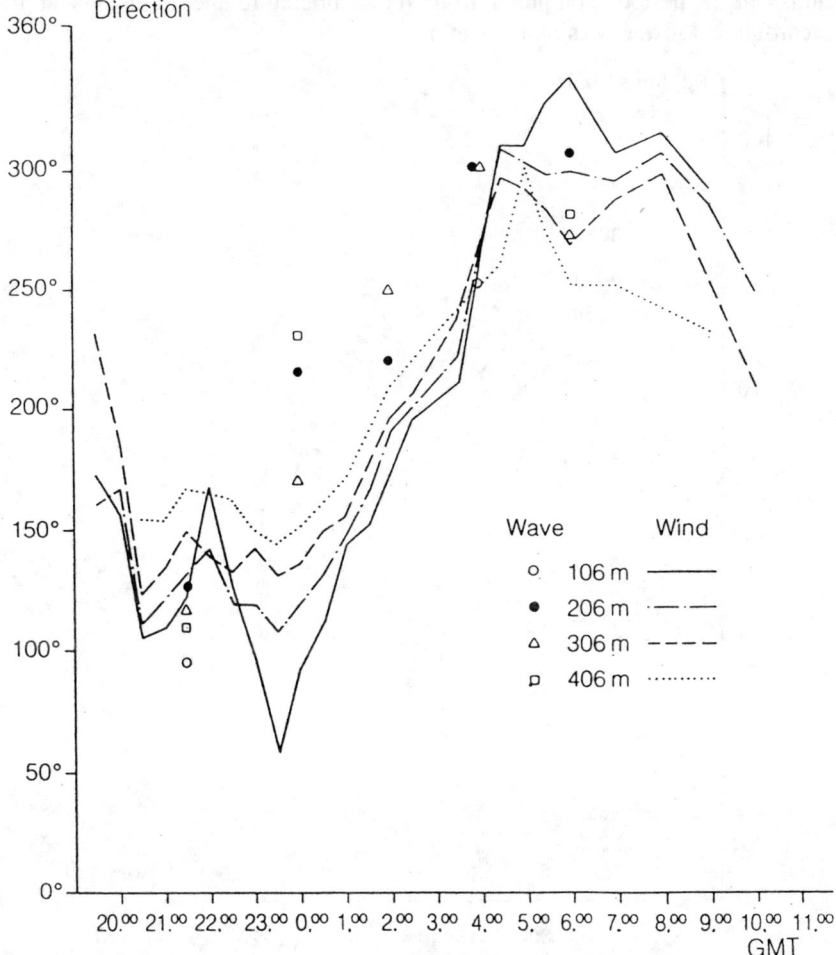

Fig. 2. Temporal variations of the wind direction at different heights and the variations of the propagation directions of the gravity waves derived from sodar network

282 Applications in the Atmosphere

to the changes of wind velocity and direction. Comparing wind and wave data, we can conclude that the waves precede about a few hours the wind changes, a fact which is not unusual for fronts.

Computations of the backscatter intensity, backscatter variance and the power spectra of backscatter intensity and vertical wind velocity for the different time ranges and heights show a large enhancement of backscatter intensity and a moderate increase of the variance and power spectral density of vertical wind velocity in the lower 300 m during midnight. In Fig. 3 the power spectral density of vertical wind velocity shows a similar behaviour for the 4 different heights with a $P(f) \sim f^{-n}$ dependence with $n = 0.97$ to 1.19 for the period range 30 min to 5 min. In all 4 heights there is a second maximum near a period of three min. The corresponding Brunt-Väisälä period, computed from the temperature measurements at the meteorological tower, was about 4 min.

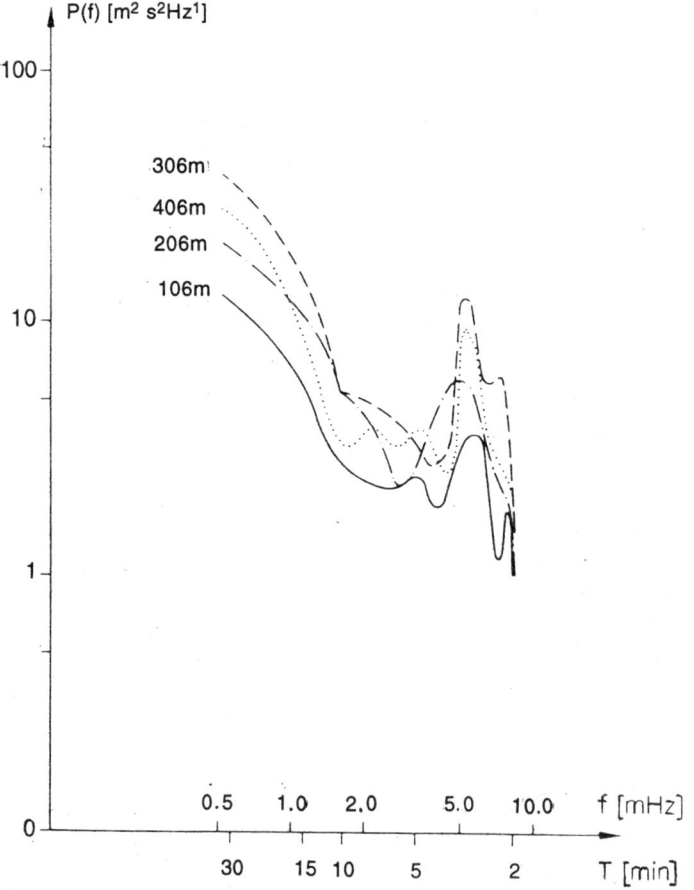

Fig. 3. Power spectral density of vertical wind velocity for 4 measuring heights. 3./4.6.1989, 23.15–01.15 GMT

The computed wavelengths for the different time ranges and heights varied mostly between 1 km and 3 km for the periods between 2.5 and 6 min. The values for the longer periods of 7.5 min to 30 min were in most cases in the range between 2 km and 20 km.

4. Computer Aided Trace Following Method and Computation of Displacement Spectra

For the investigation of undulating layers we computed time series of equidistant data of the height variations of the backscatter intensities from the time-height-matrix of backscatter intensities using a computer aided trace following the height variations of absolute or relative maxima of the backscattering intensity. At the beginning, the first height was chosen by the operator, then for each following single sodar profile a comparison with a given synthetic pattern was carried out by cross correlation within a defined height range. From these time series the power spectra of vertical displacements together with the mean heights of the oscillating layers were computed.

Table 1 shows some results from the computation of the displacement spectra of 8 cases of ground based inversions and 8 cases of elevated temperature inversions. We see that the mean height of the ground based structures is about 140 m and for the elevated inversions 300 m, the mean vertical amplitudes are for both types about 60 m with ranges from about 30 m to about 100 m. The 1/e-values of the autocorrelation functions (correlation radius) show larger values for the elevated layers, while for the mean exponent n of the $P(f) \sim f^{-n}$ relation we evaluated the same value of about 1.1.

Some examples of spectra for ground based inversions (solid lines) and elevated inversions (dashed lines) shall illustrate the variability of the displacement spectra, shown in a log-log presentation, see Fig. 4. The spectral density decreased, as

Table 1. **Properties of undulating layers**

			Ground based inversions	Elevated inversions
number of investigated events			8	8
duration of events		min	60–165	60–165
heights of undulating layers	range	m	61–243	132–568
	mean	m	140	300
vertical displacements (peak to peak)	range	m	28–81	35–102
	mean	m	55	59
1/e value of autocorrelation functions	range	min	0.1–3.7	0.9–9.5
	mean	min	2.0	3.7
displacement spectra exponent n of $P(f) \sim f^{-n}$	range		0.22–1.87	0.65–2.56
	mean		1.16	1.1

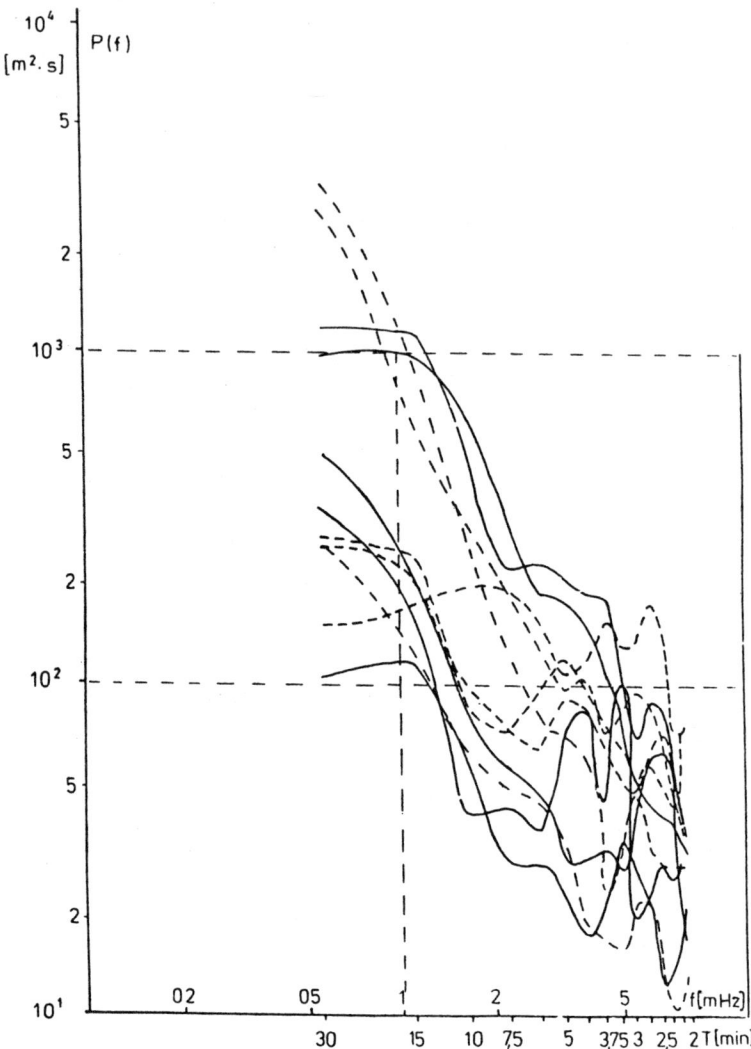

Fig. 4. Displacement spectra of undulating layers
— ground based strucures
... elevated structures

expected, with increasing frequency. But while some spectra show a rapid decrease with increasing frequency, others exhibit a flat curve with strong variations of the intensity in the different frequency ranges between 0.55 mHz and 8.33 mHz (30 to 2 min). If we evaluate a mean slope for each spectrum, we see that for the 16 events investigated there is no difference in the mean slope for $P(f) \sim f^{-n}$ of the displacement spectra between ground based and elevated structures. The investigations were carried out during the months May and June. From our gravity wave investigations with our microbarograph network we know, that in summer there are smaller gravity wave amplitudes than in winter [22].

5. Spectral Behaviour of Waves from Sodar Vertical Coherograms

For the investigation of wavelike structures in different heights in the ABL we developed a different method, which we called "sodar vertical coherogram". In the following we will demonstrate the method. From time series of sodar backscatter intensities for different heights between 50 m and 700 m, the coherence spectra for neighbouring height levels were computed using cross spectral analysis. As an example Fig. 5 shows the coherence values as function of frequency for 4 heights.

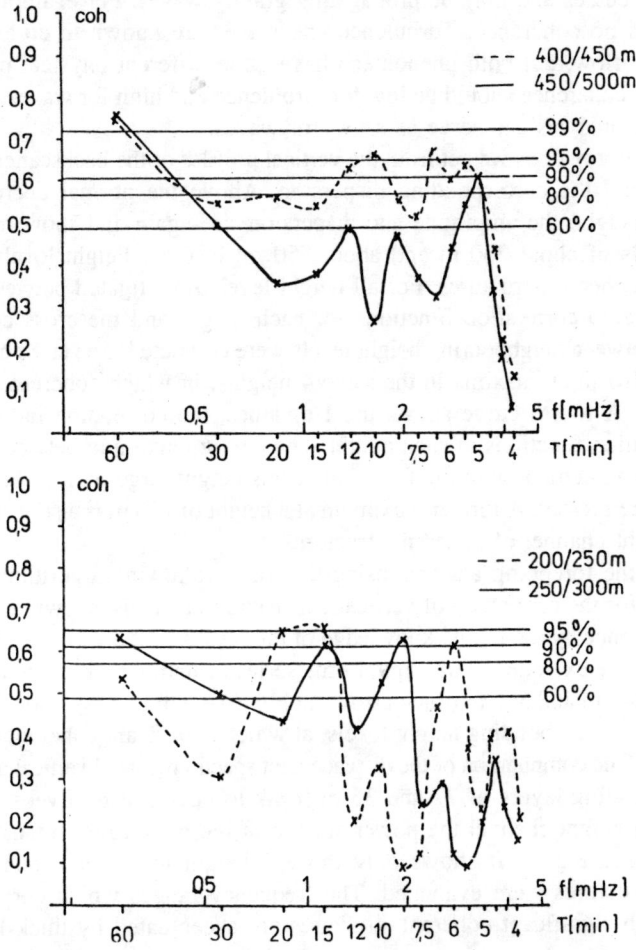

Fig. 5. Coherence spectra for neighbouring height channels

For some frequency channels the coherence values are larger than some given significance levels of 99, 95, 90 or 80%, respectively.

Fig. 6 (right part) shows for a case study the result of the coherence analysis, the ordinate corresponds to heights from 50 to 700 m, the abscissa is the frequency. The spectral analysis was carried out for frequencies between 4.17 mHz and 0.27 mHz (periods of 4 to 60 min). Only if the values of the coherence spectral values in the different height levels for certain frequencies were larger than a preset significance level of 99, 95, 90 or 80%, respectively, the corresponding isolines were drawn. We see that in 4 heights there are coherent structures with high significance for some frequency ranges, for example in about 100 and 600 m with low frequencies and in about 250 and 450 m with lower and higher frequencies.

Our hypothesis is now, that these coherent structures correspond to wavelike periodic processes and may be propagating gravity waves. For example random noise shows no coherence. Turbulence and waves are known to co-exist in an atmospheric flow, but both phenomena have quite different physical properties, for example coherence should be low for turbulence and high for waves, therefore a coherence analysis can serve as a test for waves.

Further informations we get from the vertical profiles of the backscatter intensity amplitude and the corresponding dispersion. Above the normal decrease with increasing height, the amplitude and dispersion rise again and show maxima in height levels of about 250 m and about 550 m; in these height levels we find significant coherent structures. For all height levels investigated between 50 and 700 m the auto correlation functions for each height and the cross correlation functions between neighbouring height levels were computed. The cross correlation functions also show maxima in the same 4 heights, in which coherent structures are present. A fourth curve shows the 1/e-values, the correlation radius, of the auto correlation functions in each height. The strong maximum between 550 and 600 m is in agreement with the fact, that at this height large periods of wavelike variations are present. A further maximum at a height of 250 m is also in agreement with a height channel of coherent structures.

Besides the foregoing analysis using the trace, following algorithm was also carried out for the case study of vertical coherogram-analysis, shown in Fig. 6, so that an influence by a possible knowledge of the results of the vertical coherogram analysis can be excluded. Two digital time series of undulating layers with mean heights of 242 m and 568 m (thick arrows in Fig. 6) exhibit a very good agreement with the two corresponding height levels, at which significant coherent structures were found. The computation of the displacement spectra revealed vertical amplitudes of the undulating layers of 77 and 35 m (peak to peak), respectively.

For each height channel the power spectra of the backscatter intensities were computed, see Fig. 7. It shows only those 4 height levels, in which periodic coherent structures were evaluated. The frequency ranges which correspond to periods with significant coherent structures are accentuated by thick lines. The spectral densities decrease, as expected, with increasing frequencies and with strong variations.

Sodar Investigations of Gravity Waves by Cross Spectral Analyses 287

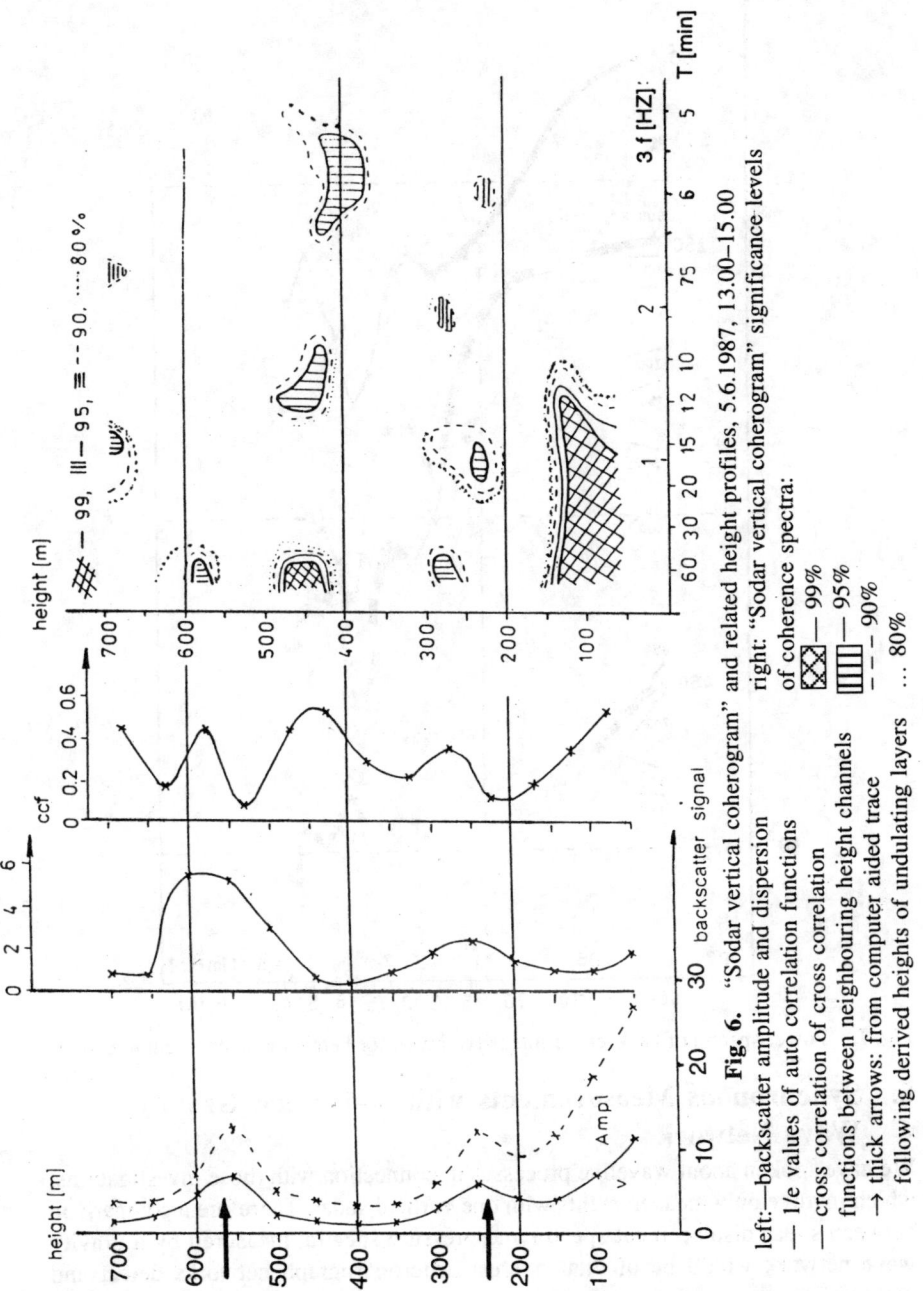

Fig. 6. "Sodar vertical coherogram" and related height profiles, 5.6.1987, 13.00–15.00
left: — backscatter amplitude and dispersion
— 1/e values of auto correlation functions
— cross correlation of cross correlation
functions between neighbouring height channels
→ thick arrows: from computer aided trace
following derived heights of undulating layers

right: "Sodar vertical coherogram" significance levels
of coherence spectra:
▨ — 99%
▥ — 95%
- - - 90%
... 80%

288 *Applications in the Atmosphere*

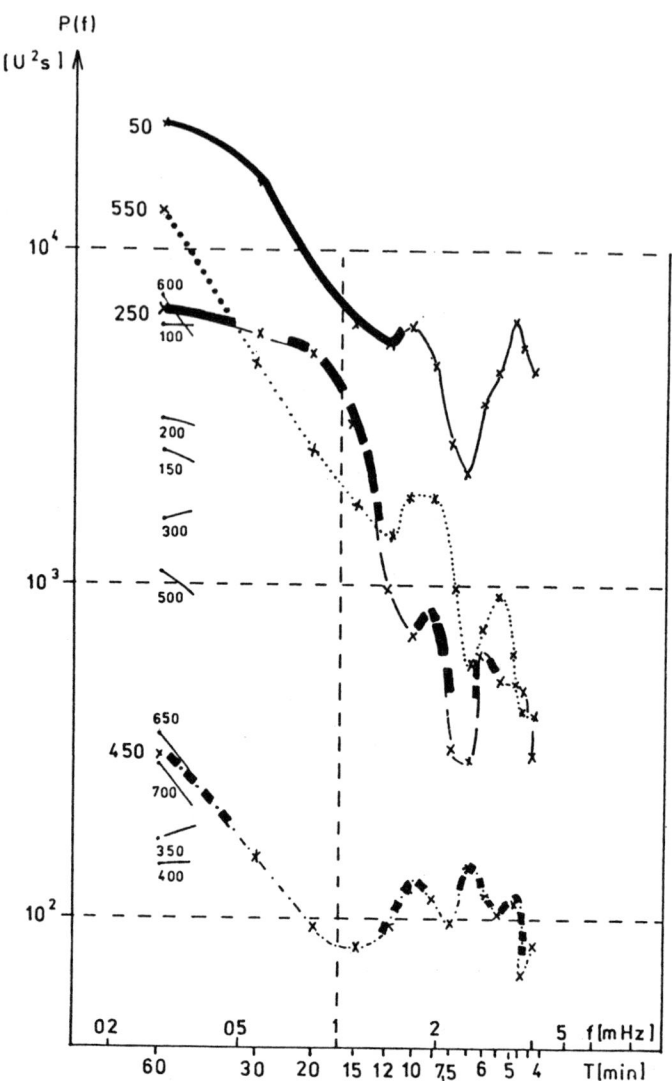

Fig. 7. Power spectra of backscatter intensity, drawn for heights with coherent structures

6. Synchronous Measurements with Sodar and Gravity Wave Network

We have spoken about wavelike processes in connection with those investigations, which involve only measurements with one vertical sodar. Therefore a comparison between sodar displacement spectra and pressure spectra, measured by a gravity wave network would be of vital interest. Microbarograph networks detect and measure pressure variations occurring with periods of a few seconds to some hours. Our microbarograph array, was basically designed to study the pressure variations associated with gravity waves with periods of some minutes to about two hours. The microbarograph-array provides data on the wave amplitudes, periods, horizontal phase speeds, directions of wave propagation and horizontal wavelengths. Comparisons bwtween microbarograph records and sodar facsimile records show that the microbarographs and sodars detect the same waves. Simultaneous measurements with sodar and microbarograph array were carried out by Beran et al, [23], Kjelaas et al, [24], Hooke et al, [24], Richner et al, [25], Cheung and Little [26], and Bull and Neisser [16], and with a vertical sodar and only a single microbarograph by [8, 9, 10].

During May-June 1987 an experiment was carried out at the Observatory Juliusruh, situated at the Baltic Sea. A vertical sodar, a microbarograph array, an instrumented 70 m-high tower with standard temperature and wind measurements as well as ultrasonic anemometers at two heights were operated in order to examine several features of the ABL structure near to the shore line, see Neisser et al, [27]. Fig. 8 shows for three successive time intervals of two hours each a cutout of the

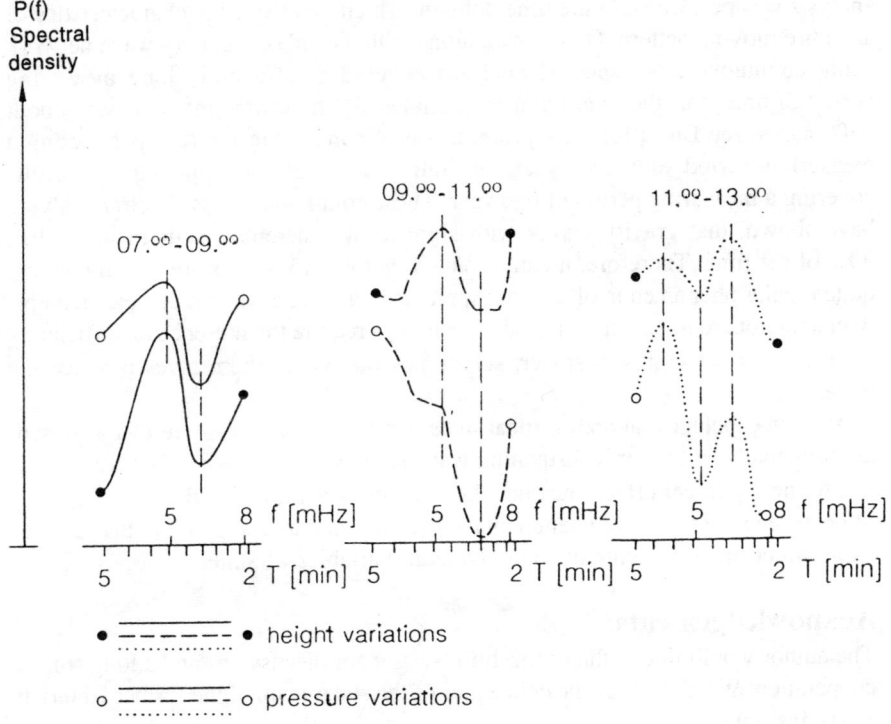

Fig. 8. Sodar-derived displacement spectra and microbarograph-derived pressure spectra, 26.05.1987, 07.00–13.00, cut-out for the period ranges

power spectra of the pressure variations of the gravity wave network and the vertical displacement spectra, evaluated with the aid of the trace following method. The spectra were computed for the frequency range 0.55 mHz to 8.33 mHz (30–2 min), Fig. 8 shows only a cutout for the frequency range 3.33 mHz to 8.33 mHz (5–2 min). For these periods, typical for processes in the ABL, the coincidence of the spectra for both methods is very good. From this comparison we can conclude that both spectra are manifestations of the same gravity wave process.

Published results about the frequency of occurrence of gravity waves and the associated wave parameters differ considerably. They depend on the method of measurements (microbarograph array, sodar, lidar, radar, wind measurements etc), on the configuration of the arrays, the method of analysis etc. Often gravity wave data were obtained and analysed as gravity wave events, consisting of a number of waves that occurred consecutively. For our wave investigations with a microbarograph array and with a sodar array we performed auto and cross spectrum analysis. The wave parameters were computed and accepted as gravity waves, only when the coherence spectra showed a coherence across the array larger than a specified significance level for any frequency band. In addition cross correlation analysis was performed in the time domain, which yielded mean characteristics of an entire moving pattern. Our investigaions with the aid of a gravity wave network using coninuous cross spectral analyses covered a sufficiently long measuring period of one year, the total number of analysed 2-hour time intervals was about 3300 cases, see Bull [21] for a presentation of comprehensive results covering a measuring period of a half year, or Bull et al, [22] for some further results covering a measuring period of one year. These continuous cross spectral analyses have shown, that gravity waves with significant coherence occur in more than 90% of the time. Therefore, it can be stated that gravity waves are a common and quite regular phenomenon of the atmospere. But, in contrast to direct measurements with a microbarograph network, sodar soundings require the presence of sufficiently intensive C_T^2 structures as tracer, sometimes the waves themselves produce the tracer.

We can conclude that both vertical sodar and Doppler sodar—carefully analysed-are principally suited for investigating wave structures in the ABL. But restrictions due to the C_T^2 tracer effect must be taken into consideration, so that the frequency of occurrence of the wave phenomena appear reduced as compared to that derived from direct measurements of hydrodynamic variables of state.

Acknowledgements

The author would like to thank Joachim Neisser for discussions and a long fruitful cooperation. We also thank the colleagues of the sodar team of the former Heinrich-Hertz-Institute.

References

1. E.E. Gossard and W.H. Hooke, "Waves in the atmosphere", Elsevier, New York, 1975.
2. F. Einaudi, D.P. Lalas, and G.E. Perona, "The role of gravity waves in tropospheric processes", Pageoph., 117, 627–663, 1979.
3. S.P. Singal, "Acoustic sounding stability studies", Encyclopedia of Environment Control Technology, Vol. 2, P.N. Cheremisinoff (Ed.), Ch. 28, 1003–1060, 1989.
4. W.D. Neff and R.L. Coulter, "Acoustic remote sensing", in Probing the Atmospheric Boundary Layer (D.H. Lenschow, Ed.), American Meteorological Society, Boston, MA, pp. 201–266, 1986.
5. E.H. Brown and F.F. Hall Jr., "Advances in Atmospheric Acoustics", Rev. Geophys. Space Phys., 16, 47, 1978.
6. S.K. Aggarwal, S.P. Singal, and S.K. Srivastava, "Sodar studies of gravity waves in the planetary boundary layer at Delhi", Mausam, 31, 373, 1980.
7. J. Neisser, G. Bull, K. Evers, M. Weimann, E. Weib, J. Keder, and I.V. Petenko, "Results of sodar investigations of the structure of the planetary boundary layer", In Proc. of the Field Exp. KOPEX-86, Czechosl. Acad. Sci, Prag, 109–141, 1988.
8. A.K. De, A. Ganguli, and J. Das, "On some characteristics of gravity wave perturbations in atmospheric boundary Layer", In S.P. Singal (Ed.), Acoustic remote sensing, New Delhi, Tata McGraw-Hill, New Delhi, 372–377, 1990.
9. R. Venkatachari and H.N. Dutta, "Acoustic sounding and microbarograph: complementary techniques to study ABL", In S.P. Singal (Ed.): Acoustic remote sensing, New Delhi, Tata McGraw-Hill, 519–525, 1990.
10. D.N. Rao, K.K. Reddy, K.S. Ravi et al, "Study of gravity waves in the atmospheric boundary layer using sodar", Proceedings of the 6th Symposium on Acoustic Remote Sensing and Associated Techniques of the Atmosphere and Oceans, University of Athens, May 26–29, 1992, Athens Isars'92, Greece, 265–270, 1992.
11. L. Eymard and A. Weill, "A study of gravity waves in the planetary boundary layer by acoustic sounding", Boundary Layer Meteorology, 17, 231, 1979.
12. M. Blez and A. Weill, "Simultaneous analysis of waves and turbulent parameters under statically stable morning conditions", In Proc. 2nd Int. Symp. on acoustic remote sensing of atmosphere and ozeans, Rome, Ed. G. Fiocco, Univ. Rome, Part 28, 1–23, 1983.
13. A. Weill, M. Blez, and F. Leca, "Observatorions of gravity waves and horizontal mixing in the atmospheric boundary layer", Annales Geophysicae, 5B, 413–420, 1987.
14. A.G. Kjelaas, D.W. Beran, W.H. Hooke, and B.R. Bean, "Waves observed in the planetary boundary layer using an array of acoustic sounders", J. Atmos. Sci., 31, 2040, 1974.
15. Fua and G. Mastrantonio, "Triaxial Doppler sodar system detecting gravity waves in the boundary layer", a case study. Proc. IInd Internat. Symp. Acoust. Rem. Sens., 27, 1–17, 1983.
16. G. Bull and J. Neisser, "Acoustic sounding of diurnal variations and gravity waves in the planetary boundary layer", Appl. Phys. B57, 3–9, 1993.
17. C.J. Nappo and R.M. Echman, "An episode of wave-generated turbulence in the stable planetary boundary layer", Proc. 10th Symp. Turb. Dif., AMS: Portland, 118–122, 1992.

18. G. Bull and Neisser, "Acoustic sounding of mesoscale structure in the planetary boundary layer", Proc. Inter. Meet. Appl. Sodar and Lidar Techn. in Air Pollution Mon. (EURASAP), Krakow, Poland, IV. 1–19, 1990.
19. H.R. Lehmann and J. Neisser, "Acoustic sounding work in the G.D.R.", In S.P. Singal (Ed.): Acoustic remote sensing, New Delhi, Tata McGraw-Hill, 490–499, 1990.
20. J. Neisser, "Über den Zusammenhang zwischen atmosphärischen Schwerewellen und hochtroposphärischen Strahlströmen", Z. Meteorol. 35, 253–262, 1985.
21. G. Bull, "A study of statistical properties of atmospheric gravity waves", Z. Meteorol. 35, 73–83, 1985.
22. G. Bull, J. Neisser und M. Weimann, "Interne Schwerewellen: Klimatologie und Zusammenhang mit instabilen Scherschichten", Abh. Meteorol. Dienst DDR Nr. 141, 175–177, 1989.
23. D.W. Beran, W.H. Hooke, and S.F. Clifford, "Acoustic echo sounding techniques and their application to gravity wave, turbulence and stability studies", Boundary Layer Meteorology, 4, 133–153, 1973.
24. W.H. Hooke, J.M. Young, and D.W. Beran, "Atmospheric waves observed in the planetary boundary layer using an acoustic sounder and a microbarograph array", Boundary Layer Meteorology, 2, 371, 1972.
25. H. Richner, W. Nater, and P.D. Phillips, "Collected papers on acooustic echo sounding, gravity waves and biometeorology", Laboratorium fur Atmosphärenphysik der Eidg. Tech. Hochschule Zürich, LAPETH-13, 113 pp, 1977.
26. T.K. Cheung and C.G. Little, "Meteorological tower, microbarograph array and sodar observations of solitary-like waves in the nocturnal boundary layer", J. Atm. Sci. 47, 2516–2536, 1990.
27. J. Neisser, Th. Foken, G. Bull u.a., "Ausgewählte turbulente und mesoskalige Prozesse in der Kontaktzone Meer-Land", Z. Meteorol. 40, 1, 38–40, 1990.

Acoustic Remote Sensing Applications
S.P. Singal (Ed)
Copyright © 1997 Narosa Publishing House, New Delhi, India

14. Sodar Monitoring of Nocturnal Boundary Layer During the Harmattan in Ile-Ife, Nigeria

J.A. Adedokun
Physics Department, Obafemi Awolowo University, Ile-Ife, Nigeria

1. Introduction

1.1 Pre-amble

Studies of the planetary boundary layer (PBL) carried out in particular over the mid-latitude and high latitude regions of the globe suggest a PBL evolution akin to that depicted schematically in Fig. 1 below. Very little information exists on both the insolation and the boundary layer depths over the tropical regions. This paper gives a preliminary report on research efforts at measuring the latter using a SODAR system. Solar radiation equipment required for measurements of the

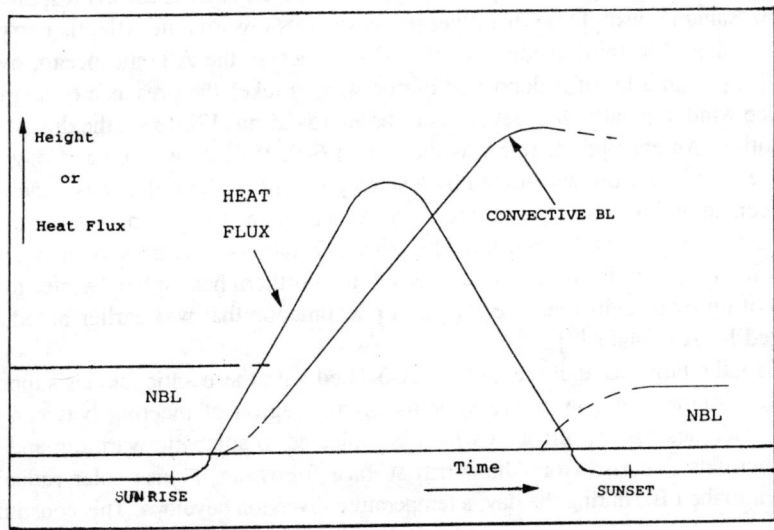

Fig. 1. A schematic sketch of the diurnal variation of the depth of the PBL.

former has also been installed at Obafemi Awolowo University (OAU), Ile-Ife, Nigeria, (Adedokun et al, [1]) venue of the current experiment (Fig. 2). Knowledge of the PBL depth is vital to the monitoring and control of atmospheric pollution, an area in which the sodar is quite reliable [2, 3, 4, 5].

Adedokun and Vaughan [6] have reported on the sodar research efforts initiated at Ile-Ife in Nigeria in the year 1988. The system has since been running and the present paper is on a study of sodar echo returns observed with the system and the height of the nocturnal boundary layer (NBL) inferred for each hour between sunset and sunrise during the harmattan months of February and March 1991 except on occasions when power cuts and other operational problems might have arisen. Also, the environmental pollution group, OAU made some simultaneous measurements of the total suspended particulates (TSP) in the atmosphere on some occasions during the period and these are compared with the relevant sodar PBL depth measurements as shown later in the paper.

1.2 The Prevailing Weather and Temperature Inversion Regions in Harmattan

The harmattan season is unique for the prevalence of the dry and dusty north-easterly winds that invade the West African region from the Lake Chad basin area [7, 8, 9, 10]. The season is often marked by spells of harmattan wind occurrence which may last between three to ten days.

The synoptic pressure pattern that characterizes the harmattan period is anticyclonic in nature. The centres of action for the circulation consist of the Saharan anticyclone which often oscillates southwards over the West African hinterland and the Azores anticyclone which oscillates eastwards towards the region [11]. The former is principally the agent behind the transport of dust from the Chad basin area right to the gulf of Guinea, while the latter is usually responsible for the Saharan dust plume that diverges westwards towards the Atlantic carrying an estimated 260 million tons of mineral dust across the Atlantic ocean, every year [12], with a lot of it deposited on the way. It takes the presence of a strong surface wind, typically, low-level jets of order 15–25 m/s [7], to set the dust storm in motion. An example of this was the March 6–9, 1991 case when a strong and turbulent surface flow was forced by a steep pressure gradient that was generated between an anticyclone over northeastern Africa and a strong depression moving over the Atlas region (Franzen et al [13]). Kalu [7] has also remarked on incidences of 'cold outbreaks' from the midlatitudes in the northern hemisphere winter in the form of upper troughs over the region, a phenomenon that was earlier noted and tracked by Adefolalu [7].

Also, the large scale descent of air associated with these anticyclones suppress the development of convective systems. At the region of meeting between the large scale descending air mass which is subjected to adiabatic warming and the hot thermals that rise from the warm surface following intense solar radiation input into the PBL during the day, a temperature inversion develops. This constitutes

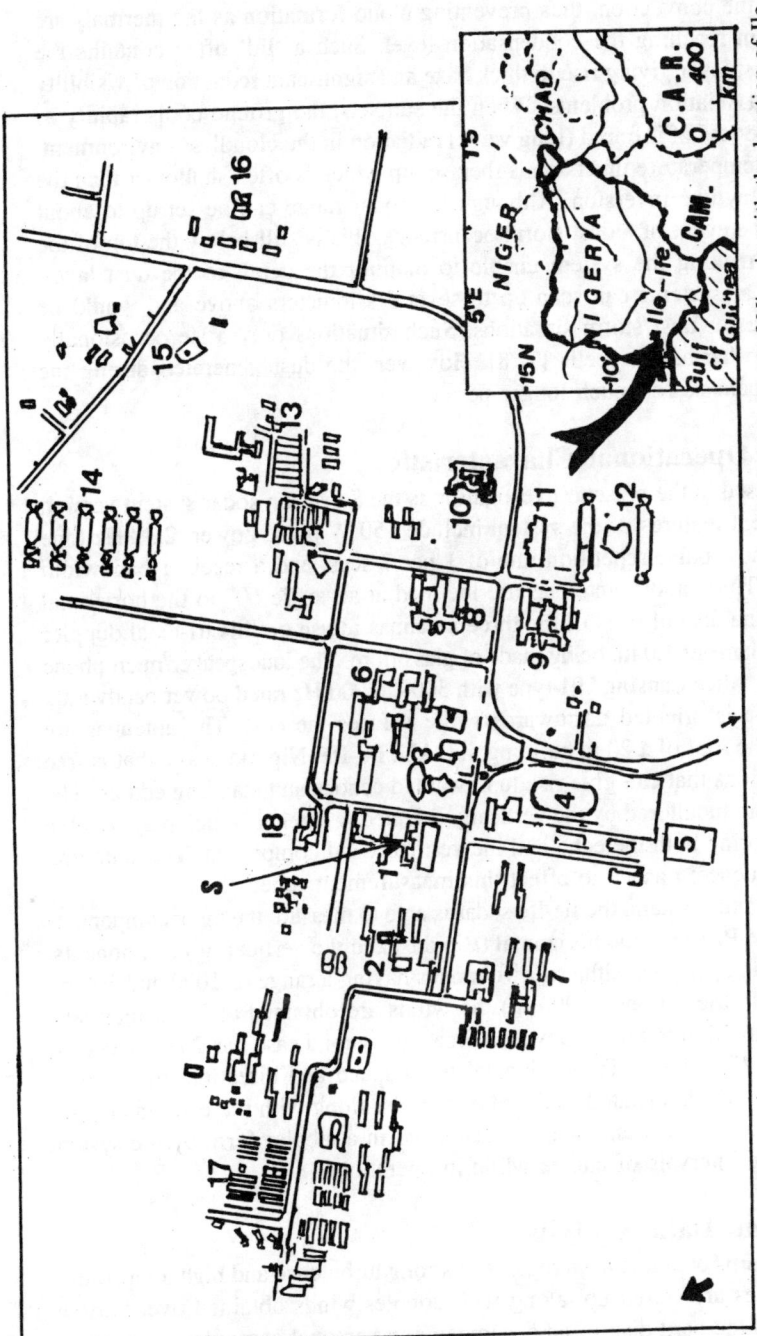

Fig. 2. Obafemi Awolowo University (OAU) site, showing the SODAR location, S. Inset is a map of Nigeria with the location of Ile-Ife. Labelled 1 to 18 on campus are: (1) Fac. of Science & Physics Building on top of which are the Sodar antennas, (2) Moremi Hall, (3) Oduduwa Hall, (4) Sport's Field, (5) Hockey Field, (6) Fac. Education, (7) Fajuyi Hall, (8) Biolog. Sciences, (9) Fac. H/Sciences, (10) Computer Blg, (11) Pharmacy, (12) Fac. E.D.M., (13) Law & Pol. Sci. (14) I.A.R.&T, (15) Conf. Centre, (16) Staff School, (17) Mozambique Hall, (18) African Studies.

a "lid" over the convection, thus preventing cloud formation as the thermals are hindered from reaching the condensation level. Such a 'lid' often contains the harmattan dust layer giving rise to thick haze and significant reduction of visibility and attendant aviation problems. When the sun sets, the ground cools rapidly as a result of the loss of infrared (long wave) radiation in the cloudless environment. A low level temperature inversion is then set up which is often shallower than the upper level daytime inversion. Although the sodar range can be set up to about 1000 m, in the event of a dust storm occurrence, this is still below the top of the dust plume, making the system unable to monitor the depth of the dust layer. However, lidar systems can scan up to several kilometers above and would be more suitable for dust storm situations. Such situations only arise occasionally giving rise to harmattan spells [7, 8]. However, the dust generated during the spells often persists for much longer periods.

2. Sodar Operational Characteristics

The system used in the present investigation is the Sensitron sodar system version 325 [6]. Salient features of the system include 150 W pulse power, 2.3 kHz tone pulse frequency, pulse repetition rate of 1 per 7 secs, and a receiver bandwidth of ±150 Hz. The u and v antennas are inclined at an angle 60° to the horizontal and the antenna dish of each of the three antennas in use on this tri-axial doppler system is of diameter 1.0 m, being made of glass fibre. The loudspeaker/microphone element is the Altec Lansing 291 type with 500–20,000 Hz rated power bandwidth. The u antenna is oriented northwards while v faces the east. The antennas are mounted on the roof of a 20 m building at OAU, Ile-Ife, Nigeria, a site that is free from obstructions that can give rise to unwanted echoes and standing eddies. The echo returns are monitored by the vertically oriented w antenna which plays double role of monitoring both the echo and the vertical wind component. The sounding follows a sequence: $uwvwu$ to effect this measurement scheme.

Being a tri-axial system, the Ile-Ife sodar is able to measure the three components of winds in the PBL, i.e. the horizontal (u and v) and the vertical (w) components. The system can operate in either of two modes having a range of 1000 and 500 m, respectively. In the former (1,000 m) the winds are obtained at 50 m intervals, beginning from 50 m minimum level whereas the second case has 25 m intervals, starting from 25 m level. The horizontal wind speed and direction, the vertical wind speed and the associated standard deviations along with the echo intensities (recorded by the vertical antenna) are displayed in a tabular form by the system at regular time intervals of choice while in operation.

3. Data and Data Analysis

Sodar echo returns obtained from regions of strong turbulence and high temperature inhomogeneities are stored up, along with doppler winds obtained over various layers aloft, in the hard disk of an accompanying personal computer. A software on the system's presentation computer is used to analyze the data off-line. The echograms, similar to the popular facsimile records, display, as a function of height and time, the temperature inhomogeneities of the atmospheric region above the sodar system, depicting regions of temperature inversions, elevated layers, shear layers, as well as convective processes going on in the atmosphere. A typical example is hereby shown in Fig. 3 where an elevated layer can be seen to have developed rather early in the morning, eroding the nocturnal surface-based inversion.

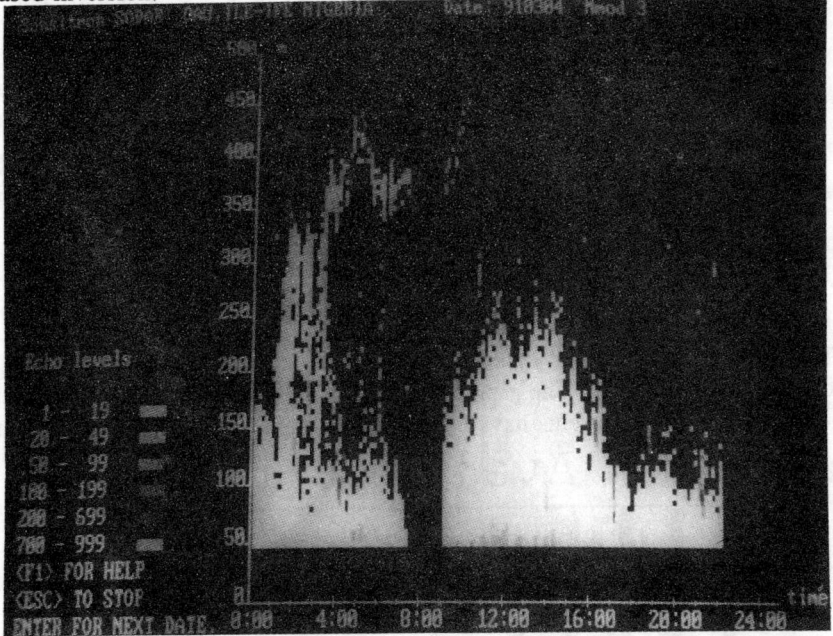

Fig. 3. An echogram of the Sodar system showing an elevated layer in the early hours of March 4, 1991.

The atmospheric phenomenon depicted here, is quite interesting. Ile-Ife, being in a tropical region, does not experience sunrise at 0400 hours (see Fig. 4 for a typical diurnal variation in solar/terrestrial radiation fluxes experienced over the station in March, although not for the same period, as our solar radiation station was only established a year later [1]). How come the thermal plumes can be seen in Fig. 3, under the elevated layer at that period? The 'causa d'etre' is the fact that the station is about 200 km from the Gulf of Guinea on the Atlantic coast, and, as the Inter-Tropical Discontinuity (ITD) is just some 1 to 2 degrees north of the station, the moist South westerly winds still get there, albeit forming a wedge under the overlying north easterly current at the ITD surface position. The various soundings available for the nocturnal cases of interest have been carefully examined and the boundary layer depth, H, determined for each hour either as the height of the ground inversion or the base of an elevated layer in the absence of the former or the maximum height of weak reflection in case of near neutral stratification.

298 Applications in the Atmosphere

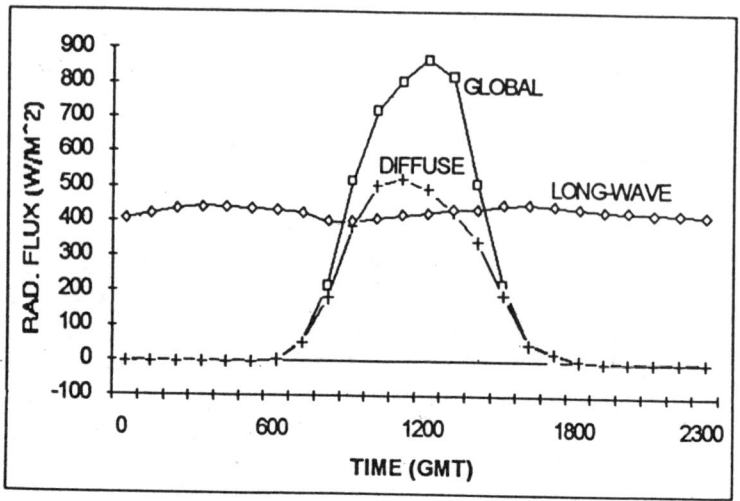

Fig. 4. A plot of the hourly means of the broad-band global, diffuse and long-wave radiation obtained at OAU, Ile-Ife on 21st March, 1992 (1).

Fig. 5 shows the number of data available in Feb. and March and the total data considered for each hour. The gap noticed in Fig. 3 is an evidence of a disruption in power supply. Power cuts have been a considerable constraint to the amount

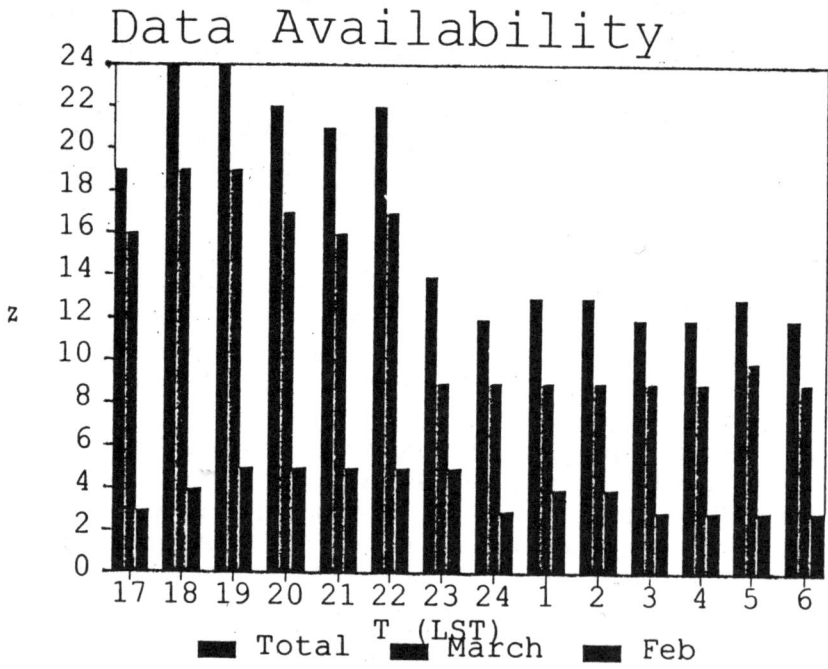

Fig. 5. A plot of the number of hourly echogram data available for the 1991 harmattan period.

of data that could be obtained. This also accounts largely for the unevenness in the number considered for each hour.

4. Results and Discussions

4.1 Variation in NBL Depth with Time

In Fig. 6, H(m) has been plotted as a function of time of the day, from 17 LST. that is, just before sunset. It can be observed that the height of the nocturnal boundary layer (NBL) first falls at 18 LST before rising again. This agrees with earlier observations made in India by Pahwa et al, [5] and others. The layer rises rather unevenly. The Feb. curve shows a greater deviation than that for March but the number of available data in March far outweigh that in February, hence, the mean agrees more with the March case. This investigation is a preliminary one and we intend to carry out a more extensive survey on this aspect later.

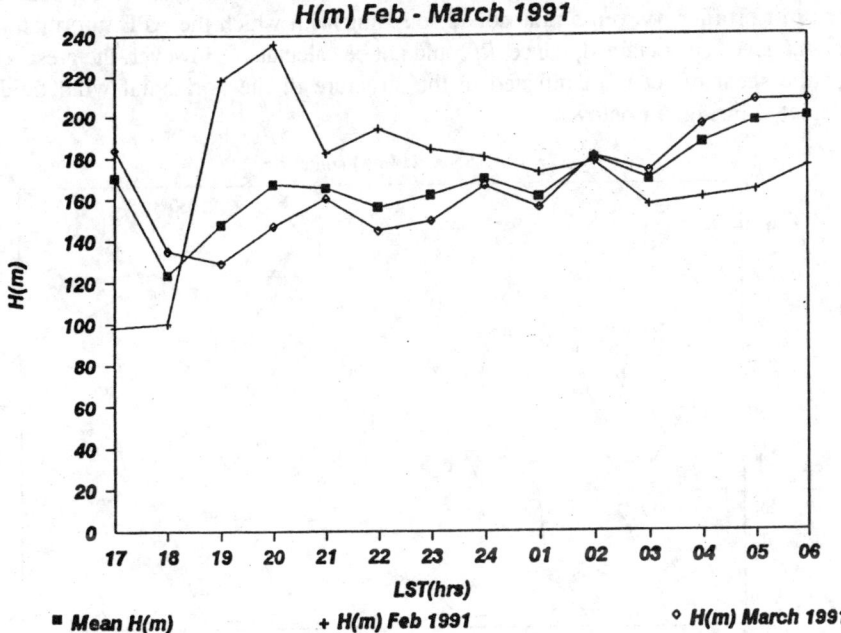

Fig. 6. A plot of the nocturnal boundar layer depth H(m) as a function of time T(LST) for the 1991 harmattan period under study.

4.1.1 The Dynamic Factor

(a) Richardson's Number
In consideration of the convective processes occurring within the PBL, an important factor to reckon with is the Richardson's number.

$$R_i \equiv -\{(g/\bar{\theta})(\partial\theta/\partial z)/(\partial v/\partial z)^2\} \tag{1}$$

where g is the acceleration due to gravity, θ is the dry bulb potential temperature, and v is the wind vector. As a matter of fact, R_i is equivalent to the ratio of the energy available to the thermals, styled 'available potential energy' (APE) to the energy due to motion, styled 'available kinetic energy' (AKE). While the temperature stratification of the atmospheric layer of interest dictates the APE, the AKE is dictated by the shear in the wind field. Hence, the larger the shear in the wind field, the larger is the AKE, the smaller is the richardson's number and the more unstable is the PBL for a given value of the APE and vice versa.

While the sodar facsimile records provide information on the APE, albeit qualitative in the present system, the associated wind fields, as observed by the system's antennae, provide information on the AKE. Sometimes, regions of small richardson's number are regarded as indicative of the depth of the turbulent stable boundary layer (14). In the experiment reported here, there was no supporting mast or profiler or even a radiosonde sounding with which the APE information might have been obtained; hence, R_i could not be calculated. However, the presence of the shear effect is manifested in the structure of the horizontal wind fields depicted in Fig. 7a below.

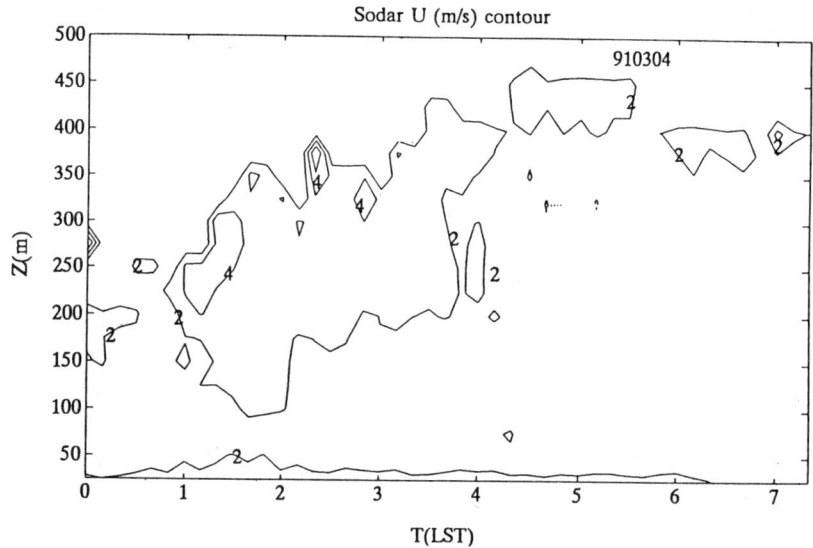

Fig. 7a. Horizontal Wind contour fields composited from Sodar wind returns obtained at 10 min. integration periods, corresponding to the first section of the event in Fig. 3. Isopleths are at 2 m/s intervals.

(b) Nocturnal Jet Structure

We have examined the horizontal wind fields associated with the elevated layer structure of Fig. 3 as shown in Fig. 7a where it can be seen that between 001 and

004 LST, a nocturnal jet developed and later disappeared over the station. This system was associated with some wind shear as could be seen in the wind field. The axis of this jet deepened with time with ocassional decrease in the jet strength as it passed over the station. When compared with the situation after the power cut (Fig. 7b), a comparatively different system could be observed as a convective layer developed with the advent of the sun inputting its radiation into the PBL.

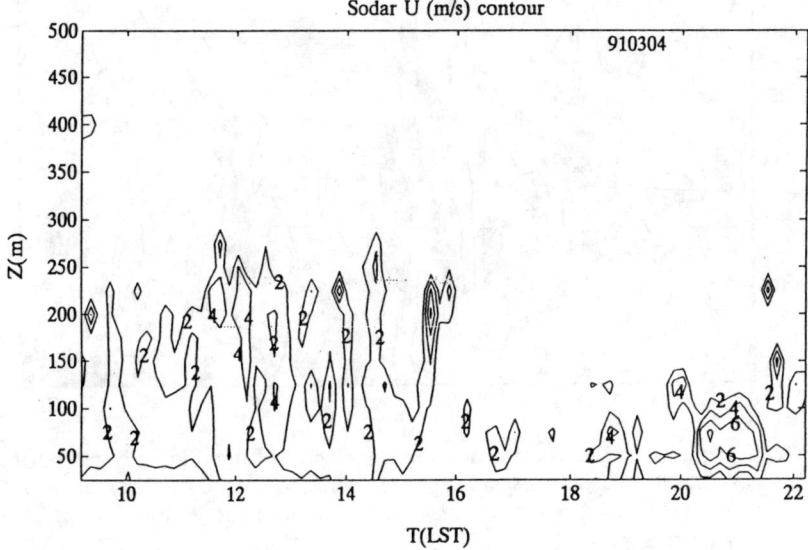

Fig. 7b. As in Fig. 7a, but for the second section of the event in Fig. 3.

(c) Convective Layer Depth

As the PBL experiences the input of solar radiation, it becomes convectively unstable and a convective layer develops. A deeper probe ino this case was attempted by compositing the sodar-observed vertical (w) wind fields, obtained, as in Fig. 7a and 7b above, at 10 minute integration time intervals. The resulting contour fields are rather complex as shown in Figs. 7c and 7d below. The regions of ascending motion, with w positive, shaded, can be seen to be more prominent in Fig. 7d than in Fig 7c indicating that more convective plumes typify the latter. Hence, with the comparatively less ascending motion observed in Fig. 7c, it can be concluded that the elevated layer noticed in Fig. 3 is due to advective processes, occasioned by the jet rather than a result of local parcel ascent that often accompanies convection.

We know that sodar capability for monitoring day time convective PBL depth is rather limited when the plumes are not capped by a stable layer (See e.g. Singal et al [15]). Hence, the value of H indicated in the convective period here can be regarded as an under-estimate. However, Singal and collaborators [15], have reported a technique for determining mixing height during daytime since about a decade

302 *Applications in the Atmosphere*

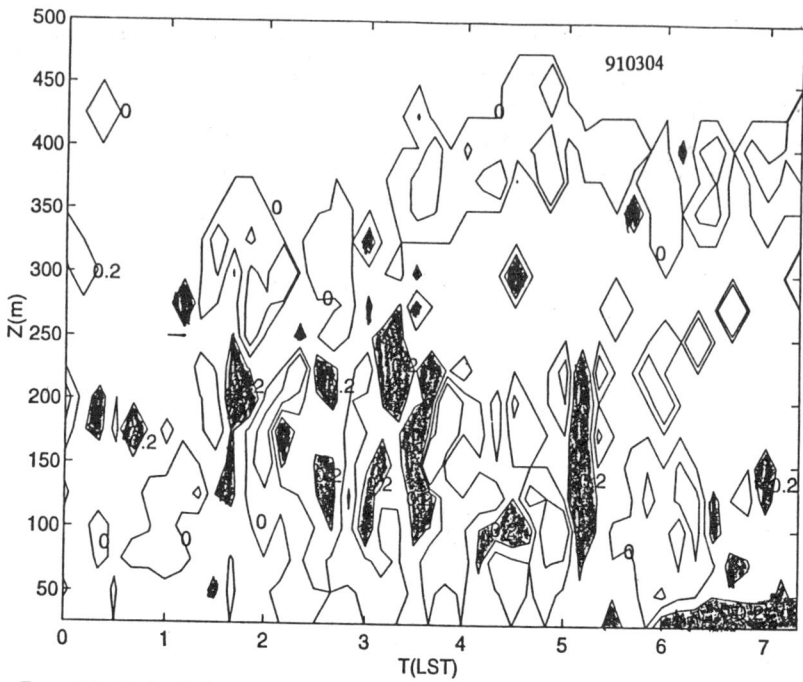

Fig. 7c. Vertical wind, w contour fields (m/s) composited from Sodar Doppler winds obtained at 10 min. integration periods, corresponding to the first section of the event in Fig. 3. Isopleths are drawn at 0.2 m/s intervals. Zones with w ≥ 0.2 m/s are shaded.

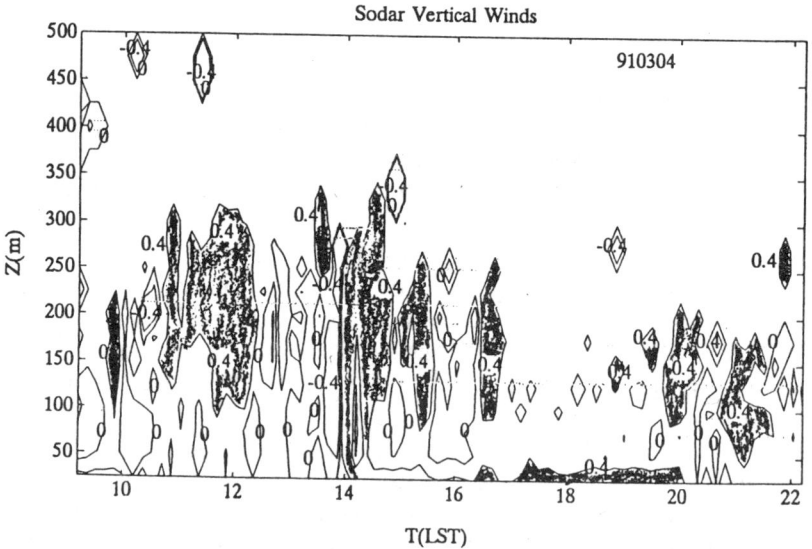

Fig. 7d. As in Fig. 7c but for the second section of the event in Fig. 3. Isopleths of w are drawn at 0.4 m/s intervals and zones with w ≥ 0.4 m/s are shaded.

ago in India. In accordance with this technique which was based on Holzworth model, the actual depth of the mixed layer H would be given by:

$$H(m) = 95 + 4.24\ H_s \tag{2}$$

where H_s is the depth of the thermal plumes measured by the sodar.

If we apply this empirical relationship to the daytime section of our observation here (Fig. 7b), we would obtain an H value of 1,155 m while the sodar indicated 250 m as the observed plume depth. This value looks more realistic. We hope to have a closer look at the relation with a consideration of many more cases of daytime sodar observations in order to effectively affirm its suitability in the region under study as it has been found applicable in several parts of India and in Poland [15].

4.2. The NBL and Total Suspended Particulates

An initial investigation has also been made regarding the manner in which the total suspended particulates (TSP) vary with H in the harmattan period. Measurements of the TSP made by the OAU environmental pollution group during Feb. and March 1991 (Obioh 1991, private communication) were compared with those of H simultaneously determined as shown in Fig. 8 where a low but significant ($r = 0.37$; $a = 110$ g/m 2, $b = -2.5$) inverse power correlation has been obtained between the two.

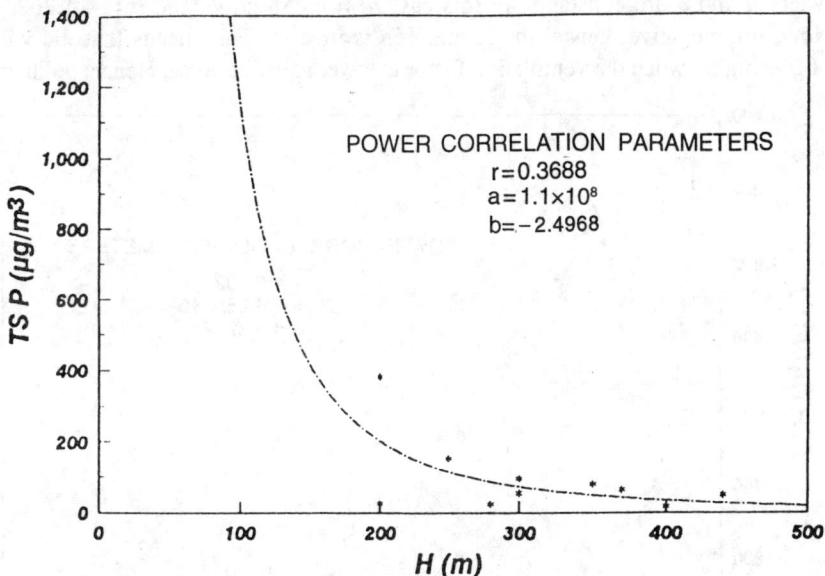

Fig. 8. A plot of The Total Suspended Particulate (TSP) concentrations ($\mu g/m^3$) against the NBL depth, $H(m)$

4.2.1 The Ventilation Factor

In our consideration of the concentration of pollutants in the atmosphere over the station, we shall examine the ventilation factor [16], given by:

$$V_F \, (m^2/s) = VH \tag{3}$$

where V is the mean wind over the atmospheric column of layer n, say, above the station of interest, given by:

$$V = \frac{1}{n} \sum_{i=1}^{n} v_i(z_i) \tag{4}$$

where z stands for the layer height.

When the mean wind values were obtained and compared with the associated TSP values, a higher power correlation ($r = 0.50$) was obtained (Fig. 9). This rightly suggests that the ventilation factor, incorporating both the wind circulation within the PBL and the planetary layer depth, not just the latter, play a part in determining the concentration of pollutants within it. This agrees with an earlier result of Pekour and Kallistratova [17]. The TSP is here found to fall with increase in V_F up to a threshold after which any rise in V_F does not affect the TSP. Hence, the TSP concentration, C, say, is given by:

$$C = aV_F^b \tag{5}$$

where a and b are constants. In this case, $a = 0.485$ g/m/s and $b = -1.262$.

Now, b is negative, hence, the relationhip is inverse. This means that the value of C is higher when the ventilation factor is lower and vice versa. Hence, pollutants

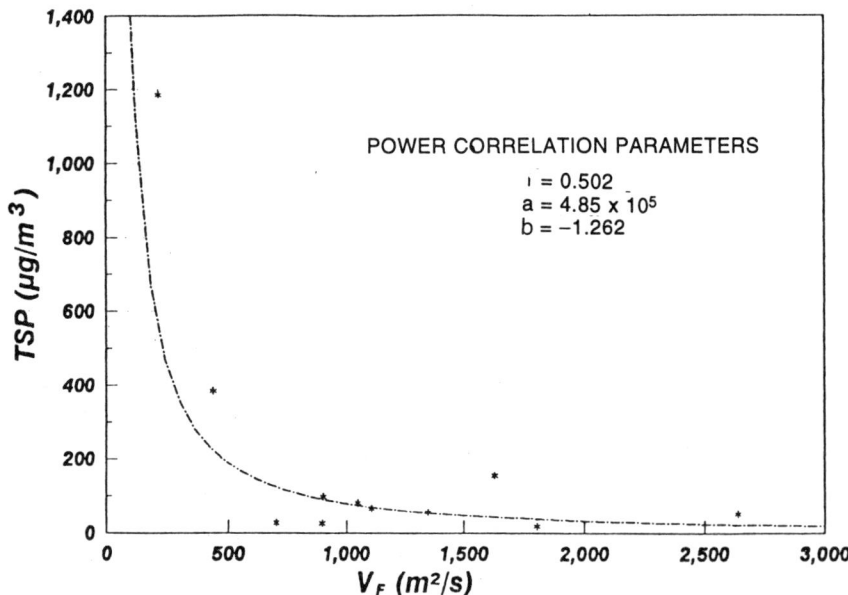

Fig. 9. As for Fig. 8 but for ventilation factor V_F (m^2/s)

accumulate over an area when there exist a lower PBL depth (engendered by low-level inversions for instance) and low wind speed.

Local sources of pollutants may also be contributing to the observed results. It needs be known whether the diurnal variation in TSP is due to the variation in the harmattan dust concentration only or local ambient sources are contributing to it. It is intended that a comprehensive measurement programme involving not only the determination of TSP and H but also, other PBL parameters would later be carried out over the station.

5. Conclusions

An evaluation has been made of sodar-derived nocturnal boundary layer depth H at Ile-Ife, Nigeria, a tropical station, during the harmattan season months of February and March 1991. While the value of H first decreases one or two hours after sunset, it rises swiftly over the next two hours before it starts to oscillate with a general upward trend. This behaviour agrees with some cases analyzed in India by Pahwa et al, [5].

On the average, the hourly mean values of H ranged between 120 and 200 m. It has risen by 76 m between 18 LST and 6 LST the following morning, that is, a period of 12 hrs. This information along with a knowledge of the sensible heat flux over the period, can be quite useful inputs into PBL models for the region. Our initial estimates show that some mid-latitude and high latitude type PBL models grossly over-estimate the region's boundary layer.

The sodar range is too shallow to be able to monitor the top of the dust layer in the event of a dust storm occurrence during the harmattan period, but a lidar system can achieve this feat as it covers several kilometers in range. However, the sodar system capability for monitoring the surface-based inversion layer along with other cases of inversion that may exist between the surface and 1 km region of he PBL makes it versatile for pollution research, monitoring and control, its inherent limitations not withstanding. Efforts along this valuable application of the system have initially shown that an inverse power law relationship exists between the total suspended particulates and the sodar-derived ventilation factor over the station. We intend to carry out more measurements along this line as well as evaluations of the data available for other periods of the year in order to make useful comparisons for the purpose of achieving a better understanding of whatever seasonal variability there may be.

6. Acknowledgements

The Ife Sodar System has been supported by International Science Programs Uppsala University, BITS, and SENSITRON AB (all of Sweden) and OAU, Ile-Ife, Nigeria. The support of Profs L. Hasselgren, B. Holmgren, E.E. Balogun, and G.O. Ajayi, and the entire members of the Atmospheric Physics group OAU are gratefully acknowledged. So also is the contribution of the environmental pollution group headed by Prof. A.F. Oluwole for the TSP data. Drs I. Obioh and Z.D. Adeyefa assisted in drawing some figures.

References

1. Adedokun J.A., Adeyefa Z.D., Okogbue, E. and B. Holmgren 1993: Measurement of solar and long wave radiation over Ile-Ife, Nigeria. Paper presented at the 3rd UN/ESA workshop on Basic Space Science. Lagos, Nigeria; Oct. 18–22, 1993. AIP Conference Proceedings 320 p. 179–190. AIP Press, N.Y.
2. D.W. Beran, Hall F.F. Jr, J.W. Wescott and W.D. Neff 1971: Application of an Acoustic Sounder to Air Pollution Monitoring. Paper presented at Air Pollution Turbulence and Diffusion Symposium. New Mexico State University, Las Cruces, New Mexico 88001. 7 pp.
3. D.W. Beran and F.F. Hall Jr. 1974: Remote Sensing for air pollution Meteorology. Bull. Amer. Met. Soc. Vol. 55 No. 9 1097–1105.
4. S.P. Singal 1990: Current Status of Air Quality Related Boundary Layer Meteorology Studies Using Sodar. In: Acoustic Remote Sensing. (S.P. Singal, Ed) Proceedings, 5th Int. Symp. on Acoustic Remote Sensing of the Atmosphere and Oceans. India. Tata McGraw-Hill Publishing Co. Ltd. New Delhi. pp. 453–476.
5. D.R. Pahwa, B.S. Gera and S.P. singal 1990: Study of the Evolution of Nocturnal Boundary Layer at Delhi by acoustic Sounding Technique. ibid. pp. 363–371.
6. J.A. Adedokun and O.O. Vaughan 1990: Sodar Doppler Wind Measurement in Ile-Ife, Nigeria. ibid. pp 395–400.
7. A.E. Kalu 1977: The African dust plume: Its Characteristics and propagation across West Africa in Winter. In: Saharan Dust (Morales, ed) New York, Wiley, Ch 5, 95–118.
8. J.A. Adedokun, W.O. Emofurieta, and O.A. Adedeji 1989: Physical, Mineralogical and chemical properties of harmattan dust at Ile-Ife, Nigeria. Theor. Appl. Climatol. 40, 161–169.
9. Adetunji J., McGregor, J. and Ong C.K. 1979: Harmattan Haze. Weather 34, 430–436.
10. E.E. Balogun 1974: The phenomenology of the atmosphere over West Africa. Proceedings of Ghana Scope's Conference on Environment and Development in West Africa. Ghana Acad. of Arts and Sciences, 19–31.
11. J.A. Adedokun 1978: West African precipitation an dominant atmospheric mechanisms. Arch. Met. Geoph. Biokl., Ser. A, 27, 289–310.
12. Shutz L. 1980: Long range transport of desert dust with special emphasis on the Sahara. Annals of the New York Academy of Sciences, 338, 515–532.
13. L.G. Franzen, J.O. Mattsson, T. Nihlen and A. Rapp 1994: Yellow Snow over the Alps and Subarctic from Dust Storm in Africa, March 1991. Ambio Vol. 23, No. 3, p. 223–235.
14. Roland B. Stull 1988: An Introduction to Boundary Layer Meteorology. Kluwer Academic Publishers. p. 178.
15. S.P. Singal, B.S. Gera and D.R. Pahwa 1994: Application of sodar to air pollution meteorology. Int. J. Remote Sensing, Vol. 15, No. 2, 427–441.
16. Richard A. Dobbins 1979: Atmospheric Motion and Air Pollution John Wiley and Sons. p. 138.
17. M.S. Pekour and M.A. Kallistratova 1993: SODAR Study of the Boundary Layer over Moscow for Air-Pollution Application. Appl. Phys. B 57, 49–55.

Acoustic Remote Sensing Applications
S.P. Singal (Ed)
Copyright © 1997 Narosa Publishing House, New Delhi, India

15. An Overview of Similarity Methods to Estimate Turbulence Quantities from Sodar Measurements in the Convective Boundary Layer

Dimitrios Melas

Laboratory of Atmospheric Physics, Physics Department,
University of Thessaloniki, 54006 Thessaloniki, Greece

Introduction

When buoyancy is the dominant mechanism driving turbulence, the boundary layer is in convective state. Convective conditions normally dominate over land in the warm period of the year during the day except when it is overcast or it is within an hour (or so) of sunrise or sunset. Under typical mid day conditions (z_i = 1000 m, $u_* = 0.4$ ms^{-1}), strong convective effects can be found with surface heat fluxes as low as ~ 25 Wm^{-2} [1].

Convective scaling was first proposed by Deardorff [2] and its ability to order field measurements of convective turbulence was demonstrated in a number of subsequent studies [3]. Dilution of pollutants is accomplished mainly by turbulence and convective scaling has proved to be an effective and concise means for modeling daytime diffusion [3, 4]. According to convective scaling theory, the governing parameters that control the turbulent structure are the buoyancy parameter, g/θ, the kinematic surface heat flux, Q_0 and the mixed layer depth, z_i. These governing parameters can be used to construct convective scales and nondimensional parameters that are convenient for parameterizing the convective boundary layer (CBL). In addition to mean wind speed and direction, it is thus essential in air quality studies to monitor continuously basic meteorological parameters, such as Q_0 and z_i. Moreover, some diffusion parametrizations include mechanical turbulence effects in the form of u_*, and a knowledge of this parameter becomes essential when studying transportation and dispersion characteristics in moderate instabilities.

Although, there is a well established practice in estimating the surface fluxes of heat and momentum and the mixed layer depth in research experiments, there still exists a lot of controversy about the determination of these quantities for operational use in air pollution monitoring. Direct turbulence measurements require research grade instrumentation and are costly to operate on a continuous basis. Estimates of mixing heights for air quality purposes are usually derived from radiosonde profiles which are often difficult to utilize because they provide a spot-like picture of the CBL which may not be representative of the Planetary Boundary Layer (PBL) depth averaged over space or time [5]. A more common problem is that radiosonde balloons are not launched frequently (usually twice daily).

Alternatively, remote sensing techniques can be used to estimate the various air quality related meteorological parameters. Singal [6] felt that urban meteorological measurements related to air pollution monitoring situations can be made up entirely of an acoustic sounder. To a large extent, however, most of Doppler sodar monitoring is restricted to mean horizontal wind and thermal pattern recognition [7] the data which can be most easily and accurately obtained. Although, sodar measured variances of vertical velocity and horizontal wind direction are also of particular importance to air pollution dispersion problems but they are contaminated with errors and should be interpreted with care [8].

This review focuses on selected similarity methods which can be used to extract additional information from sodar measurements. These methods are also referred to as indirect measurement techniques [9]. The definition of indirect measurements does not include determination of boundary layer depth using pattern recognition techniques. Instead, it means inferring one property from measurements of others, through use of empirical or theoretical relationships [9].

It is beyond the scope of this paper to present a detailed description of sodar performance. For this purpose the reader is referred to other, more detailed sources [8, 10]. However, a brief review of sodar performance will provide the basis for defining appropriate similarity methods and estimating their accuracy and is therefore presented in the next section.

The interest is focused on the convective PBL, which is particularly amenable to acoustic sounding [6] and wherein there also exists a fairly well established similarity structure. The parameters considered here are the surface fluxes of momentum and heat and the mixed layer depth. These are the basis of a wide variety of parametric relations which can be used to derive even more variables [5].

Excellent reviews about the sodar technique and applications can be found in Neff and Coulter [8], Weill and Lehmann [11], Neff [12] and Singal [7]. Parts of this review were previously published in Melas [13].

Doppler Sodar Measurements

Some of the basic boundary layer parameters are measured, more or less, directly by Doppler sodars. The list includes:

— horizontal mean wind components,
— vertical wind variance, and
— acoustic-scattering cross section.

The accuracy in sodar measurements of vertical velocities, w, and standard deviation of the horizontal wind direction, σ_θ, is not adequate for most practical applications [8, 10, 14].

Since there are many different systems with different sizes and operational characteristics and with different data reduction techniques, typical performance values presented in this section are only indicative.

Horizontal wind speed

Doppler sodars provide estimates of horizontal wind speed by frequency analysis of the returned signal. Measurements of the horizontal wind vector are relatively straightforward and compare well with in situ measurements when the wind speed is below a certain value (typically 10–12 m s^{-1} at 50 m height). Above this value the correlation is rather poor with the acoustic sounder always underestimating the wind speed. A possible explanation is that the high wind speeds are associated with an increased background noise and the signal cannot be distinctly separated from the noise within the computer of the sodar. Another likely contribution factor is the refraction of the signal by the wind. Vogt and Thomas [10] performed a long term intercomparison of wind data measured directly by sodar and tower instruments and found that the reliability and accuracy of horizontal wind speed measured by the sodar is generally good. Schwiesow [15] estimated that the accuracy for the two components of the horizontal wind is about 10%.

Wind variances

Measurements of the vertical velocity variances, σ_w^2, with Doppler sodar are subject to many sources of error (see Neff and Coulter [8] for a detailed discussion). Comparisons with in situ measurements (Finkelstein et al [16]; Keder et al [17]) show that the acoustic sounders tend to overestimate σ_w^2 at night (small σ_w^2 values) and to underestimate σ_w^2 during daytime (large σ_w^2 values). The positive biases in the σ_w^2 values at night are attributed to system errors while the underestimation of σ_w^2 during daytime is due to incomplete coverage of the vertical wind spectrum. In a recent study, Vogt and Thomas [10] concluded that the sodar can reliably measure σ_w. However, they found that the sodar underestimates σ_w when compared to a sonic anemometer, especially during daytime.

Melas [1] suggested that, after a proper calibration, the uncertainty in sodar measurements of σ_w is about 30%. Finkelstein et al, [16] compared measurements from four different, commercially available, Doppler sodars with similar measurements from in situ sensors on a 300 m instrumented tower. As a measure of the relative scatter in the data, the authors defined the percentage deviation,

$$s' = 100 \frac{(\text{RMSE}^2 - \text{BIAS}^2)^{1/2}}{(\overline{\sigma}_w)_{\text{ref}}} \quad (1)$$

where RMSE is the root mean squares difference which is sometimes called comparability, BIAS is the sample bias and $(\sigma_w)_{\text{ref}}$ is the average of σ_w from the sonic anemometer.

During daytime conditions, Finkelstein et al [16] found that s' is in the interval 27%–54% depending on measurement height and sodar system. Vogt and Thomas [10] found that, during daytime conditions, s' is ~ 28%.

Sodar measurements of the standard deviation of the cross-wind component of the wind, σ_v, and consequently the standard deviation of the horizontal wind direction, σ_θ, are of poor quality [14].

The temperature structure parameter

For monostatic sodar, the backscattered signal is proportional to the temperature structure parameter, C_T^2. The sodar equation, relating transmitted and received acoustic energy, provides means to estimate C_T^2 from sodar measurements [8]. This equation is strictly valid in the case of an isotropic and homogeneous turbulence field. In order to obtain quantitative measurements of C_T^2, the sodar has to be calibrated acoustically and the loss of acoustic energy through spreading and attenuation in the atmosphere has to be taken into account [8, 18]. It is thus questionable whether accurate C_T^2 values can be obtained routinely [19].

There has been a number of comparisons of in situ measurements of C_T^2 with estimates derived from sodar measurements [10, 17, 18 20]. The results generally show a fairly good agreement (usually within a factor of 2-3) and it is probably unrealistic to expect to obtain C_T^2 measurements better than or within a factor of two [8]. Since vertical profiles of C_T^2 range over two orders of magnitude they still can be useful.

Interpretation of sodar facsimile records

A vertically pointing antenna yields time-height facsimile records of temperature structure in the lowest few hundred meters of the atmosphere. Qualitative information about turbulent layers and their height can be obtained by interpretation of the facsimile record. Among other atmospheric features, the elevated inversion, capping a convective boundary layer, is probably the most easily recognizable feature [8]. Entrainment causes a peak in C_T^2 profiles near the base of the inversion and the thermal structure of the PBL detected by acoustic sounders provides a reliable estimation of z_i. This peak coincides with the bottom of the elevated echo layer on facsimile records.

Kaimal et al [21] found that the acoustic sounders are capable to locate the inversion base with a very good accuracy (when z_i is within the sodar range). Beyrich [22] gives a brief review on the use of sodar data to estimate mixing height. The estimated accuracy in sodar measurements of z_i is 10% [15].

Theoretical Framework

When convection is the dominant turbulence generation mechanism, the PBL has a well defined structure which may be idealized as in Fig. 1 [5]. The three major components of the PBL structure are the surface layer (the lowest 10% of the PBL, say), the mixed layer (the middle 35% to 70%) and the entrainment zone.

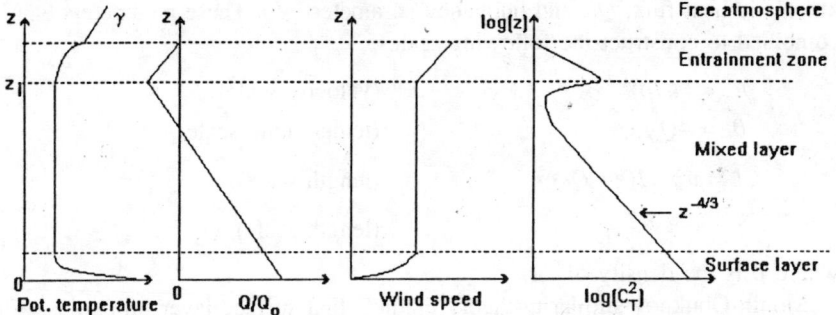

Fig. 1. Typical layers and profiles within the convective boundary layer.

Figure 2, patterned after diagrams that have been compiled by many authors [23, 24, 25], shows the scaling regions and the scaling variables in the unstable boundary layer. The axes are defined using the stability parameter z_i/L and the dimensionless height z/z_i. Following the suggestion by Olesen et al, [24], the dividing lines are of two types, the solid line and the dashed line. The dashed line is used to denote that a division is not commonly accepted in the literature.

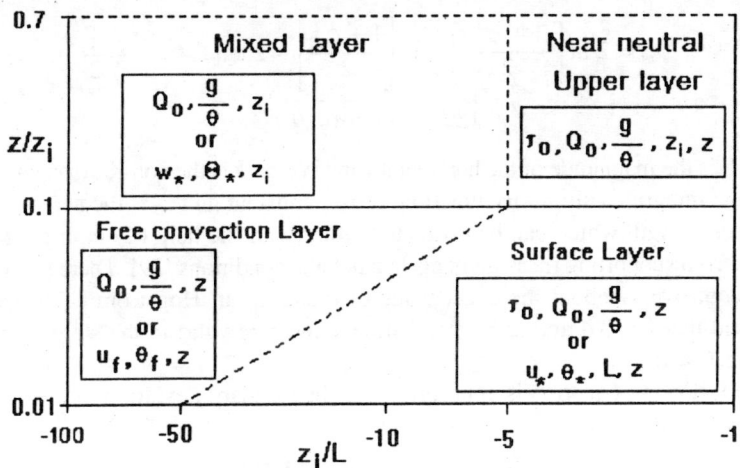

Fig. 2. The scaling regions and the scaling variables in the unstable PBL.

The similarity relationsphips listed below are strictly valid in stationary and horizontally homogeneous conditions.

Surface layer relationships

Monin-Obukhov similarity: In the surface layer, where the turbulent fluxes of heat and momentum are considered to be approximately constant, Monin-Obukhov similarity theory provides a simple conceptual framework for describing its structure. The governing parameters are height above surface, z, surface shear stress, τ_0, surface kinematic heat flux, Q_0, and buoyancy parameter, g/θ. These parameters can be combined to construct the following scales

$u_* = (\tau_0/\rho)^{1/2}$ (velocity scale),

$\theta_* = -Q_0/u_*$ (temperature scale),

$L = u_*^3\, \theta/(g\, k\, Q_0)$ (length scale),

z (length scale),

where ρ is the density of air.

Monin-Obukhov similarity theory predicts that surface layer properties, after non-dimensionalization with the above scales, are universal functions of z/L. The universal functions are determined empirically, by fitting curves to experimental data. Within the surface layer and sufficiently far from the surface, the following relationships have been proposed which relate the surface fluxes and the mean profiles [26, 27]

$$U(z) = \frac{u_*}{k}\left[\ln\left(\frac{z}{z_0}\right) - \psi_m\left(\frac{z}{L}\right)\right] \tag{2}$$

$$\psi_m = 2\ln\left[\frac{(1+\phi_m)}{2}\right] + \ln\left[\frac{1+\phi_m^2}{2}\right] - 2\tan^{-1}(\phi_m) + \frac{\pi}{2} \tag{3}$$

$$\phi_m(z/L) = (1 - \alpha_f z/L)^{1/4} \tag{4}$$

where U is the magnitude of the horizontal wind vector, k is the von Karman constant, ψ_m is the integrated similarity function, α_f is a constant and z_0 is the aerodynamic roughness length which can be estimated either with the help of empirical tables [28] or from supporting measurements in adiabatic conditions [29]. There still exists some controversy about the exact value of k and a_f but Hogstrom [27] recently proposed that $k = 0.4$ and $\alpha_f = 19.3$. These values are valid in the stability range $-1 \leq z/L < 0$.

The following equation is appropriate for the standard deviation of the vertical velocity

$$\frac{\sigma_w}{u_*} = 1.9\left(-\frac{z}{L}\right)^{1/3} \tag{5}$$

The variances of the horizontal wind components do not observe Monin-Obukhov similarity law.

Local Free convection relationships: Under light or calm winds and strong insolation the momentum flux is no longer an important governing parameter and the following scales are appropriate:

$u_f = [Q_0 z(g/\theta)]^{1/3}$ (velocity scale),
$\theta_f = Q_0/u_f$ (temperature scale),
z (length scale).

According to local free convection predictions, dimensionless groups formed with the above scales should be constants. The following similarity relationships, relevant to the present study, have been proposed

$$C_T^2 = 2.7 \left(\frac{\theta}{g}\right)^{2/3} Q_0^{4/3} z^{-4/3} \tag{6}$$

$$\sigma_w^2 = 1.4 \left(z \frac{g}{\theta} Q\right)^{2/3} \tag{7}$$

Mixed-layer relationships

The mixed layer is the middle portion of the convective PBL extending to z_i, where z_i is the height to the first inversion base. In spite of this, the similarity relationships listed below are valid approximately up to $0.7\, z_i$. Above this height, entrainment processes upset the validity of the mixed layer similarity relationships. An extension of mixed layer similarity to include entrainment effects is presented in [30]. The governing parameters in the mixed layer are Q_0, z_i and g/θ. The relevant scales in the mixed layer include the following:

$w_* = [z_i Q_0 g/\theta)]^{1/3}$ (velocity scale),
$\Theta_* = Q_0/w_*$ (temperaure scale),
z_i (length scale).

The mixed layer parametric relationships that have been proposed include the following [5]

$$\frac{z_i}{w_*} \frac{\overline{\partial u}}{\partial z} = 0 \tag{8}$$

$$\frac{\sigma_w^2}{w_*^2} = 1.8 \left(\frac{z}{z_i}\right)^{2/3} \left(1 - 0.8 \cdot \frac{z}{z_i}\right)^2 \tag{9}$$

$$\frac{C_T^2 z_i^{2/3}}{\Theta_*^2} = 2.7 \left(\frac{z}{z_i}\right)^{-4/3} \tag{10}$$

$$\frac{Q}{Q_0} = 1 - (1 - \alpha) \frac{z}{z_i} \tag{11}$$

where u is the wind component in the direction of the surface layer wind and α is the ratio of the buoyancy flux at z_i to the surface flux. The value of α typically varies between 0 to 0.4, with $\alpha = 0.2$ being a good average [5]. Alternatively, α can be calculated using a combination of sodar reflectivity data and in situ meteorological measurements [31]. This method is applicable in the morning hours when the PBL is developing rapidly and values of α may deviate considerably from the prescribed value 0.2.

Integral models

Integral methods enjoy popularity in modeling the CBL where they are known to provide realistic and stable estimates of averaged properties. Slab models are based on the assumption that mean variables are uniformly distributed in the CBL and that the surface layer and the entrainment zone have a small thickness in comparison to z_i. Under horizontally homogeneous conditions this idealization leads to the following differential equation for z_i [32].

$$\frac{dz_i}{dt} = \frac{\left(\frac{Q_0}{\gamma}\right)}{\frac{z_i^2}{(1+2A)z_i - 2BkL} + \frac{Cu_*^2 \theta}{\gamma g[(1+A)z_i - BkL]}} \qquad (12)$$

where γ is the potential temperature gradient above the PBL and A, B and C are constants approximately equal to 0.2, 5.0 and 8 respectively. It is worth mentioning that according to (12), both mechanical and convective turbulence contribute to the growth of the boundary layer.

Horizontal inhomogeneity occurs in most real life applications, a typical example being the coastal area. Under conditions where horizontal advection is the dominating forcing, the following expression is appropriate [33]

$$z_i = (1 - 0.002L)\left(\frac{2.8}{u}\frac{Q_0}{\gamma}x\right)^{1/2} + 11\ln(-L) \qquad (13)$$

Similarly, analytical expressions for other properties of the PBL are derived using some additional assumptions. Sen Gupta et al, [34] integrated the equation of conservation of sensible heat from 0 to 0.7 z_i in order to obtain a simple expression for calculation of Q_0. They obtained the following expression, for the case when advection effects are negligible:

$$Q_0 = \frac{\partial \theta}{\partial t} 0.7 z_i \qquad (14)$$

Garratt et al, [35] utilized Equation (2) together with the assumption that the wind profiles in the mixed layer are uniformly distributed (Equation, 8) and derived the following relationship:

$$k\frac{\langle u \rangle}{u_*} = \ln\left(\frac{z_i}{z_0}\right) - 0.5\ln\left(\frac{z_i}{|L|}\right) - 2.3 \qquad (15)$$

where the brackets denote layer-averaged values.

Integration of (9) from $0.1z_i$ up to the maximum sodar height, h_s, yields

$$\frac{\langle \sigma_w^2 \rangle}{w_*^2} = b \tag{16}$$

The values of b as a function of h_s/z_i are found in [36]. A good average value is $b = 0.45$.

Similarity Methods

Sensible heat flux

The C_T^2 method: Coulter and Wesely [37] presented a method for determining Q_0 from C_T^2 measurements inferred from sodar. Similar methods have been utilized by other investigators [1, 10, 38, 39]. After some rearrangement, (10) leads to

$$Q_0 = 0.48 \left(\frac{g}{\theta}\right)^{1/2} (C_T^2)^{3/4} z \tag{17}$$

Equation (17) is valid in the surface layer and in the lower and middle part of the mixed layer. The underlying hypothesis in the derivation of (17) is that the height dependence of C_T^2 approaches the free convection prediction which is often referred to as the "$z^{-4/3}$" law. In order to check the validity of the underlying assumption, it is suggested that $\log (C_T^2)$ should be plotted as a function of $\log (z)$ and the profiles that show a slope different from $-4/3$ should be rejected [37]. When applying the aforementioned procedure, it was found that the estimated surface heat fluxes compare fairly well with direct measurements except for the morning hours when (17) seriously overestimates Q_0 [10, 37]. Deviations from the "$z^{-4/3}$" law may be caused by wind shear, measurement errors (e.g. due to excess attenuation [20] or anisotropy in the temperature field [40]) and/or atmospheric conditions (when the convective requirements are not fulfilled). When chosing only data which fulfill the convective requirements ($-z_i/L > 4.5$), Melas [1] found that (17) did not perform well when compared against direct measurements. This was attributed to measurement errors.

The σ_w^2 method: It has been pointed out by many investigators that Q_0 can be alternatively estimated from sodar measurements of σ_w^2 [1, 9, 38, 39]. Short range minisodars have a good resolution within the surface layer and Monin-Obukhov similarity theory can be applied directly to extract information from sodar measurements. After some rearrangement, (5) yields.

$$Q_0 = 1.9^{-3} k^{-1} \left(\frac{g}{\theta}\right)^{-1} \frac{\sigma_w^3}{z} \tag{18}$$

As it was mentioned before, Equation (18) is valid in the surface layer.

The lowest measuring level of the acoustic sounders is usually near the top of the surface layer (35–50 m) and it is necessary to employ relationships that are valid in the mixed layer. For this purpose either (7) or (9) can be utilized. Melas [13] shows that the results of (7) and (9) are only slightly different and concluded that either of them can be used for the estimation of Q_0. The combination of (7) and (11) yields [10, 38]

$$Q_0 = 1.4^{-3/2} \left(\frac{g}{\theta}\right)^{(-1)} \frac{\sigma_w^3}{z} \left[1 - (1 + \alpha)\frac{z}{z_i}\right]^{-1} \qquad (19)$$

Applying the minimum least square method, Q_0 and z_i can be derived using the sodar measurements of σ_w^2 at the different heights.

A different expression can be derived by rearranging (16) [36]

$$Q_0 = b^{-3/2} \left(\frac{g}{\theta}\right)^{(-1)} \frac{\langle \sigma_w^2 \rangle^{3/2}}{z_i} \qquad (20)$$

The indirect technique based upon (20) has a disadvantage compared to the ones based upon (17) or (19), namely that it requires knowledge of z_i. This technique can only be used when z_i is within the sodar range, in practice mainly during morning hours, or when z_i is known from some other source. Alternatively, z_i can be estimated using a parametric relationship such as (12) or (13) [36].

The usefulness of (19) or (20) in estimating Q_0 from sodar measurements was tested by comparison to direct measurements [1, 10, 36, 38, 39]. It was found [1, 36, 38, 39] that (20) gives reasonable results under convective conditions ($-z_i/L >$ 4.5). During cases characterized by weak instability, (20) seriously overestimates the observed surface heat flux. On the other hand, a systematic investigation performed in Germany [10], revealed that sodar estimates of Q_0 based upon (19) show only a fair agreement with sonic measurements. These results are rather controversial, and indicate that further research is needed in order to produce definitive conclusions.

Finally, Q_0 can be estimated from (14) using some plausible assumption about $d\theta/dt$. Sen Gupta et al [34] assumed that $d\theta/dt$ is independent of height. The surface heat flux was then calculated by using a sodar system monitoring z_i and a thermal probe measuring the time variation of temperature. The same authors estimated that when taking into account the errors of estimating z_i and θ the above method have uncertainties ranging from 15% to 20%. The most serious restriction of this method is probably the requirement of horizontal homogeneity which is more stringent than in other methods. It can be shown that even several kilometers from a thermal discontinuity (e.g. a shoreline) the advection term in the enthalpy equation is comparable in magnitude or even larger than the local tendency term. Moreover, if the thermal probe is placed near the surface it is very accessible to local effects which might result in large errors.

Friction velocity

Unlike the heat flux, the shear stresses can be calculated directly from sodar measurements. Different techniques have been proposed for the determination of the friction velocity including direct estimates using a spectral method [11]. It is however questionable whether these estimates are accurate enough to be used for most purposes. We therefore summarize some alternative methods to calculate u_* from sodar measurements.

When the first sodar level is in the lower part of the surface layer, it is feasible to use surface layer similarity relationships. After some rearrangement, (2) yields [41]

$$u_* = Uk \left[\ln\left(\frac{z}{z_0}\right) - \psi_m\left(\frac{z}{L}\right) \right]^{-1} \quad (21)$$

It should be mentioned that (21) involves, through L, the surface heat flux. It is therefore not possible to use it as a stand alone relationship but it is necessary to combine it with an additional relationship for Q_0 (e.g. (18)).

However, most commercial acoustic sounders operate in the mixed layer and it is preferable to utilize bulk relationships, such as (15). After some rearrangement it is obtained [42]

$$u_* = k\langle u \rangle \left[\ln\left(\frac{z_i}{z_0}\right) - \frac{1}{2}\ln\left(\frac{z_i}{|L|}\right) - 2.3 \right]^{-1} \quad (22)$$

It should be mentioned that (22) involves L and z_i which must be estimated from some other source (supporting measurements or other parametric relations).

The above relationship was evaluated using observed values of z_i and values of Q_0 obtained through (20) [42]. The resulting estimates of u_* compared very well with direct measurements with no bias and a correlation coefficient $r = 0.91$. The same relationship was also evaluated using values of z_i and Q_0 estimated using (20) and (12) respectively [36]. The comparison of sodar estimates of u_* obtained through (22) with direct measurements of u_* revealed a very good agreement. The success of (22) in extracting u_* from sodar measurements of the mean wind speed and direction was attributed to the high accuracy of these sodar measurements [36]. Moreover, Melas and Kambezidis [33] found that for convective conditions, u_* is expected to show a rather weak stability dependence and is mainly influenced by $\langle u \rangle$ and z_0.

Mixed-layer depth

As it was mentioned previously, z_i can be estimated from backscatter profiles, provided that the inversion base is within the sodar range. The limited extend of sodar range, necessitates however the complement of these direct measurements with some alternative method of estimating z_i.

A rough estimate of z_i can be obtained from the σ_w^2 profile. According to (9),

σ_w^2 has a maximum at $z_m \approx 0.35z_i$. Measurements in the convective PBL, show that z_m occurs in the range $0.3z_i - 0.6z_i$ [43]. A good compromise is therefore

$$z_i \approx 2.8 z_m \qquad (23)$$

Considering the experimental range of z_m we can estimate that the uncertainties in z_i estimated from (23) are approximately 40%.

Singal et al [44] suggested a methodology to determine mixing height when the convective boundary layer grows beyond the sodar range. Mixing heights were estimated on the basis of Holzworth method and compared with the corresponding sodar measured depth of the thermal plumes and the following empirical relation was derived:

$$h = 4.24 \, z_s + 95 \qquad (24)$$

where h is the mixing height as per the Holzworth model and z_s is the depth of the sodar measured thermal plumes.

Weill et al [38] made an attempt to estimate the altitude of the convective PBL (not to be mixed up with z_i) by extrapolating the σ_w^3/z to the height where Q_0 vanishes. This level was found to be close to the inversion height. This result was however contradicted by Best et al, [45].

Enger [39] and Melas [1] proposed that (17) and (20) can be combined to yield

$$z_i = \left(\frac{2.7}{b^2}\right)^{3/4} \left(\frac{g}{\theta}\right)^{-3/2} (C_T^2)^{-3/4} \langle \sigma_w^2 \rangle^{3/2} z^{-1} \qquad (25)$$

Equation (25) is valid in the lower and middle part of the mixed-layer. Estimates of z_i based on (25) were compared with simultaneous direct measurements [1, 22, 39, 46] but the agreement between them was only fair.

It has been proposed that (12) (or (13) in nonhomogeneous conditions) can be used to estimate the development of the mixed layer, provided that Q_0 and u_* are known from some other sources. Melas and Kambezidis [36] constructed an equation set consisting of (20), (22) and (12) which requires measurements of σ_w^2, u and γ to estimate Q_0, u_* and z_i. The initial values of z_i, required for the numerical solution of (12) are easily obtained from backscatter profiles in the early morning hours. It is a far more delicate task to estimate γ which varies both in space and time. Rough estimates of γ can however be obtained from the morning radiosonde profiles. Alternatively, γ can be adapted to sodar observations in the morning hours when the mixed layer top is within the sodar range and γ is the only unknown in (12). If measurements of γ are not available, Batchvarova and Gryning [47] recommended to use $\gamma = 0.05$ (K/m) when the mixed layer is lower then 150 m, and $\gamma = 0.005$ (K/m) for mixed layer heights larger than 150 m.

It should be mentioned that there also exist an analytical solution to (12) [48]. Comparison of z_i values calculated through (12) and direct measurements of z_i

revealed that (12) is capable to produce reasonable results under typical summer conditions [22, 36] but it fails to predict z_i under complicated meteorological conditions.

Figure 3 shows the diurnal variation of mixed layer depth, friction velocity, and surface heat flux during an experimental day from the Oresund experiment [49]. The observations are shown by the solid line while the circles represent the estimates based on equation set (20), (22) and (12).

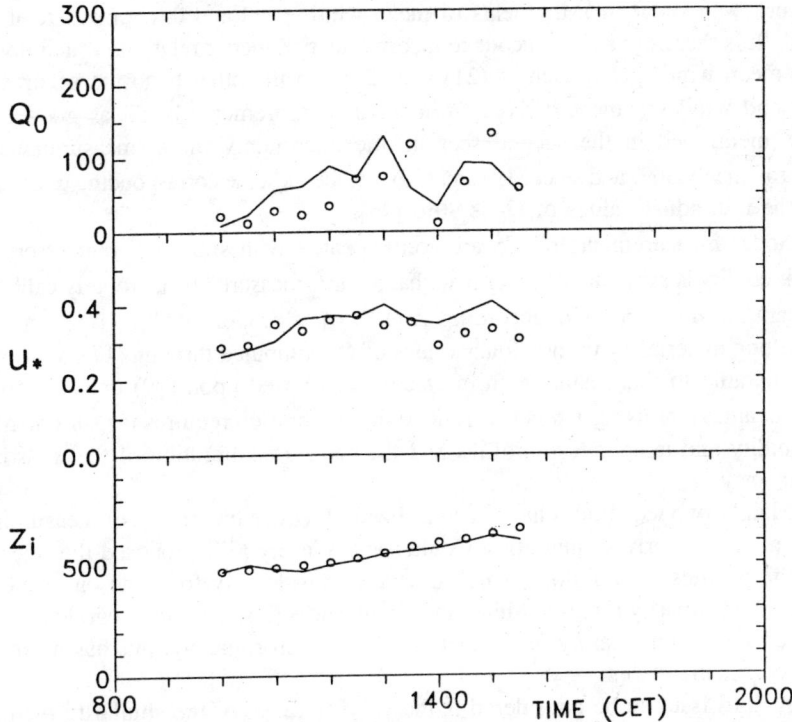

Fig. 3. The diurnal variation of the mixed layer depth, z_i (m), the friction velocity, u^* (ms^{-1}), and the surface heat flux, Q_0 (Wm^{-2}). Solid line: observations. Circles: sodar estimates based on equation set (20), (22) and (12). (From Melas and Kambezidis [36]).

Further Discussion and Conclusions

In the present paper, an attempt was made to review many of the existing similarity methods which enable the extraction of turbulence properties and mixed layer depth from sodar measurements.

A critical assessment of the existing methods should give consideration to several issues. The most obvious one is the accuracy of the underlying similarity relationships. Rather surprisingly, this is the easiest task to tackle. Most similarity methods reviewed in the present paper incorporate well established similarity relationships of the convective PBL. A major improvement can be achieved by using short range, high frequency minisodars which have a good resolution within the surface layer. In this case it is possible to use Monin-Obukhov similarity theory which is now widely accepted. In any case, similarity relations in the convective PBL are not known better than within 15–20% [9] and the accuracy of our indirect estimates can not be better than this.

A far more serious source of errors is associated with the accuracy of sodar measurements of the implied parameters. As it was mentioned previously, the accuracy in sodar measurements of mean wind speed and direction is relatively high. It is therefore advantageous to incorporate parametric relations which involve the mean wind speed, such as (21) and (22). On the other hand the accuracy of C_T^2 and wind variances derived from sodar measurements is not as good. As it was mentioned in the second section, the uncertainty in σ_w measurements is approximaely 30% and since (19) and (20) include σ_w^3, the corresponding uncertainty in the individual values of Q_0 is 90% [46].

Sodar measurements of C_T^2 are contaminated with some inherent errors and their quality is appreciably lower than that for σ_w^2 measured by a properly calibrated sodar. On the other hand, according to (17), Q_0 is proportional to $(C_T^2)^{3/4}$ and the resulting uncertainty in individual values of Q_0 computed through (17) is probably comparable to that resulting from calculations based upon (19) or (20) [46]. A disadvantage of using C_T^2 is that reflectivity calibration requires measurements of humidity and temperature profiles and the acoustic calibration of the transducer efficiency.

Mainly two techniques have been utilized to reduce the impact of measurement errors on the derived quantities. Coulter and Wesely [37] proposed the rejection of the profiles with a slope which deviates considerably from the one expected from the similarity relation. Melas and Kambezidis [36] used an integral technique to reduce the uncertainty in indirect estimates. Unfortunately, this has no impact on systematic errors.

A third issue to be considered is the validity range of the similarity methods. Strictly speaking, these methods are valid in the convective, stationary and horizontally homogeneous PBL. Although there still exists some controversy, there is some evidence [1, 25, 50, 51] that a value of $-z_i/L \approx 5$ may be sufficient to drive the PBL in convective state. Under typical midday conditions ($z_i = 1000$ m and $u_* = 0.4$ m/s), this corresponds to a surface heat flux as low as 25 W m^{-2} Thus convective conditions are expected quite often in the real PBL.

The requirement for the mixed layer scaling to hold is that, despite externally forced temporal and/or spatial changes, the turbulence is in local equilibrium. Wyngaard [50] introduced suitable time and space scales. When the convective time scale is short compared to the characteristic scales for temporal and spatial changes, then the mixed layer is in a state of local equilibrium. This is less restrictive than the requirement for stationarity and horizontal homogeneity. A more detailed discussion about the concept of local equilibrium is found elsewhere [50]. For the purpose of the present paper, it is sufficient to mention that under typical day-time conditions, time changes of mixed layer properties are slow compared to the convective time scale, except for the morning hours when the PBL is deepening rapidly by entrainment. Horizontal homogeneity is probably a more severe limitation on the validity of our equation set. Horizontal inhomogeneities occur in a wide variety of combinations but, probably, the best known example is found in the coastal urban area where the land and the sea surfaces have quite different properties in terms of temperature, roughness and wetness. Melas and Kambezidis [33] presented some calculations of the fetch downwind the shoreline required for the mixed layer to be in local equilibrium. Under light wind conditions, turbulent mixing typically dominates over horizontal advection and the mixed-layer is in local equilibrium just a couple of kilometers from the shoreline. Under conditions characterized by strong sea winds ($u > 8$ ms^{-1}), the required fetch downwind the shoreline is of the order of 10 km or more. For a somewhat more detailed discussion about the conditions under which surface inhomogeneities do not upset the validity of the parametric relations, the reader is referred to other sources [33, 36].

From the above discussion it becomes apparent that even though similarity methods are valuable tools they also have some limitations. The wide interval of the problems mentioned above makes it very difficult to recommend or reject some of the methods presented in this study. Instead it is suggested that the existing methods should be validated against direct turbulence measurements. The quality, that is the fitness-for-purpose, of these methods is expected to differ between different sodar systems, different applications and even between different system operators. This last issue is related to the operational value of the existing methods. Some of the methods are user sensitive which makes them less suitable for operational use. Since sodar measurements represent volume averages, the derived quantities are not epected to reflect local inhomogeneities in a scale of a few hundred meters or less [11]. On the other hand, turbulence instruments mounted on a tower are very sensitive to local inhomogeneities and a point-to-point comparison might lead to misleading conclusions. It is therefore necessary to perform a statistical validation against a large data set.

Unfortunately, not all similarity methods are validated against direct measurements. In addition, there has often been much reliance on a rather small data bank for validation. This might lead to an over-confidence in the ability of similarity methods to provide accurate estimations of turbulent quantities.

References

1. D. Melas, "Sodar estimates of surface heat flux and mixed layer depth compared with direct measurements', Atm. Environment, Vol **24A**, No. 11, 2847–2853, 1990.

2. Deardorff J.W. "Preliminary results from numerical integrations of the unstable boundary layer", J. Atmos. Sci. **27**, 1209–1211, 1970.
3. Briggs G.A. "Analytical parametrizations of diffusion: The convective boundary layer", J. Climate Appl. Meteorol. **24**, 1167–1186, 1985.
4. Briggs G.A. "Surface inhomogeneity effects on convective diffusion", Boundary-Layer Meteorol. **45**, 117–135, 1988.
5. R.B. Stull, "An Introduction to Boundary Layer Meteorology" (Atmospheric Sciences Library, Kluwer Academic Publishers, Dordrecht, The Netherlands) 666 pp, 1988.
6. S.P. Singal, "The use of an acoustic sounder in air quality studies", J. Scient. Ind. Res. **47**, 520–533, 1988.
7. S.P. Singal, "Monitoring air pollution related meteorology using SODAR: State of the art", Appl. Phys. **B 57**, 65–82, 1993.
8. W.D. Neff and R.L. Coulter, "Acoustic remote sensing", in Probing the Atmospheric Boundary Layer, (D.H. Lenschow Ed., American Meteorological Society, Boston) 201–239, 1986.
9. J.C. Wyngaard, "Measurement Physics", in Probing the Atmospheric Boundary Layer, (D.H. Lenschow Ed., American Meteorological Society, Boston) 5–18, 1986.
10. S. Vogt and P. Thomas, "Estimation of the sensible heat flux and the temperature structure parameter by sodar and sonic anemometer: an intercomparison", Intern. J. of Remote Sensing **15**, 507–516, 1994.
11. A. Weill and H.R. Lehmann, "Twenty years of acoustic sounding—a review and some applications". Z. Meteorol. **40**, 241–250, 1990.
12. W.D. Neff, "Remote sensing of atmospheric processes over complex terrain", in Atmospheric Processes Over Complex Terrain, edited by W. Blumen (Boston, M.A.: American Meteorological Society), pp. 173–228, 1990.
13. D. Melas, "Similarity methods to derive turbulence quantities and mixed layer depth from sodar measurements in the convective boundary layer: A review", Appl. Phys. **B 57**, 11–17, 1993.
14. J.E. Gaynor, "Accuracy of sodar wind variance measurements", Intern. J. of Remote Sensing **15**, 313–324, 1994.
15. R.L. Schwiesow, "A comparative overview of active remote-sensing techniques", in Probing the Atmospheric Boundary Layer, (D.H. Lenschow Ed., American Meteorological Society, Boston, 1986) 129–138, 1986.
16. P.L. Finkelstein, J.C. Kaimal, J.E. Gaynor, M.E. Graves and T.J. Lockhart, "Comparison of wind monitoring systems. Part II: Doppler sodars", Atmos. Oceanic Technol. 3, 594–604, 1986.
17. J. Keder, TH. Foken, W. Gerstmann and V. Schindler, "Measurements of wind parameters and heat flux with the Sensitron Doppler sodar", Boundary-Layer Meteorol. **46**, 195–204, 1989.
18. D.N. Asimakopoulos, T.J. Moulsley, C.G. Helmis, D.P. Lalas and J. Gaynor, "Quantitative low-level acoustic sounding and comparison with direct measurements", Boundary-Layer Meteorol. **27**, 1–26, 1983.
19. W.D. Neff, "Mesoscale air quality studies with meteorological remote sensing systems", Intern. J. of Remote Sensing **15**, 393–426, 1994.
20. D.A. Haugen and J.C. Kaimal, "Measuring temperature structure parameter profiles with an acoustic sounder", J. Appl. Meteorol. **17**, 895–899, 1978.
21. J.C. Kaimal, N.L. Abshire, R.B. Chadwick, M.T. Decker, W.H. Hooke, R.A. Kropfli,

W.D. Neff, F. Pasqualucci and P.H. Hildebrand, "Estimating the depth of the daytime convective boundary layer", J. Appl. Meteorol. **21**, 1123–1129, 1982.
22. F. Beyrich, "On the use of SODAR data to estimate mixing height", Appl. Phys. B **57**, 27–35, 1993.
23. S. Nichols and C.J. Readings, "Aircraft observations of the structure of the lower boundary layer over the sea, Quart. J.R. Meteorol. Soc. **105**, 785–802, 1979.
24. H.R. Olesen, S.E. Larsen and J. Hojstrup, "Modeling velocity spectra in the lower part of the planetary boundary layer", Boundary-Layer Meteorol. **29**, 285–312, 1984.
25. A.A.M. Holtslag and F.T.M. Nieuwstadt, "Scaling the atmospheric boundary layer", Boundary-Layer Meteorol. **36**, 201–209, 1986.
26. C.A. Paulson, "The mathematical representation of wind speed and temperature in the unstable atmospheric surface layer", J. Appl. Meteorol. **9**, 857–861, 1970.
27. U. Hogstrom, "Non-dimensional wind and temperature profiles in the atmospheric surface layer: Reevaluated", Boundary-Layer Meteorol. **42**, 55–78, 1988.
28. J. Wieringa, "Representative roughness parameters for homogeneous terrain", Boundary-Layer Meteorol., **63**, 323–363, 1992.
29. R.J. Barthelmie, J.P. Palutikof and T.D. Davies, "Estimation of sector roughness lengths and the effect on prediction of the vertical wind speed profile", Boundary-Layer Meteorol. **66**, 19–47, 1993.
30. Z. Sorbjan, Coulter R.L., M.L. Wesely, "Similarity scaling applied to sodar observations of the convective boundary layer above an irregular hill", Boundary-Layer Meteorol. **56**, 33–50, 1991.
31. G. Dubosclard, "A comparison between observed and predicted values for the entrainment in planetary boundary layer", Boundary-Layer Meteorol. **18**, 473–483, 1980.
32. E. Batchvarova and S.E. Gryning, "Applied model for the Growth of the Daytime Mixed Layer", Boundary-Layer Meteorol. **56**, 261–274, 1991.
33. D. Melas and H.D. Kambezidis, "The depth of the internal boundary layer over an urban area under sea-breeze conditions", Boundary-Layer Meteorol. **61**, 247–264, 1992.
34. P. Sen Gupta, P.K. Kunnikrishnan, V. Radhika and K.N. Nair, "Estimating surface sensible heat flux and surface measurements in the evolving boundary layer", Atmos. Res. **20**, 119–123, 1986.
35. J.R. Garratt, J.C. Wyngaard, R.J. Francey, "Winds in the atmospheric boundary layer" J. Atmos. Sci. **39**, 1307–1316, 1982.
36. D. Melas and H.D. Kambezidis, "A similarity method to derive turbulence parameters and mixed-layer depth from sodar measurements", Intern. J. of Remote Sensing **15**, 499–505, 1994.
37. R.L. Coulter and M.L. Wesely, "Estimates of surface heat flux from sodar and laser scintillations", J. Appl. Meteorol. **19**, 1209–1222, 1980.
38. A. Weill, C. Klapisz, B. Strauss, F. Baudin, C. Jaupart, P. Van Grunderbeeck and J.P. Goutorbe, "Measuring heat flux and structure functions of temperature fluctuations", J. Appl. Meteorol. **19**, 199–205, 1980.
39. L. Enger, "Simulation of dispersion in complex terrain. Part C. A dispersion model for operational use", Atmospheric Environment. vol. **24A**, No. 9, 2457–2471, 1990.
40. T.J. Moulsley, D.N. Asimakopoulos, R.S. Cole, B.A. Crease and S.J. Caughey, "Measurement of boundary layer structure parameter profiles by acoustic sounding and comparison with direct measuremens", Quart. J.R. Met. Soc. **107**, 203–230, 1981.

41. A. Weill, C. Klapisz and F. Baudin, "The CRPE minisodar: applications in micrometeorology and in physics of precipitation", Atmos. Res. **20**, 317–333, 1986.
42. D. Melas, "Using a simple resistance law to estimate friction velocity from sodar measurements", Boundary-Layer Meteorol. **57**, 275–287, 1991.
43. S.J. Caughey, "Observed characteristics of the atmospheric boundary layer". In Atmospheric Turbulence and Air Pollution Modelling (F.T.M. Nieuwstadt and H. Van Dop, Eds), D. Reidel, Dordrecht, The Netherlands, 107–158, 1982.
44. S.P. Singal, B.S. Gera and S.K. Aggarwal, "Nowcasting by acoustic remote sensing: experiences with the systems established at the National Physical Laboratory, New Delhi, J. Scient. Ind. Res. **43**, 469–488, 1984.
45. P.R. Best, J. Ewald, M. Kanowski, "The estimation of pollutant dispersal from Queensland power station". In Proc. of the 7th International Clean Air Conference, (Clean Air Society of Australia and New Zealand, Adelaide, Australia) edited by K.A. Webb and A.I. Smith, August 24–28, 1981, 429–448.
46. F. Beyrich, "Sodar estimates of surface heat flux and mixed layer depth compared with direct measurements: Discussion", Atm. Environment **26A**, No. 13, 2459–2461, 1992.
47. E. Batchvarova and S.E. Gryning, "Applied Model for the height of the Daytime Mixed Layer and the entrainment zone", submitted to Boundary-Layer Meteorol., **56**, 261–274, 1990.
48. D.J. Thomson, "An Analytical solution of Tennekes' equations for the growth of boundary-layer depth", Boundary-Layer Meteor. **59**, 227–229, 1992.
49. S.E. Gryning, "The Oresund experiment—A Nordic mesoscale dispersion experiment over a land-water-land area", Bulletin of the American Meteorological Society **16**, 1403–1407, 1985.
50. J.C. Wyngaard, "Lectures on the planetary boundary layer". In Mesoscale Meteorology: Theories, Observations, and Models (D.K. Lilly and T. Gal-Chen eds.), D. Reidel, Dordrecht, The Netherlands, 1983.
51. J.W. Deardorff, "Numerical investigation of neutral and unstable planetary boundary layer", J. Atmos. Sci. **29**, 91–115, 1972.

Acoustic Remote Sensing Applications
S.P. Singal (Ed)
Copyright © 1997 Narosa Publishing House, New Delhi, India

16. Sodar: A Tool to Characterize Hazardous Situations in Air Pollution and Communication

S.P. Singal, B.S. Gera and Neeraj Saxena
National Physical Laboratory, New Delhi-110012, India

Introduction

The operation of large scale industrial plants in an urban and industrial area is accompanied with the fallout of excessive smoke, toxic gases and the particulates, the so called air pollutants. Normally the mobility of the atmosphere (self purification process) is high enough to disperse these air pollutants rapidly. However, the presence of calm conditions often found associated with anticyclonic conditions, like in the coastal areas or in the valleys, can lead many a times to serious air pollution situations.

To monitor the air quality at a place, normally ground measurements of the pollutants are made at a number of sites in and around the place of interest. However, many a times due to poor diffusion and transport problems in the atmosphere, it so happens that the concentration of the effluent increases disproportionately. Further, from the ground level measurements it may not be possible to distinguish the contribution of industrial plants from other sources of urban pollution located in the same area specially when the emission is from elevated sources.

Similar to air quality problems, quality of transmission of decimetric and centimetric waves on line of sight and transhorizon paths, has been seen to depend on the structure of the lower atmosphere (atmospheric boundary layer) consisting of the different degrees of stratification and the prevailing local meteorological conditions. It is because, firstly, the radio refractive nature of atmosphere is determined by the gradient of refractivity at a place, a parameter governed by stratification and stability of the atmosphere. The atmosphere can be subrefractive (less than $-40N$), normaly refractive ($-40N$ to $-80N$), superrefractive ($-80N$ to $-157N$) and ducting (equal to or less than $-157N$). Secondly, the presence of a layering in the stable atmosphere acts as a transition zone and suppresses vertical transport of the moist air. In terms of the radio wave propagation, thin transition zones act as ducts, broad transition zones reflect, while very broad transition zones develop only random disturbances. Temperature inversions are generally associated with ducts but all observed temperature inversions do not act as ducts [1]. The ducts are responsible for anomalous propagation, extended range of the transmitter and signal enhancement, especially when the transmitter and receiver both lie within the duct.

The layer reflection or the volume scattering properties of this stable layer depend on its thickness, the scattering angle and the wavelength. If the vertically sampled wave scale for a troposcatter system is larger than the vertical outer scale of turbulence i.e. $\lambda/(2 \sin \theta) > L_v$ where λ is the wavelength, θ is the scattering angle and L_v is the vertical outer scale, then the layer will produce partial (specular) reflection, otherwise the layer is a thin volume of turbulence that produces scattering.

As per above, it is essential to make site specific measurements of the stratification structure and other meteorological parameters in the atmospheric boundary layer. In this article we shall discuss the potential of the acoustic remote sensing technique used as a tool to probe the lower atmosphere for application in hazardous situations of air pollution and microwave propagation. A brief review of the atmospheric boundary layer (ABL) is also given to appreciate the potential of the acoustic remote sensing technique.

Atmospheric Boundary Layer

Atmospheric boundary layer (ABL) is one of the most important constituent zones of the atmosphere, since it plays a crucial role in major areas of human endeavour in air pollution, aviation and communication in various ways. It can also be considered as a circulatory system of the biosphere. It is in the boundary layer the carbon dioxide and oxygen are transported to plants and animals for photosynthesis and respiration. Photochemical reactions take place in this layer which help to remove waste products and thus cleanse the atmosphere. Heat and moisture collected at the surface of the earth disperse in this layer both horizontally and vertically which effectively air condition the biosphere and provide a conduit for propagation of weather systems on all scales.

Considering technically, ABL is a part of the lower atmosphere which is influenced by friction and heating at the surface of the earth i.e. it is a layer of the atmosphere wherein there is a direct turbulent transfer of heat and momentum (Fig. 1). It extends from the ground level to the lowest few hundred meters of the atmosphere. The main characteristic of this layer is turbulence. The diurnal heating and cooling of the ground along with the terrain roughness, wind speed, wind shear and buoyancy determine the extent of this characteristic. During the day time due to the solar heating of the surface of the earth, unstable lapse (convective) conditions develop, while during night time due to the cooling of the surface of the earth, stable conditions develop.

The various phenomena, processes and constituent parts of the ABL are as follows:

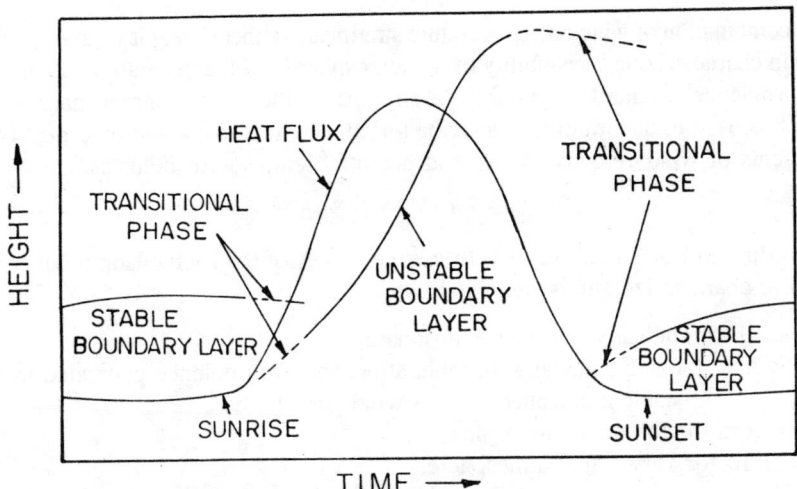

Fig. 1. Plot showing the evolution and height of the atmospheric boundary layer (both stable and unstable) in relation to sensible heat flux.

(i) Turbulence

Turbulence may be regarded as a complex assembly of locally organized but unsteady velocity patterns which interact strongly with each other as they move with the flow. Thus turbulence is a random, three dimensional state of motion which is characterized by high degree of chaotic vorticity. Unstable and stable flows tend to remain in the same state unless there is a imbalance between the paired destablising and stablising forces acting on the flow. If the net effect of destablising factor is more than the net effect of stablising factor, then turbulence will occur. It is governed mainly by the non-linear terms of the equation of motion with strong interaction between motion on different spatial and temporal scales. A linearized wave description of motion and consideration of the dynamics solely in terms of independent Fourier components is impossible from the equation of motion. Only statistical means/averages correlated in time and space describe turbulence.

Turbulence structure is considered to be made up of eddies of various sizes which interact with each other and with the mean flow. The eddy interaction cascades the energy to smaller and smaller eddy sizes until the energy is essentially lost by the action of viscosity. Thus the turbulence is dissipative and diffusive in character to a high enough degree which cannot be accounted for by the molecular diffusivities and the mean strain rate. The diameter of the eddies whose influence is predominant under any conditions roughly defines the scale of turbulence.

There are two major types of turbulence in the boundary layer:

(1) Mechanical due to the instability of the vertical wind shear and
(2) Thermal due to the buoyancy forces in the atmosphere.

328 Applications in the Atmosphere

The combination of wind and temperature stratification, therefore, plays an important role in characterizing the stability of the atmosphere and the generation and nature of turbulence. Richardson number (R_i) is one of the most common parameters which is used to quantitatively describe turbulence. It is expressed in terms of the gradients of wind velocity, $\delta u/\delta z$, and potential temperature $\delta\theta/\delta z$ as:

$$R_i = g\ (\delta\theta/\delta z)/T(\delta u/\delta z)^2$$

According to Louis et al, [2], the following values of the Richardson number are used to characterize atmospheric stability:

$R_i \leq 0$ for statically unstable atmosphere
$0 < R_i \leq 1$ for dynamically unstable atmosphere (turbulence generated in the stable atmosphere due to wind shear)
$1 < R_i \leq 10$ for stable atmosphere
$R_i > 10$ for very stable atmosphere.

Laboratory studies of the turbulence in the stable conditions have shown that in the region characterized by $0.25 < R_i < 1$, turbulence once initiated when $R_i \leq 0.25$ will persist, no turbulence will be initiated if R_i does not become less than 0.25 at any time, and existing turbulence will be suppressed i.e. the medium will become laminar in case $R_i \geq 1$.

Under dynamically unstable conditions, for critical values of shear, gentle waves begin to form on the interface. These waves grow in amplitude and eventually reach a point where each wave begins to roll up or break. This breaking wave is called Kelvin-Helmholtz (KH) wave. In this condition both static instability and dynamic instability exist and make each wave to become turbulent. This happens because the fluid reacts in a manner to undo the cause of instability which is done via turbulence. The turbulence spreads throughout the layer causing mixing of different fluids. During this process, there is a transfer of momentum between the fluids which reduces shear between the two layers. The shear may be reduced below a critical value to eliminate the dynamic instability. Thus in the absence of some external source to restore the shear, the turbulence decays in the interface ragion and the flow becomes laminar. Thus, turbulence is the mechanism whereby fluid flows tend to undo the cause of the instability e.g. in the case of static instability, it is convection that occurs and makes more buoyant fluid to move upwards, thereby stabilises the system while for dynamic instability turbulence tends to reduce the wind shear and thereby stabilises the system.

(ii) the Roughness Parameter

Roughness parameter represents the direct effect of the surface on the wind above and is thus a very useful concept. Normally, within the layer of air near the ground viscosity predominates. However, as we increase the wind speed (shearing stress), over a given surface i.e. as we depart from the ground level or as the surface becomes increasingly rough at a constant wind speed, the shearing stress becomes partly turbulent and partly viscous. Soon a stage is reached at which the pure viscous stress of the surface is outweighed by the effect of pressure forces associated with the eddying wakes from the roughness elements i.e. at this stage the viscosity ceases to influence the wind profile and thus the shearing stress primarily defines the turbulent motions. Such a aerodynamically rough stage where the flow is practically zero is a characteristic of the surface roughness and gives a measure of the roughness parameter z_0 of the surface.

Roughness length is generally an order of magnitude smaller than the actual height of the roughness elements. Above the roughness length the air is directly influenced by the local pressure forces set up by the individual roughness elements of the surface. The layer of air where the shearing stress becomes partly turbulent and partly viscous defines the inertial interval of turbulence and is thus called the layer of inertial subrange. The power spectrum in the inertial subrange has a band of wave numbers (k) defined by the following form of the energy distribution:

$$F(k) = c\varepsilon^{2/3} k^{-5/3}$$

(iii) Convective Boundary Layer

The convective atmosphere constitutes the daytime unstable boundary layer. It consists of thermal plumes i.e. updrafts surrounded by large downdrafts. They grow in the morning with the solar heating of the surface of the earth, become maximum up to a height of 1–2 km around midday and decrease in the afternoon. Solar heating of the ground generates thermal turbulence which results in the upward sensible heat flux H related to the net radiation Q^*, the driving force of the surface energy budget, as:

$$Q^* = H + LE + G$$

Here LE is the flux of energy due to evaporation ($LE > 0$) or condensation ($LE < 0$) and G is the soil heat flux. H and LE are respectively defined as:

$$H = \rho c_p \, \overline{w'\theta'} \text{ and}$$

$$LE = \rho L_n \, \overline{w'q'}$$

where ρ is the density of air, c_p the specific heat, L_n the latent heat of vaporization, w is the vertical wind velocity, θ the potential temperature, q the specific humidity and the primes represent the fluctuations in the respective parameters. On a clear day H, LE and G are positive with G having relatively small value. For characterization of turbulance in the unstable layer, the length scales of mixing height, z_i, the height above the ground, z, and the Monin-Obukhov stability length, L are important. The mixing height z_i defines the height of unstable boundary layer above the ground level wherein pollutants are mixed due to the presence of turbulence. It exends up to the top of the ground based convective layer where the lapse rate becomes almost adiabatic and is capped by a stable layer (the first elevated inversion). Monin-Obukhov length L is a length scale physically related to the height below which dynamically generated turbulence dominates buoyantly generated turbulence while above it the reverse becomes true. It is infinite in neutral conditions, negative in unstable conditions and positive in stable conditions. In terms of surface fluxes of heat and momentum, it is expressed quantitatively as:

$$L = - u_*^3/\{k(g/T) Q_0\}$$

where Q_0 is the kinematic surface heat flux ($w'\theta'$), T is the actual temperature in the bouyancy term (g/T), k is the von Karman constant (~ 0.4) and u_* is the friction velocity defined in terms of the surface momentum shear stress s, as:

$$u_*^2 = s/\rho = - u'w'$$

with u' and w' representing respectively the fluctuations in the horizontal and vertical wind velocities. The unstable boundary layer (assuming it to be horizontally homogeneous) has a vertical structure consisting of the surface layer, the free convection layer, the mixed layer and the entrainment layer.

The surface layer is confined to the region $20 z_0 < z < - L$ with $z < 0.1 z_i$ where z_0 is the surface roughness length. Surface fluxes of heat and momentum are considered to be approximately constant in this layer while wind shear plays a dominant role. A measure of the surface layer is where the shearing stress is within 90% of the surface value. The controlling parameter in this layer are the height, z, the momentum shear stress, the sensible heat flux and the buoyancy parameters. Monin-Obukhov similarity theory postulates that dimensionless local groups formed with the friction velocity u_* and scaling temperature parameter $\theta_* = (- H/(\rho c_p u_*)$ become universal functions of the stability parameters (z/L) in the layer

The free convective layer exists in the region $-L < z < 0.1 z_i$. Tennekes [3] has described the scaling velocity, u_f, and the temperature, θ_f, in this layer as:

$$u_f = (w'\theta' z g/T)^{1/3}$$
$$\theta_f = w'\theta'/u_f$$

Under light or calm winds and strong insolation, the momentum shear stress is no longer important in this layer in describing the atmospheric turbulence. The controlling parameters are the height z and the values of heat flux and buoyancy parameter in the surface layer.

The mixed layer exists in the region $0.1z < z < 0.8z_i$. The turbulence structure in this layer is insensitive to momentum shear stress parameters while the mean wind, humidity and potential temperature profiles tend to be constant due to intense vertical mixing. The height z loses its importance as a scaling parameter. The mixing height, z_i, becomes the controlling length scale. Deardorff [4] has described the scaling parameters of velocity w_* and temperature θ_m as:

$$w_* = (w'\theta'z_i g/T)^{1/3}$$

$$\theta_m = w'\theta'/w_*$$

According to Wyngaard [5] strong convective effects can be found at $z/L \cong -4.5$.

The entrainment layer exists in the region $0.8z_i < z < 1.2z_i$. In this layer the stable layer from above mixes into the developing convectively unstable region of the boundary layer. There is not only a temperature change across the convoluted interface between the deepening convectively unstable boundary layer and the capping stable layer but there is also a finite shear in the wind velocity. The distance to the inversion, $|z_i - z|$ enters as the scaling length [6] in this region.

(iv) Stable Boundary Layer

During the evening the earth starts cooling due to radiative emissions which in turn cools the air adjacent to the ground. The process continues upwards slowly from sunset to sunrise in response to slow changing surface conditions giving rise to the stable boundary layer. The height of the stable boundary layer is a couple of hundreds of meters which is an order of magnitude smaller than that under unstable conditions. It is generally measured as the height at which the heat flux (negative) falls to a certain low proportion of its surface value. The stable stratification suppresses turbulence. Turbulence in the stable boundary is driven by local shear stress which is inherently weak. It has a maximum value at the surface and decreases monotonically with height to near zero at the top of the stable boundary layer. This leads to comparably smaller eddies in the stable boundary layer. Low magnitude of fluctuations coupled with the masking of the structure turbulence by other physical processes like gravity waves, drainage and slope flows, intermittent turblence and radiative divergence raises measurement problems limiting thus knowledge about turbulence in the stable boundary layer.

Often the stable boundary layer appears to consist of a turbulent layer adjacent to the surface, capped by a zone of weak or non-existent turbulence [7]. The surface layer exists in the stable boundary layer like that in the unstable boundary layer. It is in principle lower than the Monin-Obukhov length. The scaling parameters of turbulence are the height z and the surface values of the shearing stress and heat flux as in the case of unstable boundary layer.

Above the surface layer, turbulence scaling parameters are the height z and the local values of the fluxes of shearing stress and heat. The local scaling approach is an extension of Monin-Obukhov similarity theory to the whole stable boundary layer [8]. Here the local Monin-Obukhov length, Λ is defined as

$$\Lambda = -s^{3/2}/\{k(g/T)(w'\theta')\}$$

The temperature scale for this region can be defined as $\theta_c = (w'\theta')/s^{1/2}$. Since in the local scaling region there are only two length scales, z and Λ, the turbulent quantities can be described as a function of z/Λ. For large value of z/Λ, local scaling reduces to z-less scaling, implying that z is no longer an important lenth scale for turbulence. The turbulence within the z-less region is intermittent and difficult to characterize. The background is that vertical motion is inhibited due to the stable stratification and that turbulent eddies no longer feel the presence of the surface.

Acoustic Remote Sensing—Sodar

For the study of the atmospheric boundary layer, conventional in situ techniques are often used to monitor the variability in the basic atmospheric parameters. However, the data are generally limited both in time and space and are not thus adequate to understand the meso- and microscale weather phenomena and to monitor hazardous situations of air pollution, microwave propagation and aviation etc. The necessity to use economical methods of obtaining data have resulted in a growing effort to develop remote sensing techniques which can provide data on the atmospheric boundary layer parameters continuously in both space and time with higher resolution and without disturbing the variables being measured.

Remote sensing techniques can be grouped in three principal categories according to the particular type of waves used—acoustic (SODAR), optic (LIDAR) and radio (RADAR). Out of these systems, acoustic waves have been found to be more suitable for investigating the atmospheric boundary layer. This suitability is based on the fact that interaction of sound waves with the inhomogeneities of the lower atmosphere is very much stronger than that of the electromagnetic spectrum. Moreover, the small speed of the acoustic waves also offers an advantage of longer time to handle the scattered acoustic signals thus extending investigations to very close ranges near the ground level which is normally difficult to probe with the help of electromagnetic waves because of ground clutter.

We have been working with the acoustic technique (sodar) at the National Physical Laboratory, New Delhi since early seventies. The aim was to develop the technique for use in studies of hazardous situations in air pollution and microwave propagation. In the following we discuss the various developments made to study atmospheric stability, mixing layer height, inversions, radio wave ducts, turbulence parameters, wind velocity profile, wind shear, diffusion coefficients and many other parameters relevant to characterize the atmospheric boundary layer. In this context, however, we shall presume that the reader is already aware of the acoustic remote sensing technique and sodar performance. Further that the vertically looking monostatic sodar gives information about the thermal structure, bistatic configuration gives information about wind-shear structure and Doppler sodar gives three dimensional wind information in the lower atmosphere. In this report, we mainly discuss the applications of monostatic sodar.

Sodar Pattern Recognition and Associated Meteorological Information

Various types of echograms [Fig. 2] have been seen on the monostatic sodar facsimile records. A close examination of these echograms, however, shows that

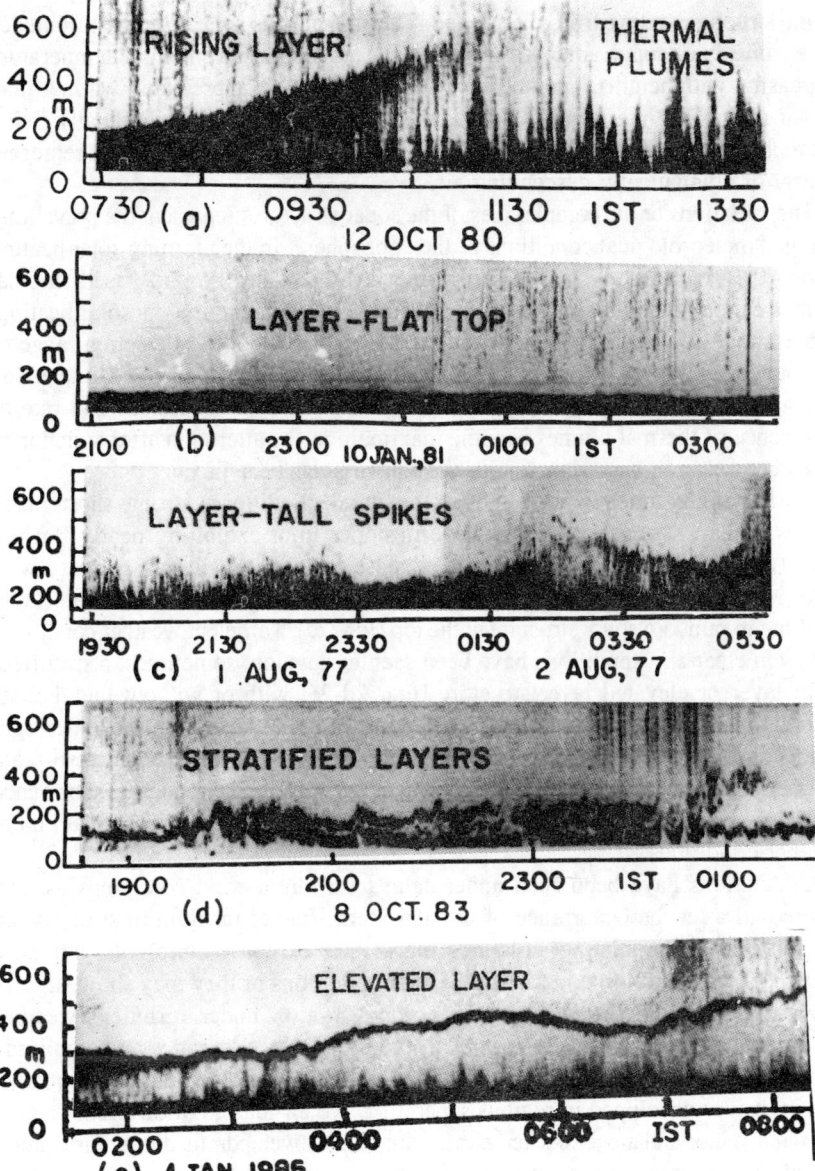

Fig. 2. Typical thermal structures observed on the monostatic sodar echograms.

basically there exist only two main categoaries—shear echoes and thermal echoes. Shear echoes tend to be horizontal and may be surface based, aloft and stratified. They are caused by turbulence in regions of static stability (potential temperature increasing with height). Thermal echoes appear in the form of stalagmites-like plume structures rising from the ground. Thermal echoes are caused by turbulence in the unstable, super adiabatic layer of the atmosphere (potential temperature decreasing with height). They occur when the surface is appreciably warmer than the air aloft. The intermittent groups of thermal echoes mark individually rising convective cells called thermals. The echo free regions between thermals represent neutral or adiabatically descending air.

The variations or the complexities of the sodar structures represent the prevailing/changed meteorological conditions of the atmosphere. In the morning solar heating erodes the nocturnal surface based stable layer forming ground based thermal plumes capped by the rising stable layer (Fig. 2a). With continuous solar heating, stable layer rises sufficiently high so that either it goes beyond detection range of the sounder or it loses sufficient turbulence to become insensitive to sodar detection making thermal plumes to pervade on the sodar echograms. Height and rate of occurrence of thermal plumes become maximum in the afternoon after which they start decreasing in accordance with the fall in solar heat flux.

During night time, under slight or no wind conditions, strong short range echoes having abrupt but almost uniform upper limit exhibiting nearly flat top layer (Fig. 2b) are formed. Thickness of these layers may slightly increase with time. Medium to strong surface winds bring in mixing within the stable layer resulting in ramdom spiky structure at the top (Fig. 2c). Turbulent weather conditions and even clear sky conditions have been seen to develop, sometimes, a stratified/multi layer or elevated layer structure [Fig. 2d-2e] with or without undulations superposed over them. Multiple layers represent stable shear structure under light wind conditions or advection of super-imposed flow due to some meso-scale weather phenomena. Elevated layers represent the presence of fog layer, subsidence inversion, marine boundary layer or some other approaching/persisting turbulent weather conditions.

Undulations have been seen under calm to medium wind conditions as also before and after the occurrence of thunder storms under medium to strong wind conditions. The undulating structures may either exhibit features of symmetric sinusoidal wave motion under clear weather conditions or they may show slightly unsymmetric rounded saw tooth type wave motion under turbulent weather conditions. These waves have periods of the order of few minutes and amplitude in the range of 100 m peak to peak. It seems that these undulations represent gravity waves developed in regions separating two air masess of different density and wind vector and are associated either with convective updrafts during fumigation period or with wind shear variations under stable conditions.

The repeatable patterns on the sodar echograms identified first by Shaw [9] have been studied by many workers in the field from time to time. Fukushima [10, 11] distinguished three general characteristic echo patterns apart from regions of no echo, wind, rain, and ambient noise. Schubert [12] proposed a system of 15 categories to identify the various phenomena of climatological interest. Clark et al, [13] proposed a numerical classification scheme of 14 different types and separated cases of surface based echoes from elevated echo layers and related them to the boundary layer meteorological conditions.

Hall [14] introduced comnputer compatible coding to the various sodar structures together with numerical notation of the height of the structure. Prater and Colls [15] developed a categorisation scheme similar to Clark et al, [13] but they identified only five major categories of sounder structures. They did not include any identification for zones of no echoes and non-recognisable noisy echoes. Maughan [16] and Maughan et al, [17] developed a numerical code to identify the various types of sodar echoes and to give information about the height of the structure and their complexity. This classification scheme was also adopted by Asimakopoulos et al, [18] to study stability frequencies in different cities in Greece. A code of eight numbers including no echo and no reliable echo conditions was introduced. Singal et al, [19] defined soder thermal structures consisting of flat top ground based layer, tall spiky top ground based layer, multi-layers, elevated layer, thermal plumes, morning eroding layer and no echoes and considered them to classify atmospheric stability.

Walczewski [20, 21] and Walczewski and Felesky-Bielak [22] considered four basic characteristic forms of sodar echoes-vertical [type 1], ground based horizontal layer [type 2], elevated horizontal layer in the absence of ground based layer [type 3] and no echo structure [type 4]. Type 1 culminated at noon, type 2 at midnight, type 3 in the morning and type 4 in the cvening around sunset time. Digital representation/coding was introduced and the sodar echograms were interpreted in terms of atmospheric stability categories.

Foken et al, [23] considered ground based structures, elevated structures, transition pattern, convective pattern, no echoes and disturbances of meteorological or technical origin as the basic sodar structures and coded them by 2-digit code figures compatible with computer pattern recognition. The patterns were represented by the divided tens-figure with the first two basic patterns sub-classified into internal [inhomogeneous] and multiple patterns for which unit figures were used to describe the necessary details. Height information of a pattern was encoded by two digits for representing the lower and upper boundary of the phenomenon and two digits were used to describe the tendency of the development of the phenomenon with the intensity of a pattern encoded by a scale.

Evers et al, [24, 25] defined the following distinct echo patterns to study the stable and unstable atmospheric conditions:

L_G — Ground based compact structures
L_E — Layer shaped elevated structures
S — Spiky structure extending from the ground with blanks on the time axis
X — Transition type between convection S and inversions L_G and L_E
C — Echo structure with considerably varying local and temporal occurrence typical of frontal passages and precipitation
NE— No echo due to missing turbulence in the $\lambda/2$ range.

Sodar Stability Classification

Atmospheric stability is one of the essential parameters for air quality studies. Giblett [26] was the first person to categorise atmospheric stability based on the behaviour of wind and vertical temperature gradients. This work was followed by Smith [27] who presented Brookhaven National Laboratory typing schemes similar to Giblett's four category scheme. Cramer [28] suggested another method of classsifying stability types. He correlated observations of the standard deviation of wind direction, azimuth and elevation angle with simultaneously measured horizontal plume spreading data to develop a four category system. This was followed by the well-known Pasquill stability classification scheme [29]. Pasquill proposed a simple scheme of turbulence types using six categories of stability conditions derivable from data of surface wind speed, insolation and night time state of the sky [Table 1]. They were listed as categories A to F in terms of increasing stability order from very unstable [A], moderately unstable [B], slightly unstable [C], neutral [D], slightly stable [E] to moderately stable [F] conditions successively.

Table 1. Pasquill Stability Classification in terms of surface wind speed, day time insolation and night time cloud conditions (Pasquill, 1961)

Surface wind speed (m/sec)	Day time Insolation			Night Time Conditions	
	Strong	moderate	slight	Thin overcast or $\geq$ 4/8 cloudiness	$\leq$ 3/8 cloudiness
< 2	A	A–B	B	—	—
2–3	A–B	B	C	E	F
3–5	B	B–C	C	D	E
5–6	C	C–D	D	D	D
> 6	C	D	D	D	D

Using measurements made up to a distance of 800 m of the dispersion of a passive non-buoyant tracer gas released near the surface, Gifford [30] developed the sigma curves to express the horizontal and vertical dispersion coefficients, σ_y and σ_z respectively, as a function of distance from the source for the various Pasquill stability conditions to estimate the concentration of pollutants downwind of a source. This system became popular as the Pasquill-Gifford (P-G) system of dispersion estimates.

Turner [13] introduced a slightly modified version of Pasquill stability scheme in terms of seven categories derivable from meteorological quantities of cloud cover, height and solar angle. Golder [32] found that the best conversion between Pasquill and Turner classes was provided by A as 1, B as 2, C as 3, D as 4, E as 6 and F as 7. Islitzer and Slade [33] established relations among Pasquill stability types, lapse rates and standard deviations of horizontal wind direction.

Carpenter et al [34] based on Tennesse Valley Authority experience developed families of sigma curves (TVA curves) using lapse rate. However, Gifford [35], Weber et al, [36] and Hanna et al, [37] found that lapse rate was practically useless as a stability indicator during unstable conditions.

Pasquill stability classes were also expressed in terms of turbulence parameters. Islitzer [38] assigned Richardson number values for Pasquill stability types. These values ranged from (–) 0.26 for type A to 0.046 for type F. The basis for this use had been the definition of flux Richardson number which is the ratio of the rate at which work is done against buoyancy forces to the rate at which turbulent kinetic energy is created by shear stresses. This definition takes into account the relative importance of turbulent transfer of buoyancy effects with respect to forced effects of shear stress. Hanna et al, [37], however, recommended to use the dimensionless parameter, S, expressed as the ratio of the height of the atmospheric boundary layer to the Monin-Obukhov length as the proper parameter to estimate stability. Monin-Obukhov length used in defining this parameter has a simple relationship with Richardson number [39] and has the same sign as that of the Richardson number. Weber [40] worked out a correlation amongst the parameters of standard deviation of horizontal wind direction, Richardson number and Monin-Obukhov length for the various stability categories as given in Table 2.

Table 2. Correlation amongst Pasquill stability classes, standard deviations of horizontal wind direction (σ_θ), Richardson number (R_i) and Monin Obukhov length (L) (Weber, 1976).

Pasquill Class	σ_θ	R_i (at 2 m)	L
A	25°	– 1.0 to – 0.7	– 2 to – 3
B	20°	– 0.5 to – 0.4	– 4 to – 5
C	15°	– 0.17 to – 0.13	– 12 to – 15
D	10°	0	—
E	5°	0.03 to 0.05	35 to 75
F	2.5°	0.05 to 0.11	8 to 35

Today numerous techniques exist to categorise Pasquill stability on the basis of meteorological measurements made close to the ground. Among others, the meteorological parameters used for the purpose are; (1) the standard deviation of the vertical wind direction, σ_φ: (2) the standard deviation of the horizontal wind direction, σ_θ: (3) the temperature difference between two levels and the horizontal wind speed, and (4) the radiation balance (i.c. insolation, cloud cover etc.) and horizontal wind speed. Since sodar echograms essentially represent turbulence in the planetary boundary layer and Doppler sodar can determine wind velocity and wind turbulence in the lower atmosphere, remote acoustic sounding can be used to determine Pasquill stability class of the atmospheric boundary layer.

In the above context, Singal et al, [19, 46–48] were the first to lay down a technique to classify Pasquill stability categories based on pattern recognition. Stability determined from simultaneously measured data of standard deviations of the horizontal wind direction were used to correlate with sodar deduced stability classification. The characteristics of the sodar echoes to classify Pasquill stability categories are given in Table 3.

Broadly the classification scheme can be described as follows.

1. Category A, representing strongly unstable conditions, is marked on the sodar echograms by well-defined families of tall plumes.
2. Category B, representing moderately unstable conditions, is marked on the sodar echograms by thermal plumes of shallow height.
3. Category C, representing slightly unstable conditions, is marked on the sodar echograms by very shallow plumes formed during late afternoon hours.
4. Category D, representing neutral conditions, is marked on the sodar echograms either by no structure or by dark bands due to strong wind induced noise.
5. Category E, representing slightly stable conditions, is marked on the acoustic sounder echograms either by a ground based layer or by a stratified layer structure of higher depth during night time.
6. Category F, representing moderately stable conditions, is depicted on sodar echograms either by shallow, firm ground based layer or by shallow stratified layer structure.

The above technique of determining Pasquill stability classification from the observed sodar echogram characteristics was used by Singal et al, to explain the distinct anomalous peaks (a weak one in the morning and a strong one in the evening) in the measured carbon monoxide concentrations on the busy traffic roads of Delhi due to vehicular traffic although the same number of vehicles on average were passing the measurement place all through the day (Fig. 3). It was found that morning weaker peak lies during the fumigation period under unstable weather conditions while the evening stronger peak was due to the presence of stable weather conditions which did not help the dispersion of carbon monoxide. Typical diurnal plots of the sodar derived stability category and simultaneously measured concentration of carbon monoxide in the atmosphere near the ground surface for a set of days having different weather conditions (Fig. 4) further showed very clearly that the prevailing stability category determined the concentration of carbon monoxide present in the atmospheric air.

This technique of stability classification was also used to determine the stability classes for any hour of the day for Ngawha Springs, New Zealand [50] from the structural details of the monostatic acoustic sounder echograms obained for the year 1983. The results when compared with the stability calsses obtained for the respective hours using the technique developed by Wratt [51] based on calculating bulk Richardson number using data from the 56 m instrumented tower operated

Table 3. Classification of Sodar Echograms in Terms of Stability Categories

S.No.	Stability class (Pasquill)	Wind direction fluctuation criteria (degrees)	Nature of Sodar echograms	Outlook
1.	Strongly unstable (A)	$\sigma_\theta \geq 23$	(i) Well defined thermal plumes	Clear sunny day with strong solar heating and light/calm winds.
2.	Moderately unstable (B)	$18 \leq \sigma_\theta < 23$	(i) Well defined thermal plumes upto shallow heights (ii) Rising layer with thermal plumes below.	Moderate solar heating and moderate winds. Bright sunny morning.
3.	Slightly unstable (C)	$13 \leq \sigma_\theta < 18$	(i) Thermal plumes upto very shallow heights	Weak solar heating, cloudy day, moderate to strong winds, and late afternoons.
4.	Neutral	$8 \leq \sigma_\theta < 13$	(i) Spiky top layer of height above 150 m. (ii) No structure (iii) Darkness due to rain or wind induced noise.	Early evening hours on clear days. After rain or storm, cloudy/windy conditions. During rain or heavy winds (storm).
5.	Slightly stable (E)	$4 \leq \sigma_\theta < 8$	(i) Flat top layer of depth more than 100 m. (ii) Surface based layer with spiky top of depth generally within 150 m. (iii) Stratified layers of depth more than 200 m.	Clear night with moderate winds.
6.	Moderately stable (F)	$\sigma_\theta < 4$	(i) Surface based layer with flat top of depth within 100 m. (ii) Stratified layers of height less than 200 m.	Clear night with strong radiative cooling and light/calm winds.

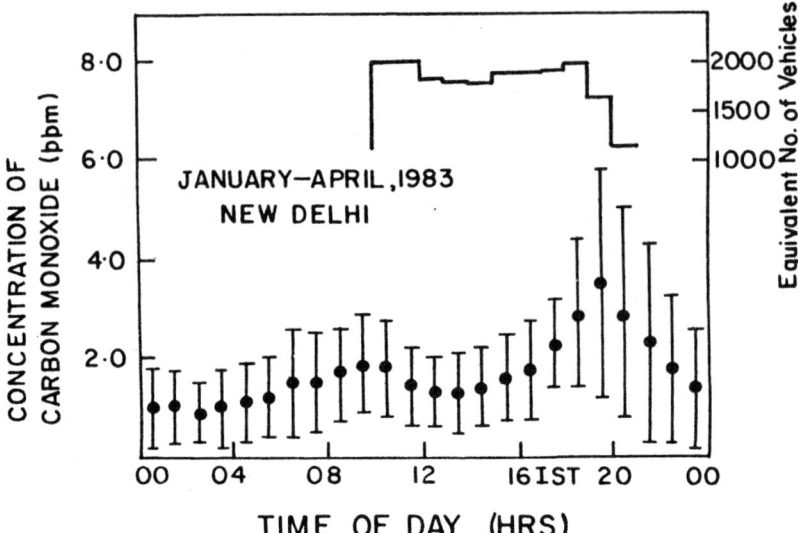

Fig. 3. Diurnal variations in the concentration of carbon monoxide and traffic density during the period January to April 1983.

simultaneously with the monostatic acoustic sounder at the same site, were seen to give a correlation for about 50% of the time. The correlation was found to become better in case limited classification of differentiating unstable, neutral and stable categories was considered. This type of result is not surprising since the two methods for stability classification differ in the sensitivity range of the meteorological parameters. While the bulk Richardson number method is sensitive mainly to conditions within the surface layer (the lowest 50 m of the atmosphere) the acoustic sounder classification scheme uses information from well up in the mixed layer up to 700 m. Further while the technique has been developed for the plains, Ngawa Springs area has a complex terrain.

Singal and associates have also correlated sodar echo patterns [52–54] obtained under various atmospheric conditions with lapse rate conditions and the associated stack plume behaviour. It has been seen [Fig. 5] that looping under strong lapse rate conditions (unstable weather) is linked with sodar thermal plume structure, coning under subadiabatic conditions is associated with evening transition from unstable to stable weather conditions, fanning is a property of the stable weather when inversion depth is more than the height of the plume stack, lofting occurs under conditions of inversion depth less than the height of the plume stack and fumigation is the period of the morning eroding of the stable layer.

Following the work of Singal et al [19], Walczewski and Felesky-Bielak [22] have also identified the stable, unstable and neutral stability classes in terms of the sodar structures. Variations in the vertical temperature gradient were used as a criteria to define the stability layer.

Evers et al, [24, 25] have also used sodar structures to identify the stability

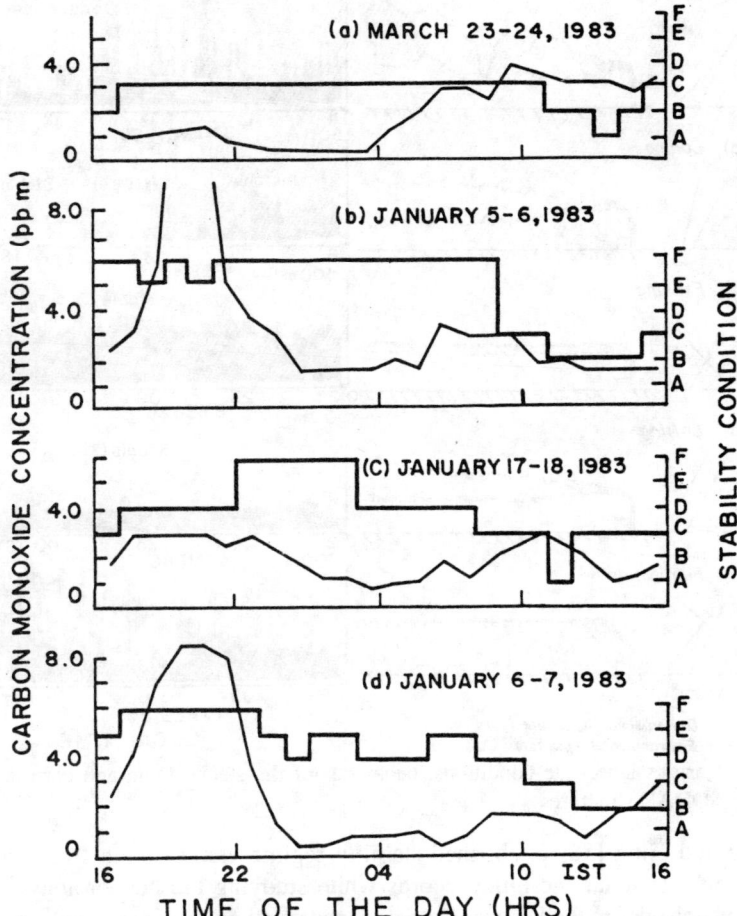

Fig. 4. Plots of simultaneous measurements of carbon monoxide (CO) concentations and sodar observed atmospheric stability obtained on various days. Thick lines represent Pasquill stability variations while thin lines represent CO concentrations.

categories of stable, unstable and quasi-neutral. Diurnal variations in temperature gradient and radiation data were used as an aid for classification. The ground based layer structures defined stable category, the spiky structures defined unstable category and the no echoes and transition echoes with small vertical extension defined quasi-neutral category. Using in addition the height information of the sodar structures, Neisser et el, [55] further defined all the seven Pasquill stability classes. However, it was found that the error rate in the determination of stability from the sodar data had a marked diurnal variation, with the discrepancy becoming largest at night. These results suggested that the use of sodar data for determining stability classes required more detailed investigations.

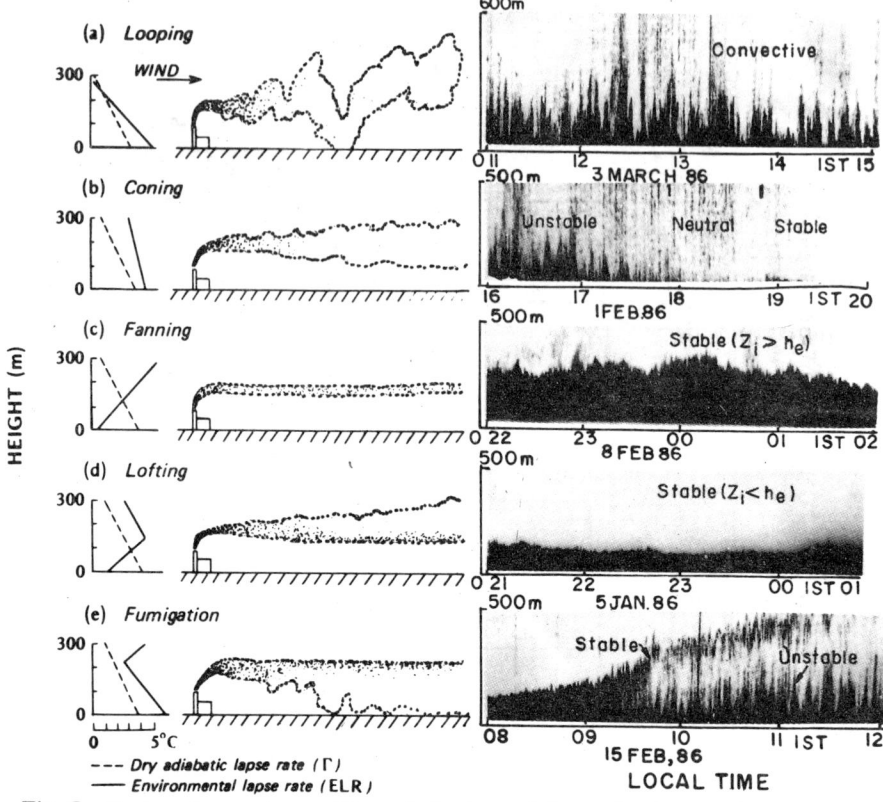

Fig. 5. Various lapse-rate conditions, behaviour of the stack plume and corresponding SODAR structures

Neff and King [56] established stability regime classification from the high resolution minisodar facsimile records while studying the meteorology of high pollution episodes in the Denver, Colorado, metropolitan area. Structures pertaining to the surface mixing layer (the lowest uniform echo layer on the sodar record), convective echoes, capping inversion, and stratified echoes were identified. A stability classification scheme was laid down which was based on the observation that conditions most conducive to limited vertical mixing were those in which there were the greatest number of fine stratified layers on the sodar facsimile records. Three classes were marked, weak, moderate and strong, with finer gradations of plus/minus designation allowed for each class.

Doppler sodar measured wind velocity has also been used to classify stability. In this respect Gland [57] and Jones et al [58] proposed a simple model of determining stability, wherein the lowest layers of the atmosphere were classified either stable or unstable depending on the value of the standard deviation, σ_w of the vertical wind speed. According to Gland if $\sigma_w < 0.45$ ms^{-1}, the atmospheric conditions were classified as stable and if $\sigma_w > 0.45$ ms^{-1}, unstable or neutral conditions were said to prevail. Jones et al, however, proposed critical value of switching atmospheric stability at 0.3 ms^{-1}. They also proposed a method of determining the top of the stable layer. According to them, the height of the successive horizontal velocity values where wind shear became maximum, was the top of the stable ground based layer.

Thomas [59] presented two schemes of Pasquill stability classification based on Doppler sodar data. According to one scheme, he used the standard deviation σ_w of vertical wind speed and the horizotal wind speed u at a height of 100 m to determine Pasquill stbility classification and according to the other scheme, he used the standard deviation σ_ϕ of vertical wind direction and the vertical profile of back-scattered amplitude both measured at a height of 100 m, to determine the Pasquill stability classifiction. He found statistical equivalence of both these schemes with the classification derived from measurements of σ_φ by a vecor vane at the 100 m level of the tower.

Best et al [60] found that the use of σ_θ for stability determination could be misleading in all but very flat and uniform terrain. For determining the stability at Stanwell (Australia), they preferred to use the turbulence parameter, σ_w/u where u is the horizontal wind velocity. Gland [61] also considered using turbulence internsity parameter, σ_w/u for stability classification. He, however, found [57] that it was leading to unrealistic results in cases of weak wind associated with strong atmospheric stability.

Application to Air Pollution

Based on the above described extensive work done on sodar capabilities at the National Physical Laboratory and elsewhere in the world, sodar is now recognized as a useful meteorological tool for air quality applications. Herein we give techniques to determine parameters related to air quality in the atmosphere using sodar data.

(i) Determination of the Mixing Depth

Mixing depth is the vertical extent of the lower atmosphere wherein turbulence is present and pollutants mix. Under unstable conditions, mixing depth is the height from the ground surface to the interface level, a potential barrier to dispersion between the elevated stable and less stable or almost adiabatic air down below. Under stable conditions, technically speaking, inversion layering exists and as such there is no mixing depth. However, in urban areas due to the presence of vertical wind shear and heat island effect, weak turbulence develops and therefore, the depth of the surface based inversion can be assumed to serve as the mixing depth. Since variations with height in the values of temperature, heat flux and momentum flux etc., give a measure of turbulence, therefore any one of the parameters can be used to determine the mixing depth.

The conventional instruments used to measure turbulence and hence the mixing depth are the radiosonde, tethersonde and instrumented aircraft etc. Since sodar also measures turbulence in the atmospheric boundary layer, it can also be used to measure the mixing depth. In the simplest form a measure of the height of the thermal plumes during daytime and of shear echoes during night time should gives a measure of the mixing depth. Reliable measurements can, however, be made only for the period when during night time ground based shear echoes are present or during the daytime an elevated capping layer is present above the range of the thermal echoes. However, the latter is only possible when the height of the structure is within the probing range of the sounder, a condition which is possible normally during the morning forenoon hours or at sites where the first elevated inversion layer is within 1 km.

The question also arises how far a measure of the height of the ground based shear echoes compares with the height of the stable boundary layer [i.e. the inversion height] as measured by the standard meterological techniques. In this context, extensive investigations have been made by many investigators [24, 62–66] to study and compare the sodar determined height of the stable boundary layer with that determined from a measure of the ground based inversion layer reported by radiosonde. Considering the fast ascent rate of the radiosonde balloon near the ground level, the slow response of the sensors and the distance of the sodar site from the radiosound site, the small discrepancies in the two results can be ignored and a good correspondence between the two results can be expected as has also been reported by many.

Von Gogh and Zib [67] compared the simultaneously measured height of the stable layer by monostatic acoustic sounder and tethered balloon borne sensors. Within the coincidence range (40–350 m) of the two systems, a good agreement between the soundings was reported in terms of the position of the statically stable zones in the atmosphere.

Russell and Uthe [68–70] made measurements of the mixing depth in the forenoon and during the nocturnal inversion periods using sodar, instrumented tower and aircraft and found that they correlated well. The slightly increased scatter in the plots (Fig. 6) was attributed to frequent occurrence of several weak inversions or isothermal layers in the evening or night time temperature profiles.

Coulter [71] compared the mixing layer heights determined by the thermal profiles, lidar and sodar observations in the forenoon under canopy conditions. He found that the overall values agreed fairly well but they had systematic differences (Fig. 7). The lidar derived values were consistently higher than the sodar derived values while the temperature profile values were consistently lower than those estimated by the other two methods. These differences were attributed to slightly different behaviour of the sensed variables near the capping inversion. Aerosols and particulate matter sensitive to lidar measurements mix to larger heights than the top of the adiabatic temperature profile, while temperature fluctuations sensitive to sodar measurements also exhibit an increase at a height above the top of the adiabatic temperature profile but it is below the maximum height of the particulate mixing.

On the basis of aircraft and acoustic sounder observations of O'Neil, Wangara,

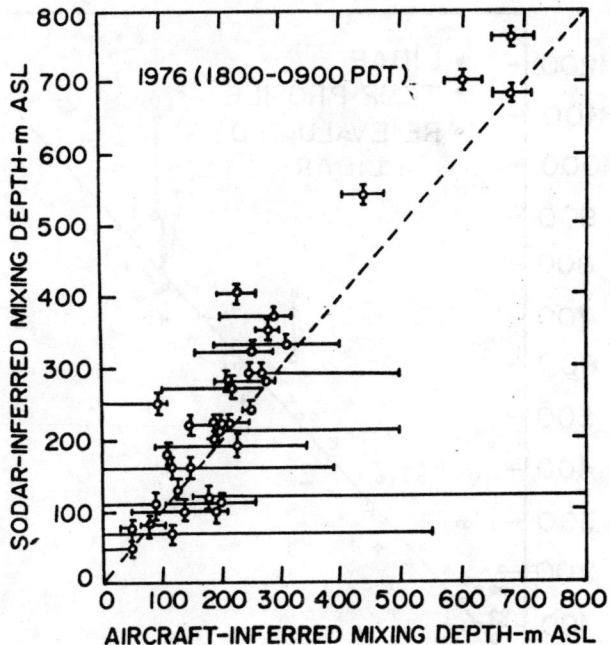

Fig. 6. Comparison of mixing depths inferred from Sodar and aircraft measurements (Russel & Uthe, 1978).

Mahrt et al, [72] found that, on the average, the top of the sodar determined boundary layer occurred just below the low level wind maximum, which, in turn, approximately coincided with the height of the maximum in gradient Richardson number. Further the temperature inversion height was often around 25% higher than the low level wind jet height.

Nieuwstadt and Driedonks [73] and Nieuwstadt [74, 75] reported that sodar determined height of the turbulent layer under nocturnal stable boundary conditions was usually smaller than that of the temperature inversion layer.

Arya [76] made detailed analysis of February 1975 Cabauw data for selected periods. A comparison of the sodar measured height with other characteristic heights of the low level wind jet and of the potential temperaure profile during nocturnal stable periods showed generally a poor correlation with the difference that for the very stable and extremely stable categories, the height of the maximum in the wind speed appeared to give a good agreement with the sodar measured height. Arya also found that sodar height was best correlated (coefficient of correlation 0.67) with the diagnostically determined height using relations proposed by Ziltinkevich [77].

Dohran et al, [78] made measurements of the sodar inversion structure heights above the city of Cologne (Germany) and reported that structures which are being classified as inversions represent only layers of temperature fluctuations and may

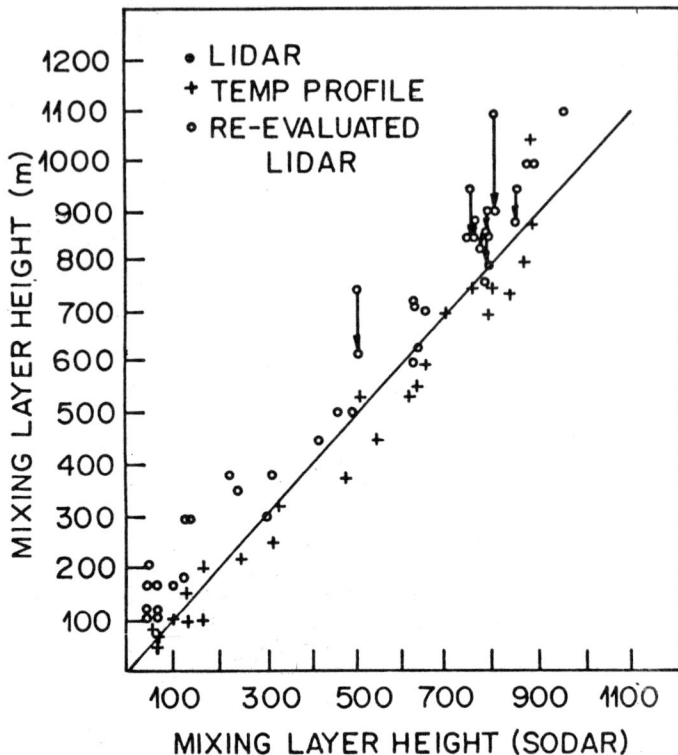

Fig. 7. Comparison of mixing layer heights determined by SODAR, LIDAR and temperature profiles (Coulter, 1979).

only belong to very small stability i.e. may not be associated with the strong temperature inversions and further that stronger vertical wind shear enhances temperature fluctuations. Thus they called the layered returns to be only associated with inversion structures and further that monostatic sodar sounding can never replace direct sounding,

Fitzharris et al, [79] described sodar measurements of inversion frequencies near Cromwell, New Zealand and found that sodar measured heights were comparable with those obtained from vertical temperature profiles from a kytoon which could measure temperatures up to 300 m above the ground.

Jones et al, [58] obtained the height of the boundary layer using Doppler sodar data. According to them the height of the successive horizontal velocity values, where wind shear became maximum, was the top of the stable ground based layer. A comparison of the mixing heights obtained from the computer model and the sodar facsimile charts showed a good correlation (Fig. 8) with a regression coefficient of 0.676.

Piringer [80] reported measurements of mixing heights in Vienna, Austria using sodar, tethersonde and radiosonde. He often found a good agreement between the

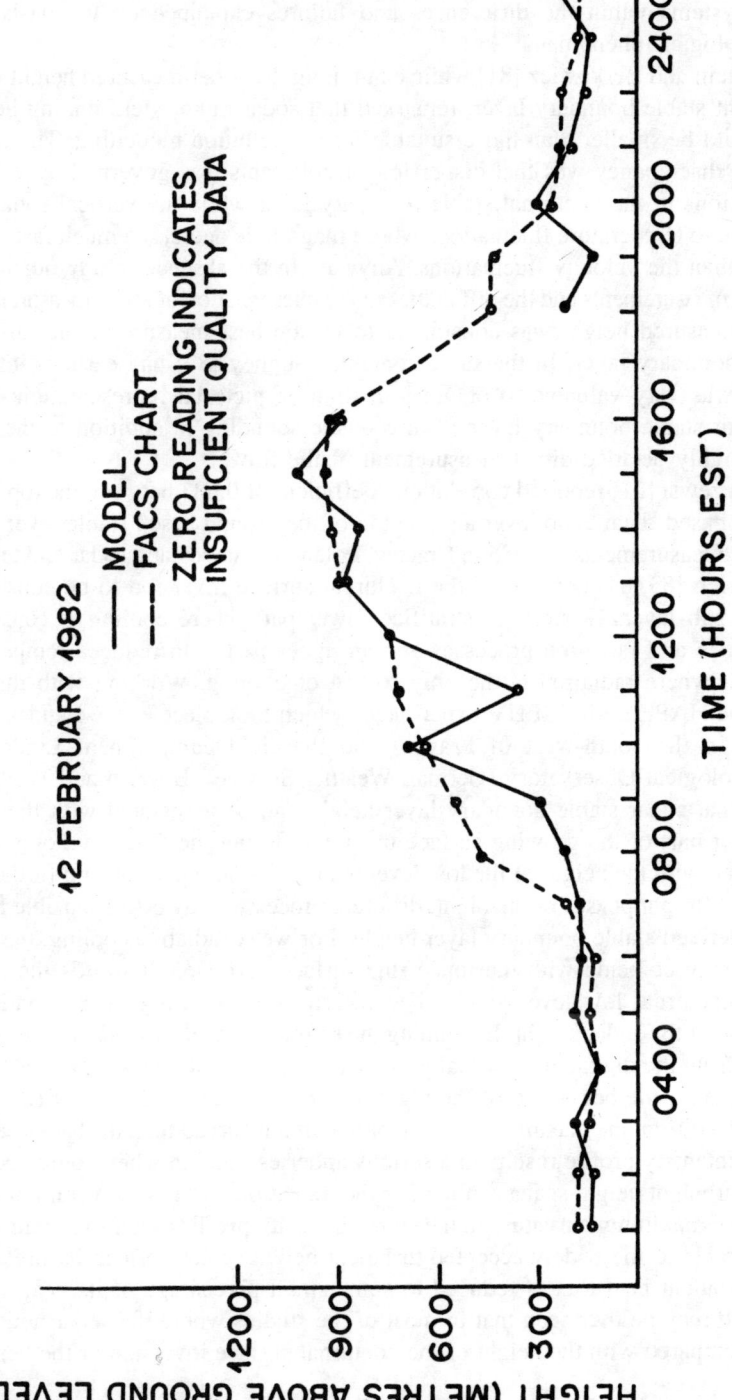

Fig. 8. Comparison between computer generated mixing height estimates using sodar wind data and estimates made from a facsimile chart recorded over the same time period (Jones et al, 1984).

three systems wihin the differences and failures explained on the basis of a meteorological phenomena.

Koracin and Berkowicz [81] while examining the sodar measured height of the turbulent stable boundary layer, remarked that sodar might yield mixing heights that could be smaller than those suitable for air pollution modelling. The reason for this discrepancy was that dispersion of pollutants was governed by velocity fluctuations in the nocturnal stable boundary layer while the vertical sodar was sensitive to temperature fluctuations whose magnitude decreased much faster with height than the velocity fluctuations. Anyway, in the absence of any other more reliable measurements and the difficulties in the interpretation of sodar measurements, sodar measured height was considered to be the best measure of the turbulent stable boundary layer. In the same context, Caughey [82] had earlier remarked that it was very valuable to obtain a qualitative pictorial representation of the turbulent stable boundary layer from acoustic sounding in addition to the more usual highly detailed direct measurement of the flow.

Walczewsi [21] reported correlation coefficient of 0.683 between the top of the ground based shear echo layer and the top of the ground based stable layer while making measurements in 1985 in Cracow, Poland by monostatic sodar and tetroon.

Beyrich [83, 84] considered the nocturnal surface inversion to be constituted of two sub layers: a strongly stratified lower part where cooling is caused by turbulence and radiation processes and an upper part with reduced temperature gradient where radiation is the only source of cooling. Working with the data from the HAPEX-MOBILHY experiment (which took place in the Landes forest region in the south-west of France) and the field campaign at Lindenberg Meteorological Observatory (German Weather Service), Beyrich and Weill [85] found that sodar stable boundary layer height can be associated with the lower turbulent part of the growing surface inversion during the first few hours of the night and with the height of the low level wind maximum later on and further that during different phases of the night, different processes may be responsible for the sodar derived stable boundary layer height. For weak radiative cooling, the sodar height may coincide with the top of the surface inversion. Towards the end of night, nocturnal low level jet can become stronger resulting in the continuous decrease of the sodar height. Examining the estimated stable boundary layer (SBL) height from the digitalized vertical profiles of sodar signal intensity, it was further found that at the beginning of the night when stratification is weak, SBL height determined from the maximum gradient or maximum curvature in the back-scattered signal intensity profile results in a serious under estimation when compared with other turbulent height scales while after the transition period of several hours, the height of maximum curvature in the signal intensity profiles seems to yield values comparable to the widely accepted turbulent height scales defined by presuming that turbulent heat flux is reduced to some small percentage of its near surface value. It may be thus seen that in most of the studies where the sodar height has been compared with the height of the nocturnal surface inversion or the height of the low level wind jet, the results are quite contradictory and cover a wide range of conclusions.

In terms of finding the day time mixing height, it has already been pointed out that direct determination of the mixing height during the daytime is limited to heights within the sodar probing range while for most of the sites, the mixing height is higher than the probing range. Different techniques to estimate the mixing height under convective conditions have been proposed using the available sodar data. In the following we try to have a brief look on these attempts.

Singal and his associates [46, 86] developed a technique to determine the mixing height during daytime when the plume is not capped by a stable layer. Studies of the convective boundary layer height on the basis of Holzworth model [87] using radiosonde data of Delhi were made. The mixing height so obtained from day-to-day was compared with the corresponding sodar measured depth of the thermal plumes and the following empirical relation was laid down to determine the mixing height from the observed height of the thermal echoes:

$$y = 4.24x + 95$$

where y is the depth of the mixed boundary layer as per Holzworth model and x is the depth of the sodar measured thermal plumes. Ths relationship has been developed for plain/smooth terrain.

Weill et al, [88] found that measurements of the Doppler velocities and their variance during the morning development of the convective boundary layer (CBL) could be very useful for studying the convective structure and the mixing height under unstable conditions. Working at Chigne (France) they found that the profiles of the mean square vertical velocity variance, $\sigma_w = \overline{w'^2}$ in the free convection had a maximum at $Z/h \approx 0.5$ (w' is the measure of the fluctuations in the vertical velocity). This result was in agreement with the findings of Deardorff [89] and could thus help to define the altitude of the convective boundary layer where heat flux vanishes.

In a subsequent paper, Weill et al, [90] investigated the relationship between the inversion height, Z_i, and the heat flux profile. They derived the following appoximate esquation for the dry convective layer:

$$\sigma_w^3/Z \cong \alpha^{3/2}(g/\theta)\ \overline{w'\theta'}$$

where α is a constant approximately equal to 1.4, θ is the potential temperature and θ' is the fluctuation in the potential temperature. Since this equation showed that the sensible heat flux, $\overline{w'\theta'}$, was proportional to σ_w^3/Z, an attempt was, therefore made to estimate the altitude of the convective boundary layer from the σ_w^3/Z profiles and the variations in the level of the capping elevated inversion layer in the convective boundary layer. The plot (Fig. 9) showed that height, h, was close to the inversion height, Z_i during the mornings and continued to remain measurable in the afternoon, offering thus an estimate of the inversion height or the mixing

350 Applications in the Atmosphere

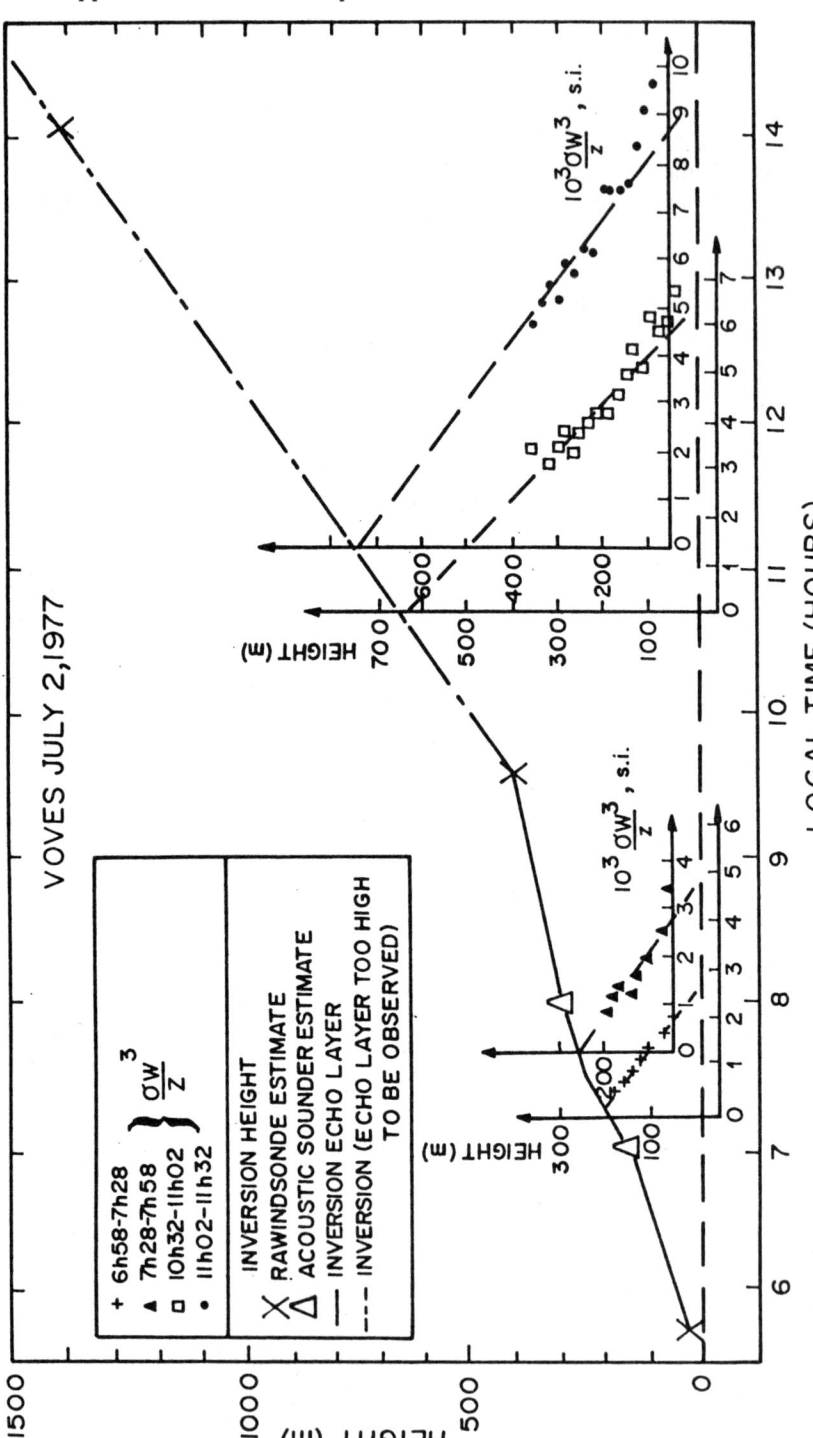

Fig. 9. Plots of the variation of the inversion height and σ_w^3/z profiles (Weill et al, 1980).

depth by the acoustic sounder in free convection. This result was, however, contradicted later by Best et al, [91].

Melas [92, 93] suggested similarity methods to measure the surface heat flux and mixing depth layer in the CBL using the capabilities of the acoustic sounder. An operational model based on Enger's work (94) was developed. It was shown that sodar measurements of the thermal structure parameter, C_T^2, and the standard deviation of the vertical wind, σ_w, could be used to estimate the CBL mixed layer depth, Z_i, and the surface heat flux, Q_0 using the following relationships:

$$Z_i = (2.7/b^2)^{3/4} \, (g/T)^{-3/2} \, (C_T^2)^{-3/4} \sigma_w^3 Z^{-1}$$

$$Q_0 = 0.48 \, (g/\theta)^{1/2} (C_T^2)^{3/4} \, Z$$

$$Q_0 = 1.9^{-3} \, k^{-1} \, (g/\theta)^{-1} \, \sigma_w^3/Z$$

where b is a constant defined as $b = \sigma_w^2/w_*^2$ and is proposed to be equal to 0.45. The thermal structure parameter C_T^2, wind variance σ_w and the mixed layer velocity scale w_* for the above computations was to be determined from the sodar data.

A critical assessment of the above approach is that similarity relations in the CBL are not known better than 15 to 20% [95]. Further, the accuracy of the C_T^2, and wind variances derived from sodar measurements is also not high. The uncertainty in σ_w measurements is approximately 30% which can, of course, be reduced by averaging while the best values of sodar determined C_T^2 are within a factor of two [96] even after knowing humidity and temperature profiles and calibrating the transducer efficiencies. Horizontal homogeneity can be a fruther factor imposing severe limitations on the validity of the similarity relations. However, since sodar measurements represent volume averages, the derived quantities are not expected to reflect local inhomogeneities in a scale of few hundred meters or less.

In a limited comparison with the direct measured values of surface heat flux and mixed layer depth, Melas [92], however, found that sodar derived estimates and direct measurements of the parameters showed a fair agreement with a correlation coeficient of 0.68 (as estimated by Beyrich [83]). However it is necessary to perform a statistical validation against a large data set keeping in view the limitations of the existing methods and the poor accuracy of the parameters measured by the sodar. Further, the applicability of the method is rather limited over complex terrain or regions with heterogenous land use or surface cover.

Beyrich [84, 97] has suggested the use of a simple mixed layer model equation derived by Batchvarova and Gryning [98]:

$$\delta h/\delta t = (1 + 2A) \, [(w'\theta')_0/\gamma h] + B \, u_*^3/\beta\gamma h^2$$

where A and B are constants with typical values of 0.2 and 5 respectively (to be adapted to sodar observations in the morning hours when the top of the rising

CBL can be easily determined from sodar data), $(w'\theta')_0$ is the surface heat flux, u_* is the friction velocity, $\beta = g/\theta$ is the buoyancy parameter and γ is the potential temperature lapse rate. The above approach is based on the studies of Driedonks [99], that parameterization of turbulence production by buoyancy and surface friction (turbulent kinetic energy budget) is sufficient to simulate most of the observed mixing height variability.

Using data of the HAPEX-MOBILHY experiment, Beyrich [97] found that model results were comparable with radiosonde data indicating that the Doppler sodar data can be reliably used to estimate and nowcast the height of a deep CBL. Application of the model, however, requires knowledge of an initial temperature profile which can be easily available from the nearest radiosounding station. Problems, of course, can arise in complex terrain.

Singal and associates [100] have reported the height of the stable and unstable boundary layer at a number of sites in India having different climatology and topography. The climate varied from moderate and humid to dry, wet, warm, cold, while the terrian varied from sea coast to hilly region and plain lands. The data have been drawn from the monostatic sodar measurements made at these sites using the approach laid down by Singal et al, (46, 86) in their earlier work. Plots of the stable and unstable boundary layer for the various locations for all the months of the year are shown in Figs. 10a and b. The following may be seen from these plots:

1. The maximum height of the unstable boundary layer at any station is 1.6 km
2. The unstable boundary layer height at Tarapore is less than 500 m. This is because Tarapore is a coastal station on the Western Ghats Range in India and thus experiences sea breeze during the day time which does not allow convective atmosphere to develop beyond a certain limit.
3. The convective boundary at Dehradun goes as low as 700 m during the monsoon months of July to September. This happens because Dehradun is located at the foot of Shivalak Hills and the sky at Dehradun remains overcast during the monsoon season.
4. The unstable boundary layer height at Jamshedpur and Jodhpur is maximum during the premonsoon (March to May) months of the year while the minimum height is during the post monsoon month of September. This is an expected behaviour from any of the plain stations. Compared to this, the maximum height of the unstable boundary layer for Delhi and Chittorgarh, the other plain stations, is not occurring during the summer season. It may be due to the blowing of the strong surface winds druing other months of the year.
5. The stable boundary layer height is not more than 250 m for any of the stations. It is generally variable throughout the year i.e it does not have a distinct maxima or minima for any month and for any one of the stations (no systematic behaviour). May be the local factors of terrain, soil characteristics and winds are responsible for this behaviour.

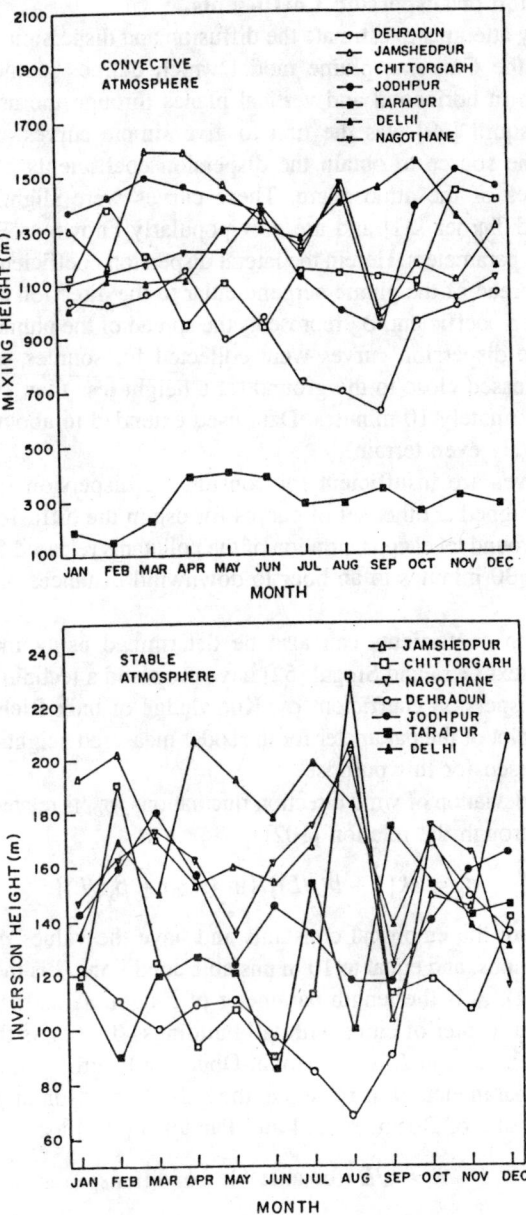

Fig. 10. Height of: (a) convective boundary layer (Day time mixing height) during maximum heating hours (1200–1400 Hrs) at different stations; (b) nocturnal stable boundary layer (nocturnal mixing height) during maximum cooling hours (0300–0500 Hrs) at different stations.

(ii) Determination of Dispersion Coefficients

Most of the early attempts to estimate the diffusion and dispersion of air pollutants were based on the Gaussian plume model which defines plume concentration distribution both in horizontal and vertical planes through the use of dispersion coefficients. Pasquill [29] was the first to give simple curves as a function of distance from the source to obtain the dispersion coefficients pertaining to the various stabilities of the atmosphere. These curves were slightly modified by Gifford [30] and Turner [31] and are now popularly known as Pasquill-Gifford (P-G) dispersion parameters. Herein the lateral dispersion coefficient, σ_Y, represents the horizontal spread of the plume perpendicular to the direction of travel and the vertical dispersion coefficient, σ_z, represents the spread of the plume in the vertical. The data for the dispersion curves were collected for sources where the tracer material was released close to the ground (at a height less than 10 m) for a time period of approximately 10 minutes. Data used extended to about 1 km of rural, open and relatively even terrain.

The P-G curves are insufficient for considering dispersion from tall stacks. Briggs [101] designed another set of curves for use in the diffusion equations for calculating the ground level concentration of the pollutants released from an elevated stack source for 30 minutes to an hour to downwind distances of approximately 10 km.

The dispersion coefficients can also be determined using monostatic sodar data. In this context Gera and Singal [52] have proposed a technique to determine the crosswind dispersion coefficient σ_Y. Knowledge of bulk Richardson number deduced from a plot of the parameter for the sodar measured height of the boundary layer has been used for this purpose.

The standard deviation of wind direction fluctuations, σ_θ, is related to the stability parameter z/L through the relation [102]:

$$\sigma_\theta = ak\{1 - b(z/L)\}/\{\ln (z/z_0) + \beta(z/L)\}$$

where a and b are the empirical constants and have the values $a = 1.5$ and $b = 1$ in stable conditions, and equal to 10 in unstable conditions, k is the von Karman's constant ($k = 0.4$), z_0 is the length parameter of roughness and is approximately equal to 1.5 m for center of cities with tall buildings, $\beta = 6$, z is the height of the stable boundary layer and L is the Monin-Obukhov length.

The stability parameter z/L is related to the bulk Richardson number R_i through the empirical results of Businger [13] and Pandolfo [104] as:

$$R_i \approx z/L \text{ (for unstable conditions)}$$

and

$$R_i/(1 - 5R_i) = z/L \text{ (for stable conditions)}$$

The use of these relationships enables the computation of σ_θ which in turn enables determination of the horizontal dispersion coefficient σ_Y through the use of the relation:

$$\sigma_\theta = \tan^{-1}(\sigma_Y/x)$$

where x is the downwind distance in meters. Thus, knowing R_i one can estimate σ_Y.

The scheme of working is to plot bulk Richardson number (from radiosonde data) at 0000 GMT as a function of the corresponding depth of the stable boundary layer (from the sodar records) as shown in Fig. 11. This plot has shown that Richardson number first increases slightly with the increasing depth of the stable boundary layer, arrives at a point of inflexion, beyond which it starts decreasing steadily with the increase in the depth of the stable boundary layer. This plot has been used to determine the prevailing value of the Richardson number for the given depth of the stable boundary layer and finally to compute σ_Y. A comparison of the computed values using the above scheme with those given by Pasquill curves for the stable conditions has shown that the two corroborate.

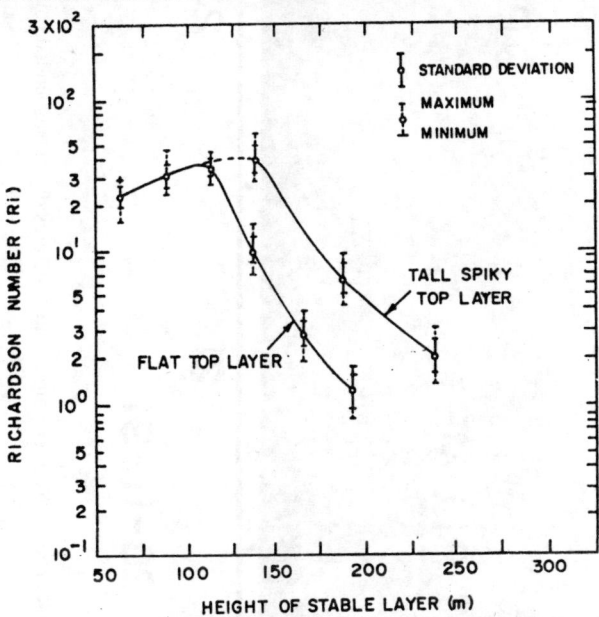

Fig. 11. A plot of the bulk Richardson number as a function of SODAR observed depth of the stable layer for flat top and tall spiky top strucures

The vertical dispersion coefficient, σ_z, can also be measured as a function of downwind distance x, in case the plume is actually passing over the sounder and can be visualized on the sodar echograms. This has been practically done for the plume echo structures observed at an elevated height on the sodar echograms (Fig. 12) on our sounder at Chittorgarh (India). These structures were seen when the stack plume consisting of hot particulate matter from a nearby cement factory passed over the sodar under favourable wind conditions (such structures have

356 *Applications in the Atmosphere*

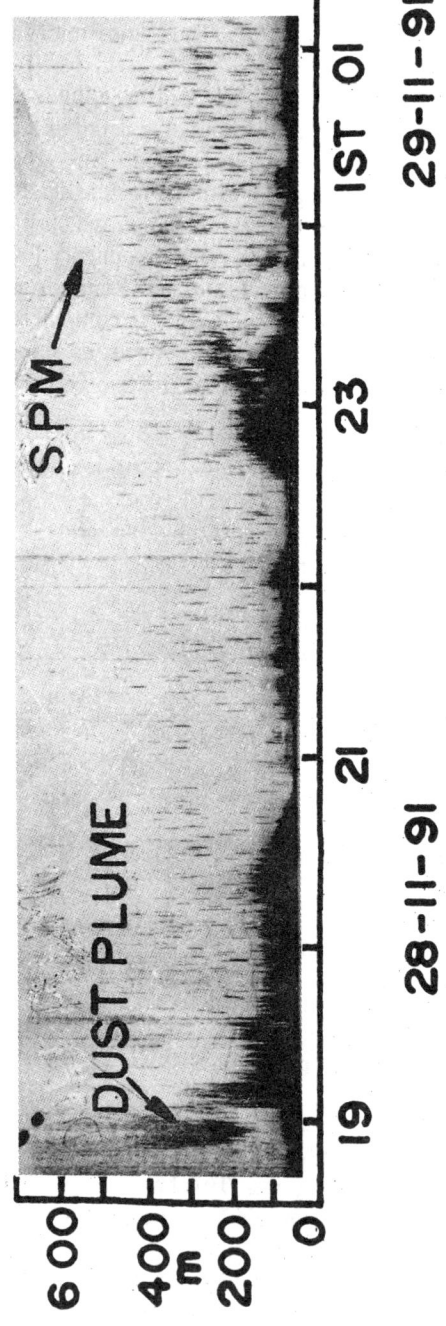

Fig. 12. A view of the SODAR echograms of elevated echo structures due to suspended particulate matter (SPM) observed at Nimbahera, Chittorgarh (India)

been seen for other stations also). A comparison of these values of σ_z under C, D & E stability categories with those obtained from P-G curves shows (Fig. 13) that the values obtained from sodar echogram data are slightly higher (10–20%). This difference in the two values is possible since the P-G curves are valid for releases at 10 m level and for open terrain while the release under consideration is at an elevated level in a complex semi-urban environment.

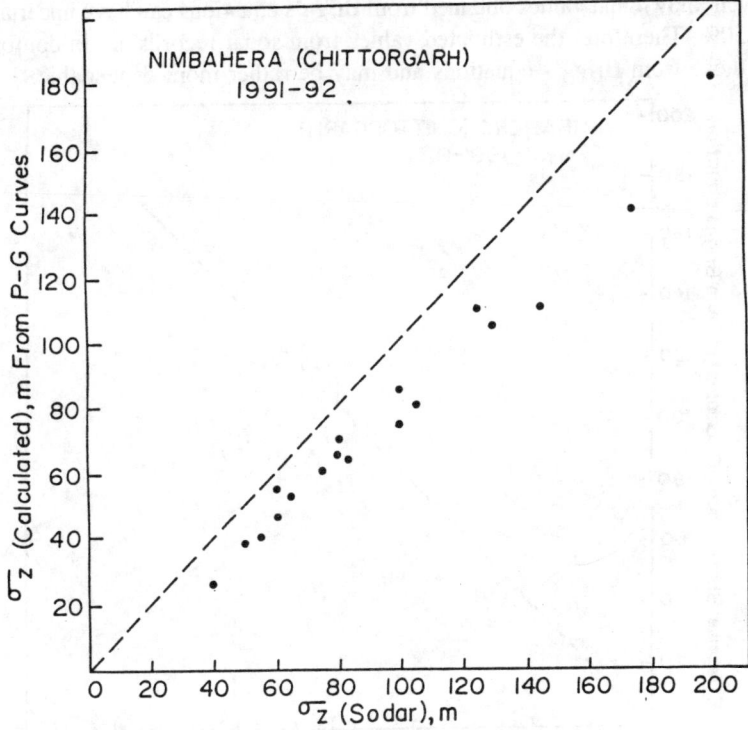

Fig. 13. A plot of the vertical dispersion coefficient, σ_z, obtained from SODAR echograms and Pasquill-Gifford curves.

In the above context, it may be mentioned that Coulter and Underwood [105] have also shown, while taking measurements of temperature structure parameters, the potential of the monostatic sodar system to estimate the relative dispersion coefficients of the cooling tower plume. They found that the width of the cooling tower plume increased linearly with distance from the tower.

(iii) Determination of the Effective Stack Height

We have already seen that on days when the stack plume passes over the monostatic sodar, vertical dispersion coefficient σ_z, can be determined. From the same sodar echograms the effective stack height, h_e, (physical stack height plus plume rise height) for the emitted plume can also be measured. The central intense part of the elevated echo gives the effective stack height. The effective stack height calculated using Brigg's equations and estimated from sodar records obtained at Chittorgarh (India) have been compared. It may be seen (Fig. 14) that effective stack height from Brigg's equations is lower by about 20% compared to that obtained from sodar records. There is nothing unusual in this discrepency since it is well known that values obtained from Brigg's equations can have uncertainties upto 50%. Therefore, the estimated values from sodar records are in conformity with those from Brigg's equations and may be rather more dependable.

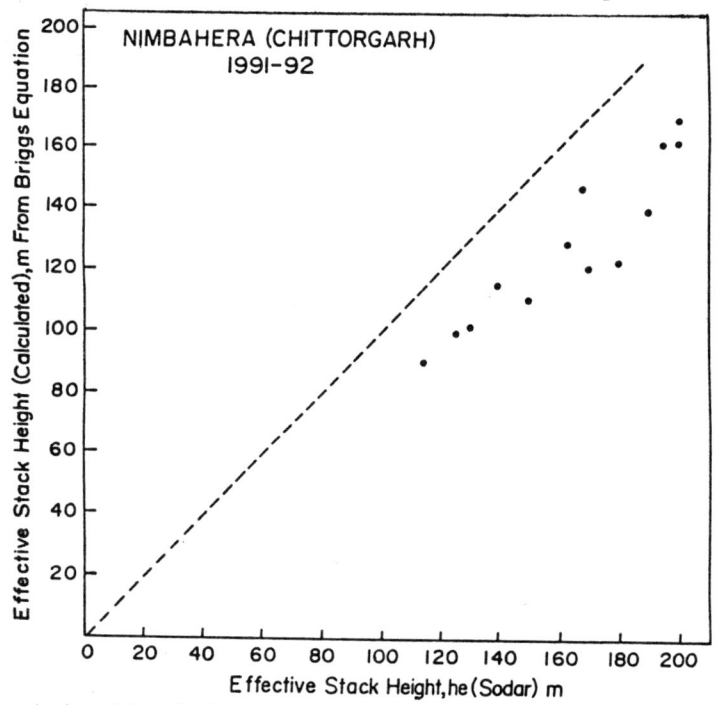

Fig. 14. A plot of the effective stack height estimated from SODAR elevated echo layer and calculated using the Briggs equation.

In case of multiple stacks of nearly equal height located at different nearby places in a factory, the effective stack height distance from the point of observation can also be estimated through geometrical configuration.

(iv) Pollution Concentration Prediction

We have seen that sodar is able to visualize the atmospheric stability, can measure wind vector in the atmospheric boundary layer and can also be used to determine dispersion related parameters, therefore sodar has a great potential to nowcast hazardous events related to air pollution. In the following we describe the efforts made to estimate the pollution concentration at a place under different stability conditions for a given source or sources of emission.

Jensen and Petersen [106] proposed a simple box model based on the height of the sodar measured mixing layer to calculate air pollution concentrations and their variation in time. Concentrations of SO_2 in a large city (Gladsaxe, Denmark) during a subsidence situation were predicted and a good agreement (Fig. 15) between model results and measurements was found.

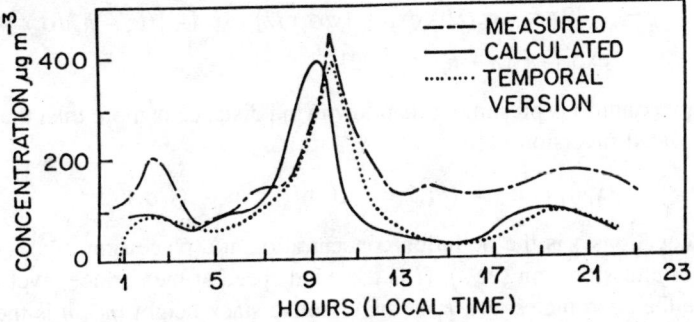

Fig. 15. Comparison of SO_2 concentration measured on 20 Feb. 1975 at Gladsax TV tower (10 minute average values) with those estimated using simple quasi-stationary box model (Jensen & Petersen, 1979).

Moulsley and Cole [107] showed that acoustic sounder measurements could be used to collect information about the size, behaviour and concentration of methane plume released in the air. For this purpose, concentration structure and wind velocity structure parameters of the methane plume were determined from acoustic scattering measurements. The following equation for the back-scattering cross-section, $\sigma(\pi)$, in terms of the concentration structure parameter, C_m^2 was derived:

$$\sigma(\pi) = 6 \times 10^{-4} k^{1/3} C_m^2 / P_0^2$$

where k is the wave number and P_0 is the total pressure. From the measurements, it was found that plume dimensions derived from sounder measurements agreed well with predictions of simple diffusion theory.

Brusasca et al, [108, 109] experimented by feeding model input data from a Doppler acoustic sounder and radio acoustic sounding system installed near the power plant. They found that the performance of the simple Gaussian model improved through the use of data from the remote sensing systems and further that it paved the way to use more complex diffusion models.

Best et al, (60, 91, 110] reviewed air pollution dispersion models for convective conditions together with a discussion of implications for power station siting, design and use of acoustic sounding data for a preferred dispersion model. It was found that for a given source location, acoustic sounder information was almost essential for the present level of understanding of convection at a complex non-ideal site.

360 *Applications in the Atmosphere*

Singal and his associates [53, 54] have worked at the National Physical Laboratory, New Delhi, India to develop a sodar based model to predict pollution concentration at a place due to emission source or sources. They have proposed to adopt the normal Gaussian dispersion model for estimating the downwind pollution concentration. The following Gaussian dispersion model equations under clear sky and canopy conditions [111] have been proposed to be used:

For clear sky

$$\gamma = Q/[2\pi\sigma_y \cdot \sigma_z U] \cdot \exp[-(y/\sigma_y)^2/2] \exp[-\{((z-h_e)/\sigma_z)^2\}/2] \cdot \exp[-\{((z+h_e)/\sigma_z)^2\}/2]$$

For canopy conditions presuming that downwind distance is more than the height of the elevated inversion base:

$$\gamma(x, y, z, h_e) = Q/[(2\pi)^{1/2} \sigma_y \sigma_z U] \exp[-(y/\sigma_y)^2/2]$$

In these equations γ is the pollution concentration at a place (gm m^{-3}), Q is the quantity of emission (gm sec^{-1}), U is the wind speed at the surface level (ms^{-1}), x, y, z are the coordinates (m), h_e is the effective stack height (m), h is the depth to the elevated inversion base (m) and σ_y and σ_z are the dispersion coefficients.

As a case study, the above approach has been applied to compute particulate matter concentration with respect to distance for a cement factory located at Nimbahera (Chittorgarh) in Rajasthan, India. The terrain is a vast valley surrounded by medium range hills, a part of the Aravali mountains. The nearest hills are at a distance of around 10 km from the site. Sodar was placed at a distance of 500 m from the stack. It has a range of 700 m and was operated for the period May 1991 to April 1992 taking sample data for about 10 days every month. The stack parameters are given in Table 4.

The Gaussian dispersion model equation used to determine the downwind concentration (surface level) of the particulate matter due to emission from a source gets slightly modified in this case and becomes

$$\gamma = Q/[\pi\sigma_y\sigma_z U] \cdot \exp[-(y^2/\sigma_y^2 + h_e^2/\sigma_z^2)/2]$$

For carrying out the concentration calculations the dispersion coefficients have been taken from the P-G curves for the respective stability category determined from sodar echograms and then suitably modified as per findings discussed in the previous sections. Calculations have been made for two locations, Gambhiri and Nimbahera, as shown in the layout map (Fig. 16). Results of the observed and calculated values for these two sites for each sampling period are given in Table 5 which also gives the sampling time of observed particulate matter. It may be pointed out that high volume sampler was used to measure the particulate matter concentrations at a few selected places and further that sampling for the time period 1700 hrs on one day to 0900 hrs on the next day for the reported values at Gambhiri was continuous. It has been seen that the model calculated values are

Table 4. Stack data for the cement factory at Nimbahera, Chittorgarh (India)

A. Kiln stacks	No.	1A	1B	2	3	4
All values at exit to atmosphere						
(i) Internal diameter [m]		1.6	1.2	2.2	2.2	3.0
(ii) Height above G.L. [m]		49.4	52.8	60.0	65.0	87.0
(iii) Stack gas temp. [°C]		140.0	140.0	150.0	150.0	300.0
(iv) Stack gas velocity [m/s]		15.0	26.0	20.0	20.0	27.0
(v) S.P.M.[a] concentration [mg/nm^3]		200.0	200.0	200.0	200.0	300.0
B. Cement mill stacks	No.		1	2	3	4
(i) Internal diameter [m]			0.6	0.6	0.6	0.9
(ii) Height above G.L. [m]			30.0	30.0	30.0	30.0
(iii) Stack gas temp. [°C]			90.0	90.0	100.0	100.0
(iv) Stack gas velocity [m/s]			12.0	12.0	14.0	12.0
(v) S.P.M.[a] concentration [mg/nm^3]			150.0	150.0	250.0	250.0
C. Other stacks			(a)		(b)	
(i) Height above G.L. [m]			60.0		40.0	
(ii) No. of stacks			6		8	
(iii) Internal diameter [m]			0.3		0.4	
(iv) Stack gas temp. [°C]			40.0		40.0	
(v) Stack gas velocity [m/s]			8.0		8.0	
(vi) S.P.M.[a] concentration [mg/nm^3]			150.0		150.0	

[a]S.P.M. — Suspended Particulate Matter

more or less comparable with the observed values. The slight mismatch is considered to be due to the possibility of local dust contribution and power failure during sampling period.

Diurnal variations in the computed concentration of the particulate matter have also been studied in correlation with the sodar determined mixing height for the above discussed two locations (Fig. 17). It may be seen that during the fumigation period (depicted on the echograms as the morning rising layer from the surface level), the model computed concentration of the particulate matter increases enormously as expected.

Pekour and Kallistratova [112] have reported a linkage between the concentration of the pollutants measured locally in Moscow, Russia with the mixing height and wind speed parameters measured by sodar during the same time period. Applying the box-model equation

$$C = l\, Q_a/(h_{mix} U)$$

where l is the linear dimension of a city, Q_a the emission per unit area, U, the mean wind velocity within the mixing region and h_{mix}, the mixing height. It has been seen that the variations in time in the pollutant concentrations (measured for soot

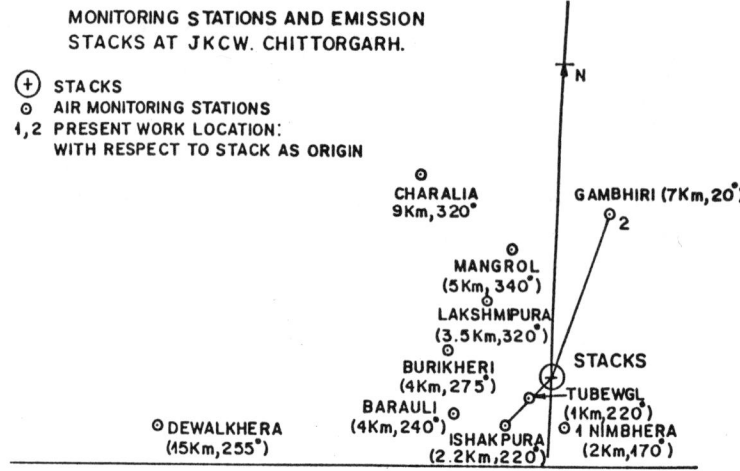

Fig. 16. Layout map of the surroundings of J.K. Cement Works at Nimbahera (Chittorgarh) marking the site of SODAR location and the two other sites where particulate matter concentrations were measured. In the brackets at each site are given the respective distance and angle held with the source at the factory site.

Table 5. Comparison of observed and calculated values of the particulate matter for the two locations at Chittorgarh (India)

Location	Date	Sampling period (hrs)	Particulate concentration ($\mu g/m^3$)		Remarks
			Observed	Calculated (NPL model)	
Nimbahera	2–12–91	09h17	610	462	Possibility of local dust contribution
Gambhiri	30-11-91	17h00	94[a]	79	-do-
	1-12-91	00h09	94[a]	67	
	1-12-91	17h00	111[a]	121	Sampler not operative for part of the time
	2-12-91	00h09	111[a]	183	

[a]Continuous sampling from 17:00 hrs on one day to 09:00 hrs on the next day

and NO_2) and the box factor $BF = 1/(h_{mix} U)$ measured using a sodar, correlate with each other as shown in Fig. 18. The positive result of the comparisons shows the usefulness of the sodar determined parameters of mixing height and stratification of the atmospheric boundary layer for air pollution meteorology.

Application to Microwave Propagation

One of the important parameters in the study of tropospheric propagation of microwaves is refractive index (n) which is a function of the temperature, pressure and humidity of the atmosphere. This parameter for propagation purposes is expressed

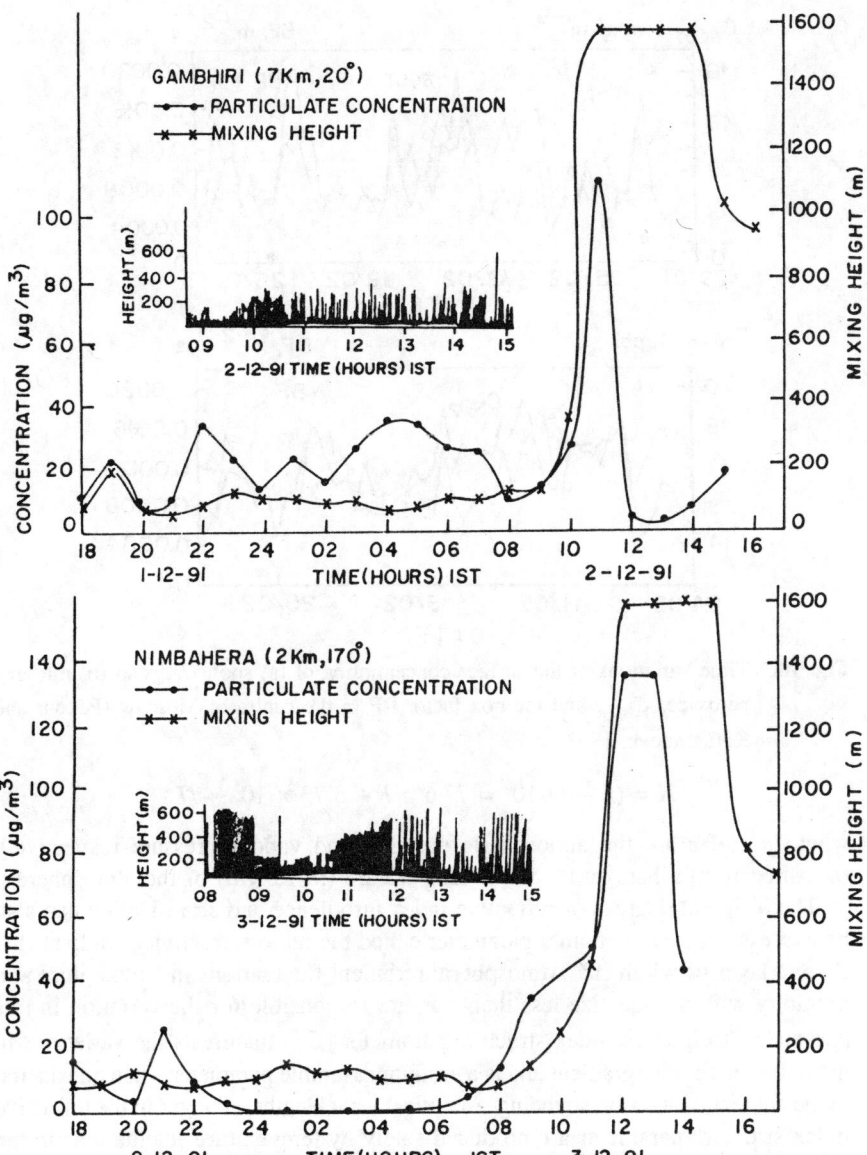

Fig. 17. Studies of diurnal variations of computed concentrations of particulate matter in relation to SODAR derived mixing height at (a) Gambhiri (Chittorgarh) and (b) Nimbahera (Chittorgarh). The overlaid diagrams are the diurnal variations in the SODAR echograms at the SODAR site near the cement factory showing the fumigation period followed by clear thermal plumes.

in the form of radio refractivity (N) which is the departure from intensity in parts per million of the refractive index and is expressed as:

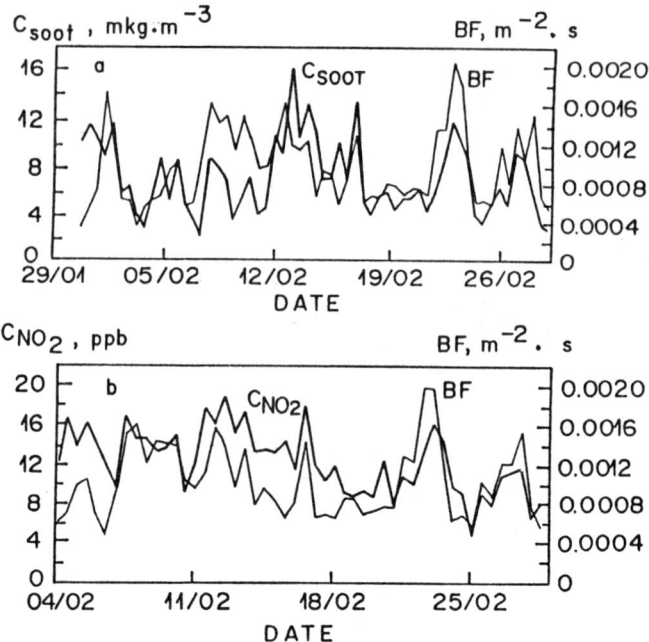

Fig. 18. Time variations of the surface concentration of (a) soot, C_{soot} and (b) nitrogen peroxide, C_{NO_2}, and the box factor BF in the centre of Moscow (Pekour and Kallistratove, 1993).

$$N = (n-1)\,10^6 = 77.6\,p/T + 3.73 \times 10^5 \cdot e/T^2$$

where p and e are the atmospheric pressure and vapour pressure respectively measured in millibars and T is the temperature (in Kelvin) of the atmosphere.

The different degrees of refractive index turbulence and stratification can also be expressed through another parameter called the radio refractivity gradient $\Delta N/\Delta z$ in a layer of width ΔZ. Atmospheric turbulent fluctuations in temperature and humidity with a scale size less than 1 m are responsible for the variation in this parameter. Refractive index structure parameter (C_n^2) manifests the variations in the radio refractivity gradient and is a easily measurable parameter when considered in the context of acoustic sounding. At optical wavelengths, changes in the refractive index structure parameter are produced solely by temperature fluctuations in the atmosphere, while for radio waves the variation in the refractive index structure parameter are also produced by atmospheric conditions of humidity in the propagating layer.

Since the structure of the atmospheric boundary layer, consisting of different degrees of stratification, determines the refractivity or refractive index structure parameter at a place and ABL stratification can be easily visualized with the help of sodar, therefore it should be possible to study microwave propagation characteristics in terms of sodar data and thus use sodar to nowcast hazardous situations of microwave propagation.

(i) Correlation of Sodar Structures and Micorwave Propagation Characteristics

Gilman et al, [113] were the first to use the acoustic sounder for studies related to the observed fadings in the tropospheric transhorizon microwave radio circuit of path length 64 km operating between New York and Neshanic, New Jersey. A correlation of the period of sodar echoes from temperature inversions and microwave radio fadings was reported.

This important work about the potential of acoustic sounding, however, remained unnoticed for quite a long time. It was only in the mid 60s that in pursuit of a similar propagation problem of finding out a linkage between troposcatter radio propagation and incursion of marine air into their experimental area in southern Australia, McAllister (114) returned to the concept of sodar. He showed that acoustic echoes from atmospheric inhomogeneities can be readily obtained to give information of the presence of temperature inversions and superimposed gravity waves. He thus rediscovered the potential use of the acoustic scattered intensity from atmospheric inhomogeneities.

. Wickerts and Nilsson [115] while studying transhorizon radio propagation over a path length of 160 km over the Baltic sea at radio frequencies between 60 and 5000 MHz later used sodar to get information of the height and width of the surface based and elevated inversion layers. They found that heights of temperature inversions are strongly correlated to the heights of duct layers and further that changes in the observed reflected signals correlated with the waves monitored by sodar on the reflecting layer. The influence of such wavy layers of high refractivity gradient on microwave propagation was also reported by Stilke [116]. According to him, microwave propagation depended on the height of the layer, terrain orography, antenna heights, amplitude and orientation of the gravity waves and antenna beam width.

Nilsson [117] while conducting studies in mid Sweden on the observed correlation of deep fadings on a 2 GHz LOS microwave link of path length 73 km with sodar monitored surface based inversion layers found that about 90% of the observed inversions at relevant heights (50 m above the line of sight) are associated with deep fadings. Mitra et al, [118] reported the use of sodar in detecting a tropospheric disturbance causing severe fadings of field strength over the LOS link between Sonepat and Delhi of a path length of 42 km.

Gera and Sarkar [119] reported correlation studies of sodar observed thermal structures and microwave propagation characteristics of the LOS path of 42 km between Delhi and Sonepat operating at a frequency of 7 GHz, with sodar placed at the receiving end at Delhi. They observed that deep fadings were correlated with stable multiple/elevated layers with or without waves superimposed over them (Fig. 19). The wavy layers were responsible for higher fadings. They further

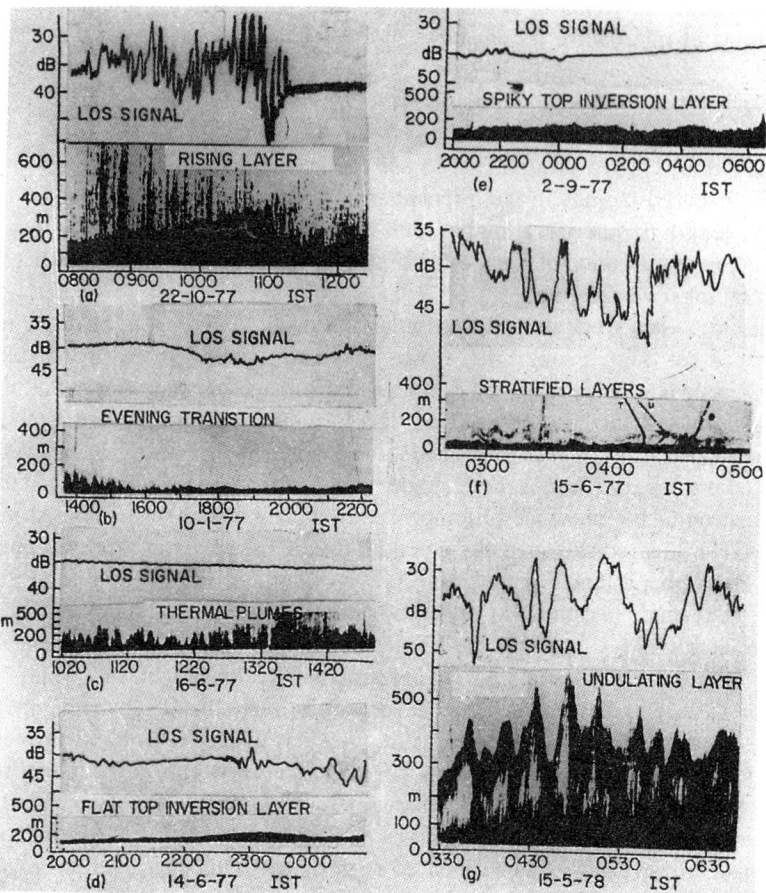

Fig. 19. Sodar structures and corresponding microwave signal characteristics. The linkage between the fading characteristics and the sodar thermal structures are self evident.

correlated the decreasing initial refractivity gradient from sub-refractive to super-refracting conditions to the increasingly stable trubulent structure of the boundary layer (Fig. 20) as observed by sodar. It was seen that for a refractivity gradient of 0 to −40 N units/km, sodar structures of the mechanical mixing type predominate, for the medium refractivity of −40 to −80 N units/km, it is the surface based inversion structures which are predominant, for the super-refracting gradient of −80 to −160 N units/km, sodar structures of multilayers and wave motions predominate and in the ducting range of −160 to −200 N units/km, sodar structures responsible for microwave fadings are predominantly perturbations and multi-layers.

Fukushima et al, [120] made studies of radio fields of UHF TV signals over a path length of 117 km between Choshi and Radio Research Laboratories Tohoku

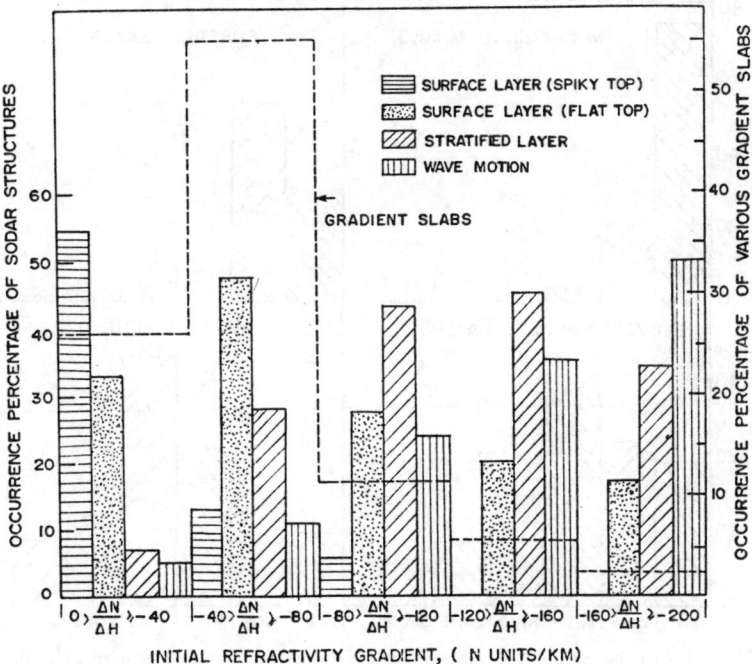

Fig. 20. Occurrence percentages of sodar structures with respect to various states of initial refractivity gradient (Gera and Sarkar, 1980).

in relation to the occurrence of atmospheric temperature inversion layers monitored by the mobile RRL Sodar over the path. They found that occasional high signals were being caused by the ducting mode propagation through a partial radio duct which occurred over the path.

Singal et al, [121] correlated the different types of stable sodar structures with the observed amplitude fadings (Fig. 21). It was found that for the first 3 dB step of amplitude fading, sodar structures of mechanical mixing predominate; in the next 3–6 dB step of amplitude fading, surface based inversion structures predominate; in the 6–9 dB level of amplitude fading, elevated/multilayers predominate; with 9–12 dB level of amplitude fading, perturbations predominate; and beyond 12 dB of amplitude fading, it is mostly the multilayers and perturbations which are responsible for the observed microwave fadings.

Singal et al, [121–123] incorporated sodar determined stable boundary layer data to evolve a ray tracing programme for the Delhi-Sonepat line of sight path obtaining phase and path difference information of the rays reaching the receiver. The computed values of the received signal strength of different values of initial refractivity gradient were found comparable to the measured values of signal strength at the receiving end (Table 6).

368 Applications in the Atmosphere

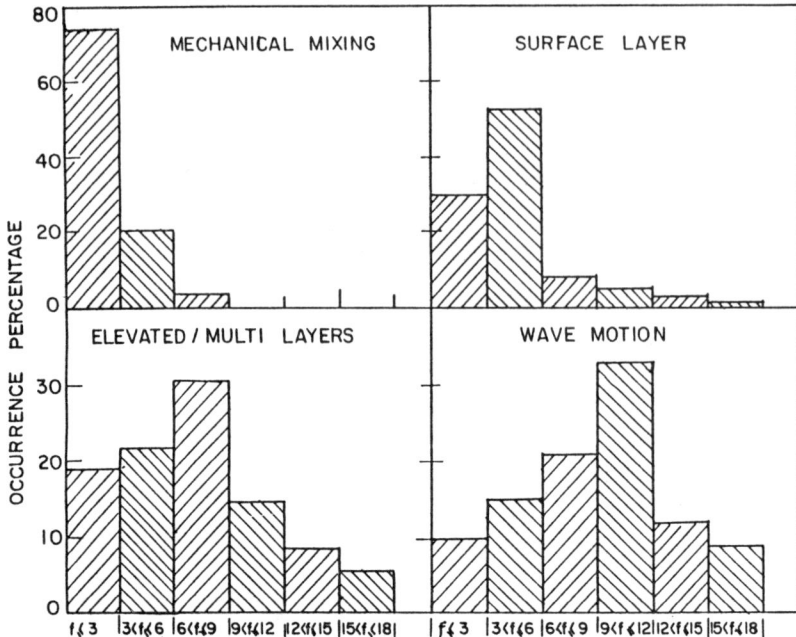

Fig. 21. Plot of the precentage occurrence of the different sodar structures under stable conditions for the various slabs of microwave amplitude fadings.

Table 6. Estimated values of various propagation parameters for a line-of-sight link under various conditions of initial refractivity gradient

Initial refractivity gradient km^{-1}	Signal received in free space P_{FS} above one uV	Propagation factor (dB) due to		Total propagation factor (dB)	Resultant signal strength (dB)	Predicted signal on NBS 101 model (dB)	Obs. signal strength (dB)
		Ground reflection	Layer reflection				
-40 N	65	-23	—	-23	42	33	38
-80 N	65	-21	—	-21	44	35	39
-157 N	65	-9	-10	-19	46	39	45
-200 N	65	-9	-8	-17	48	41	47

Singal et al, have also computed the field strength for a transhorizon path with both transmitter and receiver antennas lying within the duct layer using expressions given by Bullington [124] and Eklund and Wickerts [125] involving refractive index structure parameter. The computed values of the free space field strength and propagation factors for ground reflection and duct layer-reflection for the various transhorizon paths are given in Table 7 which also gives the propagation factor due to scattering. From this table it may be seen that sufficient field strength due to reflections can be received for a transhorizon path of 250 km in the presence of a ducting layer upto a height level of 200 m.

Table 7. Estimated field strength for a trans-horizon ground duct with both transmitter and receiver lying within the duct

Path length	Height of reflection within the duct	Free space field strength P_{FS}	Propagation factor			Received field strength
			Ground reflection	Duct reflection	Scattering	
(km)	(m)	(dB)	(dB)	(dB)	(dB)	(dB)
42	98	65	−9	−10	−60	46
90	120	56	—	−13	−68	43
140	140	52	—	−16	−73	36
190	160	49	—	−18	−80	31
240	180	47	—	−20	−84	27

Mon et al, [126] observed that propagation anomalies over an LOS link of path 44 km operating at 4.1 and 6.2 GHz near Paris were in correlation with sodar ground-based multi-layered structures. The sodar was located at 5km from one end of the LOS path. The mechanism of power fading was considered to be due to multipath and focussing and defocussing of antenna beams by sodar monitored microscale atmospheric perturbations. In another similar experiment on a LOS path of 48 km in South West France operating at 4, 7 and 11 GHz, Mon and associates [127, 128] further observed that deepest fading periods (about 70%) occurred when the top of the temperature inversion reached the mean height (within 40–100 m from the ground) of the radio link in the first Fresnel's zone, 15% of the fadings occurred when the layers were located between 100–140 m while very small number of fadings were observed when the layering was above 140 m (Fig. 22). Sodar in this experiment was located at about midpath of the LOS link where an instrumented tower (110 m) was also present.

In the PACEM experiment consisting of a LOS transmission path 36.3 km long operating at 11 GHz between Marcheville and Viabon (Eure et Loir) and of meteorological techniques of instrumented tower, tethered balloon, monostatic sodar, Doppler mini-acoustic sounder and a lidar set up at Villeau, 1.8 km off the transmission path at a distance of 7.5 km from Viabon towards Marcheville, Sylvain et al, [129] observed that stable events recorded by sodar and other meteorological instruments conform to fadings on transmission path. Claverie et al, [130] further found that sodar determined layer heights are more precise, dependable and continuous to compute refractivity profiles and periods of multipath propagation through ray tracing conforming to the observed behaviour of the propagating waves than obtained through tethered balloon system. They also observed a linkage between the sodar observed gravity waves and multipath propagation leading to anomalous microwave behaviour. The use of at least two sodars along the radiolink path was stressed.

Das et al, [131, 132] studied the behaviour of the 75 km path length, Satkhira VHF TV tower (189 MHz) at Calcutta, in correspondence with the observed

370 Applications in the Atmosphere

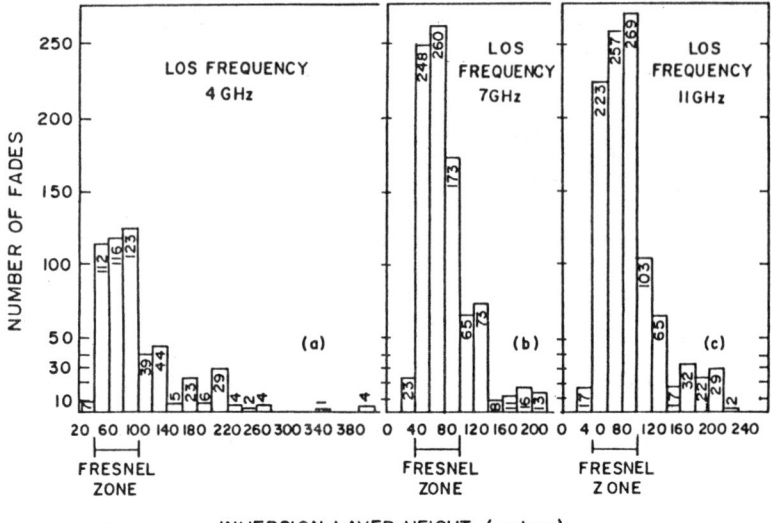

Fig. 22. Fade distribution versus inversion layer heights (Roux and Mon, 1983).

monostatic sodar thermal structures. They observed that the atmospheric refractivity and its gradient values were greatly influenced by thermal structure and stratification in the lower atmosphere. Sharp multipath fadings were observed on the days of elevated structures, morning transient periods and nocturnal surface based inversions.

(ii) Studies of Structure Parameters

Sodar provides an excellent opportunity to determine remotely the turbulence structure parameters in terms of the intensity of the back scatterd sodar echoes and the wind velocity profile. In the accurate determination of these parameters, it is, of course, essential to measure the directivity of the antennas, the acoustic power put into the main lobe, the efficiencies of the electroacoustic transducers, the acoustic absorption coefficient for sound at the operating frequency and the effects of turbulence and horizontal wind on the signal power.

Taking care of the above parameters, measurements of the temperature and velocity structure parameters (C_T^2 and C_v^2 respectively) have been made by many investigators [133–149] as functions of height and time of the day using acoustic sounding. Many of them have compared the sodar derived results with the direct measurements of the parameters and have reported them to vary within a factor of 4 (less than one for unstable conditions and within 2–4 for stable conditions) a fairly good agreement inspite of the various inaccuracies involved in the measurements. Under convective conditions, measured values of the temperaure structure parameter, C_T^2, were found to obey the following law with altitude:

$$C_T^2 = az^{-4/3}$$

while the measured values of the velocity structure parameter, C_v^2, were found to obey the law:

$$C_v^2 = b + cz^{-2/3}$$

where the terms a, b and c are constants. These laws were observed to be true well beyond the surface layer to as high as half the height of the mixed layer. This behaviour of the variations of the structure parameters with altitude corresponds to the theoretical and semi-empirical relations of Wyngaard et al, [150] based on free convection similarity theory for the surface layer upto 22 metres, extendable to higher heights on the basis of Minnesota data as shown by Kaimal et al, [151].

Brown [152] reported a small hump in the profiles of the structure parameters near the height of the capping stable layer followed by a decrease in the magnitude of the parameters. Brown and Clifford [153] had earlier predicted this behaviour on the basis of theoretical analysis (Fig. 23). The observed behaviour was attributed to the entrainment of the warmer air in the capping inversion by the rising turbulence in the thermal plumes while the inversion tended to strongly suppress turbulence.

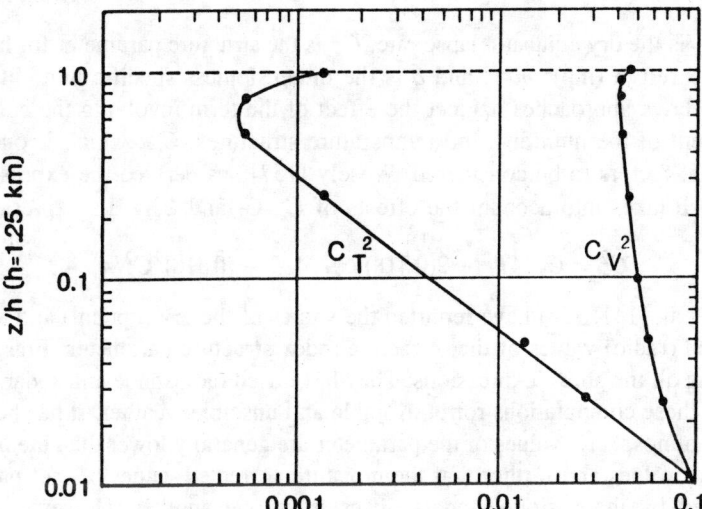

Fig. 23. Variation of the intensity of the turbulent parameters, C_T^2 and C_V^2, with altitude (Brown and Clifford, 1976).

Moulsley et al, [140] observed that the structure parameters increased greatly within the plumes compared with the level outside the plumes.

Gardiner and Hill [154] while investigating the variations in the temperature structure parameter with time at the cloud top from a sounder located at the top of the hill, found that the temperature structure parameter increases with time. This observation was interpretted to indicate an increase in the cloud top entrainment associated with a corresponding increase in temperature variations leading to the gradual dispersal of the cloud.

The refractive index structure parameter C_n^2 in dry air can be easily obtained from the temperature structure function by applying the following expression:

$$C_n^2 = \{77.6 \times 10^{-6} p/T^2\}^2 C_T^2$$

However, for radio wave propagation the effect of humidity in the atmosphere should also be included. This will modify the above expression. For this purpose, two independent approaches have been suggested by Sirkis [155] and Ottersten et al [156]. According to these approaches, the modified expressions are:

$$C_n^2 = M^2 [dT/dz + \gamma_a]^{-2} C_T^2 \quad \text{(Sirkis)}$$

with

$$M = \{-77.6 \cdot 10^{-6} p/T^2\} \{1 + 15500\, q/T\}$$
$$\{dT/dz + \gamma_a[-7800/(1 + 15500\, q/T)]\, dq/dT\}$$

and

$$C_n^2 = [77.6 \cdot 10^{-6} p/T]^2 (1 + 9620 e/pT)^2 C_T^2 + (0.373/T^2)^2 C_e^2$$

(Ottersten et al)

where γ_a is the dry adiabatic lapse rate, C_e^2 is the structure parameter for humidity 'e' measured in $(mb)^2\, m^{-2/3}$ and q is the dimensionless specific humidity term.

The above approaches neglect the effect of the term involving the correlation coefficient of the humidity and temperature structures, C_{eT}, which is one of the important factors to be considered. Wesely [157] has derived the expression for C_n^2 which takes into account the effects of C_e, C_T and C_{eT}. His expression is

$$C_n^2 = C_T^2/4T^2 + 2(0.318)\, C_{eT}/4pT + (0.318)^2 C_e^2/4p^2$$

Singal et al, [141, 158] have reported the values of the dry (optical) and moisture corrected (radio) values of the refractive index structure parameter (Figs. 24 and 25) using all the above expressions. They have used radiosonde and sodar data for making these computations for both stable and unstable weather. It has been seen that dry atmospheric values of the parameter are generally lower than the humidity corrected values and further that the moisture corrected values of the parameter computed by the various models differ from one another. However, nothing conclusive can be said since none of these derived values have been compared with the direct measurements of the parameter. Moreover, the data in itself is not large enough to come to any conclusion.

It has been thus seen that the sodar measured values of C_T^2 do not exactly correlate with the directly measured values of the parameter within the estimated mean square random error of the sodar technique of 40–50% and of the direct technique of 20–30%. Thus the sodar technique though very simple, direct, remotely applicable and continuous has certain inherent problems. The problems for the difference in results between the sodar and direct techniques have been analysed. The major

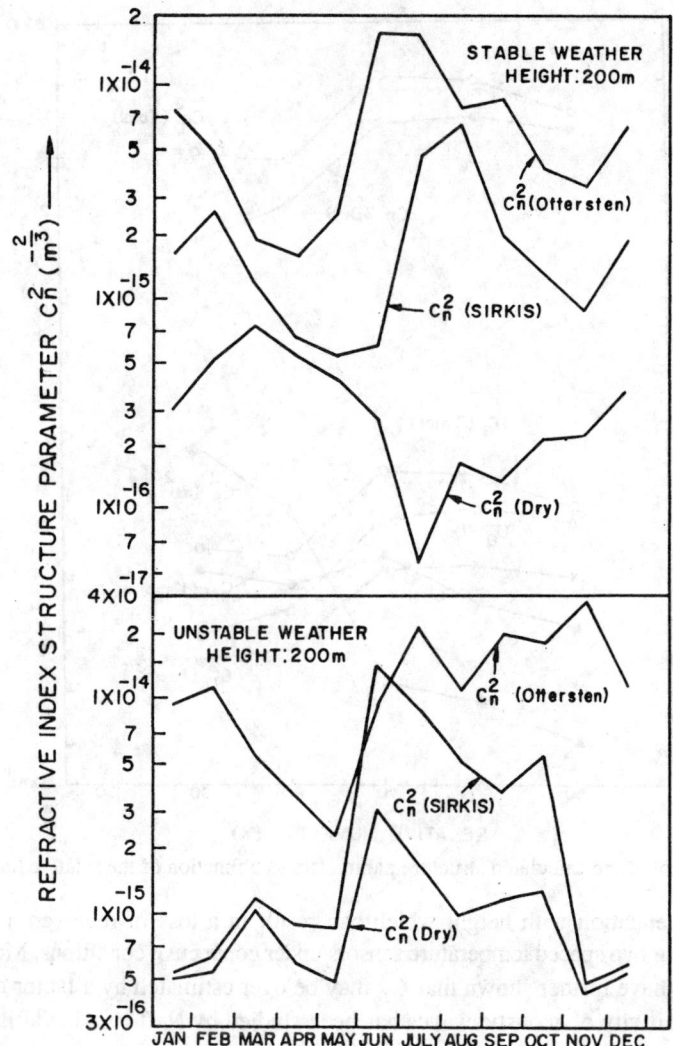

Fig. 24. Plot of mean monthly values of the refractive index structure parameter for a dry atmosphere and for a moist atmosphere following the approaches of Sirkis and Ottersten et al under both stable and unstable weather conditions.

causes for the difference are considered to be the incorrect estimation of the excess acoustic attenuation due to turbulence, beam broadening, beam tilt and wander etc., and the separation and different sizes of the volumes of the atmosphere in which the parameter is measured by the two methods. Besides, the low intensity of sodar scattering, intensity of turbulence and anisotropy can be some of the other causes for the large difference in results.

In the above context, Neff [159] and Brown and Clifford [153] have shown that beam bending by wind and beam broadening by turbulence do increase excess

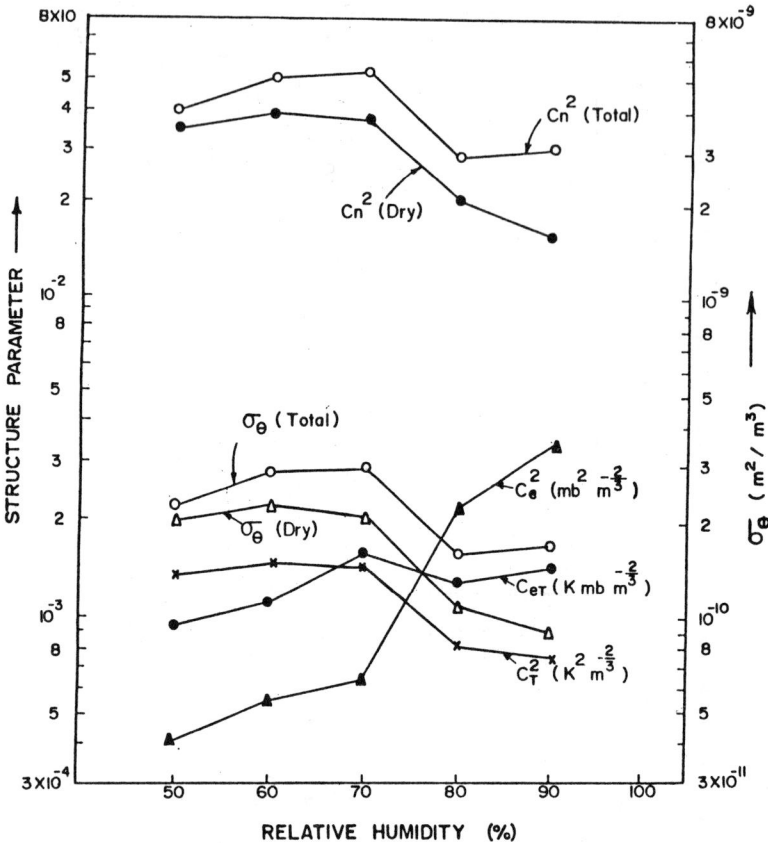

Fig. 25. Plot of the calculated structure parameters as a function of the relative humidity.

acoustic attenuation with height which can result in a loss of received acoustic power. Using two spaced temperature sensors under convective conditions, Moulsley et al, [160] have further shown that C_T^2 may be over estimated by a factor of two. Aspect sensitivity of acoustic waves has been studied by Neff [134], Kallistraova and Petenko [161] and Singal et al [162]. It has been seen that backscattering anisotropy can sometimes double the sodar measured values of C_T^2, however, on average, the effect is small and hence cannot be responsible for the systematic overestimate of the sodar measured values of C_T^2.

Irrespective of the above difference in the values of C_T^2 measured by the two techniques, the considered opinion is that the sodar technique of determining the C_T^2 values is worthwhile. It is because the values of the structure parameters vary over several orders of magnitude while the difference is only within a factor of 2 to 4.

Concluding Remarks

It has been seen that acoustic sounding technique has been extensively developed since the initial work of McAllister [114] and Little [163]. This technique is now serving as a valuable tool for remotely probing the atmospheric boundary layer. Besides the many qualitative studies of the atmospheric phenomena like thermal plumes, inversions, mixing height, stability class, gravity waves, breaking waves, sea breeze, fog and drainage winds, it provides quantitative information on many imporant atmospheric parameters like wind velocity profiles, wind shear, vertical velocity spectra and velocity and thermal structure parameters. While the information on mixing height, inversions, stability class, wind velocity spectra and velocity structure parameter is very useful in monitoring hazardous situations of air pollution, the information about the presence of gravity waves, breaking waves, stability class and elevated inversions in the atmosphere determines the extent of radio signal fadings in the line of sight propagation and the siting of the microwave antennas at a place. Further the determination of C_T^2 profiles is relevant to optical tracking as also to measure the vertical heat flux during conditions of local free convection and in transhorizon microwave propagation problems.

However, a lot of work remains to be done as yet. Investigations at various sites will have to be done to demonstrate the applicability of the sodar structures, their depth and of C_T^2 values to characterize stability, mixing height and inversion height etc., as also for their application in air pollution and microwave communication. Approach has to be worked out to obtain dependable data on variance in wind velocity not only in the vertical direction but in all other directions also. Approach to determine the true values of the structure parameters has also to be evolved. More work is also needed to be done on the aspect sensitivity of the sodar scattering. Automatic structure analysis and acquisition of the various derived parameters has also to be achieved. This will not only help in the interpretation and evolution of the objective pattern recognition scheme but may also improve the determination of the atmospheric stability, mixing height etc., from sodar data, leading ultimately to the identification of the atmospheric boundary layer models that can be used with acoustic sounder data and to the reliable extraction of key boundary layer parameters from acoustic sounding signal in most meteorological situations. No doubt, large number of careful comparative study of acoustic sounder data with in situ measurements will be needed to independently establish sodar in meteorological research.

References

1. S.C. Majumdar, S.K. Sarkar and R. Chadha, "Tropospheric surface ducting phenomena in India", J. Instr. Electr. Telecomm. Engrs., 21, 597–604, 1975.
2. J.F. Louis, A. Weill and D. Vidal-Madjar, "Dissipation length in stable layers", Boundary Layer Meteorol., 25, 229–243, 1983.
3. H. Tennekes, "Free convection in the turbulent Ekman layer of the atmosphere", J. Atmos. Sci., 27, 1027–1034, 1970.
4. J.W. Deardorff, "Convective velocity and temperature scales for the unstable planetary boundary layer and for Rayleigh convection", J. atmos. Sci., 27, 1211–1213, 1970.

5. J.C. Wyngaard, "Lectures on the planetary boundary layer", published in Mesoscale Meteorology: Theories, Observations and Models, edited by D.K. Lilly and T. Gal-Chen, D. Reidel, Dordrecht, The Netherlands, 1983.
6. A.G.M. Driedonks and H. Tennekes, "Entrainment effects in the well mixed atmospheric boundary layer", Boundary Layer Meteorol., 30, 75–105, 1984.
7. L. Mahrt, "Modelling the depth of the stable boundary layer", Boundary Layer Meteorol., 21, 3–19, 1981.
8. F.T.M. Nieuwstadt, "The turbulent structure of the stable nocturnal boundary layer", J. Atmos. Sci., 41, 2202–2216, 1984.
9. N.A. Shaw, "Acoustic sounding of the atmosphere", Ph.D. Dissertation, Department of Physics, University of Melbourne, Australia, pp. 226, 1971.
10. M. Fukushima, K. Akita and H. Tanaka, "SODAR probing of small scale temperature structures in the clear troposphere", J. Rad. Research Labs. Japan 22 No. 108, 23–43, 1975.
11. M. Fukushima "Status of SODAR observations in Japan", J. Rad. Research Labs. Japan 22 No. 109, 151–164, 1975.
12. J.F. Schubert, "A climatology of the mixed layer using acoustic methods", in Proc. of the 3rd Symposium on Meteorological Observations and Instrumentation, American Meteorological Society, Washington D.C., pp. 151–156, Feb. 10–13, 1975.
13. G.H. Clark, E. Charash and E.O.K. Bendun, "Pattern recognition studies in acoustic sounding," J. Appl. Meteorol., 16, 1365–1368, 1977.
14. F.F. Hall Jr., "Boundary layer climatologies from acoustic sounder investigations", in Proc. of the 4th Symposium on Meteorological Observations and Instrumentation, American Meteorological Society, Denver, Colo., pp. 330–332 April 10–14, 1978.
15. B.E. Prater and J.J. Colls, "Correlations between acoustic sounder disperion estimates, meteorological parameters and pollution concentrations", Atmos. Environ., 15, 793–798, 1981.
16. R.A. Maughan, "Frequency of potential contributions by major sources to ground level concentration of SO_2 in the Forth valley, Scotland: An application of acoustic sounding", Atmos. Environ., 13, 1697–1706, 1979.
17. R.A. Maughan, A.M. Spanton and M.L. Williams, "An analysis of the frequency distribution of SODAR derived mixing heights classified by atmospheric stability", Atmos. Environ., 16, 1209–1218, 1982.
18. D.N. Asimakopoulos, C.G. Helmis and D.G. Deligiorgi, "Classification of acoustic sounder facsimile records for use in air pollution experiments", in Proc. of the Intl. Meeting on Application of Sodar and Lidar Techniques in Air Pollution Monitoring (EURASAP), Krawkow, Poland, pp. 1–7, Sep. 20–28, 1990.
19. S.P. Singal, B.S. Gera and S.K. Aggarwal, "Studies of the boundary layer at Delhi using Sodar", in Proc. of the 2nd Intl. Symposium on Acoustic Remote Sensing of the Atmosphere and Oceans, Rome, Italy, pp. XXIII 1–8, Aug. 29–1 Sep., 1983.
20. J. Walczewski, "Acoustic sounding of the atmosphere for inversion layer climatology over an urban area", in Proc. of the 2nd Intl. Symposium on Acoustic Remote Sensing of the Atmosphere and Oceans, Rome, Italy, pp. XXII 1–16, Aug. 29–1 Sep., 1983.
21. J. Walczewski, "Development of SODAR and acoustic soundings of the atmosphere in Poland", Z. Meteorology, 39, 129–141, 1989.
22. J. Walczewski and M. Felesky-Bielak, "Diurnal variation of characteristic SODAR echoes and the diurnal change of atmospheric stability", Atmos. Environ. 22, 1793–1800, 1988.

23. Th. Foken, K.H. Hartmann, J. Keder, W. Kuchler, J. Neisser and F. Vogt: Z. Meteorology, 35, 348–354, 1987.
24. K. Evers, J. Neisser and E. Weiss, "Acoustic sounding of the urban boundary layer over Berlin-Adlershof in summer", Z. Meteorology, 37, 241–252, 1987.
25. K. Evers and J. Neisser, "Vertical SODAR measurements at different sites in summer," Atmos. environ. 24A, 2541–2545, 1990.
26. M.A. Giblet, "The structure of wind over level country", Meteorological Office Geophysical Memoirs No. 54, 6(4), H.M.S.O. London, 1932.
27. M.E. Smith, "The forecasting of micrometeorological variables", Meteorological Monographs, No. 4, 50–55, 1951.
28. H.E. Cramer, "A practical method for estimating the dispersion of atmospheric contaminants", in Proc. of the First National Conference on Applied Meteorology, Section C, American Meteorological Society, Hartford, Conn., pp. 33–35, Oct. 1957.
29. F. Pasquill, "The estimation of the dispersion of wind borne material", Meteorological Magazine, 90, 34–49, 1961.
30. F.A. Gifford, "Use of routine meteorological observations for estimating atmospheric dispersion", Nuclear Safety, 2, 47–51, 1961.
31. D.B. Turner, "A diffusion model for an urban area", J. Appl. Meteorol., 3, 83–91, 1964.
32. D. Golder, "Relations among stability parameters in the surface layer", Boundary Layer Meteorol., 3, 47–58, 1972.
33. N. Islitzer and D.H. Slade, "Diffusion and transport experiments", in Meteorology and Atomic Energy, edited by D.H. Slade, US AEC Report TID-24190 and US Environmental Science Services Administration, pp. 117–118, 1968.
34. S.B. Carpenter, T.L. Montgomery, J.M. Leavitt, W.C. Colbaugh and F.W. Thomas, J. Air Pollution Control Association, 21, 491, 1971.
35. F.A. Gifford, "Turbulent diffusion-typing schemes: A review", Nuclear Safety, 17, 68–86, 1976.
36. A.H. Weber, K.R. McDonald and G.A. Briggs, "Turbulence classification schemes for stable and unstable conditions", in Proc. of the Joint Conference on Applications of the Air Pollution Meteorology, American Meteorological Society, Boton USA, pp. 96–102, 1977.
37. S.R. Hanna, G.A. Briggs, J. Deardorff. B.A. Egan, F.A. Gifford and F. Pasquill, "AMS Workshop on stability classification schemes and sigma curves: Summary of recommendations", Bull. Amer. Met. Soc., 58, 1305–1309, 1977.
38. N.F. Islitzer, "Program review and summary of recent accomplishments of NRTS", in conference on AEC Meteorological Activities, May 19–22, 1964, USAEC Report BNL-914, Brookhaven National Laboratory, USA, pp. 57–64, 1965.
39. F. Pasquill and F.B. Smith, "Atmospheric Diffusion", John Wiley and Sons, New York, 3rd Edition, pp. 50–52, 1983.
40. A.H. Weber, "Atmospheric dispersion parameters in Gaussian plume modelling Pt. I : Review of current systems and possible future developments", EPA-600/4-76-0309, US Environmental Protection Agency, North Carolina, USA, July 1976.
41. D.H. Slade, "Meteorology and Atomic Energy," Report No. US AEC-TID-24190, US Atomic Energy Commission, 1968.
42. F.T.M. Nieuwstadt, "The computation of the friction velocity and the temperature scale from temperaure and wind velocities by least square methods", Boundary Layer Meteorol., 14, 235–246. 1978.

43. J.S. Irwin and F.S. Binkowski, "Estimation of the Monin-Obukhov scaling length using on-site instrumentation", Atmos. Environ., 15, 1091–1094, 1981.
44. F.T.M. Nieuwstadt and H. van Dop, "Atmospheric and Air pollution monitoring", D. Reidel Publishing Co., Dordercht, Holland, 1982.
45. Environmental Protection Agency, "On site meteorological program guidance for regulatory modelling applications", Report No. EPA-450/4-87-013, US Environmental Protection Agency, Office of Air Quality Planning and Standards, Research Triangle Park, NC 27711, USA, June, 1987.
46. S.P. Singal, B.S. Gera and S.K. Agarwal, "Nowcasting by acoustic remote sensing: Experience with the systems established at the National Physical Laboratory, New Delhi", J. Sci. Industr. Res., 4, 469–488, 1984.
47. S.P. Singal, S.K. Aggarwal, D.R. Pahwa and B.S. Gera, "Stability studies with the help of acoustic sounding", Atmos. Environ., 19, 221–228, 1985.
48. S.P. Singal, "Need for acoustic sounding (SODAR) monitoring of the atmospheric boundary layer for environmental pollution management", in Environmental Planning and Management in India, Vol. I, edited by R.K. Sapru, Ashish Publishing House, Now Delhi, India, Ch. 10, pp. 79–105, 1990.
49. S.P. Singal, S.K. Aggarwal, B.S. Gera, D.R. Pahwa and Mukesh Sharma, "Correlation of studies of surface level carbon monoxide concentration and sodar observed thermal structures", Atmos. Res., 20, 133–139. 1986.
50. S.P. Singal, E.W.D. Lewthwaite and D.S. Wratt, "Estimating atmospheric stability from monostatic acoustic sounder records", Atmos. Environ., 23, 2079–2084, 1984.
51. D.S. Wratt, "An experimental investigation of some methods of estimating turbulence parameters for use in dispersion models", Atmos. Environ., 21, 2599–2609, 1987.
52. B.S. Gera and S.P. Singal, "Sodar in air pollution meteorology", Atmos. Environ., 24A, 2003–2009, 1990.
53. S.P. Singal, "Monitoring air pollution related meteorology using SODAR: State of the art", Appl. Phys. B57, 65–82, 1993.
54. S.P. Singal, B.S. Gera and D.R. Pahwa, "Application of SODAR to air pollution meteorology", Intl. J. Remote Sensing, 15, 427–441, 1994.
55. J. Neisser, G. Bull, K. Evers, M. Weimann, E. Weiss, J. Keder and I.V. Petenko, "Results of the sodar investigations of the structure of the planetary boundary layer", in Proc. of the Field experiment Kopex-86, Czechosl. Acad. Sci., Prague, pp. 109–141, 1988.
56. W.D. Neff and C.W. King, "The use of sodars in an urban air quality study, "in Acoustic Remote Sensing, edited by S.P. Singal, Tata McGraw Hill, New Delhi, pp. 506–512, 1990.
57. H. Gland, "Qualifying test on a three dimensional Doppler sodar", Report HE/32-81.9, French Electricity Board, Department of Acquatic and Atmospheric Environment, Division of Applied Meteorology and Air Pollution, Paris, France, pp. 35, March 1981.
58. D.E. Jones, P.J. Smith, N.A. Shaw and I.A. Bourne, "Analysis of acoustic radar wind data", in Proc. of the 8th Intl. Clean Air Conference, Melbourne, Australia, pp. 337–339, May 1984.
59. P. Thomas, "Stability classification by acoustic remote sensing", Atmos. Res., 20, 165–172, 1986.
60. P.R. Best, M. Kanowski, L. Stumer and D. Green, "Convective dispersion modelling utilizing acoustic sounder information", Atmos. Res., 20, 173–198, 1986.

61. H. Gland, "Experiments with a Doppler acoustic sounder", Report HE/32-80.24, French Electricity Board, Department of Acquatic and atmospheric Environment, Division of Applied Meteorology and Air Pollution, Paris, France, pp. 35, March 1980.
62. R.J. Wyckoff, D.W. Beran and F.F. Hall Jr., "A comparison of the low level radiosonde and the acoustic echo sounder for monitoring atmospheric stability", J. Appl. Meteorol., 12, 1196–1204, 1973.
63. H.D. Parry, M.J. Sanders Jr. and H.P. Jensen, "Operational application of a pure acoustic sounding system", J. Appl. Meteorol., 14, 67–77, 1975.
64. A.K. Goroch, "Comparison of radiosonde and acoustic echo sounder measurements of atmospheric thermal strata' J. Appl. Meteorol., 15, 520–521, 1976.
65. R.B. Hicks, D. Smith, P.J. Irwin and T. Mathews, "Preliminary results of atmospheric acoustic sounding at Calgary", Boundary Layer, Meteorol., 12, 201–212, 1977.
66. S.P. Singal and S.K. Aggarwal, "SODAR and radiosonde studies of the thermal structure of the lower atmosphere at Delhi", Indian J. Rad. and Space Phys. 8, 76–81, 1979.
67. R.G. von Gogh and P. Zib, "Comparison of simultaneous tethered balloon and monostatic sounder records of the statically stable lower atmosphere", J. Appl. Meterol. 17, 34–39, 1978.
68. P.B. Russell and E.E. Uthe, "Sodar network measurements of regional mixing depth and stability patterns for an air quality model", in Preprints of the 4th Symposium on Meteorological Observations and Instrumentation, American Meteorological Society, Denver, Colo., USA, pp. 490–497, April 10–14, 1978.
69. P.B. Russell and E.E. Uthe, "Regional patterns of mixing depth and stability: Sodar network for input to air quality models", Bull. Amer. Metorol. Soc., 59, 1275–1287, 1978.
70. P.B. Russell and E.E. Uthe, "Acoustic and direct measurements of the atmospheric mixing at three sites during an air pollution incident" Atmos. environ. 12, 1061–1074, 1978.
71. R.L. Coulter, "A comparison of three methods for measuring mixing height layer," J. Appl. Meteorol. 18, 1495–1499, 1979.
72. L.J. Mahrt, R.C. Heald, D.H. Lenschow, B.B. Stankov and I. Troen, "An observational study of the nocturnal boundary layer", Boundary Layer Meteorol., 17, 247, 1979.
73. F.T.M. Nieuwstadt and A.G.M. Driedonks, "The nocturnal boundary layer: A case study compared with model calculations", J. Appl. Meteorol., 16, 115–129, 1979.
74. F.T.M. Nieuwstadt, "The steady state height and resistance laws of the nocturnal boundary layer: Theory compared with Cahan's Observations", Boundary Layer Meteorol., 20, 3–17, 1981.
75. F.T.M. Nieuwstadt, "Some aspects of the turbulent boundary layer" Boundary Layer Meteorol., 30, 31–55, 1984.
76. S.P.S. Arya, "Parameterizing the height of the stable atmospheric boundary layer", J. Appl. Meteorol., 20, 1192–1202, 1981.
77. S.S. Zilitinkevich, "On the determination of the height of the Ekman boundary layer", Boundary Layer Meteorol. 3, 141–145, 1975.
78. R. Dohrn, E. Raschke, A. Bujnoch and G. Warmbier, "Inversion structure heights above the city of Cologne (Germany) and a rural station nearby as measured with two SODARS", Meteorol. Rdsch., 35, 133–144, 1982.
79. B.B. Fitzharris, A. Turner and W. McKinley, "Cold season inversion frequencies

as measured by acoustic sounder in the Cromwell basin", New Zealand J. Science 26, 307–313, 1983.
80. M. Piringer, "The determination of mixing heights by SODAR in an urban environment", in Environmental Meteorology, edited by K. Grefen and J. Lobel, Kluwer Academic Publisher, pp. 425–444, 1986.
81. D. Koracin and R. Berkowicz, "Nocturnal boundary layer height: Observations by acoustic sounding and predictions in terms of surface layer parameters", Boundary Layer Meteorol., 43, 65–83, 1988.
82. S.J. Caughey, "Observed characteristics of the atmospheric boundary layer", in Atmospheric Turbulence and Air Pollution Modelling, edited by F.T.M. Nieuwstadt and H. van Dop, D. Reidel Publ. Co., Dordrecht, Holland, pp. 107–158, 1982.
83. F. Beyrich, "Sodar estimates of surface heat flux and mixed layer depth compared with direct measurements", Atmos. Environ., 26A, 2459–2461, 1992.
84. F. Beyrich, "On the use of SODAR data to estimate mixing height", Appl. Phys., B57, 27–35, 1993.
85. F. Beyrich and A. Weill, "Some aspects of determining the stable boundary layer depth from sodar data", Boundary Layer Meteorol., 63, 97–116, 1993.
86. S.P. Singal, S.K. Aggarwal and B.S. Gera, "SODAR (acoustic sounder) in the use of air pollution meteorology", J. Sci. Industr. Res., 39, 73–86, 1980.
87. G.C. Holzworth, "Mixing depths, wind speeds and air pollution potential for selected locations in the United States", J. Appl. Meteorol., 6, 1039–1044, 1967.
88. A. Weill, L. Eymard, M.E. Lequere, C. Klapisz, F. Baudin and P. van Grunderbeeck, "Investigations of the planetary boundary layer with an acoustic Doppler sounder", in Proc. of the 4th Symposium on Meteorological Observations and Instrumentation, American Meteorological Society, Denver, Colo., USA, pp. 415–421, April 10–14, 1978.
89. J.W. Deardorff, "Three dimensional and numerical modelling of the planetary boundary layer", in workshop on Micro-meteorology edited by D.A. Haughen, American Meteoreological Society, Boston, M.A., pp. 271–309, 1972.
90. A. Weill, C. Klapisz, B. Strauss, F. Baudin and J.P. Goutorbe, "Measuring heat flux and structure functions of temperature fluctuations", J. Appl. Meteorol., 19, 199–205, 1980.
91. P.R. Best, J. Ewald and M. Kanowski, "The estimation of pollutant dispersal from Queensland Power Station", in Proc. of the 7th Intl. Clean Air Conf., Clean Air Society of Australia and New Zealand, Adelaide, Australia edited by K.A. Webb and A.I. Smith, pp. 429–448, August 24–28, 1981.
92. D. Melas, "SODAR estimates of surface heat flux and mixed layer depth compared with direct measurements", Atmos. Environ., 24A, 2847–2853, 1990.
93. D. Melas, "Similarity methods to derive turbulence quantities and mixed layer depth from SODAR measurements in the convective boundary layer: A review", Appl. Phys. B57, 11–17, 1993.
94. L. Enger, "Simulation of dispersion in complex terrain-Part C, A dispersion model for operational use", Atmos. Environ., 24A, 2457–2471, 1990.
95. J.C. Wyngaard, "Measurement Physics", in Probing the Atmospheric Boundary Layer, ed. by D.H. Lenschow, AMS, Boston, Mass., USA, pp. 5–18, 1986.
96. W.D. Neff and R.L. Coulter, "Acoustic Remote Sensing", in Probing the Atmospheric Boundary Layer, edited by D.H. Lenschow, AMS, Boston, Mass., USA, pp. 201–263, 1986.

97. F. Beyrich, "Intercomparison of different methods for mixing height estimation in the convective boundary layer from SODAR data", in Proc. of the 7th Intl. Symposium on Acoustic Remote Sensing and Associated Techniques of the Atmosphere and the Oceans, edited by W.D. Neff, Boulder, Colorado, USA, pp. 2–19 to 2–26, 3–7 Oct. 1994.
98. E. Batchvarova and S.E. Gryning, "Applied model for the growth of the daytime mixed layer", Boundary Layer Meteorol., 56, 261–274, 1991.
99. A.G.M. Driedonks, "Models and observations of the growth of the atmospheric boundary layer", Boundary Layer Meteorol., 23, 283–306, 1982.
100. S.P. Singal, B.S. Gera and Neeraj Saxena, "Sodar studies of the mixing height at various topographical locations in India", in Proc. of the 7th Intl. symposium on Acoustic Remote Sensing and Associated Techniques of the Atmosphere and the Oceans, edited by W.D. Neff, Boulder, Colorado, USA, pp. 9–1 to 9–6, 3–7 Oct. 1994.
101. G.A. Briggs, "Plume rise predictions", in Lectures on Air Pollution and Environmental Impact Analysis edited by D.A. Haughen, A.M.S. Boston, Mass., USA, pp. 59–111, 1975.
102. K. Takeuchi, "Some studies on the fluctuations of wind direction near the ground", J. Meteor. Soc. Japan, 1, 40–51,1962.
103. J.A. Businger, "Transfer of heat and momentum in the atmospheric layer", in Prog. Arct. Heat Budget and Atmospheric Circulations, Santa Monica, Calif., Rand Corp., pp. 305–332, 1968.
104. J.P. Pandolfo, "Wind and temperature for constant flux boundary layer in lapse conditions with a variable eddy conductivity and eddy viscosity ratio", J. Atmos. Sci., 23, 495–502, 1966.
105. R.L. Coulter and K.H. Underwood, "Some turbulence and diffusion parameter estimates within cooling tower plumes derived from sodar data", J. Appl. Meteorol., 19, 1395–1404, 1980.
106. N.O. Jensen and E.L. Petersen, "The box model and the acoustic sounder: A case study", Atmos. Environ., 13, 717–720, 1979.
107. T.J. Moulsley and R.S. Cole, "The evaluation of acoustic sounder returns from a methane plume", Atmos. Environ., 14, 1063–1066, 1980.
108. G. Brusasca, G. Elisey, M. Maini and A. Marzorati, "Acoustic remote sensing for environment control in thermal power plants", in Proc. of the 2nd Intl. symposium on Acoustic Remote Sensing of the Atmosphere and Oceans, Rome, Italy, pp. XXI 1–12, Aug. 29–1 Sep., 1983.
109. G. Brusasca, F. Goppelli and G. Tinarelli, "Central intelligent unit for atmospheric control using remote sensing measurements", in Proc. of the 3rd Intl. Symposium on Acoustic Remote Sensing of the Atmosphere and Oceans, Issy-les-Moulineaux, France, Oct. 14–17, 1985.
110. P.R. Best, M. Kanowski, P. Morland and L. Stumer, "A comparison of acoustic sounder statistics at various sites on the Australian Continent", in Proc. of the 2nd Intl. Symposium on Acoustic Remote Sensing of the Atmosphere and Oceans, Rome, Italy, pp. XXXIV 1–24, Aug. 29–1 Sep., 1983.
111. D.B. Turner: Workbook of Atmospheric Dispersion Estimates: Pub. No. AP-26, Office of Air Programs, Environmental Protection Agency, USA, 1970.
112. M.S. Pekour and M.A. Kallistratova, "SODAR study of the boundary layer over Moscow for air pollution applications", Appl. Phys., B57, 49–55, 1993.

113. G.W. Gilman, H.B. Coxhead and F.H. Willis, "Reflection of sound signals in the troposphere", J. Acoust. Soc. Amer., 18, 274–287, 1946.
114. L.G. McAllister, "Acoustic sounding of the lower atmosphere", J. Atmos. Terrestr. Phys., 30, 1439–1440, 1968.
115. S. Wickerts and L. Nilsson, "The occurrence of very high field strengths beyond the horizon propagation over sea in the frequency range 60–5000 MHz", published in Modern Topics in Microwave Propagation and Air-Sea Interaction, Proc. of the NATO Advanced Study Institute, Italy, Sorrento, pp. 217–240, June 5–14, 1973.
116. G. Stilke, "Occurrence and features of ducted modes of internal gravity waves over western Europe and their influence on microwave propagation", Boundary Layer Meterol., 4, 493–509, 1973.
117. J. Nilsson, "On the correlation of deep fading of line of sight microwave link and the presence of inversion layers monitored by acoustic sounder", URSI Comission F, Open Symposium, La Baule, France, pp. 49–53, April 28–May 6, 1977.
118. A.P. Mitra, Y.V. Somayajulu, S.P. Singal, S.C. Majumdar, T.R. Tyagi, B.M. Reddy, S.K. Aggarwal, B.S. Gera, A.B. Ghosh and S.K. Sarkar, "Tropospheric disturbances of 17–21 December' 77 and its effect on microwave propagation", Boundary Layer Meteorol., 11, 103–116, 1977.
119. B.S. Gera and S.K. Sarkar, "Sodar as an indicator of microwave propagation", Ind. J. Rad. and Space Phys., 9, 86–96, 1980.
120. M. Fukushima, K. Akita and Y. Masuda, "Signal enhancement of UHF TV wave due to the occurrence of inversion layers: A case study", Bulletin Radio Research Laboratories, Denpa Kenkyusho Kiho, 16, 569–594, 1980.
121. S.P. Singal, B.S. Gera and A.B. Ghosh, "Application of sodar derived boundary layer information in microwave communication", J. Sci. and Industr. Res., 40, 765–777, 1981.
122. S.P. Singal and B.S. Gera, "Gorrelation studies of sodar structures and microwave propagation characteristics", in the Intl. symposium on Acoustic Remote Sensing of the Atmosphere and Oceans, The University of Calgary, Calgary, Canada, pp. V. 51–V. 66, June 22–25, 1981.
123. S.P. Singal, B.S. Gera and A.B. Ghosh, "Sodar echoes and line of sight microwave propagation", J. Institution of Engineers ET63, 49–54, 1982.
124. K. Bullington, "Radio propogation fundamentals" Bell System Tech. J., 36, 593–626, 1957.
125. F. Eklund and S. Wickerts, "Wavelength dependence of microwave propagation far beyond the radio horizon", Rad. Science, 3, 1066–1079, 1968.
126. J.P. Mon, A. Weill and L. Martin, "Effects of tropospheric disturbances on a 4.1 and 6.2 GHz line of sight path", Annales des Telecom., 35, 470–473, 1980.
127. J.P. Mon and S. Mayrargue, "Sodar monitoring of the lower atmosphere and prediction of propagation anomalies on line of sight path", IEEE Intl. Conference on Communications, pp. 68.6.1–68.6.4, June 1981.
128. G. Roux and J.P. Mon, "Propagation anomalies in 4, 7 and 11 GHz line of sight path", Proc. URSI Commission F, 1983 Symposium, Louvain, Belgium, pp. 57–68, June 1983.
129. M. Sylvain, F. Baudin, C. Klapisz., J. Lavergnat, S. Mayrargue, J.P. Mon, R. Nuten and M. Rooryek, "The Pacem experiment on line of sight multipath propagation" Proc. URSI Commission F, 1983 Symposium, Louvain, Belgium, pp. 13–23, June 1983.

130. J. Claverie, C. Klapisz and M. Sylvain, "The use of acustic soundings in multipath propagation studies", Presented in the 3rd Intl. Symposium on Acoustic Remote Sensing of the Atmosphere and Oceans, CNET/CRPE, Issy Les Moulineaux, France, Oct. 14–17, 1985.
131. J. Das, A.K. De, D. Dutta Majumdar, A.K. Sen and S.K. Basu Mallick, "LOS propagation characteristics in the presence of inversion layers", Ind. J. Phys., 63B, 149–160, 1989.
132. J. Das, A.K. De and D. Dutta Majumdar, "VHF propagation characterization during atmospheric disturbances", Intl. J. Remote Sensing, 10, 1227–1241, 1989.
133. D.W. Beran, W.H. Hooke and S.F. Clifford, "Acoustic echo sounding techniques and their application to gravity wave, turbulence and stability studies", Boundary Layer Meteorol., 4, 133–153, 1973.
134. W.D. Neff, "Quantitative evaluation of acoustic echoes from the planetary boundary layer", Tech. Report ERL 322-WPL 38, NOAA, Boulder, Colo., USA, pp. 34, 1975.
135. M. Fukushima, K. Akita and H. Tanaka, "Sodar probing of small scale temperature structure in the clear troposphere", J. Rad. Res. Lab. Japan, 22, 108, 23–43, 1975.
136. D.N. Asimakopolous, R.S. Coles, S.J. Caughey and B.A. Crease, "A quantitative comparison between acoustic sounder returns and the direct measurements of atmospheric temperature fluctuations", Boundary Layer Meteorol., 10, 137–147, 1976.
137. D.A. Haugen and J.C. Kaimal, "Measuring temperature structure parameter profiles with an acoustic sounder", J. Appl. Meteorol., 17, 895–899, 1978.
138. Mingyu Zhou, Naiping Lu and Yanjuan Chen, "The detection of the temperature structure coefficient of the atmospheric boundary layer by acoustic radar', J. Acoust. Soc. Am., 68, 303–308, 1980.
139. R.L. Coulter and M.L. Wesley, "Estimates of surface heat flux from SODAR and LASER scintillation measurements in the unstable boundary layer", J. Appl. Meteorol., 19, 1209–1222, 1980.
140. T.J. Moulsley, D.N. Asimakopolous, R.S. Cole, B.A. Crease and S.J. Caughey, "Measurements of boundary layer parameter profiles by acoustic sounding and comparison with direct measurements", Quart. J. Roy. Meteorol. Soc., 107, 203–230, 1981.
141. S.P. Singal, B.S. Gera and S.K. Aggarwal, "Detection of structure parameters using SODAR", Boundary Layer Meteorol., 23, 105–114, 1982.
142. D.N. Asimakopolous, T.J. Moulsley, C.. Helmis, D.P. Lalas and J.E. Gaynor, "Quantitative low level acoustic sounding and comparison with direct measurements", Boundary Layer Meteorol., 27, 1–26, 1983.
143. J.E. Gaynor, "Acoustic Doppler measurements of atmospheric boundary layer velocity structure functions and energy dissipation rates, "J. Appl. Meteorol., 16, 148–155, 1977.
144. A.E. Gur'yanov, S.L. Zubkovski, M.A. Kallistratova, G.A. Karyukin, V.P. Kukharets and I.V. Petenko. "Reliability of determination of the vertical profile of the temperature structure parameter in the atmosphere by acoustic sounding", Izvestiya, Atmos. Oceanic Phys., 17, 2, 107–111, 1981.
145. M.A. Kallistratova, G.A. Karyukin, S.N. Kulichkov, I. Keder, I.V. Petenko and N.S. Time, "Comparison of qualitative and quantitative measurements of temperature turbulence by monostatic acoustic Doppler Radar", Izvestiya, Atmos. Oceanic. Phys. 20, 2, 120–126, 1984.

146. A.E. Gur'yanov, M.A. Kallistratova, F.E. Martvel, M.S. Pekur, I.V. Petenko, N.S. Time and E.A. Shurygin, "Comparison of sodar and microfluctuation measurements of the temperature structure parameter in mountainous terrain", Izvestiya, Atmos. Oceanic Phys., 23, 9, 685–691, 1987.
147. I.V. Petenko, E.A. Shurygin, J. Neisser and Th. Foken, "Comparison of sodar and turbulent measurements", in Proc. of the Field Experiment KOPEX-86, Czechosl. Acad. Sci., Prague, pp. 37–54, 1988.
148. J. Keder, Th. Foken, W. Gerstmann and V. Schindler, "Measurements of wind parameters and heat flux with the Sensitron Doppler sodar", Boundary Layer Meteorol., 46, 195–204, 1988.
149. S.D. Danilov, A.E. Gur'yanov, M.A. Kallistratova, I.V. Petenko, S.P. Singal, D.R. Pahwa and B.S. Gera, "Acoustic Calibration of SODARS", Measurement Science Technology, 3, 10, 1001–1007, 1992.
150. J.C. Wyngaard, Y. Izumi and S.A. Collins, "Behaviour of the refractive index structure parameter near the ground", J. Opt. Sci. Am., 61, 1646–1650, 1971.
151. J.C. Kaimal, J.C. Wyngaard, D.A. Haugen, O.R. Cote, Y. Izumi, S.J. Caughey and C.J. Reading, "Turbulence structure in the convective boundary layer", J. Atmos. Sci., 33, 2152–2169, 1976.
152. E.H. Brown, "Quantitative acoustic measurements of atmospheric turbulence", in Proc. of the 4th Symposium on Meteorological Observations and Instrumentation, American Meteorological Society, Denver, Colo., USA, pp. 275–277, April 10–14, 1978.
153. E.H. Brown and S. Clifford, "On the attenuation of sound by turbulence", J. Acoust. Soc. Am., 60, 788–794, 1976.
154. B.A. Gardiner and M.K. Hill, "Acoustic sounder observations from an elevated inversion", Boundary Layer Meteorol., 36, 307–316, 1986.
155. M.D. Sirkis, "Contribution of water vapour to index of refraction structure parameter at microwave frequencies", Trans. IEEE, AP-19, 572–574, 1971.
156. H. Ottersten, K.R. Hardy and C.G. Little, "RADAR and SODAR probing of waves and turbulence in statistically stable clear air layers", Boundary Layer Meteorol., 4, 47–89, 1973.
157. M.L. Wesely, "The combined effect of temperature and humidity fluctuations on refractive index", J. Appl. Meteorol., 15, 43–49, 1976.
158. S.P. Singal, B.S. Gera and S.K. Aggarwal, "Studies of sodar observed dot echo structures", Atmosphere-Ocean, 23, 3, 304–312, 1985.
159. W.D. Neff, "Beam width effects on acoustic back-scatter in the planetary boundary layer", J. Appl. Meteorol., 17, 1514–1520, 1978.
160. T.J. Moulsley, D.N. Asimakopolous, R.S. Cole, S.J. Caughey and B.A. Crease, "Temperature structure parameter measurements using differential temperature sensors", Boundary Layer Meteorol., 23, 307–315, 1982.
161. M.A. Kallistratova and I.V. Petenko, "Aspect sensitivity of sound back-scattering in the atmospheric boundary layer", Appl. Phys., B57, 41–48, 1993.
162. S.P. Singal, B.S. Gera, M.A. Kallistratova and I.V. Petenko, "Sodar aspect sensitivity studies in the convective boundary layer", in Proc. of the 7th Intl. Symposium on Acoustic Remote Sensing and Associated Techniques of the Atmosphere and the Oceans, edited by W.D. Neff, Boulder, Colorado, USA, pp. I-19 to I-24, 3–7 Oct. 1994.
163. C.G. Little, "Acoustic methods for the remote probing of the lower atmosphere," Proc. IEEE, 57, 571–578, 1969.

17. Application of Sodar in Urban Air-Quality Monitoring Systems

Jacek Walczewski

Institute of Meteorology and Water Management
ul. Borowego 14, 30–215 Cracow, Poland

Introduction

The function of the urban air-quality monitoring systems can be described by 3 points:

(a) The assessment of air-quality, investigated through systematic measurements of concentrations and their variability in time and space of different air pollutants;
(b) Identification of cause-effect relationships to determine the status of air pollution; the knowledge on these relationships makes possible rational planning of land use, industrial development and air protection activities.
(c) Warning the population on approaching maxima of air pollution ("air pollution episodes"), to stimulate emergency air protection activities and to make possible adjustment of human behaviour and operations of urban structures for survival in severely polluted atmosphere.

The functions (b) and (c) cannot be performed in an effective way with pollution measurements only; they must include meteorological observations and analyses of meteorological data, in respect of dispersion conditions of air pollution. The dispersion conditions are partially dependant on general synoptic situation; in this part the necessary data are available from the routine meteorological service. But the direct influence on dispersion conditions is generated by the structure of atmospheric boundary layer and its variations. The phenomena occurring in boundary layer are characterized by short time and space scales as compared with synoptic processes. To record these small-scale phenomena, continuous observation is necessary of the atmospheric structures up to about 1000 m height. Sodar offers here unprecedented possibilities. One can say, that application of sodar has brought revolution in this area. Sodar characteristics of continuous automatic operation, low cost of unit of information, insignificant environmental limitations for operation, clear display of atmospheric processes in time-height cross-section, are incomparably better than for any other boundary layer sounding technique.

We will make a brief review of the forms of sodar data application in an urban air-quality monitoring system. The review has the practice-oriented character and is to a large extent based on the 15-year experience of sodar operation in urban-industrial agglomeration of Cracow, Poland, where air-pollution problems were of rather severe nature in the past decade. Now the situation is much improved, but sodar monitoring continues to play an important role in the urban informative system concerning atmosphere. Air Monitoring System in Cracow includes a central station equipped with one REMTECH PA-2 Doppler Sodar and one SAMOS-4C vertical digital sodar of Polish production. The latter enables continuous watching of Boundary Layer structures up to 1000 m height at PC Computer screen for 6 hours. Data are computer-stored and can be processed and printed.

Sodar Application in Identification of Cause-Effects Relationships Determining Air Pollution

The relationship between emission, meteorological situation and pollution concentration field, is described by dispersion models. These models have many practical applications in arrangement of air protection strategies in urban agglomeration. First of all, they are used to assess the impact on environment of the new (planned) emission sources, as well as existing sources, whose real impact cannot be determined by means of concentration measurements, because of interference of many different sources at the same urban area. Moreover, dispersion models are used for simulation studies in planning of city development and of air-protection measures.

The role of these two functions of dispersion models cannot be overestimated. The model-calculation outputs have their far-reaching consequences in the fields of economy and urban ecology. On the basis of environment impact assessment, official permissions are issued for the amount of air-pollution emissions from individual sources. On the basis of simulation studies, perspective plans are prepared for location of industrial plants, roads, heating plants and other emission sources. Future shape of the city may depend on the results of these studies.

The results of calculations are as dependable, as are the data used as input to the models. There are 4 most important model inputs, concerning meteorological data:

— atmospheric stability;
— mixing layer height;
— upper wind direction;
— upper wind speed;

These data may be required in 2 forms: as the annual or seasonal statistics (for long-range impact calculations) and as the data files for extreme conditions (for calculations of impact in "pollution episodes").

Determining of Atmospheric Stability

Many important components of dispersion model equations are depending on the categories of atmospheric stability. In classical Gaussian models this dependance refers to: dispersion parameters σ_y, σ_z; exponent m in the Sutton's formula for wind speed versus height; and formulae for plume rise. In more modern models the parameterization procedures are expressed by different formulae depending on stability categories [5, 11]. Thus, proper estimation of stability is of essential importance for the results of calculations.

For the past many years the atmospheric stability was being determined on the basis of different versions of the Pasquill scheme [3]. This scheme is still included in the official guidelines for calculations of air pollution in many countries. However it has been demonstrated in many recent publications [2, 10], that Pasquill estimations of stability categories are giving wrong description of physical processes in boundary layer in many cases. In particular, they are neglecting the "stable" category in daytime, giving, as substitute, overestimation of "neutral" category.

Sounding of the boundary layer with use of vertical sodar makes possible observation of such atmospheric structures like convective cells and inversion layers. The interpretation of sodar echoes is based on many comparisons with the results of other methods of observation and on the resulting consensus of the character of basic morphological features of the echoes of convective cells and inversion layers [14, 10]. Sodar observation enables estimation of 3 stability categories, assuming, that:

— presence of the echoes of convective cells means the negative heat flux (soil to atmosphere) and consequently *unstable* category of stability;
— precence of the ground-based layer echo means the temperature inversion near the ground, positive heat flux (atmosphere to soil) and consequently *stable* category;
— no-echo structure means adiabatic lapse rate or mechanical mixing and consequently *neutral* category.

There are also methods of distinguishing more particular types (A, B, C, D, E, F) within these 3 main categories [4, 6]. The longer series of sodar observations make possible to characterize annual, seasonal and monthly contributions of stability categories; such characteristics usually are differing from those based on Pasquill scheme (Fig. 1) and it is justified to consider them as more dependable than Pasquill data. Another important advantage of sodar estimations of stability results from the continuity of sodar observations, is to observe diurnal variations of stability and to give mean monthly, seasonal and annual diurnal characteristics of stability categories (Fig. 2). Estimations based on meteorological observations at ground level (Pasquill scheme) are sometimes made only for selected observation hours (e.g. 06, 12, 18 GMT) which does not describe adequately the diurnal variability. In general, sodar estimations of stability are dependable source of data for dispersion model calculations.

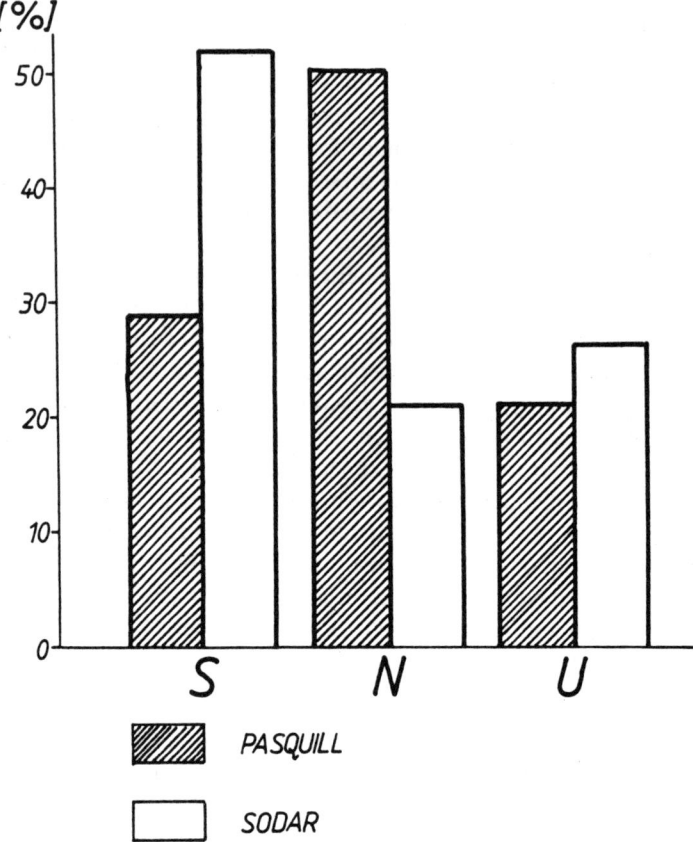

Fig. 1. Annual contribution of hours with stable (S), neutral (N) and unstable (U) stability categories in Cracow, Poland; estimations after Pasquill scheme and sodar observations.

Determining the mixing layer height

The height of layer, in which ground-borne pollutants are mixed with air due to turbulent motions, is a component of model input data, both in simple "box-models" and in more sophisticated dispersion models. Calculations of this height are giving different results depending on parameterization methods; development of dependable calculation approaches remains a problem which needs some work to be done. Sodar offers a unique possibility to observe such features of the boundary layer which are related to the mixing processes:

— in the presence of ground-based stable layer echo, top of this layer may be assumed to be the index of mixing layer height [11].
— when convective mixing layer is developed and capping inversion echo is visible in the range of sodar, the mixing height may be determined exactly as the height of the bottom of elevated layer echo from the ground level.

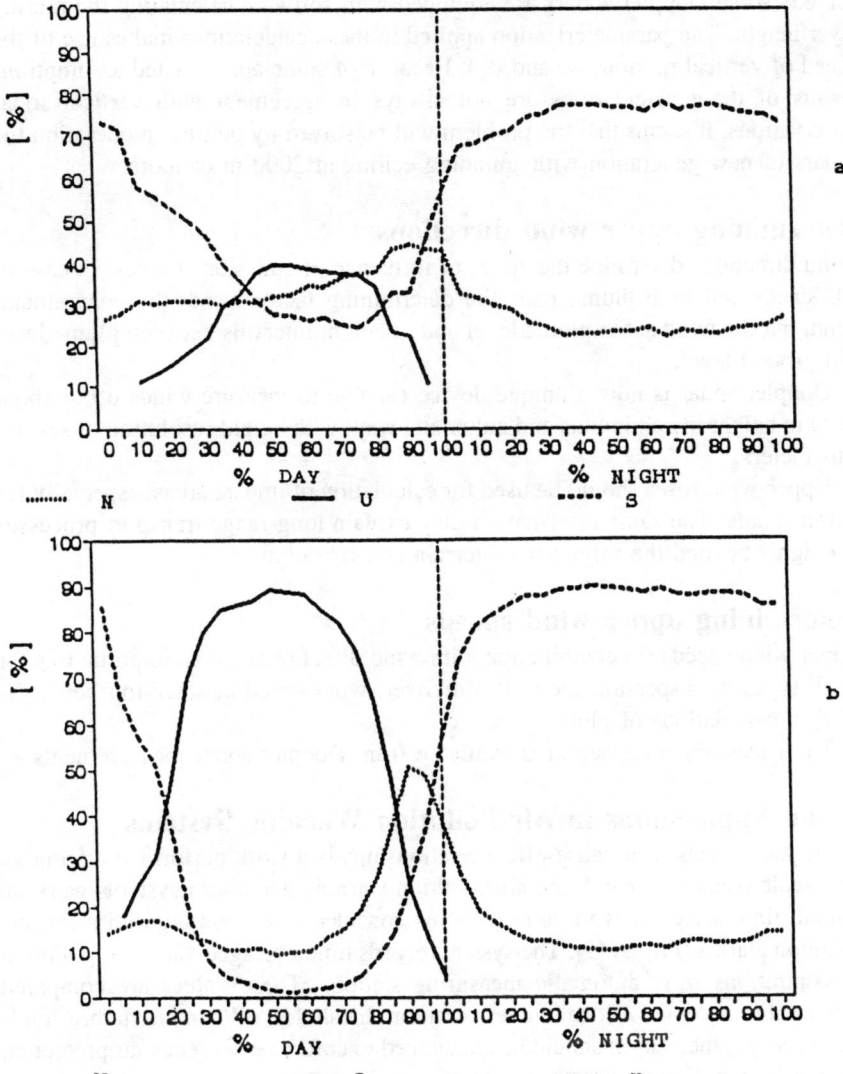

Fig. 2. Occurrence frequency of 3 main stability categories in their diurnal variability, for (a) cool season (Nov.–Feb.) and (b) warm season (Apr.–Sept.) in Cracow, Poland, in the years 1980–1992. Normalized time scale with fixed sunrise and sunset points. U—unstable,. S—stable, N—neutral.

Using the rules described above, mixing heights may be determined for majority of nights and for large percent of morning hours, on the basis of vertical sodar records. For the cases of free convection (capping inversion not visible in the range of sodar), the problem of determining the mixing height is not yet generally solved. Some Doppler sodars are equipped with software calculating the mixing layer height. The parameterization applied in these calculations makes use of the speed of vertical motions (w and σ_w). Because of some aproximated assumptions results of these calculations are not always in agreement with vertical sodar observations. It seems that the problem will be solved by putting in operation the sodars of new generation with sounding ceiling at 2000 m or more.

Determining upper wind directions

Wind directions determine the zones of influence of emission sources. In case of tall stacks and high plume rises the determining factor is not the anemometer wind, but the wind at the plume level and in height intervals between plume level and ground level.

Doppler sodar is now a unique device capable to measure winds up to about 1000 m height in continuous and automatic way, with height resolution of several tens meters.

Upper-wind roses should be used for calculation of impact areas, especially for power plants. The same information may explain long-range transport processes at heights beyond the influence of terrain configuration.

Determining upper wind speeds

Upper wind speeds (in combination with wind directions) are an important factor in all types of dispersion models. Moreover, wind speed at stack-top height, is used in calculations of plume rise.

The necessary information is available from Doppler sodar measurements.

Sodar Applications in Air-Pollution Warning Systems

While identification of cause-effect relationships is a work performed on a large time scale (seasons, years), the air-pollution warning and alarm systems work on a small time scale (days to hours). Let us consider a decision algorithm for air-pollution alarms (Fig. 3) [9]. The system records time-averaged values of pollution concentrations from automatic measuring stations. These values are compared with earilier accepted critical values—"alarm thresholds". When alarm thresholds are axceeded, then alarm should be announced to enforce emergency air protection actions. But pollution situation may change in the meantime due to change of meteorological conditions. For instance, sudden front passage with rain and strong winds may clear the air and expensive emergency actions will be purposeless.

Taking into account this possibility, a safety element has been included in the decision algorithm: meteorological diagnosis and forecast check. Here answer should be prepared for the question: will meteorological situation change in near future? Whether the poor ventilation conditions are of stable character, or they are going to become better? The answer can be given partially by routine synoptic analysis, but actual observation of boundary layer structure by sodar may say

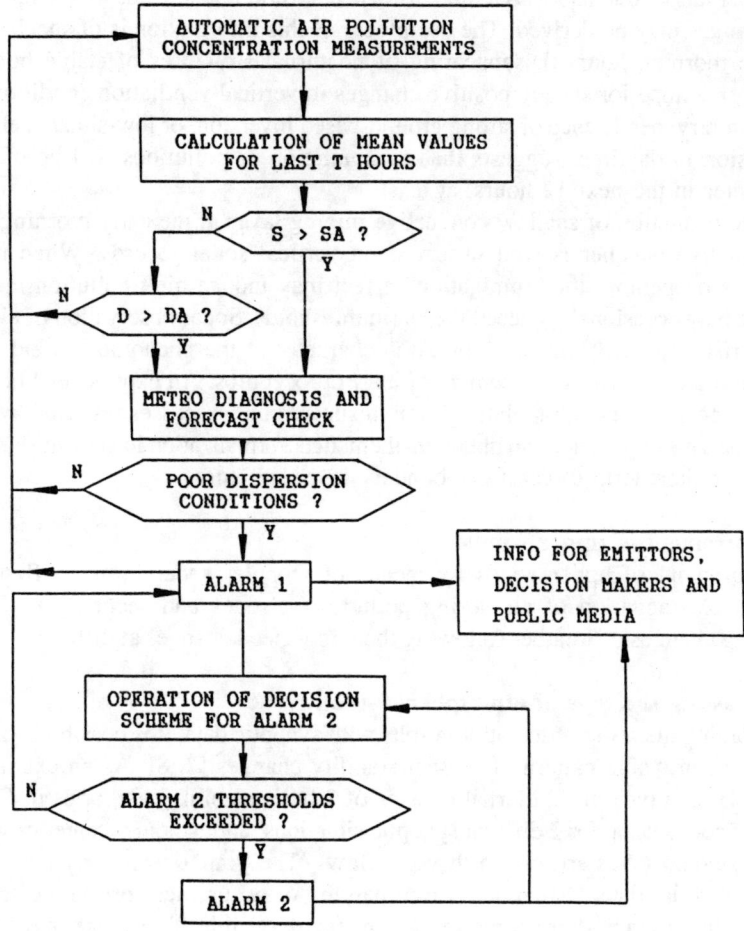

Fig. 3. Scheme of the decision algorithm for air-pllution alarms in winter. Air pollution characterized by 2 leading substances: S and D.
S, D—concentrations of pollutants S and D (means of T hours);
SA, DA—warning thresholds for substances S and D.

much more about the tendencies in a time scale of hours. In practice, both synoptic and sodar data should be used. There are several kinds of informations, available by means of sodar, which are of essential importance for verification of stability of actual pollution situation.

Detection and Location of Inversion Layers

Observation of records of vertical sodar is giving information on presence and situation of ground-based and elevated inversion layers, not only at the moment of observation, but also with a time history of several hours, from which tendencies of changes may be derived. The usefulness of this information is of special value in the morning hours. Disappearing of nocturnal structures of stable boundary layer give hope for speedy positive changes in vertical ventilation conditions. On the contrary, persistence of strong ground-based inversion, or low-situated elevated inversion in daytime, suggests that accumulation of pollutions will be of stable character in the next 12 hours, at least.

The formation of shallow convective mixing layer in the early morning hours in radiative weather is well observed on vertical sodar records. When mixing layer is deepening, the "fumigation" effect may induce high pollution maxima, which may occasionally exceed the alarm thresholds. Sodar observation of elevated layer rise can verify the short-duration character of the phenomenon and enable to avoid unnecessary enforcement of alarm procedures. An experienced observer at a given site may extrapolate with high success the processes revealed by sodar records. This experience can obtain mathematical formalization in statistical models, used for short-term forecast of boundary layer structure.

Measurement of upper winds

Measurement of upper winds by means of Doppler sodar informs, firstly the height of stagnation of air during pollution episodes and secondly to foresee coming changes of weather following the wind speed changes at different heights.

Forecasting the type of atmospheric stability

Combining the sodar observation results with synoptic data, it is possible to foresee the most probable pattern of diurnal stability changes [7, 8]. As an example, at (Fig. 4) are given mean diurnal courses of 3 basic stability categories in Cracow for the cool season for 2 differnt synoptic situations: anticyclonic (center or wedge) and cyclonic (Western or North-West flow). The graphs were prepared on the basis of sodar data. Differences of probability of occurrences of the 3 categories of stability, in their diurnal variability, are clearly visible. These data may be used as auxiliary material when considering possible changes of ventilation conditions in perspective of 24 hours.

Real-time modelling

In specific cases it is necessary to give quantitative data on air pollution in the form of a chart in a continuous way in situations of high pollution hazard, when emission data are available. This may happen e.g. in case of a big factory or power plant, exclusively (or mainly) responsible for air quality in its surroundings. In such cases coupling the computing unit with sodar station makes possible acquisition of real-time data on upper winds, mixing layer and atmospheric stability.

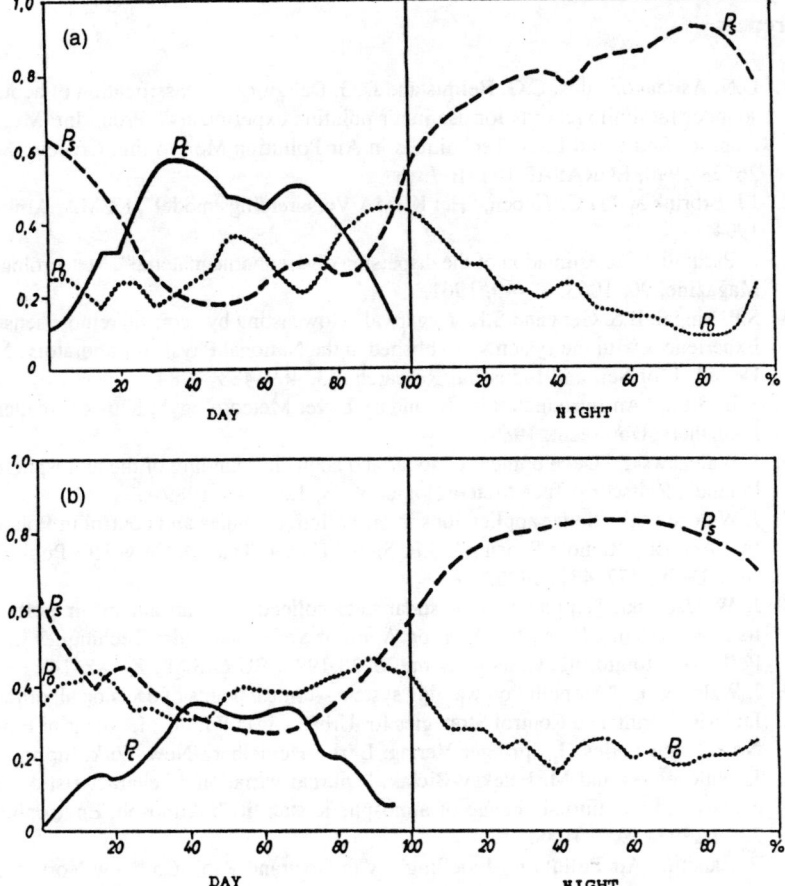

Fig. 4. Diurnal change of probability of occurrence of 3 stability categories for cool season (Nov.–Feb.) in Cracow, Poland, for selected synoptic situations: (a) anticyclonic (center or wedge) and (b) cyclonic (W or NW flow). Normalized time scale like Fig. 2.
P_s—for stable category, P_c—for unstable, P_o—for neutral.

Concluding Remarks

As it has been shown, sodar becomes necessary in operation of modern air-quality monitoring systems, which are responsible for operations of far-reaching economic and ecological consequences. Author is deeply convinced that sodars should be considered as obligatory element of big air-monitoring systems and that this fact should be taken into account in national and international regulations for meteorological and environmental services.

References

1. D.N. Asimakopoulos, C.G. Helmis and D.G. Deligiorgi, "Classification of acoustic sounder facsimile records for use in air polution experiments", Proc., Int. Mtg. on Appl. of Sodar and Lidar Techniques in Air Pollution Monitoring, Cracow, Sept. 26–28 1990, EURASAP II-1–II-7.
2. J.J. Erbrink and H.C. Tieben, "Het KEMA Verspreidings model", KEMA, Arnhem, 1994.
3. F. Pasquill, "The estimation of the dispersion of windborne material", Meteorological Magazine, 90, 1063, 33–49, 1961.
4. S.P. Singal, B.S. Gera and S.K. Aggarwal "Nowcasting by acoustic remote sensing: Experiences with the systems established at the National Physical Laboratory, New Delhi", J. of Sci, and Industrial Research, 43, 469–488, 1984.
5. R.B. Stull, "An Introduction to Boundary Layer Meteorology", Kluwer Academic Publishers, Dordrecht, 1989.
6. J. Walczewski, "Development of sodar and acoustic sounding of the atmosphere in Poland", Zeitschrift fuer Meteorologie, 39, 3, 129–141, 1989.
7. J. Walczewski, "Sodar applications in air pollution studies and control in Poland"; In: "Acoustic Remote Sensing", S.P. Singal Editor, Tata-McGraw Hill Publ. Co., New Delhi, 477–481, 1990.
8. J. Walczewski, "Application of sodar data collection as an aid in air pollution forecast system", Proc., Int. Mtg. on Appl. of Sodar and Lidar Techniques in Air Pollution Monitoring, Cracow, Sept. 26–28 1990, EURASAP, X-1-X-14.
9. J. Walczewski, "Air pollution warning system—critical points of decision algorithm", In: "Monitoring and Control Strategies for Urban Air Pollution", I. Allegrini Editor, NATO ASI Series G, Springer Verlag, Berlin-Heidelberg-New York, inprint.
10. J. Walczewski and M. Feleksy-Bielak, "Diurnal variation of characteristic sodar echoes and the diurnal change of atmospheric stability", Atmosph. Environment, 22, 9 1793–1800, 1988.
11. P. Zanetti, "Air Pollution Modelling", Van Nostrand Publ. Co, New York 1990.

Acoustic Remote Sensing Applications
S.P. Singal (Ed)
Copyright © 1997 Narosa Publishing House, New Delhi, India

18. Operational Use of Sodar Information in Nowcasting

Th. Foken[1], H.-J. Albrecht[2], K. Sasz[3], F. Vogt†

[1]German Weather Service, Meteorological Observatory,
D-15864 Lindenberg, Germany

[2]Environment Office of Brandenburg, P.O. Box 139, D-15201 Frankfurt/O Germany

[3]German Weather Service, Weather Office Leipzig, P.O. Box 32, D-04291 Leipzig, Germany

Introduction

In recent years SODAR technology has been increasingly used in industrial concentration areas, near nuclear power stations and at airports. Now there is a conspicuous tendency to include SODAR into meso-meteorological operations, i.e. forecasts and other services. Such working with SODAR should be based upon a real-time dissemination of sounding results [4]. Requirements include a specific classification and the encoding of data and, if possible, an automatic transmission. In the last decade there have been several publications on the SODAR recording techniques, i.e. the methods of recognising boundary layer phenomena which are known widely and in sufficient detail. The present paper reports on some results of a SODAR network using a special code.

Optimal Encoding of Sodar Information

There are available several reports on experimental SODAR data coding [8, 9], all of them, however, use a rigid classification of the pattern of the detected phenomena. In the present paper, a flexible coding system was used [5], which includes data from visual interpretation and from a micro-processor. These studies were mainly based on the operation of vertical-SODAR instruments designed by the former Heinrich-Hertz-Institute of the Academy of Sciences, Berlin [4].

The code was primarily designed for the coding of visual observations and manual distribution via telex or for electronic display. After several weeks of familiarisation with the systems and after classification of the basic pattern by a meteorologist (at least in difficult weather situations) the encoding can be done properly by a qualified technician. The structure of the SODAR code includes the alternative of automatic encoding of at least the basic patterns and of the height of ground based and elevated echo layers (inversions, see [2]) at stations where a computerised SODAR unit is operated. For this purpose a qualified pattern recognition is needed [3]. The code for the SODAR data includes the digits designating the station and time specific information and whether manual or machine coding was applied. The data reflecting the basic patterns contains for selected heights the measured values of the backscatter intensity, or, after a calibration, the temperature structure parameter [7]. Additionally it has been proved to be valuable to report also a parameter classifying the wind speed. This is a useful indicator of the sounding height actually reached, and of dynamic and thermal turbulence.

Generally, the basic patterns are divided into vertical and horizontal type. Distinction is made between the following basic patterns, which can be encoded by 2-digit code figures. This distinction is the most important and is given in Table 1.

Table 1. Code number for basic pattern structures (survey)

Code	Description
11	pattern touching the ground, ground based structure (horizontal type)
21	pattern not touching the ground, elevated structure (horizontal type)
30	convective pattern (vertical type)
35	transitional pattern
40	no echo (neutral mixed layer)
77	interpretation of the structure impossible
88	disturbance (of meteorological or technical origin)

There may occur several patterns at one time, which can easily be covered by the special SODAR code by including each of these patterns in sequence. The basic system of the SODAR code is given in Table 2.

Table 2. Structure of the SODAR-code

Indicator of station and time of observation	indicator of basic pattern	indicator of sub-patterns

If wave motions are observed to be associated with a pattern, the average height of the layer with waves as well as the mean amplitude of waves are reported.

In particular the significant so-called "Special Features" include

— wave pattern with periods smaller or greaer than 0.5 hrs,
— pattern continuing beyond the measurable height range,
— so-called "falling echos"; indicating destabilisation processes at the lower boundary of the "free" patterns,
— multiple patterns occurring mainly with stable stratification and usually

touching the ground, preferably at night; they are decoded as uniform patterns
— merging of a "free" and a ground based structure, which in polluted areas can cause a sudden increase of air pollution,
— disturbances caused by heavy rainfall, wind, non-meteorological echoes (smoke plumes) and by technical disturbances.

For the reporting of these parameters a priority list is needed. Two one-digit codes are reserved for the special features. Additionally, the rate of development, i.e. the intensity of a pattern can be reported. The selection of these code figures can be made if reproducible echo-grammes are available, i.e. taking into account constant parameters of the SODAR and of the recording device and excluding the possibility of variations of intensity orignating from acoustic noise.

Further, the tendency of the development of the phenomenon is described. This information, though redundant, can be beneficial for operational evaluation. The first tendency digit describes the change of the rate of development, which can be reported even if due to variable disturbances a statement on the state of development cannot be made. The second digit describes the height change (rising/lowering) in each case along with changes greater or smaller than 100 m.

Recognition and interpretation can be affected by disturbances and variable boundary layer patterns which at the time of observation do not yet show clear patterns. The interpretation of these SODAR-grams requires specialised experience of the observer, in particular regarding the data on the intensity and the recognisability and, hence, the continuous observation of the patterns. It should, however, be noted that the horizontal patterns typical of stagnation in the atmosphere are always clear and distinct, i.e. in such important situations there will be no difficulties in interpretation and coding of SODAR information.

Network of Sodar Measurements

The idea of a SODAR network was to support nowcasting of air pollution in the heavily polluted areas in the south of the former GDR and in co-operation with colleagues in northern Bohemia even in their region. Therefore the code in both territories was equal (co-author of [5]: Dr. Keder). The operational test phase with one SODAR installed at the regional weather bureau in Leipzig was in the winter 1985/86 (30.11.1985–28.02.1986). In the following years a network of 11 SODAR stations was built up. Fig. 1 shows the sites at which vertical SODAR's were operated during the winter 1989/90. These include civil stations as well as military airports. A similar SODAR network was built up in Poland [10].

Most of the vertical SODAR's were installed at regional weather stations, where a meteorologist was working around the clock, regularly. Most of the devices were located directly in the polluted areas. The organization of the network is shown in Fig. 2. Only regional weather forecast centre could use the direct output (facsimile-registration) for nowcasting. The information of all other devices

Fig. 1. The network of vertical SODAR's in the former GDR (winter 1989/90)
● : Civil station with nowcasting
○ : Civil station without nowcasting
■ : Military station with nowcasting

was distributed using the code. From the national forecast centre it was possible to distribute the code as well as maps with the inversion height etc.

After the unification of Germany and the reduction of the industrial potential in the former GDR the problem of air pollution was not as severe as in the years before. At present only at the regional forecasting centers in Leipzig (airport Leipzig-Halle) and Potsdam, a vertical SODAR is working from October to March.

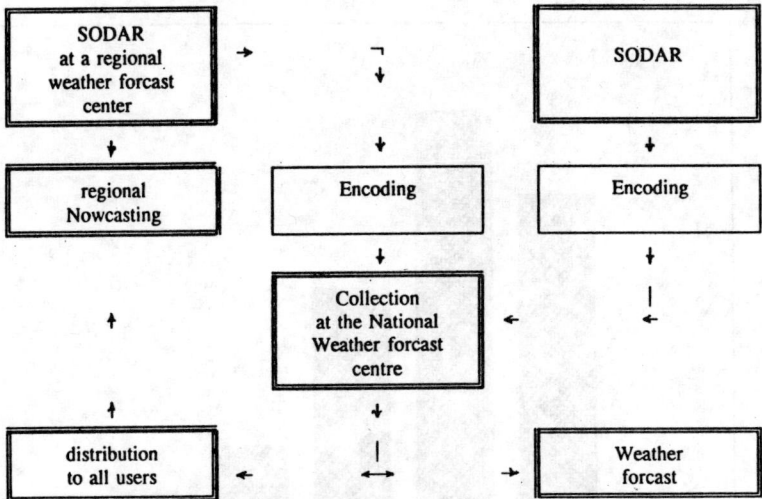

Fig. 2. Operational scheme of the vertical-SODAR network

Experiences in Operational Sodar Network

Generally it must be said that the benefit of the code information is primarily given for climatological application. For nowcasting mainly a regional overlook (for example on inversion height) is of interest. On the other hand it is necessary to use for nowcasting the full SODAR backscatter echo facsimile registration.

Climatological information

During the test phase of the vertical SODAR 1985/86 in Leipzig some climatological studies were carried out. It should be pointed out that these studies are based on the information, that had been coded operationally in real time by the different meteorologists and technicians on duty. As an example, the frequency distribution of the thickness of the ground based (code 11) and elevated (code 21) echo layer based on one-hour averaged values is given in Table 3. For the majority of cases the thickness is less than 150 m.

Table 3. Percentage of the thickness of ground based pattern (code 11) and elevated pattern (code 21)

thickness in m	00–49	50–99	100–149	150–199	200–249	250–299	300–349	350–399	> 400
code = 11	15.9	28.0	18.2	12.1	6.8	3.8	3.8	1.5	9.8
code = 21	19.7	30.3	17.8	6.6	4.0	1.3	4.0	2.6	19.7

Histogrammes of ground based echo layer height (Fig. 3) as well as of the upper height of elevated echo layers (Fig. 4) both show maxima in the range

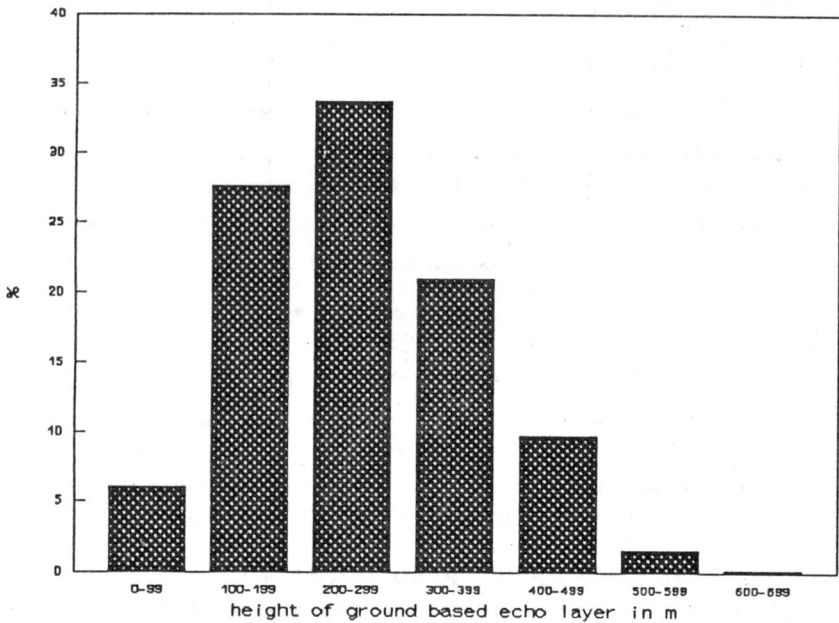

Fig. 3. Occurrence frequency of ground based echo layer heights during the test period 1985/86

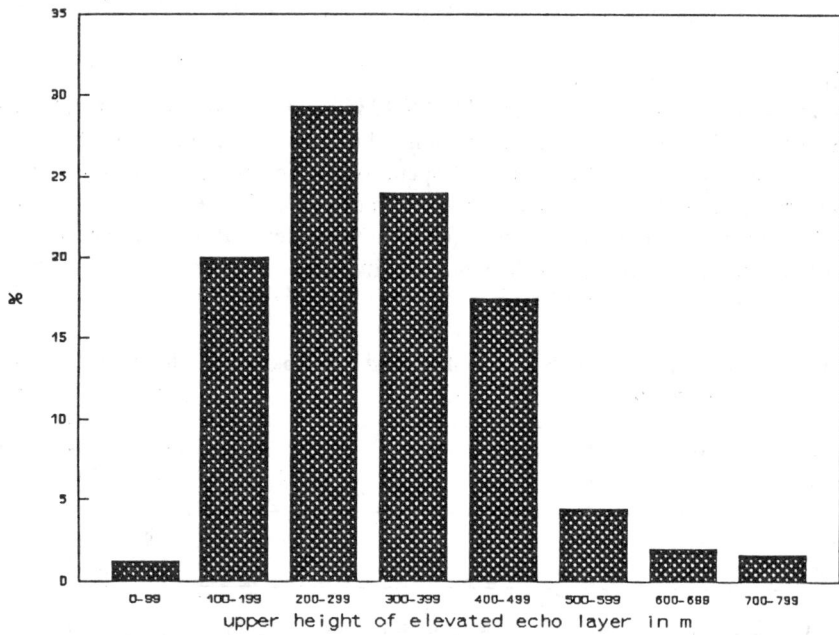

Fig. 4. Occurrence frequency of the upper height of elevated echo layers during the test period 1985/86

200-299 m for the 3-month observation period. Such limited values of mixed layer height are associated with high concentrations of pollutants.

Figures 5 and 6 show the variation of the thickness during the time period over which the structure remained for code 11 and 21, respectively. Inversions of long duration have a significant larger variation of their thickness due to the reduction at daytime and higher values in the night and morning. Of special interest is the information about the history of inversions. Table 4 shows the code number before and after an inversion in the case of ground based pattern. It is evident that an inversion is a well-developed phenomenon, that can be clearly separated from other structures. On the contrary, the code numbers 30, 35 and 40 are sometimes not easy to detect without ambiguity. Therefore an observer without much experience tends to give the code number 77 instead of 30, 35 and 40 in such situations, which often occur before and after inversions.

Such information is not only interesting for climatological investigations, but this information of code parameters can also be used for reports about meteorological situations with high air pollution or smog. For special investigations in this field, measurements with SODAR or Mini-SODAR are interesting. Therefore, the code gives a possibility to quantify the information.

Use for Nowcasting

Based on the code information or the original facsimile-registration at the regional weather stations, nowcasting of the inversion height was possible. Probability forecasts of inversion height for ground based pattern (code 11) and elevated structures (code 21) over 12 hours were made twice a day, namely at 06 UTC for the period 06 to 18 UTC and at 15 UTC for the periode 18 to 06 UTC. An example is shown in Table 5.

The code information can be used to show for instance on a display the regional distribution of the height of the ground based inversion or the first elevated inversion. Such overviews are interesting for regional forecasting or in management. Nevertheless, most helpful in nowcasting are continuous registrations of the backscatter echo. Some examples are discussed in the following.

A good forecasting of the development of an inversion is very difficult especially for the surface inversion, since the inversion phenomenon shows similar patterns on the facsimile-registration without significant differences between summer and winter and irrespective of the air masses. The generation of an inversion under clear-sky conditions starts very early in the evening. First visible structures should be used for nowcasting. But a small increase in the wind velocity or even clouds at the cirrus-level can disturb this process. By combination of all these effects the nowcasting meteorologist with some experience has good possibilities to forecast inversions. This was used with good success.

Thus nowcasting forecasts using the facsimile-registration were made for air pollution concentrations in industrial regions as well as for special prognosis for airport forcasting (METAR: $t + 2$ hours, TAF: $t + 9$ hours) and for the visibility

402 Applications in the Atmosphere

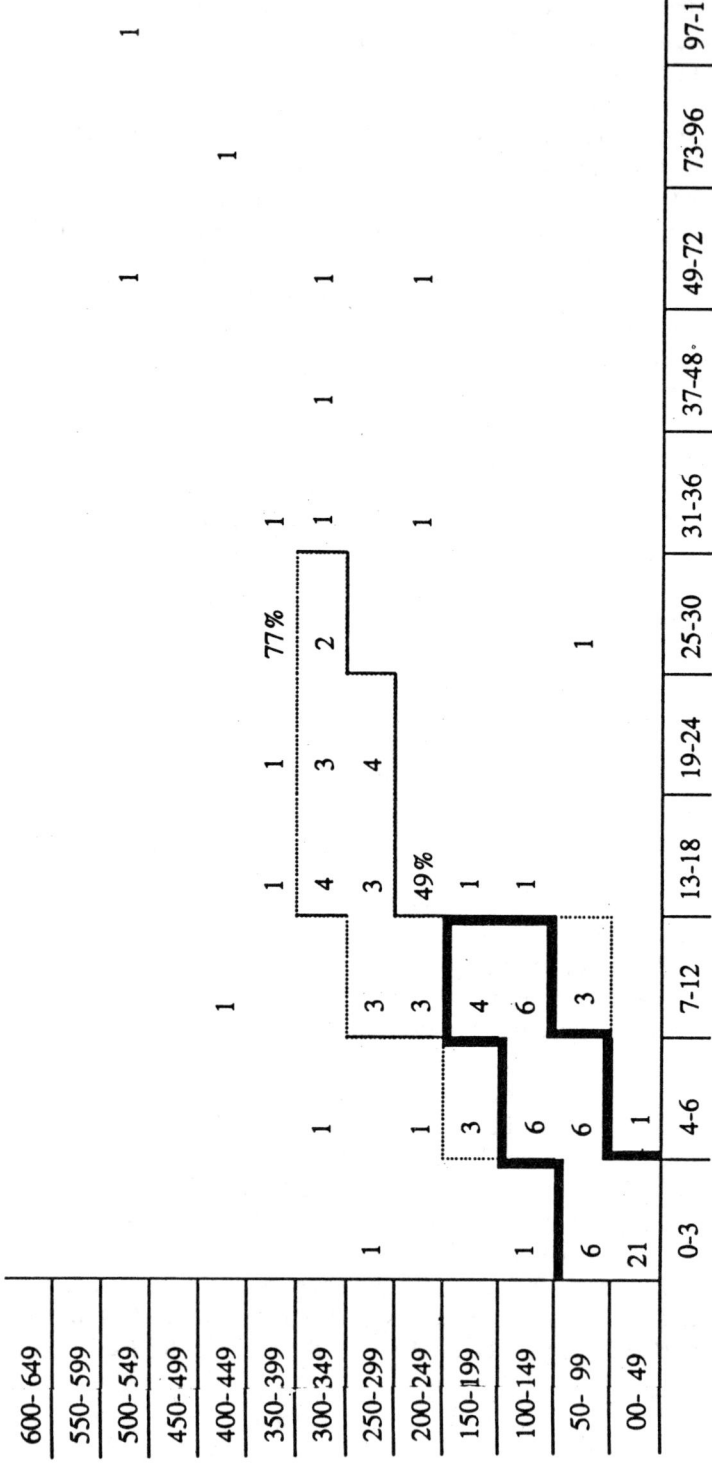

Fig. 5. Height variation (ordinate in m) of ground based echo layers dependent on their duration (abscissa in hour). The figures indicate the percentage of cases.

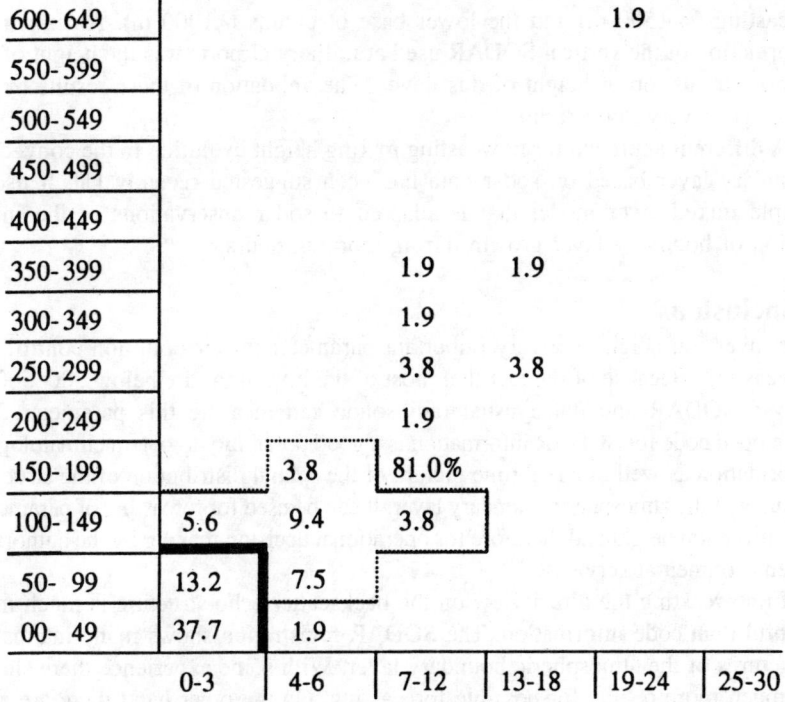

Fig. 6. Height variation (ordinate in m) of elevated echo layers as a function of their duration (abscissa in hours). The figure indicate the percentage of cases.

Table 4. Code numbers before and after the structure of code 11 (ground based pattern)

Code	21	30	35	40	77	88
Code before	9.7%	2.8%	0.0%	15.3%	29.7%	12.5%
Code after	9.7%	6.9%	4.2%	8.3%	61.1%	9.8%

Table 5. Example of a forecast of the probability of the inversion height. 99: no inversion

	code = 11				code = 21		
height	≤ 800	≤ 400	≤ 200	≤ 100	≤ 800	≤ 400	≤ 200
06–09	95	90	80	50	99	99	99
09–12	30	20	10	00	99	99	99
12–15	00	00	00	00	99	99	99
15–18	70	70	70	70	99	99	99

forcasting (< 1500 m) and the lower base of clouds (<1000 m). An important information of the vertical SODAR used at military airports was the height of fog layers, stratus top or height of dust layer. The validation of these results by the pilots gave very good result.

A different approach for nowcasting mixing height evolution in the convective boundary layer based on sodar data has been suggested recently [1]. It uses a simple mixed layer model that is adapted to sodar observations of the initial period of boundary layer growth during morning hours.

Conclusions

The inversion height is a very important parameter for air pollution control and forecasting. Because of the fact that most of the inversions are below 300–500 m, only a SODAR and not a usual radiosonde can measure this parameter. The developed code for SODAR information is a good possibility to obtain climatological information as well as a real time picture of the spatial distribution of the structure features of the atmospheric boundary layer. It can be used for forcasting of parameters like inversion height and therefore for operational decision making by the authorities of environmental services.

In nowcasting the direct view on the backscatter echo structure is much more helpful than code information. The SODAR registration shows many interesting structures of the atmospheric boundary layer. With some experience there should be much more results for possible forecasting. On the other hand there are a lot of meteorological situations (convective conditions, summer time) where the SODAR gives no new or applicable information. But during stable stratification due to missing RADAR data and with hardly usable satellite based IR pictures, SODAR gives necessary information together with other ground-based data. One problem of the operational use of a SODAR is to give information to the meteorologist only in such situations when the application of a SODAR is helpful.

Acknowledgements

The authors are much obliged to all colleagues who supported the progress of a SODAR network and its operational use and to F. Beyrich for the critical discussion of the manuscript.

References

1. F. Beyrich, "Mixing height estimation in the convective boundary layer using Sodar data", Boundary-Layer Meteorol., 74, 1–18, 1995.
2. F. Beyrich and A. Weill, "Some aspects of determining the stable boundary layer depth from Sodar data", Boundary-Layer Meteorol., 63, 97–116, 1993.
3. R. Dohrn, E. Raschke, A. Bujnoch and G. Warmbier, "Inversion structure heights above the city of Cologne (Germany) and a rural station nearby as measured with two sodars", Meteorol. Rdsch. 35, 133–144, 1982.

4. Th. Foken, "Vozmožnosti ispol'zovanija nazemnych sistem kosvennogo distancionnoogo zondirovanija atmosfernogo pograničnogo sloja v setjach meteorologičeskich slyžb", Meteorol. Isledovanija 28, 93–99, 1987.
5. Th. Foken, K.H. Hartmann, J. Keder, W. Küchler, J. Neisser and F. Vogt, "Possibilities of an Optimal Encoding of SODAR Information", Z. Meteorol. 35, 6, 348–354, 1987.
6. M. Gronak, D. Kalaβ: "Ein Vertikal-SODAR zur indirekten akustischen Sondierung der planetarischen Grenzschicht", Z. Meteorol. 36, 4, 225–228, 1986.
7. J. Keder, Th. Foken, W. Gerstmann and V. Schindler: "Measurement of wind parameters and heat flux with the Sensitron Doppler SODAR", Boundary-Layer Meteorol. 46, 195–204, 1989.
8. R.A. Maughan, "Frequency of potential contributions by major sources to ground level concentrations of SO_2 in the Forth Valley, Scotland: An application of acoustic sounding", Atm. Environm. 13, 1697–1706, 1979.
9. J. Walczewski, "Development of sodar and acoustic sounding of the atmosphere in Poland", Z. Meteorol., 39, 129–141, 1983.
10. J. Walczewski, "Dziesięć lat rozwoju Polskiego sodaru i akustycznego sondażu atmosfery w Krakowie", Wiadomości Instytutu Meteoroloii i Gospodarki Wodnej., XV, 37–46,1992.

Part THREE
Ocean Acoustics

19. Acoustic Remote Sensing of Ocean Flows

Antony Joseph K. and Ehrlich Desa
National Institute of Oceanography, Dona Paula, Goa-403 004, India

Introduction

While satellite remote sensing techniques employing active and passive optical, thermal, and microwave signals, and coastally operated remote sensing techniques employing active electromagnetic signals in the H.F. and microwave bands are used on an operational basis for remote detection and mapping of ocean surface circulation, the inability of these signals to penetrate below a very thin surface layer of the ocean surface has rendered these techniques unusable for measurement of subsurface circulation.

In addition to their physical oceanographical interest from the point of view of the circulation in the ocean layers at various depths and its dependence and possible effects on climatological conditions, subsurface currents are of some importance in marine geology because of their influence on the transportation and deposition of sediment. A knowledge of deep currents is also of biological interest because of their influence on the dispersal of organisms and the maintenance of supplies of nutrients. Regions of convergence or divergence in the horizontal movements of water mass are of particular interest because of their association with vertical movements in the form of sinking or upwelling.

The traditional means of making observations of subsurface current velocity was an indirect one, the so-called dynamical method, based on highly precise measurements of temperature, salinity, and depth, the hydrographic tables for computing density, the geostrophic equation, and an assumption regarding the 'depth of no motion'. The valuable review of Prof. K.F. Bowden [1] focusses attention on the assumptions made, and uncertainties involved in the dynamical computations regarding the depth of no motion, and the mean sub-surface current charts. In fact, the celebrated physical oceanographer Henry Stommel's letter to the editors [2] provides an indication of an almost total lack of knowledge on sub-surface currents in the early 1950's.

For lack of proper tools, direct measurements of sub-surface currents were limited to those made from current meters tethered from anchored ships and moored current meters. For a variety of reasons such limited measurements were not adequate to resolve the spatial structure of deep water motions. The extensive and remarkable developments in the field of underwater sound, made during World War II, have placed in the hands of oceanographers a means of making sub-surface current measurements by a variety of acoustic methods. Acoustic techniques have become powerful tools for measurement of ocean circulation mainly because of the ability of acoustic signals to travel long distances in water, and the inherently non-invasive nature of measurement.

Ocean flow measurement techniques can be broadly classified into three distinct groups namely; Lagrangian, Eulerian and single-probe profiling. In the Lagrangian method a synoptic view of circulation spreading over a large area is mapped. The Lagrangian method is useful in establishing large as well as comparatively small circulation routes, and in detecting and identifying gyres, meanders and jets. In the Eulerian method, flow of water past a fixed position in space is determined as a function of time. Circulation measurements in a Eulerian fashion become important when a detailed investigation is required of a particular area of interest on a long-term basis. The profiling method is employed for investigation of the depth-dependence of horizontal flows. These three methods of ocean circulation measurement using acoustic remote sensing techniques are addressed in this article.

Lagrangian Methods

In describing the circulation of the oceans, the Lagrangian method is used more frequently. Acoustic techniques used in oceanography for Lagrangian measurement of sub-surface flows are:

   (i) Tracking of sub-surface floats by ship-borne hydrophones.
   (ii) Tracking of SOFAR floats.
   (iii) Tomography.
   (iv) Reciprocal transmission.
   (v) Space-time scintillation analysis.

These five techniques are briefly discussed below.

(i) Tracking of sub-surface floats by ship-borne hydrophones

A method for sub-surface circulation measurement is the use of pinger (acoustic transmitter)–borne sub-surface floats which can be adjusted to be neutrally buoyant within a certain depth layer. After sinking to a specified depth, the float drifts with the water mass around it. A ship follows the sub-surface drifting float with the aid of acoustic receivers (hydrophones) and thus tracks the path of the sub-surface circulation along this depth layer. The possibility of using this method for measuring deep drift currents over a long period of time was first suggested by Stommel [2].

The principle used in the design of these neutrally buoyant floats (now known as Swallow floats, after their original designer J.C. Swallow) is that a solid body which is less compressible than sea water will gain bouyancy as it sinks. This is because, while the density of a solid body remains practically unchanged with increasing depth, the density of sea water increases with depth. If the float's excess weight at the surface (compared to the weight of the sea water displaced by it) is small, it may at some depth gain enough buoyancy to become neutrally buoyant. At this depth no further sinking will occur. Following the movement of such a float would give a direct measurement of the current at that depth.

An essential design criterion of a Swallow float is that besides having a sufficiently low compressibility, the float must provide enough spare buoyancy to carry the pinger, and must not collapse at the greatest working depth. In the design of the float, Swallow [3] employed aluminium alloy scaffold tubing, HE-10-WP, which has the required mechanical properties. The compressibility of a long cylindrical tube, closed at its ends, is given by the expression:

$$- (1/V) (dV/dp) = [(R_1^2/\mu) + (R_2^2/k)]/(R_2^2 - R_1^2) \qquad (1)$$

where
R_1 : internal radius of the tube.
R_2 : external radius of the tube.
V : external volume.
μ : rigidity of the tube material.
k : bulk modulus of the tube material.

In Swallow's neutral-buoyancy float, the required compressibility at the desired working depth (4500 m in the original design) was achieved by uniformly reducing (by solution in caustic soda) the wall thickness-to-mean radius ratio of the standard tube to 0.16. The required length of the tubing was 6 m. For convenience in handling, the float was made in two pieces of 3 m, one containing the pinger and battery, and the other providing buoyancy. A sketch of the Swallow float is given in Fig. 1. The two ends of the tubes were plugged with end caps and 'O' — rings. The pinger consisted of a nickel scroll resonant at 10 kHz, wound toroidally and energized by discharging a capacitor through a flash tube.

Before deployment, the mean density of each complete float and pinger was adjusted to the sea water density at the desired depth of operation and temperature by immersing it in a salt solution of the same density and temperature, and adding weights until it was neutrally buoyant. While doing the fine adjustment of the float the calculated compressibility of the float at the desired depth of operation must be taken into account. All the extra weight are put inside the buoyancy tube so that the external volume of the float remains unchanged.

The method used to track the drifting float is to lower two hydrophones over the ship's side as far apart as possible. The acoustic pulses from the neutral buoyancy float will be received at different times unless the float is located just below the ship or at an orientation perpendicular to any plane joining the positions of the two hydrophones. From the magnitude and sign of the observed time difference it is possible to estimate the bearing of the float with reference to the plane joining the positions of the two hydrophones using known values of the

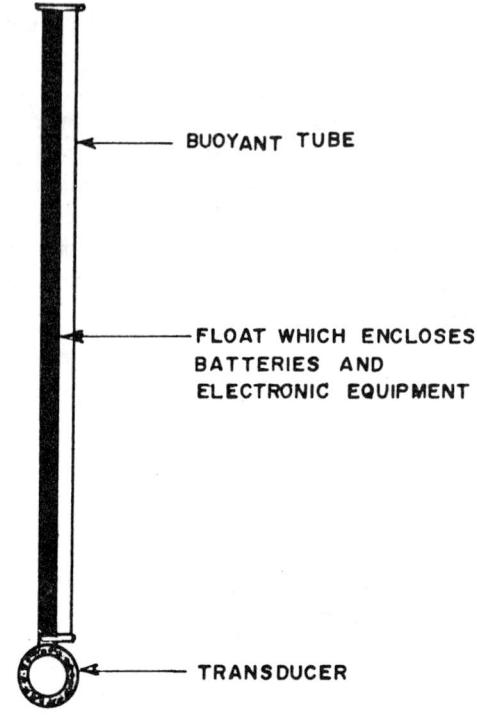

Fig. 1. A sketch of the Swallow float, after Swallow [3].

separation between the two hydrophones and the velocity of sound in sea water. To avoid errors in estimation of the bearing of the float, the hydrophones are kept fairly shallow (approximately 7 m) and are weighted to prevent their cables from straying too far from the ship's side. Detection of float position using ship-borne hydrophones is schematically shown in Fig. 2. Because the chasing ship continuously determines its own position by conventional navigational techniques, the speed and trajectory and, therefore, the direction of movement of the current-driven drift of the float at that depth can also be determined.

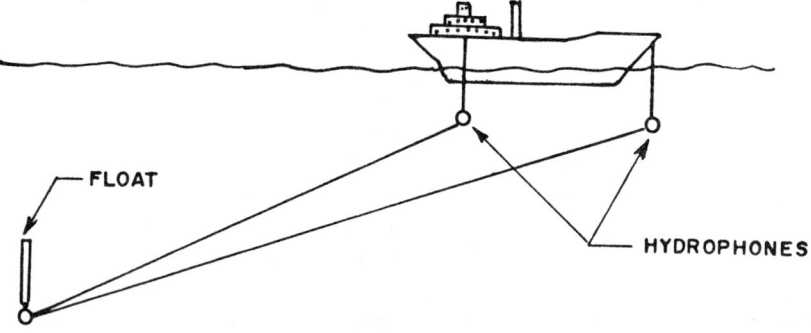

Fig. 2. Acoustic tracking of subsurface float by ship-borne hydrophones

A limitation of this method is that one ship can follow only a very small number of floats at one time. If several floats are released, e.g., at different depths, it is quite likely that they will drift off at different directions and the ship may not be able to keep track of them for long [4], [5], [6]. Further, dedicating ships to keep track of the float is too expensive for any long-term circulation monitoring programme.

(ii) Tracking of sub-surface SOFAR floats

Dr. M. Ewing's invention of SOFAR channel (Sound Fixing and Ranging) during World War II heralded a new era in under-water acoustics, paving the way for development of techniques for making continuous day by day measurements of sub-surface currents in the ocean on a permanent basis. With this development, we should eventually be able to map synoptic charts of currents at various depths in the ocean in the same way as meteorologists keep abreast of the winds. The SOFAR channel permits acoustic signalling over great distances (greater than 1000 km), provided the frequency of transmission is low (much smaller than 1000 Hz). The SOFAR channel, which is an acoustic wave guide, owes its existence to the fact that the speed of sound reaches a minimum value at a certain depth in the ocean (approximately 1000 metres). This is because the speed of sound in water is a function of pressure and temperature (and salinity to a lesser extent). As one leaves the warm surface waters of a subtropical ocean, the speed of sound decreases with increasing depth due to the decreasing temperature. Below the main thermocline, where the thermal gradients become small, the effect of pressure upon the speed of sound becomes dominant and increases with further increase in depth. The significance of the depth of minimum sound speed (sound channel axis) is that sound rays which are radiated within certain angles from the horizontal from a source located at the sound channel axis will be refracted back towards this depth. This acoustic energy, trapped in the vertical, radiates horizontally with a geometric rate of attenuation proportional to $(1/R)$, where R is the distance, instead of $(1/R^2)$ as in the case of a spherical radiator. This low rate of attenuation of acoustic signals propagating in the SOFAR channel permits acoustic signals to be detected at great distances from the source. In addition to the geometric attenuation, there is also the frequency-dependent attenuation due to scattering and absorption. Thus, acoustic signal is attenuated much more rapidly at higher frequencies. For this reason, better reception and greater ranges are obtained by transmitting at as low a frequency as possible. There are other interesting characteristics of sound propagation in SOFAR channel, which are more relevant to acoustic tomography. These characteristics are discussed in the next section on acoustic tomography.

The possibility of making use of the special characteristics of SOFAR channel to remotely explore large scale ocean circulation features in deep waters was initially suggested by Henry Stommel [2]. He suggested that the basic instrument required (for this purpose) is the oceanographic equivalent of the meteorological constant-altitude balloon: an unmanned sub-surface buoy or float, devised so as to float at a nearly constant depth, or along an isopleth of temperature or density, and equipped with a clockwork designed to drop and/or fire SOFAR charges at certain predetermined times, say once a week, for periods upto half a year. The trajectory of each buoy could be determined from the times of arrival of the explosive sound wave at three SOFAR stations. He predicted that if this method proves feasible, the way will then be clear for a major advance in physical oceanography. Encouraged by this suggestion, Rossby and Webb [7] initiated an observational program to study abyssal water motions by tracking subsurface floats which were released in the SOFAR channel, and the first Swallow float was launched in January 1968. These floats are now known as SOFAR floats.

The SOFAR floats are similar in concept to Swallow floats, which have been used with great success. However, the technology for SOFAR floats are much more demanding, in terms of reliability, economy of power, batteries with high energy density per unit weight, a low frequency (preferably less than 500 Hz) acoustic projector with low weight and high efficiency, and corrosion resistance. These special demands of SOFAR floats arise because they act as acoustic beacons for tremendous distances (more than 1000 km) which must operate over long periods of time (over a year).

The SOFAR floats have been tracked by the use of two receivers separated by a large distance. If the travel time of a pulse from the SOFAR float to these two receivers are known, then the distance between the floats and each receiver can be computed from an apriori knowledge of the mean sound velocity in that ocean basin. The position of the float will be one of the two intersection points of the two circles whose centres are the positions of the two receivers and whose radii are the two pulse-travel distances between the float and the two receivers (Fig. 3). The ambiguity can be resolved by knowing its previous location or the location

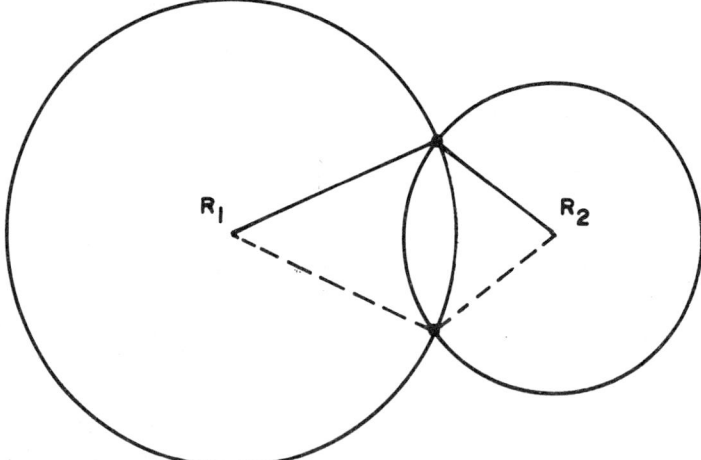

Fig. 3. Method of tracking a SOFAR float with the aid of two acoustic receivers R_1 and R_2.

where it was launched. If the pulse travel time is to be determined precisely, the time of transmission of the acoustic pulse from the float must be known precisely. For this purpose, pulses are transmitted from the float at predetermined times. This means that the clock in the float must be extremely precise. The crystal clocks are known to drift with time. To circumvent this difficulty three receivers are usually used rather than two. In this case the float position can be determined even if the clock in the float gains or loses time slightly. When three receivers are used, the difference in time of arrival between two pairs of receivers gives two hyperbolae (on a sphere), the intersection of which is the location of the float [7]. Because the 'difference' in time of arrival is used, any error due to gain or loss of time which is common to both of the received signals gets cancelled out. To get sufficient data for this purpose, a set of many pulses (usually 100) are transmitted 1 minute apart every few hours. Because the deep currents are usually small this methodology of tracking is not expected to give rise to serious errors in the position determination of the float.

In the receiver electronics some precautions are needed to keep the influence of noise level to a minimum. This is because the acoustic signal traversing a long distance (1000 km or more in the horizontal), pick up noise in various parts of the frequency spectrum, generated by ships, earthquakes, waves, rainfall etc. Any slight non-linearity of the signal amplifiers in the receiver side can generate harmonics, and this will raise the effective noise level of the system. This problem can be minimized by filtering out the noise components in the received signal and amplifying only the frequency band of interest. Usually, the received signals are detected in an approximately 1 Hz wide bandpass filter centred at the transmission frequency.

The SOFAR float is a tightly engineered Swallow float with an acoustic transducer which transmits c.w. pulses according to a precisely timed schedule. The SOFAR float constructed by Rossby and Webb (Fig. 4) consisted of a spherical housing (in two hemispheres) of adequate stiffness and resistance to hydrostatic pressure. Each hemisphere of the float was spun from aluminium alloy 6061–T6, had an internal diameter of 0.93 m, a minimum wall thickness of 25 mm and a hydrostatic collapse safety factor of two or more. All exposed external surfaces were either machined hard anodized aluminium alloy or non-conductive synthetic material. The float enclosed the required power supply and electronic equipment and supported the electro-acoustic transducers suspended about 1.5 m beneath it. The electro-acoustic transducer was a hollow cylinder of 750 mm diameter, 325 mm length, and 50 mm wall thickness, which vibrated in the fundamental radial mode. The construction was a composite of active barium titanate and passive aluminium alloy in a barrel stave configuration pre-stressed into compression by circumferential glass fibers in tension and insulated with polyurethane. The electrical connectors to the transducers were protected against fish bite with nylon tubes.

In the design by Rossby and Webb, three electro-chemical subsystems were used to supply power to the electronics and the transducer of the float. The first

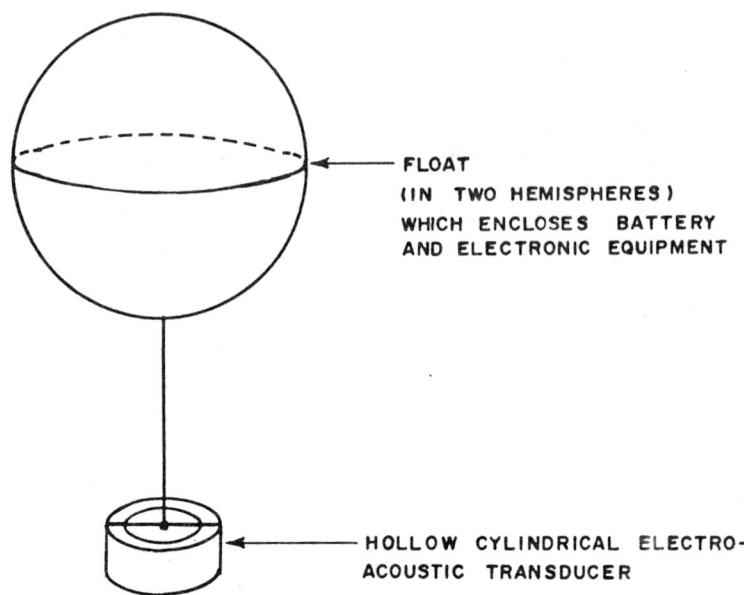

Fig. 4. A sketch of an unsuccessful SOFAR float.

subsystem consisted of 20 kilowatt-hours of energy stored in a 230 kg primary battery consisting of magnesium manganese dioxide cells. This primary battery of high energy density was used to charge a second battery consisting of 13 silver-cadmium cells of 18 A-hr rating. This second battery was able to deliver the short pulse (1.2 second long) of high current to the output power amplifier. A third battery of four cells utilizing hydrogen-copper oxide couple served to recombine any hydrogen gas evolved by parasitic corrosion of the primary magnesium anodes.

The low frequency carrier was derived from a crystal oscillator. This signal, generated in low power integrated circuits, was amplified to about 300 Watts, transformed to about 2500 volts, and powered the transducer through a series inductor. The drift rate of the oscillator was less than 9 seconds per year.

The abrupt failure of both spherical shaped floats after only a few days of operation was a cause for concern. For use in the Mid-Ocean Dynamics Experiment (MODE) at Bermuda triangle Rossby et al, [8] adopted an alternate design. This design was similar to the neutral buoyancy float originally designed by Swallow. The SOFAR floats constructed for MODE (Fig. 5) consisted of 3 cylinders of aluminium alloy.

The central cylinder was 5.2 m long, 30.5 cm diameter, with hemispherical aluminium end closures, and contained the battery and electronic equipment. The two short cylinders shown in the figure are the low frequency acoustic transducers which operate freely flooded. The upper end was open and the acoustic driver, which was a bender plate, was fitted to the lower end. These transducer tubes

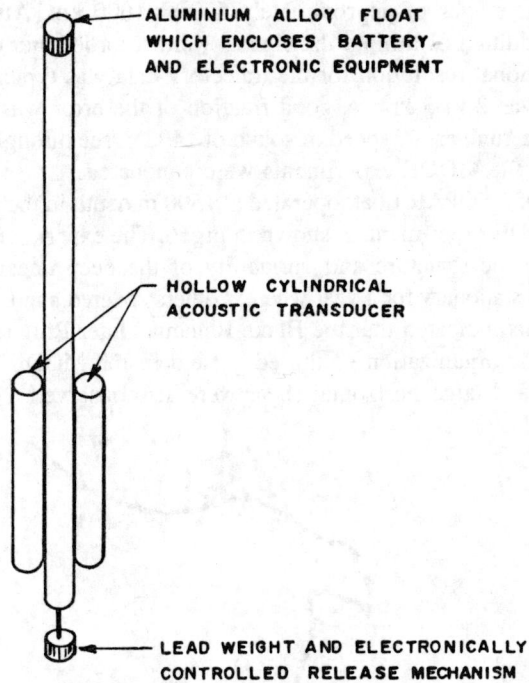

Fig. 5. A sketch of the SOFAR float used in the MODE programme.

were mounted on the main housing with 4 heavy stainless steel studs. All the 20 SOFAR floats constructed for MODE were equipped with two acoustic signaling systems: a low frequency system for long-range shore locations and a high frequency subsystem for shipboard location and command recovery. A 7.3 kg cylinder of lead was connected to the lower end through an electronically controlled release mechanism, and was jettisoned for float recovery on command from the surface. The small high frequency (10 kHz) ceramic transducer was connected to the bottom end.

All signals were derived from a temperature-compensated quartz crystal oscillator. Electrical energy was furnished by a single battery of 73 kg, 30 volt nominal, 5 kW hrs. Over 90% of the energy was used in the low frequency signal. The completed float weighed 430 kg.

For identification of all the 20 floats from the same receiver stations, the method used was to have a combination of one of three carrier frequencies (approximately 267, 270 and 273 Hz) and 7 different pulse repetition rates ranging 1437 to 1443 transmissions per day. These three differing carrier frequencies and 7 pulse repetition rates provided 21 channels, which was more than sufficient to track 20 floats.

The signals from the float were received at four widely spaced SOFAR hydrophones at ranges of approximately 500 to 1000 km. After amplification, filtering and addition of timing, the signals were recorded and used for analysis.

The navigational resolution for the trajectory data was typically ± 500 m and the accuracy was 2 to 3 km. A good fraction of the error was attributed to the assumption of a "universal" speed of sound of 1492 m/sec throughout the Bermuda triangle where the MODE experiments were conducted.

Trajectory of a SOFAR float, operated at 1500 m depth in the Bermuda triangle during the MODE experiment, is shown in Fig. 6. The experiment shed some new information on the structure and variability of the deep ocean currents. Some floats remained stationary for a year whereas others covered hundreds of kilometres. The experiments indicated that the Blake-Bahama Outer Ridge had considerable influence on the organization of the eddy field in the MODE area. Regions of sudden swirls and large horizontal shear were also observed.

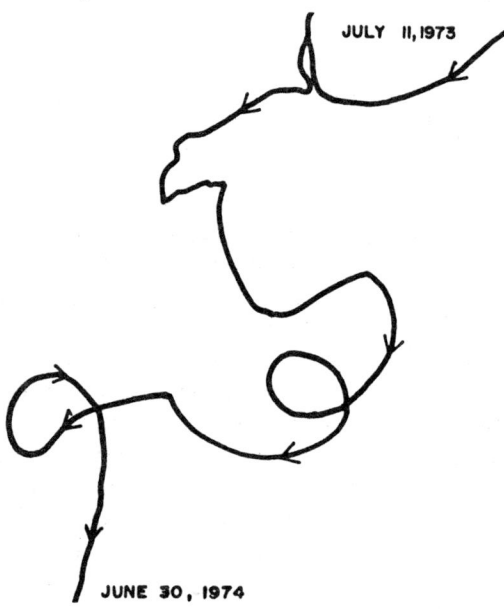

Fig. 6. Trajectory of a SOFAR float operated at 1500 m depth, after Rossby et al, [8].

In an alternate design Webb [9] used a cylindrical float (Fig. 7) as part of the POLYMODE programmes in a major study of mesoscale circulation in the mid-ocean. The two cylinders were made of corrosion-resistant aluminium alloy 6061-T6 with end closures cast in aluminium alloy A 356-T6. The main pressure resistant housing, of 5.39 m length and wall thickness of 1.9 cm for deeper floats and 1.6 cm for shallow floats, enclosed the battery and electronic equipment and provided the required buoyancy. These floats, although used without no protective coatings were reported to have performed well even after 4 years of continuous service. The signals were received at deep hydrophones from moorings and

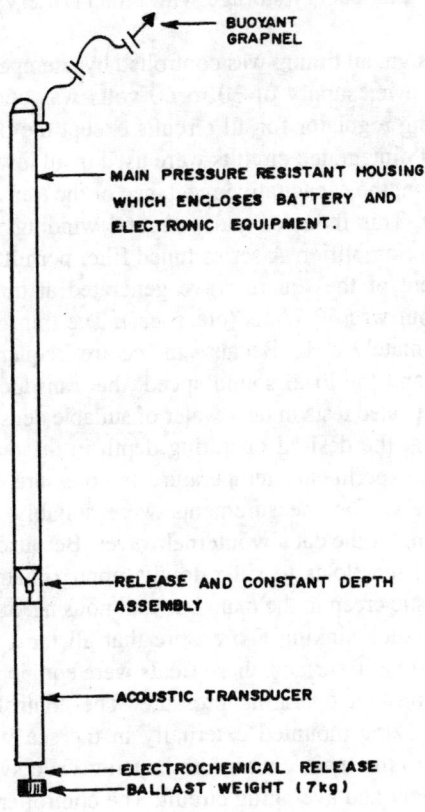

Fig. 7. Sketch of a SOFAR float used in POLYMODE programme, after Webb [9].

transmitted to land-based stations. A multiple address system using 1420 Hz phase encoded signal transmitted from a recovery ship to the float could establish communication to the float. Upon reply from the float, a recovery command sent from the ship caused the jettison of a 7 kg external ballast weight, and initiated a special fast pressure telemetry cycle to verify release and aid recovery. The jettison was effected via a simple electrochemical release discussed later. For pickup at sea a buoyant grapnel tethered to the float by a buoyant line was snared with a heaving line from the ship, and the float was hauled into its launch and recovery frame using the buoyant line.

In the electrical design of the POLYMODE SOFAR floats two alternate batteries were used successfully, either an alkaline cell pack using Eveready G cells, each pack containing 400 or 560 cells in series-parallel, the number depending on the tube wall thickness, or a lithium battery pack consisting of 72 cells of the type

660-5AS. The lithium pack is lighter and permits the use of a much shorter buoyant housing. The total energy storage with either battery was approximately 7 kW hr.

In the electronic design, all timing was controlled by a temperature compensated crystal oscillator. A power supply of 30 to 60 volts was converted to 12 volts regulated in a switching regulator for all circuits except the power amplifier. To conserve power, CMOS integrated circuits were used in all low-level applications. In the transducer section, the capacitative reactance of the transducer was matched by a parallel inductor. This inductor had a second winding which matched the impedance of the power amplifier. A series tuned filter permitted to pass only the fundamental component of the square wave generated at this power amplifier. Normal transducer input was 60 Watts (electrical). The transducer had a narrow band width of approximately 2 Hz. Because the centre frequency is a function of the transducer length and the local sound speed, the transducer was trimmed to the correct length by repeated tests in tank water of suitable density and temperature corresponding to that at the desired operating depth in the ocean.

In the POLYMODE experiments, temperature and pressure were also measured as additional parameters. The measurements were suitably averaged to avoid contamination or aliasing of the data by internal waves. Because early observations revealed a tendency of the floats to sink slowly, approximately 0.5 m per day, apparently due to inelastic creep in the main buoyant housing, adequate precautions were taken to prevent such sinking. To ensure that all the POLYMODE floats drifted over the same isobaric surface, these floats were equipped with a controller which maintained a constant operating pressure. The controller consisted of a block of anode quality zinc mounted externally in the sea water and could be electrically connected to the main aluminium housing via a switch controlled by the pressure measurement and averaging circuit. The controller works as follows: when the circuit is open, the elecro-chemical couple is quite inactive. When the circuit is closed, under control of the pressure measurement circuitry, a small salt water battery is formed. When this happens zinc goes into solution in the sea water and the whole instrument begins to rise. When the float has risen to the required level the switch is again opened so that dissolution of zinc is arrested. The regularly telemetered pressure signal indicated that all the POLYMODE floats operated within ± 5 decibar of their normal equilibrium depth.

An excellent description of the careful preparations needed before ballasting may be found in Webb [9]. This information will be very useful to those involved in the deep-sea circulation measurement using SOFAR floats.

(iii) Tomography

Acoustic tomography is employed in oceanography mainly to remotely detect mesoscale circulation features such as cold/warm-core eddy structures (known as the ocean weather) that are superimposed on a generally sluggish large-scale circulation (known as the ocean climate). The ocean mesoscale eddy field is closely analogous in character to weather systems in the atmosphere, but the two differ in sizes and life times. While the oceanic eddy structures are several hundred kilometres in diameter and have life times of a few months, the weather systems in the atmosphere are several thousand kilometres in diameter and have life times of a few days.

Detection of mesoscale eddy structures in an ocean basin using acoustic tomography relies on the measurement of travel time fluctuations induced by changes in the acoustic field within the ocean by acoustic transmission along many diverse paths. The feasibility of studying and ultimately monitoring the oceans by measuring acoustic transmissions between moorings over large distances was originally proposed by Walter Munk and Carl Wunsch [10]. The method was called 'ocean acoustic tomography' in analogy with computer-assisted tomography (CAT) used in medical diagnosis. The sound speed in the ocean is predominantly a function of temperature, and to a lesser extent salinity and depth. Thus, a cold eddy within the observation region will delay the arrival of any transmission through the eddy; and a warm eddy will cause faster arrival of the acoustic transmission at the receiver. For example, in the Bay of Bengal an observed cold-core eddy of 5° temperature drop which brings about a reduction in the ambient sound speed by 10 m/sec delays the travel times by 100 to 200 ms for a mesoscale range [11].

The ocean acoustic tomography (OAT) technique involves two aspects, namely (i) the "forward" problem of finding the behaviour of sound transmission in an ocean basin over distances of order 100 km, and (ii) the "inverse" problem of determining the interior structure of this ocean basin from travel time measurement of sound transmitted through the basin. In the OAT technique the variable acoustic travel times between all source-receiver pairs of an array of underwater acoustic sources and receivers, moored at spatial intervals of usually more than 100 km, are used to construct the three-dimensional (time-variable) eddy field, using inverse theory. The great advantage of the OAT technique over conventional point measurements or ship surveys is that the number of data points grows geometrically as the product of the number of sources (S), the number of receivers (R), and the number of resolved acoustic paths (P) as compared with the sum ($R + S$) for conventional spot measurements. This is demonstrated in Fig. 8. Additionally, path integration reduces the noise from local fine structure and internal waves which contaminate spot measurements. This superb ability of the OAT technique is achieved essentially because of the transparency of ocean to acoustic signals, and the presence of an acoustic wave guide known as SOFAR channel. In the ocean there is a minimum sound-speed axis around 1 km depth. The sound speed above this axis gradually increases upwards because of increase in temperature, and it increases downwards because of the effect of increasing pressure. This gradually increasing sound speed with reference to the acoustic axis corresponds to a gradually decreasing refractive index of the sound channel because of which the acoustic ray within this channel travels long distances with minimal loss, in

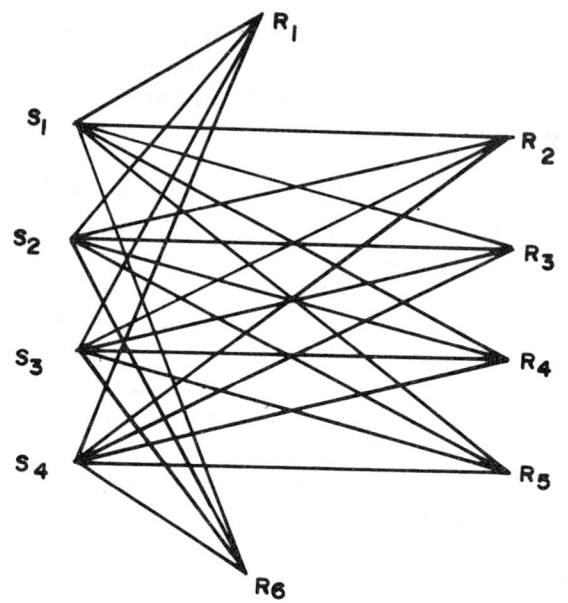

Fig. 8. A sketch of OAT implementation.

a manner similar to the propagation of light rays along an optical wave guide (Fig. 9). Because the acoustic ray paths oscillate about the axis of the sound channel (i.e., the axis of minimum sound speed) these rays, depending on their inclination, can scan long distances in the vertical plane. It is interesting to note that the steep rays which traverse a large distance of the ocean in the vertical plane arrive at the receiver faster than those flat rays which traverse a much lesser distance in the vertical. Nevertheless, this is an essential peculiarity of acoustic rays traversing in an acoustic wave guide.

The presence of an acoustic wave guide in the ocean helps essentially in the following two ways to simultaneously scan a large ocean volume within a very short time [12]. (i) The acoustic rays that are refracted back towards the SOFAR

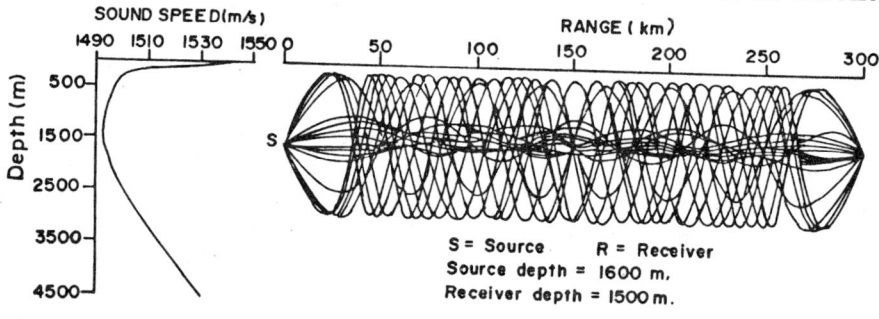

Fig. 9. Propagation of acoustic rays, about the axis of minimum sound speed, in a SOFAR channel in the Bay of Bengal, after Prasannakumar et al, [13].

axis before reaching the surface and bottom of the ocean basin lose little energy through the boundaries and, therefore, can be detected over several thousand kilometres; (ii) steep rays which sample the entire water column and generally arrive early can be distinguished from flat late rays which remain nearer the axis, and in this way information can be gathered about the depth dependence of the mesoscale eddy field desired to be detected. Different ray paths give different weights of the water column, and this permits study of vertical eddy structure. Once the travel times of various acoustic rays are obtained (i.e., direct solution is obtained), the inverse theory is applied to construct a three-dimensional map of the sound speed field in the scanned ocean basin from an apriori knowledge of the unperturbed sound speed field (reference sound speed field) in the scanned region.

Instrumentation: The acoustic transmission loss is both range and frequency dependent; this leads to an operating frequency in the 200–400 Hz range. For this reason the transmitted signal frequency employed in OAT measurements generally lies in the range 200–400 Hz and usually has a bandwidth of 2 to 100 Hz.

The source is a resonant tube approximately one-quarter wavelength (in water) long, driven at one end and open at the other. The source usually consists of a cylindrical tube of approximately 0.3 m diameter tuned for resonance at the desired transmission frequency. It is driven at the closed end by a flat circular plate of piezo-electric material (lead zirconate titanate) operated at the fundamental tube resonance. The actual tube length is about 10–20 per cent greater than one-quarter wavelength due to finite edge effects [14]. The efficiency of the source is roughly proportional to the tube cross-sectional area. The acoustic source transducers are essentially of organ pipe design derived from SOFAR float programmes and have lengths of approximately a metre. This type of transducer has severely limited bandwidth and is marginally suitable for tomograpy because a limited bandwidth of the source transducer will limit the sharpness of the processed received signal thereby limiting the multipath resolution of the receiver. However, presently there are no alternative broadband transducers available for tomographic measurements.

Because travel time of acoustic signals travelling over a distance of a few 100 km are to be measured with precision, an essential requirement of the transmitted pulse is that it must contain sufficient power (to traverse long distances) and its width must ideally be very small (so that multipath travel times can be adequately resolved). Because these two conflicting requirements are difficult to be met with a narrow pulse, an ingenious technique used in OAT is to transmit a signal (sequence) pattern of large width (so that sufficiently large power can be transmitted) whose autocorrelation function has a very sharp peak having very low sidelobes. The width of the correlation peak establishes the achievable multipath resolution of the system. For many purposes the covariance peak can be regarded as if it had in fact been the transmitted pulse. This pulse compression technique permits, therefore, pumping sufficiently high power by the transmitter without sacrificing the multipath resolution of the system. In the acoustic tomography experiment conducted by the Ocean Tomography Group in 1982 [12], a single transmission consisting of 24 consecutive sequences lasted for nearly 192 seconds. Each sequence was a 127-digit maximal length shift register. Thus transmission of each sequence required 8 seconds, and each pulse was of width approximately 63 ms. This meant that each transmission was equivalent to the transmission of 127 pulses of 63 ms width, at intervals repeated 24 times. The carrier was 224 Hz having a band width of 20 Hz, and the transmitted power level was approximately 14 Watts. At the receiver the arrival time structure of the multipath field between each source and receiver pairs was obtained by cross-correlating the coherently received incoming signal with a stored replica of the transmitted signal. Some receivers perform this cross-correlation in-situ while others store signal samples for later on-shore processing.

The validity of the interpretation regarding mesoscale eddy fields using OAT techniques depends to a large extent on the precision with which the acoustic pulse travel time is measured. For this reason, an essential requirement of OAT instrumentation is the incorporation of a precision time-base. Because the time-base derived from quartz crystal oscillators cannot provide the required long-term precision and repeatability, rubidium atomic frequency standards must be used. However, the power requirement of rubidium atomic frequency standards is too high for continuous use in a long-term moored instrumentation such as that of OAT measurements. To meet the stringent requirement of high precision in time-base and low-power consumption, a technique usually employed is to switch on the highly stable rubidium atomic frequency standard only periodically (so that power consumption is reduced) and compare the frequency of the considerably less stable (but low power consuming) crystal oscillator clock frequency with the rubidium standard to get the frequency offset of the crystal oscillator clock. The periodically measured frequency offsets of the crystal oscillator clock are then integrated to yield time corrections.

The receivers are usually equipped with a vertical array of hydrophones separated at a suitable spacing so that the ray inclinations of various multipath arrivals can be computed. The pulse arrival pattern for each ray is predicted using ray theory from an apriori knowledge of the sound speed in the ocean basin being surveyed. The measured deviations are attributed to perturbations in sound speed along the unperturbed ray paths, due to the presence of eddies, meanders etc. The three-dimensional shape and nature of the travel time perturbation can be obtained from analysis of multipath arrivals from many combinations of transmitter/receiver pairs in the horizontal array, moored near the axis of the sound channel (approximately 1 km depth). The measured travel-time perturbations are then used to generate the sound speed pattern in the ocean basin and construct a tomographic picture of the perturbations within the array (Fig. 10). From daily snap shots of the perturbations it is possible to identify cold- and warm-eddies and determine their growth, change in shape, speed and directions of their movement, and their ultimate decay.

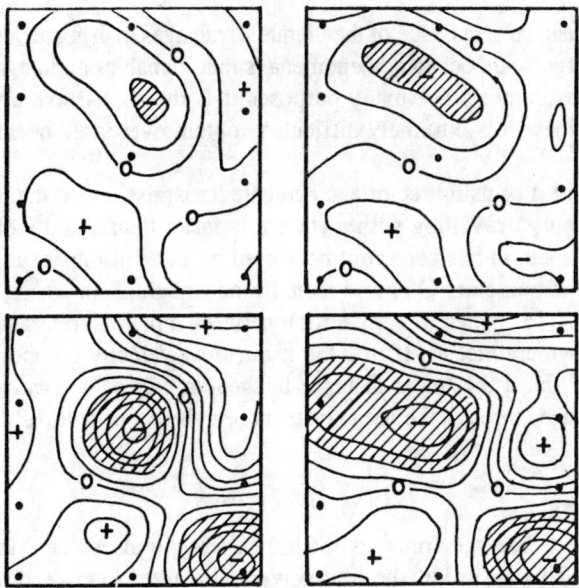

Fig. 10. Mosaic of sound speed anomaly charts in a 300 × 300 km square south-west of Bermuda, obtained from tomographic inversions. (−) indicates a cold-core eddy region. (+) indicates warm water (higher sound speed), after Behringer et al, [12].

Several tomographic experiments in the 1980's have indicated that mapping of mesoscale gyres in the ocean can be performed tomographically over large areas. The integrating properties of long range sound transmission are now permitting true basin-scale measurements without being perturbed by local small-scale influences. The proponents of OAT to monitor large ocean basins for mesoscale fluctuations have envisaged real- or near real-time use of the data for operational ocean monitoring programmes. The spatial resolution could be greatly improved by well-positioned autonomous offshore listening stations and data transmitted periodically, via telecommunication links. With the advent of miniature transmitters and receivers such data transmissions would be possible.

(iv) Reciprocal transmissions

An approach to the problem of remote measurement of large scale oceanic motion using reciprocal acoustic transmission method was advanced by Stallworth [15] followed by Rossby [16]. The central element of this scheme is that the line integral of fluid velocity along an acoustic ray between two points in a flow field is proportional to the difference in travel times of two acoustic signals simultaneously transmitted from these two points in opposite directions. By doing so, the effects of ocean currents on acoustic propagation can be separated from the effects of sound speed structure. Reciprocal acoustic transmissions can, therefore, be used to measure ocean currents. One of the unique advantages in using acoustic techniques to measure large-scale oceanic phenomena is that it enables an integral or spatially averaged measurement. For many purposes it is the spatial averages that are of interest, and these are extremely difficult to obtain over large ocean areas in any other way.

The basic idea of using reciprocal acoustic transmissions to measure current is that a sound pulse travelling with a current is faster than one travelling against a current. This method has been routinely used by meteorologists as early as 1960 in acoustic anemometers [17] and later by oceanographers for Eulerian current measurements [18], where the separation between a pair of transceivers are of the order of a few centimetres. If v is the mean flow velocity component along the acoustic path, and l the linear distance between a pair of transceivers, and c the mean sound speed through the ocean water at rest, the acoustic travel time difference is given by:

$$| (t_{ij} - t_{ji}) | = 2lv/c^2 \qquad (2)$$

where t_{ij} and t_{ji} are the transmission times from transceiver i to j and j to i respectively. To first order, the transceiver separation l between a transceiver i and another transceiver j is given by [19]:

$$l = (1/2)\ c(t_{ij} + t_{ji}) \qquad (3)$$

An average sound speed profile of the ocean basin at which the reciprocal acoustic transmissions are made, appropriate for the time during which the experiments are conducted, is usually constructed by combining data from a series of expendable bathythermograph (XBT) casts and salinity measurements taken along the lines joining different moorings of the transceiver array. Measurement of mean flow components along two or more different axes permits estimation of mean flow vector. In an ocean basin-scale measurement scenario using acoustic techniques, where many pairs of transceivers are usually deployed in a large array, it is possible to estimate the mean flow in an ocean basin using this technique.

When reciprocal acoustic transmission technique is applied for measurement of large-scale oceanic circulation, the difference in travel times of the oppositely travelling pulses are interpreted in terms of ray-averaged currents. In the simplest approach, it is assumed that sound speed and current depend only on the horizontal coordinate along a straight line ray between a given source-receiver pair. Preliminary investigations by Worcester [19] suggested that differences in travel times of oppositely travelling pulses can be interpreted in a preliminary manner as ray-averaged currents. Since a number of distinct ray paths with a variety of turning depths exist for each source-receiver pair, and each ray represents a different depth-weighted average of the ocean, the ocean-basin current estimated using this technique represents baroclinic (depth-dependent) spatially averaged current fields. However, methods now exist to estimate basin-scale current fields in several horizontal layers of the ocean depth, which yield barotropic (depth-independent) spatially averaged current fields.

The method of generation of transmission signals and processing of the received signals in the reciprocal acoustic transceivers are similar to those used in acoustic tomography experiments. Before deployment, the source-receiver pairs are usually connected in the same loop for clock synchronisation purposes. The transmission signal is usually a pulse stream consisting of period repetitions of a phase-coded linear maximal shift-register pseudorandom sequences. The advantage of such a transmission signal code is that it can be processed at the receiver to yield an output waveform which has minimum sidelobes (Fig. 11). It has the additional advantage that the processing can be implemented on a microprocessor using only adds. This enables the signal processing to be performed in-situ in the instrument itself, thus conserving memory space.

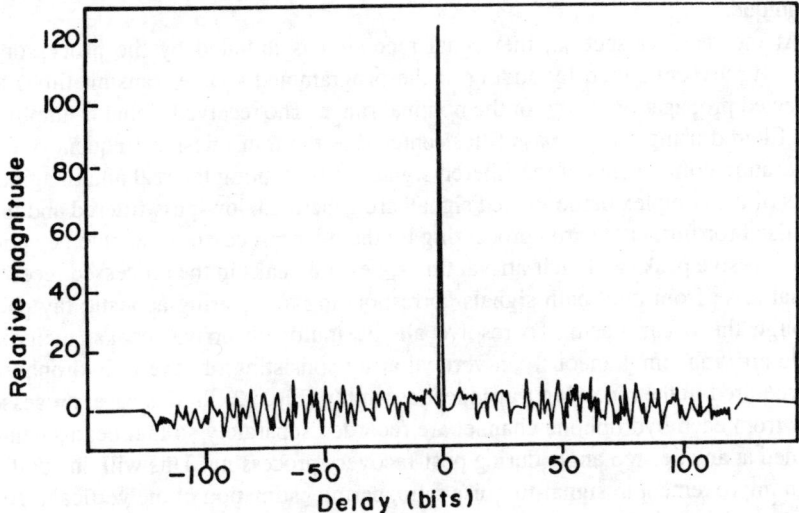

Fig. 11. Autocorrelation of pseudorandom n-bit shift register sequences used in reciprocal transmission measurements. The autocorrelation exhibits a triangular peak at zero lag, after Menemenlis et al, [25].

In the reciprocal acoustic transmission experiment conducted by Worcester et al, [20], the transmission signal code consisted of a carrier frequency of 400 Hz. The transmission length was 122.64 seconds consisting of 24 sequences of 5.11 seconds. The phase-coded 400-Hz digital signal is amplified usually by a constant power amplifier. After necessary filtering and impedance-matching, the signal is impressed upon the drivers of the acoustic transducers. The transducers are resonant tubes, driven by a pair of flat piezoelectric elements inserted at the mid-point of the tube, with an effective length of one-fourth the acoustic wave length.

In any typical ocean-basin experiment, the expected differential travel-time is only a few millisecond. For this reason it becomes necessary to have a clock with nano-second precision for the several month duration of the experiment, requiring an oscillator accurate to better than one part in 10^{10}. In the basin-wide experiment of Worcester et al, [20], a two-oscillator system was employed to achieve this precision at a reasonable level of power consumption. In this system, a low-power (10-mW) temperature-compensated crystal oscillator (TCXO) ran continuously to drive the clock. A high power (13-W) rubidium (Rb) atomic frequency standard, which returned to its previous frequency to within two parts in 10^{10} within 10 minutes after power is applied, was turned on at 6-h intervals. The frequency offset between the Rb oscillator, after permitting its warmup, and the TCXO was used in a feedback circuitry to readjust the TCXO frequency. This feedback technique increased the effective stability of the TCXO by approximately one order of magnitude. Any left-over frequency offset which still existed was measured over a 2-minute interval with a precision of one part in 10^{10} using a phase comparison technique.

At the receiver section, the signal reception is initiated by the processor at preset times, computed by adding to the programmed source transmit times the expected propagation delay for the nominal range. The received signal is amplified and filtered using a band-pass filter centered at the transmission frequency. Two quadrature components of the filtered signals, representing the real and imaginary parts of the complex demodulated signal, are generated, low-pass filtered and then digitised for further "sharp" processing by the microprocessor to enable detection of successive peaks and their arrival times. Several peaks in the processed received signal arise from multipath signals corresponding to differing acoustic ray paths through the ocean basin. To resolve all the multipath arrival peaks, including those arriving simultaneously, a vertical array consisting of several hydrophones are required in the receiving transducer assembly (Fig. 12). In this case, processed data from each hydrophone channel are recorded separately so that beams can be formed at any desired angle during post-recovery processing. This will, in addition to an improvement in signal-to-noise ratio, permit estimation of the vertical arrival angles of the several acoustic ray paths which impinged on the receiving hydrophone array, thus assisting multipath ray identification by separating simultaneous arrivals from different angles. In order to perform an inversion of travel-time data, each arrival must be associated with a particular ray path. Furthermore, ray identification is useful in performing inversions to convert travel-time differences to ocean current structure as well. In this sense, use of a vertical array of several hydrophones rather than a single hydrophone, assumes special significance. Usually, all signals arrive within ± 15° of the horizontal if mooring motion is negligible. Fortunately, mooring motion does not seriously affect the two-way travel times in reciprocal acoustic transmissions (velocity tomography), although mooring motion correction is most important in the one-way travel-times used in acoustic tomographic measurements (density tomography). This is because, in the case of reciprocal transmissions, the differential travel time is directly proportional to the ray-averaged current with respect to the mean motion of the transceivers [19]. One probable source of error in the inversion of differential travel times to obtain currents is the fact that current shear causes ray paths to differ with and against a current. Sound

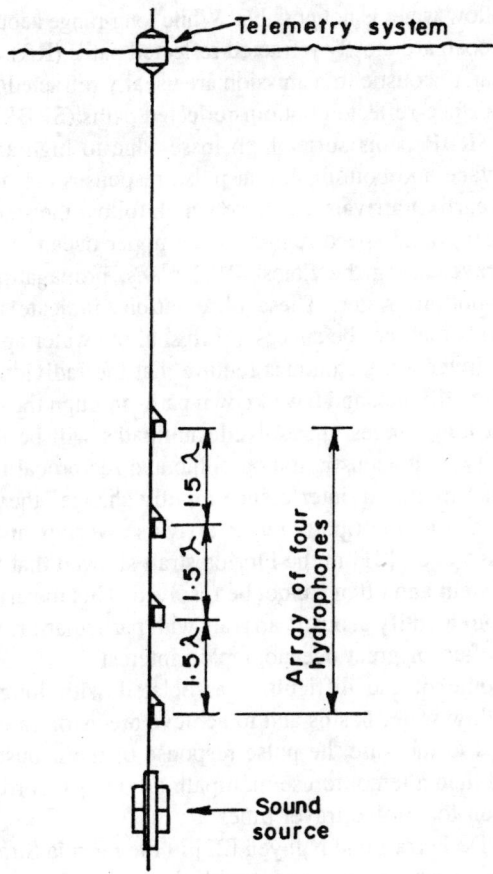

Fig. 12. Acoustic source and receiver-arrays of a typical transceiver system used in reciprocal measurements, after Worcester et al, [20].

pulses travelling in opposite directions, therefore, do not sample precisely the same part of the ocean [20]. Fortunately, by virtue of Fermat's principle, travel time is unchanged to first order in small perturbations in the ray paths. Effects associated with the non-reciprocity of ray paths are expected to be small if the sound-speed gradient and current gradient are comparable.

It has been shown that in the deep ocean the acoustic arrivals are stable and that the received acoustic pulses can be resolved and identified with ray paths. However, acoustic transmission in shallow water regions is quite different from that in deep oceans. While an axis of minimum sound speed exists in deep oceans, such an axis does not exist in shallow waters where sound speed typically decreases with depth so that ray paths are refracted downward, and propagation to moderate ranges (of the order of a few kilometres) necessarily involves bottom bounces [22]. Further, because of the loss caused by bottom interactions, long range acoustic transmission in shallow water is not possible. While long-range acoustic transmission paths in the deep ocean are mostly refracted/reflected paths (RR), the transmission paths in shallow water acoustic transmission are usually refracted/bottom-reflected paths (RBR) and surface-reflected bottom-reflected paths (SRBR). Those arrivals travelling via the SRBR paths suffer high losses due to high-angle interactions with the ocean surface and bottom so that pulse responses are dominated by the RBR arrivals. The earliest arrivals are those which follow the steepest RBR paths and reach the higher sound speed region of the upper ocean. The latest arrivals are those which travel along the flatest RBR paths, propagating in the slower sound speed near-bottom water. These observations indicate that deep ocean tomographic methods may not be successful in shallow water applications. Deep ocean tomography inversion techniques require that the individual eigen rays be separable in time, identifiable, and have known paths through the ocean. However, in shallow areas at long ranges, unresolved multipaths will be the rule [23].

A major difficulty with acoustic tomographic and reciprocal measurements in shallow water is that multipath interference rapidly changes the pulse shape and phase. Further, acoustic multipath pulse arrivals overlap and form groups. Measurements by Ko et al, [21] in the Florida strait showed that these groups are generally not consistent and often cannot be resolved. This means that, in shallow water, it is difficult to identify acoustic arrivals with particular ray paths. However, shallow areas are often of great oceanographic interest.

To surmount some of the difficulties associated with long-range acoustic propagation in shallow water basins and to achieve precision in measurement, an approach adopted is to measure the pulse response of the acoustic channel with high resolution and then attempt to use multipath groupings, correlation methods, or phase information to resolve travel times.

Experiments by De Ferrari and Nguyen [22] in the Florida Strait indicated that a convenient depth average is associated with those rays forming the late peak, and the arrival time of the late peak is a measurable and consistent feature of the data. The resulting depth-average is determined by the source and receiver depths. Thus, information about the depth-dependence of current is possible with a single parameter inversion by employing a vertical array of hydrophones. While the purpose of having such a hydrophone array in deep ocean tomographic measurements is to select the acoustic arrivals, the purpose of this array in shallow water acoustic measurements is to select average height of the ray, thus controlling the depth and extent of averaging the current. In shallow water acoustic pulse propagation, the signals arriving earlier than the late pulse generally are less stable and of lower amplitude. Even with long-time averaging, it is difficult to identify characteristic features which can be tracked and used for precise measurement of travel time. However, a consistent feature of the pulse response of a channel is a sharp peak (Fig. 13) associated with the late arriving RBR paths. The purpose of averaging several pulse responses is to smooth out interference effects. The stable and sharp cut off of the late peak is used to align the pulses prior to their averaging. The

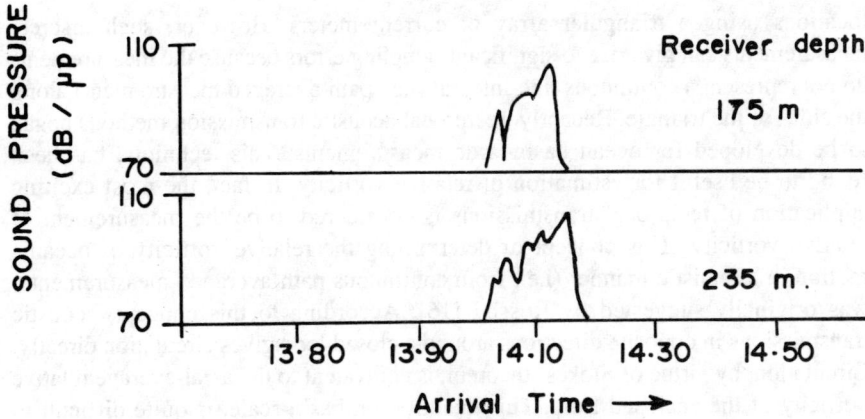

Fig. 13. Pulse responses depicting the characteristic of acoustic propagation in shallow water, wherein a sharp peak is associated with the late arriving RBR paths, after De Ferrari and Nguyen [22].

arrival time of the pulses is estimated using 'threshold cross time' method, where the arrival time of the lagging edge is estimated as the time when the edge crossed an intensity threshold. This threshold is usually an average of several thresholds which are 10 dB or more below the maximum intensity of the late peaks. The aligned pulse responses are averaged with a moving window over several records (sliding average method). Ko et al, [21] found that the pulse responses lined up very well after being aligned and averaged. Because the time corresponding to the maximum intensity of the peak is an imprecise estimate, the centroid estimation is usually used to estimate the travel time of the arrival peak. The time corresponding to the centroid of the late peak is computed using several points around the maximum value. The time variation of the late peak has been found to be a good indicator of the travel time variation for the latest, and thus the flattest, near-bottom RBR rays. Details of data analysis methods may be found in De Ferrari and Nguyen [22].

Vorticity: Circulation in an ocean basin is often associated with vorticity. Because solid earth is associated with spinning, a water mass which resides on it will suffer an additional vorticity associated with the spinning earth. Thus, the vorticity measured by an observer on earth is not absolute vorticity, but is 'relative vorticity'. Estimates of relative vorticity are necessary to determine the long-term balance of energy and momentum in an ocean basin. Muller et al, [23] pointed out the importance of the vorticity mode of motion that must co-exist with internal gravity waves at small scales. Because relative vorticity and horizontal divergence are difficult to measure using conventional instruments, the vortical mode of motion has traditionally been ignored. Muller et al, [24] made an attempt to measure potential vorticity at small scales in the ocean from measurements at three discrete locations using a triangular array of current meters. However, such discrete measurements can give rise to significant sampling errors because the measurements do not represent a continuous line-integral (i.e., path-averaged measurement) along the sides of the triangle. Recently, reciprocal acoustic transmission methods began to be developed for ocean basin-wide measurements. This technique has been found to be useful for estimation of relative vorticity. In fact, the most exciting application of reciprocal transmissions is considered to be the measurement of relative vorticity. This concept of determining the relative vorticity of oceanic motion in a realistic manner (i.e., from continuous path-averaged measurements) was originally suggested by Rossby [16]. According to this concept, acoustic transmissions in opposite directions around a closed loop gives circulation directly. Circulation, by virtue of Stokes' theorem, is equivalent to the areal-average relative vorticity of the enclosed fluid. This quantity, on basin-scale, is quite difficult to measure using traditional approach. In addition, its unprecedented sensitivity allows detection of phenomena not observable with traditional sensors.

Estimation of relative vorticity of fluid motion in an ocean basin requires measurement of path-averaged fluid velocity around a closed loop (Fig. 14). This, in reciprocal acoustic transmission method, is equivalent to the measurement of travel time differences of acoustic propagation along this closed loop. Travel time difference Δt_{ij} along an acoustic ray path, of range l_{ij}, joining points P_i and P_j is given by:

$$\Delta t_{ij} = (2v_{ij} * l_{ij})/(c_{ij})^2 \qquad (4)$$

where c_{ij} and v_{ij} are the path-averaged sound speed and fluid velocity respectively along the ray path joining P_i with P_j. Thus,

$$v_{ij} = [\Delta t_{ij} * (c_{ij})^2]/2l_{ij} \qquad (5)$$

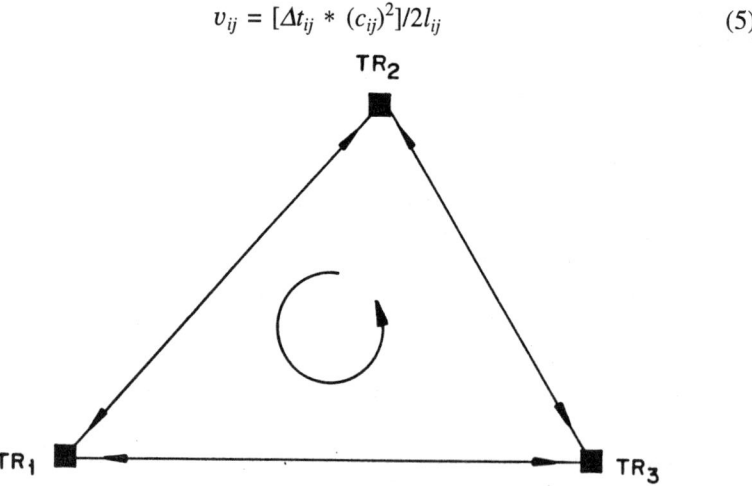

Fig. 14. Arrangement of acoustic transceivers TR_1, TR_2, and TR_3 (minimum configuration) for measurement of vorticity.

The line-integral of the fluid motion between points P_i and P_j can, under the assumptions made above, be written as:

$$\int_{P_i}^{P_j} v \cdot ds = v_{ij} * l_{ij} = [(c_{ij})^2 * \Delta t_{ij}]/2 \tag{6}$$

where v is the velocity vector and s is a unit vector along the acoustic propagation path. Estimation of vorticity, based on the method proposed by Rossby [16] requires estimation of the line-integral of fluid motion around a closed loop. Because a minimum of 3 lines are required to form a closed loop, a minimum configuration of instrumentation required for estimation of relative vorticity is a set of acoustic transceivers placed on the vertices of a triangle. Thus, the line-integral of fluid motion around a triangle $P_1P_2P_3$ becomes:

$$\oint v \cdot ds = (v_{12} * l_{12}) + (v_{23} * l_{23}) + v_{31} * l_{31})$$

$$= \{[(c_{12})^2 * \Delta t_{12}] + [(c_{23})^2 * \Delta t_{23}] + [(c_{31})^2 * \Delta t_{31}]\}/2 \tag{7}$$

According to Stokes' theorem, the circulation around a closed loop is equivalent to the surface integral of the normal component of vorticity of motion within the loop. If the triangle lies on a horizontal plane, circulation is the product of the average vertical component of relative vorticity z and the area of the triangle A. Thus,

$\oint v \cdot ds = z * A$, so that $z = (\oint v \cdot ds)/A$. By substitution,

$$z = \{[(c_{12})^2 * \Delta t_{12}] + [(c_{23})^2 * \Delta t_{23}] + [(c_{31})^2 * \Delta t_{31}]\}/(2A) \tag{8}$$

Assuming that c_{12}, c_{23} and c_{31} are nearly the same and equal to c, we get:

$$z = (c^2/2A) * (\Delta t_{12} + \Delta t_{23} + \Delta t_{31}) \tag{9}$$

This method has been used by Menemenlis and Farmer [25] to measure path-averaged horizontal current and relative vorticity of water mass in the sub-ice boundary layer of the eastern Arctic.

(v) Scintillation analysis

Acoustic scintillation technique of flow measurement is similar in principle to the well-known tracer technique where a dye, a chemical, or a radio-active isotope is injected into the flowing fluid, and its transportation over a known distance is timed to estimate the flow velocity, using the relation that velocity equals the distance travelled in unit time. The essential difference between the tracer method and the acoustic scintillation method is that while a foreign substance is injected into the flow field in the tracer method, the 'tracer' is normally provided in the latter method by some acoustically detectable random fluctuations already existing naturally in the flow field. The random fluctuations can take many forms such as velocity turbulence, presence of a second phase, temperature fluctuations etc. Internal waves, layering, and turbulence create inhomogeneities of sound speed in the ocean, and induce fluctuations in ocean acoustic transmission. Through both refractive and diffractive effects, sound waves from a point source are perturbed from the simple spherical or plane wave geometry into more complicated phase fronts [26]. A wave propagating through a random medium, such as a turbulent flow field, undergoes distortions in its amplitude and phase. Initially, only the phase of the wave is distorted as it propagates through the turbulent medium. These phase distortions or wrinkles in the wave front redirect the wave's energy, eventually producing both amplitude and phase distortions and finally intensity fluctuations (i.e., scintillation) due to interference across some distant receiving plane [28]. This results in a complicated diffraction pattern, similar in nature to the stellar scintillations (twinkling of stars) as observed near the focal plane of an astronomical telescope. In analogy, similar pattern at the receiver of an acoustic transducer is known as acoustic scintillation. The pattern of irregular intensity (scintillation pattern) in a receiving plane, perpendicular to the acoustic propagation axis, evolves with time. The evolution of the scintillation pattern occurs because of advection and/or decay of the density fluctuations or eddies that produce the wave perturbation. If the transit time of the scintillation pattern across the detectors is short compared to the eddy life times, then it is possible to derive quite accurate estimates of the intervening transverse flow from a statistical analysis of the scintillation pattern.

Scintillation techniques have long been employed to elucidate the fine scale structure and motion of the turbulent media. The acoustic scintillation analysis technique is, in fact, a spin-off from the applied optics wherin the properties of the intervening medium such as ionospheric, solar, and atmospheric winds were determined from an analysis of turbulence-induced stellar scintillations. Because of the unique spatial distribution of medium irregularities for each of these applications, (i.e., concentrated at one position along the path or uniformly distributed), different techniques have been devised to estimate flow information. However, in acoustic scintillation analysis only three methods have so far been applied to estimate oceanic flows.

The acoustic scintillation technique has so far been used for measurement of water flows in straits and channels only. The minimum configuration (Fig. 15) of a flow measuring device which employs this technique consists of a single acoustic transmitter and two acoustic receivers located in such a way that the acoustic transmission path is transverse to the flow direction. The separation L between the transmitter and the receiver ranges from a few metres to a few km. The two acoustic receivers are placed at a known distance apart in the flow field so that the receiving planes of these two hydrophones is normal to the acoustic path. The scintillation pattern at the receiver plane drifts with the flow. Upon reception, the acoustic signal from one receiver is compared with that from the other receiver

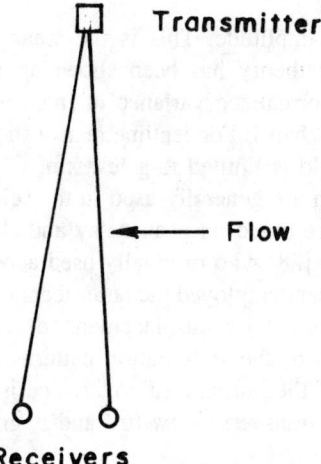

Fig. 15. Sketch of minimum configuration of an acoustic scintillation device.

to find the time lag T at which the scintillation patterns at the two receivers have maximum similarity or a best fit in waveform (i.e., correlation). From the known receiver separation and the estimated time lag, the flow velocity is computed.

Flow measurement using scintillation analysis technique is based on Taylor's frozen flow hypothesis, according to which the fluctuations in the refractive index at any point result solely from the advection of frozen refractive index field past that point by a mean flow. This hypothesis is reasonably well satisfied by atmospheric and oceanic turbulence over short distances, but is violated by internal waves. This is the reason why the method has not been used for measurement of flows in the open ocean where internal waves may be present. However, in straits and tidal channels, the flow is mainly induced by tide, and acoustic scintillation arises only from turbulence. In these environments Taylor's hypothesis is well satisfied and the acoustic scintillation technique can be applied for measurement of flows. As the refractive index irregularities are advected across the acoustic propagation path, the structure in the scintillation pattern drifts across the receiving plane with a speed proportional to the flow speed if weak scattering restriction is satisfied. The value of the proportionality constant depends on the type of incident wave. For example, the pattern drift-velocity V_p for an incident spherical wave is given by $2V_t$, where V_t is the transverse flow velocity. The factor of 2 arises because of the diverging nature of the spherical wave field and the sensitivity of the measurement to refractive eddies at mid-path [27]. For an incident plane wave, the pattern drift velocity V_p is equal to V_t.

It has been noted that acoustic scintillation analysis technique is applicable for flow measurement if weak scattering restriction is satisfied. Over short ranges, characterized well by a single transmission path from a source to any receiver, the wave front may be subjected to effects which are weak enough that the variation of signal amplitude in a plane transverse to the transmission path is a small fraction of the average amplitude. This is the weak scattering or unsaturated regime. Weak scattering theory has been shown to accurately describe wave transmission when the normalized variance of intensity (some times called the scintillation index) is less than 1. For centimetre acoustic wave length, the acoustic path length in a flow field is limited to a few km.

Two techniques which are generally used in the scintillation analysis method of flow measurements are based on peak delay and slope of covariance at zero lag. Clifford and Farmer [28] who originally used acoustic scintillation analysis for ocean flow measurement employed the latter technique. In this method a time-lagged covariance function (C) at displacement (d) is constructed to study the spatio-temporal variation of the scintillation patterns at the two receivers. The slope at zero time-lag of the normalized covariance is proportional to the path-weighted average of the transverse flow (u), and is given by [28]:

$$m_N = (1/L) \int_0^L W_{(y)} \, U_{(y)} \, dy \qquad (10)$$

where y is the constant plane, the refractive index fluctuations along which have given rise to the scintillation drift, L is the acoustic path length, and $W_{(y)}$ is a weighting function. In general, U can be expressed as $U = R * m_N$ where R is a calibration factor in metres. R takes different values for different turbulence spectral power laws and spacing $B = (d/R_f)$ where R_f is the Fresnel radius given by the square-root of the product of the acoustic wave length with the path-length. Values of the calibration factor for different values of B and spectral power laws are given in [28].

In the first experiment [28] conducted to test the validity of the scintillation technique for ocean flow measurement the acoustic transducers were mounted on two masts which were rigidly mounted on the leading edge of a research barge which could be towed at various speed steps (Fig. 16). The transducers were maintained at a depth of 2.1 m below the water surface. The transmitter-receiver separation was 12.4 m. The separation of the two receiving hydrophones, oriented along a plane perpendicular to the acoustic transmission path, was set at 15.7 cm which was approximately one-half of a Fresnel radius for the given experimental setup. The experiment was conducted in Saanich Inlet in 1982. Both the projector (P) and the receiving hydrophones (R) consisted of single hexagonal elements of 6 cm diameter. The projector was driven at approximately 214 KHz with 5 cm pulses once every 100 ms. For a transmitter-receiver separation of 12.4 m the receiving hydrophones detected only the direct pulse, without contamination by multipath propagation. The signals received at the hydrophones were amplified, detected with an rms-to-dc converter, filtered, digitized and recorded.

When the flow was zero, the signals received at both the hydrophones displayed a low-frequency fluctuation associated with slowly moving and changing patterns in the refractive index along the acoustic path. With increase in flow the signal frequency greatly increased as anticipated on theoretical grounds.

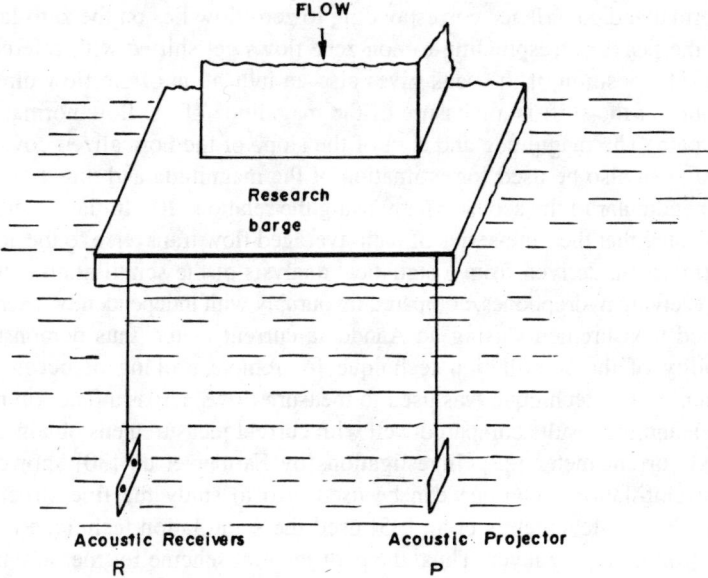

Fig. 16. A sketch of the initial experimental arrangement to test the practicability of acoustic scintillation technique of flow measurement.

Data analysis begins with breaking the edited time-series of data received at the two hydrophones into 1-minute segments, followed by log-normalization and high-pass filtering. For each segment, the cross-covariance is calculated together with the first central difference estimate of the cross-covariance slope at zero lag. Three examples of cross-covariance functions, corresponding to zero flow, and flows in two opposite directions, are shown in Fig. 17. It can be seen that the peak

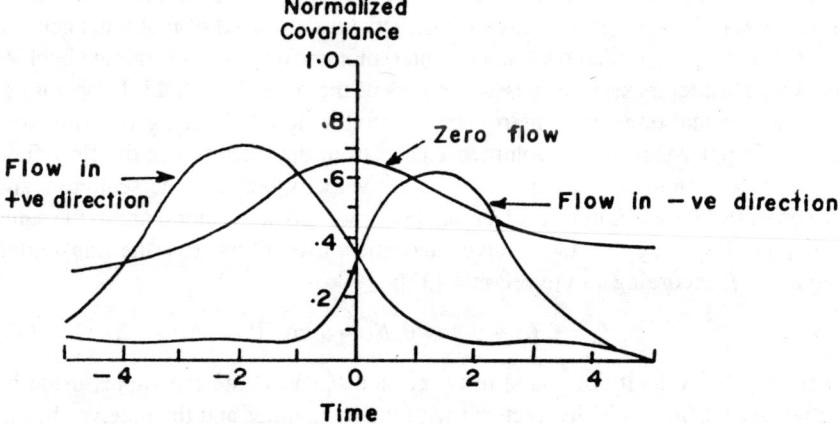

Fig. 17. Cross covariance as a function of lag for still water, and flows in opposite directions, after Clifford et al, [27].

of the normalized covariance corresponding to zero flow lies on the zero-lag axis whereas the peaks corresponding to non-zero flows get shifted with reference to this axis. The position of the peak gives also an indication of the flow direction. The amount of the shift is a measure of the magnitude of the flow normal to the acoustic path. The magnitude and sign of the slope of the normalized covariance at zero-lag can also be used for estimation of the magnitude and direction of the flow perpendicular to the acoustic path, using the relation (10). Initial experiments have indicated that the time-series of path-averaged flow transverse to the acoustic propagation path, derived from a statistical analysis of the scintillation pattern at the two receiving hydrophones, compared favourably with independently determined flow speed measurements using an Aanderaa current meter, thus demonstrating the viability of the scintillation technique for remote probing of ocean flows. Subsequently, this technique was used to measure flows across a 0.66 km path in a channel, and the results compared well with current measuremens obtained from a moored current meter [29]. Investigations by Farmer et al, [30] showed that acoustic scintillation technique can be used also to study the fine structure of turbulent flows. Menemenlis et al, [25] used the scintillation technique to study the ice-water boundary layer. Thus, the path-integral scheme for measurement of turbulence and mixing appears to prove a useful adjunct to the conventional flow measurement techniques. Recent investigations have shown that, by spatial filtering of acoustic scintillations from an array of hydrophones, it is possible to obtain horizontal profiles of fine-scale variability and transverse current along short (upto approximately 2 km) acoustic propagation path [32, 33].

Eulerian Method

Doppler is the only acoustic method available till date for remote measurement of flows from a fixed location in the flow field. This method of current measurement relies on the mechanism of acoustic volume scattering from a cloud of scatterers in the flow field. The term 'scatterer' encompasses any kind of inhomogeneity in water including suspended particulate matter, biological organisms, minute bubbles etc. These scatterers serve as passive tracers of the mean flow field. In operation, an acoustic sonar transmits a narrow beam into the flow field in a given direction, and the Doppler shift of the volumetric echo from the scatterers in the flow field is used to determine the relative radial flow velocity between the scatterers and the sonar. If the transmitter, receiver and the flow velocity vector v are in the same plane, the frequency f_r of the received acoustic wave differs from the transmitted frequency f_t according to the relation [33]:

$$f_r = f_t (c + v \cos \theta_t)/(c - v \cos \theta_r) \qquad (11)$$

where c is the velocity of sound in water, and θ_t and θ_r are the angles made by the horizontal flow velocity vector v with the transmitter and the receiver beams respectively. For a given geometry, the Doppler shift is +ve or −ve depending on the direction of the flow. For ocean flows, the received frequency differs only slightly from the transmitted frequency.

An important factor that influences the development of a Doppler sonar is the transmission frequency f_t. For a given flow velocity, higher transmission frequencies are desired to produce larger Doppler shifts. The transmission frequency also influences the transducer design, its weight, and size. However, the acoustic attenuation increases as the transmission frequency increases, and the choice of frequency will be, therefore, a compromise.

The backscattered signal, normally of microvolts in amplitude, is composed of the random sum of the individual scattering amplitudes, each having a Doppler frequency shift associated with the radial velocity of the scatterer. Since the micro-inhomogeneities that cause the scattering are of random sizes, the received signal is subjected to amplitude modulation. Phase incoherences also arise as a result of different ranges of the individual scatterers, as well as their random motion due to turbulence within the defined scattering volume. The signal is also corrupted by an additive white Gaussian noise. Thus the received signal is a highly distorted version of the transmitted signal wherein its deterministic properties are no longer valid, rather the signal is statistical in nature with Gaussian distributions. For this reason, a realistic approach for estimation of Doppler shift information is some form of statistical methods. Most of the Doppler shift information is contained in the 1st and 2nd moments of the Doppler spectrum. One method of obtaining these moments is to Fourier transform the received echoes. These can also be obtained from estimates which do not involve the Fourier transform operation.

Instrumentation: In the initial designs of acoustic Doppler current measuring devices a bistatic configuration (i.e., the transmitter and the receiver located at different points) has been employed. These transducers were inclined to each other so that the volume of inter-section of the transmitted and the received beams defined the scattering volume. Depending on the spatial separation and the angle between the transducers, the Doppler-shifted echoes were received from a reverberation volume located at a known distance from the transceivers [34]. An advantage of the acoustic Doppler measurement technique, over conventional current measuring devices, was that the geometry could be set to receive the backscattered signals from a region sufficiently away from the body of the meter so that the velocity of flow could be non-intrusively detected from the undisturbed flow field. The basic accuracy of measurements is determined by the stability of transmission frequency, precision of mechanical alignment of the transducers, and a knowledge of the velocity of sound in water. The frequency used for transmission was in the range of 1 to 10 MHz.

In situations where a vane oriented the instrument into the direction of the flow (Fig. 18), only a single transceiver set has been used [35]. In order to avoid a vane, either two sets of mutually orthogonal transceiver systems [36] (Fig. 19) or a single transmitter and three receivers pointing to a single location sufficiently above the instrument [37] have been used (Fig. 20).

For the geometries mentioned above, continuous transmission is possible because

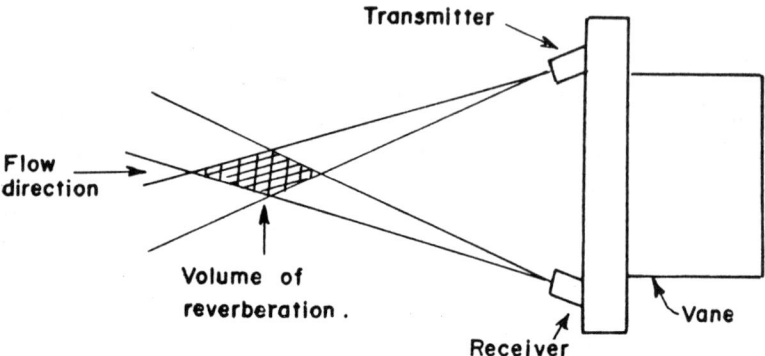

Fig. 18. Acoustic current meter geometry with a bistatic configuration.

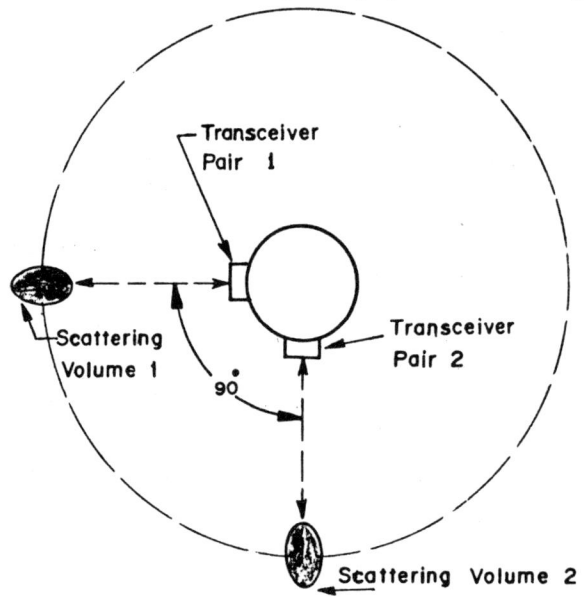

Fig. 19. Plan view of an acoustic Doppler current meter with two sets of orthogonal transceiver systems in a plane normal to the axis of the meter. Each transceiver pair consists of a transmitter and a receiver arranged as in Fig. 18.

the transmitter and the receiver are separate. In principle, precision of the velocity estimates is higher for continuous transmission than pulse transmission, but the range resolution is very poor [38]. For a Eulerian type current meter, poor range resolution is not a constraint because measurements are made from a given fixed range. A disadvantage with the continuous transmission is electrical and acoustical cross-talk between the transmitter and the receiver. One solution to this problem is the use of parametric transducers which have suppressed side-lobes and possess high directivity from a relatively small aperture [39]. In earlier designs Doppler

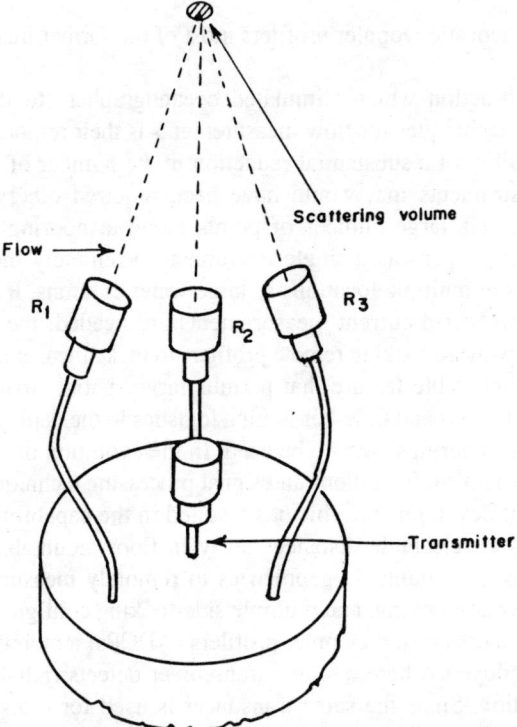

Fig. 20. Sketch of an acoustic Doppler current meter with an upward-looking transmitter at the centre and three receivers R_1, R_2 and R_3 pointing to a scattering volume far above the body of the meter.

shift of the received signals was estimated using conventional frequency measurement methods such as detection of zero-crossings of received pulse streams, or using frequency-locked-loop method. Although Doppler method permits flow measurements in a non-intrusive manner, for a long time Doppler current meters suffered from severe signal drop-out problems, arising from the inadequacy of Doppler shift estimation techniques.

Profiling Method

In the past, observations of depth-dependence of horizontal flows were made by the use of moored current meter chains. Maintaining such moorings at large depths in the ocean is difficult and rather expensive. Ideally, one might want a single package which can scan a large water column to sense the vertical distribution of horizontal motions with high vertical resolution. This requirement resulted in the development of acoustic techniques. Acoustic profilers of initial designs consisted of acoustically tracked freely sinking or rising expendable pingers. Horizontal velocity profiles were estimated from the trajectory described by the pinger and using an appropriate hydrodynamical model for the motion of the pinger. With the emergence of acoustic Doppler profilers most of the former methods have become obsolete.

A major attraction which stimulated oceanographers to the use of acoustic pulse-Doppler techniques for flow measurements is their remote sensing capability and the possibility of a substantial reduction in the number of 'single-point' flow measuring instruments that would have been required otherwise for long-term flow profiling at a large number of points along a mooring line. The acoustic Doppler technique permits a single instrument to remotely measure undisturbed flow velocities at multiple locations in large water columns. Further, in situations where only short-term current measurements are needed, the capability of hull-mountable, downward looking remote profilers to make measurements from moving vessels is a remarkable feature that permits large spatial coverage, and virtually eliminates the tedious and time-consuming logistics in the deployment and retrieval, if conventional moorings were to be used. In the evolution of this innovative new technology, system configurations and signal processing techniques have undergone many stages of development. This has resulted in the capability for flow velocity measurements at selectable distances away in floor-mountable, ship-mountable, and wall- or pillar-mountable geometries to remotely measure flows in upward looking, downward looking, and multiple side-looking configurations respectively.

In the design of acoustic Doppler profilers (ADCPs) a monostatic configuration is usually employed, where a single transceiver detects radial flow velocities in a given direction. Since the same transducer is used for transmission as well as reception, both θ_r and θ_t in equation (11) are equal (say θ). In this case the Doppler shift f_D can be reduced to a form:

$$f_D = (2f_t\, v\, \cos\theta)/c \qquad (12)$$

where $f_D = f_r - f_t$

While considering the radial flow velocity component v, the value of θ in the above equation reduces to zero so that:

$$v = (cf_D)/(2f_t) \qquad (13)$$

In commercially available acoustic Doppler profilers, three or more transceivers, inclined to each other, are often employed so that the acoustic beams are divergent from each other. The mean horizontal flow vector at a given water layer, away from the sensor, is estimated from measurements of these radial flow components in this layer (Fig. 21). The mean horizontal flow in any given layer is computed by averaging the flow 'samples' from these, often widely separated, insonified volumes along the acoustic beams.

Since radial flow components are to be measured from various layers in the water column the transceiver system is to be operated in a multiplexed mode, i.e., configuring it for transmission at one time followed by reconfiguring it as a receiver before the backscattered signals reach the transducer. In operation each transceiver emits a stream of narrow-band acoustic pulses, insonifying the scatterers

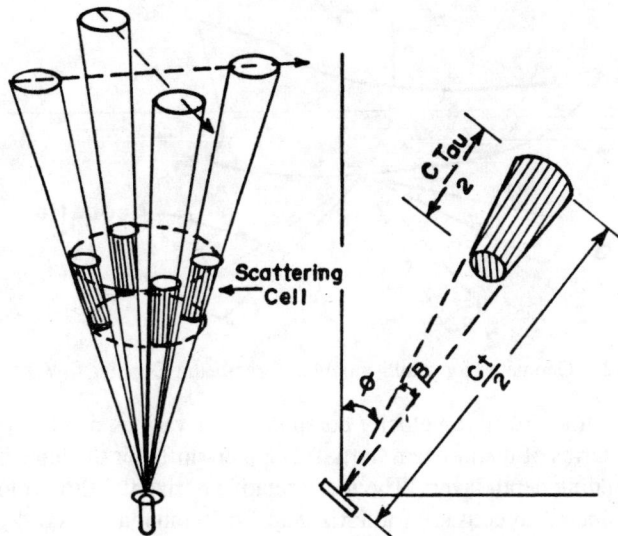

Fig. 21. (a). 4-beam configuration of an acoustic flow profiler.
(b). Expanded view of a scattering cell.

along its path, and receive the Doppler-shifted backscattered return echoes as these pulses propagate through the water column. The duration of the pulse equals the gating time T_{au}. The Doppler shift, as a function of time after transmision, yields the relative radial flow velocity as a function of the slant range in sequential layers at each of the 'range sliced' insonified volumes of the overlaying water column. Each meaurement is an averaged value over a distance into the medium along the beam, expressed in terms of the two-way travel time. If a pulse of duration T_{au} seconds, representing a time-burst, is transmitted the observation (integration) time must be equal to this duration. By adjusting the transmitted pulse length, the portion of the water column insonified at any given instant can be controlled and the desired range resolution can be achieved. In the case of stationary transducers, the Doppler-shift directly yields the absolute water velocity profiles. If the transducers are non-stationary (for example, mounted on a buoy or at the bottom of a ship), the absolute water velocity profile may be obtained from a knowledge of the motion of the transducers.

Profiler geometries: The ADCPs are of various geometries. These typically include three-, four-, and five-beam systems. The three-beam configuration is generally used only with hull-mounted systems. In a typical three-beam system, the beams are spaced 120° apart in the azimuth, and each beam points down at an angle of $\phi°$ from the vertical. One beam (say beam 1) is generally oriented forward along the ship's longitudinal axis and the other two beams (say beams 2 and 3) lie along the port and starboard quarters (Fig. 22). Based on this geometry the along-track

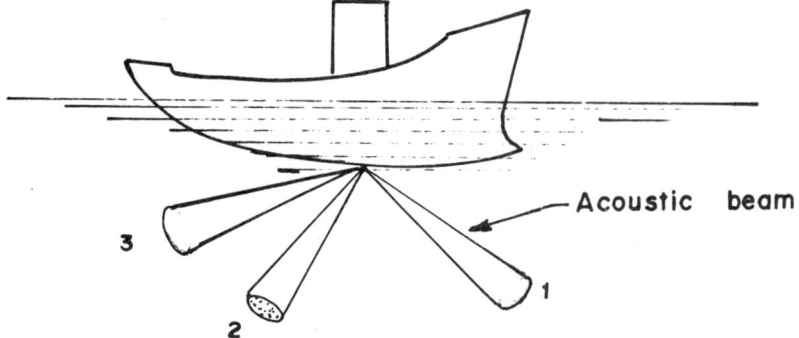

Fig. 22. Geometry for a hull-mounted three-beam Doppler flow profiler.

and the cross-track relative velocity components at various depth layers may be estimated in terms of the in-beam (radial) Doppler-shifts for the three beams from the corresponding depth layers. The mean relative horizontal flow velocity vector may be considered to consist of a horizontal flow component (say V_y) along the longitudinal axis of the ship (i.e., along its direction of motion) and an orthogonal horizontal flow component, say V_x. These components are given by the relation [40]:

$$V_y = (c/6W \sin \phi)(W_{D(2)} + W_{D(3)} - 2W_{D(1)}) \tag{14}$$

$$V_x = (c/2\sqrt{3}\,W \sin \phi)(W_{D(3)} - W_{D(2)}) \tag{15}$$

where c is the velocity of sound in water, W is the transmission frequency in angular notation, ϕ is the angle (in degrees) made by each beam from the vertical, and $W_{D(1)}$, $W_{D(2)}$, and $W_{D(3)}$ are the Doppler shifts (in angular notation) measured by the beams 1, 2, and 3 respectively. Since only three components of velocity are present, i.e., two horizontal and one vertical component, only three beams are sufficient in principle to fully describe the flow field. However, if the number of beams are more than three, the performance of the profiler correspondingly increases. For example, in a three-beam system mentioned above, the along-track horizontal flow component V_y is coupled to the Doppler shifts from all the three beams (see equation 14). Further, since there is only one beam in a given plane, the three-beam system is sensitive to pitch and roll of the vessel.

To circumvent some of the limitations inherent with a three-beam Doppler profiler system, a four-beam system has been introduced. This system employs two 'Janus configuration' consisting of a total of four beams oriented 90° apart in the azimuth, each beam making an angle of ϕ degrees to the axis of symmetry of the transducer system (Fig. 23). When mounted beneath a vessel, the system is generally operated with one pair of mutually diverging beams (say beams 1 and 2) oriented along the longitudinal axis of the ship, and other pair of beams (say beams 3 and 4) oriented along an orthogonal plane so that the profiler resolves the relative velocity field at any given depth layer into along-track and cross-track

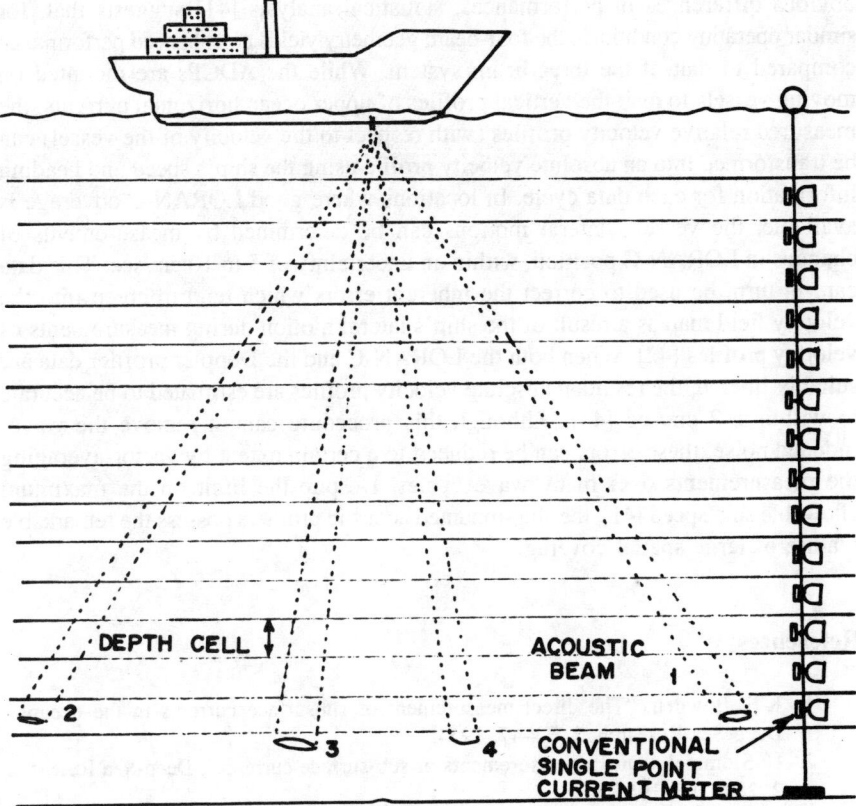

Fig. 23. Geometry for a hull-mounted four-beam Doppler flow profiler.

components. A view of the number of Eulerian current meters which can be replaced by a single profiler is demonstrated in Fig. 23. If $W_{D(1)}$, $W_{D(2)}$, $W_{D(3)}$, and $W_{D(4)}$ are the Doppler shifts measured along the beams 1, 2, 3 and 4 respectively, the relative horizontal flow component V_y, along the longitudinal axis of the ship is given by the relation [40]:

$$V_y = (c/4W \sin \phi)(W_{D(1)} - W_{D(2)}) \qquad (16)$$

and the relative horizontal flow component V_x perpendicular to the longitudinal axis of the ship is given by:

$$V_x = (c/4W \sin \phi)(W_{D(3)} - W_{D(4)}) \qquad (17)$$

The sign of W_D depends on the direction of flow relative to the axis of the transducer system. In this four-beam geometry, each horizontal flow component is estimated as a mean of the Doppler shifts from the two symmetrically oriented beams in each of the Janus configurations in the two mutually orthogonal vertical planes. It is thus evident that, in the four-beam geometry, the Janus configuration tends to minimise the sensitivity to pitch and roll of the vessel. In addition to these obvious differences in performances, statistical analysis [41] suggests that, for similar operating conditions, the four-beam geometry yields an enhanced performance compared to that of the three-beam system. While the ADCPs are mounted on moving vessels to map the vertical profiles of upper ocean horizontal currents, the measured relative velocity profiles (with respect to the velocity of the vessel) can be transformed into an absolute velocity profile using the ship's speed and heading information for each data cycle. In locations where good LORAN-C coverage is available, the vessel's lateral motions can be determined by measurements of changes in LORAN-C position, within an uncertainty of 5 to 10 cm/sec. This data can, in turn, be used to correct the inherent errors which tend to creep into the velocity field map as a result of the ship's lateral motion during measurements of velocity profiles [42]. When both the LORAN-C and the Doppler profiler data are suitably filtered, the resultant absolute velocity profiles are estimated to be accurate to within ± 2 cm/sec [43]. Although this procedure cannot remove the wave-induced noise, these errors can be reduced to a certain extent by vector-averaging the measurements over many wave cycles. Despite the limit on the maximum allowable ship speed [41], the ship-mounted acoustic profilers possess the remarkable feature of large spatial coverage.

References

1. K.F. Bowden, "The direct measurement of subsurface currents in the oceans", Deep-Sea Research, **2**, 33–47, 1954.
2. H. Stommel, "Direct measurements of sub-surface currents", Deep-Sea Research, **2**, 284–285, 1955.
3. J.C. Swallow, "A neutral-buoyancy float for measuring deep current", Deep-Sea Research, **3**, 74–81, 1955.
4. H. Weidemann, "Oceanic currents and wave recording—instrumentation", The Encyclopedia of Oceanography, **1**, 597–599, 1966.
5. W. Sturges, "Measurements of currents", Mc Graw-Hill Encyclopedia of Ocean and Atmospheric Sciences, 329–330, 1980.
6. G.L. Pickard and W.J. Emery, "Instruments and methods", Descriptive Physical Oceanography—An Introduction, Pergamon Press, 1964.
7. T. Rossby and D. Webb, "Observing abyssal motions by tracking Swallow floats in the SOFAR channel", Deep-Sea Research, **17**, 359–365, 1970.
8. T. Rossby, A.D. Voorhis and D. Webb, "A Quasi-Lagrangian study of mid-ocean variability using long range SOFAR floats", Journal of Marine Research, **33**, 355–382, 1975.
9. D.C. Webb, "SOFAR floats for polymode", MTS-IEEE Oceans'77, 44B (1)–44B (5), 1977.
10. W. Munk and C. Wunsch, "Ocean acoustic tomography: a scheme for large scale monitoring", Deep-Sea Research, **26A**, 123–161, 1979.
11. S. Prasannakumar, G.S. Navelkar, T.V. Ramana Murthy and C.S. Murthy, "Acoustical characteristics and simulated tomographic inversion of a cold-core eddy in the Bay of Bengal", Acustica-Acta Acustica, **83**, 1–8, 1977.

12. D. Behringer, T. Birdsall, M. Brown, B. Cornuelle, R. Heinmiller, R. Knox, K. Metzger, W. Munk, J. Spiesberger, R. Spindel, D. Webb, P. Worcester and C. Wunsch, "A demonstration of ocean acoustic tomography", Nature, **299**, 121–125, 1982.
13. S. Prasannakumar, T.V. Ramanna Murthy, Y.K. Somayajulu, P.V. Chodankar and C.S. Murthy, "Reference sound speed profile and related ray acoustics of Bay of Bengal for tomographic studies", Acustica, **80**, 127–137, 1994.
14. R.C. Spindel, R.P. Porter and D.C. Webb, "A mobile coherent low-frequency acoustic range", IEEE Journal of Oceanic Engineering, **OE-2**, 331–337, 1977.
15. L.A. Stallworth, "A new method for measuring ocean and tidal currents", IEEE Oceans '73, 55–58, 1973.
16. T. Rossby, "An oceanic vorticity meter", Journal of Marine Research, **33**, 213–222, 1975.
17. J.C. Kaimal, "Sonic anemometers", Air-Sea Interaction: Instruments and methods, Plenum Press, 81–96, 1980.
18. T. Gytre, "The use of a high sensitivity ultrasonic current meter in an oceanographic data acquisition system". The Radio and Electronic Engineer, **46**, 617–623, 1976.
19. P.F. Worcester, "Reciprocal acoustic transmission in a mid-ocean environment", Journal of Acoustic Society of America, **62**, 895–905, 1977.
20. P.F. Worcester, R.C. Spindel and B.M. Howe, "Reciprocal acoustic transmissions: Instrumentation for mesoscale monitoring of ocean currents", IEEE Journal of Oceanic Engineering, **OE-10**, 123–137, 1985.
21. D.S. Ko, H.A. De Ferrari and P. Malanotte-Rizzoli, "Acoustic tomography in the Florida Strait: Temperature, current, and vorticity measurements", Journal of Geophysical Research, **94**, 6197–6211, 1989.
22. H.A. De Ferrari and H.B. Nguyen, "Acoustic reciprocal transmission experiments, Florida straits", Journal of Acoustic Society of America, **79**, 299–315, 1986.
23. P. Muller, G. Holloway, F. Henyey and N. Pomphrey, "Non-linear interactions among internal gravity waves", Review of Geophysics, **24**, 493–536, 1986.
24. P. Muller, L. Ren-Chich and R. Williams, "Estimates of potential vorticity at small scales in the ocean", Journal of Physical Oceanography, **18**, 401–416, 1988.
25. D. Menemenlis and D.M. Farmer, "Acoustic measurement of current and vorticity beneath ice", Journal of Atmospheric and Oceanic Technology, **9**, 827–849, 1992.
26. T.F. Duda and S.M. Flatte, "Modelling meter-scale acoustic intensity fluctuations from oceanic fine structure and microstructure", Journal of Geophysical Research, **93**, 5130–5142, 1988.
27. S.F. Clifford, D.M. Farmer, R.J. Lataitis and G.B. Crawford, "Ocean remote sensing by acoustic scintillation techniques", Acoustic Remote Sensing, (Ed.) S.P. Singal, Tata McGraw-Hill Publishing Company Limited, 189–197, 1990.
28. S.F. Clifford and D.M. Farmer, "Ocean flow measurements using acoustic scintillation", Journal of Acoustic Society of America, **74**, 1826–1832, 1983.
29. D.M. Farmer and S.F. Clifford, "Space-time acoustic scintillation analysis: A new technique for probing ocean flows", IEEE Journal of Oceanic Engineering, **OE-11**, 42–50, 1986.
30. D.M. Farmer, S.F. Clifford and J.A. Verrall, "Scintillation structure of a turbulent tidal flow", Journal of Geophysical Research, **92**, 5369–5382, 1987.
31. G.B. Crawford, R.J. Lataitis and S.F. Clifford, "Remote sensing of ocean flows by spatial filtering of acoustic scintillations: Theory", Journal of Acoustic Society of America, **88**, 442–454, 1990.

32. D.M. Farmer and G.B. Crawford, "Remote sensing of ocean flows by spatial filtering of acoustic scintillations: Observations", Journal of Acoustic Society of America, **90**, 1582–1591, 1991.
33. C.S. Clay and H. Medwin, "Doppler effects for moving objects, sea surface and ships", Acoustical Oceanography: Principles and applications, Wiley-Interscience, 334–338, 1977.
34. F.F. Koczy, M. Kronengold and J.M. Loewenstein, "A Doppler current meter", Marine Sciences Instrumentation, **2**, 127–134, 1962.
35. M. Kronengold and W. Vlasak, "A Doppler current meter", Marine Sciences Instrumentation, **3**, 237–250, 1965.
36. E.J. Katz and W.E. Witzell (Jr), "A depth controlled tow system for hydrographic and current measurements with applications", Deep-Sea Research, **26A**, 579–596, 1979.
37. W.J. Wiseman (Jr), R.M. Crossby and D.W. Pritchard, "A three-dimensional current meter for estuarine applications", Journal of Marine Research, **30**, 153–158, 1972.
38. Pinkel, "Acoustic Doppler techniques", Air-sea interaction-Instruments and methods, (Eds) F. Dobson, L. Hasse and R. Davis, 171–199, 1980.
39. W.L. Konrad and M.B. Moffett, "Specialised drive waveforms for the parametric acoustic source", MTS-IEEE Oceans '77, 10B-1–10B-4, 1977.
40. K.B. Theriault, "Incoherent multibeam Doppler current profiler performance: Part I—Estimated variance", IEEE Journal of Oceanic Engineering, **OE-11**, 7–15, 1986.
41. K.B. Theriault, "Incoherent multibeam Doppler current profiler performance: Part II—Spatial response", IEEE Journal of Oceanic Engineering, **OE-11**, 16–25, 1986.
42. T.M. Joyce, D.S. Bitterman (Jr) and K.E. Prada, "Shipboard acoustic profiling of upper ocean currents", Deep-Sea Research, **29**, 903–913, 1982.
43. C.L. Trump, B.S. Okawa and R.H. Hill, "The characterization of a midocean front with a Doppler shear profiler and a thermistor chain", Journal of Atmospheric and Oceanic Technology, **2**, 508–516, 1985.

Acoustic Remote Sensing Applications
S.P. Singal (Ed)
Copyright © 1997 Narosa Publishing House, New Delhi, India

20. Acoustic Remote Sensing of Ocean-Atmosphere Interactions

A. Weill and H. Dupuis

CETP, 10–12 Av. de L'Europe Velizy, France

1. Introduction

Interactions between Ocean and Atmosphere are revealed by their noise as wind noise, wave breaking noise, ressac sound...Before a quantitative knowledge on acoustic waves has been acquired, sea surface motions were a fascinating interest for musicians and poets. For a long time, the science of sounds was music and scientists or philosophers (indeed both) were trying to understand how to imitate noises in the nature, tempting to reproduce natural noises. In this context Pythagoras (570–487 B.C) had analyzed vibrating strings. Until the end of the 19th century the sounds and motions of the sea were a permanent subject of pieces of poetry. We have to await Helmholtz (1863) in Germany to have a coherent theory of sound and a response to a one thousand year enigma (see Hunt (1992) for an historical survey of physics of sound).

It is possible to say that a relatively quantitative knowledge of elastic waves took place at the end of the 19th century in particular with the discovery of "Doppler" effect by Buys Ballot in 1845 on hearing sound produced by musicians playing on a train (see R. Radeau (1975)). This last effect of frequency shift was in fact understood and analyzed by a German scientist, Bergrath Doppler.

Due to navigational and military purposes early underwater acoustics was applied using bells in the water under light ships in such a way to transmit signal to ships in the vicinity using bell receivers on their hulls.

After the discovery of underwater high frequency sound sources by Fessenden in 1912, underwater acoustic sounding began to work, but the mature age of underwater acoustics and its adaptation to ocean surface studies followed the second world war. It became indeed possible to analyse the ocean bottom acoustic reflectivity, ocean depth and the particularities of dangerous obstacles: (Albers (1965), Clay and Medwin (1975)). However one of the most important problem always open was the knowledge of the ambient noise level related to sea surface state and marine animals as shrimps, fish, snapers, sharks, oceanic mammals... which makes interpretation of signal propagation difficult inside the ocean. Since the last few decades the radar technique has become widely recognized for air-sea investigations because it allows to cover large area in a short time (via aeroplanes and satellites). For example, estimations of wind speed and wave spectra are made possible using either radar backscattering at low incidence angles or using synthetic aperture imaging radar. Interpretation of the backscattering process is however not fully understood and, moreover the way it infers the wave structure at the air-sea interface still has to be improved. As sea surface noise contains geophysical informations on the sea surface, underwater acoustics can be proposed as a remote sensing tool as an alternative to radar measurements. Further spaceborne radars dedicated to rainfall measurements cannot be used for extensive studies. Hydrophones in place are intended to be used in GEWEX (Global Energy Water Experiment) to estimate the rain fall rate and compare it with radar measurements from aeroplanes. These factors explain the increasing attention paid to underwater acoustics as a remote sensing tool of air-sea interactions. In this context, recent studies have shown that many air-sea interface processes can be analyzed using underwater acoustics. The most generally used tool appears to be the inverted echosounder. In fact many parameters can be provided such as surface current by Acoustic Doppler Current profilers (Chereskin and Harding, 1993), or by acoustical current meters based on travel time measurements (Worcester, 1977), wind speed (Carey and Browning, 1988), wave breaking statistics (Ding and Farmer, 1992; Dupuis and Weill, 1993), wave spectra (Miller et al, 1993), vorticity beneath ice (Menemenlis and Farmer, 1992) or rainfall rate (Nystuen et al, 1993)... Other studies are also devoted to climatic temperature change in the ocean (Spiesberger et al, 1992; Sutton et al, 1993). In this paper we focus on wind speed, wave breaking statistics and rainfall rate estimation using passive underwater acoustics.

It is worth stating here that acoustic remote sensing of the atmosphere began before acoustic remote sensing of the ocean surface. Since McAllister (1969), a very fast develop-ment of acoustic remote sensing science using Sodar devoted to continental boundary has taken place (see (Brown and Hall (1978), Weill (1981), Singal (1989)) to quote a few). Different techniques and methods have been developed to study wind profile in the lower atmosphere (Chong (1976), Little (1969)), backscattering intensity profile: (Neff (1975)), turbulent dissipation rate: (Gaynor (1977), Weill et al, (1978)), heat flux estimate: (Coulter et al, (1983), Weill et al, (1980)). Analysis of micrometeorological phenomenae have been undertaken: thermal plumes: (Taconet end Weill (1982), (1983)), gravity waves; (Hooke et al, (1978), Eymard and Weill (1979), Blez and Weill (1983), Weill et al, (1987)), frontal system; (Gera and Weill (1987)), mesoscale turbulence; (Masmoudi and Weill (1988)), complex terrain study; (Neff and King (1987), Weill et al, (1986)).

Considering the interface between ocean and atmosphere a few fundamental questions are needed to be debated: are we able to extend methods and concepts associated with the continental boundary layer to the Maritime Atmospheric Boundary Layer (MABL) and what kind of acoustic remote sensing techniques are available in the problem of air-sea interactions study. To that end it will be useful at first to describe in part 2 what is the MABL and in part 3 what are the Sea Surface properties. Then we shall be able to discuss in part 4 acoustic remote sensing of the MABL making a partition between coastal MABL and ice-sea interactions as antartica MABL and to present in part 5, the state of art in acoustic remote sensing of the sea surface properties of surface wind, roughness and precipitation. A prospective look toward the future of acoustic remote sensing of ocean atmosphere will conclude this paper in part 6. This paper may be considered as an introduction to acoustic remote sensing of Ocean-Atmosphere interactions and does not intend to be a comprehensive review.

2. The MABL Properties

The Marine Atmospheric Boundary Layer has very different properties from A.B.L. (Atmospheric Boundary Layer) over continents. Figure 1 shows schematically some differences.

(1) The surface layer is less developed than the continental surface layer and the roughness is indeed ten to one hundred time smaller than for the continental boundary layer if we do not take into account boundary layer over snow or ice.

(2) The sea surface is moving associated with currents which have generally a very low speed of a few cm/s but they can be more intense reaching 1 m/s.

(3) The turbulence is often smaller than the continental turbulence if one takes into account the thermal and dynamic turbulence. This indicates that C_T^2 and hence the acoustic reflectivity is generally smaller than for the emerged surfaces.

(4) The sea surface is moving and evolving due to wave motions and wave breaking which characterize interactions with the wind and therefore energy exchange between the sea surface and the atmosphere is a very important feature of the MABL.

(5) The dynamic surface layer is also less developed than the ABL surface layer. This is related to the smallness of the friction velocity. If the surface layer depth can be estimated as 0.01 U^*/f where U^* is the friction velocity, f the Coriolis parameter; for $U^* \approx .2$ m/s at middle latitudes one gets a surface layer depth close to 20 m. For the ABL, 100–150 m depth is more frequently expected. This empirical depth of the surface layer is probably better described by ten times the ten meters height wind speed. Indeed, this last rough estimate is available everywhere in the world since it is independent of the Coriolis parameter.

(6) During the night, the continental boundary layer cooling often induces the formation of a ground based inversion layer. Due to sea surface properties, it is different upon the oceans and inversion layers if they appear upon the sea are due to different mechanisms as advective air-mass, frontal systems...

(7) Coupling of surface mixing layer with inversion layer is a typical mechanism observed in the MABL and decoupling is rather scarce except probably when the convective activity is inhibited in large convective boundary layers as in tropical boundary layers (see Eymard and Weill (1988)). Decoupling between the mixing

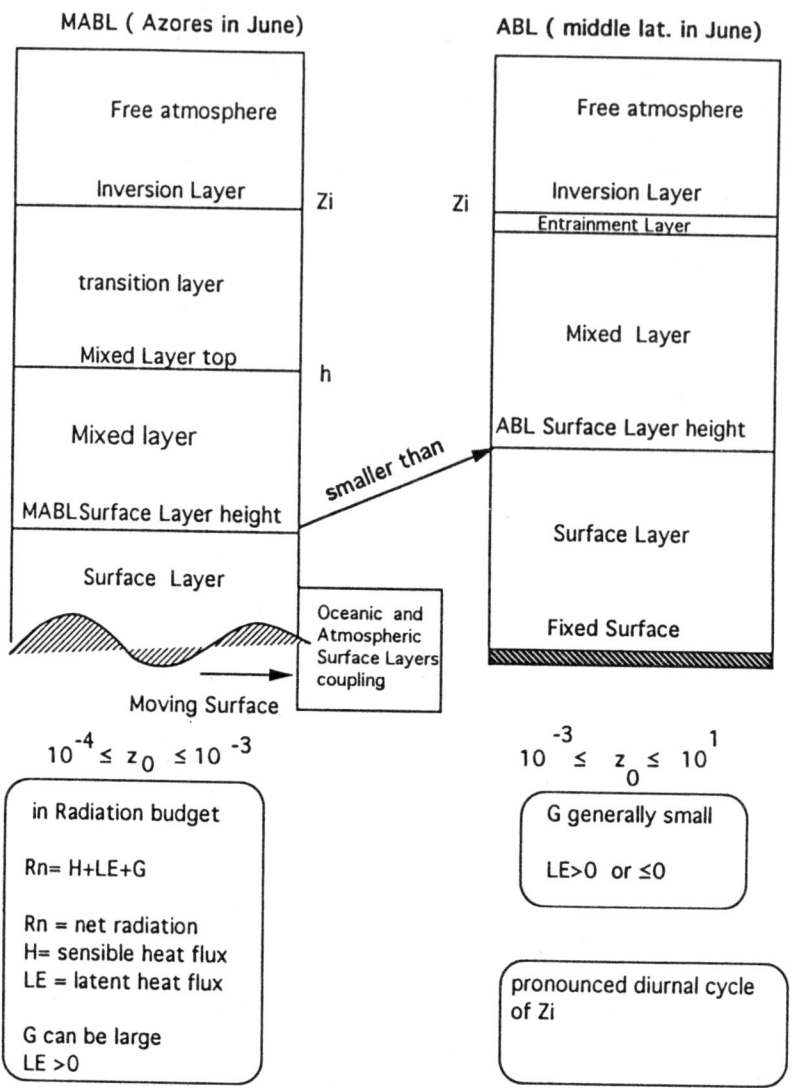

Fig. 1. Schematic difference between MABL and ABL. Notice that the main differences are a lower roughness and more frequent decoupling between the inversion layer and the surface for the MABL.

layer or well mixed layer and inversion layer can often be observed upon the sea (see Fig. 2). This indicates that parameterization of the MABL would include more probably mixed layer depth rather than inversion layer depth.

(8) For acoustic remote sensing purpose, the MABL is noisy due to sea surface noise and we do not analyze here ship noise and wind noise on the structures. The MABL noise is mainly related to sea surface noise due to wave breaking.

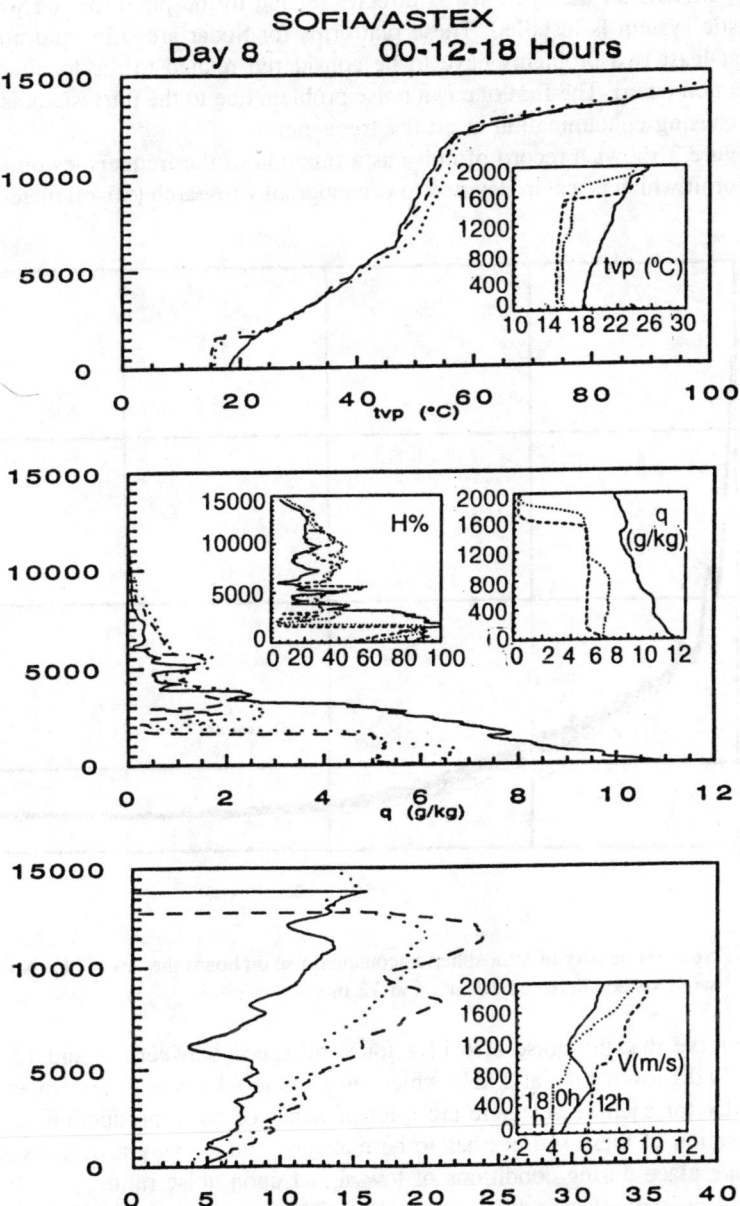

Fig. 2. Example of Radiosounding on the open sea during SOFIA/ASTEX for June 8th. In the plot are presented at the top of the figure the virtual potential temperature tvp profile, in the middle the specific humidity q profile and at the bottom the wind speed V profile for three hours 00–12–18 hours GMT. At the right of each profile a zoom corresponding to profiles for the boundary layer is shown and in the middle of the figure on the left relative humidity U% profiles are also plotted. One notices that the mixed layer depth is much smaller than the inversion height.

(9) MABL on the open sea is directly related to the platforms on which the acoustic system is installed. These platforms for Sodar are ships and not buoys and at least two problems have to be considered related to the location of the system on board. The first one is a noise problem due to the intrinsic noise of the ship causing contamination at all the frequencies.

Figure 3 shows a record of noise as a function of the frequencies on the Ship Le Noroit which is a ship devoted to oceanography research (50–60 meters long).

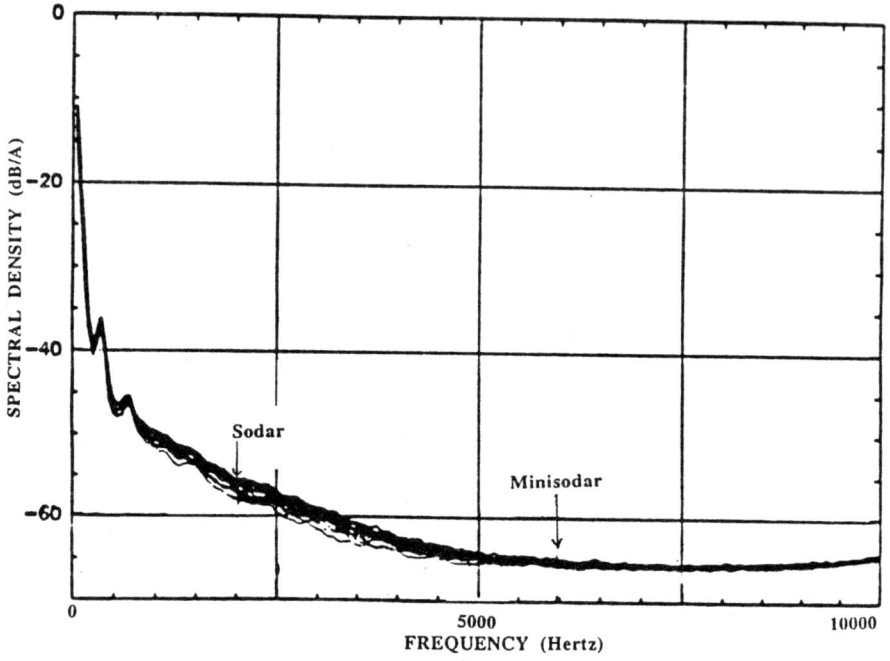

Fig. 3. Spectral density of atmospheric acoustic noise on board the vessel "Le Noroit" for wind speeds between 10 m/s and 12 m/s.

We observe that the noise at 6 kHz for wind speed between 10 and 12 m/s is about 20 dB lower than at 2 kHz which suggests to take care of the isolation. As generally for each vessel there are a lot of zones of noise production, a careful examination of these sources has to be managed and anyway acoustic sounding will take place during conditions of low signal upon noise ratio.

The second problem is due to atmospheric perturbations related to the platforms. In the case of a ship; (see Fig. 4 for Le Suroit another research ship similar to Le Noroit), if the acoustic system is in the front of the ship and the ship is moving facing the wind, a reduction of the wind speed and subsequently a modification of the turbulence will result. In fact it would affect the whole wind profile below 30 m.

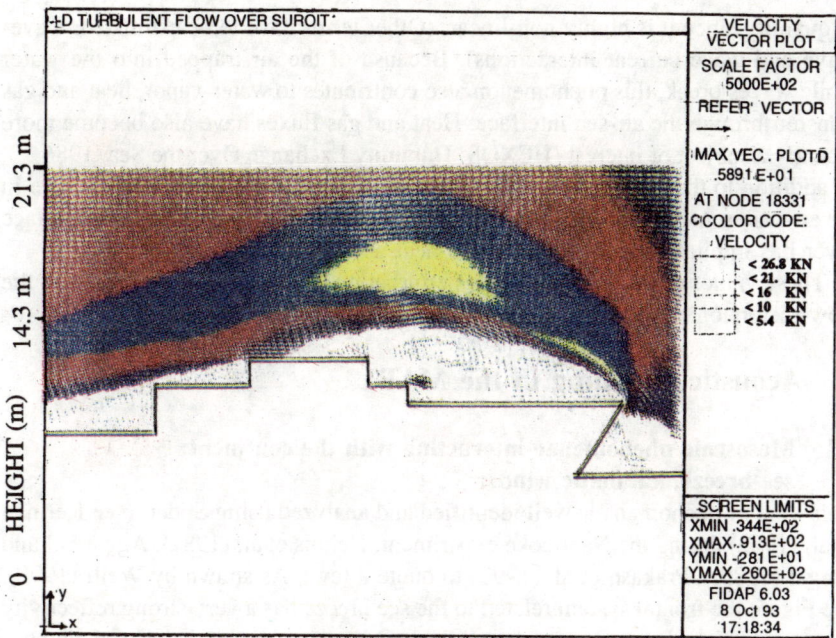

Fig. 4. A 2D simulation of perturbations generated by a 20 knots wind upstream on the ship Le Suroit. One observes a palette of four colours: (only black and white copy here)

white	below 0.27 upstream speed
dark blue:	below 0.5 upstream speed
red:	below 0.8 upstream speed
sky blue:	below 1.05 upstream speed
yellow:	below 1.34 upstream speed

We thus observe here that the use of acoustic sounder for the study of the MABL on the open sea will be different from the ABL study. Let us analyze now the problem of Sea Surface properties. Air-Sea interaction studies involve energy and flux transfer knowledge across the surface of the ocean and therefore a description of all the mechanisms involved in the exchanges.

3. The Sea Surface Properties

The main characteristics of the sea surface properties have been mentioned in the previous section, they can be detailed as follows:

The sea surface roughness induced by the ocean gravity and capillary waves is of great complexity. Among the terms to be evaluated, the energy transfer between wind and waves have been of great interest and many experimental programs have been developed to measure and parametrize the momentum flux and the ocean wave spectra in the past decades (JONSWAP: Joint North Sea Wave Project, 1973, SWADE: Surface Wave Dynamics Experiment, 1990–1991…). Understanding the wave energy balance also supposes exhaustive analysis of dissipation through wave breaking. The wave breaking term is however difficult to quantify since it is highly non-linear. (Other terms have to be estimated: wave-wave and wave current interactions). Because of the air trapped into the water while waves break, this phenomenon also contributes to water-vapor, heat and gas transfer through the air-sea interface. Heat and gas fluxes have also become more recently a subject of interest (HEXOS: Humidity Exchange Over the Sea, 1986...). In addition to these wind-wave processes, rainfalls also play an important role in the air-sea exchanges. Indeed, rainfall-induced turbulent mixing of the ocean surface layer has implications for momentum, heat and gas exchange.

These values overestimate the true perturbation but indicate the locations of the flow distorsion. Over 30 m the perturbations can be considered to be small.

4. Acoustic Sounding in the MABL

4.1 Mesoscale phenomenae interacting with the continent: see-breeze, katabatic wind.

This kind of phenomena is well identified and analyzed using Sodar, (see Kerman et al, (1982) during the Nanticoke experiment, Helmis et al, (1987), Aggarwal and Singal (1980), Prakash et al, (1992) to quote a few). As shown by Weill (1990), see Fig. 5, the frontal system related to the see breeze has a very strong reflectivity signature which allows to compute all the quantitative characteristics of the breeze

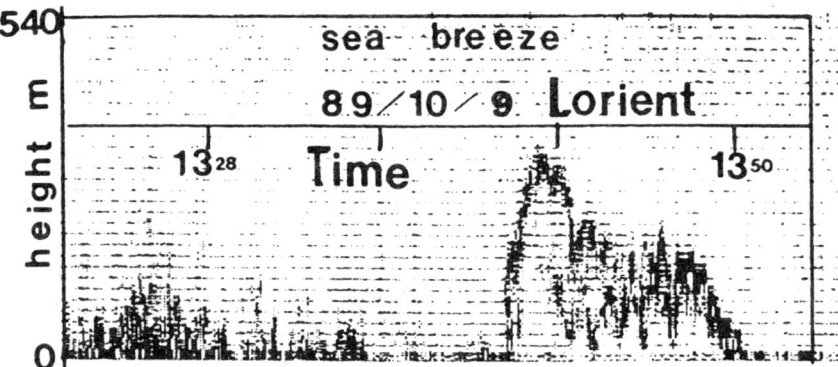

Fig. 5. A numerical facsimile record of a breeze front observed on a small peninsula.

structure as wind speed profile, turbulence level, frontal depth, vertical velocity profile. In Weill et al, (1988) it is shown how to determine at different scales all the characteristics of a maritime mesoscale disturbance related to North Sea environment during COAST experiment. See Fig. 6 showing how the turbulent spectrum can change as a function of the disturbance passage. Similarly, using acoustic sounder; ice boundary layer and interaction with sea surface have been considered in the South Pole by Neff and Hall (1976). One of the characteristic events was gravity waves theoretically considered by Fua et al, (1982).

Investigation of Katabatic wind at south pole has also been undertaken by

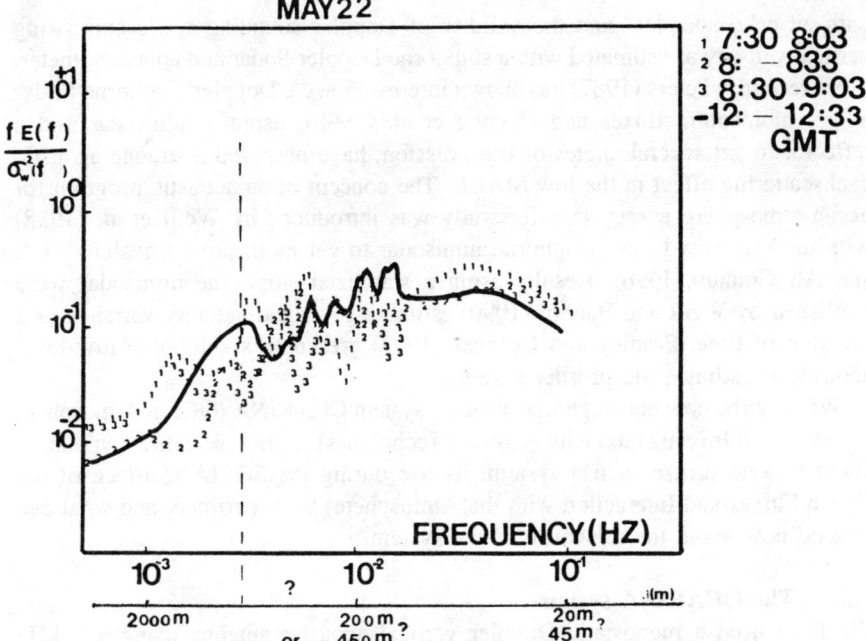

Fig. 6. Example of spectral density of vertical velocity variation during COAST experiment related to a stability change associated with a breeze phenomenon. Reprint from BLM 1988, 44, 359–371.

Kobayashi (1983) showing the same kind of structures as observed by Neff (1976). More recently more systematic studies of ice-land interactions were introduced by Fiocco et al, (1990), Argentini et al, (1990) and they have been able to document strong wind circulation and to analyze different wind regimes. They have analyzed gravity wave structures in relation with sea wind. What is remarkable in their studies is the possibility to estimate wind to noisy ambient atmosphere and the problem is probably similar to acoustic sounding in open sea. In fact as observed in their facsimile records the atmospheric flow is dynamically unstable and statically stable involving a large reflectivity level, hence although the noise is large the signal is measurable; this is not generally the case if one observes the MABL upon the open sea surface which is more generally associated with low heat fluxes and momentum production if one examines middle Atlantic regions in the spring.

4.2 The MABL on the Open Sea

The idea of shipborne acoustic sounding of the MABL has been tested in the past by several scientists (Ottersten (1973), Mandics and Owens (1975), Mandics and Hall (1976) Gaynor and Ropelewski (1979)) and their studies concerned mainly with simple Sodar often accompanied with Doppler capabilities.

Mandics and Hall (1976) have shown "the value of probing the marine atmosphere with an echo sounder" and the validity of Doppler sounding while comparing vertical velocity as estimated with a shipborne Doppler Sodar and an anemometer. More recently, Peters (1987) has shown interest to use a Doppler shipborne Sodar to get momentum fluxes and Cheung et al, (1990), using a minisodar and a reflector to get several angles of transmission, have observed a strange acoustic backscattering effect in the low MABL. The concept of an acoustic program for ocean-atmosphere energy transfer study was introduced by Weill et al, (1988) with the way how to use shipborne minisodar to get momentum transfer (Weill and Ali Goulam, 1990). Results using a stabilized shipborne minisodar were published by Weill and Baudin (1990) estimating vertical velocity variance as a function of time. Bradley and Georges (1992) presented a shipborne low level acoustic sounding wind profiler system.

We describe now our shipborne acoustic system OCARINA (OCean Atmosphere Research and Investigation with Acoustic Techniques), a priori scientific hypothesis about how to utilize such a system, its use during the SOFIA (Surface of the Ocean Fluxes and Interaction with the Atmosphere) 92 experiment and what can be said now about the relevance of the system.

4.2.1 The OCARINA system

We have used a monostatic Doppler vertical acoustic antenna using a 2 kHz frequency equilibrated with a double cardan system and stabilized with a four gyroscopic device (3000 rpm) to ensure one degree stabilization. Another three antennas 6 kHz Doppler minisodar was added (see Weill et al, 1986), supported by a stabilized platform. Each antenna of the minisodar combined a transmitter and a receiver close to each other in order to eliminate problems of reverberation inside the receiver and to be able to receive geophysical information in the lowest gates.

Roll, pitch, vertical acceleration were recorded on board the platform and these data were used to force the stabilization with a PC computer dedicated to stabilization. With this system a 0.3 degrees stabilization was awaited.

A numerical program of data processing was established to record on the same magnetic tape the minisodar and the sodar data with the ship navigation, residual roll and pitch after stabilization and vertical acceleration on the two system (Sodar, minisodar). On the tape were recorded spectrum momentum.

4.2.2 Ideal scientific use and first results

Following results obtained during GATE by Gaynor and Hall (1979), we have elaborated a strategy of MABL observations based on the use of the OCARINA system:

— The vertical stabilized antenna would be used to monitor the inversion layer and to estimate parameters as vertical velocity and vertical velocity variance.
— The minisodar would be used to estimate the horizontal wind profile,

analysis of surface layer structures related to the observed reflectivity and computation of turbulent parameters as turbulent dissipation rate.

Both the instruments Sodar and minisodar have been thought as an help to analyze parametrizations of surface fluxes either in the surface layer or in mixed layer. Figure 7 gives a schematic view about our scientific hopes and Fig. 8 gives a first raw estimate of friction velocity using vertical velocity variance from the vertical antenna.

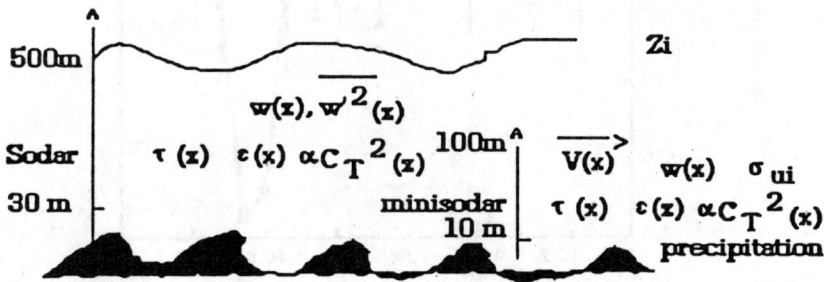

Fig. 7. Schematic sketch of acoustic sounding in the MABL (for conditions of low ship noise and low level inversion layer).

From the Sodar could be obtained a priori the inversion height, the vertical velocity variance and inversion layer entrainment.

From the minisodar could be analyzed wind speed parameterizations and relationships between wind speed profiles and Reynolds stress components.

4.2.3 Results during SOFIA 1992

During SOFIA 92 in the Azores region the Ocarina system was used in the front of the ship and a tethered balloon equipped with wind, temperature, humidity profile measurement and C_T^2 estimate was installed behind the superstructure on the bridge of the ship.

Due to a low level turbulence and strong noise at two kHz the vertical antenna was not usable contrary to what was observed during GATE (see Gaynor and Hall 1979). This was due to the fact that the wind was generally low smaller than 10 m/s and the inversion layer was high (at about 1250 m height). As observed with the tethered balloon, considering data over 40 m height, there was a very well mixed layer and its depth determined with radiosounding was generally observed between 400 m up to 800 m as in Fig. 2. At about 1000 m was found the cloud base; the inversion layer being at the cloud top.

These conditions are indeed different from the classical ABL and therefore the minisodar was able to systematically observe only the surface layer. Figure 9

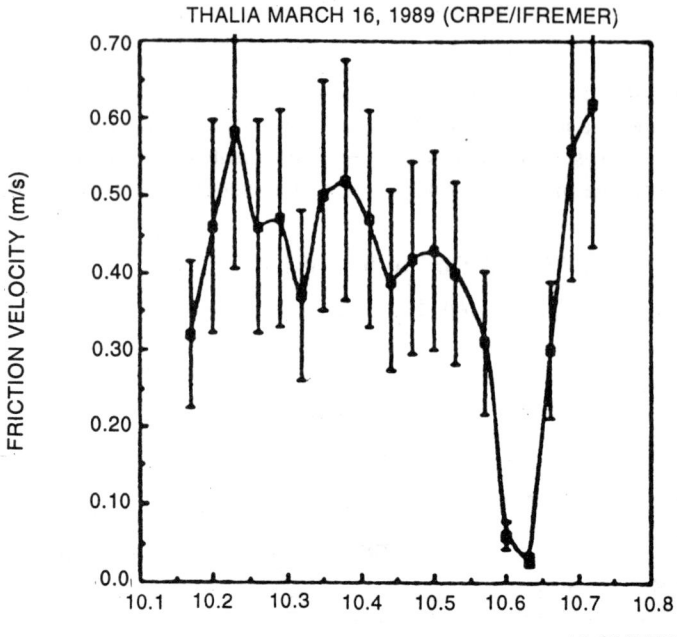

Fig. 8. Friction velocity estimated from the vertical velocity standard deviation during neutral conditions. The low values are due to an aerodynamic effect related to a small island.

shows what kind of reflectivity signature was met during SOFIA with the minisodar. In Figure 10 is also shown a comparison between a minisodar wind profile and the tethered balloon profile. Notice that the difference between the two instruments in the first 40 meters height is due to ship effect corresponding to a reverse flow behind the superstructures but in the tethered balloon profile over 40 m the differences between the two measurements are generally smaller than 1 m/s (i.e. compatible with the instrumental uncertainty).

With the tethered balloon data is computed the bulk Richardson number RG determined over 40 m since below 40 meters height, data are aerodynamically perturbed:

$$RG = gZ(\theta'v - \theta s)/(\theta'v(U - Us)^2) \qquad (1)$$

where Z is the height of measurement of the virtual potential temperature, $\theta'v$ is the vertically and temporally averaged virtual potential temperature, U is the wind speed, θs is the Sea Surface temperature, and Us the surface current.

There is no reason to compute RG at the level Z if Z is not in the surface layer and the minisodar is used to compute H the surface layer height. If $Z \leq H$, fluxes are computed with the logarithmic profiles. If Z is not in the surface layer either the tethered balloon data vertical profile over 40 m presents a well mixed layer

Acoustic Remote Sensing of Ocean-Atmosphere Interactions 461

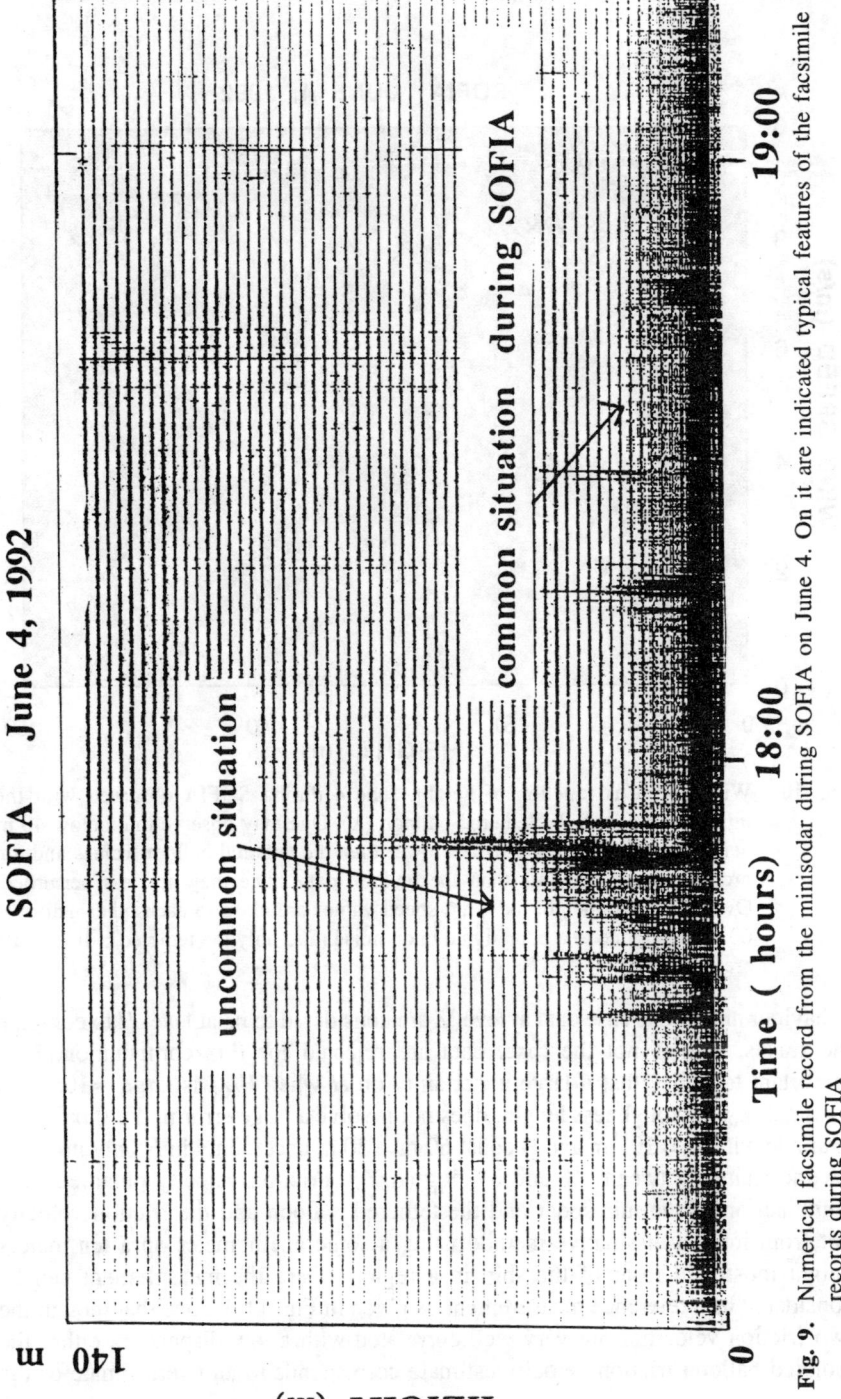

Fig. 9. Numerical facsimile record from the minisodar during SOFIA on June 4. On it are indicated typical features of the facsimile records during SOFIA

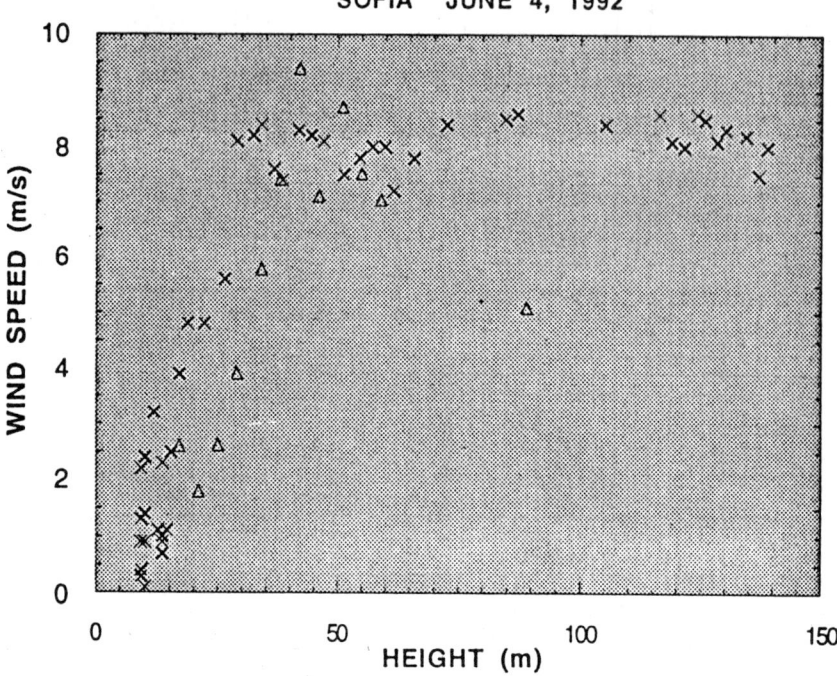

Fig. 10. Wind speed as function of height, June 4 during SOFIA observed with the minisodar (Δ) and the tethered balloon (X). One may observe that below 40 m height the minisodar speeds are lower than the tethered balloon values and are probably more accurate than the balloon data since they are less perturbed. Over 40 meters height the wind speed values are close to each other and over 60 m height due to a poor signal upon noise ratio only the tethered balloon data are available.

behaviour and in this case data at level Z are considered to be at level H to compute the fluxes, if it is not the case, data are rejected for flux computation. It is important to notice that a large uncertainty on H when H/Z_0 is large ($\approx 10^5$–10^6) for small Z_0, the roughness length, does not involve a large error in the fluxes. For example with $Z_0 = 10^{-4}$ m and H equal 60 m, a 30 m uncertainty in H only involves an uncertainty between 3% and 8% log (H/Z_0). Friction velocity obtained by a combination of the minisodar and the tethered balloon and the friction velocity got from inertial-dissipative method using a sonic anemometer on a ten meters height mostly in front of the ship in a region where the measurement can be considered as a reference measurement, is given in Fig. 11. Notice that though the two friction velocities are very well correlated with a low dispersion, either the tethered balloon friction velocity estimate corresponds to an overestimate or the

Acoustic Remote Sensing of Ocean-Atmosphere Interactions

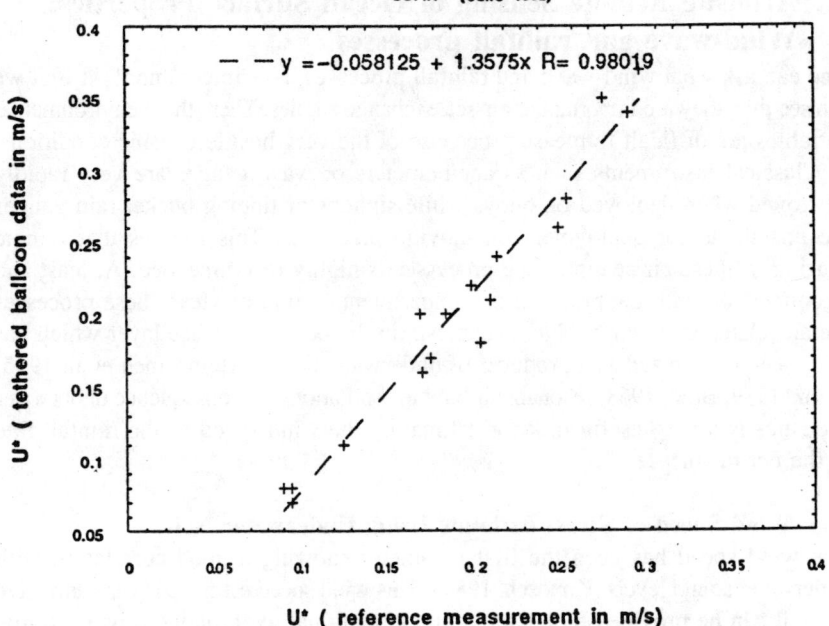

Fig. 11. Correlogram showing the friction velocity obtained from the tethered balloon data with a bulk Richardson number method (H the surface layer depth being estimated with the minisodar data) and the reference friction velocity. The reference friction velocity is obtained from a ten meters height mast in front of the ship.

reference friction velocity is underestimated. At this step we do not have enough elements to give a relevant diagnostic.

What is remarkable is also the information on the thermal turbulence got from the minisodar as described by Fig. 9 showing, as observed during the whole SOFIA experiment, a ground based turbulent layer corresponding to very small plumes which have been found to be coherent to a very small sensible heat flux, smaller than 40 W/m^2 on the average during the experiment. In Fact the wind speed and the deviation of the air temperature from the sea surface temperature SST were small.

It is too early to derive all the scientific information from the OCARINA system during SOFIA but the schematic idea of what can be obtained using Sodar technique at sea does not seem to be too far from the schematic idea depicted on Fig. 7 at least for the minisodar.

One conclusion is that during conditions of low level sensible heat flux the 2 kHz system is not a relevant technique but the minisodar at 6 kHz remains a convenient instrument.

This conclusion is indeed relative and depends drastically on the noise level of the vessel used.

5. Acoustic Remote Sensing of Ocean Surface Properties: Wind-wave and rainfall processes

One can ask what wind-wave and rainfall processes have in common! At first we can see that, they are important to air-sea exchange studies. Then, these environmental variables are difficult to measure because of the very hostile oceanic conditions.

Classical instruments such as anemometers or wave gauges are very rapidly destroyed when deployed on buoys while siphons or tipping bucket rain gauges are unsuitable for deployment on moving platforms. This implies that remote sensing of these air-sea interface processes is highly recommended. At least and in connection with the practical and experimental point of view, these processes are all related to an air bubble mechanism in the oceanic surface layer which has long been recognized as a producer of underwater sound (Heindsman et al, 1955; Franz, 1959; Bom, 1968) resonant air bubble oscillations. In consequence underwater acoustics is a very useful task for estimating the wind speed or the rainfall rate at the ocean surface.

5.1 Wind Speed or Stress Estimate Using Underwater Noise

The wind speed has been the first parameter showing a good correlation with underwater sound levels (Knudsen, 1984). This wind speed estimate is very attractive since it can be provided by a simple omnidirectional hydrophone (low cost, low energy supply, not subject to deterioration due to sea surface motion...). A "calibration" period is however necessary since the relationship depends on the experimental site conditions (celerity profiles, bottom properties and hydrophone depth...). Although the relationship is found in a large frequency range (from a few Hz to several tens of kHz), a compromise has to be found between the low frequencies ($f < 5$ kHz) whose spectral levels are easily contaminated by long distance shipping (Hamson, 1993) and the high frequencies ($f > 10$ kHz) for which a saturation is observed over high wind speed, due to bubble plume attenuation (Lemon and Farmer, 1984). Figure 12 shows an example of ship noise spectra as measured at a distance of 1 m of the ship Le Suroit (the two top curves). The ship noise levels are much higher than the wind-dependent spectra. Underwater sound propagation modelling (Dupuis et al, 1993) can be useful to predict the distance at which shipping noise is negligible.

Here in Fig. 12 corresponding to the SOFIA/ASTEX experiment, it can be seen from data and simulations that at 20 kHz, a distance of 5 km apart from the ship is widely sufficient to be able to neglect the ship noise, while distance is not sufficient at 500 Hz.

The noise level (NL) for a specific wind noise process is generally related to a power n of the wind speed (U) according to the relationship:

$$20 \log (NL) = B (20 \log U)^n \qquad (2)$$

Where B and n are two constants depending on the experimental conditions.

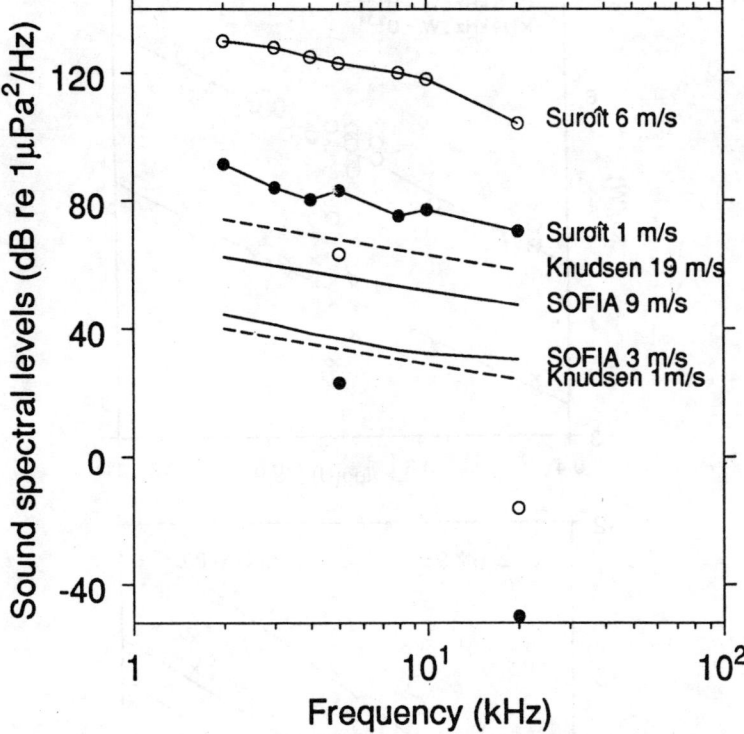

Fig. 12. Underwater sound spectra. Solid lines with filled and unfilled circles correspond to sound spectra measured at a distance of 1 m from the ship "Le Suroit", the speed of which is 1 and 6 m/s respectively. The symbols without lines indicate the ship noise predicted by a sound propagation model (Dupuis et al, 1993) at a distance of 5 km. Solid lines are wind-dependent underwater sound spectra measured during the SOFIA/ASTEX experiment, which can be compared with the Knudsen ambient noise spectra (dashed lines). From Dupuis et al, 1993.

Figure 13a shows the correlation in equation (2) applied to the case of the SOFIA/ASTEX experiment (Dupuis et al, 1933).

Two wind noise processes are generally proposed (Medwin and Beaky, 1989; Prosperetti et al, 1993), both related to wave breaking:

— the resonant oscillation of air bubbles ($f > 100$ Hz);
— the collective oscillation of the bubble clouds ($f < 1$ kHz).

Other interpretations have been proposed (nonresonant oscillations: Kerman, 1984), especially for low wind speeds (below 5–6 cm/s) during conditions where wave breaking appearance is questionable. Several authors (Carey and Browning, 1988, Kewley et al, 1990) therefore suggested two regimes of wind speed behaviour, with different powers n in equation (2). Carey and Browning (1988) reviewed the experimental results found for equation (2).

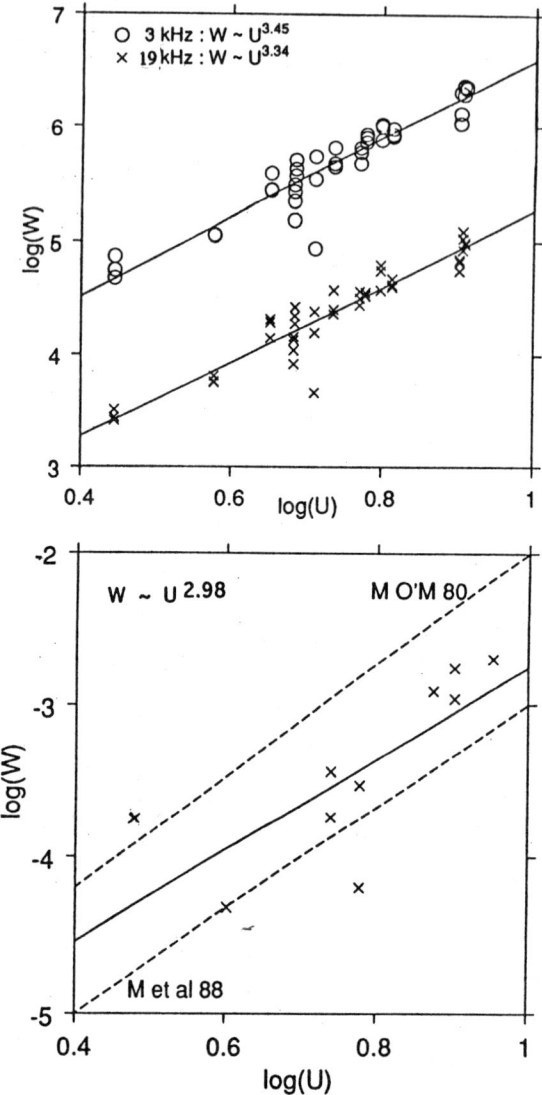

Fig. 13. a: Spectral sound levels at 3 and 19 kHz (averaged over 30 minutes) as a function of the wind speed measured on the ship Le Suroit. b: whitecap coverage as a function of the wind speed. The dotted lines correspond to the regression lines described in Monahan et al, (1988) and Monahan and O'Muircheartaigh (1980). In both panels, the power coefficient of the equation corresponds to the coefficient n in equation (2). From Dupuis et al, 1993.

As a remote sensing tool, hydrophones provide a spatial integrated measurement of the wind speed which depends on the experimental configuration (frequency, water depth...). One has however to be aware that underwater sound leads to an indirect measure of wind speed through the wave breaking process.

5.2 Wave-breaking Studies

Passive acoustics thus allows to study the entrapped air bubbles while waves break. Such transient oscillating bubbles (forming bubble clouds) have to be distinguished from the nonoscillating, non radiating bubbles (forming bubble plumes). A similarity can be done with the surface evidences of wave breaking: the whitecaps. Indeed Monahan and O'Muircheartaigh (1980) distinguish young and mature whitecaps.

5.2.1 Using a single hydrophone to derive wave breaking statistics

Farmer and Vagle (1988) first proposed to develop a model of underwater sound due to wave breaking. The idea was to describe the sources of noise at the sea surface in a coherent approach and to use it together with an acoustic propagation model to predict underwater noise. The simplest source description is as follows:

A source corresponds to a bubble cloud, acoustically radiating as a dipole. Its temporal and spatial distribution is uniform over the time and space domain of the analysis (a periodicity can however be included). The space domain is defined by the listening radius which delimits the radius of the sea surface disk from which 95% of the received ambient noise is transmitted (Fig. 14). The underwater sound propagation model provides estimations of these listening radii R1 (Farmer and Vagle, 1988), as a function of frequency and experimental conditions (water and hydrophone depth, bottom properties...). Dupuis and Weill (1993) showed how the model can be inverted to estimate mean spacing between the breakers as well as the power laws between the coverage, the size and the spacing of whitecaps and the wind speed (or any other parameter). A simple statistical analysis including mean and variance of spectral levels is needed. In fact the power law obtained in Equation (2) can also be interpreted in terms of wave breaking. Indeed, since the sound levels are supposed to be proportional to the product of the number of sources and their mean characteristic size (i.e. the coverage), the coefficient n in Equation 2 also represents the power law between the whitecap coverage and the wind speed.

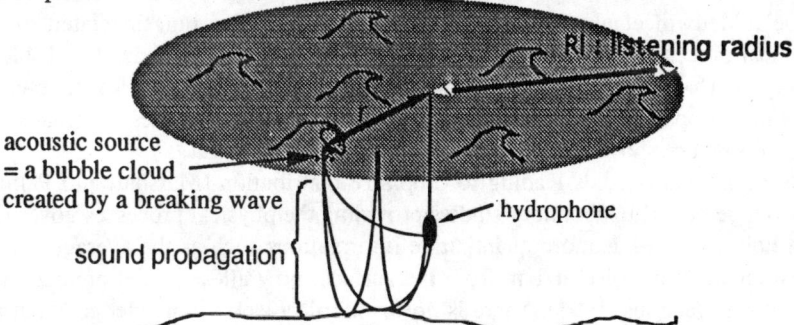

Fig. 14. Schematic drawing of the noise received at an omnidirectional hydrophone as the sum of the contributions of sources (breaking waves) inside a listening area defined by the listening radius R1. The sound propagation model allows for multiple reflections on the sea and bottom surface.

468 *Ocean Acoustics*

A comparison between power laws of the whitecap coverage using a hydrophone and a photographic system shows a good agreement between both methods (see Dupuis et al, 1993). An example is given in Fig. 13 where the coefficient n is obtained from the acoustic data (panel a) and from the photographic data (panel b) during the SOFIA/ASTEX experiment. Nevertheless, this study, conducted in the Azores region under moderate wind speed conditions seems to show that the characteristic surface of whitecaps (in m^2) increases more rapidly with the wind speed than the density of whitecaps (in m^{-2}) in contrast with other studies (Wu, 1992; Thorpe and Humphries, 1980).

5.2.2 Using an array to derive a wave breaking mapping

Ding and Farmer (1992) have proposed an acoustical system which consists of a small broad band hydrophone array of span 8.5 m and a 5 kHz bandwidth. They use a generalized cross-correlation method which estimates the time delay between signals at two hydrophones. This system can be used to measure some spatial and temporal characteristics of breaking waves, such as frequency of breaking, lifetime, velocity, and spatial distribution.

5.3 Rainfall Studies

The sound production mechanism by rainfalls has been studied in laboratory conditions using individual rain drops falling in the tank with hydrophones below the surface. Franz (1959) conducted the first laboratory experiment. However, only more recent studies (Meadwin and Beaky, 1989, Nystuen et al, 1992) ensure applicability to rainfall, because they examine individual water drops for the complete size range at their terminal velocity. It has been shown that the two principal sound production mechanisms by a water drop splash are the initial impact and radiation from any bubble trapped into the water. The damped oscillation of bubbles has been shown to be much strongly radiating than the short duration impact. Meadwin et al, (1992) recently distinguished two different processes of bubble creation related to the drop size. In their "anatomy of underwater rain noise", Medwin et al, (1992) established four drop size ranges related to the physical processes governing sound production, that are summarized in Table 1.

Figure 15 gives examples of the broad band spectral peak at 15 kHz, created by small drops. Flat spectra obtained for different rainfall rates dominated by large drops are shown in Fig. 16.

Experimental studies leading to drop size distribution (Marshall and Palmer, 1948) together with laboratory studies providing the physical processes governing rain noise, as well as more quantitative informations such as the average energy densities of drop noise at 1 m from the surface, now allow model prediction of rain noise (Jacobus, 1991). There is however still a lack of knowledge about the dependence of large rain drops upon the ocean roughness. Indeed it is well known that a surface is flattened during rainfall. And this phenomenon has been pointed out to interpret an hystereris observed in sound levels over a rain event: indeed

Table 1. Spectral sound properties as a function of drop terminology

drop terminology	drop size range	rainfall terminology	rain fall rate	sound production mechanism	rain noise spectra characteristics
minuscule	D < 0.8 mn	fog and drizzle		very weak impact noise	almost undetectable
small	0.8 < D < 1.1 mn	drizzle		impact and damped oscillations	broad spectral peak at about 15 kHz
mid-size	1.1 < D < 2.2 mm	(light) and moderate rainfall	(< 2.5 mm/h) > 2.5 mm/h < 7.6 mm/h	impact	almost undetectable
large	D > 2.2 mm	moderate and heavy rainfall	> 7.6 mm/h	impact and bubbles	flat spectra

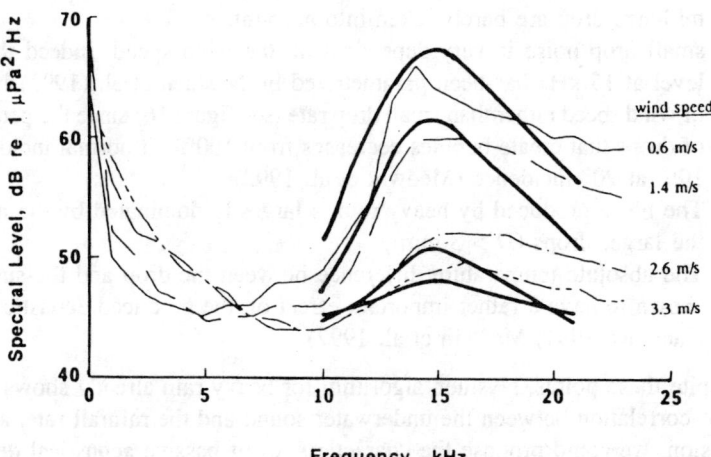

Fig. 15. Underwater sound spectra produced by light rain/drizzle. In each case the total rainfall is 0.6 mm/h and no large drops are present in the rain. The bold lines show predictions of the effect of wind on the light rain acoustic signal (from Nystuen et al, 1993).

Nystuen et al, (1993) proposed an algorithm to estimate rainfall rate during heavy rainfall although they show a different behaviour at the beginning compared to the end of the event. Other possible explanations for the hysteresis phenomenon are the influence of the bubble plume created along the event, the change of drop size distribution or the change in sea surface temperature. Furthermore a few other points have to be considered in the inversion of the rainfall rate:

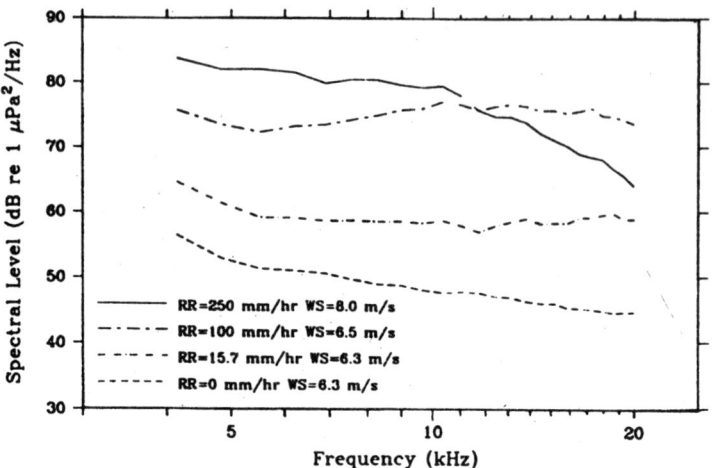

Fig. 16. Instantaneous spectra recorded during heavy rainfall. A "no rain" spectrum is shown for comparison (from Nystuen et al, 1993).

— mid-size drop are barely taken into account;
— small drop noise is very dependent on the wind speed. Indeed the peak level at 15 kHz has been parametrized by Nystuen at al, (1993) in terms of wind speed rather than small drop rate (see figure 16) since the percentage of drops that create bubbles decreases from 100% at normal incidence to 10% at 20° incidence (Medwin et al, 1992);
— The noise produced by heavy rain is largestly dominated by the noise of the larger drops ($D > 3$ mm);
— The absolute temperature difference between the drop and the surface is shown to have a rather important effect on the produced acoustic energy (Jacobus, 1991; Medwin et al, 1992).

Despite these points, Nystuen algorithm for heavy rain already shows a good level of correlation between the underwater sound and the rainfall rate, and as a conclusion, we could propose the characteristics of passive acoustical device to estimate rainfall rates at the ocean as follows: it should at least contain two frequency bands centered around a few kHz (detection and quantification of heavy rain) and around 15 kHz (detection of light rain). A few frequency bands could be added to improve the detection of a rain event (i.e. to discriminate with high wind speed events) which is done by an analysis of the spectral slope. Tentatively light rain rate quantification could be attempted using the wind speed estimate by the lower frequency (the noise level prior to rainfall could also be used to estimate the wind speed) to distinguish the effects of drop rate and of incident angle of the drop impacts. Because of the use of frequencies below 10 kHz a directional pattern is recommended to avoid ship or animal contamination due to long distance propagation.

If a sea surface temperature sensor is added it might even be possible to predict the level of the clouds from which rain has fallen as proposed by Medwin et al, (1992) because of the effect of the absolute temperature difference between the drop and the surface.

6. Conclusions

At this step one may remark that acoustic remote sensing of the MABL and some ocean surface properties (surface roughness, wind, precipitation) is mature and a lot of progress has been made. However as all these techniques are affected by human activity noises (ship noise and platform noise) and marine life noises (fishes...) and platform airflow deformation, they are therefore difficult to apply. Further, for MABL acoustic sounding stabilized platforms are necessary since platform motions in case of severe conditions have to be eliminated.

Concerning MABL Sodar sounding, considering the literature it appears that during very unstable conditions, in case of low level inversion layer, though the ship noises are often strong, one can use 2000 Hz sounding; but it was observed during SOFIA/ASTEX 92 where these last conditions were not prevailing that only high frequency (6000 Hz) minisounding was possible.

We suggest in the future to use multifrequencies in the high frequency minisodar and to combine the minisodar data with a boundary layer high frequency radar to take benefit of RASS (Radio-Acoustic) capabilities and to obtain temperature profiles in the surface layer.

For acoustic remote sensing of the surface, as it is possible (after calibration) to estimate wind speed at different scales, sea surface roughness informations related to the breaking waves and rain rate, we suggest the idea of operational drifting buoys development with a directional hydrophone, a meteorological station and computing facilities on board. Hence estimating these last parameters with an "internal" calibration with the wind speed on board, then transmitting the elaborated information via ARGOS system would be an excellent mean for intercomparison and intercalibration with satellite estimates of the same variables.

Acknowledgments

Acoustic sounding at (CRPE/CNRS, now CETP/CNRS) was supported by INSU (CNRS), DRET. SOFIA experiment has received support from INSU (CNRS), DRET, Météo-France, IFREMER. We gratefully Acknowledge their funding.
CRPE: Centre de Recherche en Physique de l'Environnement
CETP: Centre de Recherche en Environnement Terrestre et Planétaire
CNRS: Centre National de la Rechereche Scientifique
INSU: Institut National des Sciences de l'Univers
DRET: Division des Recherches et Etudes Techniques
IFREMER: Institut Français de Recherches sur la Mer

References

Aggarwal S.K., S.P. Singal, R.K. Kapoor and B.B. Adiga. A study of atmospheric structures using sodar in relation to land and sea breezes, Boundary Layer Meteor., 18, 361–371, 1980.

Albers M. Underwater Acoustics, Handbook 2, Pennstate University Press, 356 pp., 1965.

Argrentini S., R. Ocone, G. Fiocco and G. Mastrantonio. Statistical analyses of katabatic winds, In Proceeding of the 5th International Symposium on Acoustic Remote Sensing of Atmosphere and Oceans, New Delhi, S.P. Singal (Editor), MacGraw Hill Cie, 283–293, 1990.

Blez M. and A. Weill. Simultaneous analysis of waves and turbulent parameters under statically stable morning conditions, In Proc., Second Symposium on Acoustic Remote Sensing of the Atmosphere and Oceans, Rome, Italy, 29/8-1/9, Instituto Dell' Atmosfera, XXXVIII (1–23), 1983.

Bom B. The effect of rain on underwater noise levels, J. Acoust. Soc. Am., 45, 150–156, 1969.

Bradley S. and K. George. Low level acoustic wind profiling from a small sea going vessel, 6th International Symposium on Acoustic Remote Sensing and Associated Techniques of the Atmosphere and Oceans, Athens, Greece, 249–254, 1992.

Brown E.H. and F.F. Hall, Jr. Advances in atmospheric acoustics, Rev. of Geophys. and Space Phys. Vol 16, 44–110, 1978.

Carey W.M. and D.G. Browning. Low frequency ocean ambient noise, measurement and theory, in Sea Surface Sound, edited by B.R. Kerman (Kluwer Academic, Boston), 301–376, 1988.

Chapman N.R. and J.W. Comish. Wind dependence of deep ocean ambient noise at low frequencies, J. Acoust. Soc. Am., 93 (2), 782–789, 1993.

Chereskin T.K., and A.J. Harding. Modeling the performance of an acoustic Doppler current profiler, J. Atmos. Oceanic Technol., 10, 41–63, 1993.

Cheung T.K., C.G. Little and H.E. Ramm. The acoustic scattering layer observed in the low marine boundary layer, J. Atmos. Sciences, 47, 21, 2537–2545, 1990.

Chong M. Mesure des profils de vent par Sodar Doppler, Note Technique N° 22, 103 pp., 1976.

Clay C.S. and H. Medwin. Acoustical Oceanography: Principles and Applications, Wiley Interscience Publication, 544 pp., 1977.

Coulter R.L., M.L. Wesely, T.J. Martin and K.H. Underwood. Surface heat flux derived from sodar amplitude and frequency data. A comparison, In Proc., Second Symposium on Acoustic Remote Sensing of the Atmosphere and Oceans, Rome, Italy, 29/8–1/9, Instituto Dell'Atmosphera, XXXVI (1–18), 1983.

Ding L. and D.M. Farmer. Coherent acoustical radiation from breaking waves, J. Acoust. Soc Am., 92 (1), 397–402, 1992.

Dupuis H. and A. Weill. A model to estimate the density, the characteristic surface, and the coverage of whitecaps using underwater sound, J. of Geophys. Res., 98 (C10), 18, 213–18, 219, 1993.

Dupuis H., J.P. Frangi and A. Weill. Comparison of wave breaking statistics using underwater noise and sea surface analysis conducted under moderate wind speed conditions during the SOFIA/ASTEX experiment, Ann. Geophysicae 11, 960–969, 1993

Eymard L. and A. Weill. Dual Doppler investigation of the tropical boundary layer. J. Atmos. Sci., 45, 853–864, 1988.

Eymard L. and A. Weill. A study of gravity waves in the planetary boundary layer by acoustic sounding, Boundary Layer Meteor., 17, 231–245, 1979.

Farmer D.M. and D.D. Lemon. The Influence of Bubles on Ambient Noise in the Ocean at High Wind Speed, Journal of Physical Oceanography 14, 1762–1774, 1984.

Farmer D.M. and S. Vagle. On the determination of breaking surface wave distributions using ambient sound, J. Geophys. Res. 93, 3591–3600, 1968.

Fiocco G., G. Mastrantonio, R. Ocone and S. Argentini. Acoustic sounder experimentation in Victoria, Antartica, In Proceeding of the 5th International Symposium on Acoustic Remote Sensing of Atmosphere and Oceans, New Delhi, S.P. Singal (Editor), McGraw Hill Cie, 272–282, 1990.

Franz G. Splashes as sources of sound in liquids J. Acoust. Soc. Am., 31, 1080–1096, 1959.

Fua D., G. Chimonas, F. Einaudi and O. Zeman. An analysis of wave-turbulence interaction. Journal of Atm. Science, 39, 2450–2463, 1982.

Gaynor J.E. Acoustic Doppler measurement of atmospheric boundary layer structure functions and energy dissipation rates, J. Appl. Meteorol., 16, 148–155, 1977.

Gaynor J.E. and C.F. Ropelewski. Analysis of the convective modified Gate boundary layer using in situ and acoustic sounder data. Monthly Weath. Rev., 107, 985–993, 1979.

Gera B.S. and A. Weill. Doppler analysis of the frontal friction in relation to the frontal slope. J. Climate Appl. Meteorol., 26, 885–891, 1987.

Hamson R.M. Vertical array response to shipping noise: Model/measurement comparisons for range-dependent Mediterranean sites, J. Acoust. Soc. Am., 1, 386–395, 1994.

Heindsman T.E., R.H. Smith, and A.D. Arneson. Effect of rain upon underwater noise levels, J. Acoust. Soc. Am., 27, 378–379, 1965.

Helmholtz H. Théorie physiologique de la musique, translated in French by G. Gueroult, Paris, 1868.

Helmis C.G., D.N. Asimakopoulos, D.G. Deligiorgi and D.P. Lalas. Observation of sea-breeze fronts near the shoreline, Boundary Layer Meteor., 38, 395–410, 1987.

Hooke W.D., F.F. Hall Jr. and E.E. Gossard. Atmospheric waves observed in the planetary boundary layer by acoustic sounding, Boundary Layer Meteor., 2, 371–380, 1973.

Hunt F.V. Origins in acoustics, ASA publication, 186 pp, 1992.

Jacobus P.W. Underwater sound radiation from large rain drops, M.S. thesis, Naval Postgraduate Scool, Monterey, CA 93943, 1991.

Kerman B.R., R.E. Mickle, R.V. Portelli, N.B. Trivet and P.K. Misra. The Nanticoke shoreline diffusion experiment, June 78, (II internal boundary layer structure), Boundary Layer Meteor., 16, 423–437, 1982.

Kerman B.R. Underwater sound generation by breaking wind waves. J. Acoust. Soc. Am. 75, 149–165, 1984.

Kewley D.J., D.G. Browning and W.M. Carey. Low frequency wind generated ambient noise source levels, JASA 88, 1894–1902, 1990.

Knudsen V.O., R.S. Alford and J.W. Emling. Underwater ambient noise, J. Marine Res. 7, 410–429, 1948.

Kobayashi S., N. Ishikawa. Observation of an atmospheric boundary layer at Mizuho station using an acoustic sounder. Proc. Fifth Symp. Polar Research (Tokyo), 37–49, 1983.

Lalas D.P., D.N. Asimakopoulos, D.G. Deligiorgi and C. Helmis. Sea breeze circulation and photochemical pollutions in Athens, Greece, Atmos. Environment, 16, 531–544, 1983.

Little C.G. Acoustic methods for the remote profiling of the lower atmosphere, Proc. IEEE, 57, 571–578, 1969.

Mandics P.A. and E.J. Owens. Observations of the marine atmosphere using a shipmounted echosounder, J. Appl. Meteorol., 14, 1110–1117, 1965.

Mandics P.A. and F.F. Hall, Jr. Preliminary results from the Gate acoustic echo sounder, J. Appl. Meteorol., 14, 110–1117, 1976.

Marshall J.S. and W. Palmer. The distribution of raindrops with size, J. Meteorol. 5, 165–166, 1948.

Masmoudi M. and A. Weill. Atmospheric mesoscale spectra and structure functions of mean horizontal velocity fluctuations with a Doppler sodar network, J. Climate Appl. Meteorol., 27, 864–873, 1988.

McAllister L.G. Acoustic sounding of the lower atmosphere J. Atmos. Terr. Phys., 30, 1439–1440, 1968.

Medwin H., J.A. Nystuen, P.W. Jacobus L. H. Ostwald, and D.E. Snyder. The anatomy of underwater rain noise, J. Acoust. Soc Am., 92 (3), 1613–1623, 1992.

Medwin H. and M.M. Beaky. Bubble Sources of the Knudsea Sea Noise Spectra, J. Acoust. Soc. Am. 86, 1124–1130, 1989.

Menemenlis D. and D.M. Farmer. Acoustical measurement of current and vorticity beneath ice, J. Atmospheric and oceanic technology, 9, 827–849, 1992.

Miller J.H., J.F. Lynch, V Chiu, E.L. Westreich, R. Hippenstiel and E. Chaulk. Acoustic measurements of surface gravity wave spectra in Monetery Bay using mode travel-time fluctuations, J. Acoust. Soc. Am., 94 (2), 974, 1993.

Monahan E.C. and I.G. O'muircheartaigh. Optimal power-law description of oceanic whitecap coverage dependence on wind speed, J. Phys. Oceanogr., 10, 2094–2099, 1980.

Monahan E.C., B. Wilson and D.K. Woolf. Hexmax whitecap climatology: Foam crest coverage in the North Sea, 16 October–23 November 1986, in Proc. Nato Advanced Workshop, Humidity Exchange Over the Sea Main Experiment (HEXMAX), Analysis and Interpretation, Eds. W.A. Oost, S.D. Smith and K.B. Katsaros, Univ. of Washington, Seattle, Techn. report, 105–115, 1988.

Neff W.D. and C.W. King. Observations of complex terrain flows using acoustic sounders/ experiment topography and winds, Boundary Layer Meteor., 28(8), 705–710, 1987.

Neff W.D. and F.F. Hall. Acoustic echoes of the atmospheric boundary layer at south pole, Antart. J. US., 11, 143–144, 1976.

Neff W.D. Quantitative evaluation of acoustic echoes from the planetary boundary layer, Tech. Rep. ERL 322–WPL 38, 34, pp., 1975.

Nystuen J.A., C.C. McGlothin, and M.S. Cook. The underwater sound generated by heavy rainfall, J. Acoust. Soc. Am., 93(6), 3169–3177, 1993.

Nystuen J.A., L.H. Ostwald Jr., and H. Medwin. The hydroacoustics of a raindrop impact, J. Acoust Soc. Am., 92(2), Pt. 1, 1017–1021, 1992.

Ottersten H.M.K., K.R. Hardy, G. Stilke, B. Brümmer and G. Peters. Shipborne sodar measurements during Jonswap 2, J. Geophys. Res., 79, 5573–5584, 1974.

Peters G. Wind profiling by FM-CW-Radar: In Proceeding of the A. Wegener Conference on ground based Remote Sensing Techniques for the Troposphere, Hambourg August 25–28, 217 pp., 1986.

Prakash J.W.J., R.R.N. Nair, K.S. Gupta and P.K.K. Kunhikrishnan, 1992. On the structure of sea breeze fronts observed near the shoreline of Thumba, India, Boundary Layer Meteor., 59, 111–124, 1992.

Prosperetti A., N.Q. Lu, and H.S. Kim. Active and passive acoustic behavior of bubble clouds at the ocean's surface, J. Acoust. Soc. Am., (6), 3117–3127, June 1993.

Radeau R. L'acoustique, Hachette Cie, Bibliothèque des Merveilles, 327 pp., 1875.
Singal S.P. Acoustic sounding stability studies, Encyclopedia of Environment Control Technology, Vol. 2, P.N. Cheremismoff ed., 1003–1061, 1989.
Spiesberger J.L., K. Metzger and J.A. Furgerson. Listening for climatic temperature change in the northeast Pacific 1983–1989, J. Acoust. Soc. Am., 92(1), 384–396, July, 1992.
Taconet O. and A. Weill. Vertical velocity field in the convective boundary layer as observed with an acoustic Doppler Sodar, Boundary Layer Meteor., 23, 133–151, 1982.
Taconet O. and A. Weill. Convective plumes in the atmospheric boundary layer as observed with an acoustic Doppler Sodar, Boundary Layer Meteor., 25 143–158, 1983.
Weill A. Sodar micrometeorology, in proceeding of the first International conference on Acoustic Remote sensing of the Atmosphere and Oceans (T. Mathews ed.), 1–60, Calgary, Alberta, Canada, 1981.
Weill A. and F. Alli Goulam. Momentum fluxes measurement with sodar and minisodar. In Proceeding of the 5th Inernational Symposium on Acoustic Remote Sensing of Atmosphere and Oceans, New Delhi, S.P. Singal (Editor), MacGraw Hill Cie, 161–166, 1990.
Weill A., F. Baudin, C. Mazaudier, G. Desbraux, C. Klapisz, A.G.M. Driedonks, J.P. Goutorbe, A Druilhet, and P. Durand. A mesoscale shear convective cell observed during the COAST experiment: acoustic sounder measurements. Boundary Layer Meteor., 44, 359–371, 1988.
Weill A. and F. Baudin. A shipborne minisodar and Sodar, Proceeding of the AMS conference (January, New-Orleans), 419–420, 1990.
Weill A., F. Baudin, J.P. Goutorbe, P. Van Grunderbeeck and P. Leberre. Turbulence structure in temperature inversions and in convective fields as observed by Doppler Sodar, Boundary Layer Meteor., 15, 375–390, 1978.
Weill A., M. Blez, F. Leca. Observation of gravity waves and horizontal mixing in the atmospheric boundary layer, Annales Geophysicae, 5B, 5, 413–420, 1987.
Weill A., C. Klapisz, F. Baudin, C. Jaupart, P. Van Graunderbeeck and J.P. Goutorbe. Measuring heat flux and structure functions of temperature fluctuations with an acoustic Doppler Sodar, J. Appl. Meteor., 19, 199–205, 1980.
Weill A., C. Klapisz, F. Baudin. The CRPE minisodar: applications in micrometeorology and physics of precipitations, Atmos. Res., 20, 317–335, 1987.
Worcester P.F. Reciprocal acoustic transmission in a middle ocean environment, J. Acoust. Soc. Am., 62, 895–905, 1977.

Acoustic Remote Sensing Applications
S.P. Singal (Ed)
Copyright © 1997 Narosa Publishing House, New Delhi, India

21. Range-Average Inversions in Ocean Acoustic Tomography

R. Michael Jones

Cooperative Institute for Research in Environmental Sciences (CIRES)
University of Colorado/NOAA, Boulder, Colorado 80309-0216, U.S.A.

1. Introduction

Experiments over the last decade verify that ocean acoustic tomography [1, 2, 3, 4, 5, 6, 7, 8, 9] can successfully map subsurface ocean temperatures over regions of 1000 km or so [10, 11, 12]. As is well known, this technique maps the sound speed by measuring the travel times of acoustic pulses traversing the tomographic region along various paths; temperature is then inferred from the known empirical dependence of sound speed on temperature, using independent information for the salinity. Depth dependence of sound speed (and therefore temperature) is determined from the multiple acoustic paths with different turning-point depths that connect each source-receiver pair within a vertical slice through the ocean. Including further independent information (such as using climatology to give the deep ocean temperature or expressing the sound speed in terms of ocean dynamic modes) allows the determination of the sound speed (and the temperature). The usual inversion method, the stochastic inverse [11], is based on linear optimal estimation theory, a generalization of objective mapping. Adding reciprocal transmissions allows the determination of currents in addition to temperature [12]. The vertical resolution of the inversion is determined mostly by the vertical spacing of turning points; the horizontal resolution is determined mostly by the spacing of vertical slices.

For long-term monitoring of subsurface ocean temperature, especially of regions the size of ocean basins (e.g., Munk and Wunsch [13]), however, dense tomographic networks might be too expensive to deploy. Long-term basin-scale acoustic monitoring, therefore, may have to be limited to a few long acoustic paths such as the 3000 km and 4000 km paths that have been used in the Pacific [14, 15, 16, 17] and the very long paths used in the Heard Island Feasibility Test [18, 19, 20]. However, the actual lengths of paths that may be used in a program to monitor subsurface ocean temperature may depend on how difficult it is to interpret the signals for long paths.

Because most information about horizontal dependence of sound speed comes from the dependence of travel time on the horizontal position of vertical slices, that information would not be available in a configuration that has only one, or a small number, of vertical slices. In that case, information about the horizontal variation in sound speed would be limited mostly to what is available from individual vertical slices, thus putting a premium on obtaining the maximum amount of information form vertical-slice inversion.

But even with an optimum vertical-slice acoustic measurement configuration, single-slice inversion is underdetermined (as is well known) because acoustic travel time in the ocean sound channel is sensitive to only range averages [3, 4, 9, 11, 21] of some aspects of the sound speed within the vertical slice, and is insensitive to other aspects. Although including independent (nontomographic) information about the vertical asymmetry of the sound channel and range dependence assist a vertical-slice inversion [22], this is not sufficient to uniquely determine the sound speed in the vertical slice. Except for providing a range average along each vertical slice, the travel-time measurements provide little information about horizontal scales of sound-speed structure larger than a double-loop ray length (about 50 km) [23]. Although climatology could provide estimates of the magnitude of energy associated with these undetermined scale sizes, the phase in that part of the spectrum would remain unknown.

That acoustic travel time is most sensitive to a range average of the sound channel suggests using an inversion method formulated in terms of range averages, at least for situations where measurements are available for only one or only a small number of vertical slices. In developing such a method, we might follow the lead of Wunsch [24] who argues that except for the complications of the partial loops at the ends of the raypaths, the inverse problem of singal-slice tomography is best formulated directly in terms of the action J because J can be computed directly from the measurements. An inversion method formulated in terms of J suggests using Abel transforms as part of the inversion.

The Abel transform method for vertical-slice inversion developed by Munk and Wunsch [25] is an appropriate starting point for such an inversion method. The similarity of sound-channel propagation in the ocean to seismology [26] allows the application of seismological methods such as the Abel transform to vertical-slice inversion in the ocean.

Jones et al, [27] extended the method of Munk and Wunsch [25] to apply to an asymmetric sound channel. Munk and Wunsch [28] extended inversions based on Abel transforms to include adiabatic range dependence, but as a perturbative method using a three-parameter model for the sound-speed profile. Jones et al, [29, 30, 31] further developed Abel transform inversions for adiabatic range dependence as a nonperturbative method for arbitrary displacement of the sources and receivers above or below the sound channel axis.

Ocean acoustic tomography that is based on acoustic pulse travel time along raypaths is called ray tomograpy. Similarly, tomography that is based on pulse travel time for modes ducted in the sound channel is called modal tomography. Sometimes modal tomography is more applicable than ray tomography. For example, near-axial ray pulse arrivals are often unresolved, whereas the corresponding low-order mode pulse arrivals can be resolved if the frequency is low enough (Munk and Wunsch [25], Fig. 6, ambiguity diagram). Shang [32] proposed a tomography inversion method based on measured travel times of sound channel modes instead of travel times of rays. Because that method is also perturbative, the recovered profiles will depend on the choice of the background profile when nonlinear effects are important. Munk and Wunsch [25] outlined a method for nonperturbative modal tomography based on Abel transforms. Jones et al, [33] and Shang et al, [34] extended their method to a range-dependent environment.

Nonperturbative modal tomography is similar to nonperturbative ray tomography, but differs significantly in two ways:

- Modal tomography gives no information about the antisymmetric part of the sound channel (the vertical displacement of contour pairs), whereas ray tomography gives some information about the vertical displacement of the sound channel, but only on the periphery of the tomographic region (where the sources and receivers are).
- In ray tomography, the horizontal phase velocity is determined by a differentiation of travel-time measurements, whereas in modal tomography, it is determined by an integration.

We notice that it is more accurate to think of ocean acoustic tomography as a combination of tomography and seismology. That is, although horizontally it is tomography in the usual sense because the measurements represent integrations through the medium in various slices, vertically it is more like seismology because various rays or modes penetrate the sound channel to different distances from the sound-channel axis.

The nonperturbative inversion methods specifically take into account this view of ocean acoustic tomograpy (as tomography combined with seismology) by using tomographic inversion techniques in the horizontal and Abel transforms (a seismic technique) in the vertical.

The nonperturbative ray and modal tomographic inversion methods illustrated here satisfy the requirement that vertical-slice inversion be separated from horizontal-slice inversion and formulated in terms of range averages. Section 2 shows how range averaging enters naturally into ocean acoustic pulse travel-time measurements, for both rays and modes. Section 3 points out that pulse travel-time measurements are only weakly dependent on the asymmetry of the sound channel. Section 4 discusses the use of independent (nontomographic) information to augment inversions. Section 5 demonstrates range-average inversions for both ray and modal simulated measurements. Section 6 demonstrates a range-average inversion for measurements from the Slice 89 experiment. Section 7 discusses the role of range-average inversions in horizontal-slice tomography and discusses inversion difficulties caused by nonmonotonic profiles, breakdown of the adiabatic-invariant approximation, aliasing from large wavenumber sound-speed structure, and curve-fitting errors.

2. Range Averaging in Ray and Modal Tomography

Both ray and modal propagation depend significantly on range averaging through the modal phase travel time,

$$t_\phi = \int_{\text{source}}^{\text{receiver}} \hat{S}(x)\, dx$$

$$= x\bar{\hat{S}}$$

$$= \phi/\omega \tag{2.1}$$

In (2.1),

$$\phi = \int_{\text{source}}^{\text{receiver}} k(x)\, dx \tag{2.2}$$

is the unambiguous phase (including the number of cycles between the source and receiver), ω is the frequency,

$$\bar{\hat{S}} \equiv \frac{1}{x} \int_{\text{source}}^{\text{receiver}} \hat{S}(x)\, dx \tag{2.3}$$

is the range average of the horizontal component of the phase slowness,

$$k = \omega\hat{S} \tag{2.4}$$

is the horizontal wave number, $\hat{S}$ is the horizontal component of sound slowness (equal to the sound slowness at a turning point), and sound slowness is the reciprocal of sound speed.

Because sound propagation averages slowness rather than speed in (2.3), inversion naturally furnishes range averages of sound-slowness profiles rather than sound-speed profiles. Therefore, "range-averaged profiles" in this chapter will imply "range-averaged sound-slowness profiles" unless specifically stated oherwise.

Range-average inversions in both ray and modal tomography depend on being able to determine for each ray or mode the modal phase travel time t_ϕ in (2.1) or, equivalently, $\bar{\hat{S}}$, the range average of the horizontal component of sound slowness in (2.3). The methods for determining t_ϕ or $\bar{\hat{S}}$ from tomographic measurements differ for rays and modes.

In addition, it is necessary to determine for each ray or mode the action,

$$J = \oint S_z\, dz = \oint [S^2(z) - \hat{S}^2]^{1/2}\, dz \tag{2.5}$$

which is an adiabatic invariant for each ray or mode. In (2.5), S is the local sound slowness, the reciprocal of the sound speed. Again, the methods for determining J from the tomographic measurements differ for rays and modes.

The determination of t_ϕ (and therefore $\bar{\hat{S}}$) and J for each ray or mode allows the determination by interpolation of the function $J(\bar{\hat{S}})$. This allows the evaluation by Abel transform (Appendix A) of the range-averaged equivalent symmetric sound channel.

Combining the range-averaged equivalent symmetric sound channel with tomographic measurements of sound channel asymmetry (Section 3) and independent (nontomographic) information (Section 4) allows an estimate of the range-averaged sound channel plus some range-dependent information.

2.1 Ray Tomography

As is well known, ray theory provides accurate estimates of pulse travel time even when amplitude estimates are inaccurate, such as near caustics. Within the ray theory approximation, the travel time along a ray can be represented as an integral of the sound slowness S (the reciprocal of the sound speed C) along a ray that connects the source to the receiver. (Because travel time is proportional to sound slowness but inversely proportional to sound speed, using slowness instead of speed removes an unnecessary nonlinearity in ocean acoustic tomography inversion.) The pulse travel time along a ray is given by

$$
\begin{aligned}
t &= \int_{\text{source}}^{\text{receiver}} \frac{dl}{C(x,z)} = \int_{\text{source}}^{\text{receiver}} S(x,z)\, dl \\
&= \int_{\text{source}}^{\text{receiver}} \hat{S}(x,z)\, dx + \int_{\text{source}}^{\text{receiver}} S_z(x,z)\, dz \\
&= x\bar{\hat{S}} + \int_{\text{source}}^{\text{receiver}} S_z(x,z)\, dz \\
&= t_\phi + \int_{\text{source}}^{\text{receiver}} S_z(x,z)\, dz
\end{aligned}
\qquad (2.6)
$$

where x is the horizontal coordinate, z is the vertical coordinate,

$$\hat{S} = S \cos\theta \qquad (2.7)$$

is the horizontal component of sound slowness (equal to the sound slowness at a ray turning point),

$$S_z = S \sin\theta \qquad (2.8)$$

is the vertical component of sound slowness, θ is the angle between the ray and local horizontal, and $\bar{S}$ is the range average of the horizontal component of sound slowness defined in (2.3). In (2.3), x outside the integral is the total horizontal range between the source and the receiver, usually determined by tracking both source and receiver mooring motion using a pulsed acoustic navigation technique [35].

The second term on the right-hand side of (2.6) is the ray action contribution; it depends on which of the four types of rays we are considering. There are four types of rays because the ray can be going up or down at the source, and it can be going up or down at the receiver. For a ray going up at both the source and receiver, for example, (2.6) becomes

$$t_N^U = t_{2n}^U = x\bar{S} + \sum_{i=1}^{n} J_i + J_{TU} - J_{RU} \qquad (2.9)$$

where $N (= 2n)$ is the number of turning points in the ray (an even number for this case),

$$J_{TU} = \int_{\text{source}}^{\substack{\text{first upper} \\ \text{turning point}}} S_z(x, z)\, dz \qquad (2.10)$$

$$J_{RU} = \int_{\text{receiver}}^{\substack{\text{first upper turning point} \\ \text{past the receiver}}} S_z(x, z)\, dz \qquad (2.11)$$

and

$$J_i = \int_{i\text{th upper turning point}}^{\substack{i + 1\text{th} \\ \text{upper turning point}}} S_z(x, z)\, dz \qquad (2.12)$$

Equation (2.9) gives the measured pulse travel time in terms of properties of the sound channel. Typically, for each source-receiver pair, there will be several pulse arrivals, one for each eigenray that connects the source with the receiver. Each eigenray is identified by the number of turning points in the ray and whether the ray was launched upward or downward at the source. Experience at ranges of about 1000 km in the 1987 Reciprocal Acoustic Transmission Experiment north of Hawaii and in the Slice 89 experiment shows that historical data, such as in Levitus [36], are adequate to give unambiguous ray identifiers when the range between the source and receiver is known [11, 37, 38, 39, 40, 41, 42].

Only the equivalent symmetric sound-speed profile is determined by tomographic travel-time measurements (except on the periphery of the tomographic region, where the sources and receivers are located). Section 3 considers ways to determine the asymmetry of the sound channel.

A major step in nonperturbative vertical-slice tomography is to determine the separate terms in (2.9) for each eigenray connecting each source-receiver pair for which there is a measured pulse arrival time. To do that, we first consider the range of values for the various terms in (2.9).

Typically for the deep ocean, measured pulse travel times for a 1000 km path are about 670 s and have a spread of about 3 s for the various eigen rays (axial rays arrive last). Both the total and the spread are proportional to the source-receiver range. Thus, the measured travel time on the left side of (2.9) is proportional to source-receiver range.

Typically, the sound speed in the deep ocean is about 1500 m s^{-1} and varies about 3% from its minimum at the sound-channel axis (about 1 km deep) to the largest values near the surface (because the water is warmer) and near the bottom (because the water is under high pressure). Thus, the first term on the right side of (2.9) can vary from about 670 s for axial rays to about 650 s for rays that have turning points near the surface for a 1000 km path. Both the minimum and maximum travel times are proportional to range.

Each term in the sum in the second term on the right side of (2.9) varies from zero for axial rays to about 1 s for rays whose turning points are near the surface. The number of terms in the sum varies from about 20 for axial rays to about 15 for rays with turning points near the surface for a 1000 km range. The number of terms in the sum is proportional to range. Thus, the second term on the right side in (2.9) varies from about zero for axial rays to about 15 s for rays with turning points near the surface for a 1000 km path, and this term is also proportional to range.

The final two terms (from the partial loops at the source and receiver) each vary from near zero to a maximum of about 0.25 s, depending on the source or receiver depth relative to the turning-point depths. These terms are not proportional to range.

Before considering how the individual terms are estimated from the travel-time measurements, we simplify the notation in (2.9) by defining an average action

$$\bar{J} \equiv \frac{1}{n} \sum_{i=1}^{n} J_i \qquad (2.13)$$

and rewriting (2.9) as

$$t_N^U = t_{2n}^U = x\bar{\hat{S}} + n\bar{J} + J_{TU} - J_{RU} \qquad (2.14)$$

The superscript U in (2.9) and (2.14) signifies rays that are going up at the source. The subscript is the number of turning points in the ray, which is even for a ray that is going up at both the source and the receiver. This notation follows that of Munk and Wunsch [25].

It is usual to make the adiabatic-invariant approximation (e.g., Wunsch [24]) that the action integral J remains constant along the ray because it is an adiabatic invariant. In the notation of (2.12) and (2.13), the adiabatic invariant approximation is expressed as

$$\bar{J} = J_i \equiv J \text{ for } i = 1 \text{ to } n \tag{2.15}$$

Under the adiabatic-invariant approximation, J is the action for each double loop in the raypath because all double loops have the same action, even for a range-dependent sound channel. The action J is determined for each path from the tomographic measurements by [31, 43]

$$J(N) = 2t'(N) \tag{2.16}$$

where N is the number of turning points in the path, the derivative is estimated by curve fitting to measurements at discrete values for N, and only paths of the same type (e.g., upward propagation at source and receiver) are used for each of the four $t(N)$ functions.

2.2 Modal Tomography

Unfortunately, modal phase measurements do not furnish the integrals in (2.1) or (2.2) because of 360° phase ambiguities. Instead, the phase measurements furnish a discrete set of possibilities for the integral in (2.1) spaced by the wave period. Although knowledge of the ocean can limit the possibilities for the integral in (2.1) to a specific range of values, it cannot usually determine a unique value for that integral.

Fortunately, modal tomography measurements do determine the pulse travel time

$$\begin{aligned} t(m) &= \int_{\text{source}}^{\text{receiver}} S_g(m, x)\, dx \equiv x\bar{S}_g(m) \\ &= \int_{\text{source}}^{\text{receiver}} \hat{S}(m, x)\, dx + J(m) \int_{\text{source}}^{\text{receiver}} \frac{dx}{X(m, x)} \\ &= x\bar{\hat{S}}(m) + x\overline{X^{-1}}(m) J(m) \\ &= x\bar{\hat{S}}(m) + n(m) J(m) \\ &= t_\phi(m) + n(m) J(m) \\ &= \frac{\phi(m)}{\omega} + n(m) J(m) \end{aligned} \tag{2.17}$$

Thus,

$$\bar{\hat{S}}(m) = \frac{t_\phi(m)}{x} = \frac{t(m) - n(m) J(m)}{x} = \frac{t(m) - \frac{2\pi}{\omega}\left(m + \frac{1}{2}\right) n(m)}{x} \tag{2.18}$$

where $n(m)$, defined in (2.17), is the number of double loops (not an integer) for the ray that has the same turning-point depths as the mode m, and the double-loop range X is also called the modal interference distance. For modes, the action $J(m)$ is quantized as

$$J(m) = (2m + 1)\frac{\pi}{\omega} \qquad (2.19)$$

Equation (2.19) is a valid approximation except when a turning point is too close to the surface or to the bottom. The function $n(m)$ is determined by integrating a formula related to (2.16) [33, 34]. Specifically, from the definition for n, (A2), (2.17), and the derivative of (2.19), we have

$$n(m) \equiv \int_{\text{source}}^{\text{receiver}} X^{-1}(m, x)\, dx = -\int_{\text{source}}^{\text{receiver}} \frac{d\hat{S}(m, x)}{dJ(m)}\, dx$$

$$= -\frac{d}{dJ(m)} \int_{\text{source}}^{\text{receiver}} \hat{S}(m, x)\, dx$$

$$= -x\frac{d\tilde{S}(m)}{dJ(m)} = -\frac{1}{\omega}\frac{d\phi(m)}{dJ(m)} = -\frac{1}{2\pi}\frac{d\phi(m)}{dm} \qquad (2.20)$$

Taking the derivative of (2.17) gives

$$t'(m) = \frac{\phi'(m)}{\omega} + n'(m)\, J(m) + n(m)\, J'(m)$$

$$= -\frac{2\pi}{\omega} n(m) + n'(m)\, J(m) + n(m)\frac{2\pi}{\omega} \qquad (2.21)$$

$$= n'(m)\, J(m)$$

where we have used (2.20) and the derivative of (2.19). Equation (2.21) can be integrated to give

$$n(m_1) = n(m_0) + \int_{m_0}^{m_1} \frac{dt/dm}{J(m)}\, dm$$

$$= n(m_0) + A(m_1) - A(m_0) \qquad (2.22)$$

where

$$n(m_0) = [\omega t(m_0) - \tilde{\phi}(m_0)]\Big/\left[2\pi\left(m_0 + \frac{1}{2}\right)\right] - M(m_0)\Big/\left(m_0 + \frac{1}{2}\right) \qquad (2.23)$$

$t(m_0)$ is the measured pulse travel time for mode m_0, $\tilde{\phi}(m_0)$ is the measured phase (between 0 and 2π) for mode m_0, and $M(m_0)$ is an integer.

Equation (2.23) can be rewritten

$$\left(m_0 + \frac{1}{2}\right) n(m_0) = \frac{\omega t(m_0)}{2\pi} - \frac{\tilde{\phi}(m_0)}{2\pi} - M(m_0)$$

$$= N(m_0) + \left[\frac{\omega t(m_0)}{2\pi} - \frac{\tilde{\phi}(m_0)}{2\pi}\right]_{\mathrm{mod}\ 1} \quad (2.24)$$

where $N(m_0)$ is an integer to be determined by independent information. For m_0 equal to zero,

$$t_\phi(m_1) = t(m_1) - (2m_1 + 1)\, [t(0) - \tilde{\phi}(0)/\omega]$$

$$+ 2M(0)\, J(m_1) - (2m_1 + 1) \int_0^{m_1} \frac{dt/dm}{2m+1}\, dm \quad (2.25)$$

Equation (2.25) corrects an error in (30) of Jones et al, [33]. The integration in (2.20) and (2.25) can be carried out numerically. Notice that it is necessary to measure the phase at only one mode [mode zero in the example in (2.25)].

3. Sensitivity to Channel Asymmetry

Ocean acoustic tomographic measurements do not determine the asymmetry of the sound channel within the tomographic region because acoustic travel time is insensitive to it. Only at the acoustic sources and receivers is it possible to detect a significant effect of the sound-channel asymmetry on measured pulse travel times. Thus, it is possible to estimate the asymmetry of the sound channel on the periphery of the tomographic region (where the sources and receivers are), but not in the interior.

The asymmetry of the sound-speed profile is given by an Abel transform [25]:

$$\frac{z_+(S) + z_-(S)}{2} = z_A - \frac{1}{2\pi} \int_{S_A}^{S} \frac{X_+(\hat{S}) - X_-(\hat{S})}{(\hat{S}^2 - S^2)^{1/2}}\, d\hat{S}$$

$$= z_A + \frac{1}{2\pi} \int_{S_A}^{S} \frac{J'_+(\hat{S}) - J'_-(\hat{S})}{(\hat{S}^2 - S^2)^{1/2}}\, d\hat{S} \quad (3.1)$$

where z_A is the height (the negative of the depth) of the sound-channel axis, X_+ is the range of a single upper loop, X_- is the range of a single lower loop, J_+ and J_- are the corresponding single-loop actions, and S_A is the sound slowness on the sound-channel axis. Jones et al, [31] and Jones and Georges [43] give the detailed formulas needed to calculate the asymmetry at the source and receiver from travel-time measurements.

4. Use of Independent Information

A priori knowledge or independent information is used to correctly identify pulse arrivals in both perturbative and nonperturbative inversions. In that respect, the two methods do not differ.

Independent information can be used to enhance the inversion for nonperturbative as well as for perturbative methods. In nonperturbative inversions, the independent information is used at the end instead of at the beginning of the inversion, and this can be done in several ways:

- The simplest use of independent information for nonperturbative inversions is to estimate the profile below the sound-channel axis using climatological data [44], which is illustrated in Sections 5 and 6. This use of independent information plays the same role as the choice of a background profile in objective mapping inversion, except that it is used only below the sound-channel axis, where the profile is more predictable.
- Another use of independent information for nonperturbative inversions is similar to its use in objective mapping inversion. That is, the final range-averaged profile can be expressed as a background profile plus a sum of empirical orthogonal functions (EOFs) or a sum of ocean-dynamic modes. Truncating the series allows one to eliminate noise. This can be done in two ways. In the first method, one first constrains the profile below the sound-channel axis and then fits the upper profile to a background plus a sum of EOFs. In the second method, one first constructs a background equivalent symmetric profile and a set of basis profiles. If profile measurements are used in the construction, then one first calculates the equivalent symmetric profile for each measured profile and then calculates the ensemble average and the EOFs in the usual way. If ocean-dynamic modes are used for the basis functions, then only the symmetric (about the sound-channel axis) parts are used for the basis functions. In either case, one fits the equivalent-symmetric profile to the background symmetric profile plus a sum of symmetric basis profiles. This determines the range averages of the coefficients for the symmetric basis profiles. Because the background symmetric profile corresponds to an asymmetric background profile, and each symmetric basis profile corresponds to an asymmetric basis profile, we can construct the corresponding range-averaged profile. Although all these calculations can be done in terms of sound speed or sound slowness, using sound slowness avoids an unnecessary nonlinearity.

In either the first or the second method of fitting the profile found by nonperturbative inversion to a background plus a sum of basis profiles, one is able to determine a nonmonotonic profile when appropriate, even though the measurements are insensitive to the nonmonotonicity just as in perturbative inversion.

5. Example Inversions from Measurement Simulations

Here, we present examples of range-average inversions using simulated measurements. We present one example of an inversion for simulated ray pulse travel-time measurements, and two examples of inversions for simulated modal travel-time and phase measurements.

5.1 Ray Tomography

We simulated the pulse-arrival sequence shown in Fig. 1 using the HARPO ray tracing program [45] and its associated eigenray program [46]. Jones et al, [31] show the raypaths for the eigenrays. Although the sound channel used to simulate the pulse-arrival sequence in Fig. 1 was range independent, we make no use of that knowledge in performing the inversion. In fact, we determine independently from the inversion that the simulated measurements are consistent with a horizontally uniform sound channel. When we apply nonperturbative tomography to range-dependent cases, we use independent information to provide the sound speed below the sound-channel axis.

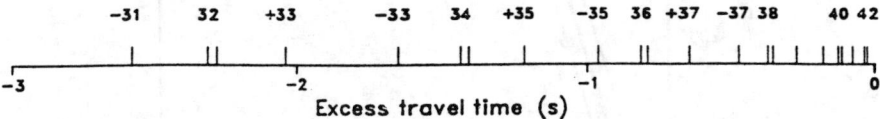

Fig. 1. Pulse-arrival sequence simulated by ray tracing [45, 46] for a horizontally uniform sound channel having the same canonical sound-speed profile as used in Munk and Wunsch [28] and Munk [48]. The travel times are relative to that for an axial ray from the source to the receiver. The numbers above the pulse arrivals indicate the number of ray loops or turning points (upper plus lower). Those with plus signs have one more upper loop than lower. Those with minus signs have one more lower loop than upper. For the pairs of pulse arrivals having an even number of turning points, the downwardly transmitted ray arrives earlier. The depths of the source, receiver, and sound-channel axis are 1200 m, 1100 m, and 1300 m, respectively.

The first step in the inversion is to display pulse travel time from Fig. 1 as a function of the number of ray turning points for each of the four kinds of rays as shown in Fig. 2. At this point, we have a choice between two inversion methods described by Jones et al, [31] and Jones and Georges [43]. The first, described in the previous sections, is advantageous for short paths. The second method, which we illustrate here, is advantageous for longer paths where the rays contain many turning points.

Instead of calculating $\bar{\tilde{S}}(J)$ for each of the four kinds of rays separately, as in the first method, we combine the measurements for them to estimate $\bar{\tilde{S}}(J)$ for a fictitious family of rays that have no partial loops at the ends, corresponding to a source and receiver on the sound-channel axis in the equivalent-symmetric sound channel.

We make the calculation by expanding in a Taylor series (through second order) the travel-time functions for the measured raypaths in terms of the fictitious raypaths. Taking appropriate combinations of the measurements then gives formulas

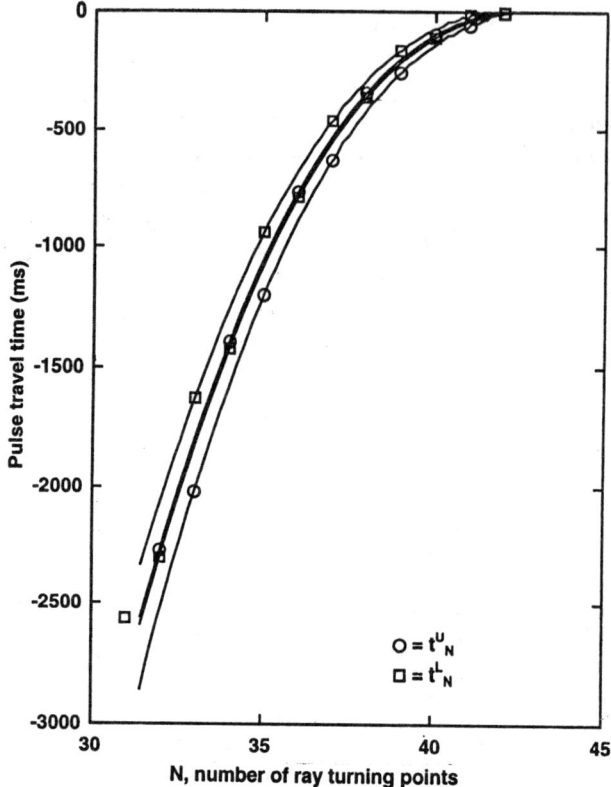

Fig. 2. Simulated measured travel times t_N^U (circles) and t_N^L (squares) from Fig. 1. The solid lines show a comparison calculated from single-loop raypath calculations. The measurements determine four functions of N because the functions of N are different for even and odd N, and are different depending on whether propagation is upward or downward at the source. There is a general increase of travel time with the number of turning points, a general characteristic of the ocean sound channel where axial rays arrive last and have the shortest ray loops [25]. The difference in travel time for upward and downward propagation for the same number of turning points is due to the asymmetry of the sound channel combined with the displacement of the source and receiver from the sound channel axis.

that can be used to estimate the individual terms in the traveltime formulas, as before. We present here the resulting formulas. The derivation is given by Jones et al, [31].

First, we write the formula for the travel time of the fictitious rays that have no partial loops at the ends. This is

$$t(N) = (N/2)\, J(N) + x\bar{\bar{S}}(N) \tag{5.1}$$

where N is the number of turning points in the ray. The function $t(N)$ is different for odd and even N.

Jones et al, [31] and Jones and Georges [43] give the formulas to calculate the function t as

$$t(2n) = \frac{t_{2n}^U + t_{2n}^L}{2} - \frac{1}{2}\frac{t''(2n)\,g'(2n)^2}{t''(2n)^2 - g''(2n)^2} \tag{5.2a}$$

$$g(2n) = \frac{t_{2n}^U - t_{2n}^L}{2} - \frac{1}{2}\frac{g''(2n)\,g'(2n)^2}{t''(2n)^2 - g''(2n)^2} \tag{5.2b}$$

$$t(2n+1) = \frac{t_{2n+1}^U + t_{2n+1}^L}{2} - \frac{1}{2}\frac{t''(2n+1)\,f'(2n+1)^2}{t''(2n+1)^2 - f''(2n+1)^2} \tag{5.2c}$$

$$f(2n+1) = \frac{t_{2n+1}^U - t_{2n+1}^L}{2} - \frac{1}{2}\frac{f''(2n+1)\,f'(2n+1)^2}{t''(2n+1)^2 - f''(2n+1)^2} \tag{5.2d}$$

where the functions f and g are defined by

$$f(N) \equiv [J_{TU}(N) - J_{TL}(N) + J_{RU}(N) + J_{RL}(N)]/2 \tag{5.3}$$

and

$$g(N) \equiv [J_{TU}(N) - J_{TL}(N) - J_{RU}(N) + J_{RL}(N)]/2 \tag{5.4}$$

Equation (5.2) uses particular combinations of the measured pulse arrivals from Fig. 2. These are shown in Fig. 3. Equation (5.2) is calculated from the measurements by iteration, which we begin by neglecting the last term in each equation in (5.2). The first and second derivatives were determined numerically by fitting a smooth curve to the discrete samples provided by the simulated measurements. Jones et al, [31] give details of the procedure. Results of the iteration are shown in Figs. 4–6 for t, f, and g. The agreement with direct ray tracing calculations is good.

We can estimate the action J for each of the fictitious raypaths. We have

$$J(N) = 2t'(N) \tag{5.5}$$

Also, we can estimate the range average of $\hat{S}$ for the fictitious raypaths. From (5.1) and (5.5), we have

$$\bar{\hat{S}}(N) = [t(N) - (N/2)\,J(N)]/x$$
$$= [t(N) - Nt'(N)]/x \tag{5.6}$$

Fig. 7 shows the range average of $\hat{S}$ versus J from (5.5) and (5.6). The range average of $\hat{S}$ on the sound channel has been subtracted from the ordinate. The amount to subtract is determined by the requirement that an extrapolation of the curve in Fig. 7 passes through the origin.

We also have the double-loop range for the range-averaged sound channel. This is

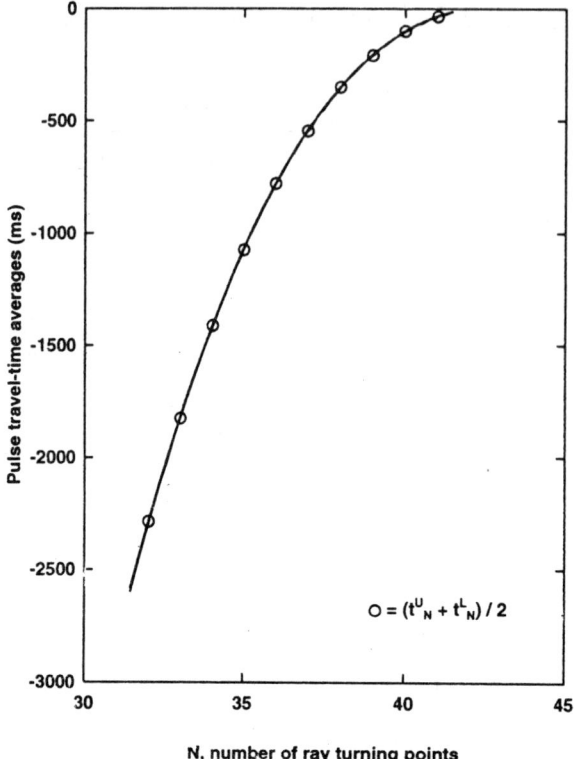

Fig. 3a. Averages of the travel times for upwardly and downwardly transmitted rays (circles) from Fig. 2. The solid lines show a comparison calculated from single-loop raypath calculations. Although two independent functions of the measurements are shown here for even and odd loop number N, it is difficult to distinguish them. Except for second-order effects (included in the inversion), these data determine the range average of the equivalent symmetric sound-slowness profile within the vertical slice.

$$X = -J'(\bar{\hat{S}}) = -J'(N)/\{d/dN \, [\bar{\hat{S}}(N)]\} = 2x/N \quad (5.7)$$

Fig. 8 shows X from (5.7) versus the range average of $\hat{S}$.

The range average of the equivalent symmetric sound-speed profile is found by applying the Abel transform (A1) to the results in Fig. 8. Because we have so far neglected the curvature of the earth, applying the Abel transform to the results in Fig. 8 determines a sound-speed profile that would give the simulated pulse-arrival sequence in Fig. 1 if the earth were flat. To convert that profile to one that could give the simulated pulse-arrival sequence in the presence of earth curvature, we use the transformation from Appendix B. The resulting profile is shown in Fig. 9. the agreement with the equivalent symmetric profile for the profile used to simulate the measurements in Fig. 1 is good.

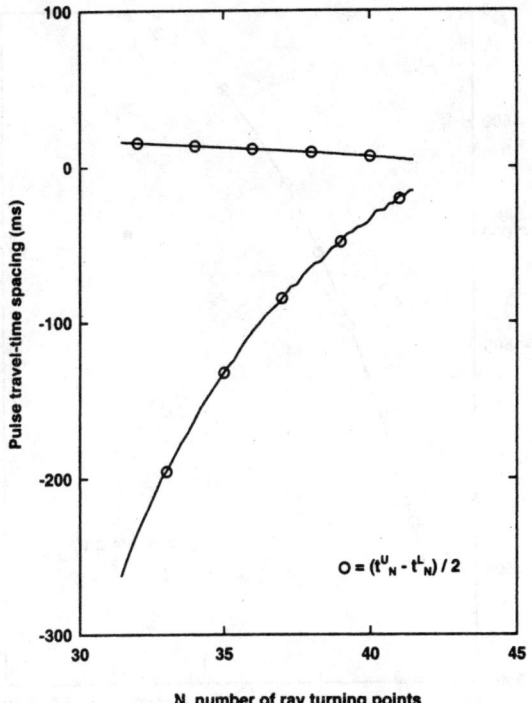

Fig. 3b. Differences of simulated measured travel time for upwardly and downwardly transmitted rays (circles) from Fig. 2. The solid lines show a comparison calculated from single-loop raypath calculations. The functional dependence on loop number N is clearly different for even and odd N. The four functions of loop number N determined by the measurements and depicted in Figs. 3a and 3b give a complete representation of the tomography travel-time measurements. Except for second-order effects (included in the inversion), these data, when combined with knowledge of the equivalent symmetric profiles at the source and receiver, determine the asymmetry of the sound channel at the source and receiver.

Determining the asymmetry of the sound-speed profile at the source and receiver is more complicated. Jones et al, [31] and Jones and Georges [43] give the details.

Fig. 10 gives the results for the profile at the source. The agreement with the profile used to generate the simulated pulse-arrival sequence in Fig. 1 is good. The profile at the receiver found by inversion also agrees with the profile used to simulate the pulse-arrival sequence in Fig. 1, and is indistinguishable from the profile at the source [31]. That the profiles at the source and receiver found from the inversion are the same within the inversion error implies that the pulse-arrival sequence in Fig. 1 is consistent with a horizontally uniform sound channel. This information is given by the inversion, which can always tell if the pulse-arrival sequence is consistent with a horizontally uniform sound channel. However, there are infinitely many other (range-dependent) sound channels consistent with the

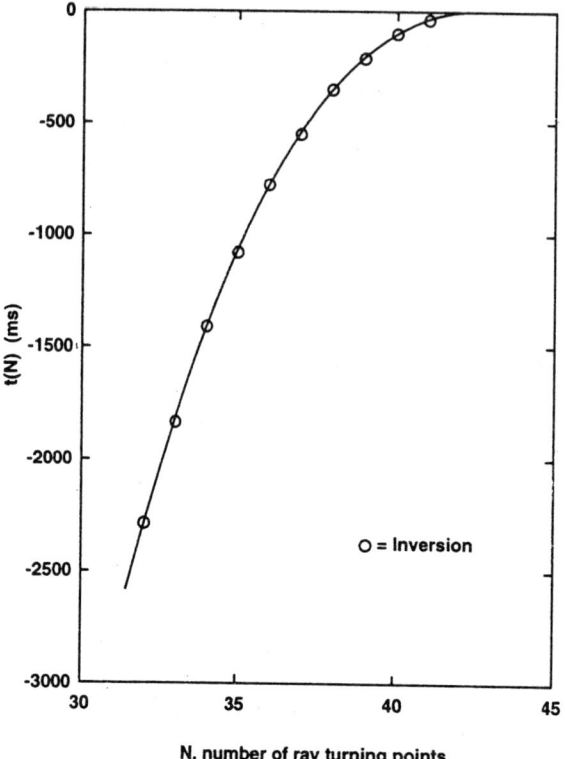

Fig. 4. Travel time, $t(N)$, as determined by iterating (5.2) for the measurements in Fig. 3 (circles). $t(N)$ is the travel time that would have been observed (relative to that for an axial ray) for an integral number of ray loops for a horizontally uniform symmetric sound channel that is the range average of the equivalent symmetric sound channel for which the measuremens were simulated. The solid line shows a comparison calculated from single-loop raypath calculations.

pulse-arrival sequence because only the range average of the equivalent symmetric profile is determined by the measurements, and because the pulse-arrival sequence gives no information about the asymmetry of the sound channel except at the source and receiver.

5.2 Modal Tomography

Shang et al, [34] simulated modal tomography measurements for an acoustic frequency of 30 Hz using KRAKEN [47] with a canonical sound-speed profile [48]. Table 1 shows the resulting pulse travel time and phase for a range of 1000 km and phase for a range of 1010 km as a function of mode number. Fig. 11 shows the modal dependence of pulse travel time.

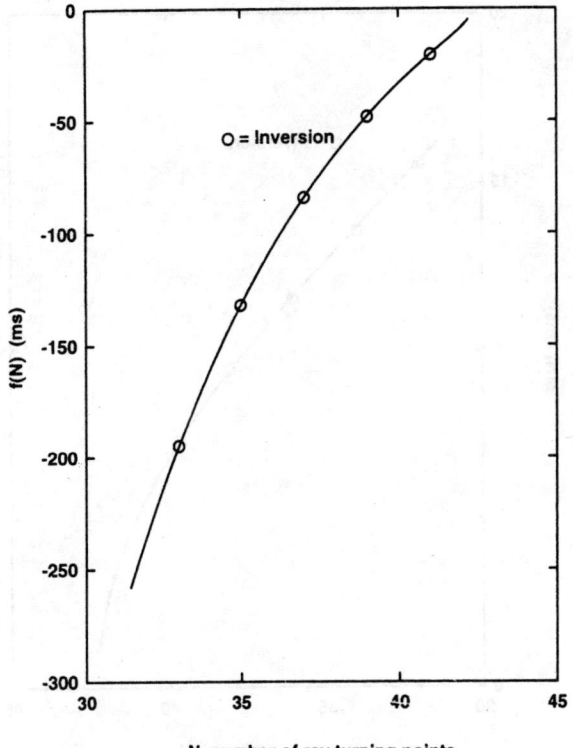

Fig. 5. $f(N)$ [defined in (5.3)] as determined by iterating (5.2) for the measurements in Fig. 3 (circles). Notice that the inversion evaluation is for odd values of the loop number N only; however, interpolation is used to evaluate f and its derivatives for other values. The solid line shows a comparison calculated from raypath calculations.

Table 2. shows the first part of the inversion. The second column gives $A(m)$ calculated from the integral in (2.22). Fig. 12 shows the functional dependence of $A(m)$.

The first guess for $n(0)$ and $n(1)$ came from an assumption that we knew something about the climatology to give a rough guess for the total phase (including the phase ambiguity) for modes 0 and 1. Taking the difference in phase between the two modes gives an estimate of 23.6 for $n(0.5)$ using (2.20). That is,

$$n\left(m + \frac{1}{2}\right) = -\frac{1}{2\pi}\frac{\partial \phi}{\partial m}\left(m + \frac{1}{2}\right) \approx \frac{1}{2\pi}[\phi(m) - \phi(m+1)] \qquad (5.8)$$

Assuming that to be equal to the average of $n(1)$ and $n(0)$ and knowing that $A(1)$ gives their difference leads to the first estimate of $n(0)$ and $n(1)$ in column 3. The rest of the values of $n(m)$ in column 3 comes from using $A(m)$ in (2.22).

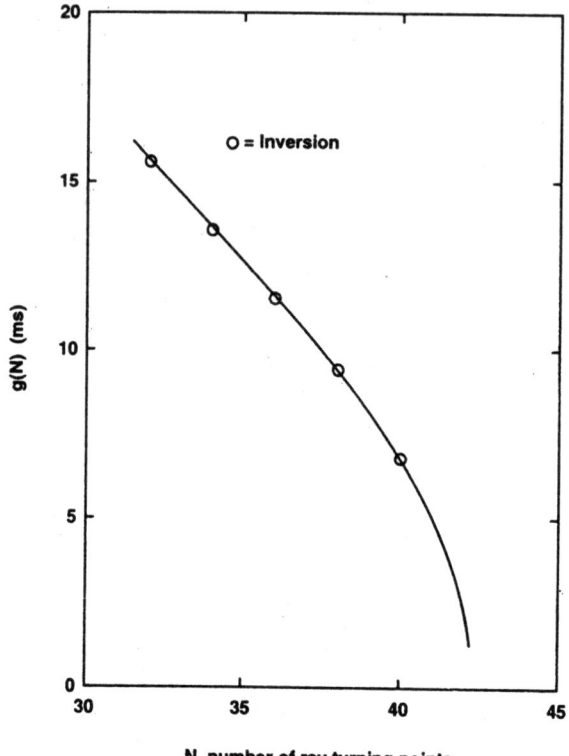

Fig. 6. $g(N)$ [defined in (5.4)] as determined by iterating (5.2) for the measurements in Fig. 3 (circles). Notice that the inversion evaluation is for even values of the loop number N only; however, interpolation is used to evaluate g and its derivatives for other values. The solid line shows a comparison calculated from raypath calculations.

An alternative method for estimating $n(0.5)$ uses the dual-range method, which assumes we have measured phase for two modes at two close ranges. We start with the approximation (5.8), and make a further approximation,

$$\phi(m, x) = x\bar{k}(m, x) \approx xk(m, x) = x\frac{\partial \phi(m, x)}{\partial x} \approx x\frac{\phi(m, x_2) - \phi(m, x_1)}{x_2 - x_1} \quad (5.9)$$

then combining (5.8) and (5.9) gives an estimate for $n(m)$ [34],

$$n\left(m + \frac{1}{2}, x\right) \approx \frac{x}{x_2 - x_1}\left[\frac{\phi(m, x_2)}{2\pi} - \frac{\phi(m, x_1)}{2\pi} - \frac{\phi(m+1, x_2)}{2\pi} + \frac{\phi(m+1, x_2)}{2\pi}\right]$$

$$\approx \frac{x}{x_2 - x_1}\left[\frac{\tilde{\phi}(m, x_2)}{2\pi} - \frac{\tilde{\phi}(m, x_1)}{2\pi} - \frac{\tilde{\phi}(m+1, x_2)}{2\pi} + \frac{\tilde{\phi}(m+1, x_1)}{2\pi}\right]_{\mathrm{mod}\,1}$$

$$(5.10)$$

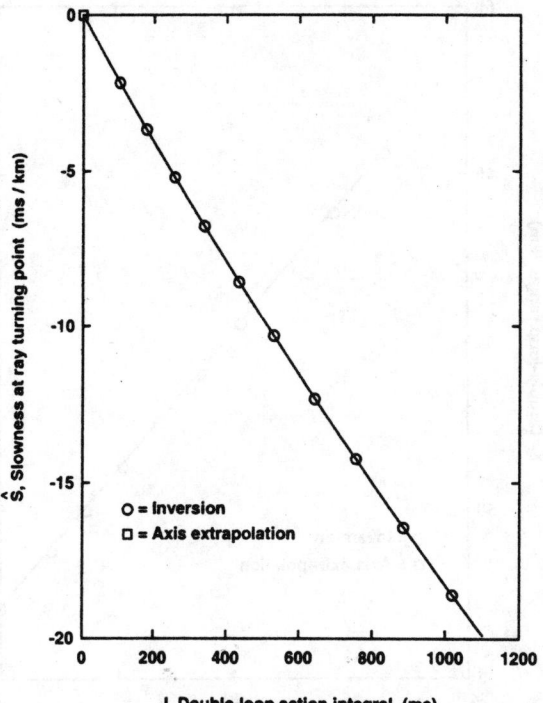

Fig. 7. Range average of $\hat{S}(J)$ calculated from (5.5) and (5.6) (circles). The value for $\hat{S}$ is relative to that assumed for the average of the sound slowness along the sound-channel axis (1.492^{-1} s km^{-1}). The error in that average as indicated by the measurements is shown by the intersection of an extrapolation of the measured points with the ordinate. That extrapolation also gives the lower limit for the integral in (A1). Thus, the measurements indicate directly the average of $\hat{S}$ along the sound-channel axis. The solid lines show a comparison calculated from single-loop raypath calculations.

For $m = 0$, $x = x_1 = 1000$ km, and $x_2 = 1010$ km, we have

$$n\left(\frac{1}{2}, x\right) \approx \frac{x}{x_2 - x_1} \left[\frac{\tilde{\phi}(0, x_2)}{2\pi} - \frac{\tilde{\phi}(0, x_1)}{2\pi} - \frac{\tilde{\phi}(1, x_2)}{2\pi} + \frac{\tilde{\phi}(1, x_1)}{2\pi} \right]_{\text{mod } 1} \quad (5.11)$$

$$= \frac{1000}{10} [0.813 - 0.934 - 0.963 + 0.320]_{\text{mod } 1} = 23.6$$

agreeing with the previous method. (The phase measurements are from Table 1.) The advantage of the dual-range method is that it requires no knowledge of climatology. The disadvantage is that it requires modal phase measurements to be made at two locations.

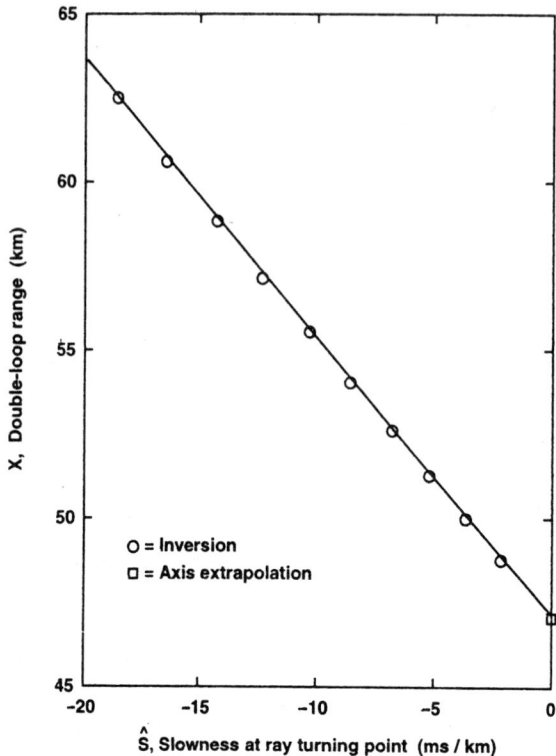

Fig. 8. Double-loop range $X(\hat{S})$ for a range-averaged sound channel calculated from (5.7). An Abel transform of this function gives the range average of the equivalent symmetric sound-speed profile. The solid line shows a comparison with single-loop raypath calculations.

The fourth column is simply the third column multiplied by $m + 1/2$. Notice that even in the absence of dual-range measurements, the only number in Table 2 not determined by the measurements is the "11" in "11.895" in the second row of column 4. It could take on any integer value, but for a 1000 km path it would not be very different from 11. Thus, $n(0)$ could be 21.79, 23.79, or 25.79, but not 22.79, for example. In practice, if climatology can provide a reasonably accurate sound-speed profile below the sound-channel axis, then only one value for the integer can give physically realistic results, as we show later. Using "10" instead of "11" gives a sound-speed profile that extends more than 100 m above the sea surface. Using "12" instead of "11" gives a profile that ends more than 100 m below the surface. Fig. 13 shows the plot of $n(m)$.

Table 3 continues the inversion process, showing first the double-loop range $X(m)$ $[= 1/\overline{X^{-1}}(m)]$, calculated from $n(m)$ using (2.17). Fig. 14 shows the plot of $X(m)$. The third column in Table 3 shows $\tilde{\hat{S}}(m)$ calculated using (2.18). Fig. 15

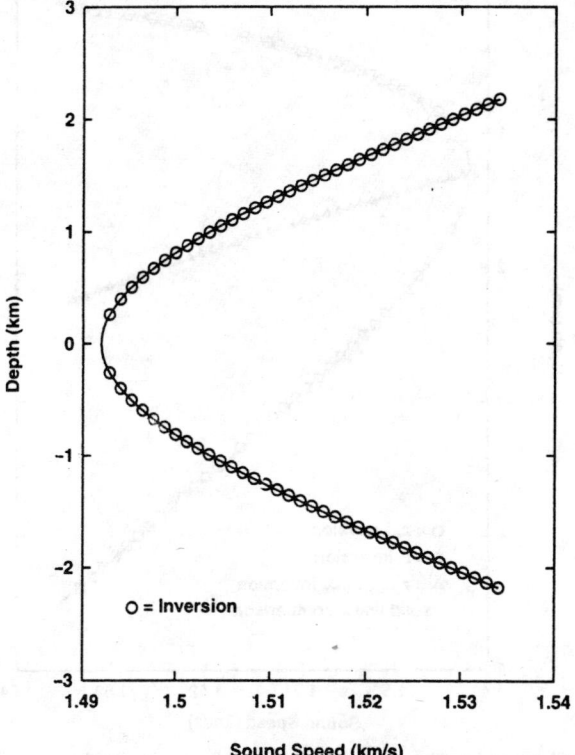

Fig. 9. Range average of the equivalent symmetric sound channel found by numerical integration of the Abel transform in (A1) (circles). The solid line gives the equivalent symmetric sound-speed profile for the profile used for the simulation of the travel-time measurements for comparison.

shows the plot of $\bar{\hat{S}}(m)$ and Fig. 16 shows the corresponding plot of $X(\bar{\hat{S}})$. The fourth column in Table 3 shows the correct $\bar{\hat{S}}(m)$ for comparison. The first row in Table 3 represents an extrapolation to the sound-channel axis. This extrapolation is necessary to perform an Abel transform.

Fig. 17 shows the range average of the equivalent symmetric sound channel found by Abel transform of $X(\bar{\hat{S}})$. We then assumed we knew from climatology the correct sound-speed profile below the sound-channel axis. Combining that with the equivalent symmetric profile in Fig. 17 gives the profile in Fig. 18. If we had used 21.79 for $n(0)$, we would have gotten a profile that extended above the surface of the ocean by more than 100 m. If we had used 25.79 for $n(0)$, we would have gotten a profile that stopped more than 100 m below the surface. Thus, it is not difficult to determine the $n(0)$ even with a small amount of independent information.

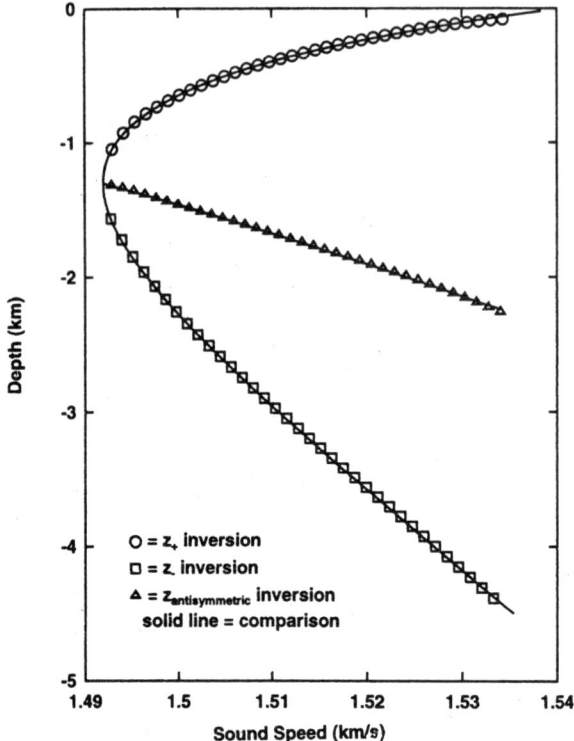

Fig. 10. Sound-speed profile at the source found by tomographic inversion. Asymmetry of the sound channel at the source (triangles) found by numerical integration of the Abel transform in (3.1). The solid line gives the asymmetry of the sound-speed profile used for the simulation of the travel-time measurements for comparison. The total sound-speed profile at the source, found by combining the asymmetry of the profile with the equivalent symmetric profile from Fig. 9, is shown by circles and squares. The solid line gives the sound-speed profile used to simulate the travel-time measurements.

Shang et al, [34] also simulated modal tomography measurements of a range-dependent case. Table 4 gives the simulated travel-time and phase measurements. Travel-time measurements are available at one range (1000 km) and phase measurements are available for two modes at two ranges (1000 km and 998 km). As with the range-independent case in Table 1, we have assumed travel time is given to 1 ms and phase to one-thousandth of a cycle (about a third of a degree). Using realistic (larger) errors for phase does not degrade the inversion, however. Fig. 19 shows the modal dependence of travel time for the range-dependent case.

Table 5 gives independent (nontomographic) data for the range-dependent case. Specifically, these are background wave numbers (to simulate climatology) for the two lowest modes, and the local wave numbers (which could have been

Table 1. Simulated measurements of modal pulse travel time and modal phase for a range $x_1 = 1000$ km (and modal phase for a range $x_2 = 1010$ km) for a range-independent example with $\omega/2\pi = 30$ Hz

Mode Number m	At $x_1 = 1000$ km		At $x_2 = 1010$ km
	Pulse Travel Time $t(m)$ (sec)	Phase $\tilde{\phi}(m)/2\pi$ (cycles)	Phase $\tilde{\phi}(m)/2\pi$ (cycles)
0	666.668	0.934	0.813
1	666.655	0.320	0.963
2	666.629	0.104	
3	666.592	0.273	
4	666.544	0.810	
5	666.488	0.697	
6	666.425	0.919	
7	666.359	0.456	
8	666.286	0.288	
9	666.218	0.389	
10	666.154	0.728	
11	666.096	0.275	
12	666.045	0.001	
13	665.996	0.882	
14	665.949	0.903	
15	665.900	0.052	
16	665.843	0.327	
17	665.780	0.729	
18	665.709	0.262	
19	665.629	0.930	

determined by measuring the sound-speed profile at the receiver array) also for the two lowest modes. These data are used to help remove the ambiguity in $n(0)$.

Table 6 shows how climatology can help remove the ambiguity in $n(0)$ [33, 34]. First, the integral $A(m)$ has been calculated from the simulated travel-time measurements in Table 4 using (2.22). The background profile for the range-dependent model has been taken to represent climatology, and the wavenumbers for the first two modes have been calculated. From those, using (2.20), $n(0.5)$ has been estimated and used with $A(1)$ to estimate $n(0)$ and $n(1)$. Then $(m + 1/2) n(m)$ is calculated and then adjusted to have the same fractional part as the quantity in column 5, using (2.24). As a consistency check, we notice that for the adjusted values, $n(1) - n(0)$ is close to $A(1)$.

Table 7 illustrates an alternative method to estimate $n(0.5)$ using a locally measured sound-speed profile at the receiver array [34]. (It is usually necessary to know the local sound-speed profile to estimate the modal eigen functions so that modes can be separated in any case.) The method of estimating $n(0.5)$ is the same as in Table 6, except that local modal wavenumbers are used instead of

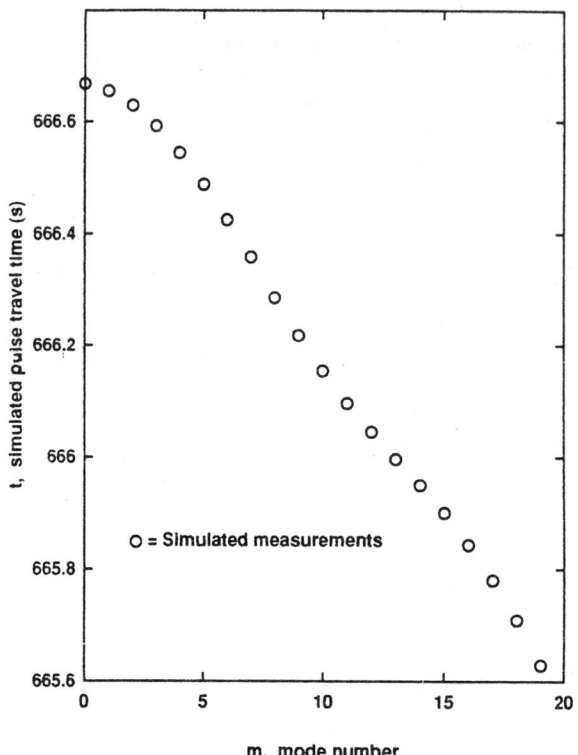

Fig. 11. Simulated modal pulse travel time versus mode number for the range-independent case in Table 1.

modal wavenumbers that estimate averages for the whole region. In this case, the adjusted value of $n(0)$ is the same, but $n(1)$ is different. In applying the consistency check, we find that $n(1) - n(0)$ is significantly different from $A(1)$, indicating an inconsistency. In actual practice, without other independent information, we would not know whether $n(0)$ or $n(1)$ were correct, but we could guess that $n(0)$ was correct because it is easier to estimate. The assumption would lead to the same adjusted (and consistent) values for $n(0)$ and $n(1)$ as in Table 6. Another possibility would be to iterate, starting with the average of the adjusted $n(0)$ and $n(1)$ for the estimate of $n(0.5)$ (= 20.351), which would also lead to the same adjusted (and consistent) values as in Table 6.

Table 8 illustrates a third method for estimating $n(0.5)$, the dual-range method [34]. As in the example for range-independent case, this uses (5.10) to estimate $n(0.5)$, but now we have

$$n\left(\frac{1}{2}, x_2\right) \approx \frac{x_2}{x_2 - x_1}\left[\frac{\tilde{\phi}(0, x_2)}{2\pi} - \frac{\tilde{\phi}(0, x_1)}{2\pi} - \frac{\tilde{\phi}(1, x_2)}{2\pi} + \frac{\tilde{\phi}(1, x_1)}{2\pi}\right]_{\text{mod } 1} \quad (5.12)$$

Table 2. Inversion of Table 1 data, using $N(0) = 11$

Mode Number m	$A(m)^1$ (s)	$n(m)$	$\left(m + \frac{1}{2}\right)n$
−0.5		23.998[3]	0.000
0	0.00	23.79[2]	11.895
1	−0.39	23.40[2]	35.100
2	−0.77	23.02	57.550
3	−1.14	22.65	79.275
4	−1.50	22.29	100.305
5	−1.83	21.96	120.780
6	−2.14	21.65	140.725
7	−2.43	21.36	160.200
8	−2.69	21.10	179.350
9	−2.92	20.87	198.265
10	−3.11	20.68	217.140
11	−3.27	20.52	235.980
12	−3.40	20.39	254.875
13	−3.52	20.27	273.645
14	−3.62	20.17	292.465
15	−3.72	20.07	311.085
16	−3.82	19.97	329.505
17	−3.94	19.85	347.375
18	−4.05	19.74	365.190
19	−4.18	19.61	382.395

[1] $A(m) \equiv \int_0^m \frac{dt/dm}{J(m)} dm = \frac{\omega}{2\pi} \int_0^m \frac{dt/dm}{m + 1/2} dm$

[2] Using the condition that $[n(1) + n(0)]/2 = n(1/2) = -1/2\pi \, d\phi/dm \approx 1/2\pi \, [\phi(0) - \phi(1)] \approx 23.6$ and $n(1) - n(0) = A(1)$.

[3] Extrapolation to sound-channel axis at $m = -.5$.

$$= \frac{1000}{2} [0.681 - 0.268 - 0.094 + 0.722]_{\text{mod } 1} = 20.5,$$

where the simulated measured phases at the two ranges come from Table 4. This time, the process leads to the same (consistent) values for $n(0)$ and $n(1)$ as in Table 6. Shang et al [34] use a variation of (5.12) to estimate only the integral part of $n(0.5)$.

Table 9 shows the estimates of $A(m)$ for all of the modes, and Fig. 20 shows the corresponding plot. The adjusted value of $n(m)$ are calculated for the rest of the modes using $A(m)$, and are shown in Table 9 and Fig. 21. The values of $X(m)$ are then calculated from $n(m)$ and (2.17), and are shown in Table 9 and Fig. 22. $S(m)$ calculated from (2.18) are shown in Table 9 and Fig. 23. Table 9 also shows a comparison with the true range-averaged values of $S(m)$ for m equal to 0 and 1.

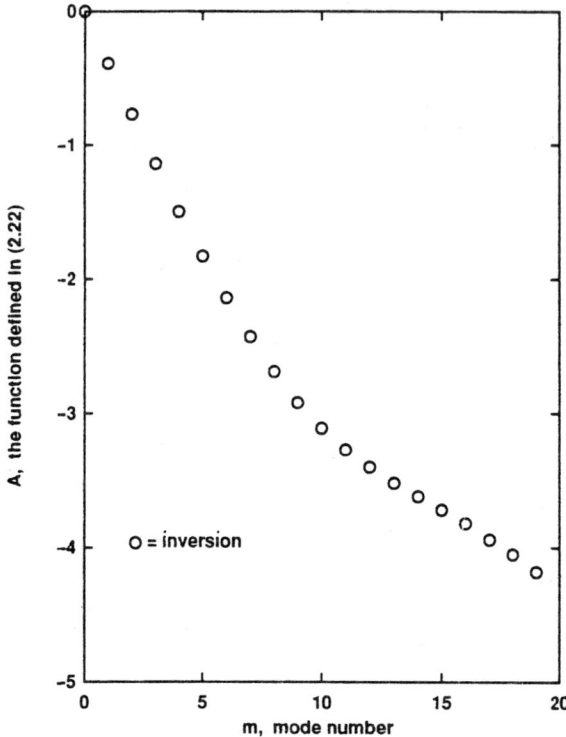

Fig. 12. $A(m)$, the function defined in (2.22) for the simulated modal pulse travel-time measurements in Table 1.

The agreement is good because $n(0)$ and $n(1)$ were adjusted to match the measured phase, and we have resolved the phase ambiguity correctly.

Fig. 24 shows a plot of $X(\bar{s})$ taken from Table 9. Applying an Abel transform (Appendix A) to Fig. 24 gives the equivalent symmetric sound-speed profile shown in Fig. 25. Constraining the sound-speed profile below the sound-channel axis to match simulated climatology gives the range-averaged sound-speed profile in Fig. 26.

Goncharov and Voronovich [49] give an example of modal tomographic measurements in the Norway Sea for a short (105 km) path.

6. Example Inversion from the Slice89 Experiment

A nearly optimum single vertical-slice experimental configuration was used in the Slice89 experiment in the northeastern Pacific Ocean in July 1989 [40, 42, 50, 51]. Pulse travel time was measured by recording, at a long vertical array of hydrophones, acoustic transmissions from a source located near the sound-channel axis 1000 km away. These measurements provide an ideal illustration for vertical-

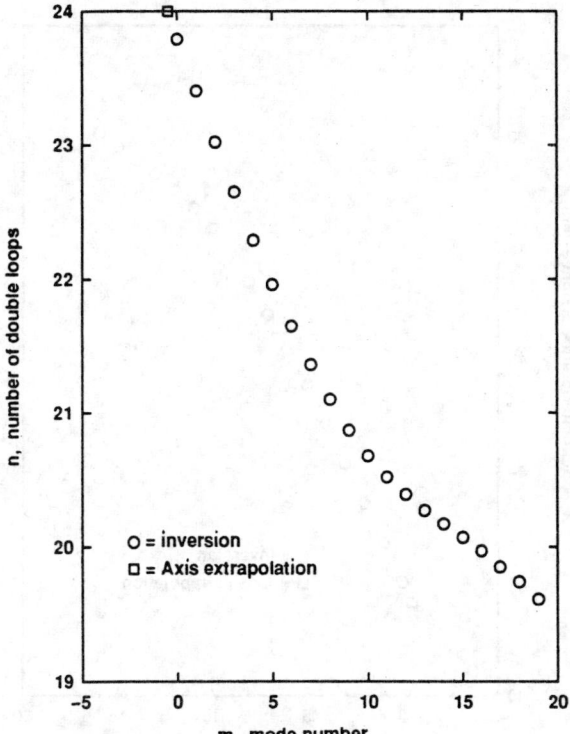

Fig. 13. $n(m)$, the number of double loops for the ray that has the same turning-point depth as the correspondence mode, from Table 2, determined from (2.22)

slice inversion because they represent nearly the maximum amount of acoustic information that can be obtained in a single vertical slice. In particular, the long vertical receiver array allows unambiguous identification of nearly all the received pulses and permits nearly the maximum vertical and horizontal resolution possible for a vertical-slice inversion.

The inversion presented here calculates the range average of the equivalent symmetric sound-slowness profile from the acoustic travel-time measurements. We use climatological data [36] to estimate the sound speed below the sound-channel axis. Combining the two gives the range average above the sound-channel axis. We compare the inversion with sound speed calculated from measurements of temperature and salinity along the great circle path connecting the source and receiver during the experiment.

The acoustic source was moored at 804 m below the surface, near the sound channel axis, at 32°0′N, 150°26′W in the eastern North Pacific Ocean. A long receiver array was suspended from R/P *FLIP*, moored near 34°0′N, 140°0′W, 1001 km to the east-northeast. R/P *FLIP* was in a tri-moor, so that the experimental geometry varied by only a few hundred meters during the experiment. Howe et

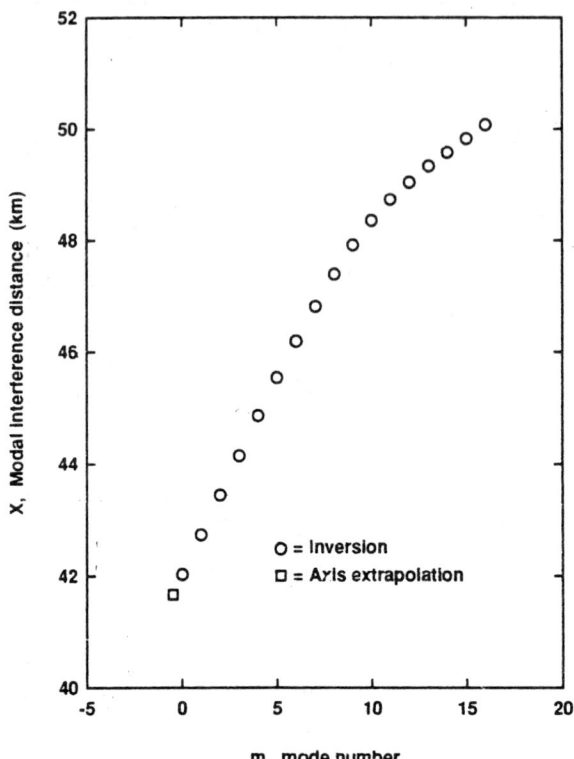

Fig. 14. $X(m) \equiv 1/\overline{X^{-1}}(m)$, the average double-loop range (modal interference distance), from Table 3, determined from $n(m)$ and (2.17).

al, [40], Duda et al, [42], Cornuelle et al, [50], and Worcester et al, [51] describe the experiment in more detail.

The source transmitted a phase-modulated signal with a center frequency of 25 Hz and 83.3 Hz bandwidth [35]. The phase modulation was encoded using a linear maximal shift register sequence containing 1023 terms. The phase-modulated digits in the transmitted signal contained three cycles of carrier each. Each transmitted pulse lasted 135.036 s. There were 400 transmissions over 14 days. The receiving array [52] was suspended with one end near the surface.

Relative motions of the source and receivers were tracked using long-baseline acoustic navigation systems, with acoustic transponders on the sea floor. The acoustic transponder positions and the mooring anchor position were determined using the NAVSTAR Global Positioning System (GPS). Absolute source positions are known to about 30 m. Relative motion of the source during the experiment was measured with an error of about 2 m. Absolute receiver positions are known to about 100 m. The best relative tracking accuracy for the receiver was about 2 m rms.

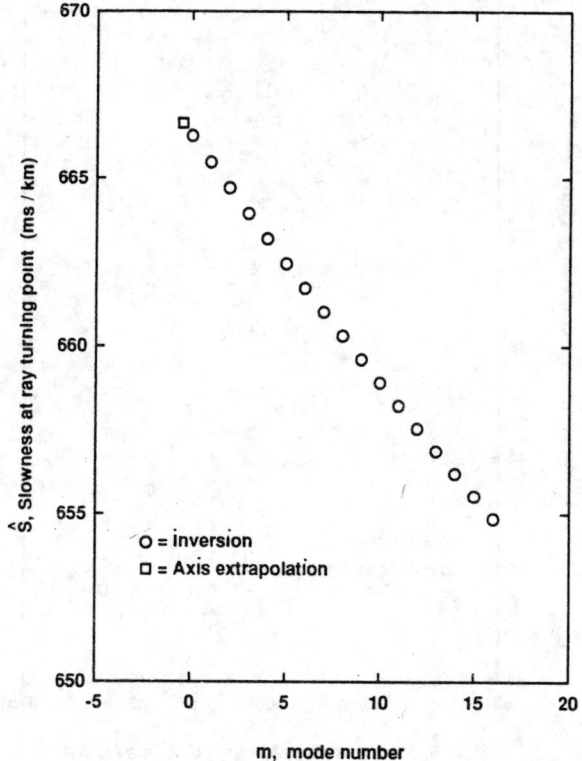

Fig. 15. $\bar{\hat{S}}(m)$, the sound slowness at the turning point, from Table 3, determined from (2.18) and (2.19).

Extensive conductivity-temperature-depth (CTD), expendable bathythermograph (XBT), and air-launched expendable bathythermograph (AXBT) measurements were made during the experiment along the great circle path connecting the source and receiver.

Fig. 27 shows the pulse arrivals for one of the 400 transmissions. We use some of these pulse arrivals to perform a vertical-slice inversion to obtain a range-averaged sound channel. Although the pulse arrivals also contain some information [23] about smaller scales equal to a ray double loop length (roughly 40–60 km), we ignore that information here.

Relative pulse-arrival identification (at least for the early arrivals) in Fig. 27 is straight forward because such patterns have been simulated by ray tracing (e.g., Munk and Wunsch [1]. Although independent information is required to give the correct absolute loop number, experience at ranges of about 1000 km in both the 1987 Reciprocal Acoustic Transmission Experiment north of Hawaii and this experiment shows that historical data, such as in Levitus [36], are adequate to

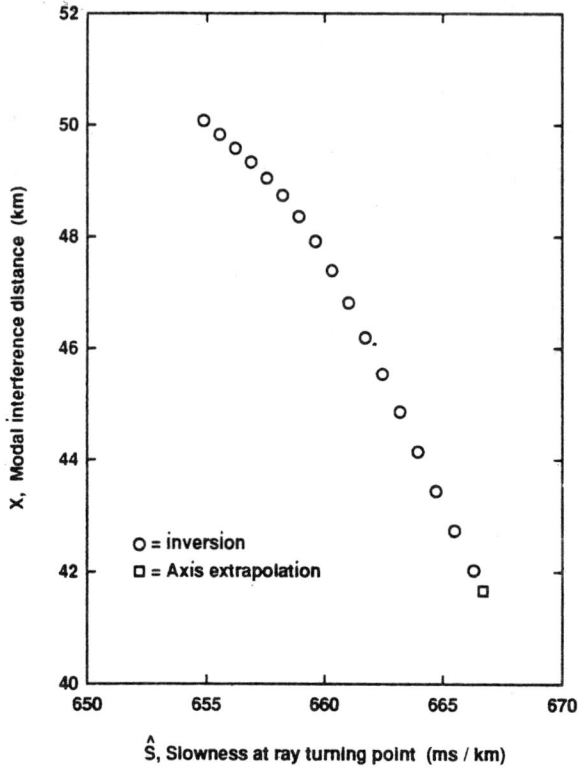

Fig. 16. Double-loop range (modal interference distance, $X(\hat{\bar{S}})$, from Table 3.

identify absolute loop numbers when the range between the source and receiver is known [11, 37, 38, 39, 40, 41].

Because using ray tracing to identify the pulse arrivals might put into question whether this inversion method was really nonperturbative, we used an independent method to determine the correct absolute loop number. Without independent information, the loop numbers indicated in Fig. 27 could have a multiple of two added or subtracted from them. To verify that the loop numbers indicated in Fig. 27 are correct, we also performed an inversion using loop numbers that are 2 less than those indicated in Fig. 27. This gave a sound-speed profile that continued about 190 m above sea level when we used climatology for the sound speed below the sound-channel axis, a clear contradiction. We also performed an inversion using loop numbers that are 2 greater than those indicated in Fig. 27. This gave a sound-speed profile whose highest point (corresponding to the upper turning points of the earliest ray arrivals) was about 250 m below sea level. Because we know that the earliest arrivals in Fig. 27 correspond to rays that have turning points near the ocean surface, we have another contradiction. Thus, the correct

Table 3. Inversion of Table 1 data (continued)

Mode Number m	$X(m)$ (km)	$\hat{S}(m)$[1] (ms km^{-1})	$\hat{S}(m)_{true}$ (ms km^{-1})
− 0.5	41.67[2]	666.65[2]	666.6666667
0	42.031	666.272	666.2644686
1	42.735	665.485	665.4773360
2	43.440	664.711	664.7034589
3	44.150	663.950	663.9424224
4	44.863	663.201	663.1936637
5	45.537	662.462	662.4565609
6	46.189	661.734	661.7306239
7	46.816	661.019	661.0152039
8	47.393	660.308	660.3096074
9	47.916	659.609	659.6129576
10	48.356	658.916	658.9242522
11	48.733	685.230	658.2425014
12	49.044	657.549	657.5666945
13	49.334	656.875	656.8960803
14	49.579	656.200	656.2300975
15	49.826	655.531	655.5683965
16	50.075	654.860	654.9109025

[1]$\hat{S}(m) = \left[t(m) - \left(m + \dfrac{1}{2} \right) 2\pi\, n(m)/\omega \right] / x$, $x = 1000$ km

[2]Extrapolation to sound channel axis at $m = -0.5$.

identification of the pulse arrivals in Fig. 27 can be accomplished without using ray tracing.

Identification of the late-arriving (near axial) pulses is more difficult because they are not resolved, as can be seen from Fig. 27. This difficulty complicates the inversion close to the sound channel axis.

In addition, there are often several neighbouring pulses that would be classified as having the same pulse identification. The source of this fine splitting may be internal waves [53] or other ocean variability. Whether it is appropriate to choose one of these neighbouring pulses or an average for the inversion is still not clear, but the effect is a second-order one.

Although the travel-time measurements in Fig. 27 contain information about both large-scale and small-scale structure in this vertical slice through the ocean sound channel, they do not determine the sound-speed structure uniquely in the vertical slice. In this inversion, we neglect the information about the small-scale structure and estimate the range average of the equivalent symmetric sound channel [25]. This allows the use of vertical-slice tomography as an acoustic thermometer [14, 16, 17].

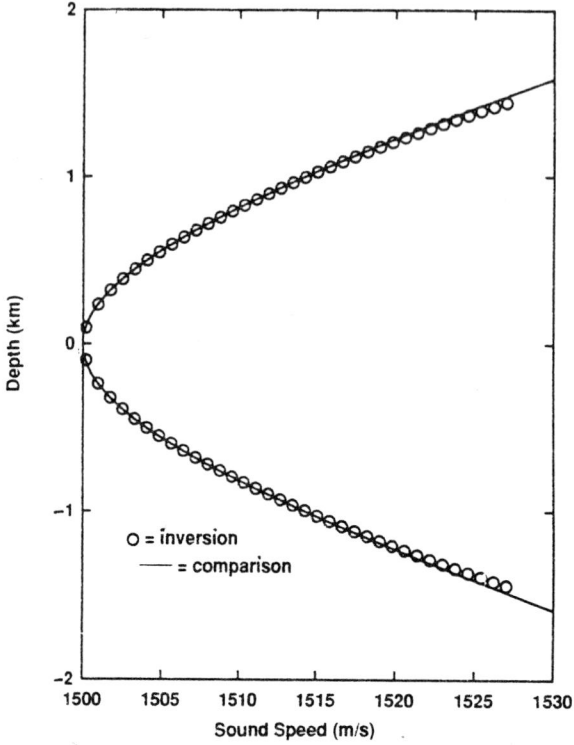

Fig. 17. Range average of the equivalent symmetric sound channel found by numerical integration of the Abel transform in Appendix A for the simulated modal pulse travel-time measurements in Table 1.

We use the pulse arrivals from only three of the receivers (at 660, 960, and 1200 m depths). Although using more receivers might increase the vertical resolution, it would require unambiguously separating vertical small-scale information from horizontal small-scale information.

Fig. 28 shows the dependence of travel time on the number of turning points (from Fig. 27) for the receiver at 660 m depth. The behaviour for other receiver depths is similar. The four functions of the number of ray turning points in Fig. 28 contain nearly all the acoustic information available from a single source-receiver pair.

To first order, averaging travel times for upward and downward propagation at the source cancels the effects of the asymmetry of the sound channel and the partial ray loops at the source and receiver [31, 43]. There are second-order corrections in (5.2), which we evaluated for Fig. 28 and determined that they were negligible for this case. Fig. 29 shows the average of pulse travel times for upward- and downward-propagated pulses—approximately the same as we would expect for rays consisting of complete ray loops for a sound channel with a

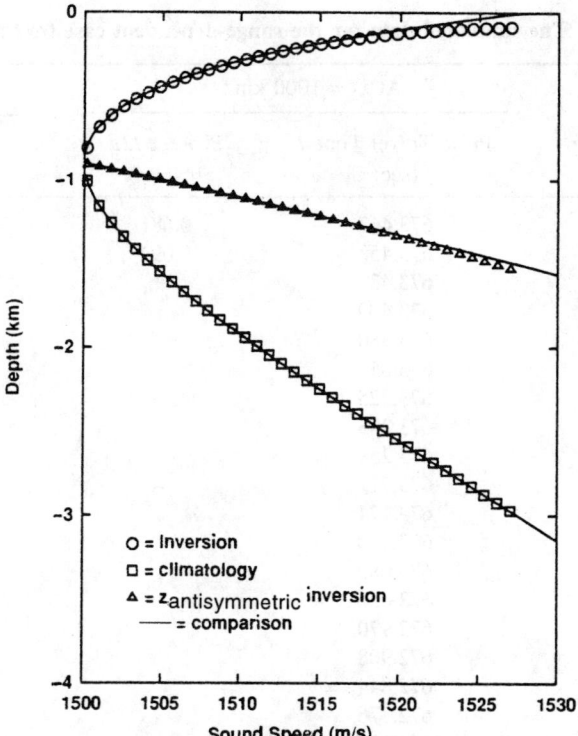

Fig. 18. Range-averaged sound channel, from combining the equivalent symmetric sound channel in Fig. 17 found by inversion with simulated climate data for the part of the sound channel below the sound channel axis.

symmetric sound-speed profile. In the adiabatic-invariant approximation, the travel time for such a raypath (e.g., Wunsch [24]) is given by (5.1).

Fig. 30 shows the difference between the travel time for upward- and downward-propagated pulses. Fitting smooth curves to the data in Fig. 30 determines functions that give the asymmetry of the sound channel at the source and receiver. Section 5.1 of this chapter demonstrates the procedure using a simulated pulse arrival sequence that was generated by ray tracing. Although the data in Fig. 30 are similar to the corresponding data in Fig. 3b, the variability of the measured data in Fig. 30 requires a more robust curve-fitting method. We plan to pursue inversion of these data after we have acquired a more robust curve-fitting algorithm.

Fig. 31 shows the result of fitting a smooth curve to the points in Fig. 29. Fig. 32 shows the derivative of the curve fit in Fig. 31 which is used to estimate J. Normally, we apply the second-order correction in (5.2) to derive Fig. 31 from Fig. 29, but an estimate showed that the correction was negligible in this case.

Fig. 33 shows the resulting plot of the action J versus the range average of sound slowness at the ray turning point (horizontal component of sound slowness).

Table 4. The simulated data for the range-dependent case ($\omega/2\pi = 30$ Hz)

Mode Number m	At $x_2 = 1000$ km		At $x_1 = 998$ km
	Pulse Travel Time t (sec)	Phase $\tilde{\phi}/2\pi$ (cycles)	Phase $\tilde{\phi}/2\pi$ (cycles)
0	673.467	0.681	0.268
1	673.458	0.094	0.722
2	673.439		
3	673.409		
4	673.380		
5	673.356		
6	673.325		
7	673.293		
8	673.258		
9	673.217		
10	673.174		
11	673.130		
12	673.080		
13	673.026		
14	672.970		
15	672.908		
16	672.844		
17	672.776		
18	672.703		

The action J in Fig. 33 is calculated from Fig. 32 using (5.5). The range average of sound slowness in Fig. 33 is calculated from Figs. 31 and 32 using (5.6).

We can use Fig. 33 to estimate the range average of the sound slowness on the sound-channel axis. Because J is zero for an axial ray, extrapolating Fig. 33 to $J = 0$ will yield the range average of sound slowness on the sound-channel axis. This gives about 675.85 ± 0.085 ms km^{-1}. The error includes both extrapolation error and the uncertainty in transmitter-receiver range.

Extrapolating to $J = 0$ also gives about 1479.69 ± 0.18 m s^{-1} for sound speed on the sound-channel axis, including an extrapolation error of ± 0.1 m s^{-1} and an error of ± 0.148 m s^{-1} due to the uncertainty in the transmitter-receiver range of ± 100 m. This does not agree with the average of the sound speed along the sound-channel axis of 1478.604 ± 0.5 m s^{-1} rms from CTD and deep XBT casts made during the experiment. The disagreement is probably not caused by contamination of the range average by aliasing from sound-speed structure on scales of the double-loop ray length [43].

The average of the sound speed along the sound-channel axis above is not the same as the minimum sound speed of 1478.548 m s^{-1} of the range-averaged sound-speed profiles because the former averages the minimum sound speed for each profile, whereas the latter averages the sound speed at the same depth. Although

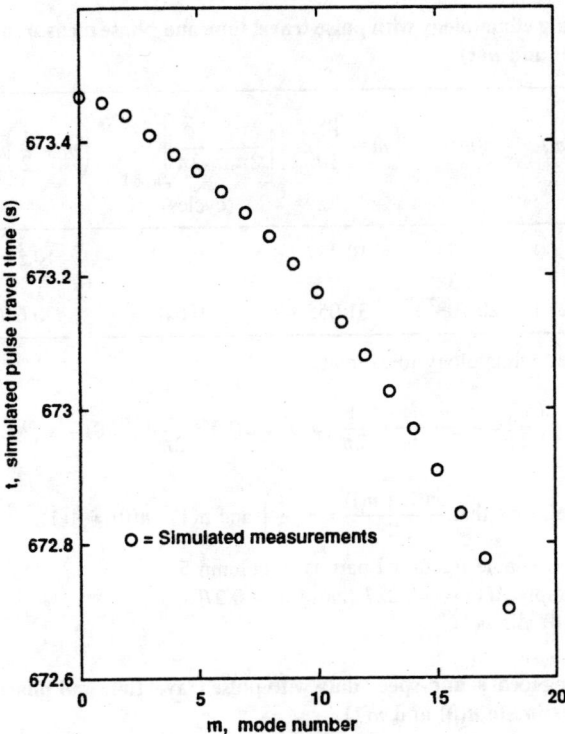

Fig. 19. Simulated modal pulse travel time versus mode number for the range-dependent case in Table 4.

Table 5. Independent (nontomographic) data for the range-dependent case

Mode Number m	Background[1] Wavenumber $k^{(b)}$ (km^{-1})	Local[2] Wavenumber at $x = 1000$ km $k^{(L)}$ (km^{-1})
0	126.866570	126.960241
1	126.735640	126.834891

[1] To simulate climatology.
[2] To simulate modal wavenumbers determined by measuring the sound-speed profile at the receiver array.

the inversion agrees better with the latter, the former is the correct one with which to compare an acoustic inversion.

Although the range average of the minimum sound speed is not the same as the reciprocal of the range average of maximum sound slowness, in this case the sound-speed variation is small enough that they agree within 1 mm s^{-1}.

The extrapolation error could be reduced by extending Fig. 33 closer to the

Table 6. Using climatology with pulse travel time and phase measurements to estimate $n(0)$ and $n(1)$

Mode Number m	$A(m)$	$n(m)$	$\left(m+\frac{1}{2}\right)n$	$\left[\frac{\omega t}{2\pi} - \frac{\tilde{\phi}}{2\pi}\right]_{\text{mod }1}$ (cycles)	$\left(m+\frac{1}{2}\right)n_{\text{adjusted}}$	n_{adjusted}
0	0.000	20.973[2]				20.658[4]
0.5		20.838[1]	10.487	0.329	10.329[3]	20.545[5]
1	−0.270	20.703[2]	31.055	0.646	30.646[3]	20.341[4]

[1] From simulated climatology to estimate

$$n\left(\frac{1}{2}\right) = -\frac{1}{2\pi}\frac{d\phi}{dm} \approx \frac{1}{2\pi}[\phi(0) - \phi(1)] = \frac{x}{2\pi}[k^{(b)}(0) - \kappa^{(b)}(1)].$$

[2] Using the conditions that $\dfrac{n(0) + n(1)}{2} = n\left(\dfrac{1}{2}\right)$ and $n(1) - n(0) = A(1)$.

[3] Adjusted to have same fractional part as in column 5.
[4] Final values imply $A(1) \approx -0.227$ (close to -0.270).
[5] Average of $n(0)$ and $n(1)$.

Table 7. Using local sound-speed data with pulse travel time and phase measurements to estimate $n(0)$ and $n(1)$

Mode Number m	$A(m)$ (sec)	$n(m)$	$\left(m+\frac{1}{2}\right)n$	$\left[\frac{\omega t}{2\pi} - \frac{\tilde{\phi}}{2\pi}\right]_{\text{mod }1}$ (cycles)	$\left(m+\frac{1}{2}\right)n_{\text{adjusted}}$	n_{adjusted}
0	0.000	20.085[2]				20.658[4]
0.5		19.950[1]	10.043	0.329	10.329[3]	20.211[5]
1	−0.270	19.815[2]	29.723	0.646	29.646[3]	19.764[4]

[1] From simulated local sound-speed measurements to estimate

$$n\left(\frac{1}{2}\right) = -\frac{1}{2\pi}\frac{d\phi}{dm} \approx \frac{1}{2\pi}[\phi(0) - \phi(1)] = \frac{x}{2\pi}[k^{(L)}(0) - k^{(L)}(1)].$$

[2] Using the conditions that $\dfrac{n(0) + n(1)}{2} = n\left(\dfrac{1}{2}\right)$ and $n(1) - n(0) = A(1)$.

[3] Adjusted to have the same fractional part as in column 5.
[4] Final values imply $A(1) \approx -0.894$ (not close to -0.270). [Thus, the resulting $n(0)$ and $n(1)$ are not consistent with $A(1)$.]
[5] Average of $n(0)$ and $n(1)$.

Table 8. Using dual-range data with pulse travel time and phase measurements to estimate $n(0)$ and $n(1)$

Mode Number m	$A(m)$ (sec)	n	$\left(m+\frac{1}{2}\right)n$	$\left[\frac{\omega t}{2\pi} - \frac{\tilde{\phi}}{2\pi}\right]_{\text{mod 1}}$ (cycles)	$\left(m+\frac{1}{2}\right)n_{\text{adjusted}}$	n_{adjusted}
0	0.000	20.635^2	10.318	0.329	10.329^3	20.658^4
0.5		20.5^1				20.545^5
1	-0.270	20.365^2	30.548	0.646	30.646^3	20.431^4

[1] From simulated dual-range measurements to estimate

$$n\left(m+\frac{1}{2}\right) \approx \frac{x_2}{x_2 - x_1}\left[\frac{\tilde{\phi}(m, x_2)}{2\pi} - \frac{\tilde{\phi}(m, x_1)}{2\pi} - \frac{\tilde{\phi}(m+1, x_2)}{2\pi} + \frac{\tilde{\phi}(m+1, x_1)}{2\pi}\right]_{\text{mod 1}}$$

$$n\left(\frac{1}{2}\right) \approx \frac{x_2}{x_2 - x_1}\left[\frac{\tilde{\phi}(0, x_2)}{2\pi} - \frac{\tilde{\phi}(0, x_1)}{2\pi} - \frac{\tilde{\phi}(1, x_2)}{2\pi} + \frac{\tilde{\phi}(1, x_1)}{2\pi}\right]_{\text{mod 1}}$$

$x_1 = 998$ km, $x_2 = 1000$ km

[2] Using the conditions that $\frac{n(0) + n(1)}{2} = n\left(\frac{1}{2}\right)$ and $n(1) - n(0) = A(1)$.

[3] Adjusted to have the same fractional part as in column 5.
[4] Final values imply $A(1) \approx -0.227$ (close to -0.270).
[5] Average of $n(0)$ and $n(1)$.

sound-channel axis by including pulses with loop numbers higher than 47, if they could be correctly identified.

Fig. 34 shows double-loop range X as a function of sound slowness at the ray turning point. Taking the Abel transform (A1) of the function in Fig. 34 (and using 1479.69 m s^{-1} for the sound speed on the sound-channel axis) gives Fig. 35 for the equivalent symmetric sound channel.

The estimate of the range average in (A1) gives a warm bias, just as with perturbative inversion [28]. The contribution to the bias from each ray is proportional to the variance of $\hat{S}$ in range for that ray. Specifically, the range average of (A1) including the next order correction is

$$\bar{z}_s(S) \approx \frac{1}{2\pi}\int_0^{J_s}\frac{dJ}{[\bar{\hat{S}}(J) - S^2]^{1/2}} - \frac{1}{2}\bar{z}_s''(S)\overline{[\hat{S}(J_s, x) - \bar{\hat{S}}(J_s)]^2}$$

$$+ \frac{1}{2\pi}\int_0^{J_s}\frac{3\overline{\alpha^2}/8 + 3\overline{\beta^2}/8 + \overline{\alpha\beta}/4}{[\bar{\hat{S}}(J)^2 - S^2]^{1/2}}dJ, \qquad (6.1)$$

Table 9. Inversion of Table 4 data for the range-dependent case

Mode Number m	$A(m)$	$n(m)$[1]	$X(m)$ (km)	$\bar{\hat{S}}(m)$ [2] (ms Km^{-1})	$\bar{\hat{S}}(m)_{\text{true}}$ (ms km^{-1})
−0.5		20.764[3]	48.16[3]	673.471[3]	
0	0.000	20.658	48.407	673.123	673.122712
1	−0.270	20.431	48.945	672.436	672.436457
2	−0.555	20.146	49.638	671.760	
3	−0.855	19.846	50.388	671.094	
4	−1.073	19.628	50.948	670.436	
5	−1.217	19.484	51.324	669.784	
6	−1.372	19.329	51.736	669.137	
7	−1.509	19.192	52.105	668.495	
8	−1.640	19.061	52.463	667.857	
9	−1.777	18.924	52.843	667.224	
10	−1.906	18.795	53.206	666.596	
11	−2.026	18.675	53.548	665.971	
12	−2.151	18.550	53.908	665.351	
13	−2.275	18.426	54.271	664.734	
14	−2.395	18.306	54.627	664.122	
15	−2.519	18.182	54.999	663.514	
16	−2.639	18.062	55.365	662.910	
17	−2.759	17.942	55.735	662.310	
18	−2.881	17.820	56.117	661.714	

[1] Based on $n(1) = 20.431$ from Tables 6 and 8 and $n(m) = n(1) - A(1) + A(m)$ for $m > 2$.

[2] $\bar{\hat{S}}(m) = \left[t(m) - \left(m + \frac{1}{2}\right) 2\pi n(m)/\omega \right]/x_2$, $x_2 = 1000$ km.

[3] Extrapolation to the sound-channel axis at $m = -0.5$.

where
$$\bar{\hat{S}}(J_s) - S = 0, \tag{6.2}$$

$$\alpha = \frac{\hat{S}(J, x) + \hat{S}(J_s, x)}{\bar{\hat{S}}(J) + \bar{\hat{S}}(J_s)} - 1, \tag{6.3}$$

and
$$\beta = \frac{\hat{S}(J, x) - \hat{S}(J_s, x)}{\bar{\hat{S}}(J) - \bar{\hat{S}}(J_s)} - 1. \tag{6.4}$$

As with perturbative inversions, the contribution to the variance from horizontal wavelengths smaller than a double-loop range (about 50 km) can sometimes be estimated, but the contribution to the variance from larger scales cannot be found from single-slice measurements. The warm bias in (6.1) from scales larger than

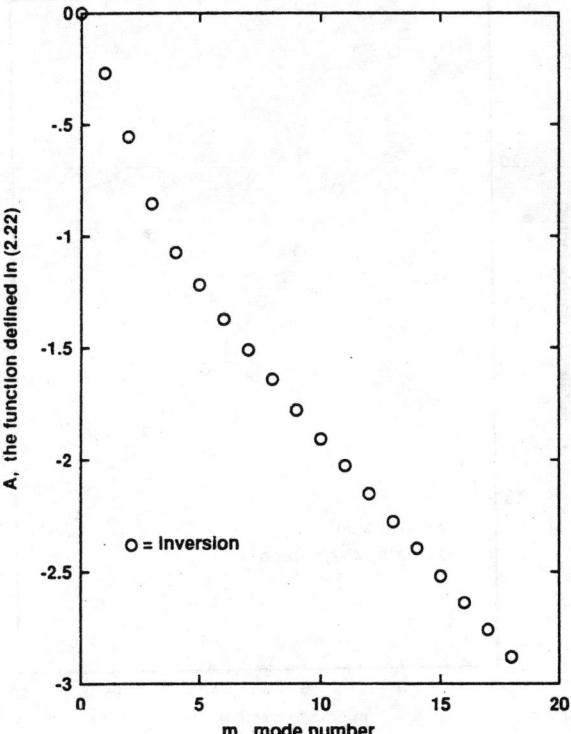

Fig. 20. $A(m)$, the function defined in (2.22) for the simulated modal pulse travel-time measurements in Table 4.

a double-loop range is probably small, but the bias from internal waves may not be negligible.

Using the range average of sound slowness in (A1) to estimate the range average of sound speed in Fig. 35 results in a cold bias. Specifically, the inverse of the range average of sound slowness underestimates the range average of sound speed by

$$\frac{1}{\bar{S}} - \bar{C} \approx -\frac{\overline{S^2} - \bar{S}^2}{\bar{S}^3} \approx -\frac{\overline{C^2} - \bar{C}^2}{\bar{C}} \leq 0 \qquad (6.5)$$

The inversion in Fig. 35 gives only the equivalent symmetric sound channel. To estimate the complete profile, we use the knowledge that the lower part of the sound channel does not vary much. That allows us to estimate the lower part of the sound channel from Levitus [36] climatology for the slice region using the Del Grosso [54] equation for sound speed. We assume a range-independent profile below the sound-channel axis between the source and receiver. Combining the equivalent symmetric profile from Fig. 35 with climatology for the part of the sound channel gives the complete sound-speed profile shown in Fig. 36.

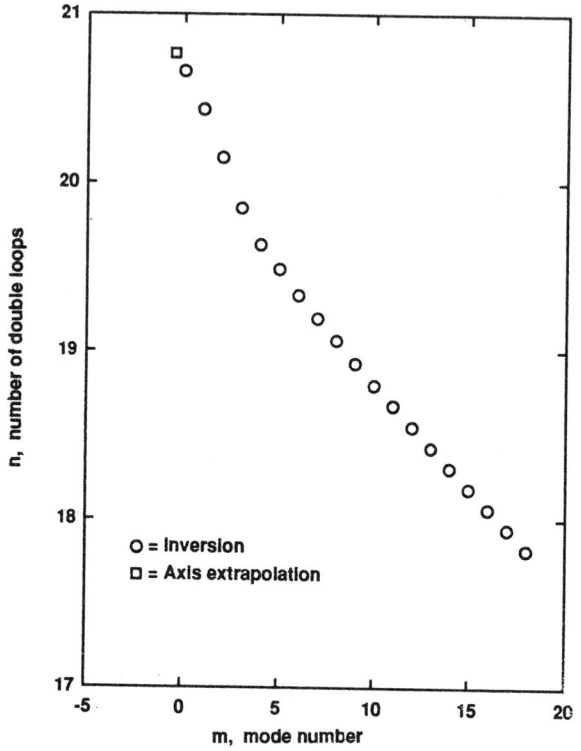

Fig. 21. $n(m)$, the number of double loops for the ray that has the same turning-point depth as the correspondence mode, from Table 9, determined from (2.22).

Notice that inversion of the measurement in Fig. 30 would give an acoustical estimate of the asymmetry of the sound channel (at the source and receiver only). Combining that with the Levitus [36] climatology for the lower part of the sound channel would give an estimate of the sound speed above the sound-channel axis at the source and receiver.

Fig. 36 also shows the range average of sound-speed profiles calculated from CTD measurements made during the experiment for comparison. Fig. 37 gives the upper 800 m of Fig. 36. The agreement is good for depths between 50 m and 300 m. The disagreement below 300 m may be caused by range dependence of the sound speed between 800 m and 2000 m. The disagreement near the surface may be caused by the difficulty in estimating the slope of the curve in Fig. 31 at the left endpoint.

A more likely cause for the disagreement is bias from internal waves, since internal waves are known to cause significant wander in the travel time of acoustic pulses. Errors should be expected for inversion of measurements from a single transmission such as considered here. Averaging the pulse travel times over roughly a 12-hour period should reduce the disagreement significantly. Fig. 38 shows individual sound-speed profiles calculated from CTD measurements for comparison.

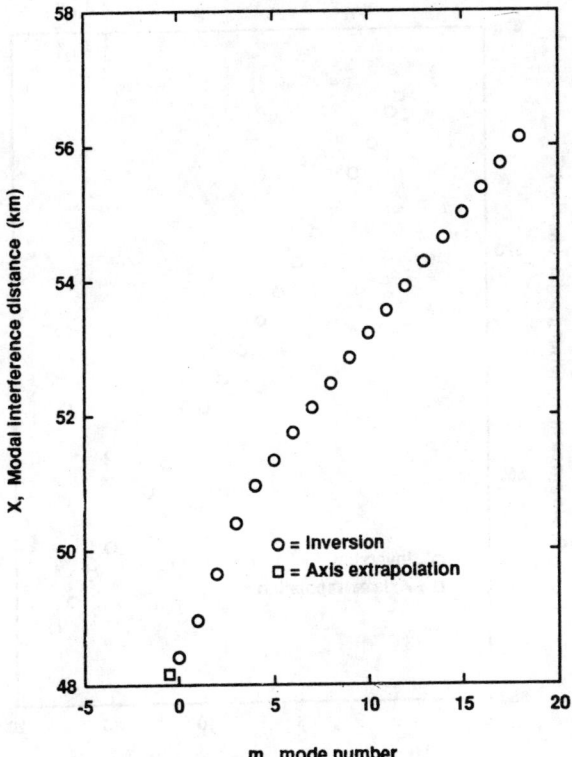

Fig. 22. $X(m) \equiv 1/\overline{X^{-1}}(m)$, the average double-loop range (modal interference distance), from Table 9, determined from $n(m)$ and (2.17).

The spread in sound speed from the individual CTD profiles allow a rough estimate of the cold bias from (6.5). This gives about 0.004 m s^{-1} at the depth of 400 m, which is quite small. Estimating the warm bias in (6.1) is more complicated, but we would expect it to be roughly proportional to the cold bias in (6.5) because it involves the variances of $\hat{S}$, which we would expect to be proportional to variance of S. Thus, we do not expect the bais in (6.1) or (6.5) to explain the disagreement between the sound-speed profile obtained from inversion with that from the range average of the CTD measurements, at least for scales on the order of the spacing between CTD profiles.

7. Discussion

The nonperturbative inversion method presented here should complement objective mapping and other perturbative methods because it gives range-averaged sound-slowness and sound-speed profiles along a vertical slice. Even for a dense network of vertical slices, separating vertical-slice inversion from horizontal-slice inversion has the advantage that standard tomographic imaging algorithms (e.g., Liebelt

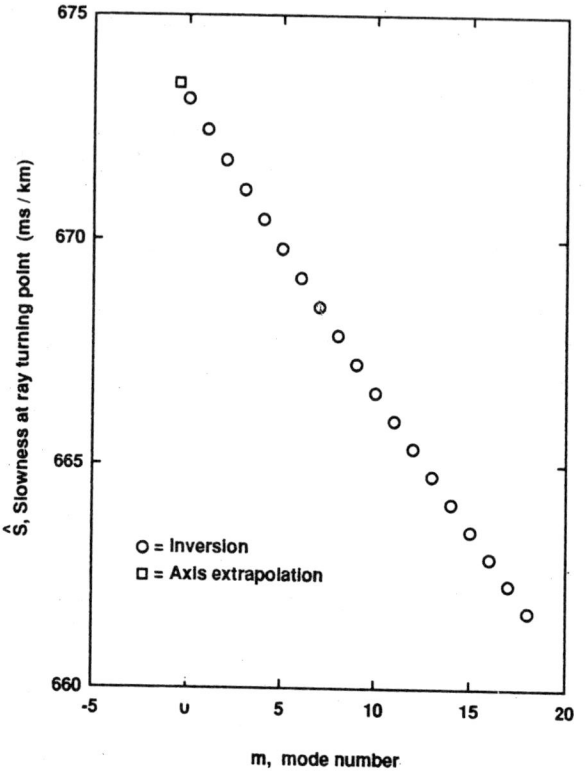

Fig. 23. $\bar{\hat{S}}(m)$, the sound slowness at the turning point, from Table 9, determined from (2.18) and (2.19).

[55], Herman [56], Menke [57], Kak and Slaney [58], Devaney [59, 60, 61, 62]) can be used for the horizontal-slice inversion because vertical-slice inversion can be formulated to yield range averages along each vertical slice. Standard tomographic imaging algorithms may not always be suitable for ocean acoustic tomography, however.

7.1 Horizontal Slice Tomography

Horizontal-slice tomography could be applied either to the range average of the equivalent symmetric profile or to the range average of the complete profile. In the former case, independent information (such as climatology) would be used to determine the profile below the sound-channel axis *after* horizontal-slice inversion; in the latter case, *before*. In either case, horizontal-slice inversion can be applied to averages of sound-speed profiles or averages of sound-slowness profiles, but the latter has the advantage of not introducing an unnecessary nonlinearity.

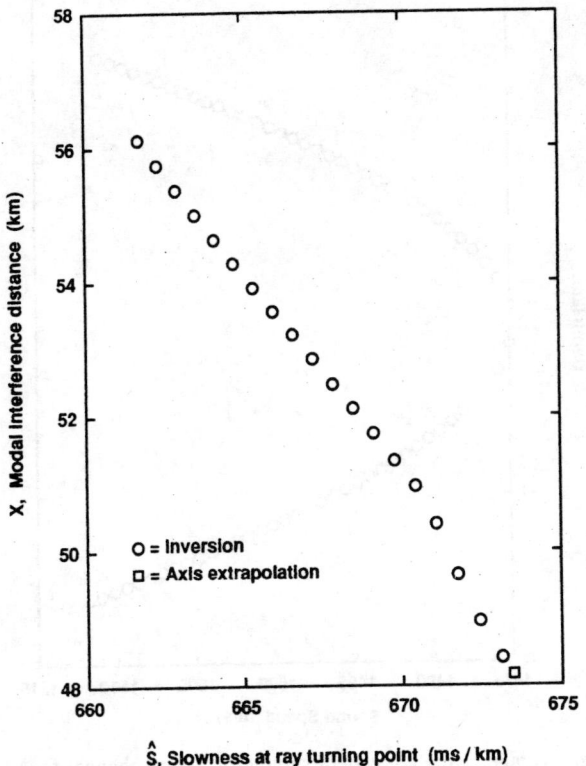

Fig. 24. Double-loop range (modal interference distance) $X(\bar{\hat{S}})$, from Table 9.

7.2 Nonmonotonic Profiles

Nonmonotonic profiles add difficulties to both perturbative and nonperturbative inversions. The difficulty (for both ray and modal tomography) results from having a range of depths for which there can be no turning points. The difficulty is explicit for nonperturbative inversions that use Abel transforms.

The difficulty is implicit for perturbative inversions, and can occur in two ways. If the background profile used for the inversion is monotonic above the sound-channel axis, then the turning-point depths for a range of background rays will differ significantly from those of the true rays. On the other hand, if the background profile is nonmonotonic, then for some rays turning-point depth can change discontinuously for small changes in the profile. In such a case, the applicability of Format's principle, upon which perturbative inversions are based, would have to be examined. An accurate inversion might require iteration.

7.3 Adiabatic Invariant Approximation

The adiabatic invariant approximation is probably valid in most of the deep oceans

520 Ocean Acoustics

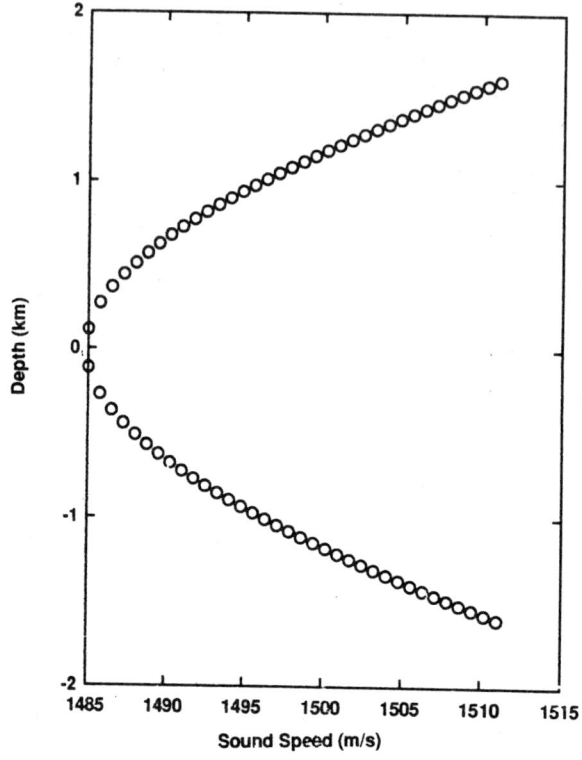

Fig. 25. Range average of the equivalent symmetric sound channel found by numerical integration of the Abel transform in Appendix A for the simulated modal pulse travel-time measurements in Table 4.

away from discontinuities such as polar fronts. But when horizontal gradients are so large that the adiabatic invariant approximation is no longer valid, the ray action will differ significantly between adjacent loops that straddle the region of large horizontal gradient. This can lead to errors in nonperturbative inversion because the range average (actually, the loop average) of the action would be used in the Abel transform instead of the correct action.

Breakdown of the adiabatic-invariant approximation can lead to errors in perturbative inversion such as objective mapping, but for a different reason. Unless the background profile used for the inversion has the appropriate horizontal gradient at the range where the adiabatic-invariant approximation breaks down, the turning-point depths calculated using the background profile may differ significantly from the correct turning-point depths. In that case, the validity of the inversion might need to be verified.

For modal tomography, breakdown of the adiabatic-invariant approximation causes significant coupling among modes. In that case, the range averaging in (2.2) and (2.3) generalizes to a phase integral in the complex range plane. Modal

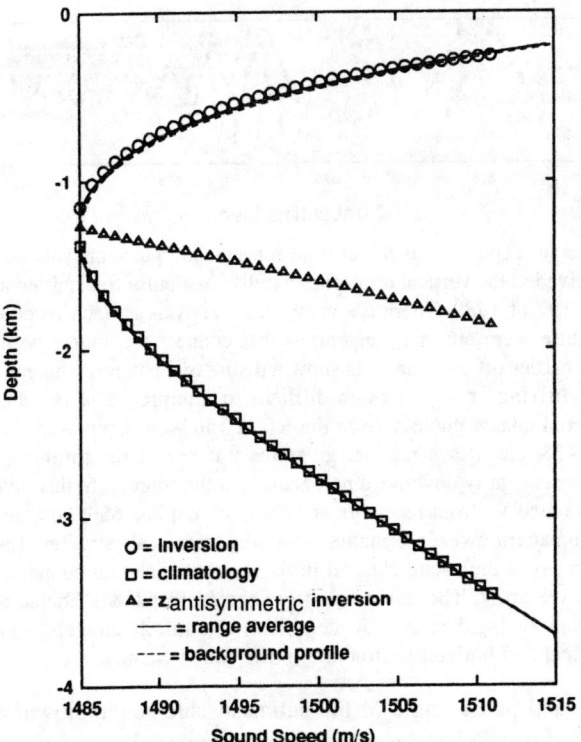

Fig. 26. Range-averaged sound channel, from combining the equivalent symmetric sound channel in Fig. 25 found by inversion with simulated climate data for the part of the sound channel below the sound channel axis.

phase for mixed-mode propagation is calculated by phase integrals around the appropriate critical coupling points in the complex range plane [63].

7.4 Aliasing from Large Wavenumber Sound-Speed Structure

The travel time measured along a ray will have a contribution that is due to aliasing from sound-speed structure having wavelengths equal to a double-loop range (about 50 km) and higher harmonics [23]. Such submesoscale sound-speed structure includes internal waves, whose amplitude decreases quickly with depth [64]. Therefore, the major effect of the aliasing will occur near the upper turning points of the ray. We can approximate the effect by taking the amplitude at the turning-point times about a third of the upper-loop range. This gives

$$x \frac{A}{3} \frac{X_+}{X} \cos \phi_A = x \frac{A}{6} \left(1 + \frac{\Delta X}{X}\right) \cos \phi_A \qquad (7.1)$$

as a contribution to the travel time from aliasing, where A and ϕ_A are the amplitude

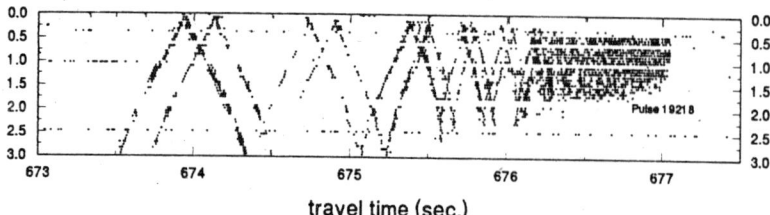

Fig. 27. Receiver depth versus travel time for acoustic pulse arrivals above a threshold received at the vertical array for an individual pulse transmitted at 0250 UTC on day 192 of 1989. There are many pulse arrivals at each receiver (each depth) because there are many eigenrays that connect the source with each receiver. The earlier off-axial arrivals show a distinctive pattern, whereas the more axial, late-arriving energy is more difficult to interpret. The ray identifiers can be ordered unambiguously from the left at 1500-m depth: −35, 36, −36, 37, −37, 38, −38, etc. (Each number gives the number of ray turning points; negative numbers signify downward propagation at the source.) In this inversion, we used pulse arrivals from receivers at only three depths: 660, 960, and 1200 m. The main pattern gives information about the large-scale structure (mostly the range average) of the sound channel in the vertical slice that contains the source and receiver array. The deviations from the main pattern (including the multiple arrivals having the same ray identifiers) give information about the smaller-scale vertical and horizontal structure in the sound channel.

(in ms km^{-1}) and phase angle of the submesoscale sound-slowness component, and ΔX is the difference in range between an upper (X_+) and a lower (X_-) loop.

Equation (7.1) assumes coherence of the submesoscale component over the entire length of the acoustic path. A more realistic assumption is incoherence at least from one turning point to the next. The effect is to divide (7.1) by $\sqrt{n}$, where n is the number of upper turning points. If the horizontal coherence length of the submesoscale component is less than $(X_+)/3$, then $(X_+)/3$ in (7.1) should be replaced by the coherence length.

The processing of travel time in (5.2a) and (5.2c) attenuates the aliasing further by a factor of about

$$| 1 - \exp\{i\pi\, \Delta X/X + 2\, i\pi\, [X_T/X + (-1)^N\, X_R/X]\} | \qquad (7.2)$$

where N is the number of turning points in the raypth, X_T is the range from the source to the nearest crossing of the raypath with the sound-channel axis, X_R is the range from the receiver to the nearest crossing of the raypath with the sound-channel axis, and X_T or X_R is negative if the source or receiver, respectively, is below the sound-channel axis.

To estimate the effect of aliasing, let us first consider a special case. For a symmetric sound channel ($\Delta X = 0$), and for the source and receiver on the sound channel axis ($X_T = X_R = 0$), the factor in (7.2) is zero, cancelling any aliasing effect. Thus, aliasing effects will be minimized for near-axial rays.

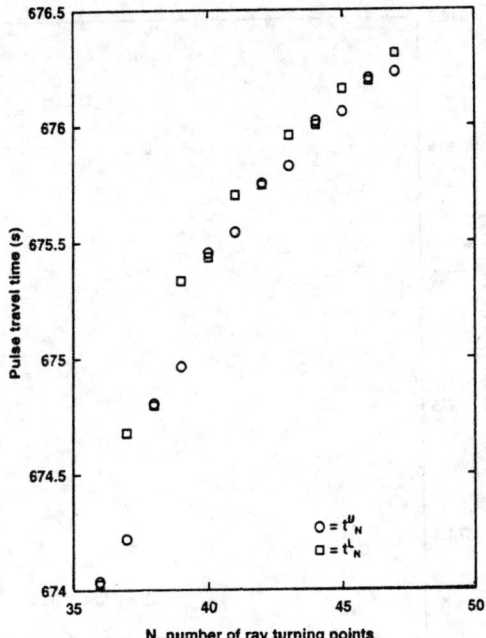

Fig. 28. Measured travel times (from Fig. 27) versus loop number (number of turning points) for a receiver depth of 660 m. The dependence of pulse arrival time on loop number for other receiver depths is similar. The circles are for upward propagation at the source. The squares are for downward propagation. For multiple arrivals having the same pulse identifiers in Fig. 27, a single pulse was chosen based on strength or continuity in time of depth. Notice a general increase of travel time with the number of turning points. This is a general characteristic of the ocean sound channel where axial rays arrive last and have ray loops with the shortest range [25]. The difference in travel time for upward versus downward propagation for the same number of turning points is due to the asymmetry and range dependence of the sound channel combined with the displacement of the source and receiver from the sound-channel axis. These data represent four functions of the number of turning points because the acoustic wave can propagate upward or downward at the source and arrive upward or downward at the receiver. These four functions contain the maximum amount of acoustic information available from a single source-receiver pair unless we include all the pulse arrivals having the same pulse identifiers.

More generally, the factor in (7.2) can vary between zero and two. To estimate the contribution to aliasing from (7.1), we consider internal waves of wavelength equal to the double-loop range of a ray (about 50 km). These internal waves having a wavenumber of about 1/50 cycle km^{-1} are near the lower end of the internal wave spectrum in the ocean (Flatté et al, [64], Fig. 3.4, p. 57). To estimate the magnitude of (7.1), we use $A = \delta S/S = \delta C/C = 5 \times 10^{-4}$, $\Delta X/X = 1/2$, and

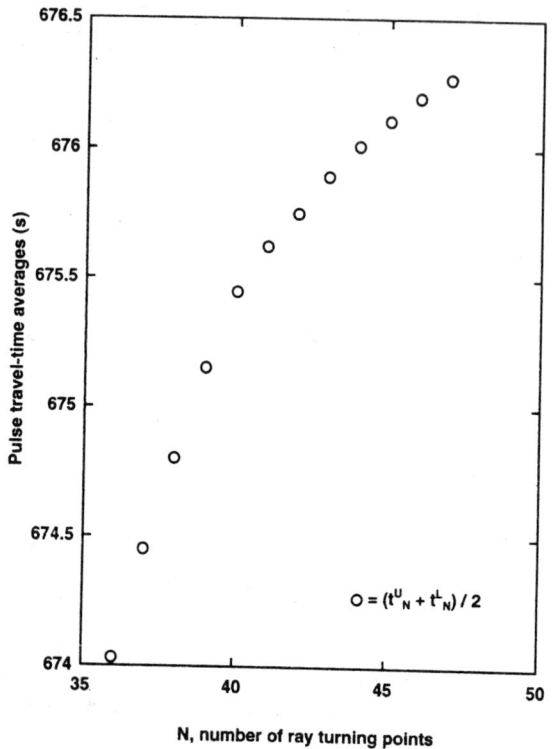

Fig. 29. Average of travel time for upward- and downward-propagated pulses for a receiver depth of 660 m from Fig. 28. To first order, averaging upward- and downward-propagated pulses cancel the effects of the partial ray loops at the source and receiver and the asymmetry of the sound channel for both even- and odd-numbered loops. That is, the travel times shown here are approximately the same as we would expect for rays consisting of complete ray loops in the equivalent symmetric sound channel, which allows measurements from even- and odd-numbered loops to be used together. To first order, these data determine the range average of the equivalent symmetric sound channel in the vertical slice containing the source and receiver. Although these data represent, in principle, two functions (for odd and even number of turning points), the two functions are indistinguishable here.

$x = 1000$ km to give about 75 ms when the raypath and internal wave are in phase. If the coherence length of the internal wave is less than 50 km, however, the aliasing effect is reduced to about 10–15 ms. This estimate of the aliasing effect is consistent with observed effects of internal waves on the travel times of acoustic waves. Because the aliasing depends on the phase of the submesoscale component relative to the raypath, the aliasing will change in magnitude and sign as the phase shifts. It is normal practice in ocean acoustic tomography to average the travel time over a period of 12 h or so to minimize this effect from internal waves.

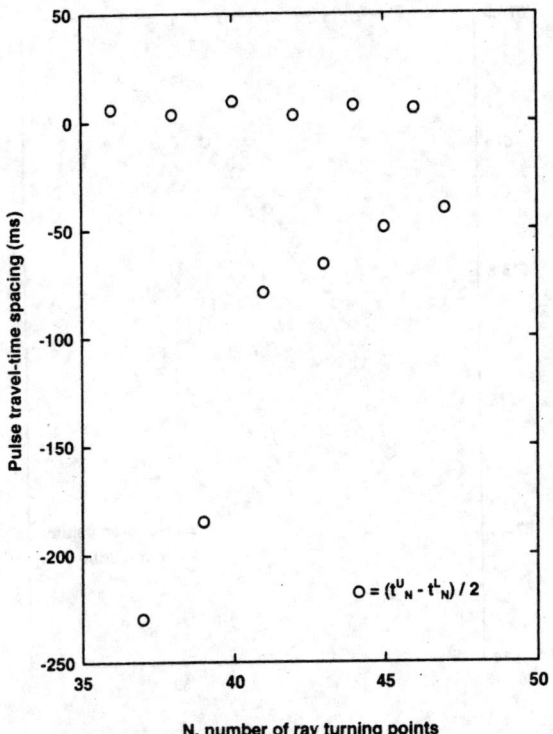

Fig. 30. Difference between the travel time for upward- and downward-propagated pulses for a receiver depth of 660 m from Fig. 28. To first order, the difference depends only on the partial ray loops at the source and receiver, and is independent of the propagation throughout most of the path. Fitting smooth curves to the data (for odd and even numbers of turning points separately) determines two functions that give the asymmetry of the sound channel at the source and receiver.

7.5 Curve-Fitting Errors

In the example using simulated measurements, we were able to do curve fitting through second derivatives to the simulated travel-time measurements so that (5.2) could be successfully iterated. One might suppose that in the presence of noise, it would be impossible to estimate first and second derivatives accurately enough. That would be true if insufficient care was taken to suppress noise in the curve fitting. It is, in fact, not difficult to estimate the slope and curvature in Figs. 3a, 3b, 29, and 30 for measured travel time, even in the presence of noise. Further, the second term on the right-hand side of each equation in (5.2) is a small correction to the first term. In the example using simulated measurements, the largest correction from the second term on the right-hand side of (5.2a) was less than 0.1 ms and for (5.2c) was 10 ms. Thus, a 10% error in estimating the curvature in Fig. 3a would result in less than a 1 ms error in Fig. 4, which would result in a relative

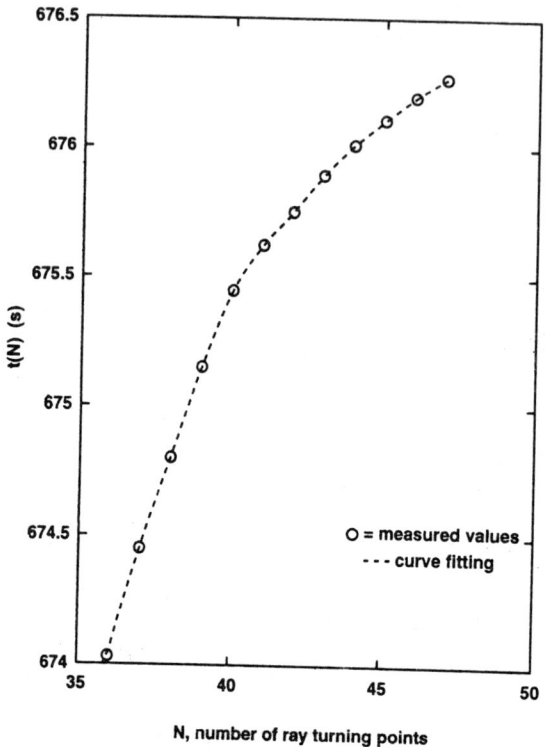

Fig. 31. Average of travel time for upward- and downward-propagated pulses for a receiver depth of 660 m, from curve fitting to the points in Fig. 29. Curve fitting is necessary to obtain estimates of first and second derivatives of travel time with respect to the number of ray turning points. The curve-fitting method used here is a cubic fit between points in which the four coefficients for the cubic are determined by the data value and the slope at each point. The slope at each interior point is determined from the parabola that fits three adjacent points. The slope of that parabola at the center data point is assigned to that data point. Assigning a slope to the points at the ends is more complicated.

error of about 1.5×10^{-6} for $\hat{S}$, and about the same relative error for sound speed. The largest correction from the second term on the right-hand side of (5.2b) was 0.03% and for (5.2d) was 0.5%. Thus a 10% error in estimating the curvature in Fig. 3b would result in an error of about 0.003% in $g(N)$ and 0.05% in $f(N)$. One millisecond of "noise" in travel time measurements would lead to an error of about 1.5 ms in the estimate of $t'(N)$ and about 0.01% error in the estimate of $\hat{S}$ for $N = 40$ (the worst case). This would lead to about the same relative error for sound speed, or about 0.15 ms^{-1}.

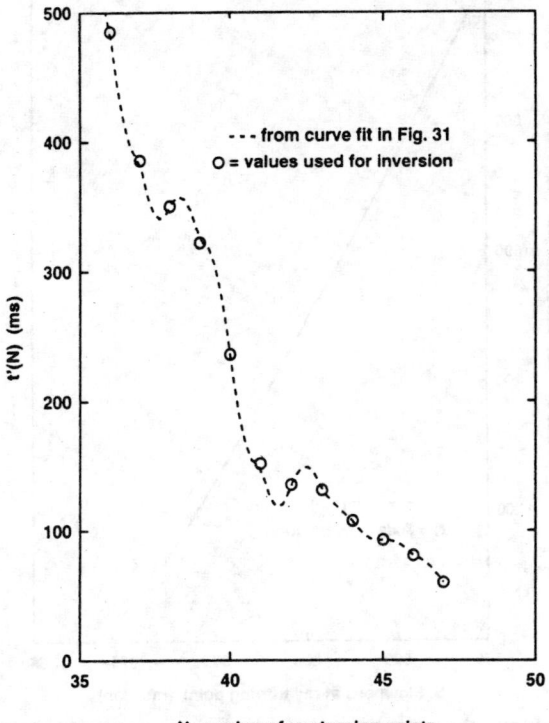

Fig. 32. Derivative of travel time with respect to the number of turning points for a receiver depth of 660 m. The derivative is necessary to estimate the action using (5.5) and the range average of ray-turning point sound slowness in (5.6). The value at each circled point in the interior is determined by fitting a parabola to three adjacent points in Fig. 31. Determining the value at the two endpoints is more complicated. This method for estimating values of $t'(N)$ (slopes of the function in Fig. 31) from the measurements in Fig. 31 is fairly robust for the interior data points. It gives less accurate estimates for the values at the two end points and between data points. This is not a problem for the present inversion, however, because the values between data points were not used in the inversion.

8. Acknowledgments

The computer program for performing the tomography inversion was written by Lloyd Nesbitt [65]. The computer program for performing the Abel transform was written by Ann Weickmann [66]. The computer program that generated all of the figures except Fig. 27 was written by Lloyd Nesbitt [67]. This research was supported by the NOAA Environmental Technology Laboratory of the U.S. Department of Commerce in Boulder, Colorado, U.S.A.

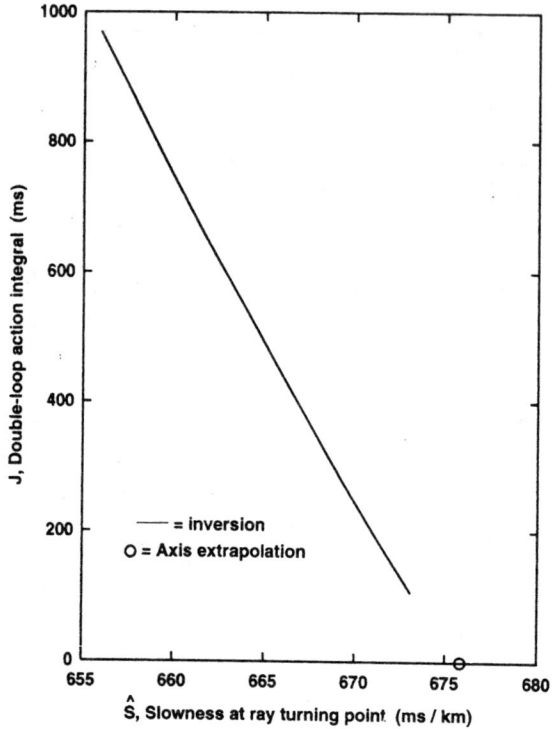

Fig. 33. Action versus the range average of ray-turning-point sound slowness. The action is calculated from Fig. 32 using (5.5). The range average of sound slowness is calculated from Figs. 31 and 32 using (5.6). Extrapolating to $J = 0$ yields the range average of sound slowness on the sound-channel axis of 675.85 ± 0.085 ms km^{-1} (1479.69 ± 0.18 ms^{-1} for the sound speed). The error includes both extrapolation error and the uncertainly in transmitter-receiver range. Results for receiver depths of 660, 960, and 1200 m are superimposed.

Appendix A. Abel Transform

The *equivalent symmetric profile* is given by an Abel transform [25],

$$z_s(S) \equiv \frac{z_+(S) - z_-(S)}{2} = -\frac{1}{2\pi} \int_{S_A}^{S} \frac{X(\hat{S})\, d\hat{S}}{(\hat{S}^2 - S^2)^{1/2}} \tag{A1}$$

$$= \frac{1}{2\pi} \int_{S_A}^{S} \frac{J'(\hat{S})\, d\hat{S}}{(\hat{S}^2 - S^2)^{1/2}} = \frac{1}{2\pi} \int_{0}^{J_S} \frac{dJ}{(\hat{S}^2 - S^2)^{1/2}},$$

where

$$X = -\frac{dJ}{d\hat{S}} = \hat{S} \oint [S^2(z) - \hat{S}^2]^{-1/2} dz, \tag{A2}$$

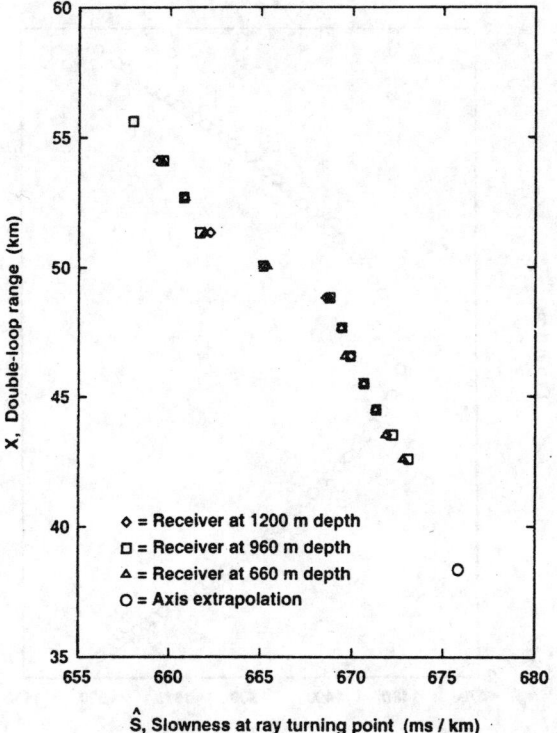

Fig. 34. Double-loop range as a function of the range average of ray-turning-point sound slowness. Double-loop range is calculated from (5.7). Results for receiver depths of 660, 960, and 1200 m are superimposed.

is the range for one double loop, $z_+(S)$ and $z_-(S)$ are the sound-slowness profiles above and below the sound-channel axis, and the upper limit in the integration is where the denominator is zero. The range-averaged version of (A1), in which S and $\hat{S}$ are replaced by their range averages, is an approximation [44] and gives a warm bias.

As is well known (e.g., Kelso [68]; Aki and Richards [26], Abel transform inversions give unique solutions for sound-speed profiles only if the profiles are monotonic above and below the sound-channel axis. Although the sound channel is monotonic below the sound-channel axis, the upper part of the sound channel often exhibits nonmonotonic structure. Thus, inversions can be unique at most from the sound-channel axis up to the first sound-speed maximum for *any* inversion method, unless supplemented by independent (nontomographic) information.

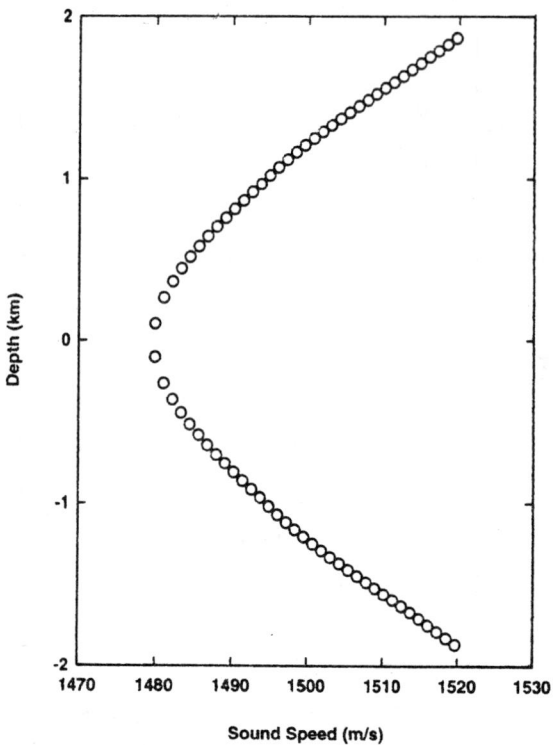

Fig. 35. Equivalent symmetric sound channel from tomographic inversion. The sound speed is the reciprocal of the range average of sound slowness. The sound speed is approximately a range average if the range dependence of sound speed is small. The profile results from taking the Abel transform (A1) of the function in Fig. 34 (and using 1479.69 m s^{-1} for the sound speed of the sound-channel axis). Using measurements for receiver depths of 660, 960, and 1200 m increases the effective vertical resolution of the profile.

Appendix B.   Curved-Earth Transformation

The flat-earth to curved-earth transformation is [69]

$$C_{\text{curved earth}} = \frac{r_0 + z_{\text{curved earth}}}{r_0 + z_1} C_{\text{flat earth}}$$

$$= C_{\text{flat earth}} \exp\left(\frac{z_{\text{flat earth}} - z_1}{r_0 + z_1}\right) \quad \text{(B1)}$$

and

$$z_{\text{curved earth}} = (r_0 + z_1) \exp\left(\frac{z_{\text{flat earth}} - z_1}{r_0 + z_1}\right) - r_0, \quad \text{(B2)}$$

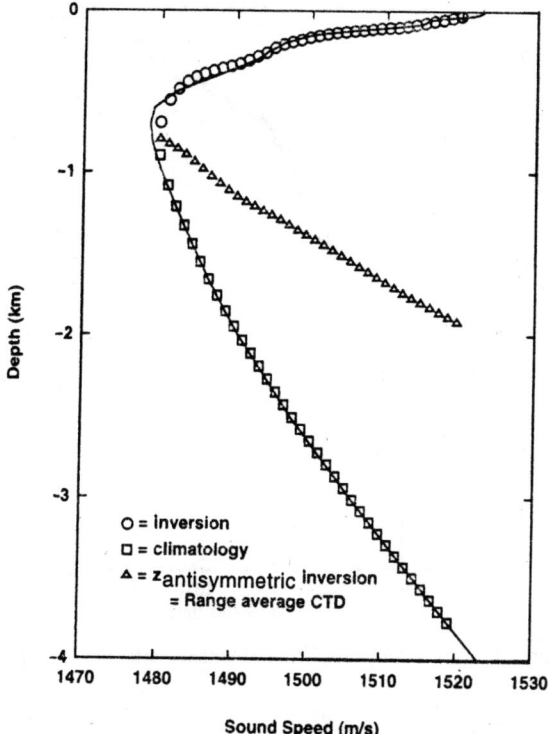

Fig. 36. Tomography inversion sound-speed profile. The circles give the sound-speed profile whose equivalent symmetric profile is that of Fig. 35 and that matches Levitus climatology below the sound-channel axis. This profile would be a range-averaged profile if the profile below the sound-channel axis were really range independent. It includes a flat-earth to curved-earth transformation [43]. The solid line gives the range average of sound-speed profiles calculated from CTD measurements made during the experiment, for comparison. The CTD measurements spanned day 187 through day 195 of 1989.

where z_1 is the height (relative to sea level) at which the flat-earth and curved-earth profiles match, and which we arbitrarily choose to be the height of the sound-channel axis z_A. The quantity r_0 is the local radius of curvature of the ocean surface in the vertical slice between the source and receiver. Because this is an exact transformation, it introduces no error.

532 *Ocean Acoustics*

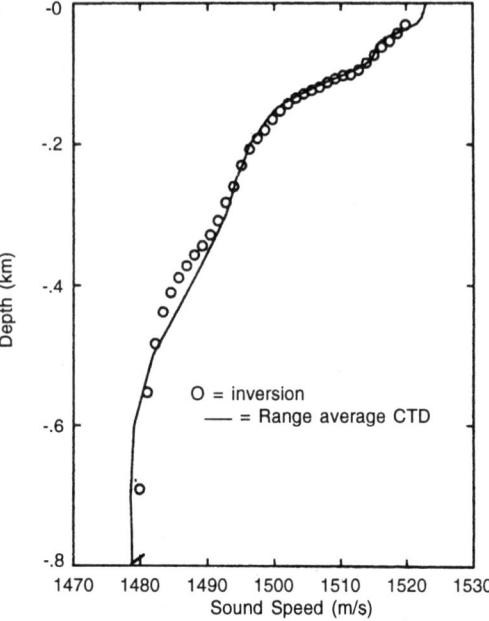

Fig. 37. The upper 800 m of Fig. 36. The agreement is good for depths between 50 m and 300 m.

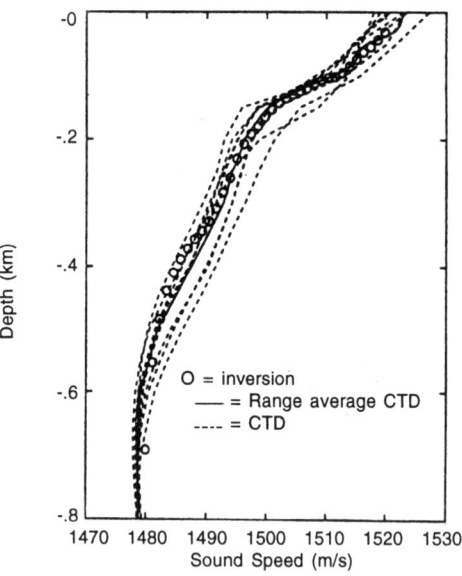

Fig. 38. As in Fig. 37, but including individual sound-speed profiles calculated from CTD measurements made during the experiment.

References

1. W. Munk and C. Wunsch, "Ocean acoustic tomography: A scheme for large-scale monitoring," *Deep-Sea Res.,* vol. 26A, pp. 123–161, 1979.
2. B.D. Cornuelle, "Acoustic tomography." *IEEE Trans. Geosci. Remote Sens.,* vol. GE-20, pp. 326–332, 1982.
3. P.F. Worcester, "Remote sensing of the ocean using acoustic tomography," in *Advances in Remote Sensing Retrieval Methods,* A. Deepak, H.E. Fleming, and M.T. Chahine, Eds. Hampton, VA: A. Deepak Publishing, 1985, pp. 1–11.
4. W.H. Munk and P.F. Worcester, "Ocean acoustic tomography," *Oceanography,* vol. 1, pp. 8–10, 1988.
5. R.A. Knox, "Ocean acoustic tomography: A primer," in *Ocean Circulation Models: Combining Data and Dynamics,* D.L.T. Aderson and J. Willebrand, Eds. Norwell, MA: Kluwer Academic Publishers, 1988, pp. 141–188.
6. Y. Desaubies, "Ocean acoustic tomography," in *Proc., Oceanographic and Geophysical Tomography,* 50th Leshouches Ecole d'Ete de Physique Theorique and NATO ASI, Y. Desaubies, A. Tarantola, and J. Zinn-Justin, Eds. Elsevier Science Publishers, 1990, pp. 159–202.
7. R.C. Spindel and P.F. Worcester, "Ocean acoustic tomography," *Scientific American,* vol. 263, pp. 94–99, 1990.
8. R.C. Spindel and P.F. Worcester, "Ocean acoustic tomography programs: Accomplishments and plans," in *Proc., OCEANS'90,* Washington, DC, September 24–26, 1990, pp. 1–10.
9. P.F. Worcester, B.D. Cornuelle, and R.C. Spindel, "A review of ocean acoustic tomography: 1987–1990," *Reviews of Geophysics,* Supplement, U.S. National Report to International Union of Geodesy and Geophysics 1987–1990, pp. 557–570, 1991.
10. Ocean Tomography Group, "A demonstration of ocean acoustic tomography", *Nature,* vol. 299, pp. 121–125, 1982.
11. B. Cornuelle, C. Wunsch, D. Behringer, T. Birdsall, M. Brown, R. Heinmiller, R. Knox, K. Metzger, W. Munk, J. Spiesberger, R. Spindel, D. Webb, and P. Worcester, "Tomographic maps of the ocean mesoscale. Part 1: Pure acoustics," *J. Phys. Oceanogr.,* vol. 15, no. 2, pp. 133–152, 1985.
12. P.F. Worcester, R.C. Spindel, and B.M. Howe, "Reciprocal acoustie transmissions: Instrumentation for mesoscale monitoring of ocean currents," *IEEE J. Ocean Eng.,* vol. OE-10, pp. 123–137, 1985.
13. W. Munk and C. Wunsch, "Observing the ocean in the 1990s, "*Phil. Trans. R. Soc. Lond.,* vol. A307, pp. 439–464, 1982.
14. J.L. Spiesberger, P.J. Bushong, K. Metzger, and T.G. Birdsall, "Basin-scale tomography: Synoptic measurements of a 4000km length section in the Pacific, " *J. Phys. Oceanogr.,* vol. 19, pp. 1073–1090, 1989.
15. J.L. Spiesberger, K. Metzger, and J.A. Furgerson, "Listening for climatic temperature change in the northeast Pacific: 1983–1989," *J. Acoust. Soc. Am.,* vol. 92, pp. 384–396, 1992.
16. J.L. Spiesberger and K. Metzger, "Basin-scale tomography: A new tool for studying weather and climate," *J. Geophys. Res.,* vol. 96, pp. 4869–4889, 1991.
17. J.L. Spiesberger and K. Metzger, "Basin-scale ocean monitoring with acoustic thermometers," *Oceanography,* vol. 5, pp. 92–98, 1992
18. W. Munk and A.M. Forbes, "Global ocean warming: An acoustic measure?" *J. Phys. Oceanogr.,* vol. 19, pp. 1765–1778, 1989.

19. R.C. Spindel, "Measuring global warming through ocean acoustics: The Heard Island experiment," *Echoes* (The newsletter of the Acoustical society of America), vol. 1, no. 4, pp. 1, 4, 5, 1991.
20. T.M. Georges, "Taking the ocean's temperature with sound," *The World & I*, vol. 7, no. 7, pp. 282–289, 1992.
21. R.B. Stoughton, S.M. Flatté, and B.M. Howe, "Acoustic measurements of internal wave rms displacement and rms horizontal current off Bermuda in late 1983," *J. Geophys. Res.*, vol. 91, pp. 7721–7732, 1986.
22. B.M. Howe, "Multiple receivers in single vertical slice ocean acoustic tomography experiments," *J. Geophys. Res.*, vol. 92, pp. 9479–9486, 1987.
23. B.D. Cornuelle and B.M. Howe, "High spatial resolution in vertical slice ocean acoustic tomography," *J. Geophys. Res.*, vol. 92, pp. 11, 680–11, 692, 1987.
24. C. Wunsch, "Acoustic tomography by Hamiltonian methods including the adiabatic approximation," *Rev. Geophys.*, vol. 25, pp. 41–53, 1987.
25. W. Munk and C. Wunsch, "Ocean acoustic tomography: Rays and modes," *Rev. Geophys. Space Phys.*, vol. 21, pp. 777–793, 1983.
26. K. Aki and P.G. Richards, *Quantitative Seismology*, vols. I&II. San francisco, CA: W.H. Freeman, 1980, 932 pp.
27. R.M. Jones, T.M. Georges, and J.P. Riley, "Inverting vertical-slice tomography measurements of asymmetric ocean sound-speed profiles," *Deep-Sea Res.*, vol. 33, pp. 601–619, 1986.
28. W. Munk and C. Wunsch, "Bias in acoustic travel time through an ocean with adiabatic range dependence," *Geophys. Astrophys. Fluid Dyn.*, vol. 39, pp. 1–24, 1987.
29. R.M. Jones, T.M. Georges, L. Nesbitt, R. Tallamraju, and A. Weickmann, "Vertical-slice ocean-acoustic tomography—Extending the Abel inversion to non-axial sources and receivers," *Colloque de Physique,* 1er Congrés Français d'Acoustique, vol. C2, pp. C2–1013-C2–1016, 1990.
30. R.M. Jones, T.M. Georges, L. Nesbitt, R. Tallamraju, and A. Weickmann, "Vertical-slice ocean-acoustic tomography inversion—Converting mullti-loop measurements to single-loop data," in *Proc. of the Fifth International Symposium on Acoustic Remote Sensing of the Atmosphere and Oceans,* S.P. Singal, Ed. New Delhi, India: Tata McGraw-Hill, 1990, pp. 239–244.
31. R.M. Jones, T.M. Georges, L. Nesbitt, and A. Weickmann, "Ocean acoustic tomography inversion in the adiabatic-invariant approximation," NOAA Tech. Memo., ERL WPL-217 (available from the National Technical Information Service, 5285 Port Royal Rd., Springfield, VA 22161), 233 pp., 1991.
32. E.C. Shang, "Ocean acoustic tomography based on adiabatic mode theory," *J. Acoust. Soc. Am.*, vol. 85, pp. 1531–1537, 1989.
33. R.M. JONES, E.C. Shang, and T.M. Georges, "Nonperturbative modal tomography inversion—Part I: Theory," *J. Acoust. Soc. Am.*, vol. 94, pp. 2296–2302, 1993.
34. E.C. Shang, Y.Y. Wang, R.M. Jones, and T.M. Georges, "Nonperturbative modal tomography inversion—Part II: Simulation", *J. Acoust. Soc. Am.*, vol. 98, pp. 560–569, 1995.
35. P.F. Worcester, D.A. Peckham, K.R. Hardy, and F.O. Dormer, "AVATAR: Second-generation transceiver electronics for ocean acoustic tomography," in *OCEANS '85 Conference Record,* 1985, pp. 654–662.
36. S. Levitus, *Climatological Atlas of the World Ocean,* NOAA Professional Paper 13, U.S. Government Printing Office, Washington, DC, GPO No. C55.25: 13

(available from the National Technical Information Service, 5285 Port Royal Rd., Sp0ringfield, VA 22161; No. PB83184093 XSP), 1982, 173 pp.
37. J.L. Spiesberger, R.C. Spindel, and K. Metzger, "Stability and identification of ocean acoustic multipaths," *J. Acoust. Soc. Am.*, vol. 67, pp. 2011–2017, 1980.
38. P.F. Worcester, "An example of ocean acoustic multipath identification at long range using both travel time and vertical arrival angle," *J. Acoust. Soc. Am.*, vol. 70, pp. 1743–1747, 1981.
39. B.M. Howe, P.F. Worcester, and R.C. Spindel, "Ocean acoustic tomography: Mesoscale velocity," *J. Geophys. Res.*, vol. 92, pp. 3785–3805, 1987.
40. B.M. Howe, J.M. Mercer, R.C. Spindel, P.F. Worcester, J.A. Hildebrand, W.S. Hodgkiss, Jr., T.F. Duda, and S.M. Flatté, "Slice89: A single slice tomography experiment," in *Ocean Variability and Acoustic Propagation, Proc. of the Workshop on Variability and Acoustic Propagation,* La Spezia, Italy, June 4–8, 1990, J, Potter and A. Warn-Varnas, Eds. Kluwer Academic Publishers, 1991, pp. 81–86.
41. P.F. Worcester, B.D. Dushaw, and B.M. Howe, "Gyre-scale reciprocal acoustic transmission," in *Ocean Variability and Acoustic Propagation, Proc. of the Workshop on Ocean Variability and Acoustic Propagation,* La Spezia, Italy, June 4–8, 1990, J, Potter and A. Warn-Varnas, Eds. Kluwer Academic Publishers, 1991, pp. 119–134.
42. T.F. Duda, S.M. Flatté, J.A. Colosi, B.D. Cornuelle, J.A. Hildebrand, W.S. Hodgkiss, Jr., P.F. Worcester, B.M. Howe, J.A. Mercer, and R.C. Spindel, "Measured wavefront fluctuations in 1000-km pulse propagation in the Pacific Ocean," *J. Acoust. Soc. Am.*, vol. 92, no. 2, pt. 1, pp. 939–955, 1992.
43. R.M. Jones and T.M. Georges, "Nonperturbative ocean acoustic tomography inversion," *J. Acoust. Soc. Am.*, vol. 96, no. 1, pp. 439–451, July 1994.
44. R.M. Jones, B.M. Howe, J.A. Mercer, R.C. Spindel, and T.M. Georges, "Nonperturbative ocean acoustic tomography inversion of 1000km pulse propagation in the Pacific Ocean," *J. Acoust. Soc. Am.*, vol. 96, no. 5, pp. 3054–3063, 1994.
45. R.M. Jones, T.M. Georges, and J.P. Riley, "HARPO, a versatile three-dimensional Hamiltonian ray-tracing program for acoustic waves in the ocean with irregular bottom, "NOAA Environmental Research Laboratories, Boulder, CO, USA, 80303, NOAA Special Report, 455 pp., 1986.
46. A. Weickmann, J.P. Riley, T.M. Georges, and R.M. Jones, "EIGEN—A program to compute eigenrays from HARPA/HARPO raysets," NOAA Environmental Research Laboratories, Environmental Technology Laborotory, Boulder, CO, USA, 80303, NOAA Tech. Memo. ERL WPL-160, 91 pp. 1989.
47. M.B. Porter, "The KRAKEN normal mode program, "SACLANT Undersea Center, Oct. 1988.
48. W. Munk, "Sound channel in an exponentially stratified ocean, with application to SOFAR," *J. Acoust. Soc. Am.*, vol. 55, pp. 220–226, 1974.
49. V.V. Goncharov and A.G. Voronovich, "An experiment on matched-field acoustic tomography with continuous wave signals in the Norway Sea," *J. acoust. Soc. Am.*, vol. 93, pp. 1873–1881, 1993.
50. B.D. Cornuelle, P.F. Worcester, J.A. Hildebrand, W.S. Hodgkiss, Jr., T.F. Duda, J. Boyd, B.M. Howe, J.A. Mercer, and R.C. Spindel, "Ocean acoustic tomography at 1000-km range using wavefronts measured with a large-aperture vertical array," *J. Geophys. Res.*, vol. 98, pp. 16, 365–16, 377, 1993.
51. P.F. Worcester, B.D. Cornuelle, J.A. Hildebrand, W.S. Hodgkiss, Jr., T.F. Duda,

J. Boyd, B.M. Howe, J.A. Mercer, and R.C. Spindel, "A comparison of measured and predicted broadband acoustic arrival patterns in travel time-depth corrdinates at 1000-km range," *J. Acoust. Soc. Am.,* vol. 95, no. 6, pp. 3118–3128, June 1994.

52. B.J. Sotirin and J.A. Hildebrand, "Large aperture digital acoustic arry, "*IEEE J. Ocean Eng.,* vol. 13, pp. 271–281, 1988.

53. S.M. Flatté, J. Colosi, T. Duda, and G. Rovner, "Impulse response analysis of ocean acoustic propagation, "in *Proc., NATO Conference on Ocean Variability and Acoustic Propagation,* La Spezia, Italy, June 4–8, 1990, J. Potter and A. Warn-Varnas, Eds. Kluwer Academic Publishers, Dordrecht/Boston/London, 1991, pp. 161–172.

54. V.A. Del Grosso, "New equation for the speed of sound in natural waters (with comparisons to other equations)," *J. Acoust. Soc. Am.,* vol. 56, pp. 1084–1090, 1974.

55. P.B. Liebelt, *An Introduction to Optimal Estimation.* Reading, MA: Addison-Wesley, 1967.

56. G.T. Herman (Ed.) *Image Reconstruction from Projections: Implementation and Applications.* Heidelberg, NY: Springer-Verlag, Berlin, 1979, 284 pp.

57. W. Menke, *Geophysical Data Analysis: Discrete Inverse Theory.* Orlando, FL: Academic Press, 1984.

58. A.C. Kak and M. Slaney, *Principles of Computerized Tomographic Imaging.* New York: IEEE Press, 1988, 401 pp.

59. A.J. Devaney, "A computer simulation study of diffraction tomography," *IEEE Trans. Biomedical Eng.,* vol. BME-30, pp. 377–386, 1983.

60. A.J. Devaney "Acoustic tomography," in *Inverse Problems of Acoustic and Elastic Waves,* F. Santosa, Y.-H. Pao, W.W. Symes, and C. Holland, Eds. Philadelphia, PA: SIAM, 1984, pp. 250–273.

61. A.J. Devaney, "Variable density acoustic tomography," *J. Acoust. Soc. Am.,* vol. 78, pp. 120–130, 1985.

62. A.J. Devaney, "The limited-view problem in diffraction tomography, "*Inverse Problems,* vol. 5, pp. 501–521, 1989.

63. K.G. Budden, "The critical coupling of modes in a tapered earth-ionosphere wave guide," *Math. Proc. Camb. Phil. Soc.,* vol. 77, pp. 567–580, 1975.

64. S.M. Flatté (Ed.), R. Dashen, W.H. Munk, K.M. Watson, and F. Zachariasen, *Sound Transmission Through a Fluctuating Ocean.* Cambridge, England: Cambridge University Press, 1979, 299 pp.

65. L. Nesbitt and R.M. Jones, "A FORTRAN program for performing nonperturbative ocean tomography inversion," NOAA Environmental Research Laboratories, Environmental Technology Laboratory. Boulder. CO. USA, 80303, NOAA Tech. Memo., ERL ETL-243, 165 pp., 1994.

66. A. Weickmann and R.M. Jones, "A FORTRAN program for performing Abel transforms," NOAA Fnvironmental Research Laboratories, Environmental Technology Laboratory, Boulder, CO, USA, 80303, NOAA Tech. Memo., ERL ETL-244, 158 pp., 1994.

67. L. Nesbitt and R.M. Jones, "A C program for generating publication-quality plots," NOAA Environmental Research Laboratories, Environmental Technology Laboratory, Boulder, CO, USA, 80303, NOAA Tech. Memo., ERL ETL-242, 160 pp., 1994.

68. J.M. Kelso, *Radio Ray Propagation in the Ionosphere.* New York: McGraw-Hill, 1964.

69. A. Ben-Menahem and S.J. Singh, "Earth-flattening transformation," in *Seismic Waves and Sources.* New York: Springer-Verlag, 1981, chap. 7.3.2, pp. 495–502.

Acoustic Remote Sensing Applications
S.P. Singal (Ed)
Copyright © 1997 Narosa Publishing House, New Delhi, India

22. Acoustic Measurements of Currents and Effluent Plume Dilutions in the Western Edge of the Florida Current

John R. Proni, Ph.D.[1] and Robert G. Williams, Ph.D.[2]

[1]Director, Ocean Acoustics Division, Environmental Research Laboratories
Atlantic Oceanographic and Meteorological Laboratory,
4301 Rickenbacker Causeway, Miami FL 33149 U.S.A.

[2]NOAA, National Ocean Service, Office of Ocean and Earth Science,
Marine Analysis and Interpretation Division, Coastal and Estuarine
Oceanography BranchNOAA/NOES 33, SSMC4 Rm 6426,
1305 East West Highway, Silver Springs MD 20910 U.S.A.

1. Introduction

The Southeast Florida Outfall Experiment (SEFLOE), which was carried out in the coastal ocean off Southeast Florida, U.S.A., between 1988 and 1993, was the most extensive application of acoustical technologies in wastewater effluent studies so far performed in the U.S.A. The general objective of SEFLOE was to develop a scientific basis for managerial regulation and rules governing effluent discharges. Physical oceanographic, biological, and chemical measurements were made both during establishment of initial effluent plume dilution and subsequent plume dilution. Acoustics played a central role in all three disciplines. In particular, acoustical systems were used to (1) detect and map the three dimensional subocean-surface effluent plume distribution, (2) guide chemical and biological sampling, (3) provide relative plume dilution measurements with space and time, (4) distinguish between effluent plumes and interfering plumes arising from other sources, (5) detect the presence of ambient biota within the water column in the vicinity of the discharge, and (6) measure three dimensional oceanic currents from near-ocean-surface to ncar-ocean-bottom every ten to fifteen minutes.

A general site map for two of the outfalls studied is shown in Fig. 1. Both outfalls have diffuser systems, which are designated by a solid circle. Note the presence of the offshore dredged material disposal site (ODMDS) and the shipping channel to the Port of Miami; interfering plumes may arise from both these locations.

An outstanding feature of the coastal ocean off sotheast Florida is the Florida

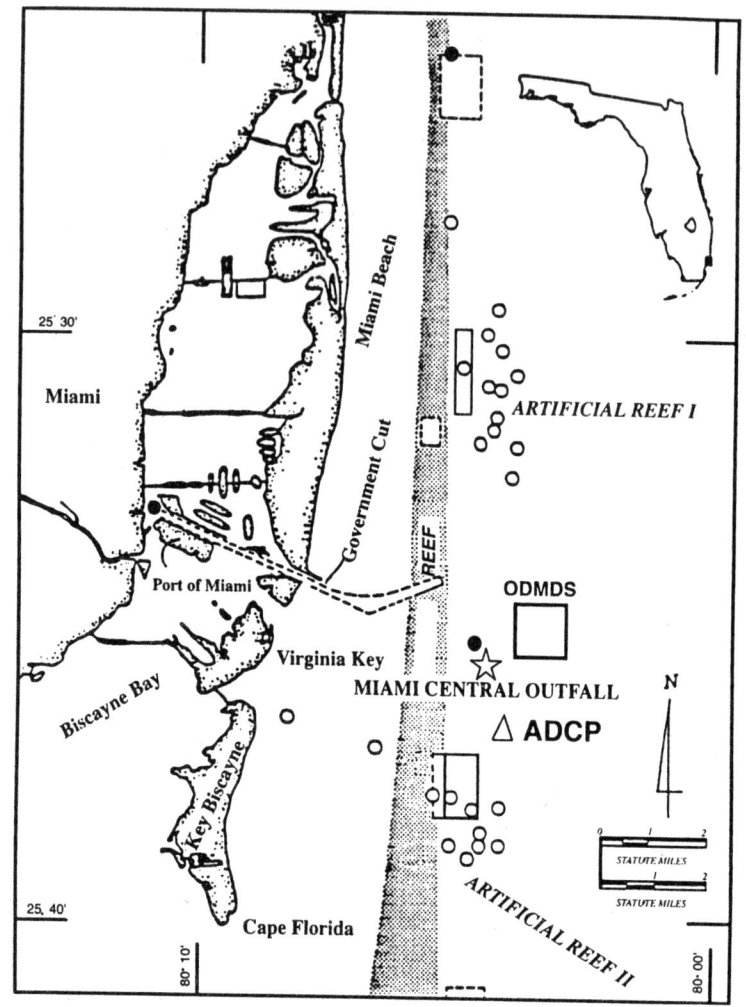

Fig. 1. Location Map showing the diffusers ●, the conventional current meter arrays ☆ and the ADCP △.

Current (as the Gulf Stream is named locally). All outfall sites are located in the western boundary region of this current, in water depths of about 30 m. Because of Wandering or meandering of the western boundary of the Florida Current, the outfalls often lay "within" the Florida Current or lay "external to" the Florida Current. Dramatically different ambient current regimes were therefore present at the sites, at different times. In this discussion, results from the measurements in the vicinity of the "Miami Central" outfall, off Virginia Key (Fig. 1) are presented.

2. Measurement Design

The overall measurement scheme for SEFLOE is discussed by Hazen and Sawyer

(1). In general, chemical and biological measurements had low sampling frequencies, e.g., biweekly, while physical oceanographic and acoustic measurements had high sampling frequencies, e.g., every 15–20 minutes. Of particular interest in the present discussion are the following instruments: (1) Savonius rotor current meters, (2) acoustic Doppler current profilers, and (3) towed acoustical backscatter systems.

A current meter mooring was installed near each outfall site. Each mooring had two Savonius rotor current meters, one near-surface and one near-bottom mounted on the mooring. At the Miami Central District, an Acoustic Doppler Current Profiler (ADCP) operating at 1228.8 kHz was installed on the bottom near the outfalls. Research ships were equipped with active acoustical systems operating at 200 kHz and 20 kHz, for the towed backscatter measurements.

In active acoustical systems a pulse of sound is emitted from a source (transducer), propagates through the water, scatters from encountered particles, and returns to the transducer. The received acoustical signal is converted to an electrical signal and recorded. The acoustic system used for SEFLOE utilized two transducers mounted in a hydrodynamically stable towbody towed alongside the research vessel at a speed of about 2 m/s. Sound pulses of 2.00×10^5 Hz frequency and 5.0×10^{-4} s duration were transmitted synchronously with pulses of either 2.0×10^4 Hz or 4.0×10^4 Hz frequency and 1.0×10^{-3} s duration at a repetition period of 2.4×10^{-1} s.

Principal system operating parameters include: frequency = 200 kHz; beamwidth = five degrees with six dB down half-angle; source level = 223.7 dB rel 1 µPa. at 1 m; receiving response of the transducer = − 183 dB re 1 V/µ Pa.

3. Observations

A. Currents

Figures 2 and 3 show current speed and direction measured near the Miami Central District outfall site by a conventional Savonius rotor current meter and an ADCP respectively for the period July 11, 1992 through August 12, 1992. The conventional current meter was at 17 meters below the ocean's surface, so an acoustic range corresponding to the 17-meter depth was selected for the ADCP. The upper panel in each figure shows the current speed and the lower panel shows the current direction. Note the similarity of data between the conventional Savonius rotor current meter and the ADCP. An extended discussion of disparities and similarities between the Savonius rotor data and the ADCP is presented in Hazen and Sawyer [1].

Between July 11–14, July 19–27, and July 31–August 12, 1992, the site was occupied by a Florida Current meander. Between July 14–19, and between July 27–31, 1992, short period, e.g., 3–8 hour current rotations were present. During these rotary current periods the Florida Current was absent from the outfall site.

A powerful capability of the ADCP is to provide nearly full water column current profiles. Figures 4 and 5 show water column horizontal current data at 18

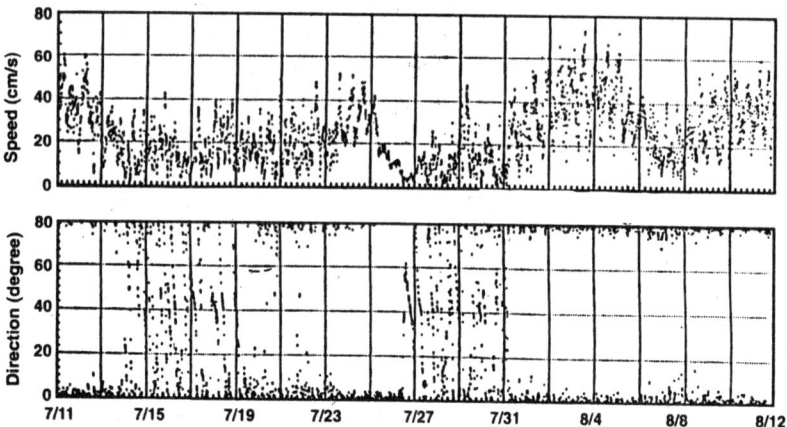

Fig. 2. Current speed and direction measured at the Miami Central outfall site for the period July 11, 1992 through August 12, 1992, at a water depth of approximately 17 meters by a conventional current meter (Savonius rotor).

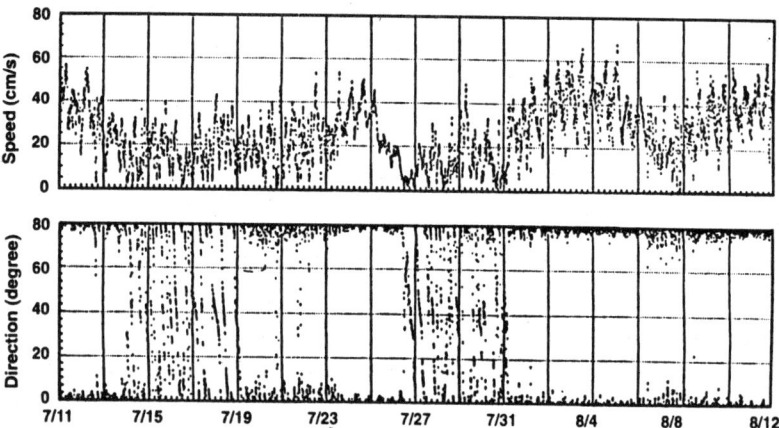

Fig. 3. Current speed and direction measured at the Miami Central outfall site for the period July 11, 1992 through August 12, 1992, at a water depth of approximately 17 meters by an ADCP.

different depths during meander present, August 3–6 and meander absent, July 26–30, 1992, periods respectively. The ambient water column current is seen to be decreasing uniformly from surface to bottom with essentially coherent rotations of ambient current direction during both times. Knowledge of the water column ambient current profile is fundamental to calculation of initial plume dilution, (Proni, Huang, and Dammann [2]). Based on acoustically derived water column current profiles, it is possible to determine the efficacy of using a single conventional current meter to estimate the vertically averaged horizontal ambient current, (Hazen and Sawyer [1]) for the Miami site.

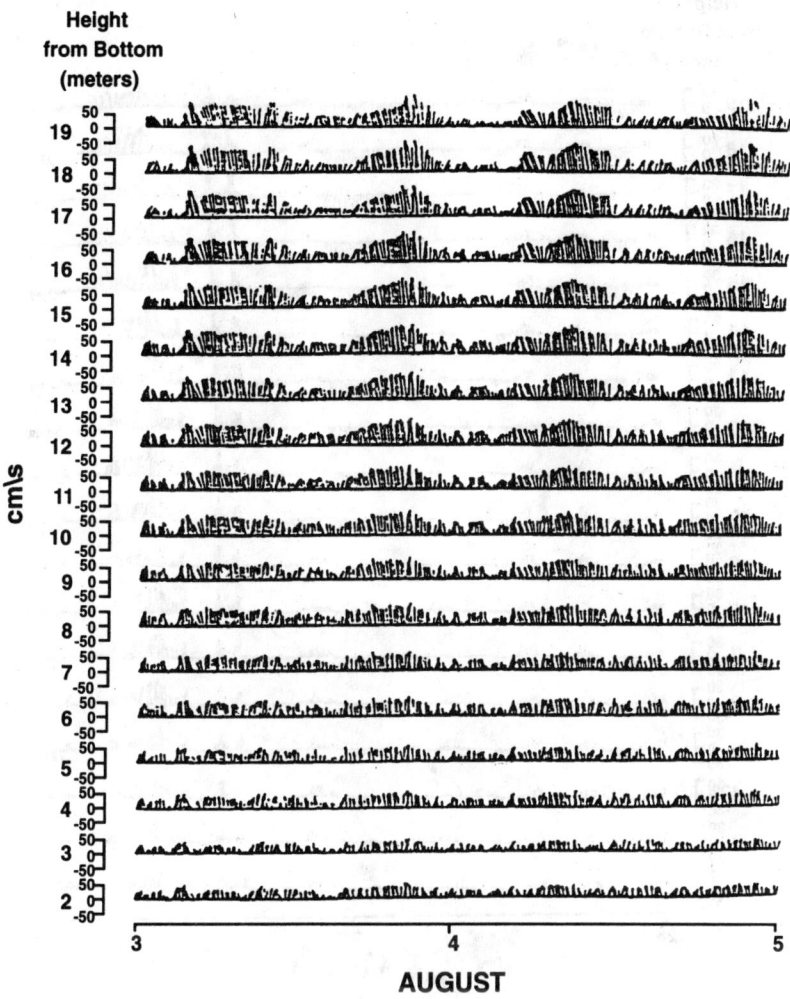

Fig. 4. Horizontal current speeds at different depths for a Florida Current meander present for August 3–5, 1992, at the Miami Central site.

B. Plumes

An example of acoustical data obtained on a near-straight line transect over the wastewater outfall (Miami Central District) is shown in Fig. 6. This transect was obtained on January 24, 1988 on a general North-South course with ships' speed at two m/s (four knots); total transect length shown is approximately 1700 meters. The data are displayed as contours of constant volume scattering strength plotted in decibels; volume scattering strength is defined as ten times the logarithm to the

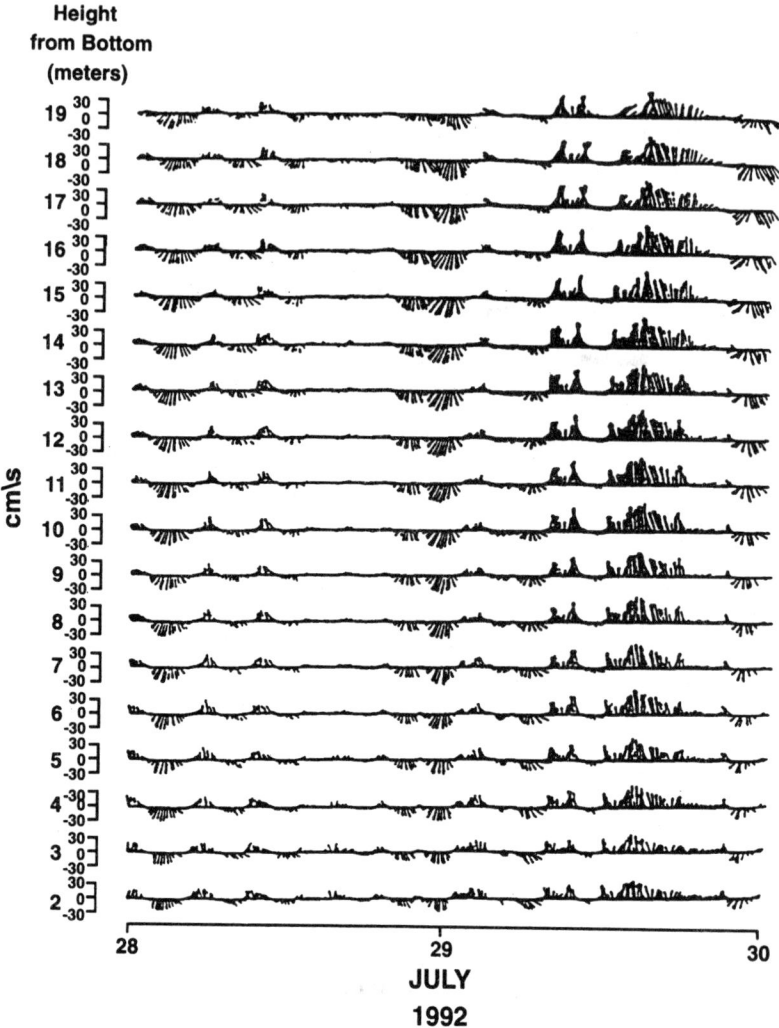

Fig. 5. Horizontal current speeds at different depths for a Florida Current meander absent for July 28–30, 1992, at the Miami Central site.

base ten of the ratio of the reflected sound intensity to the incident sound intensity and is related to equipment operating parameters through the sonar equation.

The values of the contours are actually the backscattered signal received minus a background backscattered signal which is obtained in a region of the coastal ocean close to the plume but into which the plume has not penetrated. In this way, spatially homogeneous background features (e.g., plankton of particulate horizons) are removed from the signal. The origin of the horizontal distance axis is placed

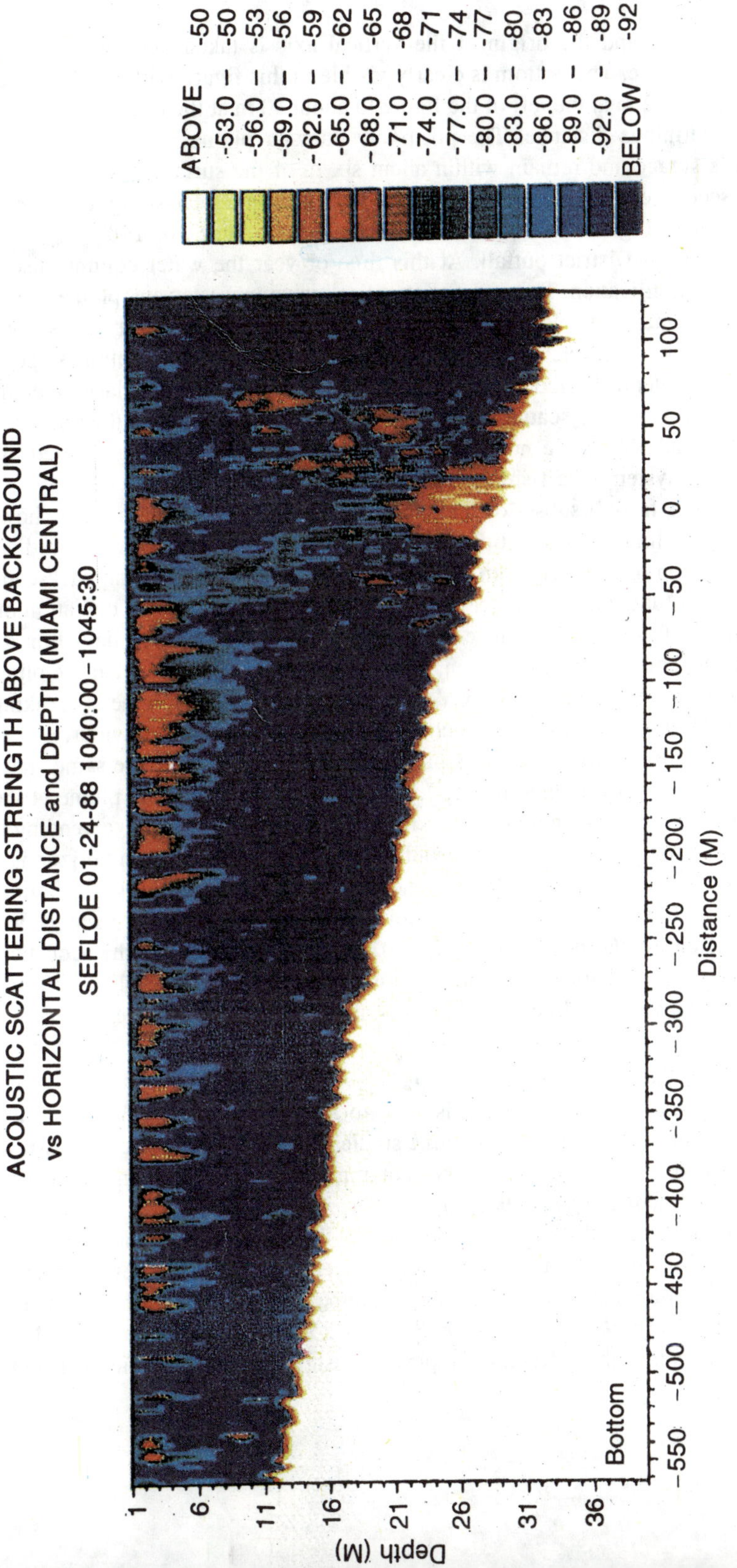

Acoustic Measurements of Currents and Effluent Plume Dilutions 543

Fig. 6. Acoustical map of Miami Central outfall made using 200 kHz system

at the outfall and the origin of the vertical axis is taken one meter below the ocean's surface. The bottom is clearly visible in this figure with depth increasing from about 12 m at the left of the figure to about 30 m at the (positive) zero meter outfall terminus distance. The positively buoyant plume is seen to rise to the ocean's surace and remain within about six m of the surface.

A second example of acoustical data obtained on a near-straight line transect is shown in Fig. 7. This transect was obtained on June 10, 1988, at the Dade County North District outfall. At this time of year the water column displayed incipient stratification. The effect of stratification upon wastewater plume formation is quite significant. Note that besides a surface plume at least one secondary plume in the 12–18 meter depth interval is visible with weaker "plumes" possible at the 7 m depth horizon and at the 24 m depth horizon. The large amount of (apparently) discrete scatterers located below 21 m and between the range values of 50 m and 450 m are not scatterers, but interference signals generated by the acoustical system of a fishing vessel operating near the outfall.

Statistics from a subset of the data (Fig. 6) were computed and are shown in Fig. 8. The following quantities are displayed: (1) the peak backscattered signal as a function of distance from the outfall, (2) the mean of the peak backscattered signal as a function of distance from the outfall, (3) the standard deviation of the signal as a function of distance and (4) the ratio of the standard deviation to the mean (the coefficient of variation, or CV). The interpretation is that the plume is continous out to about 200 m (CV < 1); beyond 200 m, the plume is breaking up (CV > 1). The CV is seen as a concise measure of plume behaviour.

A contrasting circumstance is presented in Fig. 9, showing the same statistics as above, but for the data of Fig. 7. The width of the "vertical" plume at the 15 m depth horizon is seen to be approximately 60 m (multiple ports) extending from about −20 m to +40 m. The "horizontal" plume at the 15 m depth horizon is seen to have a mean that, over the 300 meters distance shown in Fig. 9, diminishes only slightly if at all. The standard deviation is smaller than the mean (CV < 1) over the entire 260 m, (60 m to 300 m). This result shows that the acoustic data may be used to characterize both near-surface and sub-surface plumes. The sub-surface plumes would be invisible to conventional sensors at the surface.

4. Analysis

A conceptual three field model is used to help in the evaluation of the use of acoustics in wastewater outfall plume studies. The three fields are the wastewater field $F(r, \theta, z, t)$, the oceanic water column field $W(r, \theta, z, t)$ and the estimator or measurement field $E(r, \theta, z, t)$.

Cylindrical coordinates are used where r, θ, z are position coordinates and t is time. Each of these fields is comprised of many subfields, corresponding to the chemical-biological constituents of the plume. For example, for the wastewater plume field $F(r, \theta, z, t) = F \{f_1(r, \theta, z, t), ..., f(r, \theta, z, t) ... f_N(r, \theta, z, t)\}$ where f_i represents the i th subfield. Examples of wastewater plume subfields include the

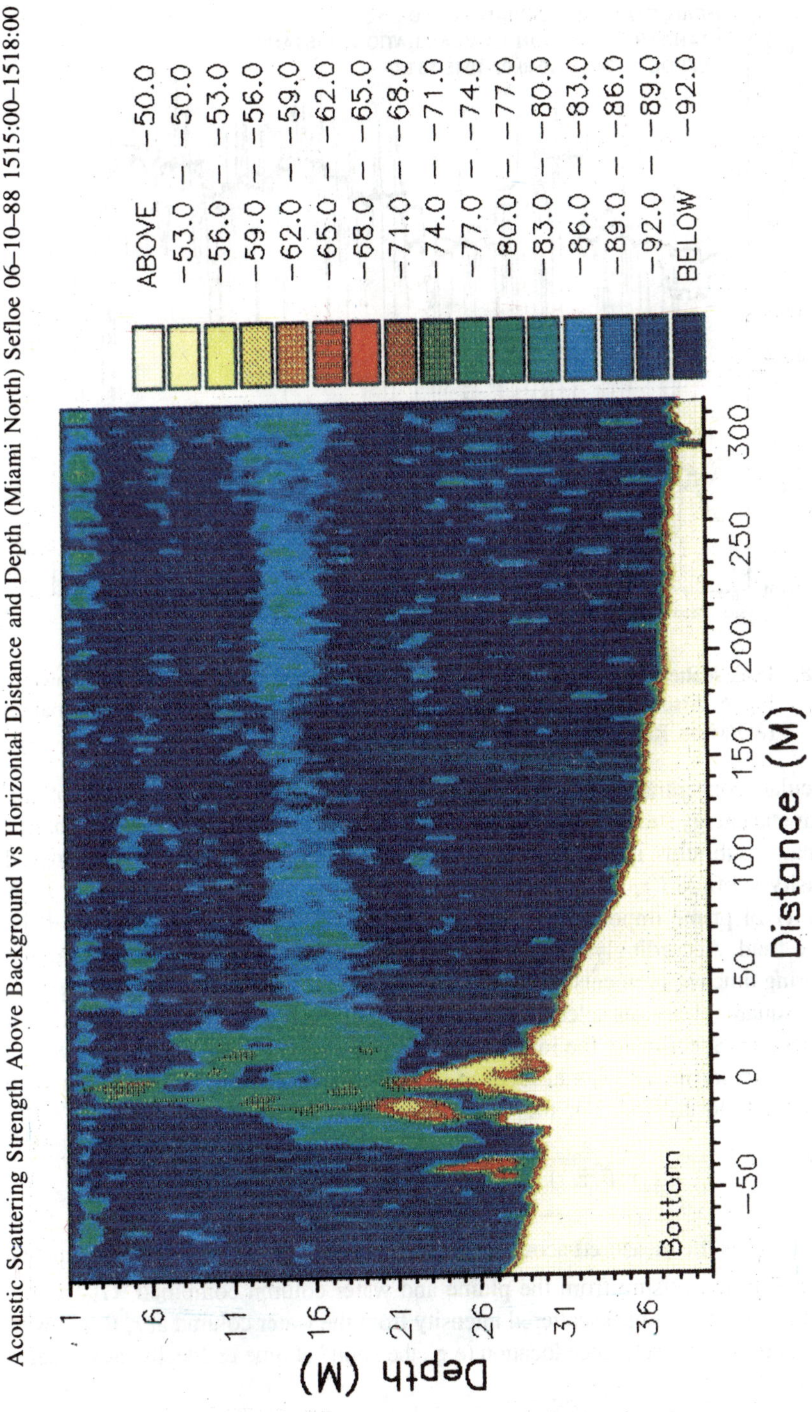

Acoustic Measurements of Currents and Effluent Plume Dilutions 545

Fig. 7. Acoustical map of Miami North outfall showing partial plume detrainment

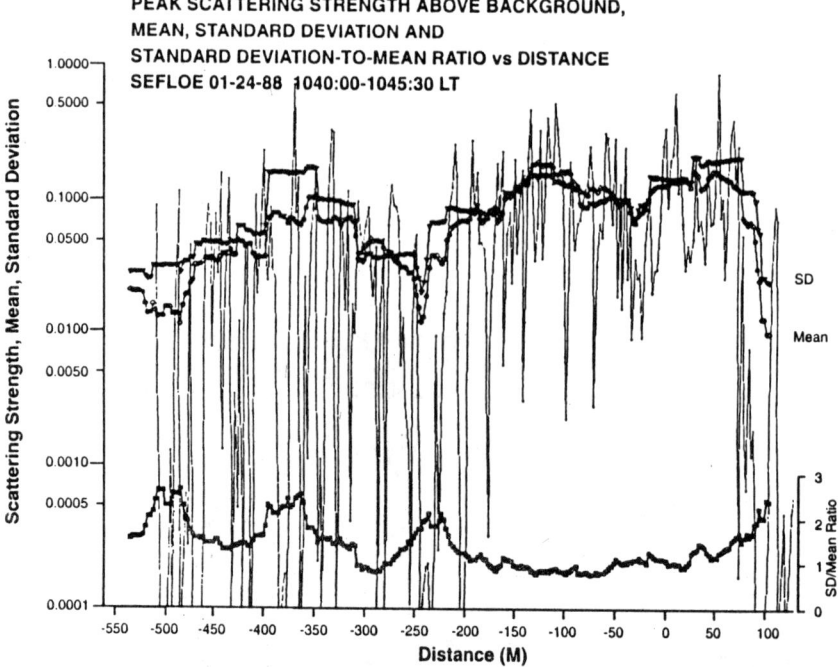

Fig. 8. Peak scattering strength as a function of distance for the acoustical data shown in Fig. 6. Also shown are the mean, standard deviation, and coefficient of variation (SD/Mean Ratio) calculated over 50 m segments.

particulate concentration subfield, the fecal coliform subfield, the nitrogen subfield, the kinetic energy subfield, the freshness subfield and so on. The two most prominent estimator subfields used in the present study are the acoustical backscattered intensity subfield $I(r, \theta, z, t)$ and the dye concentration subfield $C_D(r, \theta, z, t)$. A quantity of prime importance is the spatial rate of dilution with distance of the plume field. Accordingly, despite greatly disparate spatial resolution and sample gathering time requirements of the estimator fields, the slope of the linear regression (least-squares) line was selected as the initial test statistic for inter-estimator field comparison. Specifically the initial test statistic is the slope of the linear regression line for the normalized signal level, δ_E, versus distance from the outfall for each estimator field utilized. For the acoustical estimator field δ_A is

$$\delta_A(r, \theta, z, t) = \frac{\langle I_F(r, \theta, z, t)\rangle - \langle I_W(r, \theta, z, t)\rangle}{\langle I_F(o, o, o, t)\rangle - \langle I_W(o, o, o, t)\rangle}$$

where $\langle I_F \rangle$ is the measured acoustical backscattered intensity I at the coordinates, r, θ, z at time t arising from the plume and water column combined, $\langle I_W \rangle$ is the measured acoustical backscattered intensity from the water column at r, θ, z, t, with o, o, o, t denoting a reference location (e.g., the "boil" at time t). Ideally, as indicated

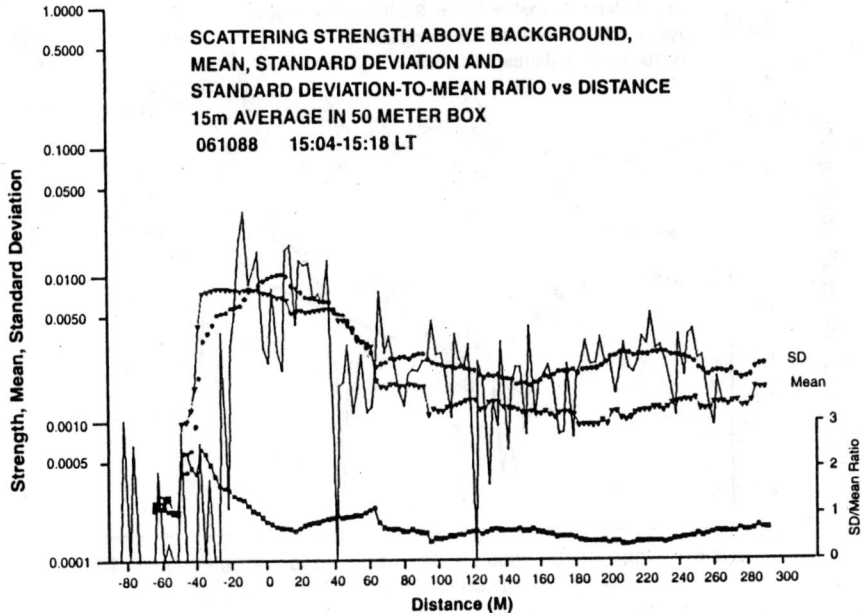

Fig. 9. Peak scattering strength as a function of distance for the acoustical data shown in Fig. 7. Also shown are the mean, standard deviation, and coefficient of variation over (SD/Mean Ratio) 50 m segments.

in the preceding equation, a point-by-point differencing of the backscattered intensity of the plume-plus-background and background alone, i.e., water column without plume, is desired. Thus far, in practice, this ideal circumstance is approximated by differencing a nearby "average" water column backscatter profile. The accuracy of this process depends on the spatial homogeneity (in r) of the water column and is analogous to the taking of a "standard" chemical water bottle sample. The logarithm of the mean normalized peak scattering strength, i.e., $\bar{\delta}_A$ (derived from the data shown in Fig. 6 and 7); the logarithm of the mean normalized dye concentration measurements, i.e., log $\bar{\delta}_D$; and the logarithm of the mean normalized fecal coliform concentration measurements, i.e., log $\bar{\delta}_{F.C.}$, as a function of distance from the surface "boil" location is shown in Fig. 10. The overbar denotes mean, where the mean used is a running mean calculated in 50 m increments of range. The "boil" value of each quantity is taken as the reference for normalization. A typical initial dilution measured for the outfall in the boil and location is 25:1. As can be seen from Fig. 10, the slopes of log $\bar{\delta}_A$, log $\bar{\delta}_D$, log $\bar{\delta}_{F.C.}$, are all on the order of 10^{-3}. The differences among the numbers 0.5, 1.1, and 1.3 are not considered statistically significant.

Figure 11 shows ambient temperature and density profiles corresponding to the data shown in Figs. 6 and 7. From these profiles it may be seen that significant water column stratification is present in June as compared to January. It is noted

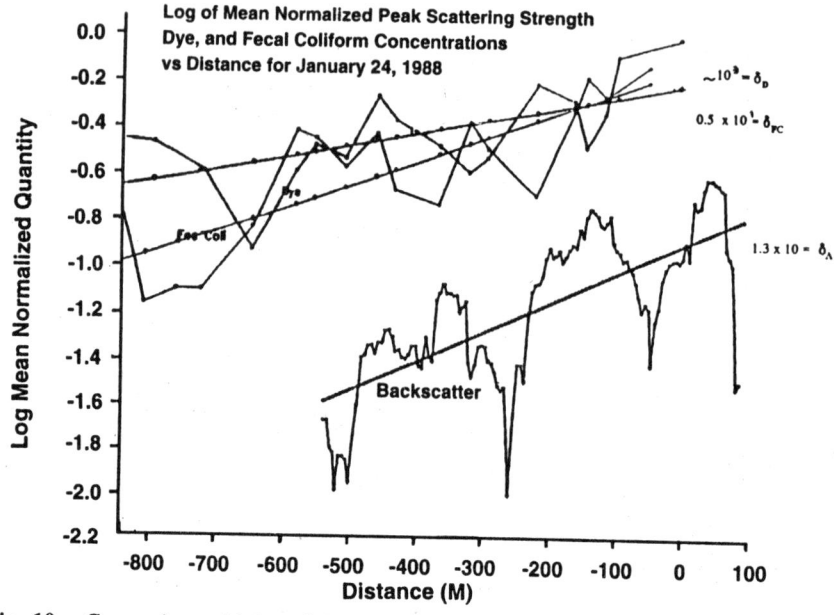

Fig. 10. Comparison of injected dye, fecal coliform, and acoustic backscatter intensity levels with distance from the Miami Central outfall for the transect shown in Fig. 6.

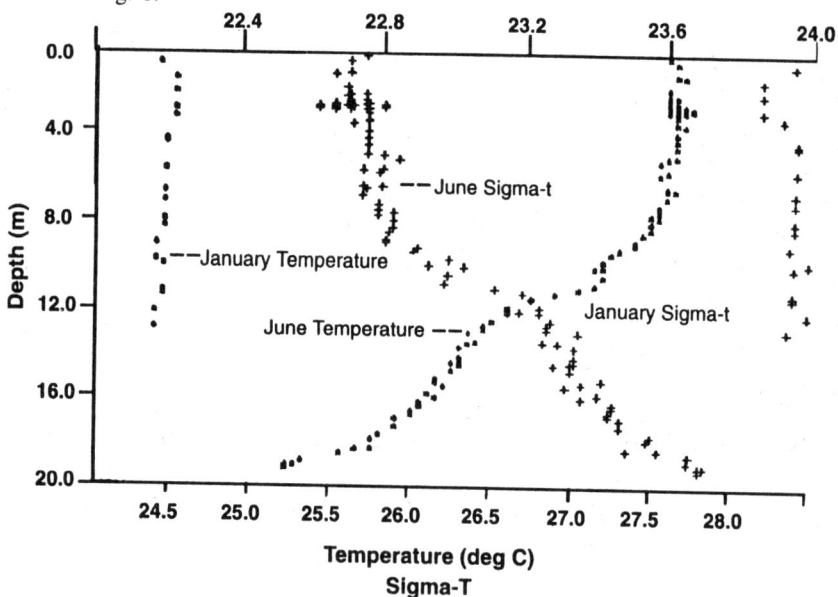

Fig. 11. Temperature and density (sigma-t) as a funtion of depth for January and June, 1992, for the Miami central outfall site.

that the "horizontal" plume portion of Fig. 7 corresponds to a depth just below a depth at which the density gradient undergoes a significant increase as depth diminishes, i.e., from 6×10^{-7} gm/cm^3/cm to 2×10^{-6} gm/cm^3/cm over the depth interval from 12 m to 9 m in the water column. A density reduction of about .0006 gm/cm^3 is seen to occur.

It is speculated that at the edges of the rising plume, entrainment of oceanic water gradually reduces the density deficit of the plume (or, equivalently, gradually increases the density of the edges of the plume) so that the "outer" portion of the vertically rising plume does not have sufficient positive buoyancy to overcome the density decrease onsetting at 12 m depth. Thus the outer portion of the plume is "peeled-off", or detrained, at a "ceiling" within the pycnocline.

5. Summary and Conclusions

The SEFLOE project has demonstrated the utility of acoustical methods for studying effluent plumes arising from wastewater outfalls in the coastal ocean. Before the SEFLOE study, basically two current regimes were expected to be present at the study sites; the first arising from the presence of the Florida Current and the second arising from large, several days in duration, eddies (Lee, [3]). Utilizing the water column ambient current data from the ADCP, a third current regime was discovered (Proni and Williams, [4]; Williams and Proni, [5]); namely, relatively short period, e.g., 3–8 hour rotary currents. The coherence of these rotary currents over the water column was demonstrated by conventional data as well. A full discussion of these rotary currents is given in Proni and Williams [4]. The poorest wastewater effluent dilutions (Hazen and Sawyer, [1]) were observed during rotary currents.

Acoustical backscatter data revealed the existence of a detrainment phenomenon in which part of the rising wastewater plume was trapped near mid-water depth. At that depth the trapped plume dispersed horizontally. In many previous studies (e.g., Proni and Hansen, [6]), it has been noted that naturally occurring organic material accrues in the coastal ocean at certain density horizons. Clearly acoustical systems may serve not only to map the subsurface deffluent plume but also to map the naturally occurring organisms thereby providing an ideal instrument for finding the joint probability of exposure of organisms to contaminants.

We have seen, based on limited case studies, that reduction of acoustical backscatter level with distance can serve as a surrogate for the reduction of effluent constituents such as fecal coliform (if present). Similarly we see that acoustics is ideally suited for guiding biological and chemical sampling both with respect to depth, range, and spatial location.

In summary, acoustical remote sensing of wastewater effluent plumes, particularly when used in combination with non-acoustic measurements, is a valuable tool for assessing the impact of effluent discharge in the marine environment.

6. Acknowledgements

The authors wish to acknowledge the support of the Miami Dade Water and Sewer Department, the City of Hollywood Utilities Department, the Broward

County Office of Environmental Services, the Florida Department of Environmental Regulation, the U.S. EPA, NOAA, and all others associated with SEFLOE.

References

1. Hazen and Sawyer. SEFLOE II Final Report and Appendices. SEFLOE Technical Advisory Committee (Editors), Hollywood, FL. June 1994. Unpublished Report. Available from: Hazen and Sawyer, 4000 Hollywood Boulevard, Hollywood, FL. 38021. Tel (305) –987–0066.
2. Proni, J.R., Huang, H., and Dammann, W.P. "Initial Dilution of Southeast Florida Ocean Outfalls. *J. Hydr. Eng.* ASCE 120 (12) 1409–1425, 1994.
3. Lee, Thomas N. "Florida Current Spin-Off Eddies" *Deep-Sea Research.* Vol. 22 pp 753–765, 1975.
4. Proni, J.R., and Williams, R.G. "Short Period Florida Current Variations" to be published 1996.
5. Williams, R.G. and Proni, J.R., "Acoustic Remote Sensing of Wastewater Outflow". *The Proceedings of the 7th International Symposium on Acoustic Remote Sensing.* Editor W.D. Neff. Boulder Colorado. October 3–7, 1994.
6. Proni, J.R., and Hansen, D.V. "Dispersion of Particulates in the Ocean Studied Acoustically: the Importance of Gradient Surfaces in the Ocean" *Ocean Dumping of Industrial Waters* Edited by Ketchum, Bostwick H., Kester, Dana R., and Park, R. Kilho. Plenum Publishing Corporation, 233 Spring St, New York, NY 10013, 1981.

23. Acoustic Propagational Characteristics and Tomography Studies of the Northern Indian Ocean

S. Prasanna Kumar, Y.K. Somayajulu and T.V. Ramana Murty

*Marine Acoustics & Acoustic Oceanography Section,
National Institute of Oceanography, Dona Paula, Goa-403 004, India.*

1. Introduction

Of late, acoustic monitoring of oceanic features is becoming more and more relevant in understanding the energetics and dynamics of meso-scale phenomena. The existence of SOFAR channel in the ocean makes possible propagation of acoustic pulses over long distances without much energy losses. It is this property of the ocean which led to the development of acoustic remote sensing techniques viz., Ocean Acoustic Tomography (Munk and Wunsch, 1979) and more recently the Acoustic Themometry (Munk and Forbes, 1989; Spiesberger and Metzger, 1991) to monitor the global warming.

Ocean Acoustic Tomography (OAT) is a technique by which the real time oceanic fields could be obtained synoptically by acoustic means. It makes use of the multipath propagation of acoustic pulses between pre-fixed sources and receivers in the ocean volume. The sequential arrival of these pulses depend mainly on the acoustic characteristics of that region. The travel time perturbations between the measured arrivals and those obtained from the model could be used to reconstruct the temporally evolving oceanic (sound speed and to a first approximation temperature) fields through apropriate inverse techniques.

The concept of OAT, triggered off a number of investigations towards the development of operational tomographic system (Spiesberger et al, 1980; Munk and Wunsch, 1982; Cornuelle et al, 1985; Worcester et al, 1985). Worcester and Cornuelle (1991) provided a lucid review of the developments towards this direction.

In India, Acoustic tomography studies were initiated, at the National Institute of Oceanography, during mid nineteen eighties under the project "Development of Acoustic technique for Remote sensing of Oceans" and were mostly confined to simulation aspects (Ramana Murty et al, 1989, 1990, 1992; Prasanna Kumar et al, 1988, 1994; Somayajulu, Somayajulu et al, 1993; 1994).

Herein, we present the results of the acoustic tomographic studies in India in the following five sections. Section 2 deals with the characteristics of the sound speed field of the northern Indian Ocean comprising of the Arabian Sea and the Bay of Bengal, followed by the simulation results (section 3), using a ray acoustic model and reference sound speed profile to delineate the acoustic waveguide characteristics in the context of tomography. In section 4, we discuss the generalised inversion model and its adequacy in reconstructing the layer-averaged as well as box-averaged sound speed anomaly in the case of a winter profile in the Arabian Sea and an eddy profile in the Bay of Bengal. Results from a medium range, short duration acoustic tomography experiment in the eastern Arabian Sea during summer 1993 are presented in section 5. Finally, section 6 outlines the Heard Island experiment, which tested the concept of Acoustic thermometry, and India's contribution to this joint programme.

2. Sound Speed Structure in the Northern Indian Ocean

An apriori knowledge of the statistics of the acoustic field is a pre-requisite for any simulation using acoustic propagation models. In view of this, the sound speeds were computed following Chen and Millero (1977), from the quasi-synthetic environmental data of Levitus (1982) for the annual mean values of temperature and salinity at 33 standard depths. The mean, minimum and maximum values of sound speed at each of the standard depths were obtained (Fig. 1).

The annual mean sound speed profile shows uniform values in the upper 40 m of the Arabian Sea while an increase with depth is seen in the Bay of Bengal, indicating the presence of a surface duct with weak gradients. An interesting feature common to the Arabian Sea and the Bay of Bengal is the depth limited

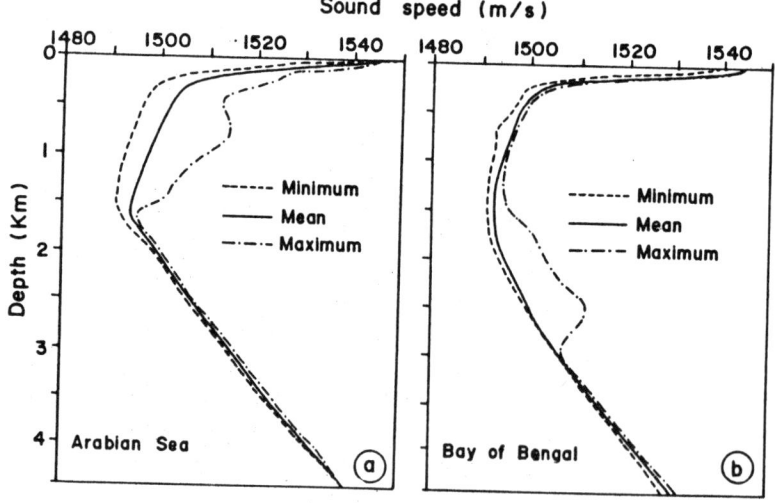

Fig. 1. Annual mean (___) sound speed profiles in the Arabian Sea and the Bay of Bengal along with the minimum (- - -) and the maximum (._._._) values.

nature of the profile, i.e. surface sound speeds exceed the near bottom values. This has an important implication in the sound propagation. Acoustic rays in an ocean with depth limited profile will propagate as surface refracted, bottom reflected (RBR) rays. In the Arabian Sea, sound speed shows maximum spatial variations between the depth levels 200 to 1500 m. This is due to a combined effect of variations in salinity (to a larger extent) and temperature. In the Bay of Bengal, spatial variations of sound speed are marginal as seen from the mean profile and the associated minimum and maximum values. However, between 1500 m and about 2800 m, the observed variations between the sound speed profiles in the Bay of Bengal is due to the presence of warmer water in the Andaman Sea. In general, at any depth sound speed values in the Arabian Sea are higher than the corresponding values in the Bay of Bengal.

Considering the importance of channel parameters such as axial sound speed, channel axis depth, conjugate depth (depth below the sea surface at which the sound speed equals the near bottom values) on acoustic propagation, the spatial distributions of these parameters are examined (Prasanna Kumar et al, 1994).

The depth of sound channel axis varies between 1450 and 1850 m in Arabian Sea as compared to 1100–1750 m in the Bay of Bengal (Fig. 2a). In general, the depth of the channel axis increases towards the northern latitudes in the Arabian Sea, while it decreases in the Bay of Bengal. The mean depth of the Channel axis in the Arabian Sea and the Bay of Bengal are 1750 m and 1500 m respectively. The sound speed at the axis shows variations up to 4 m/s and 2 m/s respectively (Fig. 2b). The eastern Bay and the Andaman Sea, are characterised by low axial depth and high axial sound speeds. The sound speed profile in the Arabian Sea shows the presence of large gradients above the axis of SOFAR channel in contrast to that in the Bay of Bengal and also relatively higher sound speed values at any given depth indicating a strong waveguide. The conjugate depth shows variation from 75 to 300 m in the Arabian Sea and 100 to 300 m in the Bay of Bengal (Fig. 3). These variations of over 200 m in conjugate depth arise primarily due to the spatial variations in bottom depth. As a result, the effective sound channel lies much below the sea surface.

3. Acoustic Ray Model

Having analayed the sound speed field, we now discuss the acoustic ray model which was used to carry out numerical experiments (simulations) to study the acoustic waveguide characterestics for use in OAT.

Ray theory, despite its short comings, remains one of the most powerful means to compute sound propagation in a complex environment. It is unique in identifying which part of the ocean is sampled by the sound signal corresponding to the arrival of each pulse. With the aid of modern computers, it is easy and economical to simulate measurements prior to the conduct of any field experiment. In view of this, ray theory has been adapted for the numerical simulation for computing the trajectory of an acoustic signal based on the assumption that (1) the wavelength

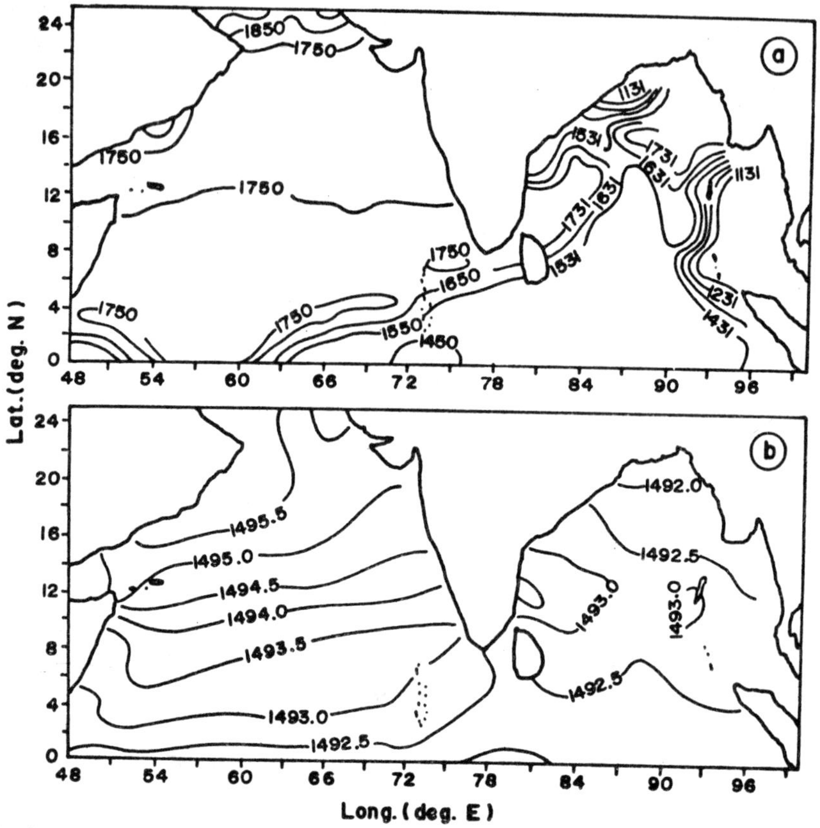

Fig. 2. Spatial distribution of SOFAR Channel axis depth (m) and sound speed (m/s) at the SOFAR axis in the northern Indian Ocean.

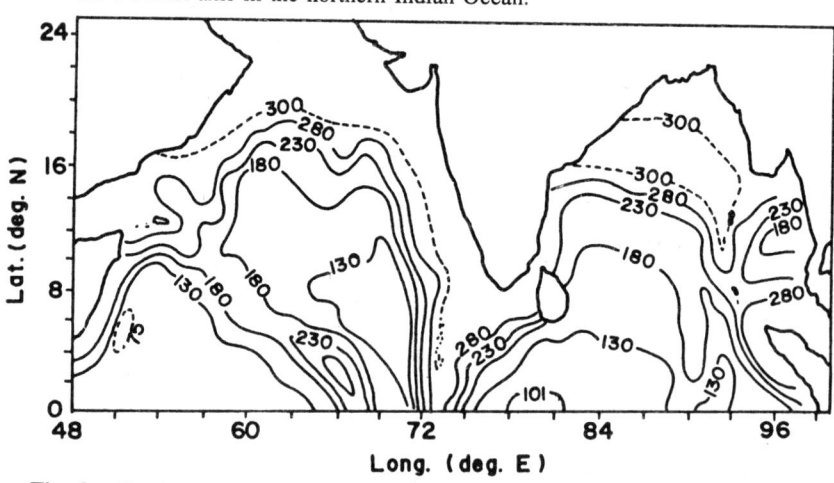

Fig. 3. Spatial variation of conjugate depth (m) in the northern Indian Ocean.

of sound is much less than the water depth, and (2) the change of sound speed is negligible over several wave lengths.

Ray geometry is governed by the following equations (Arthur et al, 1952):

$$dx = \cos \theta \, ds \tag{1}$$

$$dy = \sin \theta \, ds \tag{2}$$

$$d\theta = \frac{1}{C} [\sin \theta \, \delta c / \delta x - \cos \theta \, \delta c / \delta y] \, ds \tag{3}$$

In the above equation x and y represent horizontal and vertical co-ordinates (see Fig. 4), θ is the angle of the ray with respect to the horizontal and s is the arc length along the ray which is chosen to sample the sound speed and its gradient adeqately for determining the ray parameters, C is the sound speed and $\delta c/\delta y$ is the vertical sound speed gradient. The above equation is integrated numerically using a Runge-Kutta technique for determining the eigen rays (Mahadevan et al, 1989).

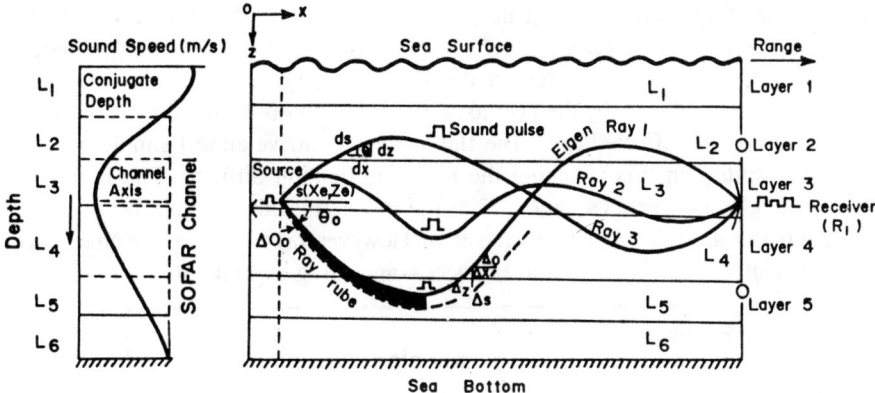

Fig. 4. Schematic presentation of acoustic rays from source to receiver, L1, ..., L6 stand for layers.

Source to receiver travel time is estimated using the equation

$$T_i = \int [C_o (s)]^{-1} \, ds \tag{4}$$

Integration is carried out along the path and $i = 1, N$ where N is the number of rays and $C_o(s)$ specifies sound speed along the ray.

Acoustic intensity calculations are based on the assumption that changes in the intensity occur only due to contraction or expansion of the ray tube — defined by two adjacent rays separated by 1/100 of a degree- and is conserved. The acoustic intensity at any point along the ray is given by (Moler and Solomon, 1970)

$$\text{Intensity} = \frac{I(x)}{I_o} = \frac{\sin \theta_o}{x \cos \theta [\Delta \theta_o / \Delta y]} \tag{5}$$

where the terms are: I_o the reference intensity at the source, θ the eigen ray angle at the point of measurement, x the distance from the source to the point of measurement, Δy the vertical distance at the point of measurement between the eigen ray and the ray in its immediate vicinity, $\Delta \theta_o$ the angle between the eigen ray and the ray in its immediate vicinity at the source. The computed intensity is converted to the practical acoustic unit decibel (dB) by the formula $10 \log (I/I_o)^2$.

The annual mean sound speed profile (reference profile) for the Arabian Sea and the Bay of Bengal were used for the computation of ray parameters (Mahadevan et al, 1989). Ray tracing was carried out for launch angles from $+10°$ to $-10°$, over a range of 300 km for a hypothetical source at the channel axis.

The ray arrival patterns (Fig. 5) obtained from ray tracing reveal that in the Arabian Sea rays with emergence angles between $(+/-)$ $6.5°$ and $(+/-)$ $7.5°$ reach the receiver first followed by the flat angle near axial rays and the surface refracted bottom reflected (RBR) rays. In the Arabian Sea, the purely refracted steep as well as flat angle rays are adequately resolvable in time. This feature makes ray identification—the crucial aspect of tomography—easier. In the Bay of Bengal also the first arrivals are the refracted rays with emergence angles between $(+/-)$ $6.0°$ and $7.0°$ followed by the near axial flat angle rays ($-4°$ to $+4°$) and the steep refracted bottom reflected rays. The flat angle rays arrive almost simultaneously (<10 ms spread in time) making the ray identification difficult. The rays with launch angles between $+4°$ and $+7°$ are purely refracted and have better travel time spread with well defined arrival pattern. However, rays with angle of emergence higher than $+7°$ undergo bottom bounces—appearing as kinks in the curve and

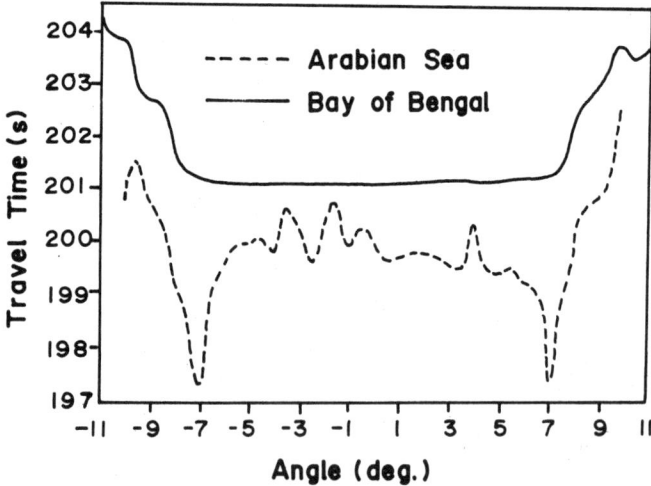

Fig. 5. Launch angle versus travel time of acoustic ray over 300 km range for the Arabian Sea and the Bay of Bengal.

lead to enhanced loss of energy at the bottom boundary. Due to the depth limited nature of the acoustic profile, the steep angle rays undergo reflection at the sea floor (lower boundary) and refraction below the sea surface (the upper boundary).

The observed arrival pattern of the acoustic pulse was analysed in the light of the prevailing oceanographic features in this region. For example, the early arrival of +6.5° to +7.5° rays in the Arabian Sea is due to the frequent scanning of the warm, high saline waters in the depth range of 500–1200 m by these rays. Since the sound speed in this layer is more, the travel times of the above rays are reduced. The warm, high saline waters of the Red Sea origin intrude into the Arabian Sea at depth interval of 500–800 m. The early arrivals of the axial rays in the Bay of Bengal result from the characteristic weak sound speed gradients prevailing within the SOFAR channel. Eigen ray computations for source at 1600 m and receiver at 1500 m show a time span of 495 ms in the Arabian Sea which is longer compared to that in the Bay of Bengal (344 ms).

3.1 Acoustic Intensity

Computations on intensity losses due to spreading for various eigen rays have been made for the Arabian Sea and the Bay of Bengal (Ramana Murty et al, 1993). The plots shown in Fig. 6a and b represent the intensity variations along steep and flat rays in the Arabian Sea and the Bay of Bengal respectively. In general, the intensity falls drastically in the first few kilometers from the source (spherical spreading), and by about 30 dB within 30 km. Beyond this range the fall in intensity is gradual. The peaks seen on the intensity curve are associated with the ray turning points where the vertical spacing between the rays constituting the ray tube will be minimum due to ray convergence. The intensity losses for different rays show variations of about 80 dB to 110 dB (Fig. 6) over 300 km range. It is also evident that the steep angle rays suffer less loss compared to flat angle rays. Greater refraction of steep angle rays results in less loss of energy.

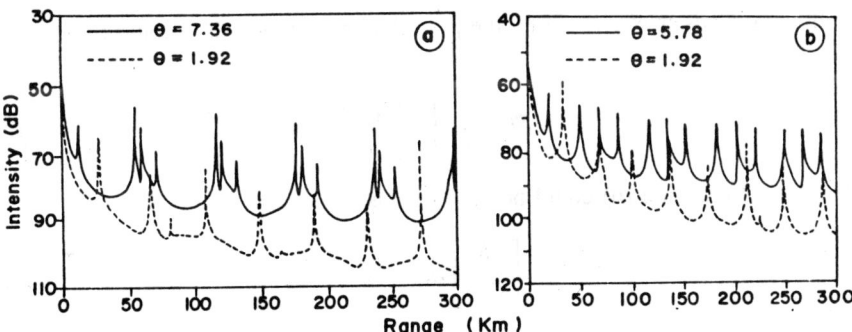

Fig. 6 Intensity loss over 300 km range (a) Arabian Sea, (b) Bay of Bengal.

3.2 Chemical Absorption

Acoustic absorption losses at two source frequencies 10 and 0.4 kHz indicated that the contributions due to boric acid and magnesium sulphate respectively predominate at these frequencies. The vertical distribution of the total losses obtained by adding contributions of all these components at 10 kHz and 0.4 kHz are shown in Fig. 7a and b respectively. The computed values for 10 kHz are useful for SONAR operations while the values for 0.4 kHz are important for OAT studies.

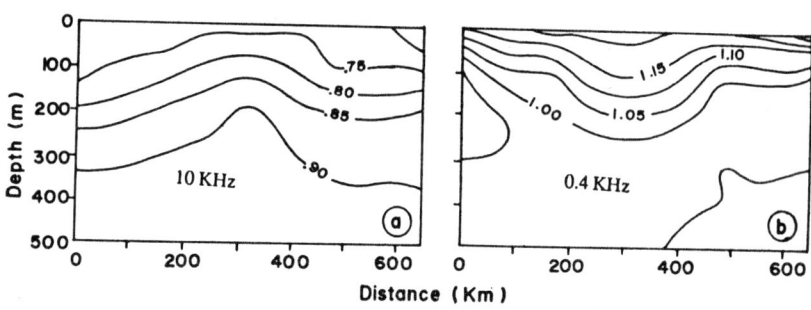

Fig. 7 Acoustic energy losses at (a) 10 kHz and (b) 0.4 kHz due to chemical absorption.

The results of computation of intensity loss due to absorption in the presence of an oceanic cold-core eddy at 0.4 kHz frequency indicated that the energy loss increased by 1.76×10^{-3} dB/km as against 1.0×10^{-3} dB/km over the ambient value. Compared to the spreading losses, losses due to absorption even in the presence of oceanic meso-scale feature like eddy, worked out to a negligible per cent (0.4%). However, for SONAR frequencies, the loss amounts to 25%. Thus, in terms of range, the meso-scale feature would limit the performance of SONAR.

4. Inverse Model

Following Chester et al, (1991), the perturbation in travel time could be written as

$$\delta T_i = - \int \frac{\delta C(x, t)}{C_o^2(x, t_o)} \tag{6}$$

The model equation (6) could be rewritten as

$$\left[\frac{-R_{ij}}{C^2_o} \right] \delta C_j = \delta T_i \tag{7}$$

After parameterising the model, the above equation can be expressed in martix-vector notation as

$$\delta T = A \delta C \tag{8}$$

where δT is the travel time difference between the measurements and those obtained by the ray model; and $A = -R_{ij}/C_o^2$ where R_{ij} is the path length of ray i in layer j and C_o is the reference sound speed. The travel time perturbation is related to sound speed perturbation through matrix A. The system of equations can be separated into (i) over-determined (ii) even-determined and (iii) under-determined problems. Each system has different type of solution by taking suitable minimization of norms. But generalised inverse solves the mixed-determined problems which usually occur.

The inversion procedure used here is based on singular value decomposition (SVD) wherein the generalised inverse operator has been constructed following the eigen value technique (Menke, 1984). The SVD provides a simple frame work for determining how well the inverted model parameters fit the data and how close model parameter estimates the true value.

According to Jackson (1972), the natural inverse operator A_g^{-1} always exists and is given by

$$A_g^{-1} = VT^{-1} U^T \qquad (9)$$

where matrices V and U are obtained by the following coupled eigen value problem:

$$AA^T U_j = \lambda_j^2 U_j \qquad (10)$$

$$A^T A V_j = \lambda_j^2 V_j \qquad (11)$$

In the above equations U_j and V_j are column vectors of matrices U and V; λ_j^2 are eigen values of martix A arranged in decreasing order. The eigen vectors corresponding to the largest eigen value indicate that large scale factors can be determined. The selection of the number of eigen values is decided by the closeness ratio approach. The solution of equation (9) is given by

$$\delta C = [VT^{-1} U^T] \delta T \qquad (12)$$

Calculations of the model resolution matrix and data resolution matrix are needed for the assessment of the resulting sound speed model.

Model Resolution: The model resolution of generalised inverse (10) is given by

$$R = A_g^{-1} A = V_p V_p^T$$

here p indicates number of factors used in SVD which is less than or equal to rank of the matrix A. The model parameters will be perfectly resolved if V_p spans the complete space of the model parameter i.e.

$$V_p V_p^T = I$$

Data Resolution: The data resolution matrix is given by

$$N = AA_g^{-1} = U_p U_p^T$$

The data are perfectly resolved if U_p spans the complete space of data. This usually tells the convergence of rays in the given domain.

4.1 Reconstruction of Winter Profile in the Arabian Sea

As the Arabian Sea undergoes seasonal changes in accordance with the changes in the atmospheric forcings, the tomographic reconstruction is applied to model these changes through numerical experiments.

Computation of data kernel: Using the annual mean and winter mean sound speed profiles, seventeen eigen rays have been identified for inversion analysis. For the reference state, ray path lengths in six tomography layers have been estimated to form the data kernel. The acoustic rays scan the vertical water column between 180 m (upper limit) and 3300 m (lower limit). In this study, a uniform water depth of 4000 m over the 300 km range and the refracted rays alone are used in the analysis as the reflected rays undergo significant intensity losses due to multiple bottom reflections. The vertical water column has been divided into six layers viz., 0–400 m, 400–800 m, 800–1200 m, 1200–2000 m, 2000–3000 m and 3000–4000 m.

The data kernel consisting of ray path lengths in the pre-set layers arranged in rows and columns of matrix-A, is used to construct the generalised inverse operator to operate on travel time perturbations (between the simulated and measured travel times) to predict the model parameter in real time. It is essential to remove the noise present in the original data kernel (A). This can be done by approximating the data kernel by the filtered kernel. The vertical distribution of the original data kernel (A) is shown in Fig. 8a. The contour pattern in this diagram indicates maximum anomaly in the presence of the sound speed minimum layer. This is due to the nature of the SOFAR channel where ray bending takes place towards the sound speed minimum forming an envelope. Fig. 8b presents the filtered data kernel obtained through reconstruction using five energetic modes belonging to the activated spaces of model and data matrices after removing the noise vector. This represents rich-information of the original signal of the data kernel (A). Both the plots are close to each other indicating that the information contained by many rays can be expressed in terms of a few modes. Thus, in the case of the waters of the Arabian Sea, it is sufficient to consider six layers and five energetic modes for the construction of generalised inverse operator (Fig. 8c) defined by equation (12) to operate on travel time perturbation for predicting the model parameter.

Travel time data: The predicted travel times of eigen rays computed using the mean sound speed profile reference state and the assumed profile (winter season mean) have been used to generate possible perturbations in travel times (T). The array of the ray travel time deviations between these two show variations from 0.7 to 117 ms. These travel time perturbations are operated by the generalised inverse

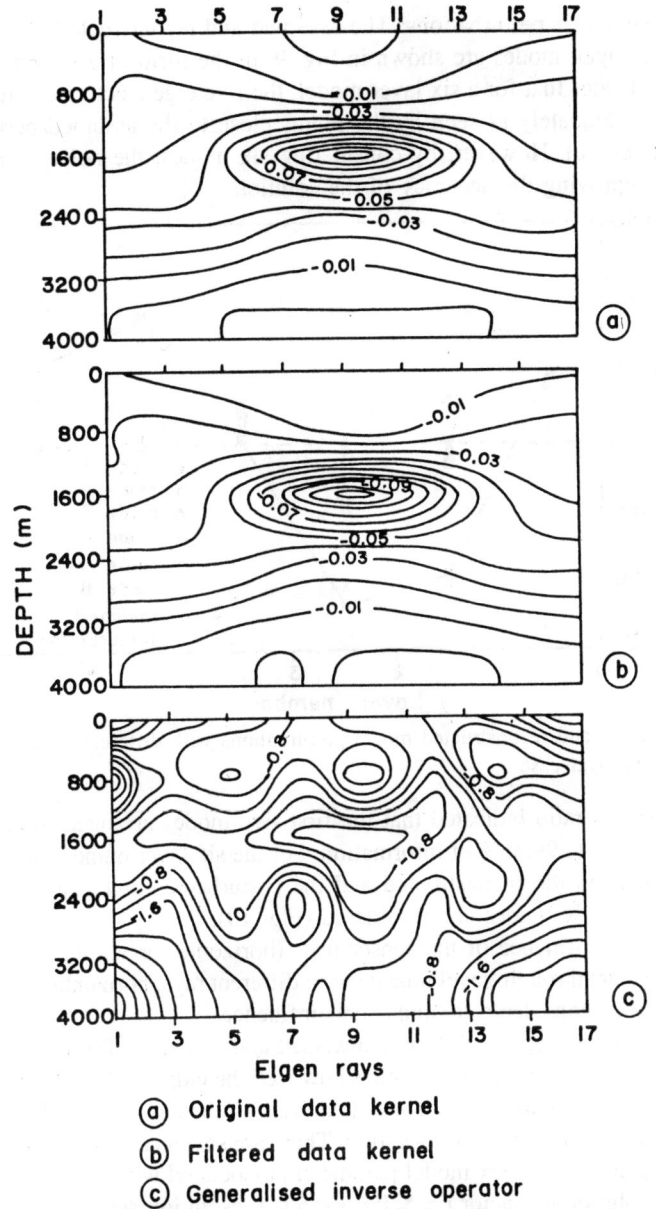

Fig. 8. Plots showing (a) original data kernel, (b) filtered kernel and (c) generalised inverse operator.

operator (Fig. 8c) to get the model parameter perturbation in different layers. The travel time data discussed above is noise free in nature representing ideal case. In real cases, errors in travel time measurements prevail due to the dynamics of the medium as said in the previous section.

Model parameter perturbations: The assumed and reconstructed perturbations for different eigen modes are shown in Fig. 9. in the form of a scatter diagram. This plot indicates that for a six layer model, the five eigenvectors can reproduce the features adequately as seen by the points close to the assumed perturbation (continuous curve). However, by considering six modes, the noise is amplified instead of improving the accuracy of the solution.

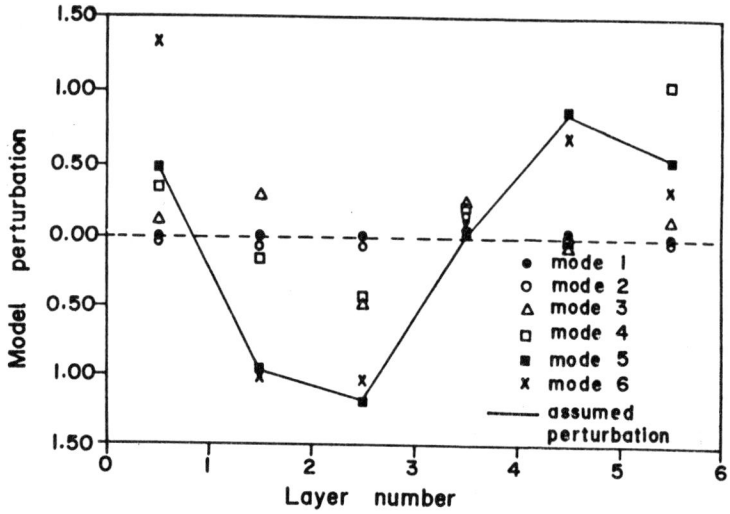

Fig. 9. Assumed and reconstructed model perturbations with different energetic modes for six layer case.

The closeness ratio indicated that the first five modes arranged in decending order, giving rise to 99.86% of information. For the six layer numerical study, the departure between the reconstructed and the perturbed/assumed profiles is less when five energetic modes are considered (Fig. 10).

To study the influence of the sensor drift (horizontal or vertical or both) on model parameters through travel time error in different rays, horizontal and vertical drifts have been considered to study their influence.

The model parameter resolution is analysed and shown in Table 1 for factors 1 through 5. The model resolution depicts how well the individual model parameters are resolved. Any value below or less than unity indicates less resolution while unit value corresponds to good resolution. There are seventeen data measurements along the eigen rays and six model parameters associated with six layers leading to perfect resolution for factor $r = 5$. For $r < 5$ the resolution deteriorates. A better resolution is expected when the sixth factor is considered but it also leads to the deterioration of resolution as the noise also gets amplified with the signal. This is due to the influence of a small positive eigen value in the activated space instead of increasing the dimension of null space.

The diagonal elements of the data resolution UU^T are shown in Table 2 for

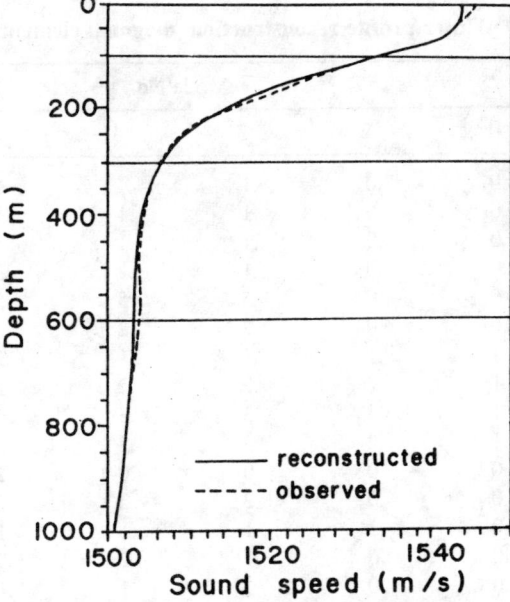

Fig. 10. Assumed and reconstructed profiles for six layer model.

Table 1. Winter profile reconstruction, diagonal elements of VV^T

Layer No.	Mode No.					
	1	2	3	4	5	6
1	.0	.0	.0	.1	.1	1.0
2	.0	.1	.5	.6	.9	1.0
3	.0	.1	.7	.7	.9	1.0
4	.7	.9	.9	.9	.9	1.0
5	.1	.6	.6	.6	.9	1.0
6	.0	.0	.0	.8	.9	1.0

factors 1, 2, ..., 5. A value of unity indicates contribution of information which is completely independent of the other observations. This table enables one to infer which of the rays present poor information resolution (resulting from sampling nearly the same ocean region) as seen from the values.

Following the analysis of the model and data resolutions, the present study indicates that the partial information contained in the data space (i.e. the solution is non-zero prediction error) is adequate to reconstruct the model parameters perfectly i.e. $VV^T = I$ which is a case of overdeterminacy and reveals the usefulness of natural generalized inverse in the least square sense.

4.2 Reconstruction of Eddy Profile in the Bay of Bengal

The sound speed profile associated with a subsurface cold core eddy, centred at

Table 2. Winter profile reconstruction, diagonal elements of UU^T

Ray No.	Mode No.					
	1	2	3	4	5	6
1	.0	.0	.1	.4	.4	.4
2	.0	.0	.1	.1	.1	.2
3	.0	.0	.1	.1	.1	.2
4	.0	.0	.1	.2	.3	.4
5	.0	.0	.5	.5	.5	.6
6	.0	.0	.2	.2	.2	.2
7	.1	.1	.1	.1	.2	.3
8	.1	.2	.2	.2	.2	.2
9	.1	.3	.3	.3	.4	.4
10	.1	.2	.2	.2	.2	.2
11	.0	.0	.0	.0	.2	.2
12	.0	.0	.2	.2	.2	.2
13	.0	.1	.1	.1	.1	.2
14	.0	.0	.1	.2	.3	.4
15	.0	.0	.1	.1	.1	.2
16	.0	.1	.1	.1	.3	.3
17	.0	.0	.1	.4	.5	.5

17° 40'N and 85° 19'E (Fig. 11a), observed in the northwestern Bay of Bengal (Babu et al, 1991) were used for the simmulation study.

Acoustic signature of the eddy is manifested by the upliftment of isopleths. The effect of the eddy was to reduce the ambient sound speed by 10 m/s (Fig. 11b). Associated with this, the depth of the SOFAR channel axis remained constant, which otherwise should have shown a deepening from south to north (Prasanna Kumar et al, 1994). Accordingly the sound speed at the channel axis below the eddy is reduced by 1.5 m/s.

Rays corresponding to the reference ocean profiles and the one associated with the eddy case have been properly identified and matched for obtaining the travel time data for use in the inversion. This was achieved by comparing the simulated arrivals of the mean as well as eddy cases in the time domain (Fig. 12). In all, 18 eigen rays were identified. Inversion has been carried considering different tomographic layers. For the assessment of the validity of the layering in the model used for inversion, it is essential to look at model and data resolution matrices. The 9 layer case with layer depths 0–50, 50–100, 100–150, 150–200, 200–250, 250–1300, 1300–1800, 1800–2500, 2500–3000 gives the best result, wherein the eddy profile was reconstructed.

Though the first 3 modes give 99% of information, all the 9 modes are required for best reconstruction of the perturbation in all the layers. Model parameter resolution is an indication of how perfectly and independently each model parameter is determined. The diagonal elements of the model parameter resolution matrix

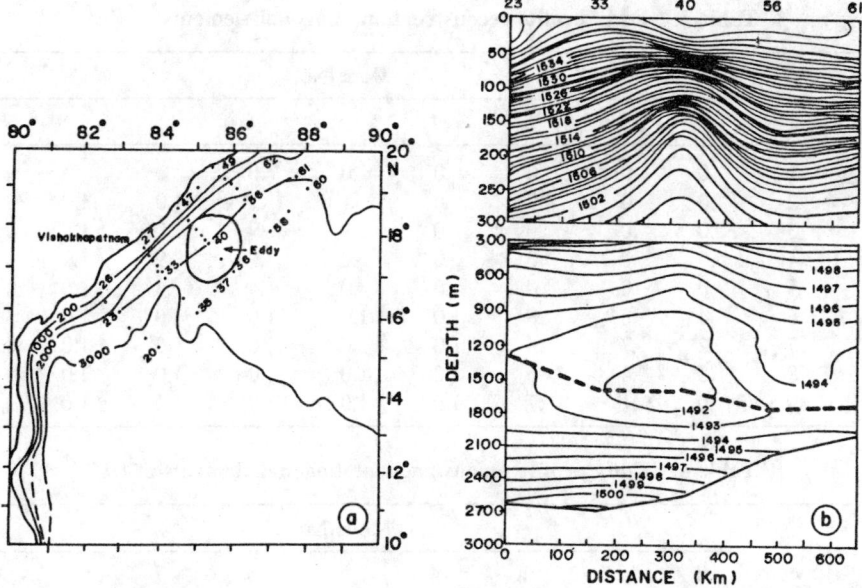

Fig. 11. Map showing the (a) location and (b) Sound speed structure across the eddy parallel to the coast. - - - - indicates the SOFAR channel axis.

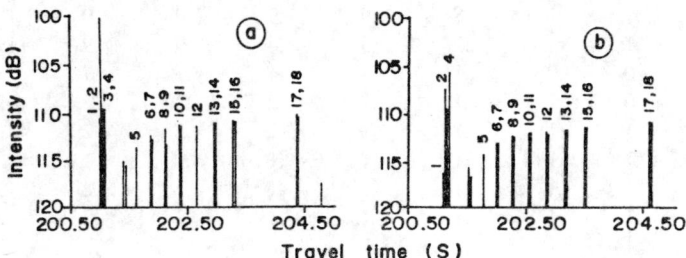

Fig. 12. Predicted travel time (a) reference profile, (b) eddy profile.

(Table 3) reveal how well the individual model parameters are resolved. A value of unity indicates a perfectly resolved parameter whereas a smaller value means inadequate resolution. It can be seen that 18 eigen rays and 9 layers with 9 eigen modes leads to a perfect resolution with unity diagonal elements. It could be noted that the resolution deteriorates as the number of eigen modes is decreased.

The data resolution matrix indicates which data contributes independent information to the solution. The diagonal elements of the data resolution matrix (Table 4) shows unity in the case of columns 3 and 4, indicating that the contribution of information by these middle order rays with launch angle $-5.55°$ and $5.62°$ respectively is exclusive. The low values observed in the rows 8 to 16 indicate poor information resolution associated with higher angle rays which scans nearly same region of the water column. The first mode, which accounts for 87.5% of

Table 3. Eddy profile reconstruction, diagonal elements of VV^T

Layer No.	Mode No.								
	1	2	3	4	5	6	7	8	9
1	.0	.0	.0	.0	.0	.0	.0	.0	1.0
2	.0	.0	.0	.0	.1	.2	1.0	1.0	1.0
3	.0	.0	.0	.0	.9	.9	1.0	1.0	1.0
4	.0	.0	.0	.0	.0	.9	.9	1.0	1.0
5	.0	.0	.0	.0	.0	.4	.4	1.0	1.0
6	.4	.8	.8	1.0	1.0	1.0	1.0	1.0	1.0
7	.2	.7	.8	1.0	1.0	1.0	1.0	1.0	1.0
8	.3	.4	.6	1.0	1.0	1.0	1.0	1.0	1.0
9	.0	.1	.7	1.0	1.0	1.0	1.0	1.0	1.0

Table 4. Eddy profile reconstruction, diagonal elements of UU^T

Ray No.	Mode No.								
	1	2	3	4	5	6	7	8	9
1	.1	.5	.5	.5	.5	.5	.5	.5	.5
2	.1	.5	.5	.5	.5	.5	.5	.5	.5
3	.1	.1	.8	.9	.9	1.0	1.0	1.0	1.0
4	.1	.1	.2	.3	.4	.9	.9	1.0	1.0
5	.0	.0	.1	.4	.4	.7	.7	.8	.8
6	.0	.1	.1	.2	.2	.2	.2	.2	.2
7	.1	.1	.1	.1	.2	.2	.2	.5	.5
8	.1	.1	.1	.1	.1	.1	.1	.1	.1
9	.1	.1	.1	.1	.1	.1	.2	.2	.2
10	.1	.1	.1	.1	.1	.2	.2	.4	.4
11	.1	.1	.1	.1	.2	.3	.3	.3	.3
12	.1	.1	.1	.1	.1	.1	.1	.2	.2
13	.1	.1	.1	.1	.1	.1	.2	.2	.2
14	.1	.1	.1	.1	.1	.2	.2	.3	.3
15	.1	.1	.1	.1	.1	.1	.3	.3	.3
16	.1	.1	.1	.1	.2	.2	.3	.4	.4
17	.1	.1	.1	.1	.3	.3	.5	.5	.5
18	.1	.1	.1	.2	.3	.3	.6	.6	.6

the variance, is mostly contributed by the near axial, flat angle and middle order rays (upto 7°).

Acoustic intensity loss along three eigen rays 12.8°, 9.95° and –2.18° computed (Moler and Solomon, 1970) using the mean sound speed profile and the one in the presence of the subsurface cold core eddy in the Bay of Bengal are presented in Fig. 13. It is evident that the loss associated with steep angle rays through the mean field is about 70 dB in contrast to the 95 dB (top panel) loss through the

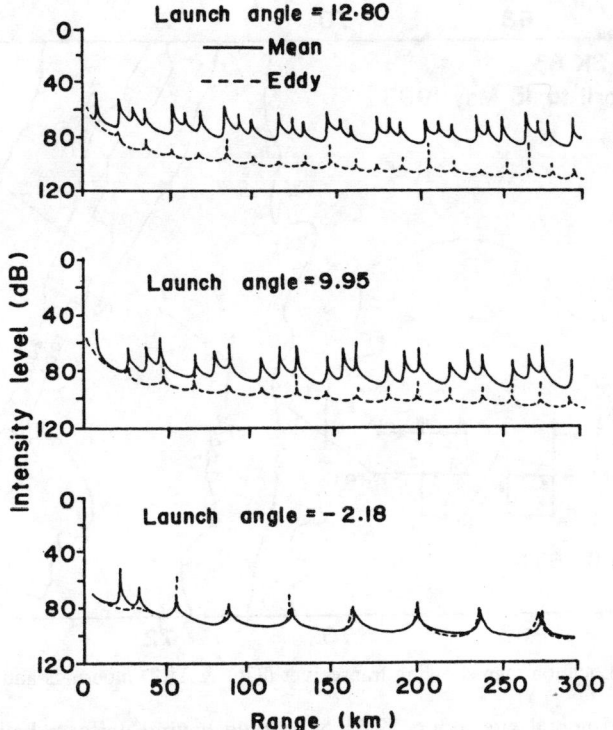

Fig. 13. Spreading loss along selected eigen rays in the presence of the subsurface cold core eddy in the Bay of Bengal.

eddy field. As the steepness of the ray decreases, the loss of intensity was also found to decrease (middle panel). The flat angle rays which do not sample the eddy undergo a loss of about 80 dB (bottom panel) in both the cases. In general it was found that the rays passing through the core of the eddy suffer an additional loss to the tune of 20–25 dB.

5. Acoustic Transmission Experiment in the Arabian Sea

Having carried out a number of simulation experiments and inversions, reconstructing the sound speed anomaly adequately, the next step was to conduct an experiment for a short duration to test the models developed and to see how close one could reconstruct the sound speed and temperature anomaly from the measured data. In view of this an acoustic tomography experiment was conducted in the eastern Arabian Sea during summer 1993 for a short duration (2–12 May) by keeping the transceivers near the sound channel axis.

5.1 Experimental Setup

Acoustic transceivers were deployed on moorings TR1 and TR2 (separated over a range of 270 km) at water depths of 4250 m and 4175 m respectively (Fig. 14).

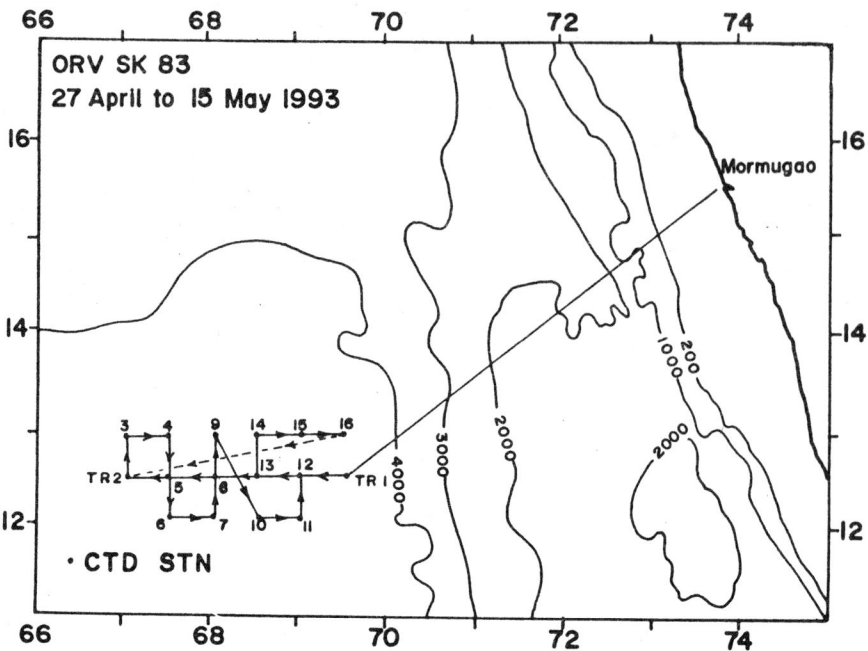

Fig. 14. Location map showing transceiver (TR1 & TR2) moorings and CTD stations.

The experimental site along 12.5° N latitude having uniform bathymetry, was chosen to minimise the bottom interaction of acoustic energy. The moorings, were designed to keep the transceivers submerged at depths of 1535 m and 1685 m respectively. The mooring assembly consisted of a marker buoy, an array of glass floats, acoustic transceiver, tilt-meters (one below the transceiver at mid-depth and the other at deeper level), acoustic releases (2 nos.) connected in parallel (for details see Acoustic Group, 1993).

The exact position of the mooring was determined by triangulation method. Two tilt meters attached to each of the moorings were used to obtain the information on mooring motion. The tilt meter has a triaxial magnetic sensor which measures the angle (tilt) in the vertical plane (ψ) and horizontal plane (ϕ). The mooring tilt data were used to estimate the drift(s) of the mooring in x-, y- and z-directions. The tilt meter also provided in-situ temperature, conductivity and pressure data.

In order to estimate the shift in position of transceiver due to the wander of the moorings during the experiment, the lean angle and direction of the mooring line were analysed. From the mooring tilt data, the magnitude of the shifts in the horizontal (x, y) and vertical (z) directions were obtained.

Acoustic transceivers used had a central frequency of 400 Hz with a bandwidth of 100 Hz. Hourly transmissions and receptions took place at 00 minutes and 30 minutes of each hour at TR1 and vice- versa at TR2. The transmitted signal consisted of 511 digit phase coded shift register sequences. The experiment started on Julian day 122 (2 May 1993) at 2000 hrs (Indian Standard Time) when TR1 transmitted and TR2 received. Four different sequences were transmitted over the duration of the experiment (Table 5). While the transmissions were going on, a hydrographic survey was carried to obtain in-situ information on the acoustic profile and the prevailing hydrographic conditions during the experiment.

Table 5. Details of transmitted signal and transmission schedule

No. of sequences transmitted	Julian day	Time (IST)	Duration (Days)	No. of sequences received
3	122	2000	1	1
17	123	2000	3	16
9	126	2000	3	8
12	129	2000	3	11

5.2 Signal Processing

The received signal is pre-amplified using in-built automatic gain controller (agc), complex demodulated, low-pass filtered, sampled and coherently averaged (Spindel, 1985). These demodulates consist of in-phase and out-phase components of the coherently averaged receptions.

The signals, thus obtained, were cross-correlated with the replica of the source signal (pulse compression) during the post processing, to generate amplitude peaks corresponding to the multipath arrivals (Fig. 15).

5.3 Simulations

Using a climatological mean sound speed profile for the Arabian Sea along 2.5° (67–69.5°E) constructed from Levitus data (1982), acoustic ray tracing was carried out using a range-independent program (Mahadevan et al, 1989) to predict the eigen ray arrival structure. For this, a flat bottom (4100 m depth) was assumed and rays were scanned for emergence angles between +12° and −12°, source and receiver at 1535 m (TR1) and 1685 m (TR2) respectively and the range at 270.92 km. This yielded 14 eigen rays connecting TR2 to TR1 (Fig. 16). Information on the acoustic travel time along with the ray path lengths in the pre-set number boxes along the source-receiver plane forms the data kernel.

The predicted times of these rays were matched with the measured travel times, after applying corrections for mooring motion. A typical ray identification is shown in Fig. 17. It was found that during 4 days of transmissions, out of the 14 eigen rays, 8 rays were stable in time (Table 6). From the travel time perturbations, sound speed perturbations (model parameter) were computed based on generalised inverse method. The 2-D inversion scheme using a box-model has been implemented to obtain sound speed anomaly in the vertical plane over the range (TR1-TR2). Range dependency has been introduced in the data kernel by using $-R_{ij}/C_{oj}^2$,

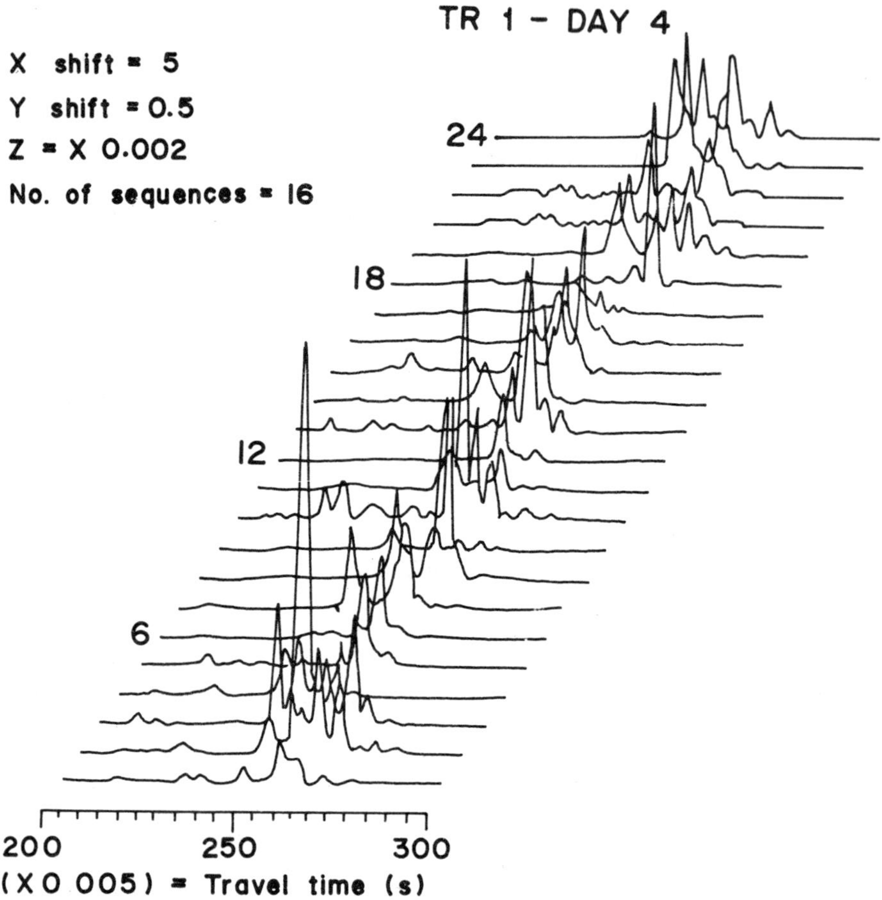

X shift = 5
Y shift = 0.5
Z = X 0.002
No. of sequences = 16

TR 1 - DAY 4

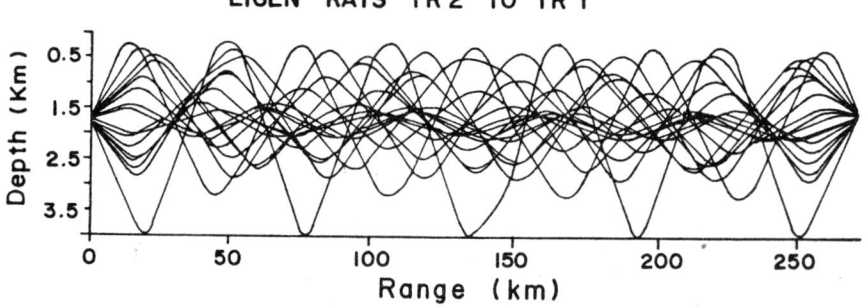

Fig. 15. Waterfall plot showing measured signal arrivals at the receiver.

Fig. 16 Eigen rays from TR2 to TR1

where R_{ij} and C_{oj} are the box-wise ray path length and reference sound speed respectively.

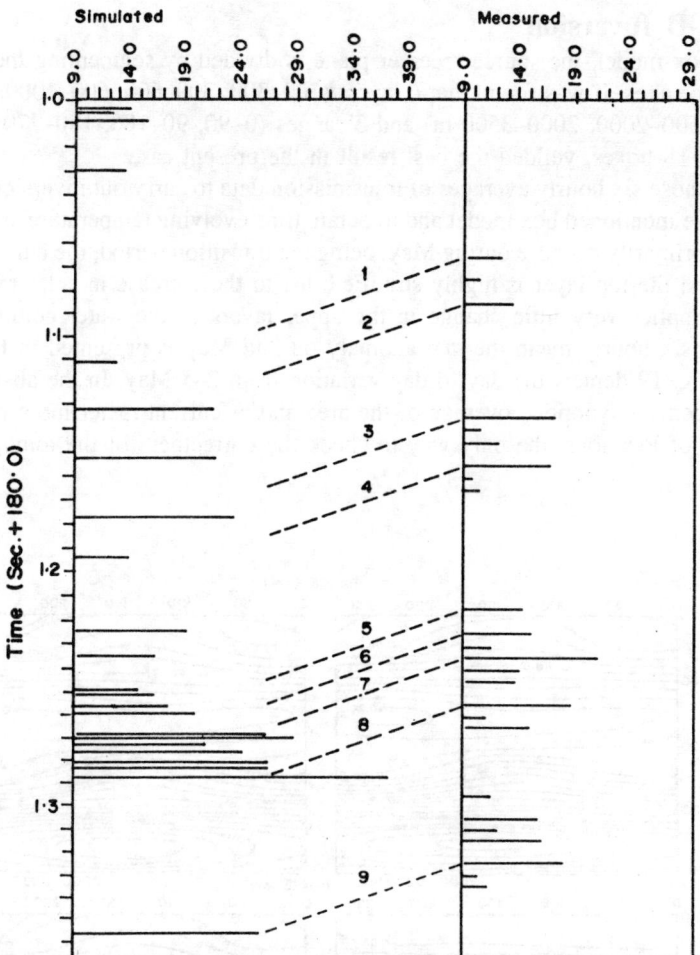

Fig. 17. Identification of eigen rays predicted with measured arrivals.

Table 6. Ray identifier for stable eigen rays

	Launch angle θ	Total turnings + N	Upper, lower turnings $(p, -p)$	Total travel time T (s)
1	8.10	9	(5, −4)	181.4543
2	6.87	8	(4, −4)	181.2587
3	5.39	8	(4, −4)	181.2166
4	4.30	9	(5, −4)	181.2865
5	2.13	11	(6, −5)	181.3595
6	−1.56	12	(6, −6)	181.3763
7	−3.14	11	(6, −5)	181.3494
8	−5.30	9	(5, −4)	181.2357

5.4 2-D Inversion

In the box-model, the source receiver plane is divided by segmenting the depth and range axes. It was found that 6 layers (100–300, 300–600, 600–1000, 1000–1500, 1500–2000, 2000–3500 m) and 3 ranges (0–90, 90–180, 180–270.9 km), forming 18 boxes, yeilded the best result in the present case.

We chose six hourly averages of transmission data to carry out inversion using the above mentioned box model and to obtain time evolving temperature anomaly. This is primarily because during May, being the transition period, the currents are weak and the top layer is highly stratified due to the increase in solar radiation which implies very little change in the upper layers of the water column. For brevity, six hourly mean thermal anomaly on 2nd May is presented in Fig. 18, while Fig. 19 depicts the day to day variation from 2–5 May. In the absence of a high density synoptic coverage of the area and/or current meter moorings at a number of locations, the only way to check the correctness of the tomographic

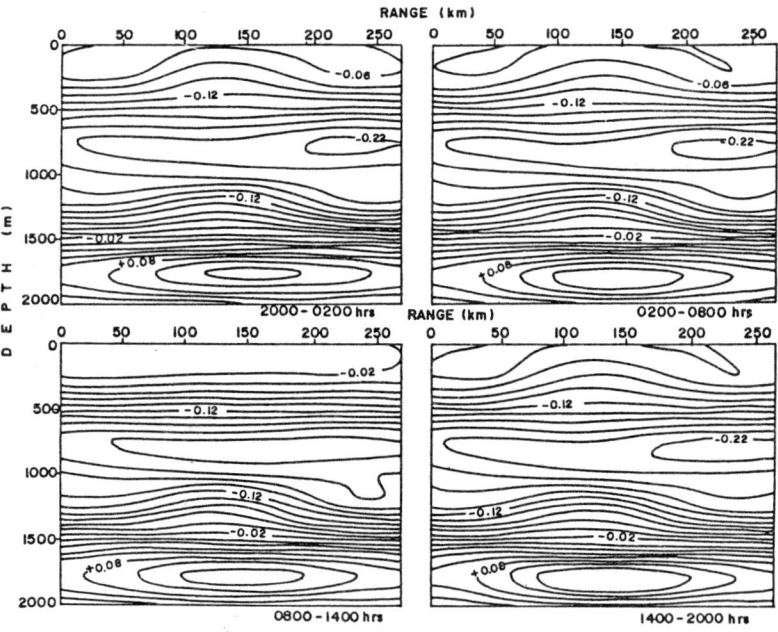

Fig. 18. 2-D temperature anomaly for 2000 hrs of 2nd May to 2000 hrs of 3rd May 1993 reconstructed through tomography.

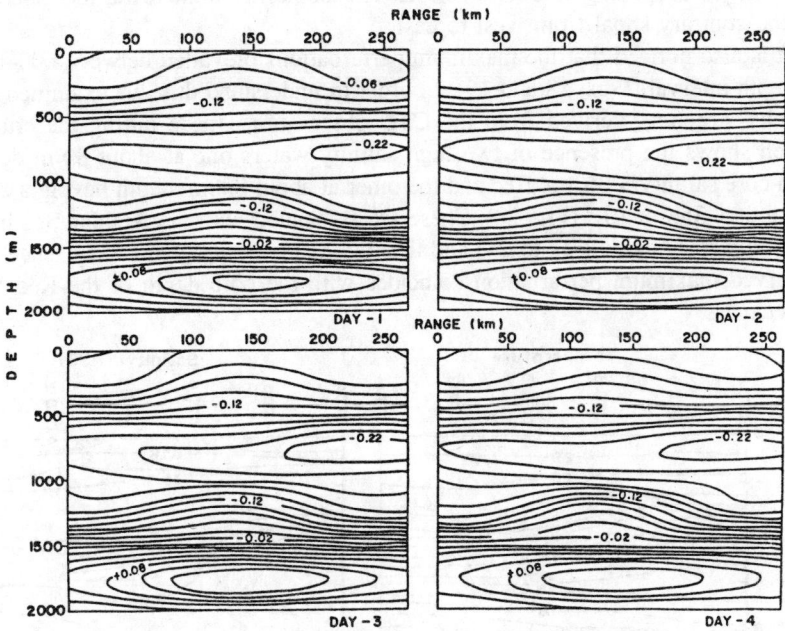

Fig. 19. 2-D temperature anomaly for 2–5 May 1993 reconstructed through tomography.

measurements is to see whether these measurements reproduce the signatures of the known ocean features.

An examination of Fig. 18 reveals changes taking place in the upper 200 m layer. The isoline –0.06, seen in the first 6 hourly average (starting from 2000 hrs of 2 May to 0200 hrs of 3 May), has been replaced by –0.04 and –0.02 isolines respectively during the subsequent 6 hourly averages (0200–0800 hrs and 0800–1400 hrs), indicating a steady warming of the top layer, while for the six hourly average between 1400 hrs and 2000 hrs, –0.04 isoline reappears indicating the cooling phase. Thus the 6 hourly average temperature anomaly essentially brings out the warming and cooling phases associated with the diurnal cycle.

Similarly, from the maps depicting variability in the thermal field over the first four days of transmissions (Fig. 19), it is discernible that the top layer exhibits gradual warming as could be noted from the changes in the isolines from –0.06 to –0.04. An examination of the SST observations (using bucket thermometer) during the same period also showed a gradual increase of 0.8°C (from 29.5°C to 30.3°C). This is to be expected, as the Arabian sea receives the maximum insolation at this time of the year. The thermal structure (Fig. 20) showed a thin mixed layer of approximately 20 m in thickness, with the isotherms in the upper thermocline (~20–27°C) deepening at around 68.5°E. The isotherms in the lower thermocline, on the contrary shoal from west to east.

It is also noticed that the maximum perturbations prevailed between 600–800 m depth interval (Figs. 18 and 19). In order to understand this, we examined the salinity structure, derived from the CTD observations made during the cruise, which shows the presence of two high salinity waters one at about 80 m depth with core salinity $> 36.1 \times 10^{-3}$ and the other at about 800 m depth having a core value of $> 35.2 \times 10^{-3}$ (Fig. 20). These are associated with the Arabian Sea high salinity water (sigma-t 23 to 24) and the Red Sea water (sigma-t 27). Hence the observed maximum perturbation coincides with the core depth of the Red Sea water mass.

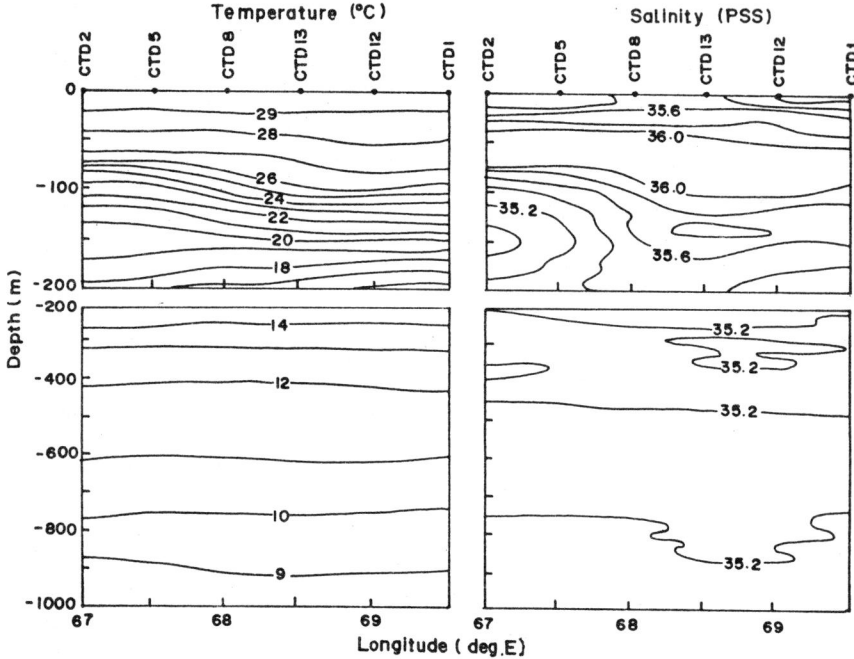

Fig. 20. Vertical section of temperature and salinity along TR2-TR1.

From the six hourly averages of the 2-D temperature anomaly signatures of diurnal variability, a gradual warming of the top layers and intrusion of Red Sea waters could be inferred from tomographic measurements.

6. Acoustic Thermometry

It is estimated that the global ocean is warming at the rate of a few millidegrees each year. Munk and Forbes (1989) put forward a novel idea of acoustically measuring the global ocean warming known as acoustic thermometry. This is based on a simple fact that the travel times of acoustic pulses between fixed source(s) and receiver(s) separated over a large range diminish (due to the increase in sound speed caused by the temperature) at a rate proportional to the warming. Continuous monitoring of travel time changes over a decade would help assess the warming of the ocean.

Heard Island experiment, a joint venture in which eight countries—U.S., Canada, Australia, Newzealand, France, South Africa, Japan and India—participated in January 1991, was a feasibility test to determine the strength and stability of acoustic signals at different listening stations spread over the four oceans. The results of the analysis of acoustic propagation paths from the source near the Heard Island site to the Indian receiving station based on the numerical simulation (Prasanna Kumar et al, 1996) and processing of received signals are as detailed.

6.1 Simulation

The climatological hydrographic data in the Indian Ocean between longitudes 69.5°–74.5°E and latitudes 55.5°S–13.5°N, at 1 degree intervals have been derived from the Levitus, 1982. Temperature and salinity values at all standard depths were used in the computation of sound speeds. In all, about 66 profiles (Fig. 21, upper panel) were used to analyse the spatial sound speed structure in the meridional direction and as input for the ray model. For the model simulation of propagation, sound speed profiles were interpolated at every 25 km range and were used as input for a range-dependent ray trace program to trace acoustic rays over a propagation range of about 7000 km.

The acoustic source, used for the experiment, was kept at site 40 nm southeast of the Heard Island (53° 15'S, 73° 40'E) in the Indian Ocean sector of Antarctica. This location was selected for the acoustic source as it provided a unique point with access to all the major oceans. The source consisted of a vertical array of transducers lowered from R/V Cory Chouest to a depth of 150 m below the surface, coinciding with the axis of the sound channel. Different types of continuous wave (CW) and phase coded signals centered at a frequency of 57 Hz were transmitted from 0000 GMT on 26 January through 2100 GMT on 31 January 1991. The transmitter schedule was fixed for one hour transmission at every three hours round the clock. Due to mechanical problems transmissions were stopped in between from 1600 GMT on 27 January to 0800 GMT on 28 January and 2200 GMT on 28 January to 2300 GMT on 29 January, 1991.

There were six different signals—a carrier wave at 57 Hz, a pentaline and 4 pseudorandom codes referred to as m-sequences. The pentaline was a 5 tone signal, the spectral lines decrease by 6 dB as one moves out from the carrier. The m-sequences were phase modulated binary codes at 57 Hz carrier. The phase shift was 45 degrees, so that for all codes half the power of the signal remains in the carrier. Codes of 255, 511, 1023 and 2047 digits were transmitted. The 255 digit code was transmitted more often than the other codes since the estimated period for the same would be sufficient to cover multipath arrival.

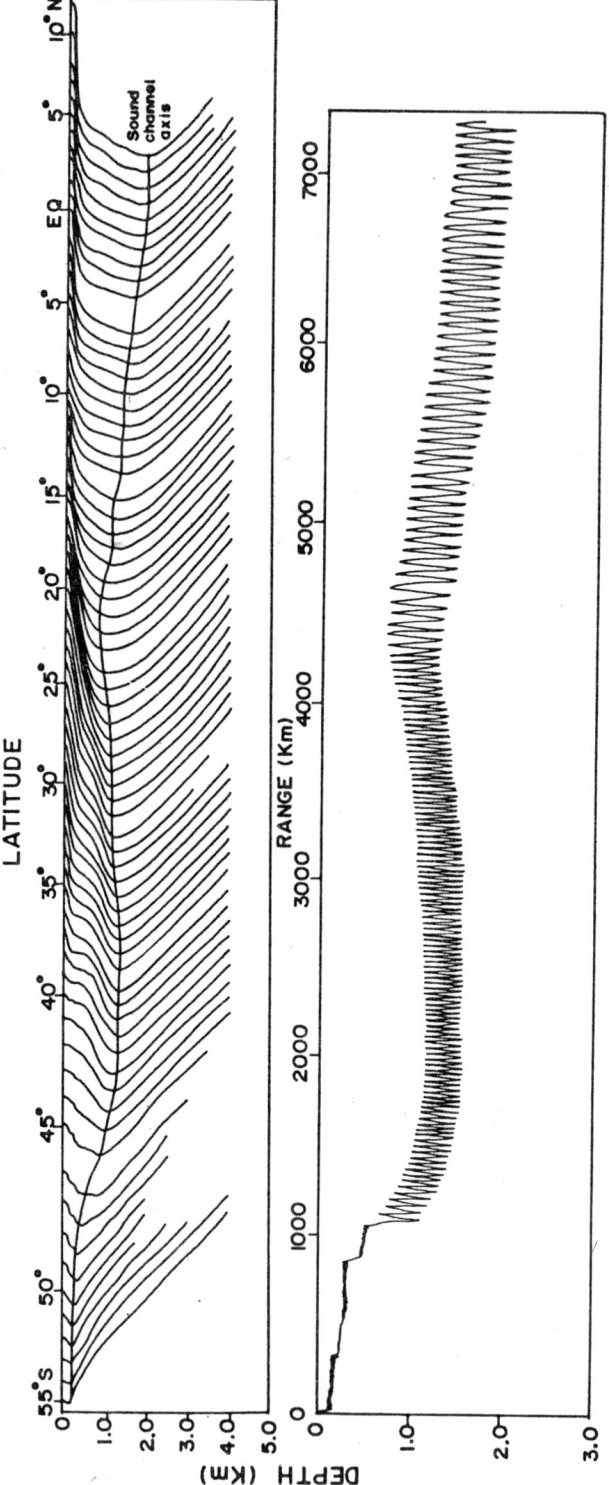

Fig. 21. Sound speed profiles (top) and a typical acoustic ray path (5° ray, bottom) simulated along 75°E meridional transect from Heard Island to Indian listening station.

The listening stations in this experiment used two types of receivers—fixed and drifting. The Indian Ocean listening station consisted of a drifting buoy with hydrophone receivers. Two types of drifting buoys—a recoverable Spartan 51-A buoy and the expendable Magnavox 41-B sonobuoy(s)-received the signals, the latter served as backups to the former.

A radio receiver onboard ORV Sagar Kanya monitored and received data from the buoy using a P.C. The received data could be monitored in real time for signal reception while the data was being received and stored on the hard disk in the background.

6.2 Acoustic Field

Sound speed structure, constructed based on the 66 profiles, along 71°E (Fig. 21, upper panel) shows significant changes in sound speed in the meridional direction which is the consequence of gradual increase of temperature from southern ocean towards the tropical region. Accordingly the depth of the SOFAR channel axis also showed variation from about 150 m near Heard Island to 1600 m in the Arabian Sea. The overall surface sound speed varied by about 80 m/s over 7000 km. The sudden change in the sound speed structure centered at 45°S (50–40°S) is associated with the Subtropical Convergence Zone, which is characterised by a rapid increase of temperature by about 8°C and salinity by 0.8 PSU towards north.

The simulated ray geometry, for launch angles +8°, +5°, +2° and −10° (Fig. 21, lower panel), depicts a gradual deepening of the ray path which is brought in by the increasing depth of the sound channel axis towards tropics. However, in the vicinity of the equator, the ray path shoals up by about 200 m, under the influence of the strong divergence. It is interesting to note that while small angle rays confine tightly to the channel axis, the large angle rays undergo surface reflection due to the combined effect of low sound speeds at the top layer and shallow axis depth.

The sudden ray jumps prominently seen for rays with launch angles 5 and 2° at a range of 1000 km is associated with the rapid changes in the sound speed structure in the vicinity of Sub Tropical Convergence. The ray travel times of these rays were of the order 1 hr 18 min. The intensity loss due to spreading was calculated to be about 150 dB while that estimated from the reception data was about 135 dB.

6.3 Signal Characteristics

On line spectral analysis, using FFT, of the signals received on board was done. Based on the FFT analysis some rough estimates were made about signal to noise ratio, and speculations as to the optimum coherent processing time. These conclusions/speculations are mainly drawn from CW and pentaline signals which contain distinct spectral peaks. During a CW reception, and assuming roughly constant ambient noise during each reception, it seems the deep sonobuoy signal is up to 20 dB stronger than the 41B sonobouy signal, for RMS spectra peaks.

This might be attributed to the deeper sonobouy's proximity to the sound channel axis and less noise in the bandpass range around 57 Hz at deeper depths. Unfortunately, only two transmissions were received on the deep sonobuoy. The best signal to noise ratio received was nominally 40 dB on the deep sonobuoy when receiving CW. On the 41B Sonobuoy, a signal to noise ratio in the low to mid 20 dB range was obtained. The m sequence receptions had visible carrier spikes, but the peak signal to noise ratio was less as only half the energy is in the carrier. A Doppler shift of −0.06 Hz to +0.08 Hz has been observed. Source ship and/or receiver drift could be a plausible cause for the observed Doppler shift. Incidentally, the receiver ship drift trajectory monitored using the GPS showed a prominent direction of drift towards southwest at the rate of 1.5 km/hr.

The received signals showed unexpected combination of phase stability and amplitude variability. When there were no transmissions the band pass filtered ambient noise of the ocean was recorded. This noise was found to be stable over the period of observation.

Off line processing of the entire data from the 11 receiving stations was done at the Cooley Electronics Laboratory, University of Michigan (Metzger et al, 1991) to determine the power, stability, time resolution and dispersions. The Indian Ocean station data has been subjected to "Frequency domain analysis" of each transmission. A hamming window was used to reduce the frequency sidelobes with 7/8 of the data, common between the adjacent windows to enhance the continuity and follow quick changes.

The post processed data for a typical reception is presented in Fig. 22. All significant FFT magnitudes, which are more than 2.5 times the median of the magnitudes of the 47 Hz to 67 Hz band, are represented by dots of same size and darkness in the plot (Fig. 22a). Thus it represents the frequency spectrogram of the signal over the reception period. The other plots represent the bin shift, magnitude and residual phase of the carrier (Fig. 22b-d). The time (minutes) versus frequency (Hz) plot (Fig. 22a) shows prominent signal at about 57 Hz indicated by the continuous dots. The dark region in the spectrogram, spread around 60 Hz is due to the noise of the ship's generators. The bin shift (Fig. 22b) indicates the presence of the Doppler shifted carrier. It could be seen that the patches with straight line are the regions with stable carrier with constant Doppler shift of about −0.1 Hz. The magnitudes of the carrier bin along with the median (Fig. 22c) shows the reception of the carrier. The fluctuations in the carrier magnitude was in part due to the multipath propagation and the variations in the number of active sources in the source array during the transmission period. The change in the phase of the carrier (Fig. 22d) elucidates the regions of constant Doppler shift with linear phase change with time. In short it is discernible from the above that the carrier with its Doppler could be traced and monitored which is required for further processing like demodulation and cross-correlation.

Thus the signals received at the Indian receiving station indicate that long range acoustic transmissions using electrically powered sources is possible with

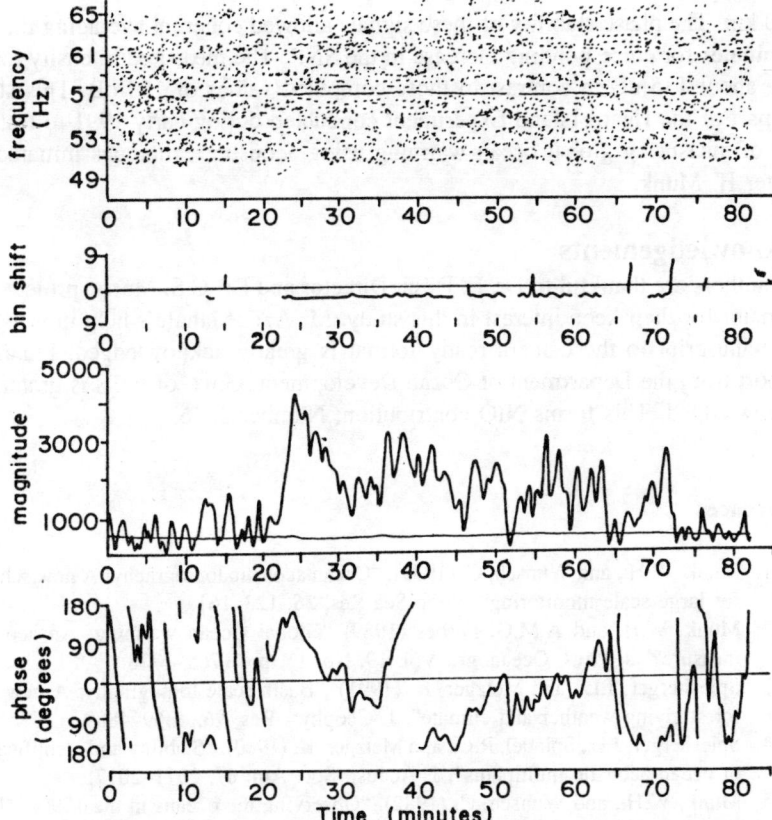

Fig. 22. Post processing of signals transmitted from Heard Island, received at Goa India station on 27 January 1991. Plots of time vs. (a) frequency dot spectrogram showing prominent signal at 57 Hz, (b) Carrier bin shift, (c) carrier magnitude and (d) carrier phase.

sufficient received signal level and stable Doppler compensated carrier frequency. Coherent averaging times of the order of 20 minutes or above were possible for increasing signal to noise ratio. Similar results were reported from the receivers located in Atlantic and Pacific (Baggeroer and Munk, 1992) proving the feasibility of acoustic thermometry for monitoring global ocean warming.

The acoustic propagations—the signal trajectories—from Heard Island in the south Indian Ocean to eastern Arabian Sea, simulated based on the climatic hydrographic data, showed a gradual deepening of the ray paths from southern ocean towards tropics associated with deepening of the axis of the sound channel. Sudden jumps were noticed in the acoustic ray paths in the vicinity of sub-tropical convergence zone. The measured acoustic signals, received at the Indian listening station through suspended hydrophones, had a signal to noise ratio (SNR) of 20 dB on an average and confirmed their detectability over distances as far as

7000 km. The phase stability of these signals allowed coherent averaging time of 20 minutes for corresponding increase in the SNR. The measured intensity losses were about 135 dB in contrast to those computed values of 150 dB. This study is a part of the Heard Island Experiment conducted during early 1991 to test the idea of measuring global ocean warming using acoustic techniques initiated by Walter H. Munk.

Acknowledgements

The authors are thankful to Dr. E. Desa, Director and Dr. C.S. Murty, project Co-ordinator for their keen interest in this study. Mr A.Y. Mahale's help in bringing the manuscript to the camera ready format is greatly acknowledged. Financial support from the Department of Ocean Development, Govt. of India is gratefully acknowledged. This forms NIO contribution: Number 2536.

References

1. Munk, W.H., and Wunsch, C. (1979). "Ocean acoustic tomography: A new scheme for large scale monitoring", Deep-Sea Res. **26**, 123–161.
2. Munk, W.H. and A.M.G. Forbes (1989) "Global Ocean warming: An acoustic measure?" J. Phys. Oceanogr., Vol. 19, No. 11, pp. 1765–1778.
3. Spiesberger, J.L., and Metzger, K. (1991). "Basin-scale tomography: A new tool for studying weather and climate", J. Geophys. Res. **96**, 4869–4889.
4. Spiesberger, J.L., Spindel, R.C., and Metzger, K. (1980). "Stability and identification of ocean acoustic multipaths", J. Acoust. Soc. Am. **67**, 2011–2017.
5. Munk, W.H., and Wunsch, C. (1982). "Observing the oceans in the 1990s", Phil. Trans. Royal Soc. London. **A307**, 439–464.
6. Cornuelle, B., Wunsch, C., Behringer, D., Birdsal, T., Brown, M., Heinmiller, R., Knox, R., Metzger, K., Munk, W.H., Spiesberger, J., Spindel, R., and Webb, D. (1985). "Tomographic maps of the ocean mesoscale: Pure acoustics", J. Phys. Oceanogr. **22**, 133–152.
7. Worcester, P.F., Spindel, R.C., and Howe, B.M. (1985). "Acoustic transmission: Instrumentation for mesoscale monitoring for ocean currents", IEEE J. Ocean Engg. **OE-10**, 123–137.
8. Worcester, P.F. and Cornuelle, B.D. (1991) "A review of Ocean Acoustic Tomography: 1987–90", Rev. Geophys., Suppl., pp. 855–865.
9. Ramana Murty, T.V., Prasanna Kumar, S., Somayajulu, Y.K., Sastry J.S., and De Figueiredo, Rui J.P. (1989). "Canonical sound speed profile for central Bay of Bengal", Proc. Indian Acad. Sci. (Earth and Planetary Sciences) **98**, 255–263.
10. Ramana Murty, T.V., Somayajulu, Y.K., and Sastry, J.S. (1990). "Computation of some acoustic ray parameters in the Bay of Bengal", Indian J. Mar. Sci. **90**, 235–245.
11. Ramana Murty, T.V., Somayajulu, Y.K., Mahadevan, R., Murty, C.S., and Sastry, J.S. (1992). "A solution to the inverse problem in ocean acoustics", Defence Sci. J., **42**, 89–101.
12. Prasanna Kumar, S., Somayajulu, Y.K., Ramana Murty, T.V., and Sastry, J.S

(1988). "Acoustic variability of central Bay of Bengal for tomography", Oceanologia **26,** 81–95.

13. Prasanna Kumar, S., Ramana Murty, T.V., Somayajulu, Y.K., Chodankar, P.V., and Murty, C.S. (1994). "Reference sound speed profile and related ray acoustics of Bay of Bengal for tomographic studies", Acustica **80,** 127–137.

14. Somayajulu, Y.K. (1993). "Some aspects of simulation of sound propagation in the Arabian Sea", Ph.D Thesis, Goa University, India, pp 125.

15. Somayajulu, Y.K., Ramana Murty, T.V., Prasanna Kumar, S., and Murty, C.S. (1944). "Some studies related to acoustic propagation in the Arabian Sea", Acoust. Lett., **17,** 173–184.

16. Chen, C., and Millero, F.J. (1977). "Speed of sound in sea water at higher pressure", J. Acoust. Soc. Am. **62,** 129–135.

17. Levitus, S. (1982). "Climatological atlas of the world ocean", NOAA Professional Paper No. 13, US Government Printing Office, Washington, D.C., pp. 173.

18. Aurther, R.S., Munk, W.H. and Issacs, J.D. (1952). "The direct construction of wave rays", Trans. Am. Geophys. Un., 33(6), pp. 855–865.

19. Mahadevan, R., Somayajulu, Y.K., Ramana Murty, T.V., Murty, C.S., and Sastry, J.S. (1989). "Tomographic forward problem: Computational details for preparation of data kernel", NIO Tech. Rept. TR-79, 1989, pp. 22.

20. Moler, C.B. and Solomon, L.P., "Use of splines and numerical integration in geometrical acoustics", J. Acoust. Soc. Am., 48, 1970, pp. 739–744.

21. Ramana murty, T.V; Y.K. Somayajulu, P.V. Chodankar and C.S. Murty (1993), "Acoustic characteristics of the waters of the Bay of Bengal", Indian J. Mar. Sci., Vol. 22, pp. 263–267.

22. Chester, D.B., Rizzoli, P.M., and De Ferrari, H. (1991) "Acoustic tomography in the straits of Florida", J. Geophys. Res. **96,** 7023–7048.

23. Menke, W. (1984) "Geophysical Data Analysis: Discrete Inverse Theory", Academic Press, New York, pp. 260.

24. Jackson, D.D. (1972). "Interpretation of inaccurate, insufficient and inconsistent data", Geophys. J. Royal Astro. Soc. **28,** 97–109.

25. Babu, M.T., S. Prasanna Kumar and D.P. Rao (1991). "A subsurface cyclonic eddy in the Bay of Bengal", J. Mar. Res., 49, 403–410.

26. Prasanna Kumar, S., Navelkar, G.S., Ramana Murty, T.V., and Murty, C.S. "Acoustic propagation in the presence of a cold core eddy in the Bay of Bengal—A case study", Proc. Pacific Ocean Remote Sensing Conf., Melborne, Australia, 1–4 March 1994, pp. 185–192.

27. Acoustic Group, (1993). "Acoustic Transmission Experiment-93", Report of Natl. Inst. of Oceanography 1993, pp. 74.

28. Spindel, R.C. (1985). "Signal processing in ocean tomography in Adaptive methods in Underwater acoustics", H.G. Urban (Ed.), D Reidel Publishing Company, 687–709.

29. Prasanna Kumar, S., Navelkar, G.S., Ramana Murty, T.V., Somayajulu, Y.K., Saran, A.K., and Murty, C.S. (1996), Numerical simulation and measurements of acoustic transmissions from Heard Island to the equatorial Indian Ocean, Indian Journal of Marine Sciences, vol. 25, pp. 173–178.

30. Metzger, K. Dzieciuch, M. and Birdsall, T.G. (1991) Heard Island Feasibility test data survey in the frequency domain, University of Michigan, 48109-2122.

31. Baggeroer, A. and W.H. Munk (1992) The Heard Island Feasibility Test, Physics Today, Sept. 1992, pp. 22–30.

Subject Index

Absorption
   acoustic 558
   coefficient 22
Acoustic
   beacons 414
   projector 414
   ray model 553
   scattering 191
   scintillation 433
   sensing limitations 20, 24
   shield 90, 96
Adiabatic
   speed 5
Air
   masses 242
   pollution 128, 343, 386, 390
   protection 385, 386
   quality 227, 308, 385
   sea interaction 455
Anisotropy
   coefficient 17
Anthropogenic 5
Aspect sensitivity 374
Atmospheric
   boundary layer 5, 326
   stratification 179
   turbulence 6, 43, 191
Attenuation
   coefficient 43, 181
Auto-correlation
   coefficient 181
   function 68

Bermuda triangle 418
Boundary layer
   marine 451
   stable 331, 392
   unstable 311, 329
Bowen ratio 193, 220, 221, 223
Box model 224, 359, 361

Bragg
   condition 6, 136, 143
   resonance 134

Calibration 192
Coding 395
Cold outbreaks 294
Communication 5, 188
Convective
   velocity scale 199
Curve fitting errors 525

Data
   acquisition system 106
   extraction technique 252
   procuring 119
Diffraction
   length 26
Diffusion 37, 224
Directional pattern 23
Dispersion
   coefficient 336, 354
Dissipation rate 192, 218
Doppler
   equation 44
   radar 137
   sodar 85, 113, 202, 308
Drainage flow 213
Dynamic range 38

Echo
   shear 333
   thermal 333
Eddy
   oceanic cold core 558
   sub surface cold core 563, 566, 567
Encoding 395
Entrainment
   layer 331
   zone 311

Fermat's principle 429, 519
Fetch 321
Filter coefficient 64
Filtering 60
First moment 107, 119
Fog capping 138
Fresnel
  radius 436
  zone 369
Friction velocity 263, 317, 330

Gravity waves 275, 278

Heat island 108
Height error 50
Holzworth model 303, 349
Hydro-phone 410, 467

Image processing 121
Intertial
  oscillations 241
  subrange 192, 329

Kolmogorov model 12

Laplace formula 5
Lee waves 265
Lump echo 206

Mixing
  depth 317, 343
  height 224, 349, 361, 388
Momentum Flux 166
Monin-Obhukov
  length 331
  stability length 329

Pattern recognition 332, 395
Phase
  array 40, 55, 197
Pinger 410
Plume model, Gaussian 227
Propagation characteristics 365

Radial
  velocity 252
  wind 98, 108
RASS equation 148
Ray model, acoustic 553

Refractivity
  gradient 366
  radio 363, 364
Resolution
  range 87
Richardson number
  bulk 354, 355
  gradient 328
Roughness
  length 329
  parameter 328

Scattering
  cross-section 10, 13, 19, 44, 86, 180, 216, 220
Sea breeze 124
Sensible heat flux 164, 193, 219, 315
Sigma curves 336
Signal
  intensity 21, 191
  processing 57, 66, 91, 106, 569
Similarity
  methods 315, 349
  theory 312, 315, 331
Sodar equation 47, 86, 185
SOFAR 413, 557
SONAR 558
Sound
  slowness 479
Spectral
  analysis 107
  distribution 218
  function 12, 180
  mean 66
  method 194
  moment 68, 70
Spectrum
  Dophler 151
  Kolmogorov's 180
  power 69
Stability
  category 338, 387
  classification 336
Stack
  height 357
Stoke's theorem 433
Structure
  coefficient 215

function 15, 43, 192
  parameter 14, 179, 191, 310, 370
Swallow floats 410

Taylor's
  hypothesis 435
Temperature error 154
Thermocline 413, 574
Thermometry, acoustic 507, 574
Tomography 420, 476
Transform
  Abel 513, 519, 520, 528
  Fast Fourier 75, 106, 120
Turbulence
  anisotropic 17
  anisotropy 22

atmospheric 5, 192
frozen 207
scale 181

Velocity measurement error 50
Ventillation factor 304
Vorticity 431

Wake 127
Wave
  breaking 467
  Herring bone 279
  Kelvin-Helmholtz 279, 328
Wind
  profiler 39, 138
  shear 238, 315, 327

OCCUPATIONAL SAFETY and HEALTH POLICY

Melvin L. Myers, MPA